T5-AFS-977

Wildflowers and Weeds of KANSAS

Wildflowers and Weeds of KANSAS

Janét E. Bare

THE REGENTS PRESS OF KANSAS
Lawrence

© Copyright 1979 by The Regents Press of Kansas
Manufactured in the United States of America

Library of Congress Cataloging in Publication Data
Bare, Janet E
Wildflowers and weeds of Kansas.
Bibliography: p.
Includes index.
1. Wild flowers—Kansas—Identification.
2. Weeds—Kansas—Identification. I. Title.
QK161.B36   582'.13'09781   78-16862
ISBN 0-7006-0176-7

Contents

List of Illustrations in Color

Plates follow page 36

Preface

This book is intended for use by beginning botanists and biologists, both those who are delving into the plant world on their own initiative without formal guidance and those who are participating in a formal high-school or college-level course in biology, elementary botany, or plant systematics. It is based upon the work *Kansas Wild Flowers* by William Chase Stevens, but the arrangement and measurements are my own. The descriptions have been rewritten for the most part. Whereas his book included both woody and herbaceous (nonwoody) plants, this revision is limited primarily to herbaceous plants, although the total number has been considerably enlarged to include most of the common herbaceous flowering plants that are native or naturalized in Kansas. The grasses and their grasslike relatives, the sedges and rushes, although they too are flowering plants, have been omitted, partially because of the sheer bulk of material which would be involved and partially because they are more difficult for the uninitiated to work with.

To a beginning botanist, the descriptions and keys in this book may at first appear overly detailed and technical and the number of species overwhelming. It has been my experience, however, that students meet with more frustration in identifying a plant when they have inadequate reference material at hand than when they have "extra" information. The latter can always be ignored if it is not needed at the moment, but the most diligent, skillful examination of a plant will not get it through a key correctly if that plant has not been included in the key or the manual. Although good photographs are of considerable assistance in checking one's identification, it sometimes happens that the important distinguishing characteristics of a species are so small that they are difficult or impossible to see in a photograph which simultaneously depicts the general aspect of the plant. In these instances, one must resort to the textual information and keys for the small distinguishing details. In all cases, it is hoped that the descriptions will encourage thorough observation of the plant structures and an awareness of the intriguing variety present in the plant world. This was also my motivation for including some of the less conspicuous plants such as the Spurge and Goosefoot families.

As in any undertaking in an unfamiliar field or new pastime, it is necessary to acquire a basic vocabulary. The descriptive terminology associated with plant parts, while it may seem foreign and discomforting at first, is in reality no more difficult than that related to automotive mechanics, sewing, office work, or sports. A relatively small amount of time spent familiarizing oneself with the terms introduced in the first sections of this book will soon pay for itself by increasing your efficiency and thereby your enjoyment in identifying a plant and will leave more time for learning about the plants themselves.

Keys and descriptions have been kept as simple and easy-to-use as possible, and nontechnical terms such as "bell-shaped" or "hairy" rather than their technical equivalents ("campanulate" and "pubescent," respectively) are used most of the time. Exceptions are cases wherein the technical terms are shorter and less cumbersome than the vernacular forms (for example, "cordate" or "ovate" for "heart-shaped" or "egg-shaped") and the terms are frequently used ones.

In the species descriptions, measurements are given in metric figures. A centimeter (cm) is equal to 2.54 inches. A millimeter (mm) is one-tenth of a cm. A decimeter (dm) is 10 cm, or one-tenth of a meter (m). A meter is equivalent to 39.97 inches, or slightly longer than a yard. In most cases a size range ("4–5 cm wide," "1–2 mm long," etc.) is given. This range indicates the size of most of the plants or plant parts a person may encounter but is not intended to be all-inclusive. Therefore, one occasionally may find a plant with measurements slightly to either side of the range given. Figures such as "(5–) 10–20 cm long" or "2–4 (–7) mm wide" indicate the occurrence of and approximate magnitude of departures from the most common size range. All measurements have been taken from Kansas specimens except in a few cases where adequate specimens were not available.

Professor Stevens was keenly aware of the fact that there is much more to be learned about wildflowers than mere names, and he brought to his readers information describing a multitude of fascinating relationships that exist among plants and between plants and other organisms. Now, with warnings of starvation and pollution rising on all sides, we are beginning to realize, as at no time in the past, the need to understand more about all aspects of our environment. As much as possible, therefore, throughout this revision I have tried to maintain Stevens's approach to botany; and I have incorporated much of his material in this book (by permission of the Regents Press of Kansas, formerly the University of Kansas Press, and of the W. C. Stevens family). For this reason there are inserted, from time to time, references to methods of pollination, use of plants by wild animals, etc., as well as the expanded sections on physiographic regions and vegetation, edible plants, poisonous plants, and weeds.

I have, however, omitted most recommendations for transplanting native species into the garden—the time has come for us to learn to appreciate our plants in their own habitats. Undisturbed woodlands and prairies are ever more difficult to find as cities enlarge and man's effects upon the environment are seen in increasingly larger areas. Some of our loveliest species are exceedingly rare and should under no circumstances be picked or dug up—among these are the yellow lady's slipper orchid and the Michigan lily. In addition to their rareness, some of these species have rather narrow ranges of tolerance for habitat conditions—conditions which are seldom duplicated in the yard—and transplants may only die. Thus the would-be wildflower gardener may have depleted the natural population by one or more plants but still has no lady's slipper in his own garden. One acceptable exception to the "do not pick" mandate is the case in which an area known to be inhabited by these species is facing imminent destruction by the bulldozer, the plow, or the cow. A happier alternative, though, would be to convince builders, urban planners, farmers, and ranchers to leave some of these areas as natural preserves to be enjoyed by present and future generations. It is possible to purchase plants or seeds of a surprisingly large number of our wildflowers from floricultural supply houses, so that one may still obtain some of our rare beauties which will grow as ornamentals without disturbing the native plant populations. On the other hand, not all of our loveliest species are rare and difficult to grow, and there are still included occasional references to various uses for plants considered to be on the "safe to pick" list.

Thanks are due to many people for assistance during the preparation of this revision. The work was supported in large part by a grant from the William Chase Stevens Fund, a fund composed of memorial gifts from friends and associates of the late Professor Stevens, and administered through the University of Kansas Endowment Association. The revision would have been impossible without the use of the University of Kansas Herbarium and Library facilities, including the Interlibrary Services Department which obtained numerous references for me. The New York Botanical Garden graciously gave permission to adapt some of the keys from *New Britton and Brown Illustrated Flora of the Northeastern United States and Adjacent Canada*. Marshall C. Johnston, Donavan S. Correll, and the University of Texas at Dallas kindly gave permission to adapt material from the key to composite genera in the *Manual of the Vascular Plants of Texas*. In some instances, keys were adapted from published journal articles by other authors. Although their keys were seldom if ever quoted exactly, references are always given to the work which served as the basis for the key in this book. Earl Robinson prepared the refreshing line drawings of leaf characters and flower structure, and Doug Carlson printed most of the black-and-white photographs, sometimes rendering miracles from negatives that left much to be desired. Debbie Baker Treanor, Teresa Snyder, and Christine King provided careful, conscientious assistance with plant measurements, distribution mapping, library research, and proofreading. Sharon Florez typed most of the manuscript and provided invaluable assistance with editing. Family members patiently gave much of their time to help with household chores and childcare and, along with other friends, offered much encouragement on days when it seemed the manuscript would never be finished.

Introduction

THE PHYSICAL GEOGRAPHY OF KANSAS

The term physical geography, or physiography, refers to the study of the present structure and phenomena of the earth's surface, including land forms, climate, and the distribution of plants and animals. Physiographic maps, as in fig. 1, are used to illustrate

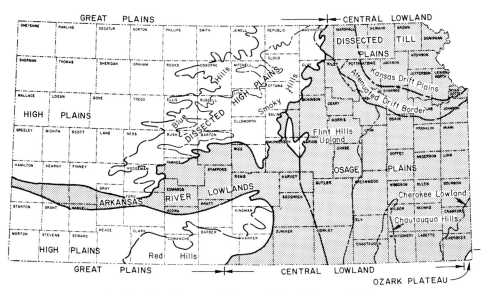

1. Physiographic map of Kansas (after Schoewe, 1949)

these features. Kansas is essentially an undulating plain which slopes gently from west to east at the rate of about 10 to 15 feet per mile. Elevation along the western border of the state is between 3,500 and 3,900 feet above sea level. This decreases gradually to about 750 feet along the southeastern border. The highest point in the state is Sunflower Mountain, located between Goose Creek and Willow Creek in Wallace County, with an elevation of 4,135 feet. The lowest point, slightly less than 700 feet, is about 3 miles south of Coffeyville in Montgomery County, where the Verdigris River crosses the border into Oklahoma. Local relief is usually less than 100 feet, although in some places it reaches 400 feet along major stream courses (1).*

Climatic patterns (fig. 2) also follow an east-west or southeast-northwest trend. The growing season in Chautauqua County is about 200 days long, while in Cheyenne

---

* A number in parentheses refers to a source listed in Literature Cited.

County it is less than 160 days. Average annual temperatures range from 58° F. in the southeast to 52° in the northwest, while average minimum temperatures range from 5° in the southeast to —20° in the northwest. Average annual precipitation varies from 42 inches in the southeast to 16 inches in the extreme west (2). In addition to receiving less rainfall, the western counties also experience higher temperatures in the summer, lower humidity, and more wind so that a large percentage of the precipitation is lost to evaporation.

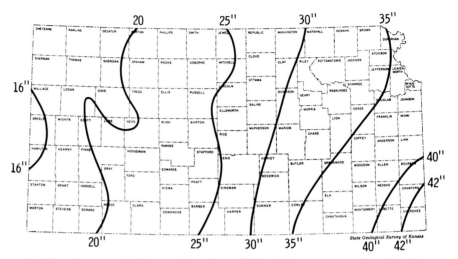

2. Climatic patterns in Kansas (after Flora, 1948): (a) Average annual precipitation

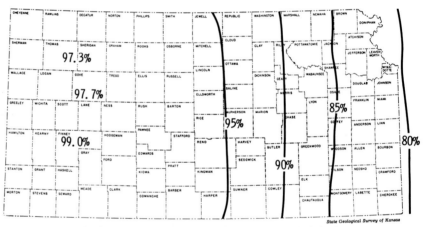

2. (b) Average annual water loss in percent of precipitation

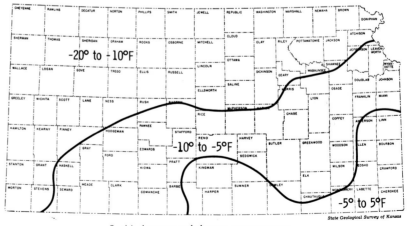

2. (c) Average minimum temperature

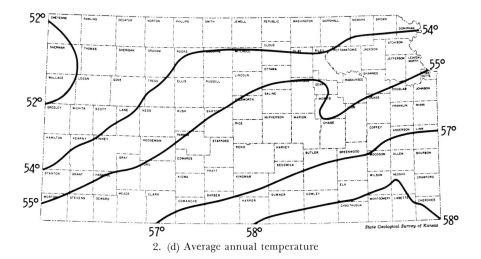

2. (d) Average annual temperature

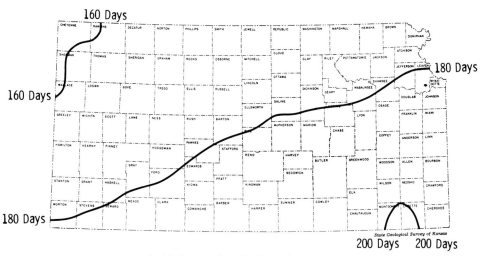

2. (e) Average length of growing season

A further significant characteristic of the Kansas climate is its variability from year to year and even from day to day. Plants must be able to tolerate a lack of dependability in the precipitation regime, as well as relatively low amounts of moisture. For plants, climatic extremes may be more important overall than the averages. Regardless of the average spring temperature or average annual temperature, for example, a single hard freeze while a plant is in flower may be enough to kill the plant or at least prevent production of mature fruits and seeds.

The surface features and vertical geologic structure of the earth's crust are the result of physical and chemical processes which have acted through an inconceivably long period of time. The continual erosive action of wind and water, combined with chemical oxidation and the alternate expansion and contraction of water freezing and thawing in cracks and crevices, gradually fragmented and pulverized the surface of the primal, pre-Cambrian rocks. These sediments eventually came to rest on the bottom of lakes and seas and provided the basic materials for sandstones, mudstones, siltstones, and similar rock types. After living organisms appeared in the waters an estimated 3.5 billion years ago, their shells, incrustations, and skeletons provided the substance for the limestone strata of subsequent periods.

The various geological strata are classified into systems according to their age. Beginning with the lowest and oldest stratum deposited on the pre-Cambrian basement rocks in Kansas, these are called Cambrian-Ordovician, Mississippian, Pennsylvanian, Permian, Cretaceous, and Tertiary. All of the formations beneath the Tertiary comprise alternate layers of shale, limestone, and sandstone. The deposition of these layers has not been continuous and uniform but has been interrupted and modified many times by periods of inundation of the land by the sea followed by periods of emergence of the land, each period continuing some millions of years. While this was going on, the replacement here and there of shallow-water forms by freshwater forms, and vice versa, and changes in climate, especially respecting temperature and humidity, profoundly affected the character of the strata. The Tertiary blankets the other strata in the western part of the state, but each of the others angles upward toward the east to outcrop over a considerable part of the state.

The history of our Tertiary system is a different story. At the close of the Cretaceous, a chain of mountains was uplifted by a compressive stress in the crust along the Rocky Mountain belt which initiated drainage eastward into Kansas. Through a long period of time, amounting to millions of years, as these mountains were being leveled by erosion, vast quantities of stream-borne gravel, sand and clay, and windborne dust and volcanic ash were spread over a large area of the Cretaceous shale, limestone, and chalk of western Kansas. Following the wearing down of this first mountain chain, there was an upthrust of the present Rockies that gave more volume and power to the eastward-flowing streams, thus greatly accelerating the distribution of sand, gravel, and silt which, together with wind-borne dust and sand, constitute the main body of our Tertiary strata.

The end of the Tertiary and the beginning of Quaternary time are placed in geological chronology at about a million years ago, when continental ice sheets originating in accumulations of snow over the High Plains west of Hudson Bay spread southward into Kansas over an area extending west to the Big Blue River and south a few miles beyond the Kansas River. When the ice finally melted, it left scattered deposits of intermingled clay, sand, gravel, and boulders which had been picked up and carried along by the glaciers as they scoured the surface of the earth. While the ice was in northeast Kansas, streams and wind were operating in the rest of the state; sand and gravel were being deposited, and extensive deposits of wind-blown soil and volcanic ash, the latter blown in from the Rockies, were laid down over a large part of the state.

The progress of the Quaternary continues before our eyes today, marked by the wearing down of the land and the deposition of part of the eroded materials in river valleys, while the remainder is carried away to lakes and seas. Winds lift clouds of dust from plowed fields and other land bare of vegetation, from the flood plains of rivers, in some places along the Arkansas River scooping up sand and depositing it in conspicuous mounds and ridges called dunes.

As you see in fig. 1, Kansas is divided into several physiographic regions. Brief descriptions of the distinguishing features of each region are given below. For more detailed information about the geology and vegetation of these units, refer to Schoewe (1), Merriam (71), and Küchler (27).

OZARK PLATEAU:

Within this small area in the southeastern corner of the state we find the oldest outcropping rocks in Kansas—cherty Mississippian limestones about 200 million years old, which contain valuable lead and zinc ores. The vegetation consists primarily of oak-hickory forest and includes many Ozarkian and southeastern species which are not found elsewhere in the state.

OSAGE PLAINS:

This region contains 4 subdivisions:

In the *Cherokee Lowlands*, the surface is an erosional plain which slopes westward at a rate of about 10 feet per mile. The total relief within the region is about 250 feet. Outcropping rocks are primarily weak shales and sandstones which are Pennsylvanian in age. The soil is mostly residual from the weathering of shale and is low in lime, phosphorus, nitrogen, and organic matter. Large areas are poorly drained, and hardpan occurs from 6 to 18 inches below the surface.

Hardpan is an accumulation of colloidal clay and of precipitated soluble carbonates washed down from the upper soil layers and deposited toward the lower limit of water penetration. The longer the process has gone on (that is, the older the geological formation), the thicker and denser is the hardpan and the more difficult it is for the roots of plants to go through it.

Farther to the north, the *Osage Cuestas* are evident as a series of northeast-southwest trending, east-facing, benchlike outcrops between which one finds flat to gently rolling plains. A large part of the soil is residual from the weathering of Pennsylvanian and Permian limestone and shale. Cement and building stone are produced here.

In the south, the *Chautauqua Hills* are distinguishable as a narrow band of hills which extend north to Woodson County. Surface rocks are chiefly thick sandstones, Pennsylvanian in age, which have weathered to produce a soil low in lime, phosphorous, and nitrogen. Local relief is nowhere greater than 250 feet. The vegetation (see color plate *1* facing page 36) is typically oak-bluestem cross timbers—the low hills mostly covered with stands of post oak and black jack oak, while lower areas and woodland openings are more suitable for bluestem prairie grasses.

To the west, the *Flint Hills Upland* (many years ago referred to as the "Kansas Mountains") extends north-south across the state and west to Arkansas City, Salina, and Clay Center and offers the greatest local relief in Kansas. The dominant feature is the range of hills with east-facing escarpments of hard Permian limestone interbedded with layers of shale and chert or flint (plates 2 and 3). Westward the Flint Hills slope gently toward the Arkansas River valley. Between the ridges the land is for the most part gently rolling, and the soils are mostly residual from the weathering of the limestones. The valleys are fertile and suited to general agriculture. Some of our more important oil and gas fields are located within this region, and the flint or chert was used by Indians as a source of materials for arrowheads, spears, knives, and other articles. The tall-grass or bluestem prairie which dominates the natural vegetational cover provides one of the finest grazing areas in the United States. The most abundant grasses here are big bluestem, little bluestem, Indian grass, and switchgrass. The woodlands, especially westward, are restricted mostly to ravines, river bottoms, and other lowlands, and the gently rolling uplands are generally devoid of trees (plate 4).

DISSECTED TILL PLAINS:

This region consists of 2 subdivisions:

The *Kansas Drift Plains* lie north of the Kansas River and are actually a northward extension of the Osage Plains but differ from the latter in having been glaciated. The cuesta topography and rock strata which prevail to the south are here overlain by a covering of glacial drift and till. Winds sweeping across this deposit spread a mantle of dust over the northeastern counties and built up the bluffs bordering the Missouri River. Soils laid down by the wind are called *loess* (from German *loesen*, to loosen). Hardpan is seldom present in these relatively young soils, and where it is apparent it is still not dense enough to keep roots from growing through it. The surface here is gently undulating except along the Missouri River bluffs, where the hills rise more abruptly. The sides of these bluffs are mostly covered with oak-hickory woodlands, but the tops of the hills often sport isolated patches of prairie, including some species such as *Lygodesmia juncea*, that are in general more western in distribution (plate 5). These western species, some of which also occur on the bluffs across the river in Missouri, are thought to be relicts of a warmer, drier period during the Pleistocene when the eastern woodlands had retreated eastward and had been replaced temporarily by more drought-tolerant prairie species.

The *Attenuated Drift Border* is a transition zone between the Kansas Drift Plains and Osage Cuestas. This region was glaciated and is covered in places with patches of glacial till and outwash, but most of the area has an erosional bedrock type of topography.

The native vegetation through the Dissected Till Plains, the Osage Cuestas, and the Cherokee Lowland is predominantly oak-hickory forest intermingled with regions of tall-grass prairie. Along the rivers and streams occur flood-plain forests dominated by trees such as the hackberry, cottonwood, American elm, and willow.

ARKANSAS RIVER LOWLAND:

This is a low-lying area which for the most part follows the valley of the Arkansas River. The surface varies from quite flat to gently rolling, with sand dunes and sand hills bordering the river or occurring sporadically at some distance from it (plate 6). The soils are chiefly loam and sandy loam from fine outwash and wind deposits of the Tertiary and Quaternary formations, but residual soils from the weathering of Permian limestone and shale predominate in the southeast section. As a whole the soils are fertile, adapted to general farming, and especially productive of winter wheat. The native vegetation is typically sand prairie or sandsage prairie. The characteristic plants are big bluestem, little bluestem, switchgrass, and sandreed, with sandhill sage becoming more abundant westward.

DISSECTED HIGH PLAINS:

This region contains 3 subdivisions.

The *Smoky Hills Upland*, which extends from the north border of the state to the Arkansas River Lowlands, is a maturely dissected, hilly region carved essentially in Dakota sandstone of Cretaceous age (plates 7 and 8). Numerous mounds and buttes account for local relief of 200 to 300 feet. Between these protruding features the surface is gently rolling. The soil is residual, principally from the weathering of the sandstone. With regard to the vegetation, this is a transition zone between the tall-grass, bluestem prairie of the east and the mixed grass or bluestem-grama prairie which prevails farther westward.

Immediately to the west of the Smoky Hills lies the *Blue Hills Upland*. The surfacing rocks here are chiefly Cretaceous also, but instead of sandstone are limestones, shales, and chalk. Relief in some places is as much as 200–300 feet per mile. Where the surface is eroded into soft shales, the land is rolling (plate *9*). Where the more resistant limestones are exposed, we find flat, plateaulike uplands with steep, rocky escarpments (plate *10*). Oil and gas are important products of this region. The dominant prairie grasses here are big bluestem, little bluestem, sideoats grama, and blue grama.

The *Red Hills Upland*, sometimes called the Permian Redbeds or the Cimarron Breaks, lies south of the Arkansas River. Tablelike plateaus, buttes, pyramids, and pinnacles carved from the red Permian shales, siltstones, and sandstones and contrasting white gypsum beds make this one of the most scenic parts of the state (plates *11* and *12*). Local relief may be as much as 300 feet, and total topographic relief is about 500 feet. In the heart of this region, bluestems and grama grasses share the prairie landscape with red cedars, our only native conifer. In Barber and Comanche counties gypsum is mined for use in plaster of paris, plasterboard, and other building materials.

HIGH PLAINS:

Westward from the Dissected High Plains stretch the High Plains, generally treeless and featureless except where traversed by rivers such as the Smoky Hill, Arkansas, or Cimarron (plate *13*). Over most of the region the land rises gradually westward, about 10 to 15 feet per mile. Across a large part of this area a thick deposit of wind-borne silt has been laid down over inwash sand, gravel, and clay from Colorado. Some of the streams have excavated their valleys through the Tertiary and deep into the chalk and chalky shale of the Cretaceous. This is strikingly exhibited along the valley of the Smoky Hill River through Trego, Gove, and Logan counties (plates *14* and *15*). South of the Smoky Hill River and its tributaries the surface is relatively unbroken as far as the Arkansas River, where there are bluffs north of the river and sand hills south of it from the Colorado line eastward into Ford County. Beyond the river to the south border, the soil is inwash material from Colorado but is mixed with relatively infertile plains marl (a composition of clay and lime) locally where stream erosion has reduced the inwash cover. A unique feature of the High Plains is the broad bed of an ancient lake between the towns Meade and Fowler in Meade County, lying below the general level and now in wet weather the site of many ponds. Vegetation in the High Plains changes from bluestem-grama prairie in the east to short-grass prairie dominated by buffalo grass and blue grama in the west. South of the Cimarron River extensive zones of sandsage prairie occur. In Gove, Logan, and Trego counties, where chalky Cretaceous formations outcrop and seleniferous soils occur, one finds chalk flat prairie (plate *16*). Here little bluestem, sideoats grama, and saltgrass become more abundant, along with seleniphiles or selenium-tolerant plants such as prince's plume, ten-petal mentzelia, wild buckwheat, and broomweed.

STRUCTURE OF THE FLOWERING PLANT

One of the most fascinating things about plants is the infinite variety in form and behavior that exists among them. The following section is designed to give the reader a general idea of how a so-called typical plant is put together and how it functions. An effort has been made to eliminate detailed descriptions and to leave only the essential basics. Although I have included some descriptions of different kinds of stems, leaf shapes, etc., it is my hope that the main body of the book and, most especially, the plants that you observe will serve as the best examples of the above-mentioned variation. The glossary should be referred to for assistance in clarifying new terms encountered in the plant descriptions and keys.

FLOWER STRUCTURE:

Flowering plants, known to botanists as Angiosperms (from Gr. *angeion*, vessel, + *sperma*, seed, referring to the location of seeds within an ovary), have their sexual reproductive parts located in structures called flowers. To be a true flower, in the botanical sense, this structure need consist of a single male part (stamen) or a single female part (pistil) only. Most flowers, however, are more complex than this, consisting of 4 concentric series or whorls of parts—sepals, petals, stamens, and pistils, arranged in that order from the outside toward the center (fig. 3). Sepals are often green and leaflike but are sometimes colored and showy, distinguishable from the petals only by their location. Collectively, the sepals are referred to as the calyx. The petals are usually white or colored, sometimes greenish, and often function to attract pollinators. The petals as a group

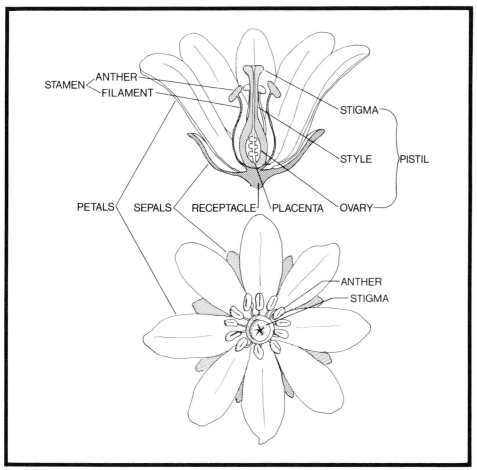

3. Flower structure

comprise the corolla. The calyx and corolla together are referred to as the perianth. A stamen consists of a stalk called the filament, atop which sits the pollen-containing portion, the anther. The stamens as a group make up the androecium. Pistils vary in shape but are frequently flask-shaped, with a swollen, ovule-containing lower portion called the ovary; a pollen-receptive stigma; and a style connecting the stigma to the ovary. A flower's set of pistils is called the gynoecium. When the sepals, petals, and stamens depart at the base of the ovary, the latter is said to be superior; when those parts depart at the top of the ovary, the ovary is said to be inferior. Sometimes the bases of the sepals, petals, and stamens are united with one another to form a cuplike structure, called a hypanthium, which surrounds a superior ovary but is not fused to it. Within the ovary are from 1 to many ovules which will eventually mature as seeds. As the seeds mature, the surrounding ovary also enlarges, and at maturity this ripened ovary containing the seeds within it is known as the fruit. Fruits are of varying sorts—some are fleshy (peach, apple, grape); some are dry (bean pod, walnut, poppy capsule). Some are dehiscent, shedding their seeds by one method or another; others are indehiscent, requiring weathering or rotting away of the fruit before the seeds can be released. Both seeds and/or fruits may come equipped with wings, plumes, hooks, or other ornamentation to aid in dispersal. Seeds of fleshy-fruited plants are frequently dispersed by birds or other animals who eat the fruits but discard the seeds.

VEGETATIVE STRUCTURE:

Most flowering plants have a vegetative body consisting of 3 parts, namely, roots, stems, and leaves (fig. 4). The roots are almost always below the ground. The stem may be of several sorts, growing either aboveground or belowground, while the leaves are always aboveground.

Roots:

Roots have 2 main functions—to anchor the plant to the ground and to absorb

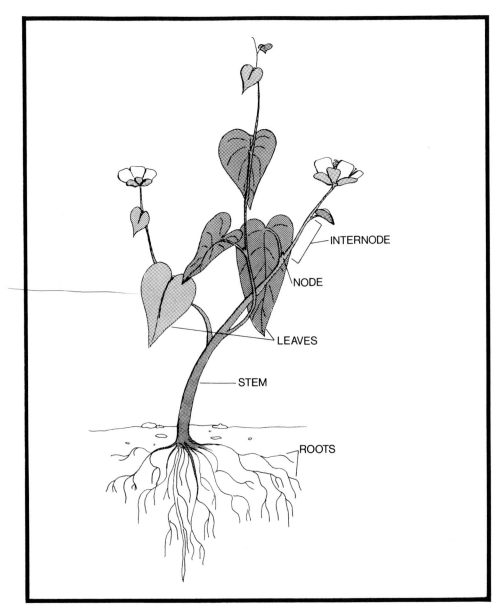

INTERNODE

NODE

LEAVES

STEM

ROOTS

4. The whole plant

water and mineral nutrients from the soil. In general, there are 2 main types of roots—fibrous roots and taproots. Fibrous roots are those such as we find in the grasses—the main roots are rather fine and stringlike, and both the main roots and the lateral roots are branched again and again. Roots of this sort are very important for holding soil particles in place, which helps to control erosion by wind and water. Taproots are those such as we find in the carrot or radish—there is 1 more-or-less enlarged main root with relatively few lateral roots coming from it. Some roots have enlarged portions for storage of food. Others, as in the case of the dandelion or the fawn lily, are contractile, serving to pull part of the plant down into the ground. A few plants, such as broomrape, parasitize other species with their roots.

Stems:

Stems function to display the leaves, flowers, and fruits, and to carry water and minerals upward from the roots and manufactured foods downward from the green parts of the plant. Points along the stem at which leaves are attached are called nodes. The section of a stem between 2 adjacent nodes is an internode. As we have already noted, stems may grow either aboveground or belowground, depending on the particular plant, and they may grow either vertically or horizontally. Examples of horizontal, aboveground

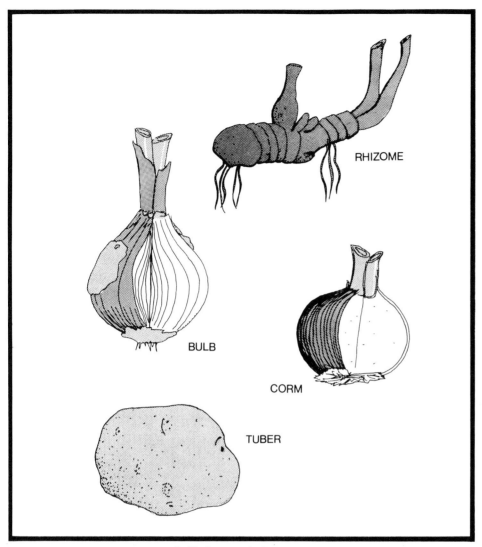

5. Underground plant parts

stems include stolons and runners. Stolons are slender, naked stems which arch across the ground and produce a new plant where they touch the ground. Runners, in contrast, creep along the surface and send out roots from the nodes. A rhizome (fig. 5) is a horizontal, underground stem. It may be distinguished from a root by the presence of nodes, buds, or scalelike leaves, all of which would be absent on a root. A corm is a short, thickened, erect, underground stem in which food is stored. Both rhizomes and corms are frequently woody. Bulbs resemble corms superficially but actually consist of a very small region of stem tissue to which are attached a number of fleshy storage scales, as in an onion. A tuber, as for example an Irish potato, is also a type of underground stem which consists largely of fleshy storage tissue. The "eyes" are actually nodes with buds. Underground stems such as those described above are usually perennial. They may produce leaves directly, or they may produce herbaceous, aboveground stems with leaves borne thereon, again depending upon the plant in question. In either case, the herbaceous parts die back at the end of each growing season.

Leaves:

Leaves function primarily in the manufacture of food for the plant. A "typical" leaf consists of 3 main parts—a broadened lamella or blade; a small, stemlike structure, called a petiole, by means of which the leaf is attached to the rest of the plant; and a pair of leaflike structures, called stipules, situated at the base of the petiole. In actuality, not all plants have petioles and stipules. Those leaves lacking petioles are said to be sessile. Stipules may be green and leaflike, or they may be modified into twining tendrils to aid in climbing or into spines or glands. Plants lacking stipules altogether are said to be estipulate.

Leaves may occur singly at a node (alternate arrangement), 2 at a node (opposite arrangement), or 3 or more at a node (whorled arrangement).

Leaves may be simple, having only a single blade, or they may be compound, having the blade subdivided into 3 or more smaller leaflets (fig. 6). Compound leaves which have all their leaflets attached to one point at the end of the petiole are palmately compound. Those with their leaflets attached in an alternate or opposite fashion along a rachis, like the pinnae of a feather, are pinnately compound. To determine whether a leaf is simple or compound, note the location of buds; buds are always situated in the axil of a leaf (i.e., on the stem, immediately above the point of leaf attachment), *never* in the axil of a leaflet.

The veins of leaves are usually arranged in 1 of 2 manners, either parallel, with the main veins running alongside one another from the base to the tip of the leaf, or reticulately, with the veins branching in many directions and forming a netlike pattern. Reticulate venation may be of either a pinnate or palmate type (fig. 7). To see the venation easily, hold a leaf toward the light.

Leaf characters—especially shape and the nature of the margin, base, and apex—are frequently used in plant identification keys or descriptions, especially at the species

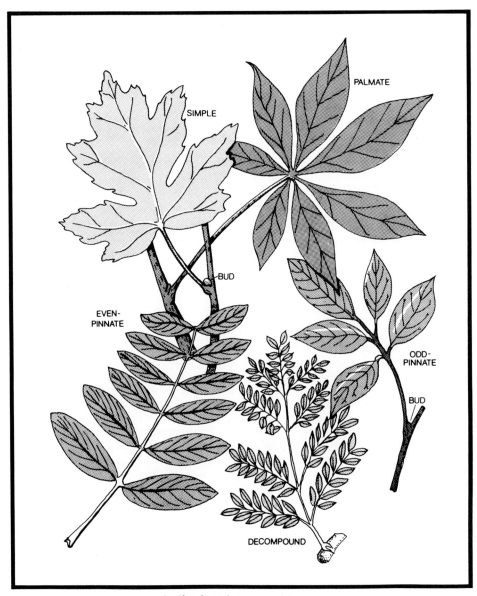

6. Simple and compound leaves

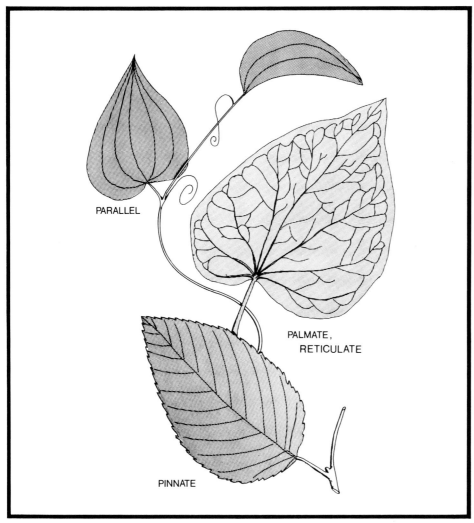

PARALLEL

PALMATE,
RETICULATE

PINNATE

7. Leaf venation types

level. The reader should refer to figs. 6–11 in conjunction with the glossary to become familiar with these terms.

PLANT NAMES

The first question we ask about an unfamiliar plant is "What is its name?" Until we know its name, we cannot look it up in the literature of plants nor conveniently impart observations about it to others. There are 2 kinds of plant names: the scientific name, which is the same in all countries; and the common name, which varies with the language. For instance, "strawberry" is *Erdbeere* in German and *frasier* in French, but in all countries of the world it has the scientific name *Fragaria*. The common name, moreover, may vary with the locality within a country, so that in one language there may be several common names for the same species; and the same common name is sometimes applied to more than 1 species that are quite different from one another. *Acer saccharinum* has 3 common names in general use: silver maple, white maple, and soft maple. *Verbascum thapsus* has approximately 140 common names. Furthermore, some plants have no common names at all.

The scientific name of a plant consists of 2 Latin or Latinized words—the first term being the name of the genus to which the plant belongs and the second term being the specific epithet. All those plants that look as if they might have sprung from the same immediate parent are classed as the same species, and those species that have the same basic characters of flowers and fruits but differ among themselves so much that we would not think it at all likely or even possible for them to have sprung from the same immediate parent are classed as belonging to the same genus. Thus the red clover (fig. 260)

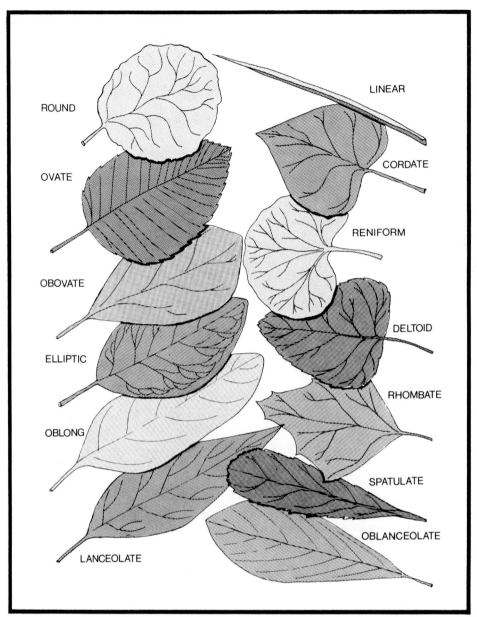

8. Leaf shapes

and the white clover (fig. 262) belong to the same genus (*Trifolium*) but to different species (*T. pratense* and *T. repens*, respectively). The genus name may be used alone when referring to a group of species, but the specific epithet is never used without the generic name before it. As in the parenthetical example above, the generic name is sometimes given merely as an abbreviation consisting of the capital letter, followed by a period, if the context makes its meaning clear.

The person (or persons) who originally published an account of and described each plant or group of plants, whether family, genus, or species, and gave it a scientific name is known as the authority of that name. The author's name may be written out, but it is usually indicated by a standardized abbreviation. For example, our common cattail was first described by Carolus Linnaeus, so the name is written *Typha latifolia* L. When the rank of a plant or plant group is changed, or when a species is transferred from one genus to another, or whenever similar changes in nomenclature are made, the name of the original author is placed in parentheses and is followed by the name of the person making the change. *Cyanotris scilloides* was first described by Rafinesque but was later transferred to the genus *Camassia* by V. L. Cory, so the name is now written *Camassia scilloides* (Raf.) Cory. Appendix 2 of this book lists full names of the authorities included here, plus brief biographical information about these men.

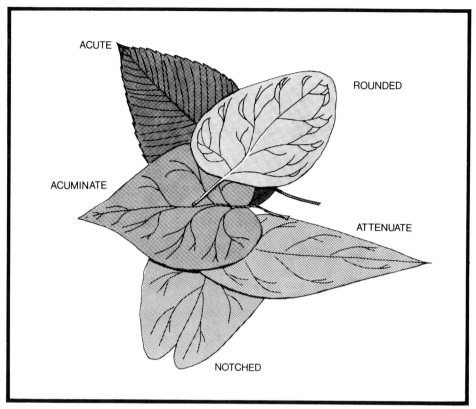

9. Leaf apexes

In this book we have tried to make the scientific name more useable and attractive by explaining its meaning and indicating its pronunciation. Accent marks are used to serve two functions—they indicate both the syllable or syllables to be stressed and the sounds of the vowels in those syllables. A grave accent mark (`) signifies the long English sound; the acute (´), the short English sound. Further aids to pronounciation are given in Appendix 1.

PLANTS AS NOTEWORTHY ACQUAINTANCES

An extra dividend that comes from familiarity with wild plant species is an open door to a host of new and interesting foods that cost only the effort of gathering and preparing them—a price that is sometimes cheap, sometimes quite dear! Wherever appropriate, the species descriptions in this book include information on edible parts and methods of preparation, and season when available for gathering. For further, more detailed discussion of edible wild plants, there are a number of popular books on the market. I especially recommend the following 4: *Edible Wild Plants of Eastern North America* by M. L. Fernald and A. C. Kinsey, revised by R. C. Rollins (1958, Harper & Row, Publishers, Inc., 452 pp., cloth); *Stalking the Wild Asparagus,* by Euell Gibbons (1962, David McKay Company, Inc., 303 pp., paperback); *Stalking the Healthful Herbs,* also by Euell Gibbons (1966, David McKay Company, Inc., 303 pp., paperback); and *Trees, Shrubs, and Woody Vines of Kansas,* by H. A. "Steve" Stephens (1969, University Press of Kansas, 250 pp., cloth and paperback). Stephens's book is especially useful in Kansas since I do not treat woody plants here.

For those wishing to sample wild foods, I include the following rules of thumb:

(1) *Be positive of identification.* Some edible plants have look-alike relatives which are deadly poisonous, so be certain you have the right plant. Familiarize yourself with the poisonous plants in this book (see index entries under "poisonous plants") so you know which plants to avoid. Also note that parts of some plants—May apple, for example—are edible while other parts are poisonous.

(2) *Do not collect in areas which have been sprayed with herbicides or insecticides.* Since roadside spraying has replaced or is used in addition to mowing in some county maintenance programs, it is wise to gather plant materials some distance back from the

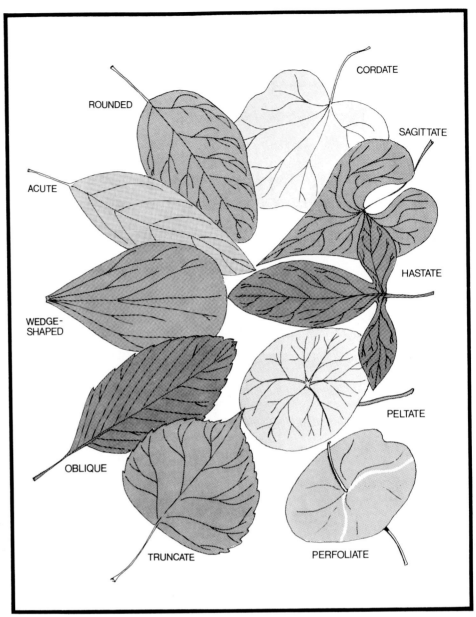

10. Leaf bases

road unless you are positive the ditches have not been sprayed. If large numbers of plants along the road are a pale green or yellow and exhibit bizarre growth forms (leaves curled, stems twisted, etc.), a herbicide such as 2-,4-D has probably been used.

(3) *Be aware of your own allergies and use good judgment.* For example, if you are allergic to broccoli and cauliflower, members of the Mustard Family, you may wish to avoid the wild members of that family as well.

(4) *Don't plan to lead a life of luxury by living off the land.* It is easy to plan a menu but more difficult (in Kansas, at least) to find many different kinds of edible plants available in quantity within a relatively small area at the same season. You may have to preserve or freeze fruits in the summer, harvest nuts and seeds in the fall, and serve your feast in the spring when fresh greens are available. Of course, you can always take advantage of whatever wild foods are in season to supplement your regular table fare.

Trouble can be avoided, whether or not you are preparing a native feast, by learning to recognize our poisonous plants. There are in Kansas over 100 species of plants, not including cultivars, which have been known to cause or strongly suspected of causing poisoning of one degree or another in livestock or humans. I would call your attention especially to poison ivy, poison hemlock, water hemlock, death camass, and annual broomweed. Our species of poison ivy (*Toxicodendron radicans* and *T. rydbergii*)

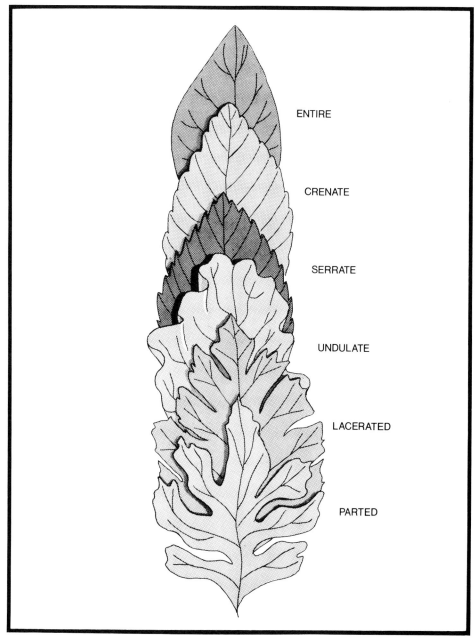

ENTIRE

CRENATE

SERRATE

UNDULATE

LACERATED

PARTED

11. Leaf margins

are woody plants and so are not included in my keys and descriptions. The plants are quite variable in habit, occurring as small subshrubs, shrubs, or woody vines which may so completely entwine a tree as to appear to be the tree itself. The leaves are 3-foliate with the leaflets ovate-lanceolate, elliptic, rhombic, or obovate, up to 20 cm long and 13 cm wide, with margins entire, toothed, or lobed. The variation in habit and leaflet shape has led to confusion among professional botanists and the general public alike regarding the number of species and/or varieties that occur. The plants produce clusters of small, greenish or white flowers in May and June and small, waxy, white or greenish berries thereafter. All parts of the plant are poisonous at all seasons of the year. Eating a leaf will *not* confer immunity but may cause serious irritation of the alimentary canal. It is possible to transfer the poison, a yellowish oil, indirectly from the plant to your person via a pet that has come in contact with the plant or from yard tools used to remove poison ivy from an undesirable location.

Should you have reason to fear plant poisoning, take the victim and an *identifiable* specimen of the suspected plant to a doctor as soon as possible—a single leaf or blossom may not be adequate. If you already know or can determine the identity of the plant in question, volunteer this information to the doctor.

Some of our poisonous plants, as well as nonpoisonous ones, have redeeming social value in that they are the source of useful drugs. Those currently of considerable interest in the medical world include *Echinacea pallida*, *Podophyllum peltatum*, and *Thalictrum dasycarpum*, to mention a few. A good many other Kansas herbs have been used by someone at some time for treating some ailment, but most are not commercially important at this time.

Included among our Kansas plants are a number which provide attractive dyes. Information regarding possible colors and suggested mordants has come in some cases from my own experience and in others from the literature. The exact color one obtains from any given plant will depend on many variables, including the specific fiber you are dying, the mordant and proportions of ingredients, and the season in which the plants were collected. While this may be cause for frustration when one is attempting to reproduce a favorite color, it adds an intriguing atmosphere of discovery to each dye session!

Finally, a few comments regarding those disreputable fellows called weeds! What is a "weed"? The definition heard most frequently is simply "a plant out of place"—a lamb's quarter in your flower bed or a dandelion in your lawn. The proper place for any given plant, however, is a matter of opinion. A more correct definition deals with the ecology of this group of plants: weeds are generally unwanted plants which thrive in habitats disturbed by man and his livestock. Frequently we find that true weeds, like crop plants, are foreigners, species not native to Kansas, perhaps not even to North America. Indeed, many are closely related to important cultivars and perhaps evolved side by side with their related crop plants as early farmers were selecting and raising desirable strains of grains or vegetables. In some instances there seems to have been an exchange of genetic material between the crop species and its companion weed species. Although the term "weed" often connotes ugliness and worthlessness, many weedy species are quite attractive and/or edible or otherwise useful.

COLLECTING AND IDENTIFYING PLANTS

For collecting plants in the field, a minimum set of equipment includes a digging implement such as a sturdy gardening trowel, a broad-bladed hunting knife, or a botanical "digger"; a cutting implement such as a pocket knife, side cutters, or small pruning shears; and some sort of container to prevent immediate desiccation of the specimens. Plastic bags are inexpensive and have almost superseded the use of a vasculum (essentially a metal box with a handle) for the last purpose. Sprinkle several teaspoons of water in the bag or add a small, moist rag to retard wilting. Although not essential in the field, a small 10X hand lens is sometimes handy.

For purposes of identification, it is usually necessary to have more of the plant than just a blossom. Frequently one is also required to know something of the fruits and of the vegetative parts—stems, leaves, and underground portions—as well. Keep this in mind when collecting your specimens.

If you do not plan to identify your plants within several hours, or if you plan to save your specimens, they should be pressed and dried. There are numerous makeshift techniques for doing this—using a dictionary or a mail-order catalogue, for example— but faster drying and more attractive specimens can be achieved by using a plant press (fig. 12). This consists of a pair of wooden frames and trunk straps, sheets of folded newsprint, absorbent blotters, and pieces of corrugated cardboard, all cut to the same size, 12″ by 18″ being standard, and with the corrugations of the cardboard running parallel to the shorter axis of the press. The wooden frames are constructed of slats of oak, ash, or hickory riveted together with copper or aluminum rivets. A press may be purchased from a biological supply house or made at home. A plant specimen should be placed between the folds of a single sheet of newsprint, carefully smoothing out the leaves and without overlapping plant parts if possible. Sandwich this combination between several blotters. Each set of 1 plant in its newsprint and surrounding blotters should be separated from the adjacent sets by corrugated cardboard. The blotters serve to draw moisture from the specimens, while the cardboard facilitates air movement through the press and speeds up the drying process. The entire stack of plants and drying materials should be strapped together firmly between the wooden frames. Weather permitting, the press and its contents may be left in the sun to dry. In rainy or humid weather, you may wish to place the press in front of a fan to speed drying, being certain to align the corrugations in the cardboard with the air flow. Various types of plant driers are used by professional botanists, one of the simplest being a rack which supports the

12. Plant press with cardboard ventilators, absorbent blotters, and folded newsprint for drying plant specimens

press or presses above a bank of hot light bulbs. Air heated by the light bulbs rises through and around the presses and speeds up the drying process somewhat. A much more satisfactory design involves the use of an electrical heating element, a small fan, and a cloth skirt to funnel all the warm air directly through the press (fig. 13).

If you plan to save your specimens, it is helpful to maintain a field notebook. Although the format of the notebook is a matter of individual preference, there are cer-

13. Portable electric plant drier with plant press

tain kinds of information which should always be included. Each specimen you collect should be assigned a collection number (exception: if you collect duplicates of the same plant at the same time at the same place, all receive the same number). This number should be recorded in the notebook *and* written on the sheet of newsprint containing that particular plant. For each specimen you should also record the date and site of collection (mileage and direction from the nearest town in the same county), habitat and soil type (dry, sandy prairie, moist woodland, etc.). Since many plants change color upon drying, a notation with regard to blossom color may circumvent difficulties later.

Equipment for identifying either fresh or dried material includes a magnifying device (an inexpensive 10X hand lens as shown in fig. 14 will suffice in most instances), a millimeter ruler, one or two dissecting needles or probes, and a plant key or set of keys such as those contained in this book. A pair of forceps is often helpful. For working with dried material, you will also need a container of warm water or a detergent solution for softening any materials you wish to dissect. A photographic wetting agent such as Kodak Photo-flo works well for this purpose; 2 drops of Photo-flo concentrate per 15 ml of water is adequate.

What is a key? A key is an aid to the identification of unknown objects, usually consisting of a series of pairs, or couplets, of contradictory descriptions. Beginning with the first pair of descriptions, you select the one which best fits your plant and are then directed to a second pair. By always selecting the correct alternative and following the accompanying directions to the succeeding step, you will eventually arrive at the name of your plant. Remember to read each alternative carefully and completely before making your choice.

There are a few plants such as the parasitic dodder, the tiny aquatic duckweeds, the milkweeds, and others which, because they are rather unusual in structure, easily confuse a beginning botanist. These plants are treated in section A of the Key to Families. You may find it handy to familiarize yourself with those plants so that you may quickly move on to other sections of the key for the majority of plants with which you will be working.

The best way to learn how to use a key is to begin with some plant you already know and run it through the key. Let us use the common morning-glory as an example: Since the morning-glory is more or less typical in floral and vegetative characters, we take the second alternative of pair number 1 and are directed to 2. The leaves are net-veined and flower parts in 5's, so we go to 3. The flowers are bisexual, having both stamens and pistils, and we are directed to 4. The perianth (consisting of sepals and

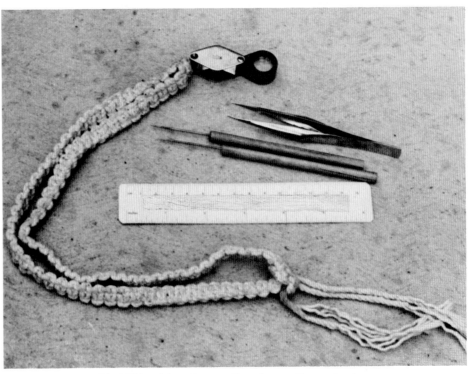

14. 10X hand lens, forceps, dissecting needles, and ruler

petals) is present, so we go to 5. The flowers have both calyx and corolla, so we go to 7. There is but a single ovary in each flower, so we go to 8. Since the ovary is superior, we are directed to 9. Stamens are of the same number as the corolla lobes (although in this case the lobes are not very prominent), and we go to 11. The corolla is of united petals, so we go to 12. The corolla is regular and the stamens of the same number as the corolla lobes, and we are directed to section L, page 29. The leaves are simple and the flowers solitary, so we take the second alternative at 1 and are directed to 2. Since the leaves are mostly cauline (that is, borne along the stem), we move on to 4. There is but a single ovary, which is not conspicuously lobed, so we go to 6. The leaves are alternate, and we go to 13. Since the leaves are entire, we are directed to 19. The flowers are lateral, rather than terminal, so we take the first choice at 19 and move on to 20. The flowers are pediceled (on short stalks), so we go to 22. The ovary has but a single stigma, so we move on to 23. Our plant is a twining vine with a large funnelform corolla, so we take the second alternative at 23 and arrive at Convolvulaceae. The next step would be to refer to the family description to see whether our plant fits within that description. If so, we may then move on to the key to genera within that family and to the key to species within the appropriate genus. After some experience working with plants, you will begin to recognize members of the more common families, such as the mustards or mints, and will be able to go directly to the family description and key to genera, thus saving yourself much time. In many cases, too, there will be only a few genera or species belonging to a given family, so that the keying process may be much shorter than in the example given above.

To check your identification, compare your plant specimen with the written description and photograph of the plant at whose name you have arrived in the key. If the description does not seem to fit your plant, go back to the key and try again, perhaps trying alternate routes through the key at points where you were uncertain of the proper choice. If you are satisfied that your identification is correct, record the name in your notebook. For those specimens you intend to keep, you should prepare a small label (fig. 15) showing the name and the collection data that you have recorded in your notebook. Then glue the specimen and the label to a fairly heavy piece of paper or lightweight posterboard. A closed box or cabinet treated occasionally with mothballs or insect repellent is recommended for storage.

Kansas Plants

Scientific name of plant

County
Mileage from nearest town in this county.
Description of habitat.
Flower color; comments about abundance of
    plants or any other pertinent info.
Date of collection.

Collector's name   Coll. no. of this
                         specimen
The University of Kansas   **Lawrence, Kansas**

---

15. Specimen label

Key to Families

1. Plants or their flowers very unusual and highly modified in structure from what is thought of as "typical" plants or flowers (if you think you may have one of these, you may find it timesaving to refer quickly to the photographs included in the sections for the families keyed out in Key A below) .... Key A, p. 21
1. Plants without any of the structures or modifications described in Key A 2
  2. Leaves usually parallel-veined, often sheathing the stem at the base; flower parts usually in 3's or 6's .... Key B, p. 22
  2. Leaves usually net-veined; parts of the flower commonly in 4's or 5's 3
    3. Flowers unisexual, containing stamens or pistil but not both .... Key C, p. 23
    3. Flowers bisexual, containing stamens and pistil 4
      4. Perianth (sepals and petals) absent .... Key D, p. 24
      4. Perianth present though sometimes quite small 5
        5. Flowers with either calyx or corolla but not with both 6
          6. Ovary inferior, the perianth apparently arising from its side or summit ....
          .... Key E, p. 24
          6. Ovary superior, appearing in the center of the flower .... Key F, p. 25
        5. Flowers with both calyx and corolla 7
          7. Ovaries 2 or more in each flower .... Key G, p. 26
          7. Ovary 1 in each flower (although it may have 2 or more styles or stigmas) 8
            8. Ovary inferior .... Key H, p. 26
            8. Ovary superior 9
              9. Stamens more numerous than the petals or lobes of the corolla 10
                10. Flowers irregular .... Key I, p. 27
                10. Flowers regular .... Key J, p. 27
              9. Stamens just as many as the petals or lobes of the corolla or fewer 11
                11. Corolla of separate petals .... Key K, p. 29
                11. Corolla of united petals 12
                  12. Corolla regular or nearly so *and* the stamens as many as the lobes of the corolla .... Key L, p. 29
                  12. Corolla *either* distinctly irregular *or* the stamens fewer than the lobes of the corolla .... Key M, p. 30

KEY A

1. Plants parasitic, usually whitish to pale orange in color, growing upon or attached to the stems of other plants and completely without connection to the soil at the time of flowering (*Cuscuta*) .... Convolvulaceae, p. 289
1. Not as above 2
  2. Plants aquatic, the entire plant body 0.3–10 mm long, consisting merely of a more or less flattened, oval or oblong, leaflike structure, sometimes with a few short roots, rarely seen flowering, usually found floating freely in the water or stranded along the shore ....
  .... Lemnaceae, p. 42
  2. Not as above 3
    3. Plants of aquatic or wet habitats, but much larger than 10 mm and definitely differentiated into leaves and stems; flowers quite small and inconspicuous, though sometimes aggregated into conspicuous spikes 4
      4. Plants with erect, more or less rigid stems; flowers unisexual 5
        5. Stout marsh plants 1–3 m tall with long, stiff, upright, swordlike leaves; flowers densely crowded into large, cylindrical terminal spikes (cattails) with male flowers above and female flowers below .... Typhaceae, p. 33
        5. Smaller plants up to 1 m tall, with leaves thinner and more grasslike than in the Typhaceae; flowers in spherical, male or female heads which are more or less sessile and evenly spaced in interrupted terminal or axillary spikes ....
        .... Sparganiaceae, p. 34

4. Plants with lax, floating or submersed stems; flowers unisexual or bisexual 6
     6. Flowers bisexual, with 4 stamens and 4 pistils, borne several to many together in loose or crowded, stalked spikes; leaves mostly alternate, the floating ones sometimes quite different from the submersed ones .... Potamogetonaceae, p. 35
     6. Flowers unisexual, the male flowers with 1 stamen, the female flowers with (1–) 4 (–5) pistils, 1 male and 1 female flower sessile and adjacent in a leaf axil; leaves opposite, very narrow, and all more or less alike .... Zannichelliaceae, p. 36
3. Not as above 7
    7. Flowers quite small, in ours consisting only of a pistil or of 2–5 sessile anthers and lacking a corolla, borne on a fleshy column (spadix) which is enclosed by a leaflike bract (spathe) .... Araceae, p. 40
    7. Not as above 8
       8. Plants with milky juice and small flower clusters resembling an individual flower, composed of a cup-shaped involucre bearing round its margin 1–5 glands, the latter sometimes with petal-like projections, or the involucre subtended by conspicuous, petal-like leaves or bracts as well; from each involucre protrude few to several stamens and one 3-lobed ovary elevated on a short stalk
       .... Euphorbiaceae, p. 203
       8. Not as above 9
         9. Plants with the individual flowers quite small but aggregated on a common receptacle into a head subtended by an involucre and in general like those of a dandelion, daisy, goldenrod, thistle, or ragweed, the whole head often resembling a single flower; individual flowers with petals united into a tubular or strap-shaped corolla (lacking in such plants as ragweed, cocklebur, and marsh-elder); stamens usually united into a cylinder through which the style protrudes; ovary inferior; calyx represented by a ring of bristles, scales, or teeth, or absent entirely .... Compositae, p. 376
         9. Plants with individual flowers easily recognized as such, but the stamens and stigma greatly modified from the normal structure and scarcely identifiable, fused to one another and forming a special compound structure in or near the center of the flower 10
           10. Flowers irregular, the lower petal differing from the other 2 in size, shape, or color; ovary 1, inferior, appearing below the sepals and petals; plants without milky juice .... Orchidaceae, p. 64
           10. Flowers regular, the 5 petals alike; ovaries 2, superior, located above the point of attachment of the sepals and petals but usually concealed; plants with milky juice .... Asclepiadaceae, p. 279

KEY B

1. Perianth none, or chaffy, scalelike, or bristlelike, never petal-like in color, texture, or size 2
  2. Plants aquatic, with floating or wholly submersed leaves (or stranded on shore at low water); flowers submersed or floating or barely raised above the water surface 3
    3. Flowers in spikes or heads .... Potamogetonaceae, p. 35
    3. Flowers axillary and solitary or a very few together, often very inconspicuous and easily overlooked 4
      4. Perianth none; stamens in each flower 1 or 2 .... Zannichelliaceae, p. 36
      4. Perianth present but small; stamens 4 .... Potamogetonaceae, p. 35
  2. Plants of land or shallow water, the leaves (under normal conditions) and the flowers completely out of water 5
    5. Inflorescence of subglobose heads .... Sparganiaceae, p. 34
    5. Inflorescence a dense, elongate spike 6
      6. Spike terminal and erect .... Typhaceae, p. 33
      6. Spikes apparently lateral (*Acorus*) .... Araceae, p. 40
1. Perianth present, at least the inner segments (and usually all of the segments) petal-like in color, texture, and/or size 7
7. Flowers unisexual; plants monoecious or dioecious 8
  8. Perianth differentiated into 3 green or greenish sepals and 3 white or pinkish petals; stamens almost invariably more than 6 .... Alismataceae, p. 37
  8. Perianth segments alike or nearly so .... Dioscoreaceae, p. 61
7. Flowers bisexual 9
  9. Ovary inferior, appearing below the perianth 10
    10. Twining vines; bisexual flowers very rarely produced .... Dioscoreaceae, p. 61
    10. Terrestrial or marsh or bog plants, not twining 11
      11. Stamens 3 .... Iridaceae, p. 62
      11. Stamens 6 .... Amaryllidaceae, p. 61
  9. Ovary or ovaries superior 12
    12. Ovaries several to many in each flower; leaves always basal or nearly so ....
    .... Alismataceae, p. 37
    12. Ovary 1 in each flower; leaves either basal or borne along the stem 13

13. Flowers irregular, the lower one much smaller than the 2 (usually blue) upper ones; fertile stamens 2 or 3 (*Commelina*) .... Commelinaceae, p. 44
13. Flowers regular 14
    14. Perianth differentiated into calyx (usually green) and colored corolla 15
        15. Leaves alternate or basal, lanceolate to linear; flowers in umbels (*Tradescantia*) .... Commelinaceae, p. 45
        15. Leaves in a single whorl of (normally) 3; flower solitary, terminal (*Trillium*) .... Liliaceae, p. 57
    14. Perianth not differentiated into calyx and corolla, its divisions essentially similar in color and texture 16
        16. Stamens 3; perianth segments 6, united below into a slender tube; aquatic or mud plants (*Heteranthera, Zosterella*) ....
        .... Pontederiaceae, p. 47, 48
        16. Stamens 6, as many as the divisions of the perianth .... Liliaceae, p. 48

KEY C

1. Leaves dissected into numerous narrow or threadlike segments; plants aquatic or sometimes found stranded on shores 2
2. Leaves pinnately dissected .... Haloragaceae, p. 255
2. Leaves palmately dissected .... Ceratophyllaceae, p. 111
1. Leaves not dissected 3
   3. Leaves compound 4
     4. Leaves 3-foliolate or palmately compound 5
       5. Flowers in a dense head on a fleshy spadix wholly or partly concealed by a large unilateral spathe (*Arisaema*, a monocotyledon with net-veined leaves keyed here for convenience) .... Araceae, p. 41
       5. Not as above 6
         6. Flowers in umbels or compound umbels (*Sanicula*) .... Umbelliferae, p. 267
         6. Flowers in spikes or panicles 7
           7. Perianth conspicuous; stamens numerous; pistils several or many (*Clematis*) .... Ranunculaceae, p. 116
           7. Perianth minute; stamens 5; pistil 1 .... Cannabaceae, p. 70
     4. Leaves pinnately compound, or ternately or pinnately decompound 8
       8. Flowers in globose heads (Mimosoideae) .... Leguminosae, p. 156
       8. Flowers solitary or in panicles 9
         9. Stem leaves alternate (*Thalictrum*) .... Ranunculaceae, p. 122
         9. Stem leaves opposite 10
           10. Stamens many; pistils several to many (*Clematis*) .... Ranunculaceae, p. 116
           10. Stamens 3; pistil 1 .... Valerianaceae, p. 367
   3. Leaves simple 11
     11. Leaves all basal 12
       12. Flowers in short or elongate spikes .... Plantaginaceae, p. 358
       12. Flowers in small open panicles (*Rumex*) .... Polygonaceae, p. 83
     11. Leaves all or mostly borne along the stem 13
       13. Leaves opposite or whorled 14
        14. Foliage densely covered with scales or branched hairs (*Croton*) ....
        .... Euphorbiaceae, p. 206
        14. Foliage glabrous or hairy but not as above 15
         15. Flowers solitary; plants of mud, swamps, or shallow water ....
         .... Callitrichaceae, p. 215
         15. Flowers in axillary or terminal clusters 16
          16. Flower clusters axillary .... Urticaceae, p. 72
          16. Flower clusters terminal 17
           17. Petals present, white or colored .... Valerianaceae, p. 367
           17. Petals absent (*Iresine*) .... Amaranthaceae, p. 96
       13. Leaves alternate 18
        18. Foliage densely covered with scales or stellate hairs (*Croton*) ....
        .... Euphorbiaceae, p. 206
        18. Foliage glabrous or hairy but not as above 19
         19. Calyx and corolla both present, the latter usually white or colored 20
          20. Plants climbing by tendrils or occasionally merely trailing; stamens 3; pistil 1; ovary inferior .... Cucurbitaceae, p. 368
          20. Plants erect or spreading or creeping; stamens more than 6; pistils 4 or more, separate or connivent .... Crassulaceae, p. 147
         19. Calyx present or absent; corolla always absent; calyx rarely petal-like 21
          21. Flowers in small axillary clusters, always individually very small 22
           22. Key for female flowers: 23
             23. Style 1, unbranched (*Parietaria*) .... Urticaceae, p. 73
             23. Not as above 24

KEY D

KEY E

8. Flowers in small terminal or axillary clusters, or solitary and pediceled in the axils; leaves entire or nearly so *(Comandra)* .... Santalaceae, p. 75

KEY F

KEY G

KEY H

KEY I

KEY J

KEY K

KEY L

1. Leaves regularly bipinnately compound; flowers in dense, peduncled axillary heads; stamens exserted far beyond the corolla (*Desmanthus*, Mimosoideae) .... Leguminosae, p. 156
1. Leaves or inflorescence not as above 2
  2. Leaves basal; inflorescence terminating a scape 3
    3. Flowers with parts in 4's, in spikes or heads; corolla membranous .... Plantaginaceae, p. 358
    3. Flowers with parts in 5's .... Primulaceae, p. 270
  2. Leaves all or chiefly cauline 4
    4. Ovary deeply lobed, appearing like 2 or 4 separate ovaries 5
      5. Leaves opposite (*Mentha*, *Pycnanthemum*) .... Labiatae, p. 317, 323
      5. Leaves alternate .... Boraginaceae, p. 300
    4. Ovary 1, not conspicuously lobed or divided 6
      6. Leaves opposite or whorled (those in the inflorescence are sometimes alternate), never compound, dissected, or deeply lobed 7
        7. Flowers in dense heads or dense short spikes; corolla 4-lobed or 4-divided 8
          8. Corolla scarious; leaves linear .... Plantaginaceae, p. 358
          8. Corolla herbaceous, white or colored; leaves broader than linear (*Phyla*) .... Verbenaceae, p. 306
        7. Flowers not in dense heads, but in some plants on short crowded racemes; corolla 4- to 12-lobed or -divided 9
          9. Stamens distinctly opposite the corolla-lobes .... Primulaceae, p. 270
          9. Stamens alternate with the lobes of the corolla (in some plants inserted so near the base of the corolla that their position is not readily ascertained) 10
            10. Lobes of the corolla 4 or 6–12 .... Gentianaceae, p. 273
            10. Lobes of the corolla 5 11
              11. Stigmas 3 .... Polemoniaceae, p. 294
              11. Stigma 1, either capitate or 2-lobed 12
                12. Ovary 1-celled; leaves entire .... Gentianaceae, p. 273
                12. Ovary 2-celled or rarely 4-celled; leaves almost always more or less dentate or angled .... Solanaceae, p. 327
      6. Leaves alternate 13
        13. Leaves conspicuously lobed, dissected, or compound (not merely hastate or sagittate) 14
          14. Twining vines, or sometimes merely trailing if a support is lacking 15
            15. Corolla funnelform .... Convolvulaceae, p. 289
            15. Corolla wheel-shaped (*Solanum*) .... Solanaceae, p. 332
          14. Erect or spreading or prostrate plants, never twining 16
            16. Corolla wheel- or saucer-shaped 17
              17. Anthers all, or all but one, connivent around the style; petals entire .... Solanaceae, p. 327
              17. Anthers separate; petals fringed (*Phacelia*) .... Hydrophyllaceae, p. 299
            16. Corolla bell-shaped to funnelform, tubular, or salverform 18
              18. Ovary 3-celled; style-branches or stigmas 3 .... Polemoniaceae, p. 294
              18. Ovary 1-celled; style-branches or stigmas 2 ...... Hydrophyllaceae, p. 297
        13. Leaves entire or serrate, or even shallowly lobed 19
          19. Flowers or flower clusters axillary or lateral 20
            20. Flowers sessile or nearly so, solitary in the axils of the leaves 21
              21. Leaves densely hairy; corolla about 1 cm wide (*Evolvulus*) .... Convolvulaceae, p. 292
              21. Leaves glabrous; corolla about 1 mm wide (*Centunculus*) .... Primulaceae, p. 271
            20. Flowers pediceled and solitary, or in axillary or lateral clusters 22
              22. Stigmas 2–4 .... Convolvulaceae, p. 289
              22. Stigma 1 23
                23. Stems spreading or prostrate to erect, never twining (in ours) .... Solanaceae, p. 327
                23. Twining vines with large funnelform corolla .... Convolvulaceae, p. 289
          19. Flowers or flower clusters terminal 24
            24. Flowers solitary and terminal 25
              25. Corolla 15 mm long or less; flowers terminating short branches ...... Boraginaceae, p. 300
              25. Corolla 7 cm long or longer; flowers in the forks of the stem (*Datura*) .... Solanaceae, p. 328
            24. Flowers several to many in terminal clusters 26
             26. Corolla wheel- or saucer-shaped 27
              27. Anthers separate; filaments, or some of them, hairy (*Verbascum*) .... Scrophulariaceae, p. 347
              27. Anthers connivent; filaments short and smooth (*Solanum*) .... Solanaceae, p. 332

KEY M

Families of Wildflowers and Weeds

TYPHÀCEAE   Cattail Family

Stout perennial marsh plants having horizontal rhizomes and scale leaves, and bearing fibrous roots and erect peduncles, or scapes, each ensheathed at its base by a group of long, upright swordlike leaves and terminating with an elongate, dense spikelike raceme of 2 contiguous or somewhat separated groups of unisexual flowers, the lower group female and the upper one male, each group subtended by a broadly linear, erect, impermanent bract or spathe. A female flower consists of a single pistil on a short stalk, or stipe, bearing a tuft of fine hairs, the stipe becoming long and threadlike later. A male flower is represented by a group of 2–7 stamens with filaments more or less united. The female flowers mature before the male ones of the same inflorescence, and cross-pollination is effected by wind-borne pollen. The fruits are tiny nuts enveloped in fluffy hairs of the stipe.

The family contains but a single genus, *Typha*.

TỲPHA L.   Cattail

NL., from an ancient Greek name, *typhe*, apparently related to *typhos*, meaning "swamp" or "marsh."

Description of the genus that of the family. The genus *Typha* contains about 15 species, 3 of which occur in Kansas. The genus is of more or less cosmopolitan distribution throughout riparian and estuarian marshes of temperate and tropical regions of the Northern and Southern hemispheres, forming vast colonies in favorable places.

The leaves have been used for thatching and for making mats and rush seats for chairs. The hair of the fuzzy fruits has been used for stuffing pillows and life jackets, and for sound-proofing and heat-insulating material; and the seeds themselves may prove to be a source of a drying oil as well as an edible oil.

Various parts of the plant may be used as food. The rhizomes may be cooked and eaten as a starchy potatolike vegetable or may be dried and pounded to provide flour. Young shoots and young spikes (before the pollen is shed) may be cooked and eaten as green vegetables, and the mature pollen produced in large quantities serves as a highly nutritious flour.

The rhizomes are also eaten by geese and muskrats, and cattail stands offer shelter and nesting cover for red-winged blackbirds, yellow-headed blackbirds, and long-billed marsh wrens.

Key to Species (3):
1. Male and female spikes normally contiguous; stigma spatulate (use high magnification) .... *T. latifolia*
1. Male and female spikes normally separated; stigma linear (use high magnification) 2
   2. Leaves plano-convex; mature spikes deep brown; widespread .... *T. angustifolia*
   2. Leaves mostly flat; mature spikes pale brown to golden brown; relatively uncommon in Kansas .... *T. domingensis*

T. angustifòlia L.   Narrowleaf cattail

L. *angusti-*, narrow, + *folius*, leaf, i.e., "narrow-leaved."
Flowering spike light brown; June–July
Marshes, borders of muddy and calcareous lakes
Scattered throughout Kansas, but less common in the southwestern part

Stems slender, 1–1.5 m tall. Leaves plano-convex, 4–8 mm wide. Male and female segments of the inflorescence usually separated 2–8 cm, the female segment 10–20 cm long

16. *Typha latifolia*

and 8–15 mm wide at maturity, light brown when flowering, becoming deep brown as the fruits mature. Fruit an achene, 5–8 mm long, subtended by copious white hairs.

T. domingénsis Pers.   Cattail

L. *-ensis*, a suffix added to nouns of place to make an adjective, in this case meaning "of Santo Domingo."
Spikes pale to golden brown; July
Edges of ponds, wet ditches, spillways below dams; primarily, but not always, in sandy soil
Uncommon; most likely to be collected in southern part

Similar to *T. angustifolia*, but taller and with more numerous, nearly flat leaves. The color of the spike resembles that of dark brown sugar.

T. latifòlia L.   Common cattail

L. *lati-*, broad, + *folius*, leaf, i.e., "broad-leaved."
Spikes dark brown or black; June–July
Marshes, muddy shores, and wet ditches
Scattered throughout

Stems stout, 1–3 m tall. Leaves flat, 1–2 cm wide. Male and female segments of the inflorescence usually contiguous, the female segment 10–15 cm long and 2.5 cm wide at maturity, dark brown. Fruit 1 cm long, with copious white hairs arising near the base. See plate *17* (color plates are between pages 36 and 37).

SPARGANIÀCEAE   Bur-reed Family

Aquatic herbs, perennial by rhizomes. Stems erect or floating. Roots fibrous. Leaves alternate, 2-ranked, linear, sessile and sheathing the stems, erect or floating. Flowers unisexual, arranged in globose heads with male heads above the female ones. Perianth of 3–6 minute membranous scales. Male flowers with 3 or more stamens. Female flowers with a single superior ovary; style simple or forked; stigma(s) simple and unilateral. Fruit a hard bony achene.

A single cosmopolitan genus, *Sparganium*, with about 20 species, is limited mostly to temperate and frigid regions of the Northern and Southern hemispheres. Two of these are in Kansas.

SPARGÀNIUM L.   Bur-reed

Probably from the Gr. *sparganion*, derivative of *sparganon*, swaddling-band, in allusion to the ribbonlike leaves.

See description of the family. Tubers produced from the creeping rhizome have been used as food by the Klamath Indians, but they are difficult to secure in quantity. The seeds are eaten by birds and waterfowl, and muskrats eat the entire plant.

Key to Species (3):
1. Stigma 2; achenes broadly obpyramidal, sessile .... *S. eurycarpum*
1. Stigma 1; achene fusiform, stipitate .... *S. americanum*

S. americànum Nutt.

"Of America" or "American."
Flowers minute, lacking a showy perianth; May–early September (mostly June–July)
Mud or shallow water of marshes and ponds
Douglas County

Stout, erect perennials, 3–10 dm tall. Leaves usually thin and flat, up to 8 dm long and 5–12 mm wide. Inflorescence simple or branched; male heads 3–10 on the main axis, 1–5 on the branches; female heads 2–4 on the central axis, 1–3 on the branches, 2 cm wide at maturity. Stigma 1. Stipe of achenes 2–3 mm long, slightly constricted, the beak straight, 3–4.5 mm long.

S. eurycárpum Engelm.

From the Greek *eurys*, broad, + *karpos*, fruit, i.e., "broad-fruited."
Flowers minute, lacking a showy perianth; May–August
Mud or shallow water
Collected from widely scattered stations, primarily in the northern part of the state

Stout, erect perennials, 5–12 dm tall, arising from a persistent creeping rhizome. Leaves erect, stiff, up to 8 dm long and 6–12 mm wide. Inflorescence branched, the central axis often with male heads only, branches with 1–2 female heads or 0–6 male ones; female heads 2–2.5 cm wide when mature. Stigmas 2. Achenes sessile, obpyramidal, 6–8 mm long, angular, with the tip truncate, retuse, or abruptly rounded to the stout 3 mm beak.

POTAMOGETONÀCEAE   Pondweed Family

Perennial rooted aquatic herbs with slender rhizomes. Stems simple or branched. Leaves alternate (or the uppermost leaves nearly opposite) with stipules sheathing the stem. The plants are usually rooted under water and produce very narrow leaves on the submersed part of the stem; if part of the stem reaches the surface, some species produce larger, leathery floating leaves of a different shape. Flowers bisexual, regular, borne in small spikes or heads with 1 to several approximate or well-separated whorls, the inflorescence mostly raised above the surface of the water. Perianth of 4 tepals. Stamens 4, each consisting of a sessile anther situated directly in front of a tepal. Pistils 4, superior, each with 1 cell and 1 ovule; the style none or very short. Each pistil produces an ovoid or subglobose achene, usually with a thin or spongy pericarp, drupelike when fresh, often short-beaked.

A single genus with about 90 species, distributed in fresh-water habitats of the North Temperate Zone.

POTAMOGÈTON L.   Pondweed

An ancient name, composed of the Gr. *potamos*, a river, + *geiton*, a neighbor, from the place of growth.

Description of the genus that of the family.

When collecting specimens of this genus for identification, it is necessary to secure fruiting material and a section of the stem bearing submersed leaves, as well as the more conspicuous floating leaves which are found in some species.

Kansas has 9 species. Because the pondweeds are rather difficult for inexperienced botanists to identify and because they are infrequently collected by amateurs, only the 4 most commonly collected species are included in the following key.

The seeds and vegetative portions of these plants are eaten by waterfowl, the seeds by marsh and shore birds, and the plants by muskrats.

Key to Species (3):
1. Submersed leaves 4 mm wide or wider .... *P. nodosus*
1. Submersed leaves less than 4 mm wide 2
  2. Stipules free, not fused to the base of the leaf .... *P. foliosus*
  2. Stipules fused to the base of the leaf, the summit projecting as a free ligule 3
    3. Achene strongly flattened, often pitted, with a conspicuous dorsal keel; floating leaves commonly produced .... *P. diversifolius*
    3. Achene turgid, not flattened; leaves all submersed .... *P. pectinatus*

P. diversifòlius Raf.   Pondweed

L. *diversus*, turned different ways (from *diverto*, to turn different ways), + *-folius*, -leaved, literally "different-leaved," referring to the narrower submersed leaves and wider floating leaves.
Flowers green, inconspicuous; May–August
Small ponds, margins of lakes
Eastern half

Stems up to 5 dm in length, usually freely branched. Submersed leaves narrowly linear, 2–5 cm long and 0.5–1.5 mm wide; stipules fused to the blade for about one-third their length, becoming fibrous in age. Floating leaves petiolate, elliptic, with 5–8 parallel veins, 1–3.5 cm long and 4–17 mm wide. The lower spikes subglobose in shape, the upper ones cylindric, up to 12 mm long. Fruits suborbicular or somewhat reniform, flattened, with a conspicuous dorsal keel and 2 sharp, narrow lateral wings (these not very obvious in immature fruits), sides depressed in the center, about 1.5 mm across.

P. foliòsus Raf.   Pondweed

L. *foliosus*, leafy, from *folium*, a leaf.

Flowers green, inconspicuous; June–July
Small ponds, marshes, margins of lakes
Mostly in the eastern half

Stems up to 1 m in length, usually freely branched. Leaves all narrowly linear, grasslike in appearance, mostly 3–4 cm long and 0.5–1.5 mm wide; stipules fused to one another about one-half their length, soon splitting and disappearing. Floating leaves not formed. Spikes subcapitate, 3–8 mm long, with 3–10 flowers. Fruits obliquely obovoid, slightly flattened, the sides depressed in the center, about 1.5 mm across, the dorsal keel narrow, sharp, often undulate.

Compare with *Zannichellia palustris*.

17. *Potamogeton nodosus*

P. nodòsus Poir.   Pondweed

L. *nodosus*, full of knots, knotty, probably in reference to the coarse cluster of fruits.
Flowers green, inconspicuous; June–early October
Shallow to deep water of ponds and lakes
Mostly in the eastern two-thirds

Stems up to 2 m in length, branched. Submersed leaves lanceolate, sometimes tapering gradually into a long petiole, 7–13 cm long and 1–1.5 cm wide; stipules 3–9 cm long, those on submersed portions of the stem soon decaying. Floating leaves petiolate, the blades elliptic or lanceolate, rounded or nearly acute at the base and apex, mostly 5.5–9.5 cm long and 1.5–4 cm wide, the petiole 3–16 cm long. Spikes cylindric, 2.5–5 cm long, the flowers crowded close together. Fruits plump, obovoid, with a sharp, narrow dorsal keel and 2 lateral ridges, brownish or reddish at maturity, 2.5–3 mm across.

Compare with *Zannichellia palustris*.

P. pectinàtus L.   Pondweed

L. *pectinatus*, comblike, from L. *pectinis*, a comb, in reference to the foliage, which nevertheless to this observer resembles hair more than it does a comb.
Flowers green, inconspicuous; June–August
Shallow water of ponds or lake margins
Scattered locations throughout

Stems up to 1 m long, many times branched and rebranched. Leaves all submersed, narrowly linear, often threadlike, mostly 2–4 cm long and 1 mm or less wide; stipules fused to the blade for about three-fourths of their length. Spikes with 2–6 unequally separated whorls of flowers, hence resembling somewhat a string of widely separated beads. Fruits plump, obovoid, sometimes obliquely so, with a very narrow dorsal keel and sometimes with 2 inconspicuous lateral ridges or wrinkles, 2–3 mm across.

ZANNICHELLIÀCEAE   Horned Pondweed Family

Submersed, perennial aquatic herbs with slender rhizomes. Stems simple or branched. Leaves alternate, opposite, or whorled, with stipules sheathing the stem. Flowers minute, bisexual or unisexual, borne singly or in clusters; when unisexual the plant monoecious or dioecious. Perianth of 3 small scales or entirely lacking. Stamens 1–3. Pistils 1–9, superior, each with 1 cell and 1 ovule; style 1; stigma prominent. Fruits sessile or stalked, beaked by the persistent style.

A small family of 3 genera, 1 of which occurs in Kansas.

ZANNICHÉLLIA L.   Horned pondweed

Named in honor of Gian Girolamo Zannichelli, a Venetian botanist (1662–1729).

Submersed aquatic herbs, branching freely from slender rhizomes, stems very slender and fragile. Leaves mostly opposite, simple, entire, 1-veined, 3–10 cm long and 0.5 mm wide, with sheathing membranous stipules. Flowers unisexual, male and female flowers adjacent in the same axil, each subtended by a minute membranous sheath. Perianth absent. Male flower of a single anther on a long filament. Female flower of 4 (1–5) separate ovaries, each containing a single ovule; style short and slender; stigma peltate. Fruit a sessile or short-stipitate achene terminated by the beaklike persistent style.

One species in Kansas.

Z. palústris L.   Horned pondweed

L. *palustris*, swampy, marshy, in reference to the habitat.

1. A portion of the Chautauqua Hills, northwest of Fredonia, Wilson County

2. Aerial view of the Flint Hills near Emporia, Lyon County

3. Flint or chert exposed in a roadcut through the Flint Hills

4. A view from I-70 showing a typical tablelike portion of the Flint Hills Upland and a wooded valley

5. Missouri River bluffs, Doniphan County, with prairie patches along the ridges

6. Sand hills of the Arkansas River Lowland, Hamilton County

7. Dakota sandstone outcropping near Kanopolis Reservoir in Ellsworth County

8. Aerial view of the Smoky Hills Upland between Kanopolis Reservoir and Osborne, Osborne County

9. Aerial view of the Blue Hills Upland, vicinity of Osborne

10. Limestone and shale exposure, vicinity of Osborne

11. Aerial view of the dissected Red Hills Upland near Medicine Lodge, Barber County

12. Red cedars in the Red Hills near Medicine Lodge

13. Wagon train encampment on the High Plains of Gove County, Kansas

14. Cretaceous chalks and shales exposed along the Smoky Hill River west of Hays

15. "Breaks" along the Smoky Hill River

16. "The Sphinx" and other chalk remnants exposed in Gove County and surrounded by chalk flat prairie

19. Tradescantia ohiensis

17. Typha latifolia

18. Sagittaria latifolia

20. Zosterella dubia

21. Androstepheum coeruleum

22. Camassia scilloides

23. Erythronium albidum

24. Lilium michiganense

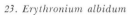

25. Trillium viridescens

26. *Hypoxis hirsuta*

27. *Sisyrinchium compestre*

28. *Cypripedium calceolus*

29. *Habenaria leucophaea*

30. *Orchis spectabilis*

31. *Spiranthes lacera*

32. *Rumex altissimus*

33. *Mirabilis nyctaginea*

34. *Mirabilis nyctaginea*

35. *Claytonia virginica*

36. Talinum calycinum

37. Nelumbo lutea

38. Anemone caroliniana

40. Aquilegia canadensis

39. Anemonella thalictroides

41. Clematis dioscoreifolia

42. Isopyrum biternatum

43. Podophyllum peltatum   *44. Sanguinaria canadensis*

45. *Dicentra cucullaria*

46. *Stanleya pinnata*
(courtesy Ralph Brooks)

47. *Amorpha canescens*
(courtesy Stephen Weiss)

48. *Astragalus crassicarpus fruits*

49. *Baptisia australis*

50. *Baptisia leucophaea*

51. *Petalostemon purpureum*

52. *Oxalis violacea*

53. *Polygala sanguinea*

54. *Impatiens pallida*

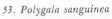

55. Hibiscus trionum

56. Viola pedata var. *pedata*

57. Viola pedata var. *lineariloba*

58. Viola pratincola

59. Viola pubescens

60. Coryphantha vivipara

61. Lythrum californicum

62. Ludwigia alternifolia

63. Oenothera macrocarpa

64. Oenothera speciosa

65. *Eryngium leavenworthii*

66. *Osmorhiza longistylis*

67. *Dodecatheon meadia*

68. *Gentiana puberulenta*

69. *Sabatia campestris*

70. *Asclepias incarnata*

71. *Asclepias purpurascens*

72. *Asclepias syriaca*

73. *Asclepias tuberosa*

74. *Asclepias viridiflora*
    with Monarch butterfly larva

75. *Convolvulus arvensis*

76. *Convolvulus sepium*

77. *Cuscuta* sp.

78. *Ipomoea hederacea*

79. *Ipomoea leptophylla*

80. *Phlox divaricata*

81. *Phlox pilosa*

82. *Phacelia hirsuta*

83. *Lithospermum carolinense*

84. *Verbena bipinnatifida*

85. *Verbena canadensis*

86. *Verbena stricta*

89. *Datura stramonium*

87. *Salvia pitcheri*

88. *Scutellaria ovata*

91. *Castilleja coccinea*

92. *Collinsia violacea*

90. *Buchnera americana*

93. *Linaria canadensis*

94. Pedicularis canadensis

95. Penstemon albidus

96. Penstemon buckleyi

97. Penstemon cobaea

98. Penstemon grandiflorus

99. Verbascum blattaria

100. Verbascum thapsus

101. Phryma leptostachya

102. Dipsacus sylvestris

103. Cucurbita foetidissima

104. Cyclanthera dissecta

105. Campanula americana

106. Triodanis perfoliata

109. Microseris cuspidata

107. Lobelia cardinalis

108. Lobelia siphilitica

112. Eupatorium purpureum

113. Liatris aspera

110. Arctium minus

111. Cirsium altissimum

114. *Liatris hirsuta*

115. *Liatris pycnostachya*

116. *Achillea millefolium*

118. *Senecio obovatus*

119. *Gaillardia pulchella*

117. *Chrysanthemum leucanthemum*

120. *Aster fendleri*

121. *Aster praealtus*

122. *Solidago rigida*

123. *Xanthocephalum sarothrae*

124. Coreopsis grandiflora

125. Coreopsis palmata

126. Helianthus annuus

127. Helianthus hirsutus

128. Helianthus rigidus

129. Melampodium leucanthum

130. Ratibida columnifera

132. Silphium laciniatum (courtesy Stephen Weiss)

131. Rudbeckia hirta

Flowers minute; late May–October
Shallow water of marshes, ponds, slow-moving streams, and backwater and pools along rivers
Mostly in the western half

Plants with a grasslike appearance. Leaves linear, 2–10 cm long and 0.5 mm or less wide. Mature achenes more or less oblong-cylindric, somewhat flattened and keeled, with a narrow membranous wing along the angles, about 4 mm long.

Compare with *Potamogeton foliosus*.

ALISMATÀCEAE

Water Plantain Family

Herbs, mostly perennial by a rhizome or rootstock, rooted in mud under water or along shores. Flowers bisexual, or monoecious, or dioecious, borne in whorls. Sepals 3, herbaceous. Petals 3, white or pink. Stamens 6 to many. Pistils superior, few to many, on a flat or convex receptacle, each pistil containing, in most species, a single ovule. Fruit an achene.

A family of 14 genera and about 60 species, of wide distribution, but mainly in temperate and tropical regions of the Northern Hemisphere.

18. *Alisma plantago-aquatica*

Key to Genera:
1. Flowers unisexual (monoecious or dioecious) or unisexual and bisexual on the same plant (polygamous) .... *Sagittaria*
1. Flowers all bisexual 2
   2. Pistils in 1 ring on a small flat receptacle .... *Alisma*
   2. Pistils in several series covering a convex receptacle .... *Echinodorus*

ALÏSMA L.

Water plantain

Ancient Gr. name for the European water plantain.

Herbaceous perennials, erect, or lax in deep-water forms, from a rootstock which is partly ensheathed by the expanded bases of the leaves and bears a tuft of fibrous roots below. Leaves all basal, simple, long-petioled, with 1 main vein and 4 or 6 other prominent "parallel" veins arising at the base; blade entire, elliptic to broadly ovate, rounded to subcordate at the base. Flowers bisexual, regular, borne in 1 or more much-branched scapes sometimes up to 3 feet tall but usually considerably less in our area, the ultimate branches of a scape terminating in whorls of slender pedicels bearing the small flowers. Sepals 3. Petals 3. Stamens 5 or more (6–10). Pistils 10–25, simple, superior, arranged in a ring on a small flat receptacle. Fruit a flattened achene.

Two species in Kansas. *A. subcordatum* Raf., not included here, is considered by some merely a variety of *A. plantago-aquatica*.

A. plantàgo-aquática L.

Water plantain

From *Plantago*, L. generic name for the dooryard plantain, because of the similarity of the leaves, + *aquatica*, of the water, referring to the habitat of the plant.
White to rose; June–September
Rooted in muddy banks or muddy bottoms of shallow waters
Primarily in the southeastern half (i.e., south of a diagonal line drawn from the southwest corner to the northeast corner)

See description of the genus. At the base of the stamens, a nectar-bearing tissue encircles the receptacle, attracting hover flies and short-tongued bees that assist in pollination.

According to Steyermark (4), plants belonging to this species are sometimes dug in autumn or spring, when the roots and solid base of the stem are filled with starch, and then cooked as a starchy vegetable. Some Indian tribes dried the roots thoroughly before cooking, reportedly to dispel an acrid taste not completely removed by boiling.

ECHINÓDORUS Rich.

Burhead

NL., from Gr. *echinos*, hedgehog, + Gr. *doros*, leather sack, because the clustered, sharp-beaked achenes of a single flower somewhat resemble a rolled-up hedgehog, and the pericarp of the achene suggests the sack.

Aquatic or marsh plants, perennial from a short rhizome. Roots fibrous. Leaves basal, simple. Flowers bisexual, regular, borne in whorls of 3–8 on an erect or repent scape. Sepals 3, persistent. Petals 3, white, deciduous. Stamens 6–20; pistils numerous, simple, superior, aggregated into a globose head; style broad at the base, tapering toward the apex to form a beak on the achene, stigma minute. Fruits seedlike ribbed achenes.

Three species in Kansas. *E. parvulus* Engelm. is known from Harvey County only.

Key to Species (3):
1. Scape erect, stamens 12 .... *E. rostratus*
1. Scape repent or prostrate, stamens 20 or more .... *E. cordifolius*

E. cordifòlius (L.) Griseb.

NL., from L. *cor, cordis*, heart and L. *folium*, leaf, some of the leaves being heart-shaped.
White; July–September
Shallow water and shores of ponds, banks of slow-moving creeks

Prostrate or creeping aquatics. Leaves erect, broadly ovate, with a rounded apex and a cordate base, 5–20 cm long and 3–15 cm wide, the lower surface with conspicuous raised cross-veins. Scape prostrate, creeping by rooting at the nodes, 6–12 dm long, often producing leaves at the nodes of the inflorescence. Flowers bisexual, in whorls of 5–15; bracts narrow, long-acuminate, 1–2.5 cm long. Sepals ovate to obovate, 6–12 mm long. Stamens 15–20, filaments slender, glabrous. Fruiting head 6–8 mm wide at maturity, achenes 1.8–2.2 mm long, evenly ribbed, the beak erect to horizontal, less than 0.5 mm long.

E. rostràtus (Nutt.) Engelm.

L. *rostratus*, beaked.
White; June–July
Shores of ponds and lakes, and along streams
Widely scattered in the eastern half, and primarily in the northeastern fourth

A fragile plant, 12–30 cm in height. Leaves long-petioled, blades entire, broadly ovate to lanceolate, usually cordate at the base, 2–20 cm long and 0.5–12 cm wide. Flowering scape simple or branched; flowers borne in whorls of 3–8, each whorl subtended by narrow pointed bracts. Sepals ovate, 4–5 mm long, remaining attached as the head matures. Petals white, broadly ovate, 5–10 mm long, deciduous. Stamens 12. Fruiting head about 8 mm wide at maturity; achenes 2.5–3.5 mm long, evenly ribbed to sharply ridged, with an erect or oblique beak about 1 mm long.

This species and the previous one are sometimes used in tropical fish aquaria, being sold under the name "poor man's lace plant."

SAGITTÀRIA L.   Arrowhead

NL., from L. *sagitta*, arrow, with reference to the shape of the leaf.

Rooted perennial aquatics, usually emersed, with rhizomes bearing apical buds that swell with stored food in late summer to early fall (then commonly referred to as tubers); roots fibrous. Normal leaves basal, simple, the blades unlobed or sagittate, with petioles usually as long as the water is deep. In deep or swiftly moving water, submersed forms are produced; these often have ribbonlike, bladeless phyllodia (actually flattened petioles). Plants monoecious, dioecious, or bisexual. Flowers regular, borne in a branched racemose inflorescence. Sepals 3, reflexed or accrescent in fruit. Petals 3, white or with a purple base, deciduous. Stamens 7 to many. Each flower with numerous, simple, superior pistils borne in several series and aggregated into a subglobose head; style 1, persisting as a beak on the fruit; stigma minute. Fruits flattened, beaked achenes with winged margins.

Eight species are known in Kansas. Five of the more common species are included in the following key.

Key to Species:
1. Sepals large and conspicuous, appressed to and surrounding the mature fruit; lower flowers perfect; fruiting pedicels very thick .... *S. montevidensis*
1. Sepals not large and conspicuous, spreading or reflexed at maturity; lower flowers either all male or all female; fruiting pedicels not conspicuously thick 2
2. Leaves not arrowhead-shaped nor with tail-like lobes at the base 3
3. Papery bracts at base of each whorl of flower stalks 9–15 mm long; southeastern Kansas .... *S. ambigua*
3. Papery bracts at base of each whorl of flower stalks 2–8 mm long; most apt to be found in sand-hill region south of the Arkansas River in central Kansas ...... *S. graminea*
2. Leaves arrowhead-shaped or with tail-like lobes at the base 4
4. Papery bracts at base of lower whorls of flower stalks more than 15 mm long .. *S. cuneata*
4. Papery bracts at base of lower whorls of flower stalks less than 15 mm long .... *S. latifolia*

19. *Echinodorus cordifolius*

20. *Echinodorus rostratus*

21. *Echinodorus rostratus* fruiting

WATER PLANTAIN FAMILY
38

S. ambígua J. G. Smith   Arrowhead

L. *ambiguus*, doubtful; reason for application uncertain, possibly because the first collected speci-
mens were misidentified as members of a previously named species.
White; June–July.
Borders of ponds, wet ditches
Southeastern part

Plants usually emergent, 3–8 dm high. Leaf blades lanceolate, 12–20 cm long and
4–10 cm wide. Flowering scape as long as the leaves or longer. Flowers in whorls of 2–12,
the upper ones usually male, the lower ones female; bracts lanceolate with acuminate
tips, 8–15 mm long, fused at the base. Sepals oblong, 6–8 mm long. Petals white, ovate,
8–10 mm long. Stamens 18–25, with slender, glabrous filaments usually longer than the
anthers. Fruiting head 12–15 mm in diameter; achenes obovate, 2 mm long, with a
prominent facial wing and with the marginal wings procurrent into the minute beak.

22. *Sagittaria graminea*

S. cuneàta Sheldon   **Wapato**

L. *cuneatus*, wedge-shaped, from the shape of the achenes.
White; late May–September
Margins of ponds and wet ditches, especially in sandy soil
Scattered, mostly in western half

Emersed or submersed. Leaves long-petioled, sagittate to lanceolate or phyllodial,
the blade when present 6–18 cm long and 1–10 cm wide, the terminal lobe deltoid-
orbicular, the basal lodes triangular, pointed, usually much shorter; phyllodia, when
present, leaflike, 0.5–2 dm long and 0.5–1.8 cm wide, or petiolelike and 3–5 mm wide.
Scape 0.5–2 dm; flowers in whorls of 2–10, the upper male or bisexual, the lower usually
female; bracts 1–4 cm long, lanceolate, fused at the base. Sepals ovate, acuminate, 5–8 mm
long. Petals white, ovate, 8–10 mm long. Stamens 10–18, filaments dilated at the base,
usually shorter than the anthers. Fruiting heads 0.8–1.3 cm wide; achenes cuneate, 2–3
mm long, the beak erect, less than 0.5 mm long.

S. gramínea Michx.   Arrowhead

L. *gramineus*, grasslike.
White or pink; late May–September
Margins of ponds, and wet ditches
Uncommon, most frequently encountered in the Arkansas River sand-hill region, especially in
Harvey County

23. *Sagittaria latifolia*

Erect and emersed, or lax and submersed. Leaves bladeless, flat, phyllodial, or
with lance-linear to broadly elliptic-ovate blade, rarely hastate or sagittate, 4–30 cm
long and 0.5–10 cm wide. Scape simple, 3–50 cm long; flowers in whorls of 2–12, the
upper ones usually male, the lower ones female; bracts 3–7 mm long, sometimes fused
at the base. Sepals ovate, 3–6 mm long. Petals white or pink, 1–2 cm long, often as wide.
Stamens 12–20, filaments broadest at the base, roughened with minute scales. Fruiting
heads 4–8 mm wide; achenes narrowly obovate, 1.5–2 mm long, winged on the margins
and with one or more facial wings, the beak minute (less than 0.75 mm long).

S. latifòlia Willd.   **Common arrowhead, wapato**

L. *latifolius*, broad-leaved.
White; July–September
Shallow water of ponds, ditches, and slow-moving streams
Throughout

Plants erect. Leaves extremely variable in shape, ranging from sagittate to rarely
lanceolate or phyllodial, blade 5–40 cm long and 2–25 cm wide, lobes varying from
broadly ovate with acute apexes to linear-lanceolate with acuminate apexes. Scape 1–12
dm tall; flowers in whorls of 2–15, the upper ones male or bisexual, the lower ones usually
female; bracts 1 cm long or less, acute or acuminate. Petals white, broadly ovate, 1–2 cm
long. Stamens 25–40, filaments slender, glabrous. Mature fruiting heads about 2 cm
wide; achenes obovate, 2.5–4 mm long, winged on both margins, the beak usually
horizontal, 1–2.5 mm long.

The large starch-filled tubers produced at the tips of the rhizomes by this species
and *S. cuneata* have been used by various Indian tribes of North America as an important
part of their diet. The members of the Lewis and Clark Expedition, sponsored by Thomas
Jefferson (1804–1806), were witnesses of this as they traveled westward across the continent. In the journal kept by them we find the statement, under the date of November 5,
1805, that on being invited into an Indian lodge in the vicinity of the Columbia River

WATER PLANTAIN FAMILY
39

they were fed with *Sagittaria* tubers (called *wapato* by the northwestern Indians) roasted in embers until they became soft, which the explorers pronounced agreeable in taste and a good substitute for bread. The women, they said, collected the tubers by going into the water breast-high and digging them out with their toes, freeing them from clinging mud until they could float to the surface, whereupon the women tossed them into a light canoe which they pushed before them, keeping this up for hours at a stretch.

The various tribes of prairie Indians—Dakota, Pawnee, Omaha, etc.—ate the tubers either boiled or roasted. The Algonquin Indians also used these tubers, which they called *katniss*.

This species and *S. cuneata* are of considerable value to wildlife, although the tubers of the former are often too large and too deeply buried to be of use to ducks. Some birds and waterfowl also eat the seeds.

See plate *18*.

S. montevidénsis Cham. and Schlecht.   Arrowhead

"From Montevideo."
Petals white, with a basal purple spot; late June–September
Shallow water and muddy margins of ponds and sluggish streams
Scattered, mostly in eastern half

Plants erect, or lax in deep-water forms. Leaves long-petioled, typically emersed, with hastate or sagittate blade 4–60 cm long and 1–30 cm wide, the terminal lobe linear to deltoid-orbicular, the basal lobes pointed, divergent. Scape stout, 1–15 cm tall; flowers in whorls of 2–10, the upper ones male or bisexual, the lower ones usually female (or with a ring of fertile stamens); bracts ovate to narrowly lanceolate, united at the base. Sepals obtuse, appressed to the fruiting head, 4–12 mm long. Petals broadly ovate, white, frequently with a purple basal spot, 7–15 mm long. Stamens 9–15, filaments roughened with minute scales. Fruiting heads 0.7–1.6 cm wide; achenes cuneate-oblanceolate to cuneate-obovate, narrowly winged on the margins, 2–3 mm long, the beak horizontal to oblique, about as long as the achene is wide.

ARÀCEAE   Arum Family

Herbaceous perennials with leaves and stalk of the inflorescence springing from a corm or rhizome. Flowers without a perianth, except in some cases consisting of a few scales, monecious, dioecious, or bisexual, borne on a fleshy axis called a spadix, which is subtended and sometimes surrounded by a large specialized foliaceous bract called a spathe. Fruit a berry or utricle.

A family of about 105 genera and 1,400 to 1,500 species, mostly tropical or subtropical in distribution, but many in temperate regions.

Two species of *Arisaema* are native to our state, and *Acorus calamus* is naturalized in a few localities. Cultivated members of this family include the calla lily (*Zantedeschia* spp.), elephant's-ear (*Colocasia antiquorum* Schott), and several species of *Philodendron*.

The leaves and roots of all members of this family contain minute crystals of calcium oxalate which are poisonous if eaten in large quantities and cause dermatitis in some people. This peppery, acrid principle may be broken down by combined heating and drying, although boiling is apparently ineffective.

Key to Genera:
1. Leaf blades simple, long and narrow .... *Acorus*
1. Leaf blades compound .... *Arisaema*

ÁCORUS L.   Sweet flag, calamus

L. name of an aromatic plant.

Herbaceous perennials arising from a stout rhizome. Leaves erect, linear, midvein usually off-center. Spadix diverging laterally from a tall 3-angled scape and surmounted by an erect, green leaflike spathe. Flowers perfect, regular, densely crowded on an elongate spadix. Perianth of 6 tepals. Stamens 6. Ovary superior, 2–4 celled; style 1, short; stigma 1, depressed-capitate. Fruit berrylike, obpyramidal, hard and dry but gelatinous within, 1–3 seeded.

One species in Kansas.

A. cálamus L.   Sweet flag, calamus

Old name for a reed.
Flowers yellowish-brown; May–July
Marshes, seepy areas, wet ditches
Known from scattered sites in Brown, Republic, Neosho, and Meade counties

Leaves crowded at base, erect, linear, mostly 0.5–1 m long and 8–25 mm wide. Scape resembling the leaves, prolonged into an erect green spathe 2–6 dm long; spadix 5–10 cm long and 1 cm wide at anthesis, 2 cm at maturity. Early introduced from Eurasia; naturalized in North America but seldom producing mature fruits. Plant, especially the rhizome, sweetly aromatic.

This plant is the source of the aromatic drug calamus which is used in toilet powders, perfumes, flavorings, and medicine. The roots may be candied, as you would orange or grapefruit peels, and the tender interior tip of the young leafy shoots is recommended by some native-food buffs as fresh spring-salad material.

ARISAÈMA Mart.

NL., from the L. *aris*, arum, + Gr. *haima*, blood, in reference to the red blotches on the spathe of some species

24. *Arisaema dracontium*

Cormose perennials. Leaves compound, long-petioled, leaflets with margins entire and veins parallel from the midrib to the marginal vein, connected by many anastomosing veinlets. Flowers monoecious or dioecious, regular, covering the basal part of a fleshy spadix and subtended by a green or purple-brown spathe. Perianth none. Male flowers above the female, composed of 2–5 subsessile anthers opening apically. Female flowers consisting of a 1-celled superior ovary with a very short style and a broad stigma. Fruit a cluster of globose red berries maturing in late summer.

Two species in Kansas.

Key to Species (3):
1. Leaflets 3; spathe expanded above, arched over the blunt spadix .... *A. triphyllum*
1. Leaflets 5–15; spadix protruding, long-acuminate .... *A. dracontium*

A. dracóntium (L.) Schott   Green dragon

Named after the tropical genus *Dracontium*, from Gr. *drakon*, the mythical dragon, sometimes represented with clawed fingers and toes—the leaves of these plants having several finger-like divisions.
Flowers yellowish; (April–) May–June
Moist, rich woods, mostly in wetter ground than the following species
Eastern third

25. *Arisaema triphyllum*

This species differs principally from *A. triphyllum* in having the leaf divided into 5–15 oblanceolate segments, the spadix whiplike above and long-extended beyond the spathe, and the flowers more frequently monoecious.

The starchy corm of this species is edible when prepared in the manner described for *A. triphyllum*.

A. triphýllum (L.) Schott   Jack-in-the-pulpit, Indian turnip

NL., *tri*, 3, + *phyllon*, leaf, the compound leaf having 3 leaflets.
Flowers green or purple; April–June
Rich, moist woods or low bottom thickets
Eastern one-fifth

Plants erect, mostly 2–5 dm tall. Corm globoid, up to 6 cm across, bearing a circle of roots around its apex and in early spring producing 1 or 2 long-petioled, trifoliolate leaves and an erect peduncle or scape with a remarkable large, green, often purple-striped bract called a spathe. Leaflets mostly ovate, with an acuminate apex and acute base. The spathe is convolute in cup-form below and tapering and overarching above. The extension of the peduncle into this enclosure is called the spadix. The latter is club-shaped and sterile above, but basally it bears a cluster of minute monoecious or dioecious flowers without calyx or corolla. When the flowers are monoecious, the male flowers are situated above the female flowers. When the flowers are dioecious, the spadix bears either male or female flowers, and only plants with female flowers bear fruit. A male flower has 4 stamens with anthers opening by apical slits, while a female flower has a single pistil with a 1-celled, 5- to 6-ovuled ovary, and a capitate stigma. Gnatlike flies are attracted to the flowers and seem to effect cross-pollination. Each ovary ripens into a shining red berry, the many berries forming a conspicuous ovoid cluster which is still held aloft on the scape after the spathe and leaves have withered away.

The corm is replete with starch associated with a ferociously pungent substance (calcium oxalate), so that biting into it or chewing a bit of it results in a harrowing, unforgettable experience. However, after the corms are thoroughly dry the hot pungency is gone, possibly by oxidation of the acrid substance. The Indians were aware of this and used to grind the dried corms into meal which they made into gruel and cooked with venison.

Another recommended method of preparation is to cut the corms into very thin slices and allow them to dry for several weeks until they are crisp and mild. Ground to flour, they then have a mild suggestion of cocoa flavor.

The bright red fruits are eaten by wildlife to a limited extent.

LEMNÀCEAE   Duckweed Family

Small, floating or submersed aquatic perennials. Plant body reduced to a small or minute oval, oblong, flat or globose, leafless thallus, with or without 1 to few short, unbranched roots. Often reproducing asexually by buds, overwintering in temperate regions by producing buds that sink below the surface into the substrate. Flowers but rarely produced, in small pouches at the edge of the thallus or on its upper surface, the inflorescence consisting of 1 or 2 female flowers and 1 or 2 male flowers. Perianth absent. Male flowers consisting of a single stamen. Female flowers composed of a simple, superior flask-shaped ovary with a single style and a concave stigma. Fruit a utricle.

A family of 4 genera distributed in fresh-water habitats throughout the world. Three genera in Kansas.

NOTE: Magnification with at least a 10X hand lens is helpful when working with these tiny plants. They are also easier to identify while fresh, rather than dried and pressed.

Key to Genera:
1. No roots present .... *Wolffia*
1. One or more roots present 2
   2. Roots 1 to a plant; plant green underneath .... *Lemna*
   2. Roots 2 to several to a plant; plant purplish underneath .... *Spirodela*

LÉMNA L.   Duckweed

Name of a water plant mentioned by Theophrastus (370–285 B.C.), a student of Plato and author of the oldest known botanical work, now known as the Father of Botany.

Thallus flat to strongly convex beneath, rotund, ovate, obovate, or stipitate, with 1–5 obscure veins radiating from a nodal point near the base. Root 1, developing opposite the nodal point, without vascular tissue. Reproduction as in *Spirodela*; ovules 1–6.

All 4 duckweed genera are used as food by wildlife. However, about nine-tenths of this use consists in feeding on common duckweed (*Lemna minor*). Kansas has 4 species of *Lemna*. *L. valdeviana* Philippi is known from a few counties only and is not included here.

Key to Species (3):
1. Thallus oblique or falcate .... *L. perpusilla*
1. Thallus symmetrical or nearly so 2
   2. Thallus 3-nerved, convex on both sides .... *L. minor*
   2. Thallus 1-nerved, flat on the lower side .... *L. minima*

L. mínima Philippi   Duckweed

L. *minimus*, smallest.
Flowers minute, rarely present
Pools
Known from Atchison, Douglas, Pottawatomie, Wyandotte, and Meade counties; collection record
    incomplete

Thallus oblong to elliptic, symmetrical or nearly so, convex above, flat beneath, 1.5–3 mm long and 1–2 mm wide, obscurely 1-nerved, comparatively thin at the margin, rounded or subacute at the base, solitary or in groups of 2–4.

L. mìnor L.   Common duckweed

L. *minor*, smaller.

Flowers minute, rarely present
Quiet waters
Known from scattered sites across the state; collection record incomplete

Thallus rotund to elliptic or obovate, symmetrical or nearly so, convex on both sides, 2–4 mm long and 1.5–3 mm wide, obscurely 3-nerved, comparatively thick at the margin, rounded to acute at the base, solitary or in colonies of 2–8 plants.

Widespread in both the Old World and New World.

L. perpusílla Torr.   Duckweed

L. perpusilla, very tiny.
Flowers minute, rarely present
Ponds and streams
Known from scattered sites in the eastern half of the state; collection records incomplete

Thallus oblong to obovate-oblong, asymmetrical, usually falcate, 2–4.5 mm long and 0.7–1.9 mm wide, veinless or obscurely 1-nerved, rounded to acute at the base, solitary or in groups of 2–6 plants.

SPIRODÉLA Schleiden   Big duckweed

From the Gr. *speira*, a cord, and *delos*, evident, because of the presence of rootlets.

Thallus flat or somewhat convex beneath, usually asymmetrical, marked with 3–15 obscure nerves radiating from a nodal point near the base. Roots 3–16, descending from the nodal point, each with a single vascular strand. Reproducing from 2 pouches, 1 on each lateral margin opposite the nodal point, usually vegetatively multiplying by budding, the young plants remaining attached for some time, forming small colonies. Flowers in the reproductive pouches, rare. Ovules 2.

One species in Kansas.

S. polyrhíza (L.) Schleiden   Big duckweed

NL., from Gr. *polys*, many, + *rhizo*, root, because of the number of rootlets present.
Flowers minute, rarely present
Ponds and margins of slow-moving streams
Collected from numerous sites throughout the state

Characters of the genus. Thallus broadly oval to ovate, purplish beneath, 3–10 mm long.

WÓLFFIA Horkel   Watermeal, ducksmeal

Named for Johann Friedrich Wolff (1788–1806), who wrote on *Lemna* in 1801.

Thallus minute, thick, globose to ellipsoid or ovoid, or flattened on the upper side. Without veins or roots. Reproductive pouch 1, near the basal end. Reproduction almost entirely by budding, the young plants soon detached. Flowers rarely produced, breaking through the upper side of the thallus. Ovule 1 per ovary.

The species of this genus are the smallest flowering plants known. As in *Lemna*, the species are very difficult to distinguish and must be studied under a high-powered or binocular microscope.

The 2 species most frequently collected in Kansas are *W. columbiana* and *W. papulifera*.

Floating mats of *Wolffia* provide food and cover for fish.

Key to Species (3):
1. Thallus rounded above, not punctate, bearing about 3 papillae .... *W. columbiana*
1. Thallus flattened above, the upper surface brown-punctate, with 1 prominent papilla near the center .... *W. papulifera*

W. columbiàna Karst.   Watermeal
NL., "of Columbia."
Flowers minute, rarely present
Still waters
Recorded from Barber, Cherokee, Douglas, Meade, and Ottawa counties, but more frequent than collections indicate

Thallus broadly ellipsoid to globose, asymmetrical, floating just below the surface, 0.3–1 mm long, the upper side convex, not dotted with brown spots, bearing usually 3 papillae along the median line.

26. *Spirodela polyrhiza* intermixed with *Wolfia* spp.

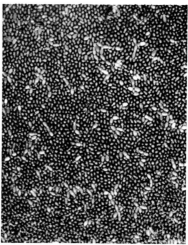

27. *Wolfia punctata* (larger plants) intermixed with *W. columbiana*

W. papulifera C. H. Thompson   **Watermeal**

NL., from L. *papula*, pimple, bearing a small papilla.
Flowers minute, rarely present
Stagnant waters and slow-moving streams
Cherokee County

Thallus broadly ovate, slightly asymmetrical, 1–1.5 mm long, rounded beneath, the upper side floating above the water, punctate with brown pigment cells, elevated near the center into a conic papilla.

COMMELINÀCEAE   Spiderwort Family

Upright or spreading or procumbent, somewhat succulent perennials or annuals. Stems slightly swollen at the nodes. In the case of perennials, the stems and coarse fibrous roots spring from short rootstocks, or tubers, while in annuals the underground system is of fibrous roots only. Leaves alternate, simple, and entire, grasslike in venation and in the sheathing habit of the leaf bases. Flowers bisexual, borne singly or in cymose axillary and/or terminal inflorescences. Perianth regular or irregular, of 3, mostly blue petals and 3 green sepals; the sepals persistent, the petals shrinking together and deliquescing after a few hours. Stamens hypogynous, usually 6, though some may be sterile. Ovary superior, 3-celled; style 1; stigma simple or at most 3-lobed. Fruit a few-seeded loculicidal capsule.

A family of largely tropical and subtropical plants, represented by 37 genera and about 600 species distributed over the warmer parts of the earth. Kansas has 2 genera and 8 species.

Key to Genera:
1. Petals all alike; 6 perfect stamens .... *Tradescantia*
1. 2 petals larger than the third; 3 perfect and 3 imperfect stamens .... *Commelina*

COMMELÌNA L.   Dayflower

Dedicated to three early Dutch botanists on account of the 2 showy petals and 1 less-conspicuous petal, Linnaeus referring to the 3 botanists named Commelin, 2 of whom—Jan (1629–1692) and Kaspar (1667–1731)—were conspicuous botanists, while the third died before becoming well known.

Herbaceous annuals or perennials with slender fibrous roots, thick perennial roots, or rhizomes, and succulent stems often rooting at the nodes. Leaves opposite, linear to lance-ovate, contracted at the base into sheathing petioles. Flowers perfect, irregular, borne in small cymes, closely subtended by a folded, cordate spathe from which the pedicels protrude. Sepals 3, herbaceous, somewhat unequal, 2 usually united at the base. Petals 3, the upper 2 blue, ovate to reniform, clawed at the base, the lower one smaller, typically white, or wanting. Fertile stamens 3, 2 with oblong anthers, 1 incurved with a longer anther. Sterile stamens 3, smaller than the fertile, with imperfect cruciform anthers. Filaments not hairy. The ovary superior, 3-celled. Fruit a 2- to 3-celled loculicidal capsule.

All species of this genus may be cooked and eaten as a green vegetable.

Key to Species (3):
1. Margins of the spathe free at the base 2
   2. Lower median petal white; anthers 6 .... *C. communis*
   2. Lower median petal blue; anthers 5 .... *C. diffusa*
1. Margins of the spathe fused to one another at the base 3
   3. Lower median petal blue, scarcely smaller than the 2 upper ones; sheaths ciliate with dark-brown hairs, not auriculate; fruit 5-seeded .... *C. virginica*
   3. Lower median petal white, distinctly smaller than the 2 upper ones; sheaths ciliate with white hairs, usually prolonged into distinct rounded auricles; fruit 3-seeded .. *C. erecta*

C. commùnis L.   Common dayflower

From L. *communis*, common.
Upper petals blue, lower petal white; (May–) July–October
Cultivated and waste ground, moist alluvial soils, and in low woods and thickets
Mostly in the eastern fourth; intr. from Asia

Herbaceous perennial with a short rhizome and fibrous shallow roots. Stems delicate, green, glabrous, ascending, decumbent or creeping, 3 dm or more in length, often rooting at the nodes. Leaves lanceolte, sheathing the stem. Flowers borne at or near the end of the stem, subtended by an ensheathing spathelike bract, its lower margins free. Sepals 3. Petals 3, the upper 2 blue and the lower 1 white. Stamens 6, 3 of them dwarfed and nonfunctional. Fruit a 3-celled capsule.

This immigrant from Asia is widely distributed in waste places and is often seen in dooryards, flourishing best in partial shade. The flower buds, of different stages of development, ensconced in the axil of the spathelike upper leaf, open (1 each morning) by erecting the upper petals and exposing the 3 fertile stamens and the slender style, all extended forward in a position to touch the bees that come for nectar held at the bottom of the flower. After a few hours, the chance for cross-pollination having been given, the petals collapse into a moist mass around the stamens and style, favoring self-pollination in case cross-pollination has not taken place.

28. *Commelina communis*

C. diffùsa Burn. f.   **Dayflower**

From the L. *diffusus*, spread out, diffuse, referring to the habit of the plant.
Sky-blue or paler; early July–October
Wooded bottom land, marshes, and waste places
Southeastern corner

Herbaceous annuals, stem succulent, diffusely branched and rooting from the lower nodes, up to 1 m in length. Leaves lanceolate, sheathing the stem. Bract subtending the flowers cordate, folded, glabrous or finely ciliate, its margins free. Petals sky-blue or paler; upper ones 6–8 mm, lower ones smaller. Anthers 5. Fruit a 3-celled loculicidal capsule, the lower 2 cells each 2-seeded, the upper one 1-seeded. Seeds black, reticulate.

C. erécta L.   **Dayflower**

From L. *erectus*, erect, referring to the habit of the plant.
Upper petals blue, lower petal white; late May–October
Roadsides, sand dunes, stream banks, especially in sandy soil
Primarily south and east of a diagonal line running from Morton County to Marshall County

29. *Commelina erecta*

Herbaceous perennials arising from a cluster of thickened fibrous roots. Stems erect or ascending, up to 1 m in length, usually branched. Principal leaves linear to lanceolate, the sheaths ciliate with white hairs, somewhat prolonged into rounded, often-flaring auricles. Spathes single or in small clusters, short-peduncled, broadly semideltoid (when folded), often with conspicuous radiating cross-veins, the margins fused together one-third of their length. Upper petals blue, 10–25 mm long; lower one white, much smaller. Stamens 6. Fruit a 3-seeded loculicidal capsule. Seeds brown, smooth.

According to Steyermark (4), "the fleshy roots of some varieties of *C. erecta* contain considerable starchy carbohydrates and have been suggested as being edible as a type of cooked vegetable."

C. virgínica L.   **Dayflower**

"Of Virginia."
Blue; July–September
Moist or wet woodland
Cherokee County

Perennial from a rhizome. Stem stout, erect or nearly so, up to 12 dm long, often widely branched. This is the most robust of our native species of *Commelina*. Leaves lanceolate, long-acuminate, scabrous above and glabrous or finely hairy beneath. Sheaths pilose-ciliate with dark-brown hairs, not prolonged into distinct auricles. Spathes usually clustered toward the summit, sessile or short-peduncled, the nerves connected by numerous cross-veins, the margins fused together for about one-third of their length. All petals blue, the lower 1 scarcely smaller than the upper 2. Stamens 6. Fruit a 3-celled loculicidal capsule, the lower cells each 2-seeded, the upper 1-seeded. Seeds reddish-fuscous, smooth or hirtellous.

TRADESCÁNTIA L.   **Spiderwort**

Named in honor of John Tradescant, gardener to Charles I of England, who assembled a wealth of plant materials and published his researches in 1658. "Wort" comes from AS. *wyrt*, meaning "herb" or "root." "Spider" comes into the picture when, the stems being broken and pulled apart, the copious mucilaginous slime in them is drawn out into thin threads suggesting spider web.

Mucilaginous perennials with a rhizome or coarse fibrous roots. Stems erect and elongated. Leaves keeled, sheathing the stem at the nodes. Inflorescence a terminal or axillary cluster of several to many flowers subtended by 2 leaflike bracts. Flowers regular. Sepals 3, herbaceous and persisting. Petals 3, blue or rose, rarely white, soon fading. Stamens 6, filaments bearded with moniliform hairs. Pistil frequently with a tuft of glandular hairs arising around the base of the style; stigma capitellate. Fruit a 2- to 3-celled loculicidal capsule with 1–2 seeds in each cell. The seeds are scurfy gray, oblong with 1 end truncated and interestingly sculptured. A linear funicular scar runs longitudinally along the adaxial surface of the seed, while on the abaxial surface a conical protuberance occupies the "floor" of a deep pit which is at first covered by a thin membrane, later exposed. From the edge of this pit, coarse ridges radiate along the sides of the seed.

For identification, magnification of hairs on the sepals is necessary. It is also most helpful to have the underground portions of the plant. Interspecific hybrids with intermediate characters occur in this genus and may cause the field botanist some consternation.

30. *Tradescantia occidentalis*

31. *Tradescantia ohiensis*

Key to Species:
1. Sepals without glandular hairs 2
  2. Sepals glabrous or merely bearded at the tip or sparsely covered with slender hairs; leaves glaucous; pedicels glabrous, 0.7–2.5 cm long .... *T. ohiensis*
  2. Sepals with large thick hairs, pubescent their entire length; leaves not glaucous; pedicels pilose, 3.5–6 cm long .... *T. tharpii*
1. Sepals with glandular hairs 3
  3. Plants green, not glaucous; pedicels 1.5–3 cm long at anthesis, heavily viscid-pubescent; sepals about 1 cm long; petals 1.8–2 cm long .... *T. bracteata*
  3. Plants glaucous; pedicels 1–2 cm long at anthesis, sparsely pubescent to glabrous; sepals 7–8 mm long; petals 1.2–1.7 cm long .... *T. occidentalis*

T. bracteàta Small   Bracted spiderwort

L. *bracteatus*, in its botanical sense "conspicuously bracted," in reference to the long, leaflike bracts subtending the flowers.
Blue, purple, or pink; April–May
Prairies, roadsides
Primarily in the eastern half

Perennial from a long rhizome. Stem usually rather stout, straight, rarely branched, 2–4 dm high at anthesis, glabrous or with minute soft hairs. Leaves glabrous or sparsely pilose at base, the larger ones 8–15 mm wide, recurved. Inflorescence usually solitary and terminal, rarely with an additional lateral one present. Pedicels heavily viscid-pubescent, 1.5–3 cm long at anthesis. Sepals about 1 cm long, with rather long, lax glandular hairs and usually with some eglandular hairs, at least in a tuft at the tip. Petals 1.8–2 cm long.

T. occidentàlis (Britt.) Smyth   Prairie spiderwort

L. *occidentalis*, western, from *occidens*, the quarter of the setting sun; so named because this species occurs only west of the Mississippi River.
Blue to rose-colored; June–August
Prairies and plains
Primary in the western half

Stems occurring singly, slender, straight, and upright, often branched, 2–6 dm tall, glabrous, and glaucous. Some roots slender, others thick and fleshy. Bracts and leaves linear, under 1 cm wide. Inflorescences solitary and terminal, or with another one peduncled from an upper node. Pedicels sparsely pubescent to glabrous, 1–2 cm long at anthesis. Sepals mostly 7–8 mm long. Both sepals and pedicels with short, stout glandular hairs, rarely glabrous or with only a few hairs near the base. Petals mostly 1.2–1.7 cm long.

T. ohiénsis Raf.   Ohio spiderwort

NL., compounded of "Ohio" and the suffix *-ensis*, signifying the place where the species is native (here actually reflecting the place from which the species was first described, although it is native to a larger area).
Blue, sometimes lilac, rose, rarely white; May–August
Prairies, along roads and railroad right-of-ways
Eastern half

Stems slender, straight or nearly so, often branched, 4–10 dm tall, glabrous, and glaucous. Roots thickened and fleshy, with some persistent root hairs. Leaves narrowly linear, under 1 cm wide, glabrous, glaucous, conspicuously dilated into the sheath. Inflorescences solitary, terminating the stem and branches. Pedicels glabrous, 0.7–2.5 cm long. Sepals glaucous, sometimes red-margined, glabrous or occasionally minutely pilose at the tip, and rarely sparsely pilose at the base, 0.7–1.3 (1.5) cm long. Petals 1.5–2 cm long. According to E. Anderson and R. L. Woodson (5), plants with sepals sparsely covered with long, slender eglandular hairs are probably hybrids between *T. ohiensis* and *T. bracteata* or *T. occidentalis*.

See plate *19*.

T. thárpii Anderson & Woodson   Shortstem spiderwort

Named after Benjamin Carroll Tharp (1885–1964), a professor of botany at the University of
    Texas at Austin and author of numerous works dealing with the Texas flora.
Purplish-blue or rose; April–May
Rocky prairies
Republic, Washington, Ottawa, Lincoln, Ellsworth, Saline, and Cowley counties

A low plant with stems 2–6 cm high, or acaulescent, during anthesis, later lengthening to 1–2 dm, villous, with 1 or 2 nodes. Leaves firm, green, often with a reddish translucent margin, the linear-lanceolate blade 1.5–3 dm long, 0.9–2.5 cm wide, loosely pilose to glabrous. Inflorescences many-flowered, solitary, and terminal. Bracts wide-spreading. Pedicels pilose, 3.5–6 cm long. Sepals 1–1.5 cm long, somewhat petaloid, purplish, rarely greenish, uniformly covered with long, thick, whitish hairs. Petals 1.5–2.2 cm long.

PONTEDERIÀCEAE   Pickerelweed Family

Herbs, perennial by rootstocks and creeping rhizomes, rooted in the mud of shores or shallow water. Flowers bisexual, more or less irregular. Sepals and petals 3 each, all of them colored and united into a tube below. Stamens 3 or 6 in our species, inserted on the perianth. Pistil of 3 united carpels. Ovary superior, 3-celled or 1-celled, 1- to many-ovuled; style 1, stigma 1–6 lobed or 6-toothed or capitate.

A family of about 6 genera and 28 species, associated with fresh water in tropical and warm-temperate regions. At least 3 species of 2 genera native to Kansas. A third genus, *Pontederia*, is reported for the state but is rare if it does occur here.

Eichornia, the water hyacinth, was introduced into the United States as an ornamental and is now becoming a notorious pest as it sometimes develops mats thick enough to clog rivers and other waterways in warm regions.

Key to Genera:
1. Leaves narrowly linear; stamens all alike; flowers yellow, solitary .... *Zosterella*
1. Leaves lanceolate or oval to reniform; stamens of 2 kinds; flowers white to blue; inflorescence 1- to several-flowered .... *Heteranthera*

HETERANTHÈRA R. & P.   Mud plantain

From Gr. *hetera*, different, + *anthera*, anther, referring to the dissimilar anthers.

Herbs of shores and shallow, quiet water, with subterranean perennial parts which give off short, erect, branching stems bearing alternate leaves with sheathing petioles, and a peduncle with 1 to few flowers having a perianth with slender tube and narrow, nearly equal, spreading sepals and petals. Stamens 3, 2 with ovoid yellow anthers, the third with a larger, oblong, greenish sagittate anther, inserted on the throat of the perianth. Ovary entirely or incompletely 3-celled. Fruit a many-seeded capsule.

Key to Species:
1. Leaf blades ovate, oblong, or lanceolate; peduncle 1-flowered; filaments glabrous .... *H. limosa*
1. Leaf blades kidney-shaped or cordate; peduncle few-flowered; filaments with long blue hairs .... *H. reniformis*

H. limòsa (Sw.) Willd.   Mud plantain

From L. *limosus*, growing in muddy places.
Blue or white; July–September
Mud and shallow water
Scattered at various locations across the state, especially in the sand-hill regions of central and
    south-central Kansas

Stems suspended in the water or creeping in the mud and rooting at the nodes. Leaf blades ovate to lanceolate or oblong, mostly 2–9 cm long and 1–3.5 cm wide; apex acute or rounded; base rounded, truncate, or somewhat cordate; petioles up to 2 dm long. Flowers borne singly. Perianth tube 2–3.5 cm long; the limb 1–2 cm long, the segments lance-linear, nearly alike, blue with a white spot at the base (or rarely all white). Filaments glabrous. Stigma 3-lobed, each lobe hairy at the tip.

H. reniformis R. & P.   Mud plantain

NL., from L. *renes*, kidneys, + *forma*, shape, in reference to the leaf blade.
White or pale blue; July–September
Mud and shallow water
Known from a few widely scattered locations in the eastern half

32. *Zosterella dubia*

Similar to the previous species but more robust. Leaf blades kidney-shaped or cordate, mostly 2–5 cm long and 2–5 cm wide; petioles mostly 0.8–1 (occasionally 2) dm long. Flowers in clusters of 3–5. Perianth tube 6–10 mm long; the limb 4–5 (occasionally 10) mm long, the inner segments elliptic, the outer ones narrower. Filaments with long blue hairs. Stigma capitate, papillose.

ZOSTERĒLLA Small   Water star-grass

A diminutive of the generic name *Zostera*, a member of the Pondweed Family; from the Gr. *zoster*, a belt, referring to the ribbonlike leaves of the original genus, + the L. diminutive suffix *-ella*, "little Zostera."

Herbaceous perennials, rooted in mud and growing with all but the inflorescence submerged in water, or stranded along the shore. Stems branching, 0.5–1 m long. Leaves alternate, elongate, linear, sessile, with thin sheathing bases. Flowers bisexual, regular, inflorescence of a single flower (sometimes 2). Perianth segments fused to form a slender, elongate, almost threadlike tube with 6 spreading yellow divisions. Stamens 3, nearly alike, with dilated filaments and sagittate anthers which coil downward after dehiscence. Ovary 1-celled with 3 parietal placentae; stigma lobed, not lifted above the anthers. Fertilization often cleistogamous. Fruit indehiscent, a 1-celled, 1-seeded utricle.

One species in Kansas.

Z. dùbia (Jacq.) Small   Water star-grass

L. *dubius*, uncertain. This species was named *Commelina dubia* in 1768 by N. J. Jacquin, director of the Botanical Garden in Vienna, before the genus *Zosterella* had been established. Jacquin's species, with its perianth undifferentiated into calyx and corolla and formed into a tube below, its herbage grasslike and submerged in the water, was indeed a doubtful *Commelina*.
Yellow; August
Shallow water and muddy shores of lake margins
South-central

See description of the genus. Leaves 2–6 mm wide, up to 15 cm long. Perianth tube 2–7 cm long. Fruit narrowly ovoid, 1 cm long. This species is sometimes included in the preceding genus as *Heteranthera dubia* (Jacq.) MacM.

Water star-grass is sometimes eaten by waterfowl.

See plate *20*.

LILIÁCEAE   Lily Family

Mostly herbaceous perennials with solid corms, scaly or coated bulbs, rhizomes, rootstocks, or woody caudex; very rarely annuals or biennials. Leaves basal or cauline, alternate or whorled, simple, the venation mostly parallel. Flowers mostly bisexual. Inflorescence various. Perianth segments mostly all petaloid and referred to as tepals (in *Trillium*, differentiated into 3 green sepals and 3 variously colored petals). Stamens 6, hypogynous or fused to the perianth, the filaments free or fused to one another. Ovary superior, 3-celled; styles 1 or 3; stigmas 3 or 1 with 3 lobes. Fruit a septicidal or loculicidal capsule or a berry.

A family of approximately 240 genera and 4,000 species distributed over most of the vegetated land areas of the earth. Over 160 genera belonging to this family are of economic importance, including the onion (*Allium* spp.), tulips (*Tulipa* spp.), lilies (*Lilium* spp.), hyacinths (*Hyacinthus* spp.), and numerous other ornamentals. *Asparagus officinalis* L., the garden asparagus, a tall perennial with dainty, filiform "leaves" (actually small branches functioning as leaves); small, greenish, bell-shaped flowers; and red berries is often found as an escape in the eastern half of the state.

Kansas has 15 native genera, including *Smilax*, woody vines not treated here.

Key to Genera (3):

1. Flowers or flower clusters produced laterally on the stem or on its branches .... *Polygonatum*
1. Flowers or flower clusters terminal or scapose 2
   2. Leaves 3, in a single whorl at the summit of the stem; flower solitary; sepals and petals differentiated .... *Trillium*
   2. Leaves in several whorls, or alternate, or basal; inflorescence various; perianth segments alike 3
      3. Principal leaves cauline, alternate or whorled, never crowded toward the base of the plant 4
         4. Stem branched, each branch bearing normal foliage leaves .... *Uvularia*
         4. Stem not branched below the inflorescence 5
            5. Perianth segments more than 30 mm long .... *Lilium*
            5. Perianth segments less than 25 mm long 6
               6. Perianth segments about 2 mm long; inflorescence rarely more than 1 dm long; stem more or less arched .... *Smilacina*
               6. Perianth segments 4 mm or longer; inflorescence 2–6 dm long; stem erect .... *Melanthium*
      3. Principal leaves at or near the base of the plant, the flower or flower cluster borne on a scape or scapelike stem which is either completely leafless or bears reduced bractlike leaves only 7
        7. Flower solitary, nodding; leaves 2 .... *Erythronium*
        7. Flowers several to many (sometimes only 1 in *Androstephium*); leaves more than 2 .. 8
           8. Flowers in an umbel or a short, irregularly branched cluster 9
               9. Flowers with a corona, pale blue .... *Androstephium*
               9. Flowers without a corona, white to pink or pale purple 10
                  10. Plants with the taste and odor of onion or garlic (except *A. perdulce*) .... *Allium*
                  10. Plants without such a taste and odor .... *Nothoscordum*
           8. Flowers in a raceme, panicle, or a spikelike raceme 11
              11. Style 1, short or elongate, or stigma nearly sessile 12
                 12. Stems arising from a woody caudex .... *Yucca*
                 12. Stems arising from a bulb .... *Camassia*
              11. Styles 3, 1 terminating each lobe of the ovary 13
                13. Petals abruptly narrowed into a basal claw .... *Melanthium*
                13. Petals not clawed .... *Zygadenus*

ÁLLIUM L.   Onion

The ancient L. name for garlic; possibly derived from Celtic *all*, pungent.

Strongly scented, pungent perennial herbs with leaves mostly grasslike, rarely tubular, mostly springing directly from a coated bulb. Flowers in simple umbels, terminating a naked stem (scape) or rarely a leafy stem, initially enclosed by a spathe composed of 2–4 membranous fused bracts. Divisions of the perianth all petal-like. Stamens 6, the filaments dilated at the base. The style threadlike; the stigma simple or slightly 3-lobed. Fruit a capsule splitting into 3 parts, each bearing 1–2 black seeds.

Cultivated plants belonging to this genus include the various onions (*A. cepa* L.), garlic (*A. sativum* L.), shallots (*A. ascalonicum* L.), chives (*A. schoenoprasum* L.), scallions (*A. vineale* L.), and leeks (*A. porrum* L.). Several of our wild species are also edible.

Six species and 4 varieties are native to Kansas. The 4 most common species and their varieties are included in the following key. The cultivated species occasionally escape from gardens and survive for awhile in the wild, and *A. vineale* is naturalized in some areas. *A. textile* A. Nels & MacBr. and *A. cernuum* Roth. are apparently native in the western part of the state but are very uncommon.

Key to Species:

1. Outer bulb-coat membranous; fall flowering .... *A. stellatum*
1. Outer bulb-coat fibrous and reticulated; spring flowering 2
   2. Flowers replaced wholly, or in part, by bulblets .... *A. canadense* var. *canadense*
   2. Inflorescence without bulblets 3
      3. Bulb-coat network coarse; bulbs lacking the odor of onion or garlic .... *A. perdulce*
      3. Bulb-coat network fine; bulbs with the odor of onions or garlic 4
         4. Bracts of the spathe 3- to 7-nerved (the following varieties not very distinct) 5
            5. Perianth pink or pinkish 6
               6. Pedicels filiform .... *A. canadense* var. *mobilense*
               6. Pedicels thicker .... *A. canadense* var. *lavandulare*
            5. Perianth white .... *A. canadense* var. *fraseri*

4. Bracts of spathe 1-nerved .... *A. drummondii*

33. *Allium canadense* var. *canadense*

34. *Allium canadense* var. *mobilense*

35. *Allium drummondii*

LILY FAMILY

50

A. canadénse L.   Wild garlic

L. -*ensis*, a suffix added to place nouns to make adjectives—"Canadian," in this instance; indicates the source of the plant used in the original description of the species.
White or pink, sometimes replaced by small bulblets; mid-April–June
Prairies, roadsides, and occasionally in open wooded areas
For distribution, see discussion of varieties below

Bulbs ovoid-conic, 0.8–3 cm in diameter, with a fibrous-reticulate outer bulb coat and whitish inner coats; bulbs with a strong smell of onion or garlic. Leaves concavo-convex in cross section, 1–3 dm long and 1.5–5 mm wide. Flowering scape 1.5–5 dm tall; spathe separating into 3, 3- to 7-nerved segments. Umbels with 15 to many flowers or pedicels. Flowers when present white or pink on pedicels 1–3 cm long, perianth broadly campanulate. Fruit lacking apical projections. Seeds black, shiny, alveolate, each alveolus with a minute pustule in the center.

As indicated in the key, this species is represented in Kansas by 4 varieties. *A. canadense* var. *mobilense* is known from Cherokee County only. *A. c.* var. *canadense* is distributed from Cherokee County west to Sumner and north to Wabaunsee and Jefferson counties. *A. c.* var. *lavandulare* overlaps the distribution of the previous 2 species but occurs west as far as Barber, Reno, and Marshall counties. *A. c.* var. *fraseri* occurs primarily in a triangular area enclosed by a line drawn from Seward County to Republic and Cowley counties and back to Seward County.

Few fertile flowers are produced by *A. canadense* var. *canadense*. The bulblets produced in the inflorescence serve instead to propagate the plants vegetatively—each bulblet being capable, once it has fallen to the ground, of developing roots and leaves and carrying on its existence as an independent individual.

Indians used this and other species of *Allium*, either raw, as a salad, or cooked with meat or in soups. When cows crop parts of the wild onion plants along with grass, the taste of onions can be detected in the milk; and when bees gather nectar from the flowers, the honey is tainted, though the taint disappears with long ripening of the honey.

A. drummóndii Regel   Wild onion

Named after James Drummond, English botanist who in the second quarter of the 19th century explored the Hudson highlands, Texas, and Louisiana.
Pink or white; late April–mid-May.
Prairie
Western two-thirds

Bulbs ovoid, 1–2 cm in diameter, with a mild smell or taste of onion or garlic; outer bulb coats brown, finely meshed, the reticula usually remaining closed. Leaves channeled, concavo-convex in cross section, 8–20 cm long and 1–4 mm wide. Flowering scape 1–2.5 dm tall; spathe separating into 2 or 3, 1-nerved segments; umbel with 10–20 flowers. Perianth rotate-campanulate, pink or white, the segments usually with a dark-rose midrib, 5–8 mm long; pedicels 0.5–2 cm long. Fruit lacking apical projections. Seeds black, shiny, finely alveolate, each alveolus usually with a minute pustule in the center.

A. perdúlce S. V. Fraser   Fraser's onion

NL., from L. *per*-, very, + *dulcis*, sweet, in reference to the fragrant flowers.
Rose; April–early May
Prairie, frequently on sandy soil
Central Kansas

Bulbs ovoid, 1–1.6 cm in diameter, lacking the typical onion or garlic odor; outer bulb coat dark brown, coarsely fibrous, with open reticula. Leaves channeled, concavo-convex in cross section, 10–15 cm tall; spathe separating into 2 or 3 mostly 5-nerved segments; umbel mostly with 3–5 (–20) flowers. Perianth urn-shaped, rose-colored, and quite fragrant, 7–8 mm long; pedicels 0.8–0.9 cm long. Fruit lacking apical projections. Seeds dull black, the alveoli without minute pustules.

A. stellàtum Ker.   Prairie onion

L. *stellatus*, starry, in reference to the cluster of flowers which somewhat resembles a starburst.
Rose-tinted; August–early October
Prairies, roadsides, and open woods
Primarily in the eastern two-fifths

Bulbs ovoid, 1–2.5 cm in diameter, with the odor of onion or garlic; outer bulb coats membranous, gray or brown. Leaves channeled, V-shaped in cross section, 2–3 dm long and 1–5 mm wide. Flowering scapes mostly 2–3 dm tall; spathe usually separating into 2, 7-nerved segments; umbel with 20 to many flowers. Perianth stellate, deep pink or rose, 5–8 mm long; pedicels 1.2–2 cm long. Fruit with 6 conspicuous projections at the apex. Seeds dull black, finely alveolate, the alveoli minutely roughened.

ANDROSTÉPHIUM Torr.   Funnel lily

From the Gr. *aner, andros*, man, + *stephanos*, crown or crest, the toothed upper margin of the corona giving the appearance of a crown.

Herbaceous perennials. Flowering stems scapose, arising from a corm. Leaves basal, few, grasslike. Perianth funnelform, the segments united below, spreading or ascending above. This genus is unusual among the Liliaceae in having the stamen filaments partly united into a tube (corona) as in *Narcissus*, this with short lobes or teeth between the anthers. Fruit a 3-celled capsule, obtusely 3-angled.

The genus, containing 3 species, appears to be native only to Mexico, Texas, Oklahoma, Kansas, New Mexico (?), Arizona (?), and California. One species in Kansas.

A. coerùleum Greene   Blue funnel lily

L. *coeruleus*, dark blue, referring to the color of the corolla (which nevertheless is sometimes
  pale blue).
Blue; mid-April–mid-May
Prairie
Central and south-central Kansas

Corm fibrous-rooted, residing about 8 cm below the surface of the soil. Foliage glabrous, gray-green. Flowering scape erect, 1–2.5 dm high. Leaves linear, up to 3 dm long. Flowers regular, in umbels, 1–9 per plant (increasing in number with age of the plant), about 2 cm long, fragrant, bright to pale blue, funnelform, with the limb slightly spreading or ascending. Lobes of the corona oblong, exceeding the anthers. Mature capsule subglobose to obovoid, 10–15 mm long, beaked. Seeds large, about 8 mm long, black, flattened, and somewhat wrinkled.

Compare with *Nemastylis geminiflora*.

The first year in the life of this plant there is a tiny bulblet bearing a single leaf and a few rootlets. Then follows in succeeding years the production of a fleshy turgid taproot, stored mostly with sap. The corm ultimately resides about 8 cm below the surface, being brought to that depth by the contraction of the turgid root, which is annually renewed until the corm arrives at its proper depth. Examining an entire plant at the time of its blooming in April, one finds that the corm consists of an upper and a lower segment, the latter producing a ring of roots around its equatorial circumference (not at the bottom, as portrayed in some manuals), the upper segment bearing the leaves and inflorescence. The following April, the bottom segment will have disintegrated, the top segment will have become the bottom segment, bearing a new top segment and a new turgid taproot, if the normal depth of the corm has not yet been reached. Thus the plant never grows old, no part of it ever living beyond its second year. The perennial nature of this plant is inherent in its growing point which each year, while producing a new corm segment with its leaves and inflorescence, is at the same time replacing every cell of its structure by new cells.

See plate *21*.

CAMÁSSIA Lindl.   Camassia

NL., from the Indian name for this plant, *kamas*, meaning "sweet."

Perennial herbs arising from a tunicated bulb. Leaves all basal, linear. Flowers bisexual, regular, arranged in a loose, many-flowered, bracted raceme. Sepals and petals essentially alike, separate, spreading to erect. Stamens 6, hypogynous. Ovary 3-celled, style 1, stigma 3-lobed. Fruit a short 3-angled loculicidal capsule producing several black, roundish seeds per cell.

One species in the state. The later blooming individuals with larger inflorescences and larger flower parts are segregated by some authors as a distinct species, *Camassia angusta* (Engelman & Gray) Blankinship, but I am including them all in *C. scilloides* here.

C. scillòides (Raf.) Cory   Atlantic camassia, wild hyacinth

NL., from L. *scilla*, squill, + Gr. suffix *-oides*, similar to.
Light blue, lilac, or white; April–May

36. *Allium stellatum*

37. *Androstepheum coeruleum*

38. *Androstepheum coeruleum*
fruiting; note corms

LILY FAMILY

51

Prairie meadows
Primarily in the east fourth

Bulbs 1–3 cm thick, residing about 8 cm below the surface of the soil. Leaves 4–16 mm wide and 2–3.5 dm long, grasslike, appearing in April about 2 weeks before the flowers. Scape to about 5 dm tall, stout. Pedicels 8–15 mm long, subtended by membranous bracts. Perianth blue to pale lilac or white, 7–14 mm long, withering and persistent at the base of the capsule.

See plate 22.

39. *Camassia scilloides*

ERYTHRÒNIUM L.   Fawn lily, dog-tooth violet

NL., from Gr. *erythros*, red (see further comments under *E. albidum* below).

Herbaceous perennials arising from a solid corm situated about 2 dm below the surface of the ground, reproducing vegetatively by means of subterranean rhizomes, buds, or droppers. Leaves 1 in immature forms, 2 in flowering forms, tapering into petioles which sheathe the base of the flowering scape, most of which is belowground, the leaves thus appearing basal. A single nodding, lilylike flower with 6 recurved or spreading tepals. Stamens 6, hypogynous, filaments awl-shaped, anthers often dimorphic. Ovary 3-celled with several to many ovules; style slender below, thickened above to the 3 short stigmas. Fruit an oblong to obovoid loculicidal capsule.

As to the distribution of the species of *Erythronium*, the Old World has only 1 species, while our own country has about 15, the choicest occurring in the Far West—Mount Rainier National Park presenting the ravishing avalanche lily, *E. montanum*, whose large flowers are pure white with orange bases; and Glacier National Park spreading before its summer visitors sheets of the golden-flowered glacier lily, *E. grandiflorum*. Kansas has 3 species.

Key to Species (6):

1. Perianth segments yellow; under side of leaves not glaucous .... *E. rostratum*
1. Perianth segments white or pale pink; under side of leaves glaucous 2
   2. Perianth segments reflexed in full bloom; leaves mottled; mature fruits held off the ground .... *E. albidum*
   2. Perianth segments spreading or at most only half-reflexed in full bloom; leaves not mottled; mature fruits resting on ground .... *E. mesochoreum*

40. *Erythronium albidum*

E. álbidum Nutt.   White fawn lily

L. *albidus*, white. The curious combination of red in the genus name and white in the species name requires explanation. In naming the genus in 1753, Linnaeus latinized the Greek name of a red-flowered orchidaceous plant, *Erythronion*. The species Linnaeus used in establishing the genus he named *Erythronium dens-canis* meaning "dog's tooth." Since this plant has rose-purple flowers and a corm shaped somewhat like a dog's tooth, the name Linnaeus gave it seems sensible. But the red in the genus name does not fit the yellow- and white-flowered American species. So, also, the common name, "dogtooth violet," most frequently used in this country, does not make sense, for the corms are too thick to resemble a dog's tooth, and the genus belongs to the Lily Family, far removed from the Violet Family. However, it may be recalled, early English writers sometimes gave the name "violet" to many kinds of dainty flowers.

It was John Burroughs who suggested "fawn lily" for the eastern yellow-flowered *E. americanum*, which has leaves mottled with brown, like a fawn's back. "Its two leaves stand up like a fawn's ears, and this feature together with its recurved petals, gives it an alert, wide-awake look." In general, these characters fit our Kansas species well enough to justify the use of his common name for them as well.

Pinkish-white; March–April
Moist woods
Primarily in the eastern fourth

Leaves lanceolate to ovate-lanceolate, abruptly attenuated, flat to half-folded, mottled on both sides with purplish-brown or light green, glaucous on both sides. Tepals usually reflexed in full bloom, with a yellow spot at base of the inner 3, white, often tinged on the under side with pink, pale blue, or lavender, 2.5–5.5 cm long. Stamens 6, filaments opposite the inner tepals slightly longer than those opposite the outer ones, anthers opposite the inner tepals maturing before those opposite the outer ones. Style not persistent on the fruit; stigmas with 3 long, slender, divergent lobes. Fruit obovoid to ellipsoid, held erect and 1.5–2 cm long at maturity.

Sterile (nonflowering) individuals propagate vegetatively by means of buds which develop on the corm, producing more corms, or by means of horizontal subterranean stems which grow laterally and produce a new corm at the tip. Flowering individu-

41. *Erythronium albidum* fruiting

LILY FAMILY

may also form buds on the corms.
See plate *23*.

E. mesochòreum Knerr

Midland fawn lily

NL., from Gr., *mesos*, in the middle, + *choros*, place, this species being of midcontinental
distribution.
White, sometimes tinged with lavender; March–April
Prairies; pastures; dry, open woods
Eastern third

Similar to the previous species. Leaves lanceolate to linear-lanceolate, gradually
attenuated, conduplicate or occasionally half-folded, usually not mottled, glaucous on
both sides. Perianth segments spreading or at most half-reflexed in full bloom, (1.3–)
2–4.5 cm long. Fruit obovoid to subglobose, resting on the ground at maturity, 1.2–1.7
cm long.

Sterile (nonflowering) individuals propagate vegetatively by means of buds, as in
the previous species, or by means of vertical subterranean stems called droppers, which
grow *downward* from the old corm and form a new corm at the tip. Flowering individuals may also form buds on the corm.

42. *Erythronium mesochoreum*

We shall use this species to illustrate the generic habits of flowering and seed
dispersal: Before the leaves come up, the flower bud, erect on its stalk, is tightly enclosed
by them, the thickened turgid tip of one leaf capping the other leaf tip and forming a
spearhead for piercing the soil. As soon as the leaves appear, the flower bud is pushed
out and the peduncle bends, inverting the flower, so saving the nectar from dilution by
rain, and better rewarding insects that effect cross-pollination. It seems that there is a
chance for self- as well as cross-pollination, for before the bud opens, the 3-lobed stigma
lies among the 3 longer, opened anthers and picks up pollen from them, as one can see
by opening a bud; but as the flower opens, the style lengthens and presents the stigma
beyond the anthers, in position to receive pollen from bees and butterflies coming from
other flowers.

After fertilization the peduncle grows tall and straightens its crook, and the ovary
gets fat with seeds till the stalk is bent over by its weight, and the ovary, now the ripened
capsule, rests upon the ground at some distance from the parent plant. Now the capsule
breaks into 3 parts corresponding to the 3 cells of the ovary, which then open and deposit
a little heap of seeds upon the ground.

The seedling plant begins as a single grasslike leaf, a tiny shallow corm, and a
few fibrous roots, but year by year the corm gets deeper, by a short downward extension
at its base which produces a new fascicle of roots, until the normal depth is reached.

E. rostràtum Wolf

Yellow fawn lily

L. *rostratus*, having a beak, from L. *rostrum*, beak, in reference to the persistent style on the fruit.
Yellow; March–mid-April
Moist woods
Cherokee County only

Leaves at flowering time strongly mottled on the upper side with purplish-brown
pigment, not glaucous. Tepals 2.5–5 cm long, the outer 3 with intense purplish-brown
specks on under side; inner 3 with well-developed auricles at the base clasping the opposite filaments. Stigma lobes swollen, short, and erect; style persistent and forming a
prominent beak on the capsule. Fruit an ellipsoidal capsule erect at maturity.

LÍLIUM L.

Lily

L. *lilium*, lily.

Tall herbs perennating from a scaly bulb. Stems simple, with numerous alternate
or whorled leaves. Flowers bisexual, regular, 1 to several in number. Perianth of 6
colored tepals, funnelform or campanulate, tepals spreading or recurved. Stamens 6,
hypogynous. Ovary 3-celled; style elongate or very short; stigma 3-lobed. Fruit a sub-
cylindric to ovoid capsule with numerous flat, closely packed seeds in each cell.

One species native to Kansas. *Hemerocallis fulva* L., the cultivated day lily, commonly escapes from gardens. It differs from *Lilium michiganense* in having leaves all
basal, rather than cauline and whorled, and orange flowers lacking spots.

In prizing the lily as second only to the rose in splendor, Pliny had in mind the
Madonna lily, *Lilium candidum* (Latin *candidus*, dazzling white); and when Jesus said,
"Even Solomon in all his glory was not arrayed like one of these," he was probably
referring to the Scarlet turk's-cap, *Lilium chalcedonicum* (Latin *chalcedonius*, pertaining

43. *Lilium michiganense*

44. *Melanthium virginicum*

to Chalcedon, an ancient Thracian city opposite Constantinople), native to Greece and Persia and occurring in Galilee around Lake Gennesaret. The lily of our prairie resembles the turk's-cap and is not lacking in the elegance that stamps the genus *Lilium*.

L. michiganénse Farwell   Michigan lily

Michigan + L. *-ense*, a suffix added to place nouns to make adjectives, in this case the adjectival form of "Michigan," this being the state from which the plant was first described.
Orange and orange-yellow, with brownish-purple spots; late June–early July
Moist prairies and low, wooded slopes
Eastern fourth; rare

Stem erect, 2 or more feet tall. Leaves lanceolate, sessile, mostly in whorls of 3–12, but only 1 or 2 at the upper nodes, 4.5–10 cm long. Flowers nodding on strong peduncles, solitary or several together at the end of the stem. Perianth segments orange with brownish-purple spots, strongly recurved or backward rolled, 5.5–8 cm long.

A horizontal rhizome produced annually as an offshoot of the bulb terminates with the formation of a new bulb. In this way the plant multiplies itself and forms colonies, so that a single bulb planted in a garden is the source of numerous progeny. In our prairies the plants occur only occasionally, either singly or sometimes in colonies, on the lower levels or slopes where drainage provides the most moisture.

The flowers are fitted for both cross- and self-pollination. Two things favor cross-pollination, namely, nectar secreted inside the thickened bases of the perianth parts, and, when the flower first opens, the position of the stigma outside and about the circle of stamens, brought about by the pronounced upward curvature of the long, stiff style. At this stage, the anthers open so that a hawk moth, for instance, poised on wing while sucking nectar, could get pollen on its head and proboscis and then carry it to the stigmas of older flowers which have been brought to a lower level inside the circle of anthers by the straightening of the style. But this later stage also favors self-pollination, for the stigmas are then where they can be touched by the anthers or catch falling pollen. We know that self-pollination is likely to be effective, for long ago Christian Konrad Sprengel, an 18th-century naturalist whose observations were of major importance in the field of pollination ecology, placed a gauze bag over a cluster of unopened buds of a turk's-cap lily and sewed the mouth tight to keep insects out. Yet capsules were subsequently formed and filled with good seeds. Sprengel was puzzled that a flower so evidently designed to attract insects by its large size, striking color, and provision of nectar could nevertheless produce seeds without the aid of insects; for he had been convinced by his previous observations "that the wise Author of nature has not created even a single hair without a definite object." Bewildered by his experience with the lily, Sprengel remarked that if he had been more concerned about proving his theory than searching for the truth he never would have reported the experiment.

The Michigan lily is treated by some authorities as a subspecies of *L. canadense*. See plate *24*.

MELÁNTHIUM L.   Bunchflower

Gr. *melas*, black, + *anthos*, flower, possibly from the fact that the perianth turns darker after blooming.

A tall leafy herb perennating from a stout rhizome. Leaves linear, alternate, and sheathing the stem. Stem and branches of the inflorescence scurfy. Flowers mostly bisexual, some polygamous, regular. Inflorescence a many-flowered terminal panicle. Perianth of 6 separate tepals, at first cream-colored or greenish, later turning darker, tepals clawed (abruptly narrowed) at the base and bearing a pair of glands at the base of the blade (the broadened part of the tepal). Stamens 6, filaments fused to the base of the tepals. Ovary 3-lobed, each lobe with a short style and a linear stigma. Fruit an ovoid septicidal capsule subtended by the withered perianth. Seeds several, flat, broadly winged.

One species in Kansas.

M. virgínicum L.   Bunchflower

"Of Virginia," an epithet used frequently by Linnaeus when naming plants brought to Europe from the American colonies.
Cream-colored, partially green-tinted; mid-June–mid-July
Moist meadows
Leavenworth, Douglas, and Miami counties; uncommon

A stately perennial with a basal cluster of broad linear leaves 3 dm or more in length and a stout stem up to 15 dm tall terminating in an ovoid panicle about 3 dm long.

Tepals spreading, cream-colored to green, 6–13 mm long, the blade twice as long as the claw, oblong to ovate or obovate, glandular at the base. Fruit ellipsoid, 3-lobed, 10–15 mm long. Seeds 5–8 mm long.

Examining flowers of different ages we find provision for both self- and cross-pollination. The opening flower has its 3 elongate, stiff styles spread apart and the 3 of its 6 stamens that are opposite the styles standing somewhat erect, with open, pollen-covered anthers—the stamens being on their way to becoming erect and finally inflexed, affording in the process the chance of contact between the anthers and the stigmas. While this is going on, the other 3 stamens are arched outward and downward so that the unopened anthers are held over their adjacent nectar glands; these anthers soon open, while their filaments begin to straighten and become erect and then incurved; but since these stamens stand between the styles, instead of opposite them as in the case of the first 3, their anthers do not make contact with the stigmas but are ready to supply pollen for cross-pollination by the bees, beetles, butterflies, and flies that visit the flowers for pollen or nectar, as the case may be. Finally all 6 stamens, with anthers barren of pollen and shriveled, are huddled around and over the ovary, while the styles stick out horizontally beyond them, still looking fresh, as though bidding for pollen brought from younger flowers.

If one would like to compare cases of this kind where stamens make movements favoring pollination, with others where the style makes similar movements, he may refer to *Teucrium canadense* and *Sabatia campestris*.

45. *Nothoscordum bivalve*

NOTHOSCÓRDUM Kunth   False garlic

Gr. *nothos*, false, + *skordon*, garlic.

Perennial herbs from a tunicated bulb, lacking the odor of onion. Leaves basal, linear. Flowers bisexual, regular, displayed in a terminal umbel subtended by 2 membranous bracts. Tepals 6. Stamens 6, filaments fused to the base of the tepals. Ovary shallowly 3-lobed; style straight and slender; stigma 1, minute. Fruit a 3-valved loculicidal capsule with 6–10 seeds per cell.

One species in Kansas.

N. biválve (L.) Britt.   False garlic

L. *bis*, twice, + *valva*, leaf of a folding door, in reference to the fact that the umbel is subtended,
        and in its younger stages enclosed, by 2 membranous bracts of the involucre.
Yellowish-white; April–May
Prairies, pastures, open woods
South and east of a line from Barber County to Shawnee County

Bulbs ovoid, 1–2 cm in diameter, with a tuft of fibrous roots at the base. Leaves ascending, grasslike, 1–3 dm in length. Scape slender, 1–3 dm in length, umbel with 5–12 flowers. Tepals oblong, 5–12 mm long. Capsule subglobose, about 5 mm long. Seeds black, angularly ellipsoid, 3 mm long.

This plant looks in all respects like a small edition of a flat-leaved wild onion but lacks the onion smell. It blooms at the same time and often in the company of *Sisyrinchium* and is sometimes confused with it by the casual observer, though there are several fundamental differences in the inflorescences, as indicated below:

Nothoscordum	*Sisyrinchium*
Ovary superior, stigma 1	Ovary inferior, stigmas 3
Stamens 6, free from one another	Stamens 3, united by their filaments
Involucral bracts membranous, finally reflexed and withering	Involucral bracts rigid, remaining stiffly upright

Both long-tongued and short-tongued bees come for nectar sparsely exuded between the stamens and the ovary.

POLYGONÁTUM Mill.   Solomon's seal

NL., from Gr. *polys*, many + *goney*, knee, suggested by the jointed rhizome. Regarding the common name, the circular, seal-like scars on the rhizomes are the inspiration of various legends, one to the effect that Solomon impressed the rhizomes with his seal as a voucher for their medicinal value. An old German story tells that Solomon used the rhizomes as magic in breaking asunder the rocks quarried for building his temple. Gerard's herbal (13) gives from Dioscorides (1st century) a still different interpretation: "Dioscorides writeth, that the roots are excellent good for to seale or close up greene wounds, being stamped and laid thereon; whereupon it was called *Sigillum Solomonis*, of the singular

46. *Polygonatum biflorum*

47. *Polygonatum biflorum* with stamens and pistils exposed

48. *Polygonatum biflorum* fruiting

LILY FAMILY

vertue that it hath in sealing or heading up wounds, broken bones, and such like. Some have thought it tooke the name *Sigillum* of the markes upon the roots; but the first reason seems to be more probable."

Herbs perennial from a creeping knotted rhizome. Stems usually unbranched, erect or arching. Leaves alternate, 2-ranked, nearly sessile or clasping the stem. Flowers perfect, regular, borne in clusters of 1–15 on axillary peduncles. Tepals 6, fused laterally to form a tubular perianth with 6 lobes at the summit. Stamens 6, inserted on or above the middle of the tube. Ovary 3-celled; style slender, deciduous by a joint; stigma obtuse or capitate, obscurely 3-lobed. Fruit a globular, black or blue berry with 1-2 seeds per cell.

One species in Kansas.

P. biflòrum (Walt.) Ell.   Great Solomon's seal

L. *bi*, two, + *flos, floris*, flower, the tubular flowers frequently being borne in pairs.
Flowers pale, green-tinted; May–July
Moist, wooded hillsides and banks
Eastern half, but more common in the eastern third

Perennial from a thickened, jointed, horizontal rhizome, renewed by a joint each year, which at its apex gives off a single, ascending, unbranched stem bearing leaves and flowers, the stem sloughing away from the rhizome in the fall, leaving a prominent seal-like scar. Stems 4–12 dm long. Leaves sessile, bases often clasping the stem, lance-elliptic to broadly oval, lighter colored and glaucous on the under side, with 1–19 prominent veins, those midway up the stem 3–5 cm wide and 8–14 cm wide. Tubular flowers 2 or 3 (less frequently 5) together, pendant, on slender pedicels at the end of an elongate axillary peduncle, mostly 1.5–2 cm long. Berry about 1 cm wide.

If one cuts the perianth open longitudinally (fig. 47), he finds that the stigma and 6 anthers are at the same level in the entrance of the tube, the anthers facing inward and discharging pollen over the already ripe stigma and so accomplishing self-pollination. However, cross-pollination also is provided for, as nectar secreted inside the tube at the base of the ovary attracts bumblebees and butterfles with proboscises long enough to reach the nectar. In getting the nectar these insects inevitably contact both stigma and anther and effect cross-pollination as they go from flower to flower.

Following pollination and the consequent fertilization of the egg contained in each of the several ovules of the 3-celled ovary, the perianth comes loose and slips off over the apex of the ovary, which is already proceeding to become a dark-purple berry, inedible for us, it is safe to assume, from what Hieronymus Bock says in his herbal (1577) concerning the European species, *P. multiflorum:* "Twelve of the little dark-colored berries when eaten purge from top to bottom."

The rhizomes, however, which contain a good percentage of a variety of carbohydrates, principally starch, were often cooked and eaten by our Indians and sometimes by pioneers when their customary rations ran out; and the young shoots were prepared and eaten like asparagus. Further, since the beginning of the Christian Era, at least, the rhizomes of Solomon's seal have been extracted or dried and powdered for making cosmetics, dressing wounds, and preparing demulcents and tonics. Nevertheless, we can but wonder whether a curious entry in Gerard's herbal could have been read with a straight face in his own day: "The root of Solomon's seal stamped while it is fresh and greene, and applied, taketh away in one night, or two at the most, any bruise, blacke or blew spots gotten by fals, or womens wilfulnesse, in stumbling upon their hasty husbands fists, or such like."

SMILACÌNA Desf.   False Solomon's seal

There are two slightly differing ideas as to the derivation of this generic name. W. C. Stevens (28) states that the NL. suffix *-ina* means "derived from" and in this case indicates a relationship to *Smilax* L. M. L. Fernald, in the 8th edition of *Gray's Manual of Botany*, says however that *Smilacina* is a diminutive of *Smilax*, which was an earlier name used by Tournefort (1700) for these plants and does not indicate any relationship or similarity to the *Smilax* of Linnaeus and later botanists.

Herbs perennating by means of a long creeping rhizome. Stems unbranched, erect or ascending. Leaves alternate, sessile or nearly so. Flowers bisexual, regular, arranged in a terminal raceme or panicle. Tepals 6, lanceolate. Stamens 6, hypogynous. Ovary globose, 3-celled; stigma obscurely 3-lobed. Fruit a 1- to 2-seeded globular berry.

Two species in Kansas.

The underground parts of *Smilacina*, after having been presoaked in lye and then boiled in water, have been used as a minor food source by some Indians. The young leaves and stems may be boiled and eaten as green vegetables. The berries have cathartic properties but are eaten by some birds and small mammals.

Key to Species:

1. Leaves broadly elliptical or ovate, apexes abruptly acuminate; flowers in a panicle; fruit red dotted with purple .... *S. racemosa*
1. Leaves lanceolate, sometimes narrowly so, apexes gradually tapering; flowers in a raceme; fruit at first green with blackish stripes, becoming uniformly red .... *S. stellata*

S. racemòsa (L.) Desf.   Feather Solomonplume, False Solomon's seal

L. *racemosus*, clustering, in reference to the panicled cluster of flowers. This epithet is sometimes a source of confusion, since, botanically speaking, this species actually has its flowers arranged in a panicle, while our other species, *S. stellata*, has true racemes.

White to greenish-white; May–July
Rich, wooded hillsides and stream banks
Easternmost 2 columns of counties

Rhizome fleshy, brownish, knotty, elongated, often 1 cm or more thick, continuous or barely forking. Stem usually curved-ascending, finely hairy, 4–8 dm long. Leaves spreading horizontally in 2 ranks, elliptic to ovate, rounded at the base, short-acuminate, finely hairy beneath, middle leaves on the stem 11–19 cm long and 4–6.5 cm wide. Panicle peduncled or rarely sessile, 3–15 cm long. Flowers very numerous, about 3 mm wide. Tepals greenish-white, lanceolate. Style short and thick, less than 0.5 mm long. Fruit red dotted with purple, 4–7 mm wide.

This is the more common of our 2 species.

49. *Smilacina racemosa*

S. stellàta (L.) Desf.   False Solomon's seal

L. *stellatus*, starry, in reference to the cluster of flowers or possibly to the spreading perianth.
White to greenish-white; May–June
Moist, especially sandy soil of woods, shores, and prairies (usually in shaded places)
Known from Republic and Scott counties; uncommon

Rhizomes slender, pale, freely forking. Stem ascending or usually erect, finely hairy or glabrous, 2–6 dm tall. Leaves spreading or more often strongly ascending, frequently folded, sessile and somewhat clasping, lanceolate or lance-oblong, gradually tapering, finely hairy beneath, 8.5–11 cm long and 1.5–3 cm wide. Raceme short-peduncled or nearly sessile, 2–5 cm long, with few to several greenish-white flowers 8–10 mm wide. Style one-third to half the length of the pistil. Fruit at first green with blackish stripes, becoming uniformly dark red, 6–10 mm wide.

TRÍLLIUM L.   Wake-robin, trillium

NL., *trillum*, euphonius combinaton of L. *tri-*, three, + the genus name *Lilium*, in reference to the leaves and flower parts in whorls of 3. "With me this flower is associated not merely with the awakening of the robin, for he has been awake some weeks, but with universal awakening and rehabilitation of nature."—John Burroughs.

Herbs perennating by a short, transversely marked rhizome, which dies away behind as it grows forward each year. As it advances, the rhizome produces frequent, fibrous, anchoring roots and a few contractile roots, and ends with a terminal bud for producing the leaf- and flower-bearing shoot the following spring. Aboveground stem erect. Leaves 3, in a whorl at the tip of the stem. Flowers bisexual, regular. Inflorescence a single terminal flower, either sessile or pediceled. Sepals 3, green. Petals 3, white or variously colored. Stamens 6, hypogynous. Ovary 3- to 6-angled or lobed, 3-celled with several to many ovules in each cell; styles 3 with inner stigmatic lines. Fruit a many-seeded berry.

Two species in Kansas. A third species, *T. gleasoni* Fern., has been reported from Shawnee County, but is of doubtful occurrence as a native.

Key to Species:

1. Leaves broad to nearly orbicular, abruptly narrowing to a short point; stamens half as long as the petals .... *T. sessile*
1. Leaves elliptic to ovate, very long-tapering to the apex; stamens a third as long as the petals .... *T. viridescens*

T. séssile L.   Toad trillium

L. *sessilis*, low, dwarf.
Petals purple, or purple with green borders, sepals green; April–May
Moist, rich woods

Extreme east-central and southeastern Kansas (Miami, Linn, Bourbon, Crawford, and Cherokee counties)

Stem stout, 2–5 dm tall at anthesis. Leaves broadly lanceolate to subrotund, sessile, mottled or green, apex acute or short-acuminate. Sepals widely spreading, linear-oblong to narrowly elliptic, 3–4 cm long, tip usually obtuse. Petals erect, oblanceolate to narrowly elliptic or nearly linear, narrowed toward the base but not clawed, 4–8 cm long, 3 times as long as the stamens, apex acute or acuminate.

50. *Trillium viridescens*

T. viridéscens Nutt. Wake-robin

L. *viridescens*, becoming green, the erect green sepals obscuring the petals, which are also sometimes greenish.
Petals green, yellow, or purple; April–May
Rich, moist wooded slopes
Cherokee County

Stem 1–3 dm tall at anthesis. Leaves broadly ovate to rotund, sessile, apex broadly rounded or obscurely cuspidate, not mottled, seldom over 9 cm long at anthesis. Sepals green, spreading or ascending, lance-ovate, 2–3 cm long. Petals brownish-purple, yellow, or green, ascending, narrowly to broadly elliptic, narrowed to a sessile base, equaling or slightly longer than the sepals and twice as long as the stamens.
See plate 25.

UVULÀRIA L. Big merrybells

NL., *uvularia*, from NL. *uvula*, the pendant lobe of the soft palate, in allusion to the pendant flowers.

Herbs arising from a slender rhizome. Stem erect, forked above the middle, the lower part bearing a few bladeless sheaths and 0–4 leaves. Leaves sessile or perfoliate, reaching full size after anthesis. Flowers perfect, regular, nodding, at first terminal but in fruit seemingly subaxillary through prolongation of the stem and branches. Perianth lilylike, of 6 yellow or greenish-yellow tepals. Stamens 6, hypogynous. Ovary shallowly 3-lobed, with several ovules per cell; styles 3, separate or united to beyond the middle. Fruit a 3-lobed or 3-winged loculicidal capsule. Seeds subglobose.
One species in Kansas.

51. *Uvularia grandiflora* fruiting

U. grandiflòra Smith Big merrybells

NL., from L. *grandis*, large, + *flos, floris*, flower.
Pale lemon-yellow; April–June
Rich woods
Extreme northeastern and extreme southeastern Kansas

Perennial by a short rootstock which gives rise to a dense cluster of somewhat thickened roots, and early each spring to a renewal of several upright glaucous stems up to 7.5 dm in height, bearing for the major part of their length leaves in the form of upright clasping scales, then forking and bearing above the fork several oval, elliptic or oblong, perfoliate leaves, minutely hairy below, up to 12 cm long. (The stem occasionally bears a foliage leaf immediately below the fork.) Flowers pale lemon-yellow, nodding; perianth companulate, tepals lanceolate, slightly gibbous at the base, 2.5–5 cm long. Stamens exceeding the style, anthers large, linear-oblong. Capsule 1–1.5 cm long.

The flower, pendant on a recurved terminal peduncle, is in structure remarkably like that of *Erythronium*. The similarity extends also to the straightening of the peduncle as the ovary ripens and to the arrangements for pollination, there being in both instances secretion of nectar around the base of the ovary, and retention of the nectar by adhesion to the broadened filaments and the closely surrounding base of the perianth until bees and butterflies suck it out, incidentally picking up pollen, which later they transfer to the stigmas of other flowers. In both instances, self-pollination is an alternative possibility.

YÙCCA L. Yucca, soapweed

NL., *yucca*, from West Indian *yuca*, vernacular name for the large tuberous roots of *Manihot esculenta*, of the Spurge Family, from which tapioca is obtained. The application of the name to the soapweed was mistakenly made by Gerard, who thought a specimen of it imported into his garden from the West Indies was the manihot.

Perennial from a woody caudex. Leaves linear, numerous, closely crowded at the summit of the caudex. Flowers bisexual, regular, borne in a raceme or panicle. Tepals alike, separate from one another. Stamens 6, hypogynous; filaments very stout, out-curved above; anthers sagittate, pollen sticky. Ovary 3-celled or imperfectly 6-celled;

stigmatic surfaces recessed in the cylindrical style. Fruit a large septicidal capsule with numerous flat, black, irregularly triangular seeds.

A North American genus of about 30 species. One species native to Kansas. Specimens from southeastern Kansas (Montgomery, Wilson, and Cowley counties) differ from our typical *Y. glauca* in having smaller seeds (8–10 mm long) and wider (1–2.5 cm broad), more flexible leaves that tend to be reflexed rather than erect. These plants are segregated by some authorities as *Y. glauca* var. *mollis* Engelm., or even as a distinct species, *Y. arkansana* Trel. In addition, *Y. smalliana* Fern. is commonly cultivated and occasionally escaped, especially in the revegetated strip-mined areas of southeastern Kansas. This latter species has reflexed leaves 2–4 cm wide, a much-branched inflorescence, and fruits 4–5 cm long which separate into rounded, rather than triangular, segments.

Yuccas were utilized by the Indians, especially in the Southwest, for a variety of purposes. The buds and inflorescences of some species were eaten raw or boiled, while the fleshy fruits of others were eaten fresh, roasted, dried, or ground and used as flour. A fermented beverage was made from the fruits, and the seeds were also used as food. Fiber from the leaves was employed in the manufacture of cord, baskets, mats, sandals, and cloth, and the sharp point of the leaves with attached fibers was used as needle and thread. The crushed roots, rich in saponin, were used for making a soapy lather and as a laxative.

52. *Yucca glauca*

Y. glaùca Nutt.

Small soapweed

L. *glaucus*, gray-green—the leaves being somewhat whitened by a wax film.
Greenish-white or cream-colored; May–June
Plains; dry, sandy or rocky slopes
Western three-fourths, and Wilson and Montgomery counties

An evergreen perennial arising from a stout, elongate, vertical rootstock. Eventually, short, ascending lateral branches are produced a few inches below the surface, terminating with leaf rosettes at the ground level and constituting, together with the primary rosette, a muliple crown of considerable size. Deeper down, the rootstock produces thick horizontal branches several feet in length with 1 or more ascending offshoots bearing leaf rosettes—a way of vegetatively begetting new individuals. Leaves all basal, linear-attenuate, plano-convex or somewhat 3-sided, stiffly ascending, 5–11 mm wide and 5–7 dm long, reaching higher than the lowest blossoms of the inflorescence, with needle-sharp tips and filiferous margins. Inflorescence a simple or sparsely branched raceme borne to a height of 1–2 m on a stout peduncle. Tepals 6, greenish-white or cream, broadly lanceolate to ovate, about 3–6 cm long and 1.5–3 cm wide. Stamens with thick, outward-curving filaments which hold the anthers away from the stigma (fig. 53). Ovary oblong, 3-celled, gradually tapering into the thickened style (figs. 53 and 54) which bears 3 stigmatic lines down its narrowly funnel-shaped apical cavity, this in turn leading by 3 tubes into the cells of the ovary, in each one of which are 2 lines of ovules. Capsule about 6–7 cm long. Seeds 10–13 mm long.

53. *Yucca glauca* with stamens and pistil exposed

There are 3 features of primary importance in the design of this flower: (1) the location of the stigmatic surface inside a cavity, (2) the position of the anthers remote from this cavity, and (3) the sticky pollen. These add up to the prevention of self-pollination and, hence, the necessity for some outside pollinating agent. Interestingly, a single species of insect—the Yucca moth (*Pronuba yuccasella*)—is equipped with appropriate structures and behavior patterns to do this and is on hand for pollination when the Yucca blooms. In fact, so intimate is the association of the Yucca and the moth that neither can complete its life cycle without help from the other.

Let us follow the sequence of events in this wonderful relationship. We stand before a Yucca that is to begin blooming in the twilight of a day late in May. Only a few basal flowers are ready on that evening, but others will follow, evening after evening, in a serial advance toward the apex of the inflorescence. Soon the flowers of that evening open up and we are aware of their fragrance, and not we alone, for small (about 25 mm long), silvery-white *Pronuba* moths are flying in from places of concealment during daylight hours. Watching one of these we see that she enters a flower, climbs a stamen, and deliberately scrapes pollen from the anther with a pair of specialized mouth parts called palps and packs it under her chin. She proceeds in this manner from stamen to stamen until she has a lump of pollen about the size of her head. She then flies to another flower on this plant or on another one, takes a position between 2 stamens, thrusts her ovipositor through the wall of the ovary and lays an egg in the ovary cavity where Yucca ovules are waiting to be fertilized. Now she walks to the top of the pistil and thrusts pollen into the stigmatic opening and works her head rapidly—the motion being mostly up and down

54. *Yucca glauca* flower with petals removed; note chimneylike opening to stigmatic chamber

LILY FAMILY

59

55. *Zygadenus nuttallii*

and lasting several seconds. Her tongue is used to force the pollen farther into the cavity than her palps can reach. She then takes a position between another pair of stamens, lays an egg in another cell, walks again to the top of the pistil, and inserts more pollen. This work continues, from flower to flower on this and other Yucca plants. In this way the ovules get fertilized and grow to be seeds.

But while the ovules are becoming seeds, each moth egg is hatching into a voracious little grub which eats its way through the middle of a row of seeds, progressing from seed to seed until it is fat with nutritive substances stored in its tissues. Thus being made ready for the next phase of its life cycle, it gnaws a hole through the wall of the seed pod just big enough for it to squeeze through and spins a slender thread by which it descends to the ground, then works its way 5–7 cm below the surface and spins a shroud about itself. There the creature pupates, emerging from the ground as an adult moth when the Yuccas bloom the next season.

So much for the moth, but what about the Yucca seed pod after the larva has taken its fill and departed? Fortunately, plenty of seeds remain uninjured for the plant to dispose of in its own way. The seed pods, which at first have the downward inclination of the flowers, gradually turn to an erect position, then split open longitudinally into 3 parts, exposing the broad, thin seeds to gusts of wind which may shake the seeds out and sweep them along to whatever destiny lies ahead.

For a more detailed account of the *Yucca-Pronuba* relationship, refer to C. V. Riley (7).

ZYGADENUS Michx.   Death camas

NL., from Gr. *zygos*, a yoke, + *aden*, a gland, in reference to the pair of glands at the base of the perianth parts of some species (a single gland present in ours).

Perennial herbs arising from a bulb or a thick rhizome. Stems erect. Leaves mostly basal, linear, those in the upper parts of the stem reduced or bractlike. Flowers bisexual or polygamous, regular, borne in a simple raceme, several racemes, or a panicle. Sepals and petals alike, white to greenish-white in our species. Stamens 6, free from the tepals. Ovary 3-lobed, the base sometimes fused to the base of the perianth, each lobe with a short divergent style and minute stigma. Fruit a capsule, surrounded at the base by the persistent withered perianth. Seeds angled. All species of this genus are thought to be poisonous.

One species in Kansas.

Z. nuttállii Gray   Nuttall Death camas

Named in honor of its discoverer, Thomas Nuttall (1786–1859), plant explorer and one-time professor of natural history and curator of the Botanic Gardens of Harvard.

Greenish-white; May–early June
Rocky prairie hillsides, especially in limestone soils
Eastern two-fifths of the state, especially in the Flint Hills, and more common southward

Bulbs with a papery coat, like an onion, and fibrous roots. Plants 3–5.5 dm tall, with many broadly linear, often conduplicate basal leaves 0.8–2 cm wide and 1.5–2.5 dm long, as well as smaller leaves or bracts on the upper parts of the stem. Inflorescence a raceme with widely spreading pedicels. Tepals oblong, with a short claw at the base, greenish-white, about 3 mm wide and 6 mm long, approximately the lower third of each tepal occupied by an obovate, thickened, glandular area. Capsules erect, ellipsoid, thin-walled, 3 or 4 times the length of the perianth, borne on filiform pedicels 1.5–2.5 cm long. Seeds angular.

The flowers seem to invite cross-pollination, for when they first open, the stamens spread widely away from the 3 styles, which are standing erect and close together, with stigmatic surfaces remote from the anthers. Soon, however, the anthers are gone or the stamens shriveled, and then the styles spread apart as if inviting pollination by insects coming from younger flowers.

It is important to know that all parts of the death camas are poisonous to man and livestock, although hogs seem to be immune. Also in early spring, when the basal leaves are up, there is so strong a resemblance between them and the leaves of the wild hyacinth and wild onion that even the Indians when digging bulbs of the latter two for food sometimes included by mistake bulbs of the death camas and were seriously poisoned by them. Let us, then, note some distinguishing features of the leaves: Leaves of the onion have the dependable onion taste and smell; leaves of the wild hyacinth and the death camas both lack this smell but are distinguishable by the fact that when cross sections of a wild hyacinth leaf are cut the midrib is seen to be hollow, a feature not occuring in death camas leaves.

The pollen and nectar apparently are not poisonous to insects, since the larvae of certain species customarily devour entire inflorescences.

56. *Dioscorea villosa*

DIOSCOREÀCEAE

Yam Family

Herbaceous twining vines, sometimes more or less woody basally, perennial by woody or fleshy, tuberous rootstocks. Leaves alternate or opposite, petioled, usually cordate or hastate, venation mostly palmate-reticulate with the primary veins extending to the apex. Flowers small, generally unisexual and then dioecious (sometimes monoecious), regular, in slender spikes, racemes, or panicles. Sepals 3, petals 3, all similar in form and color. Stamens 3 or 6. Ovary inferior, mostly 3-celled, ovules 2 or 1 in each cell; styles 3, distinct. Fruit a 3-angled or 3-winged membranous capsule, or a berry in some tropical species.

The family includes 10 genera and about 650 species distributed widely throughout the tropics and subtropics and extending slightly into the North Temperate Zone. Kansas has but a single species of 1 genus.

DIOSCORÈA L.

Yam

NL., in honor of Dioscorides (1st century A.D.), Greek physician and naturalist, who wrote a materia medica embracing about 400 species of plants, the most valued textbook of botany and medicine down to the 17th century.

Flowers dioecious, arranged in axillary panicles or racemes. Tepals white to greenish-yellow. Fruit a membranous, 3-winged loculicidal capsule. Seeds flat, winged.

One species in Kansas.

The fleshy rootstocks of several species of *Dioscorea*, called yams, are cooked as potatoes, especially in the tropics. Our sweet potato, *Ipomoea batatas* Lam., is also sometimes called by this name but is a distinctly different plant belonging to the Morning Glory Family, the Convolvulaceae.

D. villòsa L.

Atlantic yam, wild yam

L. *villosus*, hairy, in reference to the under surface of the leaves.
Greenish-yellow or whitish; June–July
Moist, open woods and thickets
Northeastern and extreme southeastern Kansas

Herbaceous, perennial twining vine up to 5 m in length, springing from a horizontal contorted rhizome. Leaves alternate above, rarely opposite and whorled below, petioled, cordate, 3.5–9 cm long with several longitudinal main veins prominent below and the under surface inconsistently hairy. Flowers small and inconspicuous. Male flowers each consisting of 6 tepals, 6 stamens, and a nonfunctional pistil; borne in loose drooping panicles. Female flowers about twice as long as the male ones; each consisting of 6 tepals, either 3 or 6 nonfunctional stamens, and a fusiform, inferior, 3-celled ovary with 3 styles and stigmas; arranged in drooping racemes. Mature capsules mostly 1.5–2.5 cm long.

A fluid extract of the rhizome has been used as an expectorant, diaphoretic, antispasmodic, and as an intestinal stimulant supposedly effective in cases of bilious colic.

AMARYLLIDÀCEAE

Amaryllis Family

Herbaceous plants perennating by scaly bulbs, rhizomes, or corms. Leaves basal, mostly linear or lorate. Stems scapelike. Flowers bisexual, regular (infrequently irregular). Inflorescence an umbel, raceme, panicle, or a single flower. Perianth of 6 petaloid parts borne on the hypanthium, which is adherent to the inferior ovary. Stamens 6, fused to the base of the perianth segments. The single pistil inferior (rarely half inferior or superior), usually 3-celled; style 1; stigmas 3, or 1 with 3 lobes, or capitate. Fruit mostly a dry capsule with black seeds.

The family contains 86 genera and about 1,300 species, including numerous ornamentals such as *Amaryllis* and *Narcissus*. Kansas has 2 native species—*Hypoxis hirsuta* and *Cooperia drummondii*, the latter being rather uncommon and known only from Cowley, Chautauqua, Labette, and Wilson counties.

HYPÓXIS L. Star-grass

NL., from Gr. *hypo*, below, + *oxys*, sharp, the seed pod tapering to its base.

Small herbs perennating from corms or short rhizomes. Leaves basal, grasslike, usually hairy. Inflorescence an irregular umbel of 2–6 flowers or a single flower. Flowers bisexual, regular, usually yellow. Fruit indehiscent or a loculicidal capsule.

One species in Kansas.

H. hirsùta (L.) Coville Common golden star-grass

L. *hirsutus*, bristly, shaggy, true of the outer surface of the perianth, the leaves, and the stem.
The common name is suggested by the starlike flowers and grasslike leaves.
Golden yellow; mid-April–early July
Mostly prairies, sometimes in open woods
Primarily in the eastern fourth

Corm up to 1.5 cm wide, bearing a circle of fibrous roots. Leaves several, grasslike, 1–2 dm long. Scape slender, with a few-flowered umbel at its summit, 0.8–1.8 dm tall. Perianth divisions stiffly spreading, green and hairy on the under side, golden above, 6–10 (–14) mm long. Capsule indehiscent, ellipsoid, 3–6 mm long. Seeds shining, black, sharply muricate, 1–1.5 mm across.

At the time of flowering, the new corm of the season, bearing the roots, has swollen but little and is still drawing nourishment from the basally attached parent corm of last year, which has lost its roots and will wholly disintegrate after delivering its store of reserve food to the new corm. For a similar corm habit in the Lily Family see *Andro-stephium*.

See plate *26*.

57. *Hypoxis hirsuta*

IRIDÀCEAE Iris Family

Perennial herbs with bulbs, corms, or rhizomes. Leaves mostly basal, equitant (2-ranked and overlapping, as in the ornamental *Iris*), linear or narrowly sword-shaped. Stems solitary to several or none, in the latter case the peduncled flowers arising directly from the corm. Flowers bisexual, usually regular. Inflorescence of 1 to several flowers, subtended by a spathe of 2 or more leaves or bracts. Sepals 3, petal-like in form and color. Petals 3. Stamens 3, distinct from one another or monadelphous, alternate with the petals. Pistil 1, 3-celled, inferior; style 1, with 3 branches which are filiform or flabellate, entire or lobed or fimbriate; stigmas 3. Fruit a 3-angled or 3-lobed loculicidal capsule.

The family includes about 60 genera and 1500 species. Four genera in Kansas. Our native species of *Iris* are quite rare and are not included here.

Key to Genera:
1. Filaments fused to one another .... *Sisyrinchium*
1. Filaments free from one another or nearly so 2
  2. Perianth orange-yellow with dark spots; perennial from a jointed rhizome ...... *Belamcanda*
  2. Perianth white; perennial from a bulb .... *Nemastylis*

BELAMCÁNDA Adans. Blackberry lily

The native Malayan name.

Perennial herbs arising from a jointed horizontal rhizome. Stems erect, branching. Leaves equitant, sword-shaped. Inflorescence loosely many-flowered. Sepals and petals widely spreading, all nearly alike. Stamens 3, filaments free from one another but inserted on the bases of the tepals. Style club-shaped, 3-cleft.

A unique feature of this genus is its fruit, which first becomes a fig-shaped, 3-celled capsule bearing on its apex the perianth segments twisted together; finally the valves of the capsule break away in such a manner as to leave the axile placentae standing, densely covered by the black and shiny, fleshy-coated seeds and looking like a blackberry.

B. chinénsis (L.) DC. Blackberry lily

NL., compounded of "China" + the suffix -*ensis*, signifying the place where the species is native.

Although the open pods do resemble blackberries, the plant is not a lily.
Orange-yellow, spotted with crimson-purple; June–early August
Roadsides, thickets, and open woods
Eastern fifth

Stems 0.5–1.3 m tall, bearing several stiffly upright leaves, branching above and terminating in umbels of large flowers reminiscent of tiger lilies, but without the fundamental characters of the Lily Family, for the ovary is inferior and there are 3 stamens (compare with *Lilium michiganense*). Tepals elliptic to oblanceolate, orange-yellow with crimson-purple spots when fresh, but turning pinkish when dried, 2–3 cm long. Capsule 2–2.5 cm long. Fleshy seeds about 4 mm across.

This remarkable member of the Iris Family, long since introduced from the Orient and a favorite in old-fashioned gardens, was brought into Kansas by the early settlers from the eastern states and has remained with us, here and there escaping from gardens and carrying on independently, most successfully in open woods from which cattle have been excluded. It is readily propagated by seeds and by division of the rhizomes and hence can easily be transplanted into the garden. The dried stalks with their berrylike clusters of seeds may be collected in August and September and make attractive additions to winter bouquets.

58. *Belamcanda chinensis*

NEMASTŸLIS Nutt. Prairie irid

NL., from Gr. *nema*, thread, + *stylos*, pillar or, botanically, style—each of the 3 styles being divided into 2 threadlike branches.

Herbaceous perennials arising from an ovoid or subglobose bulb. Stems erect, simple or branched. Leaves grasslike, 2-ranked, equitant. Flowers bisexual, regular, borne in clusters of 1–3 and subtended by 2 overlapping leaflike bracts. Tepals 6, nearly alike, united slightly at the base, spreading. Stamens 3, inserted on the base of the perianth segments, anthers coiling downward from the apex at maturity. Pistil inferior, 3-celled; styles 3, each with 2 threadlike branches tipped with a tiny globose stigma. Fruit a capsule opening by 6 deltoid teeth at the apex.

One species in Kansas.

N. geminiflŏra Nutt. Prairie irid

Gemini, twins, + *flors*, flowers, from the inflorescence which frequently consists of 2 flowers.
Blue to lilac-blue; mid-April–early June
Prairies
Southeastern sixth (west to Cowley and north to Anderson counties)

Bulbs 1–2.5 cm in diameter, with brittle, membranous, brown tunics prolonged upward into a collar around the base of the stem and leaves. Stems branched, 1.5–2.5 dm tall at flowering. Leaves mostly 1–3 dm long and 3–10 mm wide, at least one of the cauline leaves usually much longer than the inflorescence, the upper one much reduced. The flowers open in the morning and close again before sundown. Perianth segments nearly alike, ovate, elliptic or oblanceolate, blue, 1.8–2.5 cm long. Filaments free from one another or fused together only slightly at the base; anthers and pollen bright yellow. Capsule obovoid with a rather flattened apex, 1.5–2 cm long. Seeds angular, brown.
Compare with *Androstephium coeruleum*.

59. *Nemastylis geminiflora*

This, one of the loveliest of our prairie flowers, is not hardy, and seems to reach the limit of its northern range in southeast Kansas and central Missouri. It is like the wild hyacinth in appearing early and disappearing soon. Its leaves are up before the grasses are well started. Its flowers are blooming before April is past, and the capsules are ripening and the fibrous annual roots at the base of the bulb are dying away by the first week in June.

This plant has been listed in earlier manuals for our area as *N. acuta*.

SISYRÍNCHIUM L. Blue-eyed grass

NL., from Gr. *sisyrinchion*, name of an unidentified plant mentioned by Pliny (1st-century botanist). The common name is attractive but inappropriate, since the plant is not a grass; and as for "blue-eyed," the flowers of this genus are as often white as blue, while the center or eye of the flower is often yellow. "Prairie grass-irid" would be a name consonant with the habitat of the plant, its grasslike appearance, and its common ancestry with the irises. But one must admit the faults in vernacular usage will not be corrected by logic.

Low, perennial, grasslike herbs from fibrous roots. Leaves mostly basal, linear to lanceolate. Stem 2-edged or 2-winged. Flowers bisexual, regular or nearly so, delicate

60. *Nemastylis geminiflora*

Iris Family
63

and short-lasting, borne in umbel-like clusters from a 2-valved spathe; inflorescence simple or compound. Tepals alike, oblanceolate, spreading, the apexes retuse or abruptly aristulate. Filaments fused their entire length in our species. Style branches filiform, alternating with the stamens. Fruit a loculicidal capsule. Seeds globular, black.

Kansas has 4 species, the most common of which are included in the following key.

Key to Species:
1. Inflorescence compound with 2 or more long-peduncled spathes arising from the axil of a leaflike bract; outer bract of the spathe with its margins united for (2-) 3–4 mm above the base; fruits and foliage eventually drying dark .... *S. angustifolium*
1. Inflorescence simple, terminating an unbranched stem; outer bract of the spathe with its margins free to the base; fruits and foliage eventually drying light green or straw-colored .... *S. campestre*

S. angustifòlium Miller   **Blue-eyed grass**

L. *angusti-*, narrow, + *folius*, -leaved.
Purplish-blue; mid-May–late June
Moist woods, prairies
Extreme east

This species differs from the more common *S. campestre* principally in having the inflorescence compound, rather than simple, with several long-peduncled spathes arising from the axil of a large leaflike bract, and in having foliage that is a darker green and is not glaucous. Capsules frequently turning dark at maturity, 5–7 mm long. Seeds subglobose, dull black, about 1 mm across.

S. campéstre Bickn.   **Blue-eyed grass**

From L. *campester*, *campestris*, of or growing in a level field, probably in reference to its prairie habitat.
White to pale blue, with yellow centers; April–May
Prairies; occasionally in open woods
Eastern half

Leaves pale green, glaucous, tufted, equitant, usually shorter than the flowering scapes. Scapes slender, 2-edged, 1–4 dm tall, each terminated by an umbel of flowers with threadlike peduncles basally sheathed by 2 erect, lanceolate bracts; margins of the outer bract free to the base (not fused to one another at the base). Tepals varying in color from white to pale blue, 9–11 mm long. Capsules light green or straw-colored at maturity, 3–5 mm long. Seeds apparently frequently aborted.

The orange-colored anthers discharge pollen outwardly while still in the bud, and the few bees that come to the newly opened flowers for nectar secreted scantily inside the base of the perianth would incidentally pick up pollen from the protruding anthers and carry it from flower to flower, although self-pollination would seem to be a possibility, in the bud and after the flower closes.

See plate 27.

ORCHIDÀCEAE   Orchid Family

Herbaceous plants perennating by corms, rhizomes, or tuberous roots. Leaves basal and on the flowering stems, those higher up often reduced to scales; both leaves and scales with bases in the form of a closed sheath. Flower structure in this group is rather complex, and the remainder of this description may be more easily understood if the reader will refer from time to time to fig. 61. Flowers bisexual and very irregular, occurring singly or in racemes or spikes. Perianth consisting of 3 often nearly similar sepals and 3 petals, the 2 lateral ones often like the sepals but the middle and lowest one, called the lip or labellum, broader than the others, usually differently colored and sometimes spurred. Ovary inferior, 1-celled, with 3 parietal placentae bearing innumerable ovules; style 1 and stamens 3, all coalescing to form an extension above the ovary known as the column, which has 2 functional stigmas more or less coalesced into 1, and above those a modified sterile stigma which because of its form is called the rostellum (the Latin diminutive, meaning "little beak," from *rostrum*, beak), and above the rostellum 1, rarely 2, functional anthers—2 stamens in the first instance and 1 in the second being vestigial or much modified and infertile. The anthers have 2 lobes or pollen sacs, each one containing, rarely, loose but sticky pollen grains, but usually grains cohering in the form of a pollinium (as in *Asclepias*), which is united to a sticky disk produced by the rostellum. An insect coming to the flower for nectar secreted usually at the base of the labellum, or under the epidermis lining the cavity of the spur, if a spur is present, touches the sticky disk with its proboscis or head and on withdrawing from the flower carries away the disk with its attached pollinium, or pollinia, as the case may be. When now the insect visits another flower it presses the captured pollinia against the sticky stigma, to which they adhere so firmly that they are left there when the insect departs. Fruit a capsule, dehiscing by 3–6 hygroscopically sensitive valves which remain apically connate, the capsule opening medianly. Seeds minute, very abundant, colorless.

An orchid seed is very small, having a minute undifferentiated embryo, without food reserves stored in or around it, and so under natural conditions is unable to germinate without being nursed along by very different kinds of plants—namely, toadstools or other fungi, which for most of their lives exist in the substratum as strands and clumps of threadlike growth called mycelium (NL., from Gr. *mykes*, toadstool or fungus). Being without chlorophyll, the fungi cannot make their own food from soil, air, and water, as green plants do, and are forced to live off living or dead green plants by penetrating their roots and absorbing food from them, or by digesting their remains that form the humus of the soil. When an orchid seed gets into the humus and swells with absorbed water as the first step in growth, threads of mycelium enter the seed and penetrate the embryo as if to devour it; but the tables are turned—the hungry little embryo sucks food out of the mycelium which the latter has taken from the humus and the roots of other green plants; and the seedling orchid must keep on living as a parasite on the fungus until its roots are formed and enable it to live independently. Mycorrhizal fungi inhabit the roots and rhizomes of orchids as long as the latter live, but it has never been proved that orchids with chlorophyll require the continuance of this association. However, orchids without chlorophyll (such as *Corallorhiza*) have throughout their lives no food other than that supplied by mycorrhizas.

A family of approximately 450 genera and 10,000 species of wide distribution but most abundant in the tropics. Kansas has 9 genera and 16 species. Only those most likely to be encountered are included in the following section. Persons interested in further study of this group of plants are referred to D. S. Correll's *Native Orchids of North America North of Mexico* (72), the only completely illustrated orchid flora of the United States and Canada, and to Charles Darwin's work on the fertilization of orchids (73), a classic scientific paper of interest to the general reader in spite of its date of publication (1877).

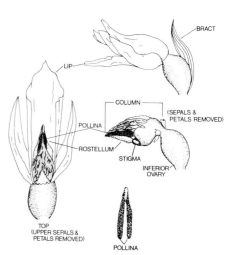

61. Flower structure in *Spiranthes*

Key to Genera:
1. Plants wholly or essentially without green color; leaves absent or reduced to scales ....
.... *Corallorhiza*
1. Plants at least partly green; most species with foliage leaves at anthesis 2
  2. Lip inflated, pouchlike; flowers yellow .... *Cypripedium*
  2. Lip not inflated; flowers white, greenish, or pinkish 3
    3. Perianth spur absent; flowers arranged in a spiral around the stem .... *Spiranthes*
    3. Perianth spur present and conspicuous; inflorescence not as above 4
      4. Lip white, sepals and lateral petals pink or lavender; lip not fringed .... *Orchis*
      4. All sepals and petals white or pale greenish-yellow; lip deeply fringed ...... *Habenaria*

CORALLORHĪZA Chat.   Coral-root

NL., from L. *corallum* or Gr. *korallion*, coral, + Gr. *rhiza*, root, in reference to the irregularly branched and thickened roots which resemble a clump of coral.

Herbaceous, yellowish, purplish, or brownish perennials, lacking chlorophyll and leaves and entirely dependent upon mycorrhizal fungi for food. Stem an erect scape with a few long-cylindrical, sheathing scales. Flowers borne in a loose terminal raceme. Sepals and lateral petals narrow, essentially alike in size, shape, and color, sometimes conspicuously veined (but not in our species); lip reflexed, usually broader, more rounded, and of a color different from other members of the perianth, and with 1 or 2 longitudinal ridges. Column short and stout, the single anther terminal with 4, soft-waxy pollen masses. Fruit drooping.

Two species in Kansas.

Key to Species:
1. Lip white dotted with purple; sepals and petals 7–8 mm long; spring blooming .. *C. wisteriana*
1. Lip white with a purple margin and 2 purple spots; sepals and petals 3–4 mm long; fall
  blooming .... *C. odontorhiza*

62. *Corallorhiza odontorhiza*

63. *Cypripedium calceolus*

C. odontorhìza (Willd.) Nutt.   Coral-root

From Gr. *odous*, tooth, + *rhiza*, root, in further reference to the underground parts of the plant.
Pink or purplish and white; September–mid-October
Oak-hickory woods
Eastern fifth (especially in the northern part of this section); rare

Similar to the following species but smaller in all respects. Scapes purple, brown, or greenish, mostly 1–2 dm tall. Receme with 5–15 flowers. Sepals and lateral petals pink or purplish (drying reddish-brown), oblong, extending forward over the column, 3–5 mm long; lip white with a purple margin and 2 purple spots, the margin entire or finely erose, not notched at the apex. Capsule 6–8 mm long.

The flowers and foliage being rather inconspicuous, this species is most frequently collected after the flowers are past their prime and the fruits have begun to enlarge, as in fig. 62.

C. wisteriàna Conrad   Coral-root

Named in honor of Caspar Wistar, American physician and author (1761–1818), after whom the genus *Wisteria* also was named.
Yellow and purple; mid-April–May
Moist, often rocky, open wooded hillsides
Eastern seventh and Chautauqua and Montgomery counties

Scape reddish or purplish, mostly 1–3 dm tall. Raceme with 10–15 flowers. Flower with the 3 sepals and 2 lateral petals yellow, spotted brownish-purple (drying reddish-brown), narrowly lanceolate, extending forward over the column, 6–8 mm long; and the broadened lower petal (lip or labellum) white with purple spots, minutely crenate on the margins and notched at the apex. The single anther 2-celled, with 2 pollinia in each cell abutting on the rostellum and adherent to its mucilaginous secretion. Capsule 9–11 mm long.

CYPRIPÈDIUM L.   Lady's slipper

NL., from Gr. *kypris*, Aphrodite, + *podion*, slipper or little foot, suggested by the shape of the labellum or lip. Some early writers give the name "Our Lady's slipper" to the English species, clearly referring to the Virgin Mary.

Perennials arising from a short rhizome with coarse fibrous roots. Stems erect. Leaves 2 to several, alternate, sheathing the stem. Inflorescence of 1 or 2 flowers. Sepals 3, spreading, broader than the lateral petals, the lower 2 separate or fused together beneath the lip. Petals 3, the lateral 2 spreading and similar in color to the sepals, the lower petal or lip a large inflated pouch. Fertile stamens 2; a third stamen transformed into a sterile petaloid projection overarching the terminal stigma; pollen granular and somewhat sticky but not united into definite clumps or pollinia. Column declined over the orifice of the lip.

One species in Kansas.

C. calcèolus L.   Yellow lady's slipper

L. *calceolus*, small shoe, again in reference to the shape of the lip.
Yellow and greenish-brown; mid-April–May
Rich, moist woods
Primarily in the northeastern corner, also in the extreme southeastern corner; rare

Stem 3–4 dm tall, terminated by a peduncle bearing 1 or 2 drooping flowers. Leaves broadly elliptic-lanceolate, ribbed by prominent parallel veins, largest leaves 15–20 cm long and 7–11 cm wide. Corolla consisting of a large, inflated, yellow and brownish-purple lip with a rounded opening on the upper side and 2 long, slender, twisted, yellowish or greenish, brownish-striped lateral petals, 4.5–6.5 cm long. Sepals broader than the lateral petals, the upper one extending over and beyond the lip, the lateral ones depressd and fused together beneath the lip, 4–5.5 cm long. Capsule 2.5–4.5 cm long. The plants in our area are recognized as var. *pubescens* (Willd.) Correll, this variety at one time being recognized as a distinct species which was more hairy than the earlier named *C. reginae*.

The design of the flower fits it for cross-pollination by bees, which enter the cavity of the lip to get nectar or other food from hairs springing from the bottom. When attempting to leave the flower the bee finds himself hemmed in by the inrolled border of the lip except at the rear where there is on each side a gap between the column and the margin of the lip. Creeping out through this gap the bee presses against the stigma and an anther, simultaneously bringing about cross-pollination and picking up a fresh supply of pollen to carry to the next blossom.

The early settlers found this orchid in the woods more frequently than we do now, because the old woods have been much chopped out, cattle and pigs have been allowed to roam the woods, and "flower lovers," whenever they run across it, inveterately transplant it to the home garden. Fortunately, the lady's slipper orchids and numerous other of our rarest plants now may be purchased from floriculture supply houses, thus enabling the wildflower enthusiast to enjoy these delicate species in his own yard without disturbing the diminishing natural populations.

See plate *28*.

64. *Habenaria lacera*

HABENÀRIA Willd.   Fringed orchid

From L. *havena*, a thong or rein, in allusion to the shape of the lip or spur of some species.

Perennial herbs arising from tubers, rhizomes, or a cluster of thickened roots. Stems erect, unbranched. Leaves basal or cauline and alternate. Inflorescence a terminal spike or raceme. Sepals and lateral petals essentially alike in form and color, the petals sometimes smaller; the lip in our species 3-lobed and deeply fringed, prolonged backward into a long spur. The single anther attached to the column by a broad base. Glands or viscid disks to which the pollen masses are attached naked and exposed, separate, sometimes widely so.

Two species in Kansas.

Key to Species (3):
1. Petals linear-spatulate, blunt, entire; lateral sepals deflected behind the lip .... *H. lacera*
1. Petals broadly obovate, toothed; lateral sepals divergent .... *H. leucophaea*

H. lácera (Michx.) Lodd   Ragged fringed orchid

L. *lacera*, lacerated, in reference to the fringed lip.
Yellowish-green or -white; late May–mid-June
Moist prairie
Southeastern corner, west to Montgomery County and north to Linn County

Stems 3–5 dm tall. Lower leaves lanceolate to elliptic or oblanceolate, mostly 9–14 cm long and 1.5–3.5 cm wide; upper leaves much reduced, narrowly lanceolate. Inflorescence a raceme of 10 to many fragrant flowers, yellowish-green to yellowish-white or dirty yellow in color, sometimes tinged with bronze or rose. Sepals more or less ovate, the 2 lower ones somewhat asymmetrical and connivent below the lip, 4–6 mm long. Lateral petals linear-spatulate, blunt-tipped, entire, nearly equaling the sepals in length; the lip deeply 3-lobed, with the 2 lateral lobes fringed more deeply than the central one; the spur at anthesis nearly equaling the ovary in length.

65. *Habenaria leucophaea*

H. leucophaèa (Nutt.) Gray   Prairie fringed orchid

NL., *leucophaea*, light brown, from Gr. *leukos*, white, without color, + *phaios*, dun-colored, dull brown, the petals turning brownish with age.
Creamy-white to whitish-green; mid–late June
Moist or wet prairies
Eastern fourth

Stems 4–7 dm tall. Lower leaves lanceolate or elliptic, 9.5–12 cm long and 2–4 cm wide; upper leaves reduced, lanceolate, sometimes narrowly so. Inflorescence a raceme of 4 to many creamy-white to whitish-green, sweetly fragrant flowers. Sepals broadly oval to obovate, 9–12 mm long. Lateral petals broadly obovate-cuneate, with toothed margins, slightly longer than the sepals; lip with 3 deep, fringed lobes, the middle lobe short-clawed and very broadly cuneate; the spur at anthesis about twice as long as the ovary.

See plate *29*.

ÓRCHIS L.   Orchis

Gr. *orchis*, testicle, from the shape of the tubers of some species.

Herbaceous perennials arising from short rhizomes and thickened roots. Stems erect, terminated by a loose, few-flowered raceme. Leaves basal, usually 2 (sometimes 1 or 3). Sepals and lateral petals similar in color but differing in size; petals connivent with the upper sepals or with all the sepals to form a concave hood over the column; lip large, turned downward and prolonged backward at the base to form a conspicuous hollow spur which is lined internally with nectiferous tissue. At the apex of the column is a single anther with 2 contiguous parallel lobes or sacs and just below this the

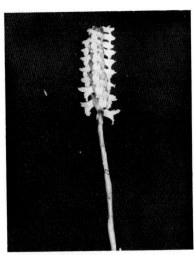

66. *Spiranthes cernua*

rostellum. Close below the rostellum is a sticky stigma. With the aid of a lens of high magnification and a dissecting needle or probe, one can see that within each anther sac is a pollinium with the pollen cohering in numerous coarse, waxy grains held together on a elastic cobwebby tissue. Each pollinium narrows basally into a small stem or caudicle which extends downward against the rostellum and is attached to it by means of a sticky disk situated just above the orifice of the spur. When an insect, in the process of gathering nectiferous tissue from the spur, touches the sticky disk with his head, the disk adheres so firmly to the insect that the pollinium is carried off to another flower.

One species in Kansas.

O. spectábilis L. Showy orchis

L. *spectabilis*, worth seeing, admirable.
Lavender or pink and white; late April–mid-May
Rich, moist woodland
Northeast corner; rare

Stem 1–2 dm tall. Leaves broadly elliptic to oblanceolate, apex rounded or acute, base narrowed to sheathe the stem, blade above the sheathing portion 1–2 dm long and 3.5–8 cm wide. Inflorescence of 2–6 (sometimes as many as 10) flowers in a loose raceme, each flower subtended by a conspicuous lanceolate or elliptic bract. Lateral petals and sepals pink or lavender, 10–18 mm long, rising together and fused to form a hood over the column; lip white, 12–16 mm long.

See plate *30*.

SPIRÁNTHES Rich. Lady's tresses

NL., from Gr. *speira*, coil or spiral, + *anthos*, flower, characterizing the appearance of the flower spike. The common name likens the spike to a braid of a lady's hair.

Herbaceous perennials with a short rootstock and a fascicle of more of less thickened roots. Stem slender, upright, variously leafy basally and scaly above. Inflorescence a spike of 1 or more spiral rows of small, mostly white flowers, each subtended by a bract closely appressed to the inferior ovary. All parts of the horizontally projecting flowers stand so close together as to give a tubular appearance. The apex of the labellum is deflexed and serves as a platform for insects. Farther back the labellum bears 2 nectar glands and at its base provides a small receptacle to receive the nectar. The short column (fig. 61) bears a rostellum at its apex and, on its upper side, close against the rostellum, an anther with a pollinium in each sac. On the under side, just beneath the rostellum, is a prominent rounded stigma. The rostellum is unusual in that it is 2-pronged at the apex, and bears between the prongs an oblong, sticky disk with 2 threads connected to the pollinia—these contrivances ensure that an insect probing for nectar touches the disk with its head or proboscis, and on backing away pulls the disk together with the pollinia from between the 2 prongs. When the insect now visits another flower the pollinia get stuck to the stigma.

There are 6 species of *Spiranthes* in Kansas, but only the 4 most common ones are included in the key to species. *S. ovalis* Lindl., an autumn-blooming woodland plant with oblanceolate basal leaves and small white flowers borne in 3 spiral rows, has been collected in Douglas, Linn, and Miami counties. *S. lucida* (H. H. Eaton) Ames has been reported from Cloud County.

Key to Species (3):
1. Flowers occurring in more than 1 spiral row .... *S. cernua*
1. Flowers occurring in a single spiral row 2
   2. Perianth segments and lip 5–8 mm long; inflorescence pubescent at least at the base
   .... *S. vernalis*
   2. Perianth segments and lip 3–4 mm long; inflorescence glabrous or nearly so 3
      3. Lip with a green median area; stem from a cluster of roots .... *S. lacera*
      3. Lip wholly white; stem from a single fusiform root .... *S. tuberosa*

S. cérnua (L.) Rich. Nodding lady's tresses

L. *cernuus*, with the face turned toward the earth; bowing forward—in reference to the orientation of the flowers.
White or creamy; mid-August–October
Prairie
Eastern two-thirds

Stems mostly 1–3 (–6) dm tall, from several fleshy tuberous roots 2–8 cm long. Leaves linear-lanceolate, mostly basal, up to 3 dm long, but usually gone by flowering time. Spike stout and blunt, the flowers borne in 2–3 spiral rows and facing downward as they open in slow progression from the bottom to the top of the inflorescence, axis of the inflorescence more or less pubescent. Perianth entirely white or creamy, 7–10 mm long; the lateral sepals narrowly lanceolate, remaining appressed to the other perianth parts. Figure 61 illustrates the morphology of the column and its relationships to the other parts of the flower.

The cut flowering stems remain fresh in appearance over a long period of time. Some plants exhale a sweet odor, of variable strength, resembling that of lilacs.

S. lácera (Raf.) Raf. — Slender lady's tresses

L. *lacera*, lacerated, irregularly torn, in reference to the margin of the lip.
White, with a greenish spot on the lip; August–early October
Prairie
Eastern half

Stems 1.5–4.5 dm tall, from several divergent fleshy roots. Leaves basal, oval to oblong, 2–5 cm long, sometimes present at anthesis. Spike very slender, the flowers borne in a single twisted row, the axis glabrous or nearly so. Sepals and petals white, 5–6.5 mm long; the lip white with a green median area. The lower 2 sepals are subulate, but the margins tend to roll inward, giving them an acicular or linear appearance, and they eventually diverge widely downward from the other perianth parts.

See plate *31*.

67. *Spiranthes lacera*

S. tuberòsa Raf. — Little lady's tresses

L. *tuberosus*, knobby, from *tuber*, a knob or swelling, in reference to the root.
Flowers entirely white; September–early October
Open oak-hickory woods
Eastern two-fifths, more common in the southern half of this area

Stems very slender and delicate, 1.5–2.2 dm tall, usually from a single tuberous root. Leaves mostly basal, oval to oblong, 2–4 cm long, usually absent at anthesis. Spike very slender, glabrous, the flowers borne in a single spiral. Perianth entirely white, 3–4 mm long; lower 2 sepals ensiform, diverging outward at the tip but otherwise lying parallel to the other perianth parts.

S. vernàlis Engelm. & Gray — Vernal lady's tresses

L. *vernalis*, pertaining to spring, for *ver*, spring. The first collections of this species came from the southern limits of its range, where they are spring-blooming.
Yellowish-white or creamy, with 2 yellow or yellow-green spots on the lip; June–August
Prairie
Eastern half

68. *Spiranthes tuberosa*

Stems rather robust, 4–7 dm tall, from a cluster of 3–7 elongate fleshy roots. Leaves basal and cauline, linear and grasslike, usually present at anthesis; basal leaves 5–15 cm long. Flowers arranged in a single spiral, the axis finely hairy. Sepals and lateral petals white or creamy, 4–8 mm long; the lip with 2 yellow or yellow-green spots. The lower 2 sepals are subulate with the margins tending to roll inward, thus giving them the appearance of being acicular, and eventually diverge somewhat from the other perianth parts.

SAURURÀCEAE — Lizard's Tail Family

Erect perennial herbs. Leaves alternate, simple, entire, stipules adnate to the petiole. Flowers bisexual (sometimes dioecious), in dense, slender, peduncled spikes or racemes. Perianth absent, but each flower subtended by a small bract. Stamens 8 or fewer, hypogynous, with distinct filaments. Gynoecium of 3–4 distinct simple pistils or one compound pistil with 3–4 carpels, superior or half-inferior. Fruit a semisucculent follicle or a fleshy capsule dehiscing by apical valves.

A small family with 3 genera and 4 species. One species, *Saururus cernuus*, occurs in Kansas.

SAURÙRUS L. — Lizard's tail

From Gr. *sauros*, a lizard, and *oura*, tail, describing the appearance of the inflorescence.

Erect succulent herbs, perennial from a creeping rhizome. Stems jointed. Leaves

69. *Spiranthes vernalis*

70. *Saururus cernuus*

71. *Cannabis sativa*

petioled, cordate or ovate, palmately net-veined. Inflorescence drooping at the tip. Stamens 6–8. Pistil 1 with 3–4 carpels united at the base; styles very short; stigmas recurved. Fruit a subglobose, somewhat fleshy, rugose, indehiscent capsule.

S. cérnuus L.   Lizard's tail, water-dragon

L. *cernuus*, with the face turned toward the earth, nodding, in reference to the drooping tip of the inflorescence.
White; (May–) June–September
Swampy woods, marshes, borders of slow-moving streams
Extreme eastern and southeastern Kansas (Cherokee and Linn counties)

Extensively creeping by means of aromatic rhizomes or by producing adventitious roots at the lower nodes. Flowering stems 5–12 dm high. Spike 1–3 dm long. Stamen filaments white, threadlike, much longer than the pistil. Fruit 2–3 mm thick, rust-colored at maturity.

CANNABÀCEAE   Hemp Family

Erect, annual or climbing perennial herbs. Leaves opposite, sometimes alternate, simple, with palmate venation or palmately compound, margins lobed and/or serrate, stipules present. Flowers unisexual and dioecious, regular. Male flowers with 5 sepals, no petals and 5 stamens; arranged in loose panicles or racemes. Female flowers with 1 sepal, no petals, and 1 superior 1-celled pistil with 2 styles and 2 stigmas; arranged in dense bracteate spikes. Fruit and achene invested by the persistent calyx which becomes dry or membranous in age.

The family contains 2 genera and 4 species of pantropical and temperate distribution. One species of each genus occurs in Kansas.

Key to Genera:
1. Stems twining, principal leaves merely lobed .... *Humulus*
1. Stems erect, principal leaves palmately compound .... *Cannabis*

CÁNNABIS L.   Hemp

L. *cannabis*, hemp.

Erect, annual, dioecious herbs. Leaves alternate or opposite, digitately compound with 5–11 leaflets. Male flowers in paniculate racemes; sepals 5, imbricate. Female flowers in leafy-bracted spikes; perianth of 1 sepal; pistil 1 and stigmas 2. Fruit a slightly flattened achene.

One species in Kansas.

C. satìva L.   Hemp, marijuana

L. *sativa*, that which is sown.
Flowers green, inconspicuous; late July–September
Roadsides, waste places; rich, fertile alluvial soil of river flood plains and valley bottoms, frequently near streams
Eastern two-thirds and scattered western Kansas localities

Tall branched annuals, up to 3.5 m tall. Leaves palmately divided into 3–7 linear-lanceolate leaflets with coarsely serrate margins. Larger leaflets mostly 7–15 cm long and 0.8–2.4 cm wide. Male flowers 3–3.5 mm long; in slender, loose axillary panicles. Female flowers in erect, leafy-bracted spikes, each flower consisting of a thin undivided calyx with numerous glandular hairs, closely enwrapping a sessile ovary which is surmounted by 2 slender, soon-withering stigmas. Fruit a hard ovoid achene with somewhat flattened sides, 3.5–4.25 mm long.

The plant has been cultivated from ancient times for the long and strong bast fibers in its bark, used in making thread, twine, cordage, and rope, as well as fabrics and bags. The resinous secretions of its leaves and flowers have been used from time immemorial as medicines and intoxicants, and in recent years the plant has attracted much attention in the U.S. and elsewhere because of its alleged narcotic properties. The seeds are included in some birdseed mixtures and yield a drying oil which is used in soap, paints, and varnishes, and as an illuminating oil.

HŬMULUS L. **Hops**

LL. name of Teutonic origin.

72. *Humulus lupulus*

Twining herbaceous perennial vines with rough stems. Leaves opposite, sometimes alternate, simple, usually 3–7 lobed, palmately net-veined, with serrate margins. Flowers dioecious, regular. Male flowers pendulous in an axillary panicle, a flower consisting of a 5-parted calyx and 5 stamens opening by apical pores, through which the pollen is shaken by wind and wafted to the female flowers of adjacent vines. Female flowers in axillary, scaly, conelike clusters, a pair of flowers subtended by each scale of the cone; a flower consisting of a 1-celled ovary bearing 2 papillose styles, and an undivided persistent calyx closely investing the ovary. Fruit a slightly flattened achene enclosed in the persistent calyx and covered by the accrescent bracts.

Two species in Kansas. *H. japonicus* Sieb. & Zucc., an introduced species, is rarely collected. It differs from the following species in having its leaves 5- to 7-lobed and the bracts of the female spikes conspicuously ciliate.

H. lùpulus L. **Hop**

L. *lupulus*, little wolf, from *lupus*, wolf. Name used by Pliny for the hopvine—because, as he said, when the vine is growing among willows it strangles by its close embrace, as a wolf does a sheep. Note that the vine is "hop" or "hopvine," while the conelike inflorescence is called "hops."

Flowers greenish; July–October
Waste ground, roadsides, along railroads
Eastern third

Twining or prostrate, herbaceous perennial vine which dies back in the fall to its vertical rhizome and stout perennial roots. From the rhizome strong flexible stems beset with stout recurved prickles grow out to about 6 m, twining clockwise about any suitable support or, lacking support, sprawling with intertwined stems.

The hop, native of Eurasia and North America, has been cultivated since ancient times. Pliny (1st century) tells that the Romans cultivated it in their gardens and ate the young shoots as we do asparagus and counted them good for a torpid liver. Herbals of the Middle Ages, probably following more ancient usage, recommended the hop as a laxative and blood purifier and curative of various afflictions. "Juice of the hop dropped warm into the ear," says A. Lonicerus (1564), "draws inflammation to a head and expels the abscess." The reputation of the hop did not diminish as time went on, for we read in Grieve and Leyel's *A Modern Herbal* (11): "Hops have tonic, nervine, diuretic, and anodyne properties. . . . An infusion of one-half ounce Hops to 1 pint of water will be found the proper quantity for ordinary use. It has proved of great service also in heart disease, fits, neuralgia and nervous disorders, besides being a useful tonic in indigestion, jaundice, and stomach and liver afflictions generally." All parts of the plant have been used for medicine, but the greatest potency resides in the secretions of the glandular hairs on the cone-scales and persistent calyx around the 1-seeded achenes—secretions of tannin, volatile oil, resin, and bitter substances that give hops their characteristic odor and taste. When the hop cones are shaken the glandular hairs fall off, giving the powder called Lupulin, which can be conveniently administered in the form of a pill.

Again, beer was being made with hops in the Netherlands in the 14th century and in England 2 centuries later, for improving the flavor and preserving the quality of the brew. In the 16th century Gerard's herbal was published—the renowned folio of 1,630 pages. The volume is divided into 3 books, concerning which Gerard tells us on the introductory page, "Each booke hath Chapters, as for each Herb a bed: and every Plant presents thee with the Latine and English name in the title placed over the picture of the plant. . . . Then followeth the Kinds, Description, Place, Time, Names, Natures and Vertues, agreeing with the best received Opinions."

Now here we give what Gerard says under "The Vertues" in the chapter entitled "Of Hops":

"The buds or first sprouts which come forth in the spring are used to be eaten in sallads; yet are they, as Pliny saith, more toothsome than nourishing, for they yeeld but very small nourishment. The floures are used to season Beere or Ale with, and too many do cause bitternesse thereof, and are ill for the head.

"The floures make bread light, and the lumpe to be sooner and easilier leavened, if the meale be tempered with liquor wherein they have been boiled.

"The manifold vertues of Hops do manifest argue wholesomenesse of beere above ale; for the hops rather make it a physicall drinke to keep the body in health, than an ordinary drinke for the quenching of our thirst."

73. *Boehmeria cylindrica*

Our American Indians also were aware of the "vertues" of the hop and used it in diverse ways for a variety of afflictions; the Dakotas, for instance, employed a decoction of the cones for fevers and intestinal disorders, and applied the soft mass of chewed roots to wounds, sometimes in combination with roots of *Physalis lanceolata* and *Anemone canadensis*.

Another species, *Humulus japonicus* Sieb. and Zucc., is cultivated as an ornamental and is occasionally found as an escape. Japanese hops, as it is called, has the principal leaves with 5–7 lobes instead of 3, and the bracts of the female spikes spinulose-ciliate rather than entire. The stem is also much rougher than that of *H. lupulus*.

URTICÀCEAE   Nettle Family

Annual or perennial herbs, or in the tropics some shrubs and trees. Leaves opposite or alternate, simple, serrate or entire, petioled, stipulate (except in *Parietaria*). Flowers small; monoecious, dioecious, or polygamous; regular; variously distributed on axillary peduncles. Perianth present or absent. When present, usually consisting of 4–5 sepals, either distinct or fused to one another. Stamens as many as the calyx lobes and opposite them. Female flowers with 1 simple 1-celled pistil, superior or inferior, with 1 style and 1 stigma which may be simple or tufted. Fruit an achene or drupe.

Pollination in this family is accomplished by an interesting mechanism. In the bud, the filaments are bent inward. As the flowers open, the anthers dehisce and the filaments straighten elastically, throwing out a cloud of pollen which settles down as it is dispersed by air currents. Hover flies and beetles visit the flowers and play a subordinate part in pollination.

A family of about 42 genera and 600 species of wide geographic distribution, mostly in the tropics and subtropics. Five genera in Kansas.

Key to Genera (3):
1. Plants with stinging hairs 2
  2. Leaves alternate .... *Laportea*
  2. Leaves opposite .... *Urtica*
1. Plants without stinging hairs 3
  3. Leaves alternate .... *Parietaria*
  3. Leaves opposite 4
    4. Flowers in axillary spikes; achene completely enclosed by the calyx .... *Boehmeria*
    4. Flowers in axillary panicles or glomerules; achene protruding beyond the calyx ...... *Pilea*

BOEHMÈRIA Jacq.   False nettle

Named for Georg Rudolph Boehmer (1723–1803), professor at Wittenberg.

Perennial herbs, shrubs, or trees (ours herbs). Leaves opposite or alternate, petiolate; blades 3-nerved, toothed or rarely lobed; stipules free. Flowers minute, apetalous, monoecious or dioecious, in axillary inflorescences. Male flowers with a 4-lobed calyx and 4 stamens; a rudimentary ovary also present. Female flowers with a tubular calyx, flattened to ovoid, with 2–4 minute teeth at the summit. Ovary 1, simple, superior, enclosed by the calyx; style 1, filiform, protruding beyond the calyx, persistent; stigma 1, elongate. Fruit an achene surrounded by the accrescent calyx which is narrowly 2-winged.

One species in Kansas.

B. cylíndrica (L.) Sw.   False nettle, bog hemp

NL., from Gr. *kylindrikos*, cylindrical, perhaps referring to the elongate inflorescences.
Flowers white or greenish, minute; June–October
Moist or wet soil of meadows or low woods
Mostly in the eastern half

Perennial herbs arising from a rhizome, resembling true nettles in appearance but lacking stinging hairs. Stem erect, little branched (if at all), up to 10 dm tall. Leaves opposite (occasionally alternate), simple, long-petiolate; blades ovate or sometimes lanceolate, with 3 main veins arising at the base and numerous lateral veinlets departing pinnately from these (especially from the central one); apex acuminate; base acute, rounded, slightly cordate, or truncated, sometimes oblique; margins coarsely serrate, each tooth with a tiny mucro; blades of the middle stem leaves 6–15 cm long and 3–7 cm wide. Flowers minute, white or pale green, arranged in small clusters along spikes arising from the upper leaf axils. Male and female flowers as described for the genus. Achenes ovate, shiny black, minutely winged.

Plants grown in sunny habitats have smaller, firmed, more hairy leaves than those of shady habitats and have at times been separated as a distinct species, *Boehmeria drummondiana* Wedd., or as a variety of *B. cylindrica, B. c.* var. *drummondiana* (Wedd.) Wedd.

LAPÓRTEA Gaud.   Wood nettle

Named in honor of Francois L. de Laporte, Count of Castelnau, 19th-century entomologist.

Herbs (shrubs or trees in the tropics) with large serrate leaves and axillary stipules. Stems and leaves with fine stinging hairs, as in *Urtica*, which on being touched cause stinging, itching, and a reddening of the skin. Flowers apetalous, monoecious or dioecious, clustered in loose cymes. Male flowers with 5 sepals, 5 stamens, and a rudimentary ovary. Female flowers with 4 sepals (the outer 2 or 1 of them usually minute) and 1 simple superior pistil; the single style elongate with a hairy stigma down one side. Fruit an achene.

One species in Kansas.

74. *Laportea canadensis*

L. canadénsis (L.) Wedd.   Canada wood nettle

Canada + L. -*ensis*, a suffix added to nouns of place to make adjectives—"Canadian," in this instance.
Flowers greenish, minute; July–September
Rich, moist woods
Eastern third

A herbaceous, green-stemmed perennial, growing upright, 0.5–1.3 m tall. Leaves alternate, simple, long-petiolate, with inconspicuous deciduous stipules; blades mostly ovate to broadly ovate, occasionally lanceolate, oblanceolate, or broadly elliptic, with either 1 or 3 main veins arising at the base and lateral veinlets departing pinnately from these; apex acuminate; base rounded or cuneate; margins coarsely serrate, the teeth sometimes terminated by a tiny white bristle; blades of the middle stem leaves 10–20 cm long and 6–13 cm wide. Stems, petioles, and main veins of the leaves with thick, spreading stinging hairs with swollen bases intermixed with very fine, spreading nonstinging hairs; leaves with scattered appressed hairs on the upper surface. Plants dioecious or monoecious, the minute greenish flowers arranged in loose compound clusters with spreading main divisions. When monoecious, female flowers predominate in the upper sprays and male flowers in the lower ones. Branches of the female sprays and the pedicels become flattened and winged as the fruits mature, and the terminal tips of the female inflorescences are spatulate. Male flowers of 5 sepals, 5 stamens, and a stunted nonfunctional pistil. Female flowers with 4 unequal sepals, a somewhat compressed ovary, and an awl-shaped style. Achenes black or dark brown, lustrous, flattened, D-shaped, sometimes tipped by the persistent style, 2–3 mm across, subtended by 2 persistent sepals.

PARIETÀRIA L.   Pellitory

L. *parietaria*, the Classical Latin name for the European *Parietaria officinalis*, from *L. paries*, wall, *parietarius*, of a wall, from the habit of plants of the European species, which grow in the crannies of dry walls.

Annual or perennial herbs lacking stinging hairs. Leaves alternate, with 1 prominent vein, entire, without stipules. Flowers minute, green, apetalous, monoecious or polygamous, situated in short axillary clusters and subtended by leafy green bracts. Male flowers with 4 more or less united sepals, 4 stamens, and a rudimentary nonfunctional ovary. Female flowers with a tubular 4-lobed calyx and a superior ovary, the stigma sessile, or nearly so, and tufted. Achenes shiny.
Compare with *Acalypha*.
One species in Kansas.

P. pennsylvánica Muhl.   Pennsylvania Pellitory

"Of Pennsylvania."
Flowers minute, green; June–August
Woods and prairie habitats, such as thickets and ravines, which offer shade and some moisture
Eastern three-fourths

Pubescent, stingless annuals with weak, simple or sparingly branched, ascending or reclining stems, 1–4 dm long. Leaves alternate, simple, with slender petioles and no stipules; blades lanceolate to elliptic or ovate, mostly with a single prominent vein and a few delicate lateral veinlets; apex more or less acuminate with a blunt tip; base cuneate or acuminate; margins entire; blades of the middle stem leaves 3–10 cm long and 0.5–2 cm wide. Flowers polygamous, with male, female, and bisexual ones in the same short axillary clusters. Achenes ovoid or ellipsoid, slightly compressed, brown, smooth, and very shiny, about 1 mm long or slightly longer.

75. *Urtica dioica*

PÍLEA Lindl.   Clearweed

From L. *pileus*, a felt cap worn by the Romans, in reference to the shape of the larger sepal of the female flower which, in the original species of this genus, partly covers the achene like a small cap.

Erect annual or perennial herbs lacking stinging hairs. Leaves opposite, blades 3-nerved, toothed, the stipules inconspicuous and united. Leaves and stems with numerous epidermal cystoliths appearing as minute short white lines on dried specimens (use magnification). Flowers minute, apetalous, monoecious or dioecious, situated in axillary cymes. Male flowers with 4 sepals, fused to one another, and 4 stamens. Female flowers with 3 free sepals, 3 scalelike staminodes (nonfunctional stamens), and 1 simple superior pistil with a tufted sessile stigma. Fruit a flattened achene subtended by the persistent calyx.

One species in Kansas.

P. pùmila (L.) Gray   Clearweed

L. *pumilus*, dwarf, short.
Flowers white or greenish; September–October
Moist, rich woods
Eastern eighth

Erect annuals, 3.5–5 dm tall. Leaves opposite, simple, long-petiolate; blades ovate to elliptic or lanceolate, sometimes rhombate, translucent, with 3 main veins arising from the base and delicate lateral veins departing pinnately from these; apex short-acuminate, terminated by a long fingerlike tooth; base cuneate or rounded; margins coarsely serrate or nearly crenate, the teeth sometimes tipped by a tiny mucro; blades of the middle stem leaves 3.5–9 cm long and 2.5–6 cm wide. Flowers minute, white or greenish. Branches of the inflorescence becoming flattened and winged, but the terminal branchlets not spatulate. Achenes shaped like flattened teardrops, green, sometimes with purple blotches, 1.5–2 mm long.

ÚRTICA L.   Nettle

L. *urtica*, ancient name for a stinging nettle, from *uro*, to burn.

Annual or perennial herbs with fibrous stems and stinging hairs. Leaves opposite. Flowers monoecious or dioecious, in spikes, panicles, or headlike clusters. Male flowers with a deeply 4-lobed calyx, 4 stamens, and a rudimentary ovary. Female flowers with 2 pairs of sepals, the outer pair small and inconspicuous, the inner pair larger and tightly enclosing the ovary, but lacking a corolla. Ovary 1, simple, superior; stigma sessile, tufted. Fruit a flattened, ovate achene enclosed by the persistent calyx.

Two species in Kansas. *U. chamaedryoides* Pursh, a slender annual usually branched from the base, is known from a few sites in Cherokee, Labette, and Neosho counties.

U. dioìca L.   Tall nettle, stinging nettle

NL., from Gr. *di-*, two, + *oikos*, house—this species being mostly dioecious, that is, having male and female flowers on different plants.
Flowers greenish, inconspicuous; late June–October
Low, moist ground, frequently in alluvial woods
Primarily north of a line drawn from Cherokee County to Sherman County

Unbranched or little-branched perennials, 7–13 dm tall, with angled stems arising from a rhizome. Leaf blades ovate to narrowly lanceolate, 3.5–15 cm long and 1.5–7 cm wide, with 3 main veins arising at the base and numerous lateral veinlets departing from them; apex usually acute to long-acuminate (occasionally rounded on a few leaves), the base broadly acute to rounded or slightly cordate, the margins coarsely serrate, the petioles 1–6 cm long, the veins of the lower surface quite prominent. Leaves and calyces with minute appressed hairs; stems and petioles with larger, scattered, hollow stinging hairs as well. Flowers minute, greenish, borne in loose, branched axillary clusters. Male and female flowers mostly on different plants. Achenes about 1 mm long.

This species has been listed as *U. procera* in earlier manuals for our area.

76. *Comandra umbellata* ssp. *pallida* fruiting

SANTALÀCEAE — Sandalwood Family

Herbs, shrubs, or trees, our herbs characterized by a semiparasitic habit. Leaves mostly alternate, simple, entire, leaflike or reduced to scales, lacking stipules. Flowers bisexual or unisexual, regular, in terminal or axillary inflorescences or solitary. Flowers apetalous, though with a 4- to 5-lobed calyx that may be colored like a corolla. Stamens of the same number as the perianth segments and opposite them. Pistil inferior, 1-celled and 2- to 4-ovuled; 1 style; 1 capitate stigma. Fruit a nutlike achene or drupe, composed of the coalesced hypanthium and ovary with the persisting sepals at its apex and a single maturing seed inside.

The family includes about 30 genera and at least 400 species of temperate and tropical regions. Kansas has a single genus and species.

COMÀNDRA Nutt. — Bastard toadflax

NL., from Gr. *kome*, tuft of hairs, + *aner, andros*, man—the stamens or male members of the flowers being adherent to the hairy-tufted base of the sepals.

Low perennials arising from slender branched rhizomes which produce at intervals simple or branched upright stems. Roots fibrous, some of them modified into haustoria which penetrate the roots of other plants and derive additional nourishment from them. Being green, the *Comandras* are able to make their own food and, like the mistletoe, probably rob their hosts of little more than water. Leaves alternate, simple, entire, sessile, lacking stipules. Flowers bisexual, regular, situated in terminal or axillary clusters. Sepals typically 5, sometimes 4, white (yellowish when dried). Arising from the inner epidermis of the sepals are numerous long, yellowish unicellular hairs, the upper portions of which break off to release what is thought to be a balsamlike resin. After the anthers have dehisced, the hairs frequently become attached to the anther connectives. The purpose of this development is unknown, although relationships to insect attraction and pollination have been suggested. The flowers are visited by various kinds of short-tongued bees, butterflies, beetles, and especially flies. Stamens of the same number as the perianth lobes and inserted at the base of the latter; pollen sticky and adhering in shiny yellow masses. Ovary inferior, or half-inferior with the top free from the floral tube; style filiform; stigma only slightly broader than the style. A shallowly lobed nectiferous disk protrudes from the inner surface of the perianth tube above the ovary. Fruit a 1-seeded drupe.

Kansas has 1 species.

77. *Comandra umbellata* ssp. *umbellata*

C. umbellàta (L.) Nutt. — Bastard toadflax

L. *umbellatus*, umbellate, from *umbella*, umbrella, in reference to the inflorescence, which has the flower pedicels arising from a common point, like the ribs of an umbrella.
White, drying yellowish; mid-April–late June
Prairies, infrequently in rocky open woods
Scattered throughout

Stems erect, more or less branched, flowering stems mostly 1–2 dm tall. Leaves alternate, narrowly lanceolate to lanceolate, elliptic, or oblanceolate; apex rounded, acute, or short-acuminate, sometimes with a sharp tooth on the end; base cuneate to attenuate; margins entire; 1–4 cm long. Flowers (including the inferior ovary) 4–7 mm long. Mature fruits (excluding the persistent sepals) about 8 mm long.

Much morphological variation occurs within this species, and Kansas plants are recognized as belonging to 2 subspecies—ssp. *pallida*, which occurs primarily in the west two-thirds of Kansas, and subspecies *umbellata*, which occurs in the northeast sixth and Cherokee County. Ssp. *pallida* typically has herbage gray-green, quite pale, and glaucous; leaves rather thick, firm, and somewhat succulent, and uniformly colored on both surfaces; sepals 2–3 mm long; fruits less than 6 mm wide; and rhizomes which are white or beige in cross section. Ssp. *umbellata* has green herbage; leaves thin, soft, and lighter colored on the lower surface; sepals 2.7–4.5 mm long; fruits 6 mm or more wide; and rhizomes blue or blackish in cross section. Where the ranges of the 2 subspecies abut or overlap, plants with intermediate characters occur.

ARISTOLOCHIÀCEAE   Birthwort Family

Woody climbers or low, acaulescent woodland herbs with perennial rhizomes. Leaves alternate, simple, entire, petioled, lacking stipules. Flowers bisexual, regular or irregular, arranged in racemes, axillary clusters, or solitary. Perianth composed of 3 sepals fused to one another. Stamens 3–36, mostly 6 or 12, free or fused to the style. Pistil 1, inferior or superior, 4- to 6-celled, each cell with several to many ovules; style 1, short and stout; stigmas as many as the cells. Fruit a septicidal capsule.

The family includes 7 genera and about 400 species, most of which are tropical. Kansas has 2 genera, *Aristolochia*, a woody vine not included here, and *Asarum*.

ÁSARUM L.   Wild ginger

NL., from L. *asarum*, hazelwort, wild spikenard, from Gr. *asaron*, the ancient name for the European species. The common name is suggested by the somewhat aromatic and pungent taste of the rhizome.

78. *Asarum canadense*

Perennial acaulescent herbs arising from a branched rhizome bearing coarse fibrous roots. Leaves 2, opposite, simple, petioled. The single flower bisexual, regular, borne between the petioles of the leaves on a reclining peduncle, often nearly hidden under the fallen leaves of surrounding trees. Calyx of 3 sepals fused laterally to one another about half their length. Petals absent or represented by 3 rudimentary lobes alternating with the calyx lobes. Stamens 12, epigynous, converging closely around the style. Pistil inferior, 6-celled, with a single style and a 6-lobed stigma. Fruit a somewhat fleshy globular capsule which opens irregularly or is loculicidal.

Kansas has 1 species.

A. canadénse L.   Canada wild ginger

L. -*ense*, a suffix added to nouns of place to make adjectives—"Canadian," in this case.
Brownish-purple; early April–mid-May
Moist woods rich in leaf mold
Extreme northeast and extreme southeast

Leaves broadly reniform or cordate, palmately net-veined; apex acute, short-acuminate, or merely rounded; base auriculate, with 2 deep sinuses and 2 long lobes; margins entire; the blade 4.5–7.5 cm long and 7.5–13 cm wide at flowering time. Both surfaces of the leaf or only the veins with short appressed hairs; the petioles, pedicel, and outside of the calyx tube with few to many long kinky hairs. Calyx lobes dark brownish-purple, triangular, acuminate, usually recurved, 1.2–2.5 cm long. Fruit a capsule which opens irregularly. Seeds large, more or less pear-shaped or ovoid, with a conspicuous funicular scar along one side, reddish-brown when fresh, wrinkled and gray when dried.

The dried and powdered rhizomes of this species have been used in the United States and Canada as a substitute for ginger. The European species, *A. europaeum*, is one of the plants in earliest use as a medicine. Dioscorides, Greek medical writer, and Pliny, Roman naturalist, both of the 1st century, recommended it for dropsy and sciatica, for inducing perspiration, and as a purgative. Matthiolus, personal physician of Kaiser Ferdinand I, says in his herbal of 1563: "The peasantry have no better medicine against colds or an attack of fever than to boil a handful of the leaves in wine or water, and after addition of honey to drink freely of this for several days." Even some of the modern German herbals recommend *Asarum* for intermittent fever, dropsy, sciatica, asthma, diseases of the liver or spleen, gout, indurated tumors, etc. Our wild gingers, also, were used medicinally in former times, as we learn from Griffith's *Medical Botany* (50), a warm decoction of the rhizome being employed as an aromatic stimulant and diaphoretic, and the dried and powdered leaves as snuff.

POLYGONÀCEAE   Buckwheat Family

Perennial, biennial, or annual herbs or vines, or some Central American shrubs or trees. Leaves alternate, simple, mostly entire and in our species, except those of *Eriogonum* and a few species of *Polygonum*, with a stipular sheath (the ocrea) extending from the node upward around the stem. Flowers bisexual or unisexual, regular, the inflorescence various. Sepals 3–6, herbaceous or petal-like. Petals absent. Stamens mostly 4–9, free from one another, but inserted on the base of the perianth. Pistil 1; the ovary superior, 1-celled, and 1-ovuled; styles 2–3, more or less united; stigmas capitate, 2-cleft or -toothed, or tufted. Fruit a rounded, flattened, or a 3- to 4-angled or -winged achene.

A family of about 40 genera and 800 or more species, most of which are distributed in temperate regions of the Northern Hemisphere. Cultivated members of this family include buckwheat (*Fagopyrum esculentum* Moench) and rhubarb (*Rheum rhaponticum* L.).

Three genera are represented in Kansas.

Key to Genera:
1. Plants without stipular sheaths; flowers subtended by a bell-shaped, obconical, top-shaped, or cylindric toothed involucre .... *Eriogonum*
1. Plants with stipular sheaths (present only as 2 subulate, often lacerated lobes in *P. tenue* and *P. sagittatum*); involucre absent 2
   2. Style 3-parted; stigmas tufted; achenes 3-sided .... *Rumex*
   2. Styles separate, short or obsolete; stigmas minute, capitate, 2-cleft or 2-forked; achenes 2- or 3-sided .... *Polygonum*

ERIÓGONUM Michx.   Eriogonum

NL., from Gr. *erion*, wool, + *gony*, knee or joint, in reference to the woolliness of some species, especially at the nodes.

Sometimes shrubs but mostly perennial, rarely annual or biennial, acaulescent or leafy-stemmed herbs with strong taproots. Leaves clustered at the base and alternate or opposite on the stem above, simple, entire, without stipules. Flowers bisexual or dioecious, usually several on threadlike pedicels which arise from the bottom of a bell-shaped, obconical, top-shaped, or cylindric involucre and extend beyond its toothed rim. Intermixed with the flower-bearing pedicels, one will find threadlike bracts as well as pedicels from which the flowers have already dropped. Sepals 6, united at the base. Stamens 9, inserted on the base of the calyx. Styles 2–3, threadlike, with capitate or capitellate stigmas. Fruit a rounded or 3-angled achene, rarely winged on the angles.

Kansas has 6 species, 5 of which are included in the following key. *E. effusum* Nutt. var. *rosmarinoides* Benth. in DC., a woody subshrub, is endemic to Logan, Gove, Trego, and Lane counties.

The seeds of various species of *Eriogonum* are of moderate importance as food for wildlife.

Key to Species:
1. Low acaulescent perennials arising from branched or unbranched woody caudexes 2
   2. Leaf blades elliptic or oblanceolate; calyx glabrous, quite petaloid, cream-colored ....
   .... *E. jamesii*
   2. Leaf blades oblong to narrowly elliptic or nearly linear; calyx densely pubescent on the outside, more herbaceous than petaloid, yellowish on the inside .... *E. lachnogynum*
1. Taller, leafy-stemmed annuals, biennials, or perennials, from taproots 3
   3. Flowering or fruiting involucres 3–6 mm long; calyx densely pubescent with long silky hairs .... *E. longifolium*
   3. Flowering or fruiting involucres 2–3 mm long; calyx glabrous 4
      4. Annuals or biennials; stems and upper surfaces of the leaves floccose-tomentose; involucres and lower surface of the leaves densely woolly; fruits rounded ...... *E. annuum*
      4. Perennials; stems and upper surface of the leaves with long, white appressed hairs; involucres glabrous or with only a few hairs; lower surface of the leaves glabrous except for the midrib; fruits 3-angled and winged .... *E. alatum*

E. alàtum Torr.   Winged eriogonum

L. *alatus*, furnished with wings, true of the achenes.
Flowers greenish-yellow; June–October
Dry, rocky, calcareous prairie banks
Scott and Russell counties

Erect perennials, 0.3–1.3 m tall, with a deep taproot. Stem unbranched below the inflorescence. Principal leaves spatulate to oblanceolate, 6.5–15 cm long, in a basal rosette. Stems with long, white appressed hairs; leaves with long, white appressed hairs on the upper surface, glabrous beneath except for the midrib. Flowers bisexual. Inflorescence a large open panicle; involucres obconical, 2–2.5 mm long, few in a cluster; involucres subtended by long-triangular or awl-shaped bracts, branches of the inflorescence subtended by subulate or linear bracts or reduced leaves mostly 1 cm or more long. Sepals 6, yellowish-green, fused laterally at the base, in 2 series (although this becomes less evident in older flowers); 1–2 mm long calyx lobes, oblong to ovate. Stamens 9, the anthers

79. *Eriogonum alatum*

yellow. Pistil fusiform. Achene with 3 broad, scarious longitudinal wings, mostly 4–5 mm long.

Although this species is not known from many Kansas counties, it is tall and robust and therefore quite conspicuous in the areas where it does occur. Its habit of growth is similar to that of *E. longifolium.*

E. ánnuum Nutt. — Annual eriogonum

L. *annuus,* of a year's duration, from *annus,* year.
White; mid-July–September
Dry, sandy prairie
Mostly in the western two-thirds and more abundant in the southern half of that area

Erect annual or biennial. Stem unbranched or with many ascending branches, 2.5–7.5 dm tall. Basal leaves usually gone at flowering time; cauline leaves narrowly oblong or oblanceolate, sessile or short-petiolate, with a single prominent vein; apex acute; base attenuate; margins somewhat revolute; leaves (including the petiole if present) of the main stem 3–9 cm long and 0.3–1 cm wide. Stems and upper surface of the leaves floccose; lower surface of the leaves, branches of the inflorescence, and involucres densely woolly. Flowers bisexual or dioecious. Inflorescence a flat-topped compound cyme at the apex of the stem or of its branches when present; involucres narrowly top-shaped, 2–3 mm long; involucres and branches of the inflorescence subtended by 2 or more triangular bracts fused together part of their length. Sepals 6, in 2 series, the outer 3 obovate, and inner 3 somewhat narrower, all united for a short distance at the base, white or pink, drying reddish-brown, 1–1.5 mm long. Stamens 9 or absent, filaments arising from the base of the sepals, sometimes intermingled with woolly hairs, the anthers purple. Pistil 1 or absent, obclavate. In female flowers, the pistil is usually surrounded by dense woolly hairs. In bisexual flowers, the pistils apparently reach anthesis after the stamens have released their pollen. Achenes obpyriform with a sharp apex, 1.5–2 mm long.

E. jámesii Benth. — Eriogonum

Named in honor of Edwin James, the first botanist to collect this species. James served as
    botanist and geologist with Stephen Long's expedition of 1819–1820.
Cream-colored; June
Dry, rocky prairie hillsides and bluffs
Scott and Logan counties

Erect perennials, mostly 1–1.5 dm tall, spreading close to the ground from a branched, woody, underground caudex. Leaves mostly basal, the blades elliptic to oblanceolate, light green above, with a single prominent vein; apex rounded, acute, or abruptly acuminate; base tapering gradually to the long petiole; blades 0.5–1 dm wide and 2.2–4 cm long, petioles about the same length. Stems, petioles, involucres, outside of calyx, and upper surfaces of leaves and bracts with long, white silky hairs; under surfaces of leaves and bracts with very dense, white woolly hairs. Flowers bisexual. Inflorescence a compound cyme; involucres obconical, 6–8 mm long, 1 per cluster, subtended by bracts or reduced leaves similar to the basal leaves but smaller and short-petiolate or sessile. Sepals 6, cream-colored with a green midrib, turning brownish-orange with age, fused laterally at the base, in 2 series, 5–7 mm long; calyx lobes oblong or oval. Stamens 9, the anthers pale yellow, bases of the filaments with long silky hairs. Pistil fusiform. Achenes 3-angled, shiny, about 4 mm long.

E. lachnógynum Torr. — Eriogonum

NL., from Gr. *lachneis,* woolly, + -*gynum,* female, in reference to the extremely hairy pistil of
    this species (this character, by the way, is also found in *E. longifolium*).
Yellowish; late May–October
Dry, rocky, sandy prairie
Extreme southwest corner

Erect perennials, 1–2.5 dm tall, spreading close to the ground from a branched, woody, underground caudex. Leaves entirely basal, the blades mostly oblong to narrowly elliptic, sometimes nearly linear, grayish- or silvery-green, with a single prominent vein, mostly 2–5 mm wide and 2–3 cm long, the petioles usually shorter; apex acute or acuminate; base attenuate. Stems, involucres, outer surfaces of the sepals, and upper surfaces of leaves with long, silky white hairs; under side of leaves with dense woolly hairs. Flowers bisexual. Inflorescence cymose; involucres obconical, 3–4 mm long, few in a cluster, without bracts or subtended by ovate or awl-shaped bracts 1.5–2.5 mm long;

80. *Eriogonum annuum*

81. *Eriogonum jamesii*

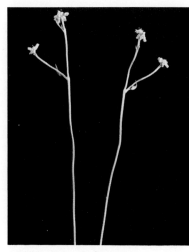

82. *Eriogonum lachnogynum*

BUCKWHEAT FAMILY
78

branches of the inflorescence subtended by narrowly lanceolate or linear bracts 3–15 mm long. Sepals 6, yellowish (although the color is obscured somewhat on the outside by the dense pubescence), fused laterally at the base, in 2 series, 3–4 mm long; calyx lobes elliptic or oblong. Stamens 9, the anthers yellow. Pistil obclavate, covered with woolly white hairs.

E. longifólium Nutt.   Eriogonum

NL., from L. *longus*, long, + *folium*, leaf, i.e., "long-leaved."
Flowers yellow or green; late June–October
Dry, gravelly prairie hillsides or sand dunes
Southwest corner and Harvey County

Robust erect perennials, 5–7 dm (or more) tall, from long taproots. Stem unbranched below the inflorescence. Principal leaves basal, oblanceolate to narrowly elliptic or linear, petiolate, with a single prominent vein; apex acute; base tapering gradually to the petiole which in turn becomes broader and somewhat winged near its base; blades of the basal leaves 0.5–1.5 cm wide and 6–14 cm long, the cauline leaves usually much smaller and sessile although similar in shape. Stems, involucres, and upper surface of the leaves covered with dense, matted gray hairs; under surface of the leaves with very dense, white woolly hairs; outside of calyces with long, silky, white, spreading or appressed hairs. Inflorescence cymose, several times branched; involucres top-shaped, 3–6 mm long; involucres and branches of the inflorescence subtended by awl-shaped bracts 3–10 mm long. Sepals yellow or green, 4–5 mm long, the calyx lobes oblong or elliptic. Stamens pale yellow, becoming orangish on drying. Pistil obclavate, covered with long, silky white hairs.

POLÝGONUM L.   Knotweed, smartweed

NL., from Gr. *polys*, many, + *gony*, knee, in allusion to the prominent nodes of stems and branches.

Annual or perennial herbs, or sometimes twining vines, frequently with conspicuously swollen nodes. Leaves alternate, simple, entire, mostly with a single prominent midvein and pinnate lateral veins, petiolate, the stipules united (except in a few species) to form a membranous sheath (ocrea; pl., ocreae) at each node. Flowers bisexual, regular, the inflorescence various but frequently a compound raceme; within the inflorescence, small clusters of flowers are ensheathed by membranous ocreolae (sing., ocrealus) similar to the ocreae found at each node of the stem. Calyx petal-like, with 4–5 (sometimes 6) lobes. Petals absent. Stamens 3–8, the filaments fused basally to the calyx. Pistil flattened or 3-angled, with 2–3 styles or style branches and stigmas. Fruit an achene, either 3-angled or lens-shaped (lenticular) in cross section.

The fruits of many species are eaten by waterfowl and other birds.

Kansas has 18 species of knotweeds, the 13 most common of which are included in the following key.

Key to Species (3):

1. Vines; leaf blades cordate to deltoid with sagittate bases 2
    2. Plants annual; mature calyx scarcely longer than the achenes .... *P. convolvulus*
    2. Plants perennial; mature calyx much longer than the achene, the outer lobes strongly
        winged .... *P. scandens*
1. Plants not vines; leaf blades various, but not cordate to deltoid 3
    3. Flowers in small axillary clusters; leaves jointed at the base 4
        4. Leaves with 2 longitudinal folds, the margins minutely spiny-toothed .... *P. tenue*
        4. Leaves flat or with the margin rolled toward the lower side, the margins entire 5
            5. Plants erect; ocreae becoming dark reddish-brown and deeply lacerated with age,
                the several prominent veins persisting as bristles .... *P. ramosissimum*
            5. Plants prostrate (or ascending, in dense vegetation), often forming mats; ocreae
                green and herbaceous at the base, with 2 or more membranous, colorless or pink,
                acuminate lobes above .... *P. arenastrum*
    3. Flowers in terminal (or also axillary) spikes or racemes; leaves not jointed 6
        6. Styles persistent in fruit, becoming hardened and more or less perpendicular to the
            body of the achene; sepals 4 .... *P. virginianum*
        6. Styles withering in fruit; sepals commonly 5 7
            7. Racemes terminal, solitary or paired (less frequently 3 or 4); achenes lens-shaped;
                plants perennial .... *P. coccineum*
            7. Racemes terminal and axillary, usually numerous; achenes and duration various 8
                8. Ocreae entire or merely lacerate 9
                    9. Racemes frequently nodding; outer sepals in fruit strongly 3-nerved, each
                        nerve ending in an anchor-shaped fork .... *P. lapathifolium*

83. *Eriogonum lachnogynum*, close-up of flowers

84. *Eriogonum longifolium*

85. *Polygonum*; note ocrea

BUCKWHEAT FAMILY

79

86. *Polygonum arenastrum*

87. *Polygonum bicorne*

9. Racemes erect; outer sepals in fruit with inconspicuous, irregularly forked nerves 10
   10. Stamens and style shorter than to about the same length as the sepals; achenes round or nearly so in outline, flattened .... *P. pennsylvanicum*
   10. Stamens or style conspicuously longer than the sepals; achenes round or round-ovate in outline, flattened or concave on one side and more or less convex on the other .... *P. bicorne*
8. Ocreae fringed with bristles 11
  11. Ocreae with spreading hairs .... *P. hydropiperoides*
  11. Ocreae with appressed hairs or glabrous; appressed hairs, when present, adnate at the base 12
  12. Calyx glandular-punctate (use high magnification) 13
    13. Glands amber, distributed on the tube and lobes of the calyx, often most abundant toward the base .... *P. punctatum*
    13. Glands whitish, confined to the inner lobes of the calyx .... *P. hydropiperoides*
  12. Calyx not glandular (use high magnification) 14
    14. Styles 2 or 3; achenes lens-shaped or 3-angled (if 3-angled, the faces slightly concave), tending to remain attached to the pedicels, the sepals weathering away from them; racemes usually less than 4 cm long, mostly rounded at the apex; marginal cilia of ocrea usually less than 3 mm long .... *P. persicaria*
    14. Styles 3; achenes all 3-angled (the faces flat), falling as they mature; racemes usually more than 4 cm long, tapering to the apex; marginal cilia of the ocrea 3 mm long or more .... *P. hydropiperoides*

P. arenástrum Lord. ex. Bor.

From L. *arena*, sandy place, perhaps applied because this species sometimes is found growing in sandy soil.
Flowers greenish or pinkish, inconspicuous; May–October
Waste ground, riverbanks, and disturbed areas in pastures (frequently in sandy soil)
Mostly in the eastern half, but also collected from scattered sites in the western half; perhaps intr. from Europe

    Smooth annuals, quite variable in aspect, often forming dense mats. Stems much branched, prostrate or, when growing in dense herbaceous vegetation, ascending, up to 8 dm long. Leaf blades mostly elliptic, occasionally lanceolate or oblanceolate; 0.4–5 cm long and 0.2–1.5 cm wide; apex acute or rounded; base acute; margins entire; ocreae green and herbaceous at the base with 2 or more membranous, colorless or pink, acuminate lobes above. Flowers borne singly or in axillary clusters of 2 to 3. Sepals 5, 1.5–2 mm long, green with wide white or pink petaloid margins. Stamens 8, the anthers yellow. Styles 3, the stigmas capitate. Achenes 3-angled, with 2 sides convex and the third (narrowest) side concave, reddish-brown, finely striated, about 2 mm long or slightly longer.

P. bicòrne Raf.   Pink smartweed

L. *bi-*, two, + *-cornis*, horned, in reference to the 2 styles.
Pink, less frequently white; July–October
Wet places such as stream and pond margins, moist ditches
Throughout

    Erect annual or short-lived perennial, up to 2 m tall. Stems much branched, the upper parts frequently with dark glandular hairs. Leaf blades mostly lanceolate or narrowly so, glandular-dotted beneath, up to 12 cm long and 2.5 cm wide; apex and base acute to acuminate; leaf margins and sometimes the midvein of the lower surface bearing rigid forward-pointing bristles; the ocreae usually lacking bristles. Flowers borne in erect terminal and axillary racemes 1–6 cm long and 1–1.5 cm wide. Calyx 3–5 mm long. Stamens 5. Styles 2. On some plants, the stamens protrude beyond the rim of the calyx, while the style is short and remains included. On other plants, the reverse is true, with the stamens being included and the 2 styles exserted. Achenes round or round-ovate in outline, flattened or concave on 1 side and more or less convex on the other, dark brown, and shiny, 2.5–3.5 mm long.

P. coccíneum Muhl.   Water smartweed, shoestring smartweed

L. *coccineus*, scarlet—in reference to the flower color.
Pink; June–October
Low, wet places along ditches, margins of lakes and streams
Throughout

Perennial, increasing and forming colonies by dark branching rhizomes, which to some people resemble shoelaces and account for its common name in some localities. Stems erect, up to 1.5 m tall. Leaf blades lanceolate or ovate-lanceolate, up to 18 cm long and 6 cm wide; apex acuminate; base acute or rounded, sometimes slightly truncated or cordate. Stems, leaves, petioles, and ocreae with rigid appressed hairs. Racemes terminal only, borne singly or in pairs, 1.5–10 cm long and 1–1.5 cm wide. Calyx 3–4 mm long. Stamens 5. Styles 2. Achenes round or nearly so in outline, flattened, biconvex, dark reddish-brown, and shiny, 2–3 mm long.

88. *Polygonum coccineum*

P. convólvulus L. — Dullseed cornbind, black bindweed

NL., from L. *convolvo*, to roll around, in allusion to the twining stem, or perhaps because of its similarity to species of *Convolvulus*.

Flowers whitish; May–August

Fields, waste places, fence rows

Throughout, but more common eastward; intr. from Europe

Twining annual. Stems smooth or minutely roughened, up to 1.5 m long. Leaf blades cordate to deltoid, dull green in color, up to 6 cm long; apex acute or acuminate; base sagittate; ocreae lacking bristles. Flowers borne in slender, interrupted, branched axillary racemes. Calyx whitish or greenish, minutely roughened, 2–4 mm long, enclosing the 7 stamens and the short style with its 3-lobed stigma and closely investing the fruit at maturity. Achenes 3-angled, black, muriform, about 3.5 mm long.

P. sagittatum L., known from a few scattered localities, is a slender reclining annual with narrowly lanceolate leaves with sagittate bases. The stems are 4-angled, with downward-pointing barbs; the ocreae are not cylindric and sheathing but consist of 2 triangular lobes.

P. hydropiperoìdes Michx. — Mild water-pepper

NL., from *hydropiper*, the specific epithet of another species of *Polygonum*, + the suffix -*oides*, -like, that is, resembling *P. hydropiper*. The name of the latter species is the latinized common name "water pepper," acquired because of the plant's pungent taste.

White or pink; June–October

Wet places

Mostly in the eastern half

Annual or perhaps a short-lived perennial. Stems with some branches decumbent and rooting from the lower nodes, up to 1 m tall. Leaf blades mostly lanceolate or narrowly lanceolate, up to 16 cm long and 2 cm wide, frequently glandular-dotted on the lower surface and with stiff forward-pointing bristles along the margins; ocreae with appressed hairs and long slender marginal bristles. Flowers borne in erect terminal and axillary racemes (2–) 4–7 cm long and 0.5–0.8 cm wide in fruit. Calyx 2–2.5 mm long. Stamens 6, enclosed by the calyx. Styles 3. Achenes ovate in outline, mostly 3-angled (although those near the apex of the inflorescence are sometimes lenticular), dark reddish-brown, and shiny, 2–2.5 mm long.

89. *Polygonum hydropiperoides*

P. lapathifòlium L. — Curltop Lady's-thumb

NL., from Gr. *lapathon*, for a kind of sorrel or *Rumex*, + L. *folium*, leaf.

White to pale rose; July–October

Wet, open places

Throughout; perhaps intr. from Europe

Erect annual. Stems simple or branched, sometimes with some branches decumbent and rooting at the lower nodes, up to 2 m tall. Leaf blades narrowly to broadly lanceolate or elliptic, glandular-dotted beneath, up to 25 cm long and 4.5 cm wide; apex long-acuminate; base acute or acuminate; margins with rigid forward-pointing bristles and the midveins and petioles with broader flattened bristles or scales; ocreae with conspicuous veins but lacking hairs or marginal bristles, usually becoming lacerated with age. Racemes frequently nodding, often paniculate, 1.5–8 cm long and 0.6–0.8 cm wide in fruit. Calyx 2–3 mm long. In fruit, the outer 3 sepals each has 3 conspicuous (under magnification) green nerves, each ending in an anchor-shaped fork. Achenes nearly round in outline, flattened, frequently with a concave depression on each side, dark reddish-brown, 1.5–2 mm long.

90. *Polygonum lapathifolium*

P. pennsylvánicum L. — Pennsylvania smartweed, pinweed

NL. "Pennsylvanian."

Pink or white; July–early October

BUCKWHEAT FAMILY

81

Wet places, waste ground, roadside ditches
Throughout

Annual from a taproot. Stems erect, branched, up to 1 m tall, the upper stems and internodes beset with tiny stalked glands. Leaf blades lanceolate to elliptic, up to 15 cm long and 3.5 cm wide; apex acuminate to acute or rounded; base mostly acuminate; margins with minute rigid forward-pointing hairs; ocreae with inconspicuous veins and lacking bristles of any sort. Racemes erect, terminal and axillary, 1–5 cm long and 0.8–1.5 cm wide. Calyx 3–4.5 mm long. Stamens mostly 8. Styles 2. Styles and stamens enclosed by the perianth. Achenes round or nearly so in outline, flattened, dark brown, and shiny, 3–3.5 mm long.

91. *Polygonum pensylvanicum*

P. persicària L. — Lady's thumb

An old generic name, said to come from the leaves resembling those of *Persica*, the peach.
Pink or white; June–October
Stream banks, lake margins, ditches, and other wet or moist places
Mostly in the eastern half; intr. from Europe

Erect annual. Stems with some branches decumbent and rooting at the lower nodes, up to about 8 dm tall. Leaf blades mostly elliptic, sometimes lanceolate or almost linear, up to 17 cm long and 3 cm wide; apex and base acute to acuminate; margins with short rigid forward-pointing bristles, the midvein and petiole with longer appressed hairs; ocreae with appressed hairs and long marginal bristles. Racemes erect, terminal and axillary, 1–4 cm long and 0.6–1 cm wide. Calyx 2–3 mm long. Styles 2 or 3. Achenes 3-angled in cross section and ovoid, or lenticular, and ovate to broadly elliptic, dark brown or reddish-brown, and shiny, 2–2.5 mm long, tending to remain attached to their pedicels and the sepals weathering away from them rather than falling freely as in many Polygonums.

P. punctàtum Ell. — Water smartweed

NL., from L. *punctatum*, meaning "pricked in," "sharp point," or "small hole," apparently in reference to the tiny, dotlike glands on the sepals.
Flowers white or greenish; June–October
Margins of ponds and streams, roadside ditches, and other wet or moist places
Mostly in the eastern half

Erect annual. Stems simple or branched, up to 1 m tall. Leaf blades lanceolate or elliptic, up to 12 cm long and 2.5 cm wide; apex acuminate, acute, or sometimes rounded at the very tip; base acute or acuminate; lower surface glandular-dotted (use magnification); margins with minute forward-pointing bristles; ocreae with tiny appressed hairs and with slender marginal bristles. Flowers arranged loosely in slender racemes 3–5 cm long and about 3 mm wide (up to 12 cm long and 5 mm wide in fruit), which may be either erect or arching. Sepals 4, glandular-dotted, 2–3 mm long. Stamens 6. Styles 3. Achenes 3-angled, dark- to reddish-brown, about 2.5 mm long.

P. ramosíssimum Michx. — Bushy knotweed

L. *ramosissimus*, very much branched.
Flowers green, sometimes tinged with pink, inconspicuous; August–early October
Waste ground, flood plains, dry prairie hillsides and ravines
Primarily in the southwest and northeast fourths

Erect annual, up to 12 dm tall. Stems slender, smooth, usually much branched above, but often with only a single main stem below. Leaf blades narrowly elliptic to oblanceolate or linear, sessile, mostly 1–2.5 cm long and 2–5 mm wide (occasionally larger), most of them falling by the time the plants are in fruit; apex acute or rounded base acute or acuminate; margins entire; ocreae becoming dark reddish-brown and deeply lacerated with age, the several prominent veins persisting as bristles. Flowers borne in clusters of 2 or 3 along the upper parts of the branches. Sepals 6, strongly keeled, green with white or pink petal-like margins, 2–3.5 mm long. Stamens 6. Styles 3, the stigmas capitate. Achenes 3-angled, dark reddish-brown, shiny, mostly 2.5–3 mm long.

P. scándens L. — Hedge cornbind, climbing false buckwheat

L. *scandens*, climbing.
Flowers whitish- or yellowish-green; August–October
Thickets and fence rows
Throughout

Smooth perennial vine, up to 2 m long. Stems annual, branched. Leaf blades cordate or deltoid, mostly 2–6 (occasionally up to 10) cm long and 1.5–3.5 (–5) cm wide; apex acuminate; base sagittate; ocreae with a few prominent veins but lacking bristles or hairs. Flowers drooping, borne in loose axillary racemes which are often branched, 1.5–6 cm long (up to 7 cm long in fruit), excluding the peduncle, and about 1 cm wide (2–3 cm wide in fruit). Calyx 2.5–3 mm long in flower, 8–10 mm long in fruit; 3 of the sepals with flat, toothed or crimped wings 1.5–3 mm wide. Stamens 8. Style very short, the stigma obscurely 3-lobed. Achenes 3-angled, black, shiny, about 4.5 mm long.

P. ténue Michx. Slender knotweed

92. *Polygonum scandens*

NL., from L. *tenuis*, slender.
Flowers greenish, sometimes tinged with pink, inconspicuous; July–October
Sandy prairie
East half

Erect annual, up to 4 dm tall. Stems erect, slender (1–2 mm wide), somewhat 4-angled and -winged, with several main branches arising near ground level. Leaves linear, up to 3 cm long and 2 mm wide; apex acute; margins entire or with minute spiny forward-pointing teeth; ocreae of 2 subulate lobes which are often much lacerated. Flowers erect, nearly sessile, borne singly or in axillary clusters of 2 or 3, mostly at the upper nodes. Sepals 5, keeled, with white or pinkish margins, 2.5–3 mm long. Stamens 8, the anthers purple. Styles 3, the stigmas capitate. Achenes 3-angled, black, shiny, about 2.5 mm long.

P. virginiànum L. Virginia knotweed

NL., "of Virginia."
Flowers whitish or greenish; July–September
Low, moist woods, especially on alluvial soil
East third

Erect or reclining annual. Stems mostly unbranched, up to about 1 m tall. Leaf blades broadly elliptic to ovate or occasionally lanceolate, mostly 8–20 cm long and 5–10 cm wide; apex acuminate; base rounded or acute; margins and upper surface of leaves with long appressed hairs, under surface with short fine appressed hairs; ocreae with long marginal bristles and appressed hairs. Flowers borne in slender terminal and axillary racemes 1.5–3 dm long and 10–15 mm wide in fruit, the flowers becoming widely spaced. Calyx 3–4 mm long, 4-parted. Stamens 4 or 5. Styles 2. Achenes ovate in outline, somewhat flattened but biconvex, reddish-brown, shiny, topped by the 2 persistent rigid reflexed styles, about 4 mm long (excluding the styles).

93. *Polygonum virginianum*

RÙMEX L. Dock, sorrel

L. *rumex*, sorrel.

Mostly coarse weedy annuals or herbaceous perennials. Leaves alternate, simple, entire, petioled, with stipules in the form of a membranous, soon fragmenting, whitish or brownish sheath (ocrea) around the stem to a short distance above the node. Flowers bisexual or unisexual, regular, borne in small verticels (whorls) aggregated into a compound inflorescence. Sepals 6, in 2 series of 3, the outer 3 green and spreading, the inner 3 larger, sometimes with the midrib enlarging into a spongy or corky tubercle (called a "grain") on the back, becoming colored and often winged and converging over the pistil. Stamens 6. Pistil 3-angled with 3 short styles, each with a branched stellate stigma. Fruit a 3-angled achene enclosed by the inner 3 sepals.

Kansas has 11 species of dock; the 7 most common ones are included in the following key.

Young shoots of most species can be eaten as cooked green vegetables, although some require several changes of water to remove their initial bitter taste. Various species have also been utilized as dye sources.

Key to Species (3):

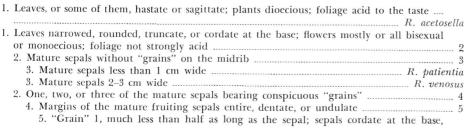

1. Leaves, or some of them, hastate or sagittate; plants dioecious; foliage acid to the taste .... *R. acetosella*
1. Leaves narrowed, rounded, truncate, or cordate at the base; flowers mostly or all bisexual or monoecious; foliage not strongly acid 2
  2. Mature sepals without "grains" on the midrib 3
    3. Mature sepals less than 1 cm wide .... *R. patientia*
    3. Mature sepals 2–3 cm wide .... *R. venosus*
  2. One, two, or three of the mature sepals bearing conspicuous "grains" 4
    4. Margins of the mature fruiting sepals entire, dentate, or undulate 5
      5. "Grain" 1, much less than half as long as the sepal; sepals cordate at the base,

94. *Rumex acetosella*

95. *Rumex altissimus*

usually dentate, 6–10 mm long .... *R. patientia*
     5. "Grains" 1–3, those best developed at least half as long as the sepal; sepals cordate
       or truncate at the base, margins various, usually less than 6 mm long 6
        6. Leaves with crisped or undulate margins, "grains" very turgid and obtuse,
          about two-thirds as wide as long, more or less ovoid .... *R. crispus*
        6. Leaves flat; "grains" half as wide as long or narrower, usually tapering to a
          point at the apex .... *R. altissimus*
   4. Margins of the mature fruiting sepals with a few long slender spinose or bristle-form
     teeth 7
       7. Plants glabrous; inner sepals 4–5 mm long at maturity .... *R. stenophyllus*
       7. Plants with minute hairs either on the petioles, lower surface of at least the lower
         leaves, or on the stem; inner sepals about 2 mm long at maturity .... *R. maritimus*

R. acetosélla L.   Red sorrel, sheep sorrel

NL., from L. *acetum*, having become sour, vinegar, + L. suffix *-ella*, in a small degree—the
     herbage being less acid than that of the garden sorrel, *R. acetosa*.
Flowers green, reddish, or purplish; late April–mid-July
Pastures, prairie, roadsides, and open woods
Eastern half, mostly in the eastern fourth; intr. from Europe

Perennial from a creeping rhizome. Stems erect, slender, mostly 1–5 dm tall.
Blades of the lower leaves variable, mostly hastate but sometimes unlobed and elliptic or
oblanceolate, 1.5–5 cm long; uppermost leaves nearly linear and without lobes. Flowers
dioecious, in erect panicled racemes, about 1 mm long at anthesis. Inner sepals ovate.
Anthers yellowish-green or reddish, on *very* short filaments. Mature achenes yellowish-
brown, lustrous, about 1.5 mm long, just equaling the inner sepals and closely invested
by them.

Also present in eastern Kansas, but uncommon, is *Rumex hastatulus* Baldw. (from
L. *hastula*, little spear, in reference to the shape of some of the lower leaves). It also has
hastate leaves and is similar in appearance to *R. acetosella*, but differs in having a taproot
rather than a rhizome and in having its sepals in fruit expanded into broad reticulate
wings much exceeding the achene.

The fast-growing, branching rhizome of *R. acetosella* makes this plant troublesome once it gets started where it is not wanted. Pliny wrote of it as being so vigorous
"that it is said that when it has once taken root it will last forever and can never be
extirpated from the soil." Sheep sorrel thrives on poor soil and is indeed an indicator
of the latter.

The young leaves of this plant are pleasantly tart and refreshing when chewed
and have often been used in salads and greens. The tartness of the leaves, due to their
content of acid oxalates, is the special quality that "sharpens the appetite, cools the
liver . . . and in the making of salads imparts a grateful quickness to the rest as supplying
the want of oranges and lemons," at least according to John Evelyn (1622–1726), English
author of *Sylva* and a famous diary. The leaves also serve to quench the thirst. Occasional individuals develop a dermatitis upon exposure to the foliage. The plant has
been used in Iceland to produce a light grayish-pink dye.

R. altíssimus Wood   Pale dock

L. *altissimus*, highest, tallest.
Flowers green; May–July
Fields, roadsides, alluvial soil, waste ground
Throughout Kansas

Perennial with a stout taproot, sometimes with heavy lateral roots as well. Stems
sturdy, erect, branched or unbranched, 0.5–1.3 m tall. Leaves flat, not undulate or
crisped; blades lanceolate, oblong-lanceolate, or narrowly elliptic, pinnately veined, those
of the larger leaves 10–25 cm long and 2.5–5 cm wide, with a single prominent midvein
and fine lateral veins; apex acute or acuminate; base attenuate, cuneate, or rounded;
margin entire, occasionally very finely undulate; petioles of larger leaves 2–3 cm long.
Entire plant glabrous. Inflorescence loose, 1–3 dm long. Flowers unisexual, monoecious,
1.5–2.5 mm long at anthesis. Male flowers with 6 elongate yellow anthers on very short
filaments. Inner 3 sepals ovate, obovate, or elliptic at anthesis, later becoming cordate
and then 4–5 mm long; at least 1 of the inner 3 with a corky "grain." Achenes brown,
lustrous, about 2.5 mm long.

R. *mexicanus* Meisn., Mexican dock, occurs only rarely in Kansas. It differs from
R. altissimus in having the fruiting sepals more triangular and with 3 large "grains"
instead of 1. *R. verticillatus* L., water dock, is known from the eastern border of the state.

It also has 3 large "grains," which are often noticeably lighter in color than the rest of the sepal, and has fruiting pedicels which are 10–15 mm long and strongly deflexed.

See plate *32*.

R. crispus L. — Curly dock, sour dock, yellow dock

L. *crispus*, curled, uneven, wavy, wrinkled—in reference to the leaf surfaces and margins.
Flowers green; May–July
Roadsides, fields, waste places, disturbed prairie
Throughout; intr. from Europe

96. *Rumex crispus*

Perennial from a heavy taproot. In its first year from seed, a strong yellow taproot is formed, crowned with a rosette of leaves which may remain green all winter. The second year and every year thereafter, the crown produces new leaves and an erect, branched or unbranched stem 2–5 dm tall. Leaves more or less undulate and margins somewhat crisped; blades lanceolate to linear-lanceolate, those of the larger leaves 10–20 cm long and 2.5–4 cm wide, with a single prominent midvein and anastomosing lateral veins which on the under side of the leaves are raised and conspicuous; apex acute; base attenuate, cuneate, rounded, or slightly cordate; petioles of larger leaves 3–12.5 cm long. Entire plant glabrous but both surfaces of the leaves covered with minute wartlike thickenings (use magnification). Inflorescence branched, mostly 2–4 dm long, becoming quite dense as the fruits mature, with several to many linear leaves intermixed with the whorls of flowers. Flowers bisexual, 1.5–2 mm long at anthesis, with 6 beige anthers on short filaments. Inner 3 sepals oval at anthesis, later becoming cordate, with rounded apex and reticulate veins, then mostly 4–5 mm long; a large ovoid corky "grain" present on 1 of the 3 inner sepals, very small "grains" sometimes present on the other 2. Achenes dark brown, lustrous, about 2 mm long.

The roots of this species may be used to produce a black dye.

R. maritimus L. — Golden dock

L. *maritimus*, growing by the sea, in reference to the habitat of this species in eastern North
    America.
Flowers green; July–September
Banks of rivers and creeks, margins of lakes and ponds
Scattered sites throughout the state except for the southeast fourth

97. *Rumex crispus*

Annual with a taproot and fibrous lateral roots. Stems erect, branched or unbranched, up to 8 dm tall. Leaf blades narrowly elliptic, narrowly lanceolate or oblong-linear, those of the larger leaves mostly 8–15 cm long and 1.5–3 cm wide, with a single prominent midvein and very fine, anastomosing lateral veins; apex acute or sometimes rounded; base attenuate or cuneate, sometimes cordate on the lower leaves; margin entire or finely erose, sometimes undulate; petioles of larger leaves 3–8 cm long. Leaves and stem with numerous short stiff erect hairs (use magnification). Inflorescence branched, 1–3 dm long, the whorls separate or contiguous, becoming more dense as the fruits mature, intermixed with numerous small leaves. Flowers unisexual and bisexual, polygamomonoecious, about 1.5 mm long at anthesis. Inner 3 sepals subulate or long-triangular at anthesis, later becoming more broadly triangular with 5 or 6 long, divergent marginal bristles, then about 2 mm long; all 3 inner sepals with a prominent "grain." Achenes brown, lustrous, about 1.5 mm long.

R. obtusifolius L., bitter dock, also occurs in Kansas and would probably be keyed out here with *R. maritimus*. The former differs conspicuously from the latter, however, in being a very tall (up to 1.5 m), robust perennial.

R. patiéntia L. — Patience dock

From L. *patientis*, enduring, patient, from *patior*, to endure.
Flowers green; April–June
Roadsides and waste ground
Primarily in the eastern half; intr. from Europe

Perennial from a heavy taproot. Stem erect, up to 1.5 m tall. Leaf blades elliptic, lanceolate, or ovate-lanceolate, those of the larger leaves 6–20 cm wide and 15–50 cm long, with a single prominent midvein and very fine, anastomosing lateral veinlets; apex acute or acuminate; base cuneate or attenuate, sometimes truncated or subcordate on the large basal leaves; margins entire, sometimes undulate; petioles of the larger leaves 4–15 cm long. Entire plant glabrous. Inflorescence branched, 2–5 dm long, the whorls becoming more dense as the fruits mature, sometimes with a few reduced leaves intermixed. Flowers bisexual or bisexual and female, 4–5 mm long at anthesis. Inner 3 sepals ovate

98. *Rumex stenophyllus*

99. *Rumex venosus*

or deltoid at anthesis, later becoming cordate with a rounded apex and conspicuous reticulate veins, then about 7–10 mm long; 1 inner sepal with a prominent "grain." Achenes brown, lustrous, about 3.5 mm long.

R. stenóphyllus Ledeb.　　　　　　　　　　　　　　　　　　　Dock

NL., from Gr. *stenus*, narrow, + *phyllis*, foliage, in reference to the width of the leaves.
Flowers green; June–August
Alluvial flood plains, sand bars, margins of lakes and ponds
Primarily in the southwest and northeast fourths; intro. from Eurasia

Perennial from a heavy taproot. Stems erect, branched or unbranched, up to 1.5 m tall. Leaf blades lanceolate, narrowly elliptic, or narrowly oblong (nearly linear, in some cases), those of the larger leaves 2–7 cm wide and 10–25 cm long, with a single prominent midvein and very fine, anastomosing lateral veins; apex acute or acuminate; base attenuate or cuneate, sometimes rounded; margins entire or finely erose, sometimes finely crisped or undulate; petioles of the larger leaves 2–10 cm long. Plants glabrous, but leaves sometimes roughened with minute wartlike thickenings (use magnification). Inflorescence branched or unbranched, 1.5–4 dm long, the whorls remote or contiguous, becoming denser as the fruits mature, sometimes with small linear leaves intermixed. Flowers bisexual, about 3 mm long at anthesis. Inner 3 sepals ovate or broadly elliptic with irregularly toothed margins at anthesis, later becoming broadly cordate or nearly orbicular, with numerous irregular sharp teeth or bristles, then about 4–5 mm long; all 3 inner sepals with a prominent "grain." Achenes brown, lustrous, 2–3 mm long.

R. obtusifolius L., bitter dock, is found in extreme southeastern Kansas and in several counties along the eastern border. It differs from *R. stenophyllus* in having minute hairs on the under surface of the leaves, cordate or truncate bases on the large lower leaves, and only 1 of the 3 mature sepals with a "grain."

R. venòsus Pursh　　　　　　　　　　Wild begonia, wild hydrangea

L. *venosus*, conspicuously veined, descriptive of the rose-colored sepals.
Flowers greenish or rose; April–mid-June
Sandy banks and dunes, roadsides
Western four-fifths

Perennial from a rhizome. Aboveground stems branched or unbranched, erect or eventually decumbent, 1.5–3 dm tall. Leaf blades ovate, lanceolate, or elliptic to oblanceolate or obovate, those of the larger leaves 3–13 cm long and 1.5–5 cm wide, with a single prominent midvein and fine, anastomosing lateral veinlets; apex acute or short-acuminate; base cuneate or attenuate, rarely rounded; margins entire, sometimes slightly undulate; petioles of larger leaves 0.5–4 cm long. Plants glabrous. Inflorescence branched, about 1 dm long at anthesis. Flowers bisexual, about 5 mm long at anthesis. Inner 3 sepals ovate, obovate, oblong, or elliptic, later becoming more or less orbicular with broad, ear-shaped basal lobes and conspicuous reticulate venation, then 2–3 cm long; all sepals lacking "grains." Achenes brown, glossy, about 6–7 mm long.

CHENOPODIÀCEAE　　　　　　　　　　　　　　Goosefoot Family

Annual or perennial herbs, shrubs, or rarely small trees. Leaves usually alternate, rarely opposite, simple, estipulate. Flowers minute, bisexual, or unisexual and then monoecious or dioecious, usually regular, borne in clusters. Sepals 2–5, fused to one another, usually persisting in fruit. Petals absent. Stamens usually of the same number as the calyx lobes and opposite them, free from one another, hypogynous or inserted on the calyx. Pistil 1; ovary usually superior, 1-celled and 1-ovuled; styles and stigmas 1–3. Fruit a 1-seeded utricle with a membranous wall, less frequently an achene.

A family of about 102 genera and 1,400 species of world-wide distribution but especially abundant in xerophytic and halophytic areas. Cultivated members of this family include the garden beet (*Beta vulgaris* L.), spinach (*Spinacea oleracea* L.), and summer cypress (*Kochia scoparia* (L.) Roth).

Kansas has 15 genera. Only the 8 most common genera are included in the following key.

The flowers in this family are so tiny that magnification is usually necessary for identification. It is also desirable to have plants with at least some mature fruits present.

Key to Genera (3):

1. Fruit free and exposed; calyx of 1–3 sepals on one side of flower only 2
   2. Leaves linear, entire; sepals 1–3, scarious, minute .... *Corispermum*
   2. Leaves, at least the main ones, with 2 large salient teeth; sepal 1, herbaceous ...... *Monolepis*
1. Fruit generally surrounded and more or less enclosed by the persistent calyx or bracteoles; calyx of (3–) 5 sepals or (2–) 5-lobed 3
     3. Fruit enclosed and concealed by a pair of rhombic, rounded, or deltoid bracteoles .. *Atriplex*
     3. Fruit surrounded or enclosed by the persistent calyx 4
        4. Foliage leaves all entire and linear to narrowly lanceolate 5
           5. Fruiting sepals, each with a flat, horizontal dorsal wing, or transversely keeled 6
              6. Leaves unarmed .... *Kochia*
              6. Leaves, especially those of the inflorescence, spiny-tipped .... *Salsola*
           5. Fruiting sepals concave, hood-shaped, vertically keeled, or spiny on the back 7
              7. Flowers few to several in glomerules, these forming a terminal panicle .... *Chenopodium*
              7. Flowers sessile in groups of 3 in the axils of the leaves .... *Suaeda*
        4. Foliage leaves either toothed or lobed or broader than lanceolate 8
           8. Fruiting calyx horizontally winged .... *Cycloloma*
           8. Fruiting calyx unchanged in form, its lobes vertically keeled or hood-shaped .... *Chenopodium*

ATRIPLEX L.   Saltbush

L. *atriplex*, ancient name for the garden vegetable orache (*Atriplex hortensis*), probably from L. atri-, black or dark, + *plexus*, network, from the conspicuous reticulate venation on the bracteoles of this species.

Annual or perennial herbs or shrubs. Foliage often scurfy or mealy. Leaves mostly alternate. Flowers minute, regular, unisexual, male and female flowers on the same plant or on different plants, borne in small dense clusters at the nodes, in the upper axils, in terminal leafless or bracted spikes, or both. Male flowers with 3–5 sepals or calyx lobes and 3–5 stamens inserted on the base of the calyx, a rudimentary ovary sometimes present. Female flowers lacking a perianth but usually enclosed by 2 broad bracteoles; pistil with 2 threadlike styles.

Kansas has 3 native species. Of these, *Atriplex canescens* L., four-wing saltbush, is a shrub 4–25 dm tall and is omitted here. In addition, *A. hortensis* L., orache, and *A. rosea* L., red orache, introduced from the Old World, are very rare in our area.

Key to Species:

1. Bracteoles enclosing the fruit at maturity herbaceous or spongy or somewhat succulent in texture, easily separable or eventually separating to the base, deltoid, triangular, or broadly rhomboidal in outline, the margins entire or with a few teeth, the backs more or less smooth or with a few tubercles .... *A. patula*
1. Bracteoles enclosing the fruit at maturity hard or bony in texture over the fruit, not easily separable, more or less orbicular in outline, the margins herbaceous and deeply toothed, the backs with several conspicuously projecting tubercles .... *A. argentea*

A. argèntea Nutt.   Silverscale

From L. *argenteus*, silvery, descriptive of the herbage of this species.
Flowers minute, greenish; July–mid-September
Dry, sandy soil, shale, or clay soil, often in alkali salt flats
Western two-thirds

Erect annual, up to 0.7 m tall, often branched from the base. Foliage, stem, and inflorescence usually silvery-scurfy throughout. Leaves alternate, graduating from petioled below to sessile or nearly so above, blades of the main leaves of the stem broadly lanceolate to ovate, rhombate, or deltoid, up to 5 cm long but usually shorter. Male and female flowers borne on the same plant, either in separate spikes or intermingled in short glomerulate axillary or terminal spikes, often with reduced leaves intermixed. Male flowers with 5 calyx lobes. Bracteoles surrounding the female flowers quite hard and bony at maturity and not separable, for that reason sometimes mistaken for the true fruit, more or less orbicular in outline, the margins deeply toothed, the backs of the bracteoles usually with several spiny tubercles.

This species and the following are reported to cause hayfever.

A. pátula L.   Spearscale

NL., from L. *patulus*, open, spread out, broad, probably in reference to the open branching habit of this species.
Flowers minute, greenish; August–October

Alkali salt flats
Western three-fifths

Erect or somewhat prostrate annual, up to 1 m tall, quite mealy or scurfy in appearance or glabrate and bright green, sometimes reddish. Leaves alternate, petioled, blades of the main leaves of the stem lanceolate to rhombate or strongly hastate, up to 7.5 cm long. Male and female flowers usually borne intermingled on the same plant in leafless glomerulate axillary or terminal spikes. Male flowers with 4 or 5 calyx lobes. Bracteoles surrounding the female flowers herbaceous, becoming more prominent as the fruits mature, then triangular, deltoid, or broadly rhombate, the margins entire or with a few teeth, the backs smooth or with a few tubercles. Seeds flattened, round to ovate in outline, black and lustrous, or appearing dull because surrounded by remnants of the translucent membranous fruit wall, about 1.5 mm long.

Our plants are recognized as var. *hastata* (L.) Gray (from L. *hastatus*, hastate, i.e., with equal, more or less triangular basal lobes directed outwards, in reference to the shape of some of the leaves).

The young growing tips of the plants may be cooked as greens and, according to Fernald and Kinsey (52), are much superior to *Chenopodium* and *Amaranthus*. The pollen is reported to be an important contributor to hayfever.

CHENOPÒDIUM L.   Goosefoot

From Gr. *chen* or *chenos*, goose, + *pous*, foot, from the leaf shape of some species.

Annual or perennial herbs. Leaves linear to ovate, deltoid, or rhombic. Flowers bisexual, in many species arranged in small round dense cymes (called *glomerules*) and these then grouped into axillary or terminal spikes, racemes, or panicles. Sepals 3–5, fused basally. Stamens of the same number as the calyx lobes. Pistil with 2–3 styles. Fruit thin-walled, closely invested by the persistent calyx, with a single lens-shaped seed. The leaves and stems, especially in young growth, and the sepals are often covered at least partially with conspicuous inflated white hairs which impart a "mealy" or farinose ("covered with a meal-like powder") appearance.

Kansas has 13 species. Because this is a rather difficult group for beginning botanists, only the 6 most commonly collected species are included in the following key.

Both the seeds and the young foliage of various species of *Chenopodium* have been used as food by Indians and white settlers in North America, as well as by prehistoric peoples in Europe. Alexander Johnston (8) reports the discovery, near a prehistoric Blackfoot archeological site in Alberta, Canada, of a cache containing 4–5 liters of cleaned *C. album* seeds. I have made palatable muffins from the seeds of *C. berlandieri*. The seeds were gathered in the fall, prior to and after several hard frosts. After being cleaned, the seeds were ground in a blender. The freshly ground flour has a sweet scent, similar to that of molasses. The most satisfactory recipe has been the following, an adaptation of a bran muffin recipe:

3 cups all-purpose flour	½ cup water
1 cup *Chenopodium* flour	2 cups sour milk
2 tsp. salt	2 Tbsp. melted butter
⅓ cup sugar	1 beaten egg
1½ tsp. soda	

Sift dry ingredients together. Add liquid ingredients and stir just enough to moisten the dry. Spoon batter into greased muffin tin or paper baking cups. Bake at 425° F. for 15–20 minutes.

The muffins have a strong flavor, reminiscent of a cross between cornbread and rye bread. Although the color is not especially attractive (the batter being the color of wet ashes), there are never leftovers when I serve the hot muffins, topped with butter and elderberry jelly, to a class of botany students.

Key to Species:
1. Under side of leaves glandular-dotted .... *C. ambrosioides*
1. Under side of leaves not glandular-dotted 2
   2. Seeds both vertical and horizontal in the same inflorescence (early stages may appear to have all horizontal seeds); sepals 3 or 4 (sometimes 5); short plants (up to 4 dm), branched from the base and having no single main stem .... *C. glaucum*
   2. Seeds horizontal, rarely and exceptionally a few vertical; sepals 5; taller plants (up to 1.5 m), often branched above but usually not from the base, a single main stem easily discernible 3
      3. Principal leaves linear to narrowly ovate or oblong, less than 1.5 cm broad, 2 to many times longer than broad, entire, 1- to 3-nerved 4

4. Leaves rather thick or leathery, silvery below and light green or yellowish-green above; flowers crowded into glomerules and these into dense spikes; fruits all maturing at about the same time .... *C. desiccatum*
4. Leaves rather thin or membranous, more nearly the same color below as above; flowers mostly somewhat separated from one another, the inflorescence open and lax; mature fruit and young perfect flowers present and intermingled on the plant at the same time .... *C. standleyanum*
3. Principal leaves deltoid, deltoid-rhombate, or ovate, 1–3 times as long as broad, pinnately veined, entire or variously toothed 5
5. Leaves thin, membranous, or leathery; flowers in nearly the same stage of development (occasionally, pistillate flowers present later, especially in late season forms); sepals enclosing mature fruit; membranous fruit wall not readily separable from mature seeds .... *C. berlandieri*
5. Leaves thin; fruit maturing very prominently mixed so that mature fruit and young bisexual flowers are present in the glomerules at the same time (except in late stages); sepals largely exposing mature fruit; membranous fruit wall usually separable from mature seeds 6
6. Leaves up to 7.5 cm long, rounded or tapering at base, the teeth short and ascending, gradually reduced to entire bracts subtending most of the inflorescence branches; membranous fruit wall readily separable; seeds 1.1–1.5 mm broad .... *C. standleyanum*
6. Leaves up to 15 cm long, rounded, truncate, or slightly cordate at base, the teeth usually long-tapering and spreading; inflorescence without bracts or the main branches subtended by rather well-developed leaves; membranous fruit wall usually separable; seeds 1.5–2.5 mm broad .... *C. gigantospermum*

100. *Chenopodium ambrosioides*

C. ambrosiòides L.   Mexican tea, wormseed

NL., from *Ambrosia*, the genus which includes our ragweeds, + *-oides*, -like, i.e., resembling *Ambrosia*.
Flowers minute, greenish; July–November
Creek banks, roadsides, margins of fields, and other waste ground
Primarily in the eastern half, especially in the eastern fourth; intr. from tropical America

Erect annual or perennial herb, up to 1 m tall, often branched above. Blades of the main leaves lanceolate or elliptic, up to 11 cm long, gradually reduced above to conspicuous leafy bracts subtending the branches of the pyramidal inflorescence, pinnately veined; apex acute; base gradually narrowed to a winged petiole; margins coarsely serrate. Under sides of the leaves and bracts, upper portions of the stem, and sometimes the calyx lobes and the fruit glandular-dotted. Flowers in any given glomerule maturing at about the same time, although newly opened flowers and mature fruits may be present on different parts of the same plant. Sepals 5. Fruit with a smooth wall separating readily from the seed, mostly enclosed by the sepals at maturity. Seeds vertical or horizontal, reddish-brown and shiny, less than 1 mm wide.

This species is cultivated in some areas, including Maryland and North Dakota, as a source of ascaridol, an anthelmintic oil contained in the resinous secretions on the leaves and in the seeds. Overdoses of the expressed oil have been fatal to humans and to livestock, but grazing animals usually avoid the plant. Needless to say, *C. ambrosioides* is not recommended for use as a potherb nor as a source of flour.

C. berlandiéri Moq.   Lamb's quarter

Named in honor of its discoverer, Jean Louis Berlandier (1805–1851), a French-Swiss physician who collected plants in Texas and northern Mexico.
Flowers minute, greenish; July–December
Roadsides, pastures, waste ground
Throughout

Erect annual, up to 1 m tall, the stems usually branched above and often with conspicuous vertical green-and-white stripes. Blades of the main leaves rhombic, occasionally deltoid or ovate, with 3 main veins arising near the base of the leaf and smaller lateral veins departing pinnately from these, up to 5 cm long; apex rounded or acute; base acute or short-acuminate; margins wavy or irregularly serrate above the widest part of the blade; petioles up to 5 cm long. Calyx lobes, under side of the leaves, and sometimes the stem with a conspicuous white mealiness. Flowers in dense glomerules, these arranged into terminal and axillary panicles, all maturing at about the same time. Sepals 5, each with a prominent green keel and wide scarious margins. Fruit with a reticulate (netlike) wall adhering strongly to the seed, completely enclosed or nearly so by the persistent calyx. Seeds horizontal, black, and shiny but often appearing dull because of the adherent fruit wall, 1–1.5 mm wide.

This is our most common species. It resembles quite closely an introduced

101. *Chenopodium desiccatum*

102. *Chenopodium gigantospermum*

European species, *C. album* L. (from *albus*, white, particularly a dull rather than a glossy white because of the mealiness which covers the young foliage), which is occasionally collected throughout our state. *C. album* differs primarily from *C. berlandieri* in having a smooth fruit wall and in lacking the prominent keel on the sepals. In the eastern fourth of the state, *C. missouriense* Aellen may also be found. Vegetatively it resembles both *C. berlandieri* and *C. album*. Like *C. album*, it has a smooth fruit wall and sepals without a prominent keel. It differs from *C. album* primarily in having slightly smaller seeds (0.9–1.2 mm wide) and in having its flowering time restricted in our area to September and October. Both *C. album* and *C. berlandieri* flower from June or July into October or November and are rarely seen in flower as late as December.

The young plants, picked early in the growing season, are delicious as cooked green vegetables. If cooked for just a few minutes in lightly salted water, they retain their bright-green color well.

C. desiccàtum A. Nels. — Narrow-leaved goosefoot

NL., from L. *desiccatus*, dried up, perhaps in reference to the dry habitat in which this plant
    often grows.
Flowers minute, greenish; May–September
Roadsides, waste ground, dry prairie
Throughout, but less common in the southeast sixth

Erect annual, up to 8 dm tall, the stem usually branched above, occasionally from the base but even then a single main stem usually discernible. Blades of the main leaves rather thick and leathery, linear, elliptic, oblong, or, less frequently, subhastate, up to 5 cm long; apex rounded to acute, often apiculate; base acute; margins entire, occasionally with 2 shallow lobes or teeth; petioles up to 1.7 cm long. Upper surface of leaves light green or yellowish-green; lower surface of leaves, upper portions of the stem, and calyx lobes silvery or white with a dense mealiness. Flowers in compact glomerules and these in terminal and axillary racemes, most flowers on the plant maturing about the same time. Sepals 5. Fruit with a smooth wall separating readily from the seed, enclosed by the sepals at maturity. Seeds horizontal, black, and shiny, 1–1.5 mm wide.

C. gigantospérmum Aellen — Maple-leaved goosefoot

NL., from Gr. *gigant-*, giant, very large, + *-spermus*, -seeded, this species having quite large
    seeds, compared to other members of the genus.
Flowers minute, greenish; July–October
Moist woods, ravines, and thickets
Throughout, but more common in the eastern half and in north-central Kansas

Erect annual, up to 1.5 m tall, branched or unbranched above. Blades of the main leaves broadly ovate, less frequently deltoid or broadly lanceolate, up to 15 cm long, pinnately veined; apex acute or acuminate; base cordate, rounded, or truncate; margins with several to 7 large, spreading acuminate teeth alternating with broad rounded sinuses; petioles mostly 1–7 cm long. Calyx and upper portions of the stem somewhat mealy; leaves glabrous or slightly mealy; herbage bright green. Flowers borne in loose terminal or axillary panicles, mature fruit and young bisexual flowers intermingled. Sepals 5. Fruit smooth, usually separating readily from the seed, sometimes adhering in places, mostly exposed at maturity. Seeds horizontal, black, and shiny, with a prominent lateral ridge or wing, 2–2.5 mm wide (rarely as small as 1.5 mm).

This species is listed as *C. hybridum* L. in some manuals.

C. glaùcum L. — Oak-leaved goosefoot

NL., from Gr. *glaukos*, bluish-green or gray, sea-colored, in reference to the foliage.
Flowers minute, greenish; June–September
Dry lake margins, sandy flood plains, pastures
Mostly in the southern half, more common westward; intr. from Europe

Erect or prostrate annual, usually branched from the base with no single main stem discernible, 1–4 dm tall. Blades of the main leaves ovate to lanceolate or oblong, up to 3 cm long, usually with but a single noticeable vein; apex rounded; base acute or acuminate; margins sinuate or coarsely toothed; petioles mostly 2–10 mm long. Under surface of the leaves densely mealy, upper surface somewhat so. Flowers in compact glomerules and these in short axillary and terminal spikes. Flowers in any given glomerule maturing at about the same time, although newly opened flowers and mature fruits may be present on different parts of the same plant. Sepals 3 or 4 (sometimes 5).

Fruit with a smooth wall separating readily from the seed, mostly exposed at maturity. Seeds vertical or horizontal, reddish-brown, and shiny, less than 1 mm wide.

C. incanum (S. Wats.) Heller (from L. *incanus*, quite gray, hoary) is a western species occasionally collected in the western third of Kansas. It resembles *C. glaucum* in size and in being densely mealy and branched from the base. It differs in having most of its leaves 3-lobed and in having 5 sepals instead of 3 or 4.

Plants of *C. glaucum* and of *C. album* have been found to contain concentrations of nitrate potentially dangerous to livestock.

C. standleyànum Aellen   Goosefoot

Named after Paul Carpenter Standley (1884–1963), American botanist who studied the floras of
  the Southwest, Mexico, and Central America.
Flowers minute, greenish; July–October
Moist woods, ravines, and thickets
Eastern three-fifths

Erect annual, up to 1 m tall, with stems usually branched above. Lower leaves lanceolate or rhombic, mostly 3–6 cm long, gradually becoming narrower above and entire; 1 or 3 main veins arising from the base and lateral veins departing pinnately from these; apex acute or acuminate; base acute or acuminate; margins entire to sinuate or coarsely serrate, sometimes with only 2 or 3 teeth, these at the widest part of the leaf. Calyx, upper portions of stem, and under side of leaves sparsely mealy; leaves bright green, rather thin and membranous. Flowers borne 1–4 in small, open glomerules, these grouped loosely in a slender, often nodding panicle, with young bisexual flowers and mature fruits conspicuously intermingled. Sepals 5. Fruit wall smooth, separating readily from the seeds, mostly exposed at maturity. Seeds horizontal, black, and shiny, about 1.3 mm wide.

CORISPÉRMUM L.   Bugseed

NL., from Gr. *coris*, a bedbug (*Cimex lectularius*), + *sperma*, a seed, the fruits of this plant
  somewhat resembling this tiny insect.

Annual herbs with much-branched stems. Leaves alternate, sessile, narrowly linear, or nearly filiform, with a single prominent vein. Flowers bisexual, regular, borne in terminal spikes, each flower subtended and hidden by a scarious-margined bract. Calyx of 1 white hyaline sepal situated between the pistil and the stem. Stamens 1–3, protruding beyond the calyx but often hidden by the bract. Fruit compressed, the inner face more or less concave and the outer convex, with a scarious wing.

Kansas has 2 species. *C. nitidum* L. (from L. *nitidus*, shining, polished, again in reference to the fruit) is rarely collected and only in the western third of the state. It differs from *C. hyssopifolium* described below in being a taller plant with filiform leaves and in having the flowers widely spaced in slender spikes (so that the bracts overlap only slightly or not at all) and the fruits smaller (2–3.5 mm).

C. hyssopifòlium L.   Bugseed

NL., from L. *Hyssopus*, a genus of the mint family, + *folium*, leaf, because of a resemblance
  between the leaves of these plants.
Flowers minute, whitish; August–September
Sand hills and dry, sandy waste ground
Western half; intr. from Eurasia

Stems much branched, up to 6 dm tall. Young foliage and bracts often with stellate hairs, becoming glabrous with age. Leaves narrowly linear, mostly 2–3 cm long and 1.5–2 mm wide. Flowers borne in compact spikes 6–10 cm wide, the bracts subtending the flowers overlapping and concealing the stem within the spike. Achenes elliptic to ovate, 3–4 mm long, with a broadly winged margin.

CYCLOLÒMA Moq.   Winged pigweed.

NL., from Gr. *cyclon*, whirling around, + *loma*, the hem or fringe of a robe, in reference to the
  winged fruiting calyx.

Much-branched annual herb. Leaves alternate. Flowers bisexual or female, regular, borne in slender, few-flowered terminal spikes. Calyx lobes 5, strongly keeled and arching over the pistil. Stamens 5. Ovary flattened, lenticular; styles 3. Fruit a utricle surrounded by the persistent calyx which by fruiting time has developed a broad, scarious, erose horizontal wing completely encircling the calyx.

Kansas has but a single species.

103. *Cycloloma atriplicifolia*

104. *Kochia scoparia*

C. atriplicifòlium (Spreng.) Coult.   **Winged pigweed**

NL., from *Atriplex*, another genus in the Chenopodiaceae, + L. *folium*, leaf, i.e., having leaves resembling those of *Atriplex*.
Flowers minute, greenish; June–September
Sandy soil
Common except in the southeast sixth

Plants up to 8 dm tall. Leaves oblanceolate, coarsely dentate, pinnately veined, up to 6.5 cm long and 1.5 cm wide. Herbage glabrous or with some white arachnoid (resembling spider web) pubescence. Seeds black and shiny, but tightly invested by the membranous, arachnoid fruit wall, about 1.5 mm wide.

KÒCHIA Roth   **Summer cypress**

Named in honor of W. D. J. Koch (1771–1849), a German botanist.

Annual or perennial herbs or shrubs. Leaves alternate or opposite. Flowers mostly bisexual, some female, regular, borne in axillary and terminal spikes. Calyx 5-lobed, each lobe eventually developing a rounded, scarious horizontal wing. Stamens mostly 5. Pistil 1; ovary superior, depressed (flattened horizontally); styles and stigmas mostly 2. Fruit a thin-walled, 1-seeded utricle, enclosed at maturity by the persistent calyx.

Kansas has a single species.

K. scopària (L.) Schrader   **Summer cypress**

NL., from L. *scoparius*, a broom.
Flowers minute, greenish; July–October
Flood plains, fields, roadsides and other waste ground
Common throughout except in the southeast fourth where it is infrequently collected; intr. from Eurasia

Erect, much-branched annual, often forming pyramidal or rounded bushes up to 1 or 1.5 m tall. Leaves alternate, linear to linear-lanceolate, up to 5.5 cm long and 8 mm wide, with either 1 or 3 prominent veins arising at the base, petioled on lower parts of the plant but becoming sessile above. Stems more or less tomentose; leaf margins with long, spreading hairs. Flowers subtended by leaflike bracts. Seeds lenticular, about 1.5 mm broad.

MONÓLEPIS Schrad.   **Poverty weed**

NL., from Gr. *mono*-, one, + *lepos*, a scale, in reference to the single sepal.

Annual herbs. Leaves alternate. Flowers bisexual, sessile, borne in dense clusters in the upper axils. Sepal 1, herbaceous. Stamen 1, situated between the sepal and the pistil. Pistil 1; ovary laterally compressed; styles 2, very short; stigmas 2. Fruit a 1-seeded utricle with a reticulate wall. Seed lenticular, vertical.

Kansas has but a single species.

M. nuttallìana (Schultes) Greene   **Poverty weed**

Named in honor of its discoverer, Thomas Nuttall.
Flowers minute, greenish; late April–early June
Prairie ravines, open woods, flood plains, roadsides, waste ground
Scattered throughout most of the state

A rather fleshy annual, much branched from the base, up to 3 dm tall. Main leaves narrowly lanceolate or oblanceolate to elliptic or rhombic, with 2 (less frequently 4 or 0) prominent teeth or lobes about halfway up the blade, mostly 1.5–3.5 cm long (including the petiole) and 0.5–1.5 cm wide; apex acute or rounded; base acuminate. Sepal 1, oblanceolate or spatulate, much longer than the pistil, fleshy and bractlike in appearance. Seeds 1–1.3 mm broad.

SÁLSOLA L.   **Saltwort**

NL., from L. *salsus*, salty.

Erect annual herbs, much branched and globular, rather shrubby. Leaves linear, somewhat succulent, becoming indurated and spiny. Flowers bisexual, regular, each subtended by a pair of bractlets, borne singly or a few together in the axils of upper leaves. Calyx 5-lobed, each lobe eventually developing a scarious horizontal wing, the tips of the lobes connivent and erect over the fruit. Stamens 5. Pistil 1; ovary horizontally compressed; styles 2, filamentous; stigmas 2. Fruit a 1-seeded utricle.

Kansas has but 1 species.

S. kàli L.
Russian thistle, tumbleweed

NL., from Arabic *qaliy*, ashes of saltwort, an old generic name.
Flowers minute, greenish; July–September
Fields, roadsides, and waste ground
Throughout except in the southeast sixth, more common westward; intr. from Eurasia

Bushy annuals, up to 1 m tall. Stems with minute spines, often with vertical red-and-green stripes. Leaves narrowly linear, spine-tipped, the lower ones nearly round in cross section, up to 4 (rarely 8) cm long and 1 mm wide, becoming shorter, flatter, and more rigid on upper parts of the plant. Calyx white and membranous, sometimes becoming pink or reddish as the fruits mature. Seeds green, top-shaped, narrowest at the base, with a depression at the upper end and a groove spiraling downward, about 1.5 mm broad.

Although this plant is considered a serious weed in areas where it is overly abundant, it has been used successfully as emergency fodder for livestock during periods of severe drought. Large specimens are occasionally spray-painted and/or flocked and used as unusual Christmas trees.

105. *Salsola kali*

SUÀEDA Forsk. ex Scop.
Sea-blite

NL., from *suwed mullah*, the Arabic name of *S. baccata*.

Somewhat fleshy annual or perennial herbs, usually branched from the base. Leaves alternate. Flowers minute, bisexual or with both bisexual and unisexual flowers on the same plant, borne 1–3 (usually 3) in the axils of upper leaves. Calyx 5-lobed, fleshy, sometimes keeled or crested. Stamens 5. Ovary somewhat flattened; styles 2–5. Fruit a thin-walled, 1-seeded utricle. Seeds lens-shaped, usually horizontal but occasionally vertical.

Kansas has 1 species.

S. depréssa (Pursh) Wats.
Sea-blite

NL., from L. *depressus*, pressed down, low.
Flowers minute, greenish; July–October
Salt flats
Central and southwestern Kansas

Stems mostly branched from the base, 1–10 dm tall. Leaves linear or subulate, somewhat rounded in cross section, mostly 1–3 cm long and 1–1.5 mm wide. Calyx lobes in fruit each with a horizontal triangular wing and a vertical hornlike projection. Fruit enclosed by the persistent calyx. Seed shiny, black, about 1.2 mm wide.

AMARANTHÀCEAE
Amaranth Family

Annual or perennial herbs, shrubs, trees, or vines. Leaves alternate or opposite, simple, estipulate. Flowers bisexual or unisexual, regular, the inflorescence various. Sepals 2–5, more or less dry and membranous. Petals absent. Stamens usually 5, opposite the calyx lobes, the filaments distinct or fused to one another part or all of their length. Pistil 1; the ovary superior, 1-celled and 1-ovuled; styles and stigmas 1–3. Fruit a utricle or nutlet, or a capsule, drupe, or berry.

The Amaranth Family includes about 64 genera and 800 species of wide distribution, most abundant in the tropics of America and Africa.

Kansas has 4 genera.

Key to Genera (3):
1. Leaves alternate; anthers 4-celled .... *Amaranthus*
1. Leaves opposite; anthers 2-celled 2
  2. Flowers in small axillary clusters; calyx not woolly .... *Tidestromia*
  2. Flowers in terminal spikes or panicles; calyx densely woolly, at least in the female flowers 3
    3. Flowers bisexual, in dense heads or spikes .... *Froelichia*
    3. Flowers unisexual, male and female flowers on different plants, borne in loose pyramidal panicles .... *Iresine*

AMARÀNTHUS L.
Pigweed, amaranth

NL., from Gr. *amarantos*, unfading, because the dry calyx and bracts do not wither.

Annual herbs. Stems erect or prostrate. Leaves alternate, petiolate. Flowers unisexual, the male and female flowers either on the same plant or on separate plants, arranged in dense axillary or terminal spikes, each subtended by 3 bracts. Sepals 2–5, free from one another. Stamens usually 5, opposite the sepals, the filaments often fused part of their length. Fruit an indehiscent or dehiscent utricle (opening horizontally near the middle in the species included here). Seeds black, shiny, nearly round to ovate in outline, biconvex.

The young foliage of our pigweeds, especially *A. retroflexus*, may be used as a potherb, and the seeds may be ground and mixed with wheat flour for use in muffins and "buckwheat" cakes. In addition, the seeds of various species are eaten by birds and mammals, including the mourning dove, rose-breasted grosbeak, slate-colored junco, pocket mouse, and kangaroo rat.

Kansas has 8 species. The 6 most commonly collected species are included in the following key. In the case of the dioecious species included here, *A. arenicola* and *A. tamariscinus*, it is difficult to distinguish between the species using only male plants, and one will do well to collect both male and female plants when possible.

Key to Species:
1. Plants prostrate .... *A. graecizans*
1. Plants erect 2
  2. Male and female flowers on separate plants 3
    3. Tepals of female flowers 5, spatulate, 1–1.5 mm wide at the widest point; inner tepals of male flowers with a spiny tip, but the dark midribs not excurrent; plants primarily occurring in the southwest fourth of the state .... *A. arenicola*
    3. Tepals of female flowers 1 or 2, elliptic or subulate, less than 0.5 mm wide at the widest point; inner tepals of male flowers with the conspicuous midrib excurrent as a spine; plants primarily occurring in the eastern two-thirds of the state ....
    .... *A. tamariscinus*
  2. Male and female flowers on the same plant 4
    4. Flowers all or chiefly in small axillary clusters, sometimes also forming a small terminal panicle .... *A. albus*
    4. Flowers chiefly in elongate, simple or paniculate terminal spikes; smaller axillary clusters or spikes may also occur 5
      5. Tepals of female flowers elliptic, about 1.5 mm long and 0.25–0.50 mm wide ....
      .... *A. hybridus*
      5. Tepals of female flowers oblong to oblanceolate, 2–4 mm long and 0.5–1 mm wide
      .... *A. retroflexus*

A. álbus L. — Tumbleweed

L. *albus*, white, particularly a dull rather than a glossy white, the stems of this species often being whitish.
Flowers minute, greenish; July–September
Sandy flood plains, pastures, roadsides, waste ground
Scattered throughout

Plants erect, up to about 1 m tall and wide. Stems whitish, much branched from the base. Blades of larger (earlier) leaves oblanceolate or elliptic, mostly 2–4 cm long and 0.5–1.5 cm wide; later leaves much smaller; apex rounded; base acute or attenuate. Flowers borne in axillary clusters; male and female flowers on the same plant. Bracts acuminate-lanceolate, sharp-pointed, 2–4 mm long. Tepals of male flowers 3, narrowly elliptic or oblong, about 1–1.25 mm long; stamens 3. Tepals of female flowers 3, linear or subulate, often reddish, about 1 mm long. Seeds about 0.75–1 mm long.

A. arenícola I. M. Johnst. — Sandhills amaranth

NL., from L. *arena*, sandy place, + *colo*, to inhabit, these plants indeed growing in sandy soil.
Flowers minute, green or pinkish; July–early October
Sandy soil of prairies, flood plains, fields, roadsides, and waste ground
Southwest fourth, primarily; also recorded from Morris and Brown counties

Plants erect, up to about 2 m tall, very similar to *A. tamariscinus*. Stems much branched, the lateral branches ascending, often strictly so. Blades of the main leaves lanceolate to elliptic or oblong, mostly 2–7.5 cm long and 0.6–2.5 cm wide; apex rounded; base acute or attenuate, often oblique. Flowers borne in globular clusters in leaf axils or contiguous along elongate terminal spikes 0.5–3 dm long; male and female flowers on separate plants. Bracts ovate to lanceolate, spiny-tipped; 1–2 mm long. Tepals of male flowers 5, lanceolate, spiny-tipped, 2–3 mm long; stamens 5. Tepals of female flowers 5, broadly spatulate and recurved or spreading, about 2.5 mm long, with broad, membranous, often pinkish margins. Seeds about 1–1.25 mm long.

A. graecìzans L.
Prostrate pigweed

According to Fernald (9), simulating *A. graecus.*
Flowers minute, greenish; late June–October
Dry soil of prairies, fields, roadsides, and waste ground
Throughout

Plants prostrate. Stems much branched from the base, 2–6 dm long. Larger (earlier) leaves with blades oblanceolate to obovate, 0.8–4 cm long and 0.5–1.5 cm wide, with a rounded or acute apex, attenuate base, and callused margins and lower veins deciduous; later leaves lanceolate to oblanceolate, elliptic, or rhomboid, 0.4–1 cm long and 2–4 mm wide. Flowers borne in axillary clusters; male and female flowers on the same plants. Bracts acuminate, 1.5–2 mm long. Tepals of male flowers 4, oblong, membranous, 2–3 mm long, with a green midrib and spiny tip; stamens 3–4. Tepals of female flowers 4–5, subulate to oblong, minutely spiny-tipped, 1.5–3 mm long. Seeds about 1.5 mm long.

106. *Amaranthus tamariscinus*

A. hỳbridus L.
Green amaranth, pigweed

NL., from L. *hybrida, hibrida,* mongrel.
Flowers minute, greenish; August–October
Roadsides, fields, flood plains, waste places
Eastern half, primarily, with scattered collections from the western half; intr. from tropical America

Plants erect, up to about 2 m tall, similar to *A. retroflexus.* Stems much branched, often reddish. Leaf blades ovate to lanceolate, rhombate, or elliptic, the main ones mostly 3.5–12 cm long and 2–6 cm wide; apex acute or rounded; base obliquely attenuate. Stems, axis of the inflorescence, petioles, and veins of under surface of leaves with minute kinky hairs. Flowers borne in axillary spikes and paniculate terminal spikes; male and female flowers on the same plant. Bracts long-acuminate, 2–4 mm long, the green midrib extended into an awn. Tepals of the male flowers 5, lanceolate to lanceolate-acuminate, 1.5–2.5 mm long, membranous, with a green midrib; stamens 5. Tepals of female flowers 5, elliptic, membranous, mucronulate, about 1.5 mm long and 0.25–0.50 mm wide. Seeds about 1 mm long.

A. retrofléxus L.
Green amaranth, pigweed

NL., from L. prefix *retro-,* back, backward, + *flexus,* bend, the tips of the bracts and tepals of this species sometimes being somewhat recurved.
Flowers minute, greenish; mid-July–October
Dry soil of fields, pastures, roadsides, flood plains, and waste ground
Throughout; intr. from tropical America

Plants erect, up to about 2 m tall, similar to *A. hybridus.* Leaf blades lanceolate to ovate or rhombate, the main ones mostly 3–11 cm long and 1.5–5 cm wide; apex acute or rounded; base acute or attenuate, often oblique. Stems, axis of the inflorescence, petioles, and under surface and margins of leaves with minute kinky hairs. Flowers borne in axillary spikes and paniculate terminal spikes; male and female flowers on the same plant. Bracts lanceolate-acuminate, 3–5 mm long, the green midrib extended into a rigid awn. Tepals of male flowers 5, elliptic to subulate, about 2 mm long, membranous, with a green midrib; stamens 5. Tepals of female flowers 5, oblong to oblanceolate, silvery-white with a green midrib, the apex rounded or truncate, and somewhat ragged, with a tiny mucro or spine, 2–4 mm long and 0.5–1 mm wide. Seeds 1–1.2 mm long.

A. tamaríscinus (Nutt.) Wood.

NL., from the genus *Tamarix,* i.e., resembling *Tamarix.*
Flowers minute, green or purplish; July–October
Flood plains, fields, pastures, roadsides, waste ground
Eastern two-thirds, primarily, with scattered collections from the western third

Plants erect, up to about 2 m tall, very similar to *A. arenicola.* Stems more or less branched, the branches ascending, often strictly. Blades of the main leaves oblong to elliptic, lanceolate, or rhombate, mostly 2.5–12 cm long and 1–5 cm wide; apex acute, rounded, or occasionally notched; base acute or attenuate, often oblique. Flowers mostly borne in elongate axillary or terminal spikes 0.5–3 dm long; male and female flowers on separate plants. Bracts lanceolate, mostly 1.5–2.5 mm long, with a heavy green midrib extended into a spine. Tepals of male flowers 5, lanceolate to elliptic, 2.5–3 mm long; stamens 5. Tepals of female flowers 1 or 2; when 2, 1 is membranous and less than 1 mm long and the other is narrowly lanceolate, about 2 mm long, with a moderately heavy midrib. Fruit with irregular longitudinal ridges or somewhat tuberculate. Seeds 0.8–1 mm long.

FROELĪCHIA Moench   Cottonweed

Named in honor of J. A. Froelich (1766–1841), a German botanist.

Our species erect annual (perhaps biennial) herbs with white woolly or silky herbage. Leaves opposite. Flowers bisexual, each subtended by membranous bracts, borne in elongate terminal spikes. Calyx tubular, 5-lobed, persistent, becoming urn-shaped or conic, hardened, and longitudinally ridged or tubercled in fruit. Stamens 5, united into a tube about as long as the calyx. Style slender; stigma capitate. Fruit an indehiscent utricle enclosed by the calyx.

Kansas has 2 species.

Key to Species (3):
1. Plants rather stout, unbranched or with 2–3 erect branches above; main leaves mostly more than 1 cm wide; mature fruiting calyx flask-shaped, 6 mm long, the lateral wings entire, rather ragged (erose), or sharply toothed .... *F. floridana*
1. Plants slender, unbranched or often branched several times at the base; main leaves mostly less than 1 cm wide; mature fruiting calyx conical, 4 mm long, with 2 longitudinal rows of flattened spines .... *F. gracilis*

F. floridàna (Nutt.) Moq.

"Of Florida."
Flowers inconspicuous, whitish; July–mid-October
Sandy prairies and roadsides
Southwestern and central Kansas, with a few collections from eastern counties

Stems stout, up to 2 m tall, unbranched or with 2–3 erect branches above. Leaves mostly on the lower half of the stem, the blades narrowly elliptic to oblong or broadly linear, 1.5–8.5 cm long and 0.4–1.5 cm wide; apex acute; base attenuate. Stem and under surface of leaves densely woolly. Flowers whitish; spikes mostly 1.5–4 cm long. Calyx at anthesis 4–6.5 mm long, the tube densely covered with long, white woolly hairs. Mature fruiting calyx flask-shaped, 6 mm long, with 2 longitudinal wings or "crests" which may be entire, ragged, or toothed.

Our plants are all recognized as belonging to *F. floridana* var. *campestris* (Small) Fern. (L. *campestris*, of the plains).

F. grácilis (Hook.) Moq.   Cottonweed

L. *gracilis*, slender.
Flowers inconspicuous, whitish; mid-June–early October
Mostly in sandy soil of prairies, roadsides, and flood plains
East half and southwest fourth, with a few records from the northwest fourth

Stems slender, up to about 5 dm tall, often branched several times at the base. Leaves mostly on the lower half of the plant, the blades elliptic to narrowly elliptic or linear, 1–12 cm long and 0.2–1.7 cm wide; apex acute; base attenuate. Stem and under surface of leaves with long, silky white hairs. Flowers whitish; spikes mostly 0.5–2 cm long. Calyx at anthesis 3–4 mm long, the tube densely covered with long, woolly or silky white hairs. Mature fruiting calyx conical, about 4 mm long, with 2 longitudinal rows of flattened spines.

IRESĪNE P. Br.   Bloodleaf

NL., from Gr. *eiresione*, a wreath or staff entwined with narrow strips of wool, in reference to the woolly flowers.

Herbs or shrubs (our single species a perennial herb). Leaves opposite, petiolate. Flowers minute, scarious; male and female flowers on the same plant, or on different plants, or intermingled with bisexual flowers; each subtended by 3 bracts; borne in open pyramidal panicles. Calyx deeply 5-lobed, often long-woolly. Stamens 5, fused to one another at the base. Style 1, short; stigmas 2–3, usually threadlike. Fruit a tiny indehiscent utricle.

Kansas has 1 species.

I. rhizomatòsa Standl.   Bloodleaf

NL., having rhizomes, from Gr. *rhizoma*, mass of roots of a tree, from *rhizousthai*, to take root, from *rhiza*, root.

Whitish, minute; late August–October
Low, moist woods
Southeast eighth

Perennial from slender rhizomes. Stems erect, 0.3–1.5 m tall, mostly unbranched except within the inflorescence. Leaf blades lanceolate to ovate (some of the lowermost occasionally nearly round), mostly 6–12 cm long and 2–7 cm wide; apex acute to acuminate; base attenuate. Flowers whitish, minute, borne in small spikelets arranged in panicles, each flower subtended by 3 minute membranous bracts (use magnification); male and female flowers on separate plants. Sepals 5, membranous, about 1.5 mm long. Mature fruiting calyx subtended by long, white silky hairs which give the female inflorescence a fluffy appearance. Seeds reddish-brown, shiny, nearly round in outline, biconvex, about 0.5 mm wide.

TIDESTRÒMIA Standl.

Named in honor of Ivar Tidestrom (1864–1956), American botanist active in the western and southwestern United States.

Annual or perennial herbs (our single species annual), the herbage often covered with stellate hairs (use magnification). Leaves opposite, petiolate. Flowers minute, bisexual, subtended by membranous bracts and borne in small axillary clusters. Calyx membranous, 5-lobed, with 2 lobes smaller than the other 3. Stamens 5, fused to one another at the base. Style 1, short; stigma capitate or 2-lobed. Fruit a smooth indehiscent utricle.

107. *Iresine rhizomatosa*

T. lanuginòsa (Nutt.) Standl.

NL., from L. *lanuginosus*, covered with short soft hair or down, from *lanugo*, down, from *lana*, wool.

Flowers minute, whitish; late June–early October
Sandy prairies and banks
Western third, primarily

Annual herbs. Stems prostrate, decumbent, or ascending, much branched, sometimes becoming reddish, up to 1.5 dm long. Leaf blades obovate to nearly round or ovate, those of the main leaves 1–3.5 cm long and 1–3 cm wide; apex rounded or broadly acute; base acute or short-attenuate, often oblique. Calyx 1–3 mm long.

NYCTAGINÀCEAE   Four-o'clock Family

Herbs, shrubs, or trees, but ours all herbs, the trees and shrubs being tropical. Leaves mostly opposite, simple, estipulate. Flowers in ours bisexual, regular. The flowers, sometimes borne singly but usually in clusters, are subtended by involucral bracts which in *Bougainvillea* are colorful and showy but in most genera are not showy. In some genera, these bracts may be fused together, thus simulating a calyx. Sepals 5, petaloid, fused to one another, the calyx tubular or bell-shaped, closely investing the ovary and abruptly constricted above it, then more or less expanding to a 4- or 5-lobed border. Petals absent. Stamens equaling or exceeding in number the lobes of the calyx. Pistil 1; ovary superior, 1-celled and 1-ovuled; style and stigma 1. Fruit a utricle with the persistent calyx base adherent to it, often termed an *anthocarp*.

The Four-o'clock Family contains 28 genera and about 250 species, mostly of the Western Hemisphere. Cultivated members include *Bougainvillea* spp., the four-o'clock (*Mirabilis jalapa* L.), and sand verbena (*Abronia umbellata* Lam.). The brilliant Bougainvilleas, native to South America, owe their rich coloring not to their flowers, which are actually inconspicuous, but to their brick-red, bright rose, or purple involucres.

Only 3 genera are native to Kansas. *Tripterocalyx micranthus* (Torr.) Hook., a southwestern species, is known only from Hamilton County, and the genus is not included in the following key. For a brief description, see comments under *Abronia fragrans*.

Key to Genera:
1. Petaloid calyx white, tubular below with a flared limb above, 2–3 cm long; flowers arranged in showy rounded umbels .... *Abronia*
1. Petaloid calyx usually pink or purplish, funnelform, 1–2 mm long; flowers borne singly or in clusters of 2–4 .... *Mirabilis*

108. *Abronia fragrans*

ABRÒNIA Juss.   Sand verbena

NL., from Gr. *abros*, graceful, delicate, in reference to the flowers.

Erect or more or less prostrate annual or perennial herbs. Leaves opposite, with one member of each pair often smaller than the other, petioled. Flowers borne few to many in a head, each head having a long stalk (peduncle) of its own and subtended by 5 scarious bracts. Calyx funnelform or salverform (that is, narrowly tubular below and abruptly expanded above into a so-called "limb" at right angles to the calyx tube), 5-lobed. Stamens 5. Anthocarp with 2–5 lobes or wings.

Kansas has but 1 species.

A. fràgrans Nutt.   Snowball sand verbena

L. *fragrans*, emitting a fragrance.
White; mid-May–August
Dry, especially sandy soil
Primarily in the southwest eighth

Perennial. Stems decumbent, branched from the base, 6 dm or more long. Main root a thickened taproot penetrating the soil to about 1 m. Blades of main leaves broadly lanceolate to ovate or narrowly triangular, 2.5–6 cm long; apex rounded or acute; base acuminate to truncate or somewhat cordate. Stems and leaves with both glandular and eglandular hairs. Bracts subtending the heads white or pale pink, ovate to obovate, 1–2 cm long, the apex rounded to acute. Flowers white, fragrant, opening at dusk. When the inflorescence is young, the bracts tightly enclose the flower buds and seem to be the calyx of a single flower. The true calyx is narrowly tubular and white above, with a swollen, purplish-pink, winged base. As stated in the description to the family, there is no corolla. As the fruit matures, the membranous ovary wall becomes adherent to the swollen base of the perianth. Anthocarp tough and leathery, 0.5–1 cm long, with 2–5 wings, the surface somewhat reticulate and glandular.

Tripterocalyx micranthus, sand puffs, is a decumbent or ascending annual with branched stems up to about 3 dm tall. The clusters of small pinkish flowers are not nearly so conspicuous as the fruits, which are approximately 1.5–2.5 cm long and have 2, 3, or 4 broad, pink, transparent, reticulately veined wings. In Kansas, this species is known only from Hamilton County.

MIRÁBILIS L.   Four-o'clock

L. *mirabilis*, wonderful, strange, the first species described for the genus being a tropical plant with large showy flowers.

Erect perennial herbs. Leaves opposite. Flowers borne in terminal leafy clusters. Involucre subtending the inflorescence 5-lobed, herbaceous. Calyx in our species funnelform or bell-shaped, 5-lobed, petal-like. Stamens 3–5. Fruit a hard anthocarp, usually 5-ribbed in cross section and often with tuberculate ornamentation. When wetted, the ridges and tubercles become quite mucilaginous. On the epidermis of the involucre and anthocarps of all our species, numerous short white lines may be seen with the aid of a hand lens. These tiny lines are enlarged epidermal cells filled with raphides (needle-shaped crystals), probably of calcium oxalate.

Key to Species (64):
1. Anthocarp glabrous 2
  2. Leaf blades linear, 2–7 mm wide; stems glabrous .... *M. glabra*
  2. Leaf blades lanceolate or broader, 10–40 mm wide; stems glandular-hairy .... *M. carletonii*
1. Anthocarp more or less hairy, sometimes only sparsely so 3
  3. Leaves all conspicuously petioled except the uppermost; blades rounded to cordate at the base and sharply differentiated from the petioles .... *M. nyctaginea*
  3. Leaves sessile or nearly so, sometimes attenuate to a short, stout petiole; blade not sharply differentiated from the petiole 4
    4. Leaf blades linear or lance-linear, 2–5 mm wide; ridges on anthocarps without tubercles .... *M. linearis*
    4. Leaf blades lanceolate or broader, 4–14 mm wide; ridges on anthocarps tubercled .... *M. albida*

M. álbida (Walt.) Heimerl.   White four-o'clock

NL., from L. *albidus*, white, the stems and under surface of the leaves often whitened with a waxy "bloom."
White, pink, or rose; June–October
Well-drained soil of prairies and open woods
East half, primarily

Stems erect, 2–10 dm tall, often whitened with a waxy bloom. Leaves very narrowly lanceolate to lanceolate or oblanceolate, pinnately veined, those of the stem below the inflorescence mostly 6–10 cm long and 0.4–1.4 cm wide, the under side usually whitened; apexes mostly acute or acuminate; bases tapering gradually so that no petiole is evident, or less frequently acute with a very short petiole. Inflorescence branched and leafy. Involucres with 1–3 flowers, 4–6 mm long at anthesis, up to 13 mm long in fruit. Calyx white to pink or rose, 8–10 mm long. Anthocarps ellipsoid or obovoid with a constriction near the basal end, 4–5 mm long, hairy, with 5 vertical ridges or rows of tubercles, more or less tuberculate between the ridges as well.

109. *Mirabilis albida*

M. carletònii (Standl.) Standl.   Four-o'clock

Named after Mark Alfred Carleton (1866–1925), an agronomist who worked for the U.S. Department of Agriculture, specializing in wheats and their rusts.
Pink; June–August
Sandy prairie
Central and south-central Kansas

Stems erect, 6–12 dm tall, often whitened. Leaves sessile, thick, lanceolate to ovate or narrowly deltoid, pinnately veined, those of the stem below the inflorescence mostly 5–7.5 cm long and 1.5–3 cm wide, the under surface usually whitened; apexes acute or rounded; bases acute to truncate. Inflorescence branched, the bracts conspicuously smaller than the leaves below. Involucres with 1–4 (usually 3) flowers, about 5 mm long at anthesis, up to 12 mm long in fruit. Calyx pink, about 7 mm long. Anthocarps narrowly ellipsoid, about 5 mm long, glabrous, with 5 rather smooth vertical ridges, tubercles present between the ridges.

M. glàbra (Wats.) Standl.   Smooth four-o'clock

L. *glabra*, glabrous, i.e., lacking hairs.
White or pink; July–September
Sand hills
Collected in the southwestern eighth and Barton and Stafford counties

Stems erect, 8–15 dm tall, frequently whitened. Leaves linear to lanceolate, those of the stem below the inflorescence mostly 4–9 cm long and 0.4–1.5 cm wide, frequently whitened beneath; apex rounded, acute, or acuminate; base attenuate to the point of attachment, or acute and with a short petiole. Inflorescence branched, the bracts within the inflorescence conspicuously smaller than the leaves on the stem below. Involucres with 1–3 flowers, about 4 mm long at anthesis, up to 8 (sometimes 10) mm long in fruit. Calyx white or pink, about 7 mm long. Anthocarp ellipsoid, about 5 mm long, glabrous, with 5 vertical ridges, tubercles present between the ridges.

110. *Mirabilis carletoni*

M. lineàris (Pursh) Heimerl.   Four-o'clock

L. *linearis*, linear, in reference to the leaves.
White, pink, or reddish-purple; June–October
Dry prairie
Primarily in the west four-fifths, more common westward

Stems erect, 2–10 dm tall, often whitened. Leaves sessile, linear, with a single prominent midvein, those of the stem below the inflorescence mostly 5–10 cm long and 2–5 mm wide, the under surface sometimes whitened; apexes acute or acuminate; bases attenuate. Inflorescence branched, the bracts within the inflorescence conspicuously smaller than the leaves below. Involucres with 1–3 flowers, about 5 mm long at anthesis, up to 10 (rarely 12) mm long in fruit. Calyx white to pink or reddish-purple, 10–15 mm long. Anthocarps ellipsoid with a constriction near the basal end, about 5 mm long, with white hairs and 5 vertical ridges that may be slightly roughened but are not at all tuberculate, some tubercles present between the ridges. Lower forms of our plants that are densely branched from the base are recognized by some authors as *M. diffusa* (Heller) Reed.

M. nyctagínea (Michx.) MacM.   Wild four-o'clock

NL., from NL. *nyctago*, an obsolete synonym for the four-o'clock genus, *Mirabilis—nyctago* being composed of Gr. *nyx*, *nyktos*, night, + L. *ago*, *agere*, which in application to plants means "to put forth" or "sprout," in reference to the habit of blooming in the late afternoon and on through the night.
Pink or reddish-purple; May–October
Prairies, open woods, roadsides, waste ground
Nearly throughout, but not much collected in the west fourth

111. *Mirabilis linearis*

FOUR-O'CLOCK FAMILY

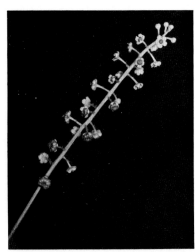

112. *Phytolacca americana*

Stems erect, sometimes whitened, 3–10 dm tall, usually several arising from a thickened perennial root. Leaves petioled, ovate-lanceolate, cordate, or triangular, pinnately veined, the blades of those on the stem below the inflorescence mostly 4–12 cm long and 2.5–10 cm wide; apex acute or somewhat rounded; base acute to truncate or cordate. Involucres with 3–5 flowers, 5–6 mm long at anthesis, up to 12 (sometimes 15) mm long in fruit, arranged in rather dense, umbelliform terminal clusters. Calyx pink or reddish-purple, 10–13 mm long. Anthocarps ellipsoid, 4–5 mm long, hairy, with 5 tubercled vertical ridges, and tubercles also present between the ridges.

See plates *33* and *34.*

PHYTOLACCÀCEAE Pokeberry Family

Herbs, shrubs, trees, or vines. Leaves alternate, simple, entire, petiolate. Flowers bisexual or dioecious (male flowers and female flowers borne on separate plants), regular, borne in racemes or panicles. Sepals 4–5, more or less fused and usually persistent. Petals absent. Stamens of the same number as the sepals and alternate with them, or more numerous. Ours with 1 pistil; ovary superior, 5- to 15-celled, with as many ovules; style absent or very short, stigmas of same number as the cells. Fruit a berry.

A family of about 17 genera and 125 species, primarily occurring in the American tropics and subtropics.

Kansas has but a single genus and species.

PHYTOLÀCCA L. Pokeweed, pokeberry

NL., from Gr. *phyton*, plant, + NL. and Italian *lacca*, for a resinous substance secreted by a scale insect (*Tachardia lacca*) which contains a red dye resembling cochineal—the crimson juice of the berry suggesting the name for this plant.

Herbs, shrubs, or trees. Flowers bisexual or dioecious, borne in racemes. Sepals 5. Stamens 5–30. Fruit a subglobose berry.

P. americàna L. Poke, pokeweed, pokeberry

NL., "of America."
Greenish-white or pale pink; May–October
Rich, moist soil of creek banks, pond margins, ditches, and waste ground
Primarily the eastern half

Smooth, erect herbs, up to 2.5 m tall, perennial from a large thick taproot. Blades of the main leaves broadly lanceolate, up to 25 cm long and 10 cm wide; apex acute or short-acuminate; base attenuate; petioles mostly 1–3 cm long. Flowers bisexual, borne on stout pedicels 0.5–1 cm long and subtended by 2 linear or subulate bracts, widely spaced in drooping racemes. Sepals 5, ovate, greenish-white or pale pink, becoming red or purple in fruit, 2–3 mm long. Stamens 10. Styles 10. Fruit a dark purple, subglobose berry, about 7 mm wide, with 10 seeds arranged in a ring. Seeds shiny, black, 2.5–3.5 mm wide.

All parts of the plant contain a saponin, phytolaccatoxin, toxic to both humans and livestock. The root especially contains a high concentration of the poisonous substance, and it is interesting to note the use of the roots in early American medicine as the source of a powerful drug for treating ulcers, parasitic skin disorders, and chronic rheumatism. The juice of the berries has been used at times to treat tremors, cancer, and hemorrhoids and as a diuretic alternative, emetic, and purgative. Current research indicates that extracts of *P. americana* may be useful as a molluscicide and thereby have some value in the control of fresh-water snails and parasitic diseases spread by these creatures.

In spite of the toxin present in pokeweed, the *young* shoots, as they first appear in the spring, have been safely used as a cooked green vegetable by boiling in 2 waters and discarding the first. *DO NOT EAT* the older herbage. There are conflicting reports regarding the edibility of the fruits. While some cases of poisoning have been attributed (although not conclusively) to the berries, they have apparently been used successfully in pies and wine and for coloring frosting and candies. The seeds and berries are eaten by a variety of birds and mammals, including the mourning dove, bluebird, catbird, mockingbird, cedar waxwing, opossum, and raccoon. The juice of the berries also yields a red dye or ink.

AIZOÀCEAE
Carpetweed Family

From the African genus *Aizoon* L.

Annual or perennial herbs, or shrubs, often succulent. Leaves opposite, whorled, or alternate, simple. Flowers bisexual (in ours), regular, borne singly in the axils of the leaves or terminally. Sepals 4–5, free from one another, free from or fused to the ovary. Petals absent. Stamens variable in number, typically twice as many as the sepals. In some genera with numerous stamens, the outer ones are sterile and petaloid, causing the flower to resemble the head of a member of the Composite Family. Pistil 1; ovary superior (rarely inferior), 1- to 5- (20-) celled, ovules usually many; style 1 or absent; stigmas 2–20. Fruit usually a circumscissile or loculicidal capsule.

A large family, especially in South Africa, with at least 100 genera and about 600 species. Representatives of about 800 genera, including *Lithops*, *Faucaria*, and *Cryophytum*, are cultivated as ornamentals. The so-called living stones belong to this family.

Kansas has 3 genera, but *Mollugo* is the only one commonly found. *Glinus lotoides* L., introduced from tropical America, is known only from Wilson County. *Sesuvium verrucosum* Raf., sea purslane, is a succulent prostrate annual or perennial herb with opposite leaves, quite similar in appearance to *Portulaca oleracea* in the Portulacaceae. It is known from Rice, Stafford, Ford, and Barber counties.

113. *Mollugo verticillata*

MOLLÙGO L.
Carpetweed

An old name for a species of *Galium* (bedstraw), perhaps used with this genus because of the similarly whorled leaves.

Annual herbs. Leaves opposite, whorled, or all basal. Flowers borne on long threadlike pedicels in terminal or axillary clusters. Sepals 5, fused to one another (at least slightly so), persistent. Stamens 3–5. Ovary superior, 3- to 5-celled; styles and stigmas 3–5. Fruit a thin-walled, many-seeded loculicidal capsule.

Kansas has 1 species.

M. verticillàta L.
Carpetweed

NL., from L. *verticillatus*, whorled, in reference to the leaf arrangement.
Flowers small and inconspicuous, greenish or white; July–October
Fields, roadsides, waste ground, and other disturbed areas
Probably throughout, although not much collected in the northwestern sixth; intr. from tropical America

Stems prostrate, much branched from the base. Leaves in whorls of 3–8; blades linear to spatulate or oblanceolate (those of the basal leaves sometimes obovate), with 1 prominent vein, up to 3.5 cm long and 2 cm wide; apex rounded or acute; base attenuate. Flowers borne at the nodes in whorls of 2–6. Sepals oblong or elliptic, fused together only slightly at the base, light green with 3 darker green veins that usually converge at the apex and white scarious margins, 2–2.5 mm long and 1 mm wide. Stamens 3. Styles 3, very short; stigmas papillose. Seeds more or less bean-shaped, reddish-brown, usually with 5–6 darker parallel ridges around the periphery (use high magnification), about 0.5 mm across.

PORTULACÀCEAE
Purslane Family

Annual or perennial herbs, or shrubs. Leaves alternate or opposite, or mostly basal, simple, more or less fleshy or succulent, stipulate (except in *Claytonia*). Inflorescence various. Flowers bisexual, regular. Sepals 2, free from one another or fused basally. Petals 4–5, rarely more, free from one another or fused basally. Stamens usually of the same number as the petals and opposite them but sometimes more, usually free from one another but occasionally borne on the corolla. Pistil 1; ovary superior or partly inferior, 1-celled, with 2 to many ovules; style 1 with 2–3 lobes or branches and the same number of stigmas. Fruit a circumscissile or loculicidal capsule.

The family comprises about 16 genera and more than 500 species, mostly of the Americas. Cultivated members of this family include the moss rose (*Portulaca grandiflora* Hook.) and jewels of Opar (*Talinum paniculatum* (Jacq.) Gaertn.).

Kansas has 3 genera.

Key to Genera:
1. Ovary partly inferior; fruit a capsule opening along a horizontal suture, the top half falling away .... *Portulacca*
1. Ovary superior; capsule opening longitudinally 2
   2. Flowers solitary or borne in racemes; leaves flat; flowers white with pink streaks; blooming in early spring .... *Claytonia*
   2. Flowers borne in long-peduncled terminal cymes; leaves round in cross section; flowers bright pink or reddish-purple; blooming in late May or early June into September .... *Talinum*

114. *Claytonia virginica*

CLAYTÒNIA L.   Spring-beauty

Named in honor of Dr. John Clayton, botanist and notable plant collector of colonial days.

Small herbs, perennial from fleshy corms. Stems unbranched below the inflorescence and bearing 2 opposite, linear, somewhat succulent, estipulate leaves. Flowers borne in loose racemes. Sepals free from one another. Petals 5, free from one another. Stamens 5, fused to the narrow base of the petals. Ovary superior. Fruit a 6-seeded capsule, opening along 3 longitudinal sutures, subtended by the persistent calyx.

Kansas has but 1 species.

C. virgínica L.   Virginia spring-beauty

"Of Virginia."
White or pale pink, with pink or red veins; late March–early May
Open woods and thickets, prairie
East half, especially the east fourth

Stems slender, 1 to several, arising from a starchy globose corm up to 3 cm wide, with thin fibrous roots, residing about 5 cm below the surface of the ground. Leaves dark green, mostly 3.5–10 cm long and 1–4 mm (sometimes up to 8 mm) wide. Sepals ovate or oval, 4–6 mm long. Petals white or pale pink, with pink or red veins, spreading, broadly elliptic to obovate, narrowed to a short claw at the base, 8–25 mm long. Capsule ovoid, about 4 mm long at maturity. Seeds black, glossy, faintly ornamented with lines of minute pits (use high magnification), more or less lenticular, with a wedge-shaped notch at 1 end, 1.5–2 mm wide.

This charming little flower is one of the earliest out in the spring. Though it occurs in the open woods, it also spreads into open land and yards with its racemes of dainty flowers. Although the plant is but meagerly equipped with leaves, these apparently photosynthesize with such efficiency that the seeds have been nursed to maturity and the corms replenished with food in time for the plant to bid its honorable adieux to spring and return to its subterranean abode before larger competing species have time to overgrow and crowd it out.

The starchy corms, gathered when they are well filled, are edible and are usually prepared by boiling in salted water. The flowers are so beautiful, though, that it seems a shame to destroy the plants merely to satisfy a curious palate.

See plate 35.

PORTULÁCA L.   Purslane

The Latin name for purslane, of unknown meaning.

Succulent annuals. Leaves alternate, mostly borne on the stem (rather than basally). Flowers borne singly or in dense clusters at the ends of branches. Sepals fused at the base. Petals usually 5. Stamens 8 to many. Ovary half or completely inferior. Fruit a many-seeded circumscissile capsule.

Kansas has 3 species. *P. retusa* (not included in the following key) is quite similar to *P. oleracea* but is rarely collected. The 2 species can be distinguished most easily on the basis of seed characters, those of *P. retusa* being larger (about 1 mm wide), iridescent when viewed under magnification, and ornamented with sharp-pointed, widely spaced tubercles. The seeds of *P. oleracea* are usually less than 1 mm wide, glossy black, and covered with crowded rounded tubercles. In addition, *P. grandiflora*, moss rose, is sometimes found as an escape from cultivation.

Key to Species (4):
1. Plants completely glabrous; leaves flat, broadest in the upper half, the larger ones 8–16 mm wide; petals yellow .... *P. oleracea*
1. Plants with hairy tufts in the axils of the leaves; leaves cylindrical or nearly so with curved or rounded sides, about the same width throughout, about 2 mm wide; petals red or rose pink .... *P. mundula*

P. múndula I. M. Johnst.   Purslane

NL., from L. *mundulus*, diminutive of *mundus*, clean, nice, neat.
Red or rose-pink; June–October
Mostly sandy soil, sometimes in shallow soil over limestone
Primarily in south-central and southeast Kansas

Prostrate or ascending succulents with taproots. Leaf blades linear, round or nearly so in cross section, up to 1.5 cm long. Leaf axils and inflorescences with tufts of long white hairs. Flowers borne 2 to several in terminal clusters. Sepals triangular or long-triangular with acute tips, sometimes pink. Petals red or rose-pink, obovate, up to 6 mm long. Stamens 10–15 (rarely more). Ovary half-interior, style branches 3–5. Seeds iridescent or glossy black, with minute rounded tubercles, about 0.5 mm wide.

P. oleràcea L.   Purslane

From L. *oleraceus*, resembling a garden vegetable.
Yellow; June–October
Fields, gardens, waste places, and disturbed ground
Primarily south and east of a line drawn from Morton County to Doniphan County, intr. from Europe or Eurasia

Prostrate succulents with a taproot and fibrous lateral roots. Leaf blades obovate to oblanceolate, up to 3 cm long; apex rounded; base acute. Herbage glabrous or with a few short straight hairs at the nodes. Flowers borne singly or a few together at the end of short lateral branches. Sepals broadly ovate or round, 2.8–4.5 mm long. Petals yellow, 3–4.5 mm long. Stamens 6–10. Ovary half-inferior; style branches 4–6. Seeds glossy black, ornamented with minute, rounded, crowded tubercles (use high magnification), usually less than 1 mm wide.

Although this plant has been much maligned as a lowly weed, it has been eaten and enjoyed by peoples of Europe and Asia for thousands of years and is still grown as a garden vegetable in those parts of the world. The herbage of purslane may be used in fresh salads, as a cooked green vegetable, or for pickling, while the seeds may be ground and used for flour or cooked as mush. Euell Gibbons (10) suggests 9 different methods for preparing various parts of the plant for the table.

The small seeds of this and other species of Portulaca are also relished by birds and rodents, including the lark bunting, horned lark, and various species of kangaroo rat.

TALÌNUM Adans.   Fameflower

NL., from the Senegalese name of a species in this group native to Western Africa.

Herbaceous perennial arising from a thickened rhizome. Leaves all basal, linear, quite succulent, and nearly round in cross section. Flowers borne in long-peduncled terminal clusters. Sepals free from one another, deciduous in fruit. Petals 5, free from one another, falling early. Stamens 5 to many. Ovary superior; stigma lobes or style branches 3. Fruit a many-seeded capsule opening along 3 longitudinal sutures.

Kansas has 2 species.

The flowers in our species open in the afternoon, about 4 o'clock, and close again in the evening to open no more. Yet the many buds open serially from day to day over a considerable period, and when flowering is done the many seed pods and the clustered foliage are still interesting features.

Cross- as well as self-pollination would seem to take place. The anthers and stigma are mature when the flowers open, but the stamens spread away from the stigma and give a chance for cross-pollination by insects. As the corolla closes, however, the stamens are brought together around the stigma, rendering self-pollination possible.

Key to Species:
1. Stamens 4–8 (usually 5) .... *T. parviflorum*
1. Stamens many (30–45) .... *T. calycinum*

T. calycìnum Engelm.   Rock-pink fameflower

NL., from L. *calyx*, cup or calyx of a flower, + the suffix *-inus*, indicating quality or attribute— the calyx in this instance contrasting with that of the following species.
Bright reddish-pink; June–mid-September
Sandy and rocky soil
Central third and southern half

Plants in our area mostly 1–3 dm tall. Leaves up to 6 cm long and 2 mm wide. Sepals ovate, mostly 6–7 mm long, often persistent even in fruit. Petals oblanceolate or

115. *Portulaca mundula*

116. *Talinum calycinum*

obovate, bright reddish-pink, mostly 10–15 mm long. Stamens 30–45, the anthers bright yellow. Capsules ovoid or nearly globose, mostly 5–6 mm long at maturity. Seeds short-reniform (somewhat bean-shaped), shiny black but with a translucent white "skin" (testa) so that to the naked eye they look gray, mostly 1–1.2 mm across.

This species transplants easily into the rock garden.

See plate *36*.

T. parviflòrum Nutt. Prairie fameflower

NL., from L. *parvus*, little, + *flos, floris*, flower.
White, pale pink, or bright pink; June–July
Rocky or sandy soil
Primarily in the eastern third

Plants mostly 1–2.5 dm tall. Leaves up to 6 cm long and 2 mm wide. Sepals ovate, mostly 2–3.5 mm long, falling early. Petals oblanceolate, pale to bright pink, mostly 5–7 mm long. Stamens 4–8 (usually 5), the anthers bright yellow. Capsules ovoid, mostly 4–5 mm long at maturity. Seeds short-reniform, shiny black but with a translucent white "skin" (testa) so that to the naked eye they look gray, 1 mm or less across.

CARYOPHYLLÀCEAE Pink Family

After NL. *Caryophyllus*, a synonym of *Dianthus*, from Gr. *Karyophyllon*, clove tree—the flowers having the odor of cloves.

Annual or perennial herbs, infrequently suffrutescent shrubs. Stems cylindric, commonly swollen at the nodes. Leaves opposite or whorled, simple, entire, usually narrow, often with the bases fused together and sheathing the stem or connected by a line, sessile or petiolate, stipules present or absent. Flowers bisexual, or rarely polygamous (bisexual and unisexual flowers on the same plant), regular, borne singly or in clusters. Sepals 4–5, separate or united into a tube or cup. Petals of the same number as the sepals or calyx lobes, or sometimes none, distinct and free, often expanded above and narrowed toward the base ("clawed"). Stamens twice as many as the sepals. Pistil 1; ovary superior, 1-celled, with 1 to many ovules; styles and stigmas 2–5. Fruit a capsule (in ours) dehiscing apically by valves or teeth, or a utricle or achene.

A family of about 80 genera and 2,100 species centered primarily in the Mediterranean region. Cultivated members of this family include sandwort (*Arenaria* spp.), easter bells (*Stellaria holostea* L.), various pinks (*Dianthus* spp.), campion (*Silene* spp.), baby's breath (*Gypsophila paniculata* L.), and others.

Kansas has 11 genera. Those most commonly encountered are included in the following key.

Key to Genera (3):
1. Fruit a 1-seeded utricle; petals none .... *Paronychia*
1. Fruit a few- to many-seeded capsule; petals usually present 2
  2. Calyx of separate sepals; leaves with or without stipules 3
    3. Petals deeply 2-cleft or 2-parted, seldom lacking 4
      4. Capsule ovoid or oblong, dehiscent by valves .... *Stellaria*
      4. Capsule cylindric, dehiscent by terminal teeth .... *Cerastium*
    3. Petals entire or emarginate, sometimes lacking 5
      5. Capsule cylindric; inflorescence (in ours) umbellate .... *Holosteum*
      5. Capsule ovoid or oblong; inflorescence not umbellate .... *Arenaria*
  2. Calyx of united sepals; leaves without stipules 6
    6. Styles 3 (in some species 4, but usually 3) .... *Silene*
    6. Styles 2 7
      7. Calyx not subtended by bracts; leaves elliptic, ovate, or lanceolate .... *Saponaria*
      7. Calyx subtended by 1–3 pairs of bracts; leaves in ours linear .... *Dianthus*

ARENÀRIA L. Sandwort

NL., from L. *arena* or *harena*, sand, in which many of the species grow.

Small annual or perennial herbs. Leaves opposite, sessile or nearly so, estipulate. Flowers bisexual, borne in terminal clusters. Sepals 5. Petals 5, entire or emarginate. Styles mostly 3. Capsule separating into 3 segments which are either entire or 2-cleft at the apex.

Kansas has 4 species. *A. hookeri* is known only from Cheyenne County and is not included here. *Sagina decumbens*, pearlwort, is known from Woodson, Neosho, Chautauqua, and Cherokee counties. This delicate, much-branched annual resembles a miniature *A. patula* but has 4–5 styles rather than 3, and attains a height of only 2–10 cm, with flowers 2–3 mm long. The sepals are ovate or broadly lanceolate, rather than narrowly lanceolate, are herbaceous with broad, white membranous margins, rather than strawlike, and lack conspicuous venation.

Key to Species:

1. Leaves ovate, (1–) 2–4 mm wide; sepals obscurely 3- to 5-veined; capsules dehiscent by 6 teeth or valves .... *A. serpyllifolia*
1. Leaves linear or linear-subulate, 1 mm or less wide; sepals strongly 3- to 5-veined; capsules splitting into 3 segments 2
   2. Tufted perennial; primary leaves with secondary leaves fascicled in their axils, imparting to the plant a superficially mosslike appearance; plants primarily of the western half of Kansas .... *A. stricta*
   2. Slender, loosely branched annuals; primary leaves only present; plants primarily of the eastern half of Kansas .... *A. patula*

117. *Arenaria patula*

A. pátula Michx. Sandwort

NL., from L. *patulus*, outspread, in reference to the growth habit of this plant.
White; late April–June (sometimes into July)
Prairies, fields, roadsides, usually in well-drained soil
Southeast sixth and Johnson and Douglas counties

Slender, loosely branched annuals, up to 2 dm tall. Leaves linear or linear-subulate, 5–15 mm long and 1 mm or less wide, sessile, each pair united basally around the stem. Flowers borne in open terminal clusters. Sepals narrowly lanceolate, 4–6 mm long, strongly 3- to 5-veined, and with a strawlike appearance when dried. Petals white, oblanceolate, 2-lobed at the apex, mostly 7–10 mm long. Styles and stigmas 3. Capsules shorter than and enclosed by the persistent sepals, splitting at maturity into 3 segments. Seeds brown (slightly iridescent under high magnification), wrinkled, about 0.5 mm long; under high magnification they resemble tiny dried apples.

A. serpyllifòlia Michx. Thyme-leaved sandwort

NL., from the genus name *Serphyllum* + L. -*folius*, leaved, i.e., having leaves like those of *Serphylum* (thyme).
White; April–June
Rocky prairie, pastures, roadsides and other waste ground, open flats in sandy soil
Eastern half

Delicate, tufted annuals up to 2 dm tall. Leaves ovate, mostly 3–6 mm long and 2–4 mm wide, sessile. Stems, leaves, and sepals minutely hairy (use magnification). Flowers borne in loose terminal clusters. Sepals lanceolate to ovate, obscurely 3- to 5-veined, the apex acuminate and margins membranous. Petals white, lanceolate, the apex acute, shorter than the sepals. Styles and stigmas 3. Capsule pear-shaped, slightly longer than the persistent sepals, splitting apically into 6 teeth. Seeds dark brown or black, iridescent, nearly round in outline, with a shallow notch on one side, about 0.5 mm wide and 0.25 mm thick, with 3 rows of radiating ridges on the circular faces and 2–4 rows of tubercles around the edge.

A. stricta Michx. Rock sandwort

NL., from L. *strictus*, very straight and upright, in reference to the growth habit of this plant.
White; May–early July
Dry, rocky or sandy prairie
Western half, with a few collections from Cowley and Crawford counties in the southeast

Tufted perennials, up to 2 dm tall. Leaves linear, with bundles of secondary leaves arising in the axils of the primary leaves and imparting to the plant a superficially mosslike appearance; primary leaves mostly 5–10 mm long and 1 mm or less wide, sessile, each pair united basally around the stem. Flowers borne in rather dense terminal clusters, usually at least 3 cm above the uppermost leaves. Sepals lanceolate, strongly 3- to 5-veined, 4–6 mm long. Petals white, oblong or oblanceolate, the apex truncate or broadly rounded, 6–7 mm long. Styles and stigmas 3. Capsule about the same length or shorter than the sepals, subtended by the persistent sepals and petals, splitting into 3 segments. Seeds dark reddish-brown to black, very finely ridged, more or less bean-shaped but sometimes resembling dried apples as in *A. patula*, about 0.75–1 mm long.

CERÁSTIUM L.   Mouse-ear chickweed

NL., from Gr. *cerastes*, horned, in reference to the slender capsule.

Small annual or perennial herbs with opposite, sessile, estipulate leaves. Flowers bisexual, regular, white, borne in terminal clusters. Sepals 5. Petals 5, the apexes 2-lobed or 2-cleft. Stamens 10 (in ours) or 5. Styles 5. Capsule cylindric, usually longer than the sepals at maturity, the apex splitting into 10 short teeth. Seeds tuberculate, somewhat resembling small tan pouches.

Kansas has 4 species. *C. glomeratum* Thuill., introduced from Europe, is known only from Cherokee County and is not included here. It will key out below with *C. vulgatum* forma *glandulosum* but differs from that plant in having sepal hairs which extend beyond the apex of the sepals.

Use of the following key requires at least a 10X hand lens, and a 20X dissecting microscope is even better.

Key to Species:
1. Sepals with short glandular hairs only; purple spot not present at the apex of each sepal 2
    2. Pedicels becoming hooked immediately below the calyx as fruits mature, 1–5.5 cm long in fruit .... *C. nutans*
    2. Entire pedicel descending as fruits mature, but not hooked immediately below the calyx, 1–8 mm long in fruit .... *C. brachypodum*
1. Sepals with long nonglandular hairs only, or with both nonglandular hairs and shorter glandular hairs; a purple spot present near the apex of each sepal (rarely absent) 3
    3. Sepals with long nonglandular hairs only .... *C. vulgatum*
    3. Sepals with both glandular and nonglandular hairs .... *C. vulgatum* forma *glandulosum*

C. brachypòdum (Engelm.) Robins.   Mouse-ear chickweed

NL., from Gr. *brachys*, short, + *podos*, foot, the pedicels of this species being conspicuously shorter than those of the closely related *C. nutans*.
White; April–mid-May
Prairies, roadsides, and open woods, often on rocky soil
Eastern half

Slender annual. Stems erect or ascending, often much branched from the base, with short glandular hairs above and long, spreading nonglandular hairs below, mostly 1–3 dm tall. Leaves oblanceolate below to elliptic or oblong above; apex acute; base acuminate on lower leaves, somewhat rounded on upper leaves; margins with glandular hairs. Bracts subtending the subdivisions of the inflorescence completely herbaceous, lacking membranous margins. Sepals lanceolate, with an acute or short-acuminate apex, white or transparent membranous margins and short glandular hairs, 3–4.5 mm long. Petals elliptic, about 4–5 mm long. Capsule slightly curved, about 1 cm long. Seeds about 0.5 mm long.

This species is considered by some authorities to be a variety of *C. nutans*.

C. nùtans Raf.   Nodding chickweed

NL., from L. *nuto*, to nod or droop, in reference to the calyces in fruit.
White; April–May
Moist woods
Extreme eastern counties

Annual. Stems erect or ascending, several times branched at the base and sometimes above as well, with spreading nonglandular hairs and shorter glandular hairs, up to 4 dm tall. Leaves oblanceolate or spatulate below to narrowly lanceolate, elliptic, or oblong above, 2–6 cm long and 0.4–1.4 mm wide, with scattered glandular or nonglandular hairs; apex acute or ending rather abruptly with a tiny sharp point; base acute or slightly rounded. Bracts subtending the subdivisions of the inflorescence completely herbaceous, lacking membranous margins. Sepals 4.5–5 mm long, lanceolate, with an acute or short-acuminate apex, white or transparent membranous margins, and short glandular hairs. Petals narrowly obovate, longer than the sepals. Fruiting pedicel hooked downward immediately below the calyx. Capsule slightly curved, 0.8–1 cm long. Seeds about 0.8 mm long.

C. vulgàtum L.   Common mouse-ear

NL., from L. *vulgatus*, common.
White; April–June
Prairie, open woods, waste ground
Easternmost 2 columns of counties; intr. from Eurasia

Short-lived perennial, rooting from the lower nodes and forming mats. Flowering stems erect, with long spreading hairs, up to 3 dm tall. Leaves oblanceolate or spatulate below to lanceolate or elliptic above, with long pointed hairs on both surfaces, mostly 1–2 cm long and 0.2–0.8 cm wide; apex and base acute or rounded. Bracts subtending the subdivisions of the inflorescence with broad membranous margins. Sepals 4–5 (–6) mm long, narrowly lanceolate, with white or transparent membranous margins, a purple spot near the apex (rarely absent), and long nonglandular hairs which do not extend beyond the apex. Plants which also have short glandular hairs on the sepals are recognized as forma *glandulosum* (Boenn.) Druce. Petals 4–5 mm long. Capsule scarcely curved, 0.5–1 cm long. Seeds about 0.6 mm long.

DIÁNTHUS L.   Pink

NL., from Gr. *Dios*, the genitive form of *Zeus*, + *anthos*, flower, Zeus's (or, using the Latin equivalent, as is commonly done, Jupiter's) own flower—the carnation (*Dianthus caryophyllus*), because of its exceptional beauty and fragrance, being the flower thus dedicated.

Ours annual or biennial herbs. Leaves narrow, opposite, estipulate. Flowers bisexual, borne singly or in clusters, each flower subtended by 1–3 pairs of narrow bracts. Calyx 5-lobed, tubular. Petals 5. Stamens 10. Styles 2. Capsule many-seeded, the apex splitting at maturity into 4 teeth.

Kansas has 1 species.

118. *Cerastium vulgatum*

D. arméria L.   Deptford pink

Probably from Celtic *ar*, near, + *mor*, sea—there being a superficial resemblance between the Deptford pink and the sea pink, *Armeria vulgaris*. Deptford pink, according to Gerard, comes from its growing "in a field next Deptford, as you go to Greenwich."

Pink; June–July
Moist prairies and waste ground
East fourth; naturalized from Europe

Erect, stiff, finely pubescent annuals or biennials from a taproot. Stems up to 8 dm tall, with decided swollen nodes. Leaves linear, grasslike, 1–8 mm wide, clustered at the base and opposite above, with bases ensheathing the stem. Flowers in crowded terminal clusters of 3–9 flowers, each subtended by 1–3 pairs of slender bracts which resemble the sepals and which may be considerably longer than the flowers. Calyx long-ellipsoid with 5 acuminate teeth, minutely hairy, conspicuously veined, 1.5–1.8 cm long. Petals pink with white spots and ragged margins, clawed, 1.7–2 cm long, with the expanded portion 5–6 mm long. Styles and stigmas 2. Capsule long-ellipsoid, 1-celled, many seeded, shorter than and enclosed by the persistent calyx and corolla. Seeds black, punctate, rather irregular in outline but more or less tear-drop shaped with a point at one end, mostly 1–1.5 mm long.

119. *Dianthus armeria*

Cross-pollination is favored by the anthers maturing somewhat ahead of the stigmas. Since the tubular calyx holds the claws of the petals together in the form of a very narrow cylinder about 1 cm long, it would seem that butterflies and moths would be best adapted to secure the nectar and effect pollination.

HOLÓSTEUM L.   Jagged chickweed

NL., from a name applied by Dioscorides to a plant of uncertain identity, from Gr. *holos*, all, + *osteon*, bone.

Small annual or biennial herbs. Leaves opposite, estipulate. Flowers bisexual, borne in terminal umbels. Sepals 5. Petals 5, somewhat jagged at the apex. Stamens usually 3–5 (sometimes as many as 10). Styles 3. Capsule ovoid, many-seeded, the apex at maturity splitting into 6 teeth.

Kansas has but 1 species.

H. umbellàtum L.   Jagged chickweed

NL., with umbels. The botanical term for this type of inflorescence comes from L. *umbella*, parasol.

White; late March–mid-May
Roadsides and waste ground
Eastern half, especially the southeast sixth; naturalized from Europe

Inconspicuous annuals. Stems erect, branched several times at ground level, up to 2.5 dm tall. Lower leaves oblong, oblanceolate, or spatulate, up to 30 mm long and 5 mm wide; upper leaves lanceolate or oblong, mostly 8–25 mm long and 3–8 mm wide.

PINK FAMILY
107

Upper portions of the stems and sometimes the margins of the upper leaves with minute glandular hairs. Umbels with 3–11 flowers. Sepals lanceolate, 3–4 mm long, sometimes purplish, the margins membranous. Petals white, lanceolate, about 5 mm long, the margins irregularly toothed near the apex. Capsules 5–6 mm long. Seeds about 0.75 mm long, tan to orange, short-oblong and flattened, with a shallow notch at one end and a short "wing" (actually the funiculus) protruding from the inner face.

Although individual plants of the jagged chickweed are very inconspicuous, one's attention may be caught by colonies along the roadside when the plants are in fruit and have turned from green to pale yellow or tan.

PARONÝCHIA Mill.   Whitlow-wort

According to M. L. Fernald (9), *Paronychia* is the Gr. name for a whitlow or felon, a disease of the nails, and for plants with whitish scaly parts, once supposed to cure it.

Annual or perennial herbs. Leaves opposite, with conspicuous membranous stipules. Flowers bisexual, borne singly in leaf axils or in terminal clusters. Sepals 5, free from one another or united slightly at the base. Petals 5, sometimes reduced to minute teeth or hairlike structures, or entirely absent. Stamens 2–5, arising from the base of the calyx. Style branches or stigmas 2. Fruit a membranous utricle.

Kansas has 4 species. *P. depressa* Nutt., a low, mat-forming perennial, is known from Morton and Cheyenne counties only and is not included in the following key. It is quite similar in appearance to *P. jamesii* but is smaller, with stems seldom more than 6 cm tall.

Key to Species:
1. Cespitose perennials with several to many stems forming a dense clump; leaves linear or needlelike, not noticeably flattened; plants of the western half of Kansas .... *P. jamesii*
1. Slender, wiry annuals with a single stem and elliptic, flattened leaves; plants of eastern Kansas 2
  2. Stems minutely hairy; sepals with a sharp-pointed tip and eventually becoming longitudinally ribbed (use magnification); plants primarily of southeastern Kansas .. *P. fastigiata*
  2. Stems glabrous; sepals with a blunt tip, not longitudinally ribbed; plants of the eastern fifth of Kansas .... *P. canadensis*

P. canadénsis (L.) Wood   Forked chickweed

L. -*ensis*, suffix added to names of place to make adjectives—"Canadian," in this instance.
Flowers minute, greenish; late June–August
Open woods
Eastern fifth

Slender, erect annual. Stems many times forked, glabrous, 1–2 (–4) dm tall. Leaves thin, elliptic, the main ones 0.7–2 cm long and 0.3–1 cm wide. Flowers borne singly or in clusters of 2–3 in the axils of leaves or leaflike bracts. Sepals oblong, elliptic, or lanceolate, slightly united at the base, 1–1.5 mm long, with membranous margins and blunt tips. Petals absent. Stamens 2. Ovary covered with papillose projections. Seeds shiny, dark reddish-brown, nearly round in outline, biconvex or subgloboid, about 1 mm long.

P. fastigiàta (Raf.) Fern.   Forked chickweed

NL., from L. *fastigiatus*, sloping to a point, from *fastigium*, slope or roof; in botanical usage, with branches erect and more or less appressed or forming a conical bundle.
Flowers minute, greenish; July–early October
Dry, rocky, or sandy places of open woods, pastures, or roadsides
Southeast eighth, primarily

Slender, erect annual, very similar to *P. canadensis*. Stems many times forked, minutely hairy (use magnification), 0.7–2.5 dm tall. Leaves thin, elliptic, with hairy margins, the main leaves 0.8–1.8 cm long and 2–4 mm wide. Flowers borne singly or in small clusters in the axils of leaves or leaflike bracts near the tips of the stem branches. Sepals narrowly elliptic or narrowly oblong, slightly united at the base, 1–1.5 mm long, with narrow membranous margins and hoodlike, sharply pointed tips, eventually becoming longitudinally ribbed (use magnification). Petals absent. Stamens 2. Ovary smooth, flat-topped. Seeds shiny, reddish-brown, biconvex, about 0.75 mm long.

P. jàmesii T. & G.

Named after Edwin James (1797–1861), a young naturalist who traveled with Maj. Stephen

Long's expedition to the Rocky Mountains (1819–1820) and who was the first botanist to collect this plant.

Flowers inconspicuous, greenish; June–mid-September
Dry, rocky or sandy prairie
Western half

Cespitose perennial with several to many stems forming a dense clump, but *not* mat-forming. Stems erect, many times forked, minutely roughened, 1–2.5 dm tall. Leaves linear or needlelike, not noticeably flattened, the main ones 1–2 cm long and 0.5–0.75 mm wide; stipules white, membranous, long-acuminate, quite conspicuous. Flowers borne in branched terminal clusters; subdivisions of the inflorescence and the individual flowers subtended by membranous bracts similar to the stipules. Sepals lanceolate, slightly united at the base, hairy, 1.5–2.5 mm long, with hoodlike acuminate tips. Petals threadlike, about as long as the 5 stamens. Seeds ovoid, less than 1 mm long.

SAPONÀRIA L.   Soapwort

NL., from ML. *saponarius*, pertaining to soap, from L. *sapo*, soap—in allusion to the fact that
       plants of this genus yield the glucoside saponin, which has the property of producing
       a soapy lather.

Annual or perennial herbs. Leaves opposite, estipulate, the bases ensheathing the stem. Flowers bisexual, borne in clusters. Calyx tubular, 5-lobed, membranous. Petals 5, "clawed." Styles usually 2. Capsule many-seeded, the apex splitting at maturity into 4 (or sometimes 6) teeth.

Kansas has but 1 species.

S. officinàlis L.   Soapwort, bouncing Bet

NL., *officinalis*, pertaining to drugs and medicines, from L. *officina*, laboratory, such as one
       where drugs are prepared for use. As a plant name, *officinalis* or *officinale* indicates that
       the plant is or was formerly kept for medicinal use by pharmacists.

Light pink or white; June–August
Roadsides and waste places
Eastern half; intr. from the Old World

Herbaceous perennial producing colonies from rhizomes. Stems smooth, erect, sparingly branched, up to 8 dm tall. Leaves sessile above, petiolate below, with 3–5 prominent parallel veins; blades elliptic or lanceolate, mostly 5–9 cm long and 1–3 cm wide; apex rounded, acute, or short-acuminate; base acute or acuminate. Fascicles of secondary leaves often present in the axils of the primary leaves. Flowers borne in dense, compound terminal clusters, the individual flowers on a short stalk subtended by a pair of spreading bracts at the end of the pedicel. Calyx tubular, slightly expanded near the base, sharply toothed above and slit down a short distance on one side, 1.8–2.2 cm long. Petals light pink or white, with long claws and spreading limbs about 1–1.5 cm long and with a bifid ligule at the base of the limb, the narrow calyx tube holding the claws together in the form of a tube which contains the nectar. Stamens 10, 5 longer than the others. Pistil fusiform; styles 2. Capsule ellipsoid, thin-walled, about 1.5 cm long. Seeds black, round, flattened, finely pustulate, about 2 mm long.

The flowers are proterandrous, the anthers of the longer stamens protruding and exposing pollen as the flower opens, the shorter anthers after a time doing likewise, while the 2 styles with opposed stigmatic surfaces are concealed within the corolla tube, but later are elongate and spread apart beyond the tube. Hawk moths busy themselves among the flowers as they open with fresh fragrance in the later afternoon, poising before a flower like a hummingbird while sucking nectar from the bottom of the corolla; and, as they go about, inevitably transferring pollen from younger flowers to the stigmas of older ones.

We find differences of opinion as to the quality of the fragrance of the soapwort. Grieve and Leyel's *A Modern Herbal* (11) says of the flowers "no odour"; yet Parkinson's *Paradisus Terrestris* (12), published in England more than 300 years earlier, says the flowers are "of a strong, sweet scent, somewhat like the smell of Jasmine flowers." To W. C. Stevens (28), the smell was like that of wild plum blossoms, modified by that of the perennial garden phlox. Parkinson says further that "the ordinary Sopewort, or Bruisewort, with single flowers is often planted in gardens, and the flowers serve to deck both the garden and the house."

Both in England and on the Continent the soapwort was esteemed for its therapeutic virtue. The ancient Greeks and Latins cultivated it as a soap substitute and for medicinal use as a diuretic, diaphoretic, blood purifier, and appetizer, and decoctions of the plant were applied externally to eczema.

120. *Paronychia jamesii*

121. *Saponaria officinalis*

PINK FAMILY

109

122. *Silene antirrhina*

123. *Silene stellata*

SILÈNE L. — Campion, catchfly

Possibly after the fabled Silenus, or Silen, companion of Bacchus, portrayed as ever drunken and driveling saliva—the stems and calyces of many *Silene* species exuding sticky secretions.

Annual or perennial herbs. Leaves opposite, sessile, estipulate. Flowers bisexual or sometimes unisexual, borne in clusters. Calyx 5-lobed. Petals membranous, variously lobed or dissected, each with a narrow "claw" and usually with a pair of scales attached at the juncture of the claw and the expanded portion. Ovary usually elevated on a short stalk (stipe); styles 3 (sometimes 4 or 5). Capsule 1-celled, or 3-celled for at least part of its length, the apex splitting at maturity into 6 teeth.

Kansas has 4 species. The 2 most commonly collected are included in the following key.

Key to Species:

1. Plants annual; petals 2-lobed, as long as or 2–3 mm longer than the calyx (sometimes rudimentary or completely absent); stems usually with dark sticky bands below the nodes .... *S. antirrhina*
1. Plants perennial; petals deeply lacerated, much longer than the calyx; stems lacking dark sticky bands .... *S. stellata*

S. antirrhìna L. — Sleepy catchfly

NL., from Gr. *anti*, like, + *rhis, rhinos*, nose, the tapering apex of the calyx tipped by the closed corolla suggesting the snout of some animals. Or, according to *Gray's Manual of Botany* (9), with leaves as in *Antirrhinum*, the genus which includes the snapdragon. Small insects crawling up the stems are sometimes caught in glutinous spots—a curious fact that seems to have no significance for the plant, though it has been suggested that small-fry insects that could creep into the flowers and get nectar without helping in cross-pollination are prevented from reaching the flowers. So the name "catchfly"; "sleepy" comes from the fact that the flowers are open only a few hours each day and are quick to close after the sun is clouded or when the plants are carried indoors.

Pale pink or purplish (or white above in forma *bicolor*); April–June
Fields, prairies, waste ground
East four-fifths

Slender, erect annual. Stems up to 8 dm tall, with several ascending branches above and dark glutinous bands below the nodes. Leaves oblanceolate below, becoming narrowly lanceolate, elliptic, or linear above, mostly 1.5–6 cm long and 0.1–1 cm wide; apex acute to acuminate; base acuminate. Flowers pale pink or purplish (white above in forma *bicolor*), slender-pedicelled, borne in loose terminal panicles. Calyx fusiform, 10-nerved, 4–8 mm long, with reddish-purple teeth, becoming campanulate-ovoid and 6–9 mm long in fruit. Petals falling early (or absent altogether in forma *apetala*), 2-lobed, as long as or longer than the calyx, the scales minute or lacking. Styles 3. Capsule 3-celled, about the same length as and closely invested by the persistent calyx. Seeds reddish-brown, tuberculate, about 0.5 mm long.

S. stellàta (L.) Ait. f. — Starry silene, starry campion

L. *stellatus*, starlike, from *stella*, star—in reference to the flowers. "Campion" is from L. *campus*, for a field of various characters, including a field where athletic contests were held, some species of *silene* having been used to make chaplets for crowning the victors.

White; mid-June–mid-August
Moist woods and thickets
East third

Erect herbaceous perennial. Stems several, up to 1 m tall, with swollen nodes. Leaves lanceolate (occasionally elliptic), mostly in whorls of 4 (in pairs toward the base and sometimes in pairs and much reduced near the inflorescence), the main leaves 4–9 cm long and 0.5–3 cm wide; apex acuminate; base rounded to acuminate. Leaves and stems finely pubescent. Flowers showy, borne in loose panicles terminating the main stems. Calyx bell-shaped, minutely pubescent, 0.8–1.3 cm long. Petals white, fringed by deep incisions, 1.5–2 cm long, lacking scales. Stamens 10, protruding, quite fragile, adding to the fringed appearance of the flower. Ovary ellipsoid; styles 3. Capsule about 7 mm long, elevated on a stipe 2–3 mm long, surrounded at least for awhile by the persistent calyx. Seeds more or less bean-shaped, minutely ornamented; black, reddish-brown, or buff; 1–1.5 mm long.

STELLÀRIA L. — Chickweed

NL., from *L. stellaris*, pertaining to a star, from L. *stella*, star, in reference to the outspread perianth.

Small annual or perennial herbs. Leaves opposite, estipulate. Flowers bisexual, borne singly in the leaf axils or in terminal clusters. Sepals 5. Petals 5 (or absent), usually 2-lobed. Stamens 10 (sometimes only 2 in *S. media*). Styles 3 (sometimes 4 or 5). Capsule ovoid or globose, several- to many-seeded, the apex splitting at maturity into twice as many teeth as there are styles.

Three species have been recorded from Kansas, but only *S. media* is common.

S. mèdia (L.) Cyrillo — Common chickweed

L. *medius*, of the middle, middling, or moderate in size—in reference to the plant as a whole. The feeding of chicks with the seeds and young shoots is an old custom. Gerard tells in his famous *Herbal* (13), "Little birds in cadges (especially Linnets) are refreshed with the lesser chickweed when they loath their meat."

White; April–June, primarily, although some plants will be found blooming throughout the summer and into fall

Roadsides, pastures, waste ground, disturbed areas

Eastern half primarily, but also known from scattered western counties; intr. from the Old World

Annual. Stems diffusely branched, prostrate or ascending, up to 3 dm tall, smooth except for a hairy line along now one side and then another. Leaves petioled below and sessile above; the blade ovate or broadly elliptic, mostly 5–20 mm long and 3–10 mm wide; apex acute; base acute; margin entire, sometimes with a few hairs. Flowers white, on slender pedicels, borne singly in the leaf axils or in loose terminal clusters. Sepals lanceolate, with sparse glandular pubescence, about 4 mm long. Petals deeply 2-parted, shorter than the sepals. Stamens 10 or fewer. Styles 3. Capsule ovoid, as long as or slightly longer than the sepals, splitting into 6 segments. Seeds orange to tan, round in outline, flattened, with 4 rows of tubercles around the sides, 0.8–1.0 mm long.

124. *Stellaria media*

CERATOPHYLLÀCEAE — Hornwort Family

Submersed perennial aquatics, usually rootless. Leaves whorled, dichotomously dissected, sessile, estipulate. Stems elongate and much branched. Flowers unisexual, regular, quite small and inconspicuous, borne singly in the leaf axils and subtended by an 8- to 15-lobed involucre or calyx. Petals absent. Male flowers with 12–16 stamens arranged spirally on a flat receptacle; filaments short and thick, free and distinct; the connective between the pollen sacks thickened and projecting apically as 2 sharp points. Female flowers with a single pistil; ovary superior, 1-celled, 1-ovuled; style 1 and stigma 1. Fruit an achene.

The family contains but a single genus and 3 species of cosmopolitan distribution. Kansas has 1 species.

CERATOPHÝLLUM L. — Hornwort

From the Gr. *ceras*, a horn, + *phyllon*, leaf, the leaves being divided into narrow, sharp-pointed segments.

Characteristics those of the family.

C. demérsum L. — Coontail

NL., growing under water, from L. *demergere*, to plunge into.

Flowers minute, green, rarely seen; July–September (?)

Lakes and ponds

Throughout

A highly variable species, with many of the differences in size, leaf shape, flaccidity, and fruit set apparently caused by differences in habitats. Leaves finely divided into 2–4 capillary or slightly flattened linear segments 1.0–1.6 mm wide, with widely spaced, spinulose teeth along 1 margin. Achene ellipsoid, wingless, terminated by the persistent style and with or without 2 basal tubercles or spines, 4.5–6 mm long excluding the style.

The achenes and leaves are eaten by ducks, and the plants provide some shelter for fish, aquatic insects, and small invertebrate animals. The plant is commonly used in aquariums.

125. *Nelumbo lutea*

NYMPHAEÁCEAE

Water Lily Family

Herbaceous, perennial (rarely annual) water plants, ours with horizontal rhizomes rooted in mud at the bottom of shallow, quiet water. Leaves alternate, simple, floating or emersed, mostly long-petioled, the length of the petioles varying with the depth of the water. Flowers bisexual, regular, borne singly on peduncles from the rhizomes and riding on the surface of the water or lifted somewhat above it. Sepals 4–6 or numerous in ours, distinct. Petals numerous, distinct, scarcely differentiated from the sepals in *Nelumbo*. Stamens 12 to many in ours. Pistils 1 to many, usually superior, although in *Nelumbo* they may appear to be inferior by virtue of their being recessed in pits in a top-shaped extension of the receptacle. Fruit a follicle, an aggregate of indehiscent nutlets, or a leathery berry.

A family of 8 genera and about 90 species, widely distributed in fresh-water habitats. The various cultivated water lilies (*Nymphaea*, *Nuphar*, and *Victoria* species), the East Indian lotus (*Nelumbo nucifera* Gaertn.), and the Egyptian lotus (*Nymphaea lotus* L.) belong to this family.

Kansas has at least 3 species in as many genera but only the following species is common.

NELÚMBO Adans.

Lotus

The native Ceylonese name for the Hindu lotus, which the Greeks called *lotos* and the Latins *lotus*.

Herbaceous aquatic perennials. Leaves circular, with the petiole centrally attached, floating on or lifted above the surface of the water. Sepals and petals many, quite similar in appearance. Stamens many. Pistils many, 1-celled, superior, situated in pits in the top-shaped receptacle; style 1, very short; stigma 1. Fruits hard, indehiscent nuts retained within the pits of the dried receptacle.

Kansas has 1 species.

N. lùtea (Willd.) Pers.

Lotus

L. *luteus*, golden yellow. It is the custom to call this the yellow water lily, but since it is not a lily the Algonquin Indian name of *chinquapin* would be better. However, for the sake of uniformity and to set this plant where it belongs beside its Asiatic relative, the Hindu lotus, we adopt the name recommended by the American Joint Committee of Horticultural Nomenclature.

Pale yellow; July–August
Ponds, lakes, slow streams
Primarily south and east of a line drawn from Barber County to Jefferson County

Perennial by a vigorous horizontal rhizome which, under favorable conditions, may increase in length by 40 feet in a single year and which gives off lateral branches as it progresses. Its habit of growth reminds one of a strawberry runner, since a very long internode is terminated by a short segment composed of very short internodes with intervening nodes from which the roots, leaves, flowers, and lateral branches originate—this segment being followed by a very long internode, and so on. Examining the composition of the short segment, one finds a unique succession of parts, i.e., a node bearing a scale with the flower springing from its axil; a short internode; a node bearing the foliage leaf, with a lateral branch of the rhizome springing from its axil; and also fascicles of roots from the nodes. The long internode has the habit of enlarging here and there in the form of a banana, and through the summer and fall stuffing these with reserve food, mainly starch, which sustains the rapid growth of the following season. By examining cross sections of the leaves, stems, and roots at different levels, we discover that large intercellular spaces are continuous from the stomata in the upper epidermis of the leaves into the remotist roots buried in the mud beneath the water, enabling all parts to breath oxygen that is taken in through the leaves.

Leaves 3–6 dm in diameter with the center depressed or cupped. Flowers pale yellow, 1.2–2.5 dm wide. Sepals ovate, 4 or more, counting transition forms between them and the oblanceolate petals. Stamens many, with the anthers dehiscing toward the inside of the flower and a hooked extension of the connective above the anther sac, 2.0–3.5 cm long. Pistils several, widely dispersed, each so deeply imbedded in the top-shaped extension of the receptacle beyond the insertion of the stamens that only its tip, terminated by a squat style and capitate stigma, is to be seen. Self-pollination is impossible, but cross-pollination is effected by wild bees coming for nectar and pollen through a brief period of early morning. In late August or September, the ovaries ripen into nuts 1.2–1.5 cm long with a thin, hard, tough shell like an acorn enclosing a single seed which occupies nearly all the space within the shell. Later the peduncle bends at its apex, causing the receptacle to hang downward, thus allowing the nuts, which are at this time free in the receptacle cavities, to fall into the water and sink to the bottom. Finally, the old receptacle breaks off and floats about, eventually disintegrating and setting free any nuts still imprisoned. The shell of the nut is at first nearly impermeable to water, and germination takes place only after the nuts have been submerged for a considerable time. They can be made to germinate as soon as ripe by cutting a small hole through the shell and placing them in water.

Mrs. Sara Robinson tells in her book *Kansas, Its Interior and Exterior Life* (14) that the Kaw Indian women often waded into an oxbow lake, some 6 miles northwest of Lawrence, to gather the nuts, estimating that there were enough for the subsistence of several families for weeks. Indians also used the mealy tubers, which they peeled and boiled with meat and hominy, using the tubers fresh or after they had been sliced and dried for long-keeping. The nuts they steeped in water and parched in sand to get out the kernels, which they ate with meat or mixed with fat in making soup. *Nelumbo* nuts are among my favorite foods from wild plants and to me have a flavor much resembling malt.

See plate *37*.

RANUNCULACEAE   Crowfoot Family, Buttercup Family

Annual or perennial herbs, herbaceous or woody vines, or occasionally shrubs or trees. Leaves mostly alternate, but may be opposite, whorled, or entirely basal, simple or compound, mostly estipulate. Flowers bisexual (unisexual in *Thalictrum* and some *Clematis* spp.), regular or irregular, the inflorescence various. Sepals 3–15, free and distinct, often soon falling, frequently petal-like. Petals about the same number as the sepals, or absent. Stamens usually many, spirally arranged. Pistils 1 to many, superior, 1-celled, 1- to several-seeded. Fruit an achene, follicle, or berry.

A family of about 35 genera and 1,500 species, distributed primarily in the cooler temperate regions of the Northern Hemisphere. Numerous ornamentals including the buttercups (*Ranunculus* spp.), clematis (*Clematis* spp.), windflowers (*Anemone* spp.), larkspurs (*Delphinium* spp.), columbine (*Aquilegia* spp.), and the peony (*Paeonia* spp.) are represented in this family.

Kansas has 10 genera. *Actea alba* (L.) Mill. and *A. rubra* (Ait.) Willd. are both quite rare and are not included here.

Key to Genera (3):
1. At least 1 of the petals or sepals prolonged backward into a long or short spur 2
  2. Each of the 5 minute sepals prolonged backward into a short spur .... *Myosurus*
  2. Not as above 3
    3. Each of the 5 large conspicuous petals prolonged backward into a long spur;
     flowers (in ours) reddish .... *Aquilegia*
    3. One of the large petal-like sepals (the upper one) prolonged backward into a long
     spur; flowers (in ours) blue or white .... *Delphinium*
1. None of the sepals or petals spurred 4
  4. Flowers deriving their color and size principally from the numerous stamens or pistils;
    sepals small, inconspicuous, usually early deciduous .... *Thalictrum*
  4. Flowers deriving their color principally from the expanded sepals or petals 5
    5. Leaves opposite or whorled 6
      6. Stems climbing or scrambling .... *Clematis*
      6. Stems or scapes erect or ascending 7
        7. Sepals 4, thick or fleshy, connivent into an urn-shaped calyx .... *Clematis*
        7. Sepals commonly 5 or more, thin, spreading 8
          8. Leaves deeply incised or parted but not truly compound .... *Anemone*
          8. Leaves ternately compound with separated leaflets .... *Anemonella*
    5. Leaves alternate or all basal 9
      9. Flowers green to yellow .... *Ranunculus*
      9. Flowers white to pink, rose, blue, or blue-purple 10
        10. Leaves simple, dissected into numerous narrowly linear segments; aquatic plants
           .... *Ranunculus*
        10. Leaves 1–3 times compound, the leaflets not linear; plants of moist woods .. *Isopyrum*

126. *Anemone canadensis*

127. *Anemone caroliniana*

ANÉMONE L.

Anemone

Gr. *anemone*, the ancient Greek name for plants of this genus, from *anemos*, wind. Why this name? Pliny says it was given because the flowers open only when the wind blows. Fernald (9) says the generic name is a corruption of *Na 'mān*, the Semitic name for Adonis, from whose blood the crimson-flowered *Anemone* of the orient is said to have sprung.

Erect perennial herbs. Leaves mostly basal and petiolate, simple and palmately lobed or dissected, or compound; some species with 1 or 2 whorls of sessile or petiolate leaves or bracts on the stem below the inflorescence. Flowers bisexual, regular, borne singly. Sepals 4–20, mostly petal-like. Petals absent. Stamens many. Pistils many, elevated on a cylindrical or subglobose receptacle; style short or elongate (depending upon the species), persistent as a beak on the fruit; stigma minute. Fruits achenes.

Kansas has 5 species, the most common 3 of which are included in the following key. For *A. cylindrica* and *A. decapetala*, see comments under *A. virginiana* and *A. caroliniana*, respectively.

Key to Species (3):
1. Achenes thinly pubescent or strigose, in a subglobose head; flowers white .... *A. canadensis*
1. Achenes densely covered and concealed by long cottony hairs, in an ovoid to cylindric head; flowers greenish, white, pink, or purple 2
   2. Involucral leaves sessile; sepals white, pink, or purple, definitely petal-like .... *A. caroliniana*
   2. Involucral leaves distinctly petioled; sepals white or greenish-white, rather herbaceous
   .... *A. virginiana*

A. canadénsis L.

Meadow anemone

L. -*ensis*, suffix added to nouns of place to make adjectives—"Canadian," in this instance.
White; May–June
Low ground and along roadside drainage ditches
Northeast sixth

Erect plants, 2–6 dm tall, with long-petioled leaves and scapose peduncle borne near the ground surface from the end or branch of a rhizome. Because of the branching rhizomes, this species is apt to occur in patches, often of large size. Leaves simple, orbicular, palmately lobed, mostly 5–15 cm wide, the margins of the lobes coarsely serrate. Leaves and stems with long appressed hairs. Flowers borne singly or 2 or 3 together, each on long slender pedicels above 1 or 2 sessile, leaflike bracts. Sepals 5, white, petal-like, somewhat unequal in size, obovate or broadly elliptic, 1–2 cm long. Fruiting head globose. Achenes flat, winged, abruptly narrowed to the persistent style, mostly 5–7 mm long (including the style).

The Indians not only admired this plant but applied the crushed rhizomes to wounds and infusions of the rhizomes to sores, including sore eyes, and they attributed to it mystic powers to amend afflictions of various kinds.

A. caroliniàna Walt.

Carolina anemone

The Latin suffixes -*anus*, -*ana*, -*anum*, indicate persons, nations, or places with which the plant named is somehow associated.
White, pink, or purple; April–May
Prairie
Eastern four-fifths

Plants with a single upright flowering stem, 0.5–3 dm tall, and 2–6 petioled basal leaves arising from a short starchy rhizome which sends out additional, more slender rhizomes. Leaf blades reniform to triangular, 1–4 cm wide, deeply cleft into 3 segments which are also lobed or parted; the lowest leaf (an overwintering leaf?) is usually purplish, and its smallest divisions are more or less oblong and up to 7 mm wide; the remainder of the leaves have linear divisions seldom wider than 2 mm; petioles 1–8 cm long. Flowers borne singly atop a long pedicel subtended by 2 sessile leaflike bracts usually situated less than half the distance up the stem. Sepals 6–20, petal-like, white to pink or purple, oblanceolate to narrowly oblong, 0.5–2 cm long. Pedicel and lower surface of sepals with long white hairs. Anthers bright yellow. The many stamens and pistils are borne on the convex receptacle which later elongates to produce an ellipsoid fruiting head 1.5–2 cm long. Achenes about 4 mm long, densely covered with long, white woolly hairs, the persistent styles projecting beyond the wool.

The flower is often mistaken for a daisy, but the daisy is very different, being a head of many tiny flowers. The anemone flowers are closed during the night and open about 8:30 A.M., closing again about 4:30 P.M. They apparently respond to a certain degree of light intensity, for they stay closed on cloudy days. No description or photograph can recapture the enchantment of these flowers for those who experienced it in pioneer days when in early April in delicate tints it overspread the prairies around homes and schoolhouses and away toward the horizon. Through all our mowing and pasturing, this anemone remains with us in many counties and sometimes in large colonies.

A. decapetala Ard., known in Kansas from Meade and Clark counties only, is similar in appearance to *A. caroliniana* but is a larger, more robust plant arising from an oblong, tuberous root. The flowers are white inside and lavender to violet on the outside. The fruiting head is cylindric, rather than ellipsoid, and usually 2–4 cm long. The achenes are woolly, as in *A. caroliniana,* but the short styles are inflexed and covered by the wool.

See plate *38.*

128. *Anemone virginiana*

A. virginiàna L. — Tall anemone, thimbleweed

"Of Virginia."
Flowers greenish or greenish-yellow; June–July
Woods and thickets
Eastern fifth

Plants with a single flowering stem and 1–7 long-petioled basal leaves arising from a fibrous-rooted rhizome, mostly 0.5–1 m tall, leaves mostly 6–20 cm wide, so deeply lobed as to appear trifoliolate, each major lobe with 2–3 smaller lobes and coarsely serrate margins. Leaves and stems with long appressed or spreading hairs. Flowers borne singly or 2–3 together, each on longer slender pedicels above 2–3 petiolate leaves or bracts. Sepals 5, greenish or greenish-yellow on the upper surface, woolly-white on the lower surface, 7–13 mm long. Styles pale or merely crimson-tipped. Fruiting head "thimble-shaped," hence one of the common names, mostly 2–3 cm long and 1.2–1.5 cm wide at maturity. Achenes about 4 mm long (including the style), densely covered with long, white woolly hairs.

A. cylindrica Gray differs from this species in having the major lobes of the leaves more deeply divided into narrower segments, involucres consisting of 5–9 (rarely 3) petioled leaves (3 of which are much larger than the others), crimson styles, and fruiting heads 2–4 cm long and 0.6–1 cm wide.

ANEMONÈLLA Spach — Rue anemone

NL., from *Anemone* + L. diminutive suffix *-ella,* hence "little anemone."

Perennial herbs with thickened tuberous roots. Leaves mostly basal, 2- to 3-times ternately compound, petiolate; several opposite or whorled sessile leaves borne beneath the inflorescence. Flowers bisexual, arrangd in 3- to 4-flowered umbels. Sepals 5–10, petal-like. Petals absent. Stamens many. Pistils mostly 8–12. Fruits fusiform achenes tipped with the persistent stigmas.

Kansas has 1 species.

129. *Anemonella thalictroides*

A. thalictroìdes (L.) Spach — Rue anemone

NL., from the genus *Thalictrum* + the Greek suffix *-oides,* similar to, in reference to the similarities in the foliage
White or pink; April–May
Open woods
Douglas, Wyandotte, Leavenworth, Labette, and Cherokee counties

Low perennial arising from a cluster of tuberous roots. Stems smooth, up to 3 dm tall. Leaves all basal, biternately compound, with wiry petioles and petiolules, the leaflets mostly 1–3 cm long, broadly elliptic to obovate, with 3 shallow terminal lobes and a cordate base. Inflorescence an umbel subtended by a whorl of leaflike, slender-stalked bracts. Sepals usually 5, white or pink, broadly lanceolate to elliptic, ovate, or obovate, 5–15 mm long. Pistils 4–15. Mature achenes about 4 mm long, strongly ribbed.

Floral adaptation to bees is similar to that in *Isopyrum.* The tuberous roots are starchy and edible when cooked.

NOTE: Compare with *Isopyrum biternatum.*
See plate *39.*

AQUILÈGIA L. — Columbine

NL., from L. *aquila,* eagle, because of the talonlike appearance of the spurred petals. Or, according to some authors, derived from L. *aqua,* water, + *legere,* to collect, from the evident fluid at the bases of the hollow spurs. The common name comes from L.

columbinus, dovelike, from *columba*, dove—the uprising spurs of pendant flowers suggesting 5 doves in a ring around a dish.

Perennial herbs arising from a stout rhizome. Principal leaves basal, long-petiolate, 2- to 3-times ternately compound. Flowers bisexual, regular, borne singly on long peduncles. Sepals 5, petal-like. Petals 5, each prolonged backward (upward as the flower is pendant) in the form of a hollow spur which serves as a nectar receptacle. Stamens many. Pistils 5, maturing into a cluster of many-seeded, erect, follicles (each essentially a dry capsulelike fruit opening along a single suture).

One species in Kansas.

130. *Aquilegia canadensis*

A. canadénsis L.   **Wild columbine**

L. suffix *-ensis* added to nouns of place to make adjectives—"Canadian," in this instance.
Flowers purplish-red outside, yellow inside; April–July
Wooded hillsides and bluffs
Eastern third

A fragile perennial, 0.3–1 m tall, arising from a short, erect underground stem with thickened fibrous roots. Leaves 2- to 3-times ternately compound, the lower leaves (including petiole) mostly 1–2 dm long, the upper leaves smaller; leaflets 1–3.5 cm long and 0.5–4 cm across, divided into 2 or 3 major lobes, these with additional, shallower rounded lobes. Sepals reddish- to greenish-yellow, ovate to broadly lanceolate, 1–1.5 cm long. Petals purplish-red outside, yellow inside, 2.5–3.5 cm long. Stamens yellow, projecting in a column around the pistils and extending beyond the corolla. Follicles 1.2–3 cm long, excluding the slender persistent styles, sparsely hairy (use magnification), and with prominent veins when dried. Seeds black, shiny, 1.5–2 mm long and 0.5–1 mm wide.

The anthers open a few at a time, beginning before the stigmas are ready to receive pollen, so that cross-pollination happens when hummingbirds and bumblebees going from younger to older flowers are sucking nectar from the deep spurs.

The American Indians, who saw a lyric and romantic expression of nature in the wildflowers, sometimes employed this species to help them in their love affairs. A young man going courting would grind the black seeds and rub the fragrant meal in the palms of his hands, after which a handshake with the girl of his choice would win a favorable response to his suit.

See plate *40*.

CLÉMATIS L.   **Clematis**

L. *clematis*, Gr. *klematis*, a classical name for various trailing or climbing vines.

Herbaceous or woody perennials, either climbing or erect. Leaves opposite, simple or compound. Flowers bisexual or unisexual (then dioecious), regular, borne singly or in clusters. Sepals 4–6, petal-like. Petals none (or sometimes present and very small). Stamens many. Pistils many. Fruits achenes tipped by the long, persistent styles.

Kansas has 4 species. The 3 most commonly encountered species are included in the following key. *C. virginiana* L., which is known from Douglas and Cherokee counties, is a vine with compound leaves, each with 3 coarsely toothed leaflets, and white flowers with spreading sepals as in *C. dioscoreifolia*. The achenes and styles are covered with extremely fine, silky white hairs.

Key to Species:
1. Leaves simple; plants erect, not climbing, growing in prairies of north-central Kansas ......
.... *C. fremontii*
1. Leaves compound; climbing plants, found in woodlands of eastern or southeastern Kansas 2
    2. Leaflets entire; sepals white, spreading; plants of southeast Kansas .... *C. dioscoreifolia*
    2. Leaflets mostly 2- to 3-lobed; sepals purple, the calyx urn-shaped; plants of the eastern
        third of Kansas .... *C. pitcheri*

C. dioscoreifòlia Lévl. & Vaniot   **Virgin's bower, sweet autumn clematis**

NL., from the genus *Dioscorea*, named in honor of the Greek naturalist Dioscorides, + L. *folium*, leaf, i.e., having leaves like those of *Dioscorea* (yam).
White; July–October
Railroad embankments, roadsides, and waste places
Cherokee, Neosho, Crawford, and Cowley counties; intr. from Japan, escaped from cultivation and became established in some areas

High-climbing perennials. Stems 2–3 (–10) m long. Leaves compound, petiolate; leaflets usually 5, mostly ovate, less frequently deltoid or elliptic, leatherlike in texture,

1.5–6.5 cm long, with a rounded apex and acute or somewhat cordate base. Flowers fragrant, white, borne in clusters. Sepals 4, oblanceolate or elliptic, spreading, 7–15 mm long. Pistils subtended by long, silky white hairs. Body of the achene 3–4 mm long, the plumose style 5–15 mm long.

For a long period the vines are covered with clusters of white flowers exhaling the fragrance of wild plum blossoms, and thereafter, far into the winter, they are decorated with clusters of purplish achenes tipped with persisting, long, silvery, plumose styles.

See plate *41*.

C. fremóntii S. Wats.   Frémont's clematis, leather flower

Named in honor of Gen. John C. Frémont, U.S. Army Corps of Topographical Engineers, who
    made 5 expeditions westward across the United States, discovering many new species
    of plants.
Flowers purple on the outside, white on the inside; May–June
Rocky prairie
North-central sixth

Erect herbaceous perennials arising from a vertical rhizome bearing many strong fibrous roots. Stems several times branched, 2–4 dm tall, sparsely covered with white cobwebby hairs. Leaves simple, sessile or very short-petiolate, usually ovate, less frequently lanceolate or elliptic, leathery, light green, with lighter, quite prominent, reticulate venation, those from the middle regions of the stem 5–11 cm long and 3–7 cm wide; apex acute or rounded; base acute; margin entire or occasionally with a few large teeth; a few cobwebby hairs present on the under surface of the leaves and along the margins (use magnification). Flowers borne singly atop pedicels 1.5–5 cm long at the tips of the stems or in lower leaf axils. Calyx urn-shaped, 2–3 cm long; sepals 4–5, acuminate-lanceolate, white on the inside and purple on the outside, with white, densely woolly margins. Fruiting head globular, mostly 4–5 cm wide. Body of the achene 5–7 mm long; style 1.5–2.5 cm long, threadlike.

Clematis fremontii is one of our rarest plants. Its entire distribution is limited to north-central Kansas, where it usually occurs in the vicinity of outcrops of Fort Hays Limestone or Smoky Hill Chalk, 3 adjacent counties in south-central Nebraska, and 3 counties in east-central Missouri.

C. pítcheri T. & G.   Leather flower

Named in honor of its discoverer, Dr. Zina Pitcher (1797–1872), U.S. Army surgeon and botanist.
Dull purple; June–August
Low woods and thickets, hedgerows
Eastern two-fifths

Herbaceous perennials with many strong fibrous roots arising from a rhizome near the surface of the ground. Stems climbing or scrambling. Leaves pinnately compound with 3–9 stalked leaflets, 1 or more of them, usually the apical 1, sometimes represented by twining tendrils by which the vine is held to a suitable support; leaflets entire or more or less deeply 2- to 3-lobed, broadly ovate to lanceolate, 2–8 cm long and 1–6 cm wide; apex rounded, acute, or short acuminate; base acute, rounded, or slightly cordate. Flowers nodding, borne singly on long, ascending pedicels arising from leaf axils. Calyx urn-shaped, mostly 2–3 cm long; sepals 5, lance-acuminate, dull purple with woolly white margins on the upper (adaxial) and sometimes the lower surface. Fruiting heads globular, 4–6 cm wide. Body of the achene 7–8 mm long and 6–7 mm wide, glabrous or with short hairs, especially near the base of the threadlike style which is about 1.5–2 cm long.

The plants are visited primarily by honeybees and bumblebees.

The plants can be propagated by sowing the achenes out-of-doors as soon as ripe, or they can be readily transplanted from the wild. It should be noted, however, that they require a deep, rich soil.

DELPHíNIUM L.   Larkspur

NL., from Gr. *delphinion*, larkspur, from *delphin*, dolphin, the unopened flowers having some
    resemblance to that graceful marine mammal. "Larkspur" is the old English name,
    from the resemblance of the spur to the long claw of the hind toe of the European
    crested lark.

Erect, sparsely branched annual or perennial herbs. Leaves alternate, palmately lobed or divided, long-petioled. Flowers bisexual, irregular, arranged in terminal racemes. Sepals 5, petal-like, the upper one prolonged backward in the form of a hollow spur.

131. *Clematis dioscoreifolia*

132. *Clematis fremontii*

133. *Clematis fremontii* fruiting

CROWFOOT FAMILY

117

Petals 2–4, relatively small, the upper 2 forming nectar-producing spurs enclosed within the spur of the calyx, the 2 lower ones, when present, covering the many stamens and the 3–5 pistils. Fruits many-seeded follicles.

Many species of larkspur contain poisonous alkaloids and have been known to cause illness and sometimes death of livestock. The plants are apparently the most toxic in their youngest growth.

Kansas has 2 native species. In addition, *D. ajacis* L. is occasionally found escaped from cultivation. This species differs from the 2 following in being an annual and having only 1 pistil per flower.

Key to Species:
1. Flowers deep blue; follicles widely spreading at maturity; plants of the eastern fourth of the state .... *D. tricorne*
1. Flowers white, sometimes tinged or spotted with blue; follicles erect at maturity; plants occurring throughout Kansas .... *D. virescens*

134. *Clematis pitcheri*

D. tricòrne Michx.   **Rock larkspur**

L. *tricornis*, having 3 horns, from *tri-*, the combining form of *tris*, 3, + *cornu*, horn, such as that of cattle or sheep—the fruits of this larkspur consisting of 3 sharp-pointed, wide-spreading follicles.
Purplish-blue (very rarely pink); April–May
Moist, rich open woods, less frequently in prairies or along roadsides
Eastern third

Herbaceous perennials with a cluster of tuberous roots. Stem erect, unbranched, easily broken, mostly 2.5–7 dm tall. Leaves ascending, long-petioled, in a scant basal rosette and a few on the stem; blades mostly 2–8 cm long, deeply dissected, the ultimate divisions oblong, mostly 2–7 (–10) mm wide. Raceme 0.8–2 dm long. Flowers purplish-blue (very rarely pink), about 3 cm long. Petals 4, somewhat prolonged forward and forming a vestibule before the spur, the 2 lower petals forming a canopy over the stamens and pistils. Follicles separate from one another and widely divergent at maturity, mostly 1.5–2 cm long. Seeds black, shiny, 1.5–2 mm long.

135. *Delphinium tricorne*

D. viréscens Nutt.   **Plains larkspur**

L. *virescens*, becoming green, and so not fully green, the fine pubescence of this species giving that effect.
White, bluish-tinged; May–June
Prairie
Throughout

Herbaceous perennials arising from a short, erect rhizome with several thickened roots. Stems erect, up to 1.5 m tall, softly pubescent. Leaves mostly basal or at least below the middle of the stem, petioled below, sessile above; blades mostly 2–7 cm long, very finely dissected, the ultimate divisions linear, mostly 1–2.5 mm wide. Raceme 1–2.5 (sometimes up to 5 or 6) dm long. Flowers white, sometimes tinged or spotted with blue, 2–3 cm long. Follicles erect and contiguous at maturity, about 2 cm long. Seeds brown, highly variable in shape, ornamented with heavy, corrugated ridges, mostly 2–2.5 mm long.

This wandlike perennial is a prominent feature of the prairies and plains in all quarters of Kansas, growing singly or in colonies among the grasses or along the undisturbed borders of cultivated fields. At the time of its blooming, it is always rendered conspicuous by overtopping the grasses.

Closely examining an open flower, we discover an interesting mechanism for effecting cross-pollination by insects. When a flower opens, and for some time thereafter, the anthers are held in front of the nectar-bearing spurs. In older flowers toward the base of the inflorescence, the anthers, after discharging their pollen, have moved aside, leaving the stigmas exposed in front of the spurs. Bumblebees—the most frequent visitors in quest of nectar—proceed from the base of a raceme toward its apex, so that on leaving a raceme with pollen from younger flowers on head and mouth parts, they deposit this pollen on the stigmas of the basal flowers of the next raceme visited.

136. *Delphinium virescens*

ISOPÝRUM L.   **Isopyrum**

NL., from Gr. *isopyron*, a name used by Dioscorides and Pliny for a plant (possibly a *Fumaria*) which cannot now be positively identified from their descriptions.

Delicate perennial herbs. Leaves alternate, 2- to 3-times ternately compound, stipulate. Flowers bisexual, regular, solitary or panicled (but not in umbels). Sepals 5, petal-like. Petals absent. Stamens many. Pistils usually 3–6. Fruits follicles, each with a "beak" formed by the persistent style.

One species in Kansas.

I. biternàtum (Raf.) T. & G.　　　　　　　　　　　　False rue anemone

NL., from L. *bi-*, the combining form of *bis*, twice, + NL. *ternatus*, consisting of 3's, the leaf being divided into 3 parts and each part into 3 leaflets. The true rue anemone is *Anemonella thalictroides*.

White; April–mid-May
Moist, rich woods
East fourth

Plants erect, with fibrous and occasionally slightly tuberous roots. Stems slender, wiry, up to 3 dm tall, bearing a few short-petioled, ternately compound leaves or leaflike bracts above and with a few basal, long-petioled, biternately compound leaves springing directly from the rootstock. Leaflets oblanceolate to obovate, circular, or somewhat reniform, 0.8–2 cm long and with 2–3 major lobes or divisions, these usually with 2–3 smaller rounded lobes. Stipules membranous lobes fused to the base of the petiole. Flowers white, borne singly in the axils of leaflike bracts. Sepals oblanceolate or obovate, 7–11 mm long. Pistils 3–6 (usually 4). Fruits divergent, smooth, 2- to 3-seeded, 6–8 mm long. Seeds smooth, brown, about 2.5 mm long.

This dainty woodland sprite is up and blooming among our earliest spring flowers —bloodroot, corydalis, toothwort, Dutchman's breeches, wild ginger, trillium, merrybells, and other occupants of the forest floor that hasten to fulfill their vegetative and reproductive functions before the spreading forest canopy is intercepting too much of the vitalizing energy of sunlight. In fact, in some seasons leaves can be seen as early as January.

The flowers open in the morning and close in the evening for several days, the anthers of the many stamens opening a few at a time each day. Cross-pollination is sure to happen because, before the anthers begin to open, the styles bend outward and present receptive stigmas to the wild bees that come to gather the abundant pollen and lick up the scant nectar secreted by glands situated between the insertions of the sepals.

This plant and *Anemonella thalictroides* are often confused, understandably so, since they are rather similar in appearance, occur in the same habitat, and bloom at about the same time. Flowers of *Anemonella* are arranged in umbels subtended by a whorl of leaves or leaflike bracts, while those of *Isopyrum* are usually borne singly. The fruits of the 2 species are also quite distinct.

See plate *42*.

MYOSÙRUS L.　　　　　　　　　　　　　　　　Mousetail

NL., from Gr. *mys*, *myos*, mouse, + *oura*, tail—suggested by the elongate fruiting receptacle.

Small acaulescent annuals with a tuft of linear basal leaves. Flowers bisexual, regular, borne singly on leafless scapes. Sepals 5, spurred at the base. Petals 5, very small and narrow, "clawed," with a nectiferous hollow at the summit of the claw. Stamens 5–20. Pistils many, borne on a slender, much-elongated receptacle. Fruits achenes, 4- or 6-angled in face view.

One species in Kansas.

M. mínimus L.　　　　　　　　　　　　　　　　Tiny mousetail

L. *minimus*, smallest.
Flowers greenish, inconspicuous; April–May (rarely collected as late as August)
Mud, and low, moist soil
Scattered throughout

Inconspicuous plants up to 1.5 dm tall. Leaves 0.5–3 mm wide. Flowers greenish, the perianth 3–8 mm wide. Sepals 3–6 mm long, including the spur, the expanded portion oblong. Elongated receptacle 0.3–2 cm long in flower, up to 5 cm long in fruit. Achenes about 1–1.5 mm long.

RANÚNCULUS L.　　　　　　　　　　　　　　Buttercup, crowfoot

L. *ranunculus*, little frog or tadpole, diminutive of *rana*, frog, the name being applied to this group of plants because several of its species inhabit wet places where frogs abound.

Annual, biennial, or perennial land or water plants. Leaves alternate, simple, variously toothed, lobed, divided, or dissected. Flowers bisexual, regular, solitary or in sparse, open clusters. Sepals usually 5, green or yellow, falling early. Petals usually 5,

137. *Delphinium virescens* leaves

138. *Delphinium virescens* fruits

139. *Isopyrum biternatum*

CROWFOOT FAMILY

119

140. *Myosurus minimus*

141. *Ranunculus abortivus*

142. *Ranunculus abortivus*

yellow (rarely white) with a nectiferous pit at the base of the blade. Stamens usually numerous. Pistils several to many, borne on a globoid or cylindrical receptacle. Achenes tipped with a "beak" formed by the persistent style.

All species of *Ranunculus*, as far as is known, contain a toxic compound, protoanemonin. The amount of the toxin contained in a given plant will depend upon certain variables, including the species, stage of growth, and conditions under which the plant grew. Ingestion of the plants will cause severe gastrointestinal upset, and severe cases of livestock poisoning resulting in death of animals have been reported. In some persons, contact of the plant juices with the skin will cause a dermatitis similar to poison ivy.

About 13 species are known from Kansas. The 6 most widely distributed species are included in the following key.

Key to Species:

1. Petals white; leaf divisions threadlike; plants aquatic .... *R. longirostris*
1. Petals yellow; leaf divisions not threadlike; plants aquatic or terrestrial 2
  2. All of the leaves crenate; plants spreading by means of stolons .... *R. cymbalaria*
  2. All or part of the leaves deeply lobed, dissected, or compound; plants sometimes with stems repent and rooting at the lower nodes, but not spreading by stolons 3
    3. Basal leaves mostly bean-shaped, usually merely crenately toothed, occasionally some of them 2- to 3-lobed; stem leaves with 3–6 leaflets or deep lobes which may vary from obovate to oblanceolate or narrowly elliptic .... *R. abortivus*
    3. All leaves deeply palmately lobed or divided 4
      4. Petals 2–5 mm long; plants usually with 1 flowering stem arising from the cluster of basal leaves, the stem glabrous, rather stout (2–35 mm wide at the base), much branched above (occasionally branched from the base), up to 6 dm tall .. *R. sceleratus*
      4. Petals 6–15 mm long; plants with several flowering stems arising from the cluster of basal leaves, the stems hairy, slender (1–3 mm wide at the base), only slightly branched, if at all, and up to 3 dm tall 5
        5. Leaflets or leaf segments with rounded apexes, lobes, and/or teeth; stems not becoming repent; some roots conspicuously thickened .... *R. fascicularis*
        5. Leaflets or leaf segments with sharply pointed apexes, lobes, and/or teeth; stems eventually becoming repent and rooting at the nodes; roots only slightly thickened, if at all .... *R. septentrionalis*

R. abortìvus L.   Littleleaf buttercup

NL., from L. *abortus*, miscarriage, incompleteness—the tiny petals giving that impression.
Yellow; April–May
Moist soil of open woods, creek banks, prairie, pastures, roadsides, and waste ground
East half

Biennial or perennial from a very short rootstock and a fascicle of slender fibrous roots. Stems erect, somewhat succulent, up to 7 dm tall. Basal leaves long-petioled with broadened bases ensheathing the stem; blades 1.2–4 (–6) cm wide, mostly cordate with crenate margins, but 1 or more with blades 3-lobed or -parted, similar to the nearly sessile blades higher up. Flowers borne singly on terminal or axillary peduncles. Sepals ovate to elliptic, 2.5–4 mm long. Petals yellow, at least the distal half turning white upon drying, lanceolate, elliptic, or rhombic, mostly 2–3 mm long. Receptacle hairy. Achenes green or straw-colored, lustrous, nearly round or ovate in outline, biconvex, about 1.5 mm long, with only a very minute remnant of the style (use magnification).

Despite its smallness and disparaging species name, the flower is complete and perfect in all its parts, even to tiny nectiferous pits at the base of the petals which are known to attract honeybees, bumblebees, various short-tongued bees, hoverflies, and beetles.

R. micranthus Nutt., recorded from Wyandotte and Cherokee counties, is similar to *R. abortivus*, but differs in having its basal leaves mostly ovate in outline; its roots fusiform thickened; its leaves, sepals, and fruiting heads smaller; its achenes dull; and the receptacle without hairs.

R. cymbalària Pursh   Seaside crowfoot

From the genus *Cymbalaria*, in the Figwort Family, which has similar leaves.
Yellow; June–September
Mud, especially in alkaline places
West half, primarily, with a few scattered collections from the northeast fourth

Erect, tufted perennials, up to 1.5 dm tall, spreading by means of stolons. Leaves mostly basal, long-petioled; blades nearly round in outline to cordate or reniform, mostly 0.5–2.5 cm wide, with a rounded apex and crenate margins. Flowers borne singly or (more frequently) in clusters of 2–10 on a scape protruding conspicuously above the basal rosette. Sepals lanceolate to elliptic, 2.5–4 mm long. Petals yellow, at least the apex and sometimes the entire petal turning white under drying (occasionally remaining yellowish), spatulate to oblanceolate, 4–5 mm long. Achenes greenish or straw-colored, thin-walled, 1.5–2 mm long, with longitudinal veins and an erect beak.

R. fasciculàris Muhl.   Prairie tufted buttercup

NL., from L. *fascicularia*, things carried in a bundle, from *fasciculus*, small bundle—in reference to the clustered stems and roots.
Yellow; April–mid-May
Open woods and prairies
Southeast eighth

A tufted, pubescent perennial with a fascicle of fibrous roots and some thickened tuberous roots mostly 1–3 (–5) cm long. Stems erect, up to 2.2 dm tall. Basal leaves long-petioled, the lowest ones with blades ovate or 3-lobed, the others with 3 leaflets—the upper one 3-cleft and stalked, the lower 2 sessile and irregularly and less deeply cleft. Upper leaves similar to the 3-foliolate basal leaves but with shorter petioles. Hairs on stems, leaves, and sepals mostly appressed. Flowers borne singly at the tips of the stems or from the upper leaf axils. Sepals yellow or greenish, often with a dark purple midvein, broadly lanceolate, 5–8 mm long. Petals yellow, the distal three-fourths of the upper surface often turning white in drying, oblanceolate to oblong, 6–15 mm long. Achenes greenish or straw-colored, the body biconvex, more or less obovate, mostly 2–2.5 mm long, with a narrow margin and a straight or curved beak.

143. *Ranunculus cymbalaria*

R. longiróstris Godr.   Longbeak buttercup, white water-crowfoot

NL., from L. *longus*, long, + *rostrum*, beak, i.e., long-beaked, descriptive of the achenes.
White; late April–June
Ponds and slow streams
Scattered in the west three-fourths

Perennial aquatic herbs with floating or creeping stems. Leaves submersed, several times palmately divided into threadlike segments, firm, 1–2.5 cm long. Flowers borne singly and lifted above the water on pedicels 2–5 cm long. Sepals ovate to lanceolate, 3–4 mm long. Petals white, oblanceolate to broadly obovate, mostly 6–9 mm long. Achenes greenish or straw-colored, transversely ridged, the body about 1.5 mm long.

The entire body of the plant must get all its material needs from the substance of the water itself, and from the gases and minerals dissolved in the water. The dissected leaf, with its permeable epidermis (the surface not waterproofed as in aerial leaves), is a direct adaptation to this situation, for such a leaf, compared to an entire one, presents a much greater absorptive surface per unit of tissue. We are reminded of the gills of fishes and tadpoles.

144. *Ranunculus sceleratus*

R. flabellaris Raf. is a robust, yellow-flowered aquatic species with some leaves similar to those of *R. longirostris* and some bean-shaped leaves. It is known from scattered sites in the northeast fourth.

R. sceleràtus L.   Cursed crowfoot

NL., from L. *scelerator*, evil-doer, perhaps because the juice of this plant is irritating to the skin or is toxic if ingested.
Pale yellow; late April–September
Moist or wet places
Scattered throughout

Annual. Stems erect, much branched above (occasionally branched from the base), rather stout (2–35 mm wide at the base), up to 6 dm tall. Basal leaves long-petioled; the blades 1–6 cm wide, bean-shaped to cordate or triangular in outline with 2–3 primary lobes, each of these with smaller rounded lobes or teeth. Upper leaves gradually becoming shorter-petioled and more deeply cleft, those near the apex being sessile and parted nearly to the base into 3 elliptic lobes with or without teeth. Flowers borne singly on ascending axillary peduncles or on peduncles terminating the branches. Sepals ovate to lanceolate, somewhat petal-like, 2.5–5 mm long, with appressed hairs. Petals yellow, sometimes turning white on drying, 2–5 mm long, "clawed," the broadened portion short-oblong to ovate or obovate. Achenes greenish or straw-colored, ovate or nearly round in outline, biconvex, about 1 mm long (or shorter), with faint horizontal ridges and a very short beak (use high magnification).

145. *Ranunculus hispidus*

146. *Thalictrum dasycarpum*, female plant

147. *Thalictrum dasycarpum,* in fruit

CROWFOOT FAMILY
122

R. septentrionális Poir.   Northern buttercup

L. *septentrionalis*, north, northern.
Yellow; April–mid-May
Moist or swampy places of woods, river banks, and roadsides
Extreme east

    Tufted perennial with thick fibrous roots. Stems at flowering time erect or ascending, 2–3 dm tall, some branches later becoming repent and rooting at the nodes. Herbage glabrous or with silky appressed hairs. Basal leaves long-petioled, trifoliolate, the terminal leaflet with 3 primary lobes, the lateral leaflets each with 2 primary lobes which are sometimes completely separated from one another on the petiolule and appear to be separate leaflets; blades of the terminal leaflets 2–4 cm long. Upper leaves similar to the basal leaves; all leaves with serrate or serrate-crenate margins. Flowers borne singly on axillary pedicels. Sepals falling early, lanceolate, 6–10 mm long, with spreading or appressed hairs. Petals yellow, the distal two-thirds often turning white on drying, broadly oblanceolate, 7–14 mm long. Achenes biconvex, thin-margined, the body 3–4.5 mm long and the beak 1.5–3 mm long.

    R. recurvatus Poir., known from Shawnee, Douglas, Leavenworth, and Wyandotte counties, is similar in appearance to *R. septentrionalis* but has leaves which are only 3-lobed, never trifoliolate. *R. hispidus* Michx. has been collected in Douglas and Cherokee counties; it has spreading, rather than appressed, hairs on the stems and petioles, and usually occurs in drier habitats than does *R. septentrionalis.*

THALÍCTRUM L.   Meadowrue

Name used by Pliny and Dioscorides, 1st century, for the meadowrue, *Thalictrum flavum.* "Rue," the name employed in the earliest English herbals for the garden or common rue, is a contraction of L. *ruta*, bitterness or unpleasantness, and *ruta* is the name used by Pliny and other medical and natural-history writers for a rank-smelling, nauseous-tasting south European plant famed for curing a large variety of diseases and discomforts—the plant named *Ruta graveolens* by Linnaeus. This is the original, bona fide rue, after which various other plants are named, because of a similarity of foliage—the meadowrue and fen rue (*Thalictrum*), rue-anemone (*Anemonella*), false rue-anemone (*Isopyrum*), and goat's-rue (*Galega*).

Perennial herbs. Leaves alternate, 2- to 3-times ternately compound, stipulate. Flowers unisexual with male and female flowers on separate plants (infrequently bisexual), regular, borne in terminal clusters. Sepals 4–5, green or petal-like, falling early. Petals absent. Stamens in male flowers many. Pistils in female flowers 4–15. Achenes with a "beak" formed by the persistent style, usually conspicuously ribbed.

    Two species in Kansas. *T. dioicum* L. is known from Doniphan County only and is not included here. It is a smaller plant than *T. dasycarpum*, seldom taller than 6 dm, has rather thin leaflets, and has the leaves of the upper and middle parts of the stem borne on long petioles.

T. dasycárpum Fisch. & Avé-Lall.   Purple meadowrue

NL., from Gr. *dasys*, thick, + *karpos*, fruit—the achenes of this species having thick walls.
Flowers greenish-white to purplish; mid-May–June
Shaded places of woods, thickets, meadows, and roadsides
Eastern half, with a few scattered collections from farther west

    Perennial from a short, erect rhizome with thick fibrous roots. Stems erect, often purple, 1–2 m tall. Stipules membranous lobes fused to the base of the *very* short petioles. Leaflets ovate to elliptic, oblong, obovate, or oblanceolate, 1–4.5 cm long and 0.5–3.5 cm wide; margins entire or with 2–3 sharp-pointed (occasionally rounded) teeth. Sepals greenish-white to purple, elliptic, lanceolate, or oblanceolate, deciduous, 2–3 mm long. Male flowers 3–8 mm long, the stamens with threadlike filaments. Female flowers 2–5 mm long; ovary angular-ovoid; style elongate and papillose on the inner surface only. Achenes ovoid, short-stipitate, with several longitudinal ridges, the body of the achene 3–5 mm long.

    Cross-pollination is effected by the wind.

    The ripening seedlike fruits have a delicate odor, which Indians employed by scattering the fruits over their clothing and rubbing them in. The Indians were open-eyed to the beautiful features of plants as well as to their usefulness. To them beauty was beneficent beyond the pleasure of seeing it, for it suggested the love and care of the Great Spirit for them. So the grace of *Thalictrum* could afford success to wooing lovers and assistance to warriors in distress when the plant was used as a charm. Thus the lover seeking to gain the affection of a maiden would crush the inflorescence of *Thalictrum* with spittle in the palm of his hand; if the desired one would accept that hand when proffered for handshaking the suitor's success was thought to be assured. Then there was the Pawnee custom of incorporating the plant with a white clay and applying this compound to the lips and nose of horses in order to increase their stamina during forced marches in retreat before pursuing enemies.

Roots and especially the fruits of this species contain the alkaloid thalicarpine, which has been approved by appropriate federal agencies for clinical evaluation as an anticancer drug. Attempts are currently being made to culture the meadowrue for its crop of fruits since, should the alkaloid become a useful drug, the quantity of roots needed would be so great that the plants would soon become rare.

BERBERIDÀCEAE   Barberry Family

From the genus *Bérberis*, the barberry, a small shrub with species native to all continents except Australia.

Shrubs or perennial herbs. Leaves alternate or basal (or a single pair of opposite leaves in *Podophyllum*), simple or pinnately compound, mostly estipulate and/or with petioles dilated at the base. Flowers bisexual, regular, solitary or borne in clusters. Sepals 6, rarely 4. Petals of the same number or more, similar to or clearly differentiated from the sepals. One or 2 whorls of often-petaloid nectaries sometimes present. Stamens as many as the petals and opposite them or twice as many, then in 2 whorls and the outer stamens opposite the petals. Pistil 1, superior, 1-celled, with 1 to few or many ovules; style short and thick or wanting; stigma capitate. Fruit a berry or a leathery capsule.

A family of about 10 genera and 200 species, primarily of north temperate regions. Cultivated members of the family include various barberries (*Berberis* spp.) and *Mahonia*. Kansas has 2 native species in 2 genera. *Caulophyllum thalictroides*, blue cohosh, is known only from the loess hills in Doniphan County and is not here included.

PODOPHÝLLUM L.   May apple

NL., from Gr. *pous, podos*, foot, + *phyllon*, leaf, the duck's foot being in mind when the name was compounded.

Erect herbs, perennial from rhizomes, with a single aboveground stem bearing 2 large opposite leaves and a single flower. Sepals 6, falling early. Petals 6 or 9. Stamens twice as many as the petals. Pistil large and ovoid with a large, thick sessile stigma. Fruit a large many-seeded berry.

Kansas has 1 species.

P. peltàtum L.   Common May apple

NL., from L. *peltatus*, armed with a shield, from L. *pelta*, shield—the form of the leaf blade and the attachment of the petiole near its center suggesting this.
Creamy white; mid-April–early May
Rich, moist woods
East fourth

Herbaceous perennials, up to 5 dm tall, with a long branching rhizome near the surface, rooting at the nodes, sometimes giving off a single leaf on a long upright petiole, sometimes a vertical stem ending with 2 long-petioled leaves and a single flower terminating the stem in the fork of the petioles. The upright stem, pair of leaves, and flower bud are all well along in their development when they emerge from the ground. An interesting sight in the April woods is their coming up, 1 leaf bent sharply downward like a closed umbrella, enfolding and making gangway for the rest. Leaves simple, orbicular, parted nearly to the petiole into 4–7 more or less obtriangular segments, these bilobed and coarsely serrate at the summit. Sepals 6, petaloid, deciduous. Petals 6 or 9, obovate or less frequently oblanceolate, creamy white, 2–3 cm long. Stamens twice as many as the petals, 1–1.5 cm long, the anthers linear-oblong. Fruit yellow or pinkish, 3–5 cm long.

The fruit of the May apple is edible when it ripens in July or August and may be eaten raw or used for jelly or pie. The flavor is rather bland, even unpleasant to some persons. Asa Gray described it as being "mawkish, eaten by pigs and boys." But no one needs to be afraid of eating it in moderation; indeed, the white immigrants soon discovered that the Indians ate it with relish.

The rhizomes, leaves, and seeds are another story. They are not edible because they contain a poisonous resinous substance, podophyllin, unpalatable but medicinal in properly regulated doses. The Indians used the rhizomes as one of their most effective purgatives, and our own medicinal practice makes use of it in the treatment of certain papillomas. According to E. P. Claus in his text *Pharmacognosy* (15), several hundred tons of the dried rhizomes and roots are dug annually to supply both domestic and export demand. Most of the commercial supplies come from the central states and from Virginia and North Carolina.

See plate *43*.

148. *Argemone polyanthemos*

PAPAVERÀCEAE   Poppy Family

Annual, biennial, and perennial herbs with a milky or colored juice, or latex, and in California, Mexico, and South America some shrub and tree forms. Leaves mostly alternate, simple with lobed or divided blades, estipulate. Flowers bisexual, regular, mostly solitary, on elongate peduncles springing directly from a rhizome or from above-ground stems. Sepals 2, rarely 3, in some species soon falling off. Petals 4–6 or 8–12, in 1–2 (rarely 3) whorls. Stamens 12 to many. Pistil 1; ovary superior, 1-celled (sometimes appearing 2- to several-celled by intrusion of the parietal placentae into the cavity), with many ovules; style 1 or obsolete; stigmas several or 1 with several lobes. Fruit a capsule opening by apical pores or splitting into 2 to several parts.

The family comprises about 28 genera and 250 species distributed primarily in the tropics and subtropics of the Northern Hemisphere. Various ornamental poppies and the opium poppy (*Papaver somniferum* L.) belong to this family.

Kansas has 2 native genera.

Key to Genera (3):
1. Petals 8 or more; flower solitary on a scape; leaf 1, basal; rare plants of moist, rich woods
   .... *Sanguinarea*
1. Petals 4 or rarely 6; leaves borne on the stem, spiny; common plants of dry, open, often sandy places .... *Argemone*

ARGEMÒNE L.   Prickly poppy

Gr. *argemone*, a name used by Dioscorides and Pliny for a plant which Pliny says had an orange-colored juice. The latter could not have been one of the Argemones of the Linnaean system of classification, for they are native only to America.

Erect, prickly annuals or biennials with a yellow juice, or latex. Leaves alternate, sessile, more or less clasping the stem, with prickly pointed teeth. Flowers showy, white or yellow (rarely lavender), borne singly on a short peduncle. Calyx of 2–3 hooded or horned sepals. Petals 4–6. Stamens many. Style very short or absent; stigma with 4–6 radiating stigmatic zones. Capsule splitting above into 3–6 parts.

Kansas has 2 species.

Key to Species (16):
1. Leaf surfaces prickly on the primary and secondary veins above and below and also minutely hispid or prickly, often closely so, between the veins; stems usually closely prickly; petals white .... *A. squarrosa*
1. Leaf surfaces prickly almost exclusively on the primary and secondary veins above and below, sometimes essentially smooth; stems usually with more widely spaced prickles or almost smooth; petals white, rarely lavender .... *A. polyanthemos*

A. polyánthemos (Fedde) G. Ownbey   White prickly poppy

NL., from Gr. *polys*, many, + *anthemon*, flower, i.e., many-flowered.
White; late May–August
Sandy soil of prairies, flood plains, and roadsides
Western three-fourths, with a few records from the east fourth

Erect annuals with a deep taproot and bright yellow latex. Stems up to 12 dm tall, sparingly prickly, the spines spreading or recurved. Leaves glaucous, variously pinnately lobed or parted, 4–18 cm long and 2–10 cm wide, with crisped and spiny-toothed margins and a few spines along the main veins of the lower surface, the upper surface usually completely smooth. Sepals prickly, 1.5–3 cm long including the long terminal horns. Petals white (rarely lavendar), broadly obovate or fan-shaped, 2.5–5.5 cm long. Stamens many (150+), yellow. Stigma purple, 3- or 4-lobed, quite conspicuous. Capsules ellipsoid,

3–4 cm long at maturity, heavily armed with spreading or recurved spines, opening and discharging seeds at the apex. Seeds dark brown, globose, about 2 mm long, with a honey-combed surface and a narrow, 2-horned crest along 1 side.

A. squarròsa Greene   White prickly poppy

L. *squarrosus*, roughened by spreading scales, scurfy, scabby. In botanical description, "squar-rose" also means that the part in question is spreading or recurved at the end. This is true of the prickles on the capsules and stems of this plant.

White; late-May–August
Dry prairie
Southwest sixth and Logan County

149. *Sanguinaria canadensis* fruiting

Erect perennials with bright yellow latex. Stems up to 12 dm tall, densely prickly, the spines slender, spreading. Leaves glaucous, bluish, more or less deeply pinnately lobed, 10–12 cm long and 3–7 cm wide, with crisped and spiny-toothed margins; spines present on the veins of both the lower and upper surfaces and minute spines or hairs in between the veins. Sepals prickly, mostly 3–3.5 cm long including the long terminal spine. Petals white, broadly obovate or fan-shaped, 3.5–6.5 cm long. Stamens many (150+), yellow. Stigma purple, 4- or 5-lobed, quite conspicuous. Capsules ellipsoid or lanceolate-ovoid, 2.5–5 cm long at maturity, heavily armed with spreading or recurved spines, the largest of which have smaller spines arising from them. Seeds about 2.5 mm wide.

SANGUINÀRIA L.   Bloodroot

NL., from L. *sanguinaria*, used by Pliny to describe *Polygonum aviculare*, employed in stanching blood, from L. *sanguis*, blood. Linnaeus gave the name to this plant because of the copious red latex in the rhizome.

Herbaceous perennial arising from a thick horizontal rhizome. A single, palmately lobed leaf and a 1-flowered scape are produced each spring. Sepals 2, falling early. Petals usually 8 (sometimes more). Stamens many. Style short and stout; stigma 2-lobed. Fruit an ellipsoid or fusiform capsule which opens along 2 sutures.

One species in Kansas.

S. canadénsis L.   Bloodroot

L. *-ensis*, suffix added to nouns of place to make adjectives—"Canadian," in this instance.

White; early April
Rich soil of woods
Extreme east

Leaf blades bean-shaped to round in outline, 3- to 7-lobed, about 3–5 cm wide at flowering time but up to 2 dm wide as the fruits mature; petioles up to 3 dm long. Petals white, oblanceolate or somewhat spatulate, about 1–3 cm long. Stamens yellow. Capsules 3–5 cm long, glaucous. Seeds reddish-brown, oblong in outline, 3–3.5 mm long, with a prominent crest along the funicular side.

This is among the first of our woodland plants to bloom in the spring, and those who love its simple beauty know they must be on the lookout at the time of its blooming, in the first 2 weeks of April at our latitude, lest they miss the sight of the flower buds peering through the covering of forest leaves, each bud wrapped round protectively by the single leaf that keeps company with it. Soon the bud shoots ahead of its companion, drops off its 2 sepals, and spreads wide its 8 or more pure-white sepals, revealing many golden stamens and a green pistil in their midst.

Bumblebees and honeybees coming to the flowers for pollen only, since there is no nectar, assist in cross-pollination.

The leaf companion now spreads out its lobed blade, but by delaying the enlarge-ment of its parts avoids concealing the flower. At night the flowers close, to reopen the following morning, but all too soon the petals and stamens drop away while the ovary steadfastly carries on and becomes a capsule full of seeds. Now the leaf grows surprisingly large, its petiole lengthening and its blade increasing in all dimensions, its food-making function becoming the dominant concern, while its products are being drawn away through the petiole into the rhizome for storage.

We must take a look at the rhizome, for there is something unusual about it. Removing one carefully from the forest mold we discover that its habit is to increase by the formation of both terminal and lateral segments, each segment capable of organizing a leaf-and-flower bud (mixed bud) at *both* ends, and providing itself copiously with long fibrous roots, and so becoming an independent individual. With reasonable care these segments can be transplanted at any time.

When broken or bruised, the rhizome yields a red liquid. This is not its sap but a special secretion, or latex—a characteristic feature of the Poppy Family. The Indians used this red latex to paint themselves as well as their wares, and they dyed articles by boiling them in water together with the rhizomes.

In the early days of the white man's medicine in this country, physicians experimented with the alkaloid sanguinarine which occurs in the latex and discovered that it must be used with great caution to avoid fatalities. Still they employed it as an emetic and expectorant and for rheumatism, ulcers, and diseases of the liver.

See plate *44*.

FUMARIÀCEAE   Fumitory Family

Named after the European *Fumaria parvifolia* in this family, from L. *fumus*, smoke—so named, according to Pliny, because the juice of it when put into the eyes causes them to water as smoke does.

Smooth annual, biennial, or perennial herbs. Leaves alternate, all basal or borne on the stem, compound and/or much dissected. Flowers bisexual, irregular, in racemes or panicles. Sepals 2, scalelike. Petals 4, 1 or both of the outer ones spurred, the 2 inner ones cohering at their tips and enclosing anthers and stigma. Stamens 6, the filaments fused in 2 sets of 3, each set with a central 2-celled anther and 2 1-celled anthers. Pistil 1; ovary superior, 1-celled, with 2 to many ovules; style 1; stigma 2-lobed. Fruit a capsule (in ours) or an indehiscent, 1-seeded nut.

The family contains about 19 genera and 425 species, including the cultivated bleeding heart (*Dicentra spectabilis* Lem.), native to Japan.

Two genera occur in Kansas.

Various species of *Corydalis* and *Dicentra* have been shown to contain isoquinoline-structured alkaloids which cause livestock poisoning. Alkaloids of this group also occur in the Papaveraceae, a family closely allied to and sometimes combined with the Fumariaceae.

Key to Genera:
1. The 2 outer petals spurred or saccate at the base; flowers white or pink .... *Dicentra*
1. One outer petal spurred or saccate at the base; flowers yellow .... *Corydalis*

CORÝDALIS Medic.   Corydalis

NL., from Gr. *korydallis*, crested lark—there being a resemblance between the long spur of the European *Corydalis cava* (L. *cavus*, hollow, bent around—the spur being bent at the tip) and the hind claw of the European crested lark.

Delicate annual or biennial herbs. Leaves bipinnately or tripinnately compound, more or less finely dissected; lower leaves petioled, upper ones nearly or quite sessile. Flowers borne in terminal racemes. Sepals membranous, triangular or obcordate, often acuminate, the margins erose or coarsely toothed. Only the upper of the outer 2 petals spurred; inner 2 thickened at the tips and united as described for the family. Capsules with 2 lines of crested seeds. Seeds black, shiny, more or less round in outline, biconvex.

Bees—the principal insect visitors—press the inner petals down while getting nectar from the spur. This causes the anthers and stigma to protrude and touch the under side of their bodies, allowing cross-pollination to occur when the bee visits another flower. Self-pollination undoubtedly also takes place while the anthers and stigma are enclosed by the inner petals, but the flowers of most of the species are reported to be infertile to their own pollen.

Five species occur in Kansas. Only the 3 most common are included in the following key. In addition, *C. flavula* (Raf.) DC. may be found in the eastern 2 columns of counties and *C. aurea* Willd. var. *occidentalis* Engelm. is occasionally picked up in the extreme southwestern and south-central counties. *C. flavula* has smooth fruits with pedicels 10–15 mm long and often pendulous. *C. aurea* var. *occidentalis* is very similar to *C. curvisiliqua* var. *grandibracteata* but differs from the latter in having the seeds only faintly ornamented with minute punctae and the outer petals lacking a well-developed winglike margin.

Key to Species:
1. Fruit covered with translucent inflated hairs; plants of the southeast sixth of Kansas
.... *C. crystallina*

1. Fruit without inflated hairs 2

    2. Lower bracts mostly 4–7 mm long; sepals 0.25–1 mm long; seeds about 1.5 mm wide, only very faintly ornamented with concentric rings of minute punctae; primarily in the east third of Kansas .... *C. micrantha*

    2. Lower bracts mostly 12–15 mm long; sepals 2–3 mm long; seeds (1.5–) 1.8–2 mm wide, conspicuously ornamented (under a hand lens) with concentric rings of punctae; plants primarily of central and south-central Kansas .... *C. curvisiliqua* var. *grandibracteata*

150. *Corydalis crystallina*

C. crystallìna Engelm.   Mealy corydalis

L. *crystallinus*, made of or resembling crystals—because of the translucent inflated hairs on the capsules.

Yellow; late April–June

Roadsides, waste ground, fields, eroded or otherwise disturbed areas of prairies and open woods

Eastern third and Saline County

    Glaucous winter annual. Stems ascending, branching, mostly 2–5 dm tall. Leaf blades mostly 3–8 cm long. Bracts ovate to ovate-acuminate, the lower ones mostly 6–10 (–15) mm long, the upper ones gradually becoming shorter toward the apex. Flowers on short pedicels. Sepals 1–1.5 mm long. Petals yellow, the upper and lower petals prominently crested on the back; spurred petal 1.5–2 cm long. Capsules upright, nearly sessile, cylindrical, 12–14 mm long (excluding the persistent threadlike style), covered with translucent inflated hairs. Seeds about 2 mm wide, with concentric rings of minute punctae (use magnification).

C. curvisilìqua Engelm. var. grandibracteata Fedde

NL., from L. *curvus*, bent, + *siliqua*, pod, the fruits of this species being curved. The varietal name is NL., from L. *grandis*, large, + *bracteata*, from *bractea*, small leaf, in reference to the conspicuous bracts which subtend the individual flowers.

Yellow; May–June

Sandy soil of prairies, flood plains, and roadsides

Central and south-central Kansas and Meade and Morton counties

    Glaucous winter annual or biennial. Stems erect or ascending, up to 4 dm tall. Leaf blades mostly 1.5–9 cm long. Bracts elliptic to lanceolate or ovate, the lower ones mostly 12–15 mm long, the upper ones gradually becoming shorter toward the apex. Flowers on short pedicels. Sepals 2–3 mm long. Petals yellow, outer petals with a well-developed winglike margin, crested along the midrib or merely keeled; spurred petal mostly 15–18 mm long. Capsules upright, sessile or nearly so, straight or curved inward toward the axis of the inflorescence, 18–30 mm long (excluding the persistent threadlike style) at maturity. Seeds (1.5–) 1.8–2 mm wide, with concentric rings of minute punctae (use magnification).

151. *Corydalis curvisiliqua* var. *grandibracteata*

C. micrántha (Engelm.) Gray

NL., from Gr. *mikros*, small, + *anthos*, flower, i.e., small-flowered.

Pale yellow; April–May

Prairies, roadsides, flood plains

East half, more common eastward

    Glaucous or green winter annual. Stems erect or ascending, up to 3 dm tall. Leaf blades mostly 2–3 (–6) cm long. Bracts elliptic to ovate, the lower ones mostly 4–7 mm long. Flowers on short pedicels. Sepals mostly 0.25–1 mm long. Petals pale yellow; spurred petal mostly 9–14 mm long. Capsules upright, sessile or nearly so, straight or curved inward toward the axis of the inflorescence, mostly 5–13 mm long (excluding the persistent style) at maturity. Seeds about 1.5 mm wide, only very faintly ornamented with concentric rings of minute punctae (use magnification).

DICÉNTRA Bernh.   Bleeding heart

NL., from Gr. *di-*, prefix signifying twofold, or double, + *kentron*, spur—the 2 outer petals being each extended backward in the form of a spur.

    Delicate perennial herbs with biternately compound leaves. Flowers flattened, more or less heart-shaped, nodding, borne on an upright scape. The 2 outer petals spurred at the base and spreading above, the 2 inner petals thickened at the tip and cohering as described for the family. The plants contain various poisonous alkaloids and, when taken in quantity by cattle browsing early-spring woodland vegetation, may cause serious and even fatal poisoning.

    One species in Kansas.

D. cucullària (L.) Bernh. Dutchman's breeches

NL., from L. *cucullus*, hood or cowl—the tips of the inner pair of petals cohering over the style and stamens.

Pale pink or white; April–May

Rich, open woods

East third

152. *Dicentra cucullaria*

Delicate perennial springing from a cluster of small tubers about 5 cm below the ground surface. Leaves much dissected, all basal, with long petioles, mostly 3–10 cm long excluding the petiole and 4–16 cm wide. Flowers pale pink, nodding, borne in terminal racemes arising directly from the underground parts, 1.5–2.5 cm long. Capsule ellipsoid or fusiform, about 1 cm long excluding the style. Seeds black, shiny, resembling a football helmet in profile, about 1.5–2 mm long.

On close examination, one sees that each spur of the outer pair encloses 3 stamens, the filament of the middle one bearing a nectar gland that delivers nectar into the spur; the anthers are held close around the stigma under the cohering tips of the narrow inner petals. The anthers discharge their pollen over the stigma, but cross-pollination may also occur; for when bees are extracting nectar from the spurs, they press aside the inner pairs of petals, thus bringing their bodies in contact with the stigmas as well as with the anthers. Then, in going from flower to flower of the same plant and of different plants, they may effect cross-pollination.

See plate *45*.

CRUCÍFERAE Mustard Family

From the cruciform arrangement of the 4 petals; the name given to the family by Linnaeus. Some authorities use the name Brassicaceae for the sake of uniformity. The International Code of Botanical Nomenclature permits the use of either name.

Annual, biennial, or perennial herbs, usually with a pungent taste. Leaves mostly alternate, simple (rarely compound), estipulate. Flowers bisexual, regular, in terminal and sometimes axillary clusters. Sepals and petals 4 each, free from one another, the petals arranged in the form of a cross and often "clawed." Stamens 6, 4 long and 2 short (rarely 2 long and 4 short or fewer than 6; usually 2 in our species of *Lepidium*), usually subtended by minute nectar glands (use high magnification). Pistil 1; ovary superior, 2-celled (rarely 1-celled), the dividing wall between the cells being referred to as the "septum"; style 1 or obsolete; stigmas 1–2, when 1 often 2-lobed. Fruit an elongate pod (silique) or a short rounded or flattened pod (silicle), often tipped with the persistent style (referred to as the "beak") and/or stigma.

A family of about 350 genera and 2,500 species, primarily of northern temperate regions, including a variety of plants cultivated in gardens, fields and greenhouses: turnips (*Brassica rapa* L.), rutabaga (*Brassica napobrassica* Mill.), Brussels sprouts (*Brassica oleracea* L. var. *gemmifer* Zenker), cauliflower and broccoli (both forms of *Brassica oleracea* var. *botrytis* L.), various mustards (*Brassica* spp.), radish (*Raphanus sativus* L.), candytuft (*Iberis* spp.), stocks (*Matthiola* spp.), sweet alyssum (*Lobularia maritima* Desv.), and many others. *Hesperis matronalis* L., mother-of-the-evening, is a conspicuous purple-flowered ornamental occasionally seen as an escape in eastern Kansas.

Kansas has 26 genera with species which are either native or introduced and naturalized. In addition to these, several others are occasionally found as short-term garden escapes. The most common genera are included in the following key.

Note to the collector: Representatives of this family are among our first-blooming spring plants, and many are conspicuous at that time in vacant lots and other waste areas near human dwellings. For these reasons, impatient botanists, eager to begin the season, often collect immature or atypically small specimens. Since the earliest-blooming individuals often differ significantly in general aspect from typical later-blooming specimens and because mature fruits are used heavily for key characters, one may avoid needless difficulty in identifying mustards by waiting for mature full-sized plants. Also, members of this family (and others) will occasionally bloom again in late summer or fall when day length and other climatic conditions approximate those of spring, and these plants may be somewhat atypical in form and size.

Key to Genera (3):

1. Fruit flattened at right angles to the septum separating the 2 cells of the fruit, the septum therefore much narrower than the width of the fruit, its position indicated by a nerve

153. *Alliaria officinalis*

154. *Arabis laevigata*

the middle, the terminal segment much the largest, acuminate, sharply serrate; flowers commonly pale violet .... *Iodanthus*
29. Lobing of the leaf not so restricted, the lateral and terminal segments about the same size and shape, or the terminal segment broad and not sharply serrate 30
    30. Cauline leaves entire or with a few low teeth .... *Arabis*
    30. Cauline leaves pinnatifid or pinnate 31
        31. Seeds narrowly winged .... *Arabis*
        31. Seeds wingless .... *Cardamine*
26. Principal leaves entire or merely toothed 32
    32. Fruit distinctly flattened 33
        33. Siliques elliptic; seeds wingless .... *Draba*
        33. Siliques linear; seeds with a membranous wing (except in *A. shortii*, which has fruits 1.5–4 cm long) .... *Arabis*
    32. Fruit round or 4-angled in cross section 34
        34. Cauline leaves auriculate-clasping (with ear-shaped lobes clasping the stem) or cordate-clasping at the sessile base 35
        35. Fruits strictly appressed; foliage more or less pubescent .... *Arabis*
        35. Fruits widely spreading; plants glabrous or nearly so .... *Iodanthus*
        34. Cauline leaves not clasping at the base, sessile or petioled 36
        36. Cauline leaves deltoid, with the odor of onions .... *Alliaria*
        36. Cauline leaves circular to ovate or oblong, lacking the odor of onions 37
            37. Cauline leaves entire or sinuate or with 1–6 low teeth on each side .... *Cardamine*
            37. Cauline leaves with many sharply projecting teeth ...... *Iodanthus*

ALLIÀRIA Heister ex Fabricius   Garlic mustard

NL., from *Allium*, the genus to which the onion belongs, from the odor of the crushed herbage.

Erect biennial or perennial herbs with the odor of onions. Leaves petiolate, triangular to heart-shaped in our species. Flowers white, borne in terminal and axillary clusters. Petals spoon-shaped, gradually narrowed toward the claw. Filaments flattened. Short stamens each subtended by a ring-shaped gland; each pair of long stamens with a 3-angled gland. Pistil cylindric. Siliques 4-angled, several-seeded.

One species in Kansas.

A. officinàlis Andrz.   Garlic mustard

NL. *officinalis*, pertaining to drugs and medicines, from L. *officina*, laboratory, such as one where drugs are prepared for use. As a plant name, *officinalis* or *officinale* indicates that the plant is or was formerly kept for medicinal use by pharmacists.
White; April–June
Low, moist woods, primarily
East third; intr. from Europe

Biennial. Stems but little branched, if at all, up to 1 m tall. Blades of main leaves mostly 2.5–10 cm long (measured from the distal end of the petiole to the apex of the blade) and 2.5–10 cm wide; apex short-acuminate to acute or rounded; base cuneate or short-attenuate; margins coarsely toothed or crenate. Sepals lanceolate to oblong, 2.5–3 mm long, the apex slightly hooded. Petals white, 3–5 mm long. Siliques 2.5–5 cm long at maturity, spreading. Seeds dark brown, more or less oblong to ellipsoid, longitudinally ridged, 3–3.5 mm long.

ÀRABIS L.   Rock cress

From Gr. *arabis*, Arabian.

Annual, biennial, or perennial herbs. Basal leaves petioled, upper leaves sessile. Flowers white, cream, or pink, borne in terminal and axillary racemes. Filaments slender. Glands various. Pistil cylindric. Siliques more or less flattened, many-seeded.

Kansas has 5 species, 3 of which are included in the following section. *A. laevigata* (Muhl.) Poir., which may have leaf bases either auricled or not, is known from Cherokee, Crawford, and Douglas counties. It differs from *A. shortii* and *A. canadensis* in having leaves glabrous rather than pubescent, and from *A. virginica* in having entire leaves and in having fruits longer than 5 cm which are usually recurved or reflexed at maturity. *A. hirsuta* (L.) Scop. is known from Republic, Marshall, Riley, and Pottawatomie counties. It differs from our other species in having the mature fruits quite erect or appressed to the axis of the inflorescence.

Key to Species:

1. Cauline leaves auricled or sagittate at the base; plants much branched from the base ......
.... *A. shortii*
1. Cauline leaves not auricled or sagittate at the base; plants branched or unbranched 2
    2. Stems much branched from the base; leaves pinnatifid, divided nearly to the midrib; petals 2–2.5 mm long; fruits less than 5 cm (mostly 1.5–2.5 cm) long, ascending ....
.... *A. virginica*
    2. Stems usually unbranched, sometimes with a few branches above; leaves entire or with dentate or serrate margins; petals 4–7 mm long; fruits 4–9 mm long, widely spreading, recurved, or reflexed at maturity .... *A. canadensis*

A. canadénsis L.   Sicklepod rockcress

L. -*ensis*, suffix added to nouns of place to make adjectives—"Canadian," in this instance.
White; mid-April–June
Wooded hillsides and bluffs
East fourth

Annual or biennial. Stems erect, up to about 1 m tall. Basal leaves obovate to oblanceolate or spatulate, those borne farther up the stem oblanceolate to elliptic or lanceolate; blade mostly 0.3–1.3 (–2.3) cm long and 0.5–4 cm wide; apex rounded or acute; base acute to acuminate; margins remotely dentate, coarsely serrate, or nearly entire. Herbage smooth or sparsely hairy. Flowers white. Sepals oblong or broadly elliptic, 3–4 mm long. Petals narrowly oblong or narrowly lanceolate, 4–7 mm long. Glands 4, triangular, situated between the long and short stamens. Pedicels spreading widely at flowering time. Siliques linear, flattened, more or less curved, 4–9 mm long, becoming widely spreading, recurved, or reflexed at maturity. Seeds orange, more or less ovate or oblong, flattened, with a broad membranous margin, mostly 3–4 mm long (including the wing).

A. shòrtii (Fern.) Gleason

Named in honor of Charles W. Short (1794–1863), medical botanist and professor at the University of Kentucky.
White; April–May (–August)
Open woods, especially in alluvial soil
East fourth, primarily

Annual. Stems 2–5 dm tall, usually much branched from the base, the branches erect to ascending. Basal leaves oblanceolate to obovate or nearly round in outline; those borne farther up the stem oblanceolate or elliptic, mostly 1.5–4.5 cm long and 0.5–1.8 cm wide, with auriculate bases and coarsely serrate margins. Stems with simple and branched hairs; upper surface of leaves with scattered simple hairs; lower surface of leaves with branched hairs (use high magnification). Flowers white, sometimes tinged with purple. Sepals lanceolate to oblong or long-triangular, 1.5–2 mm long. Petals linear or narrowly oblanceolate, 2–3 mm long. Glands minute (or absent?). Pedicels spreading or somewhat ascending in flower and in fruit. Siliques linear, straight or somewhat curved, 1.5–4 cm long. Seeds wingless, about 1 mm long.

A. virgínica (L.) Poir.

"Of Virginia."
White; April–early May
Moist places in open woods, prairies, pastures, roadsides, and waste ground
South two-thirds of the east half

Annual or biennial. Stems ascending or decumbent, branched from the base, up to 4 dm tall, with spreading hairs near the base, sparsely hairy or glabrous above. Leaves finely pinnately lobed nearly to the midrib, 1.5–5 cm long and 0.5–1.5 cm wide. Flowers white, sometimes tinged with pink or purple. Sepals elliptic, oblong, or narrowly oblong, 1–1.5 mm long. Petals narrowly oblanceolate to narrowly oblong or linear, mostly 2–2.5 mm long. Glands minute. Pedicels ascending in flower and in fruit. Siliques linear, flattened, slightly curved, 1.5–2.5 cm long. Seeds orange, flattened, more or less round or ovate in outline, about 1 mm long, with a narrow membranous wing.

This species is intermediate in characters between other species of *Arabis* and members of the genus *Cardamine*. For this reason, it is sometimes placed in a separate genus, *Sibara*, the scientific name of which is an anagram of *Arabis*.

BARBARÈA R. Br.   Winter cress

According to Fernald (9), "Anciently called the Herb of St. Barbara, the seed of *B. verna* being sown in western Europe near St. Barbara's Day in mid-December."

155. *Barbarea vulgaris*

156. *Camelina microcarpa*

Biennial or perennial herbs with bright-green foliage. Basal leaves overwintering, petiolate, pinnatifid; cauline leaves (those borne higher on the stem) entire to pinnatifid, sessile, the bases clasping the stem. Flowers yellow, borne in terminal and axillary racemes. Petals spoon-shaped to obovate. Filaments slender, that of each short stamen subtended by a semicircular gland; each pair of long stamens with a short, erect gland. Ovary cylindric, gradually tapering to the slender style. Silique round or slightly 4-angled in cross section, several-seeded.

One species in Kansas.

B. vulgàris R. Br. Common winter cress, yellow rocket

L. *vulgaris*, common.
Yellow; April–May
Roadsides and waste ground
Eastern fourth and Saline County, probably in other counties as well

Perennial. Stems erect, usually branched, up to about 8 dm tall. Basal leaves 3–15 cm long, with 1–4 pairs of small lateral lobes and a larger ovate or nearly round terminal lobe; those leaves borne farther up the stem mostly 1–12 cm long, becoming smaller and less deeply lobed toward the inflorescence, with the uppermost merely coarsely toothed rather than lobed. Flowers closely crowded in the racemes at anthesis. Sepals lanceolate to narrowly lanceolate or elliptic, 2–3.5 mm long. Petals mostly 6–8 mm long. Siliques thick-walled, spreading widely to ascending or erect at maturity, 1.5–3 cm long, terminated by the slender persistent style or "beak" 1.5–3 mm long. Seeds oblong, 1–1.5 mm long.

The leaves and flower buds of this plant may be cooked and eaten early in the spring, providing some of the earliest of our wild mustard greens. The flavor is very strong, so unless one enjoys other strong vegetables (such as turnip greens), he may wish to use several cooking waters.

CAMELÌNA Crantz False flax

NL., from Gr. *chamai*, dwarf, + *linon*, flax—suggested by the slender, ascending stem, the form of the upper leaves, and the oily seeds.

Erect annual herbs. Basal leaves spoon-shaped; upper leaves sessile, linear to lanceolate with sagittate bases clasping the stem. Stems and leaves with both simple and branched hairs. Flowers yellow, borne in terminal racemes. Petals spoon-shaped. Filaments slender. Short stamens each with a pair of semicircular glands. Silicle obovoid, slightly flattened, with a narrow thickened margin, 8- to 24-seeded, terminated or "beaked" by the persistent style.

Two species in Kansas. *Camelina sativa* (L.) Crantz, a European cultivar, is rarely collected and is not included here.

C. microcárpa Andrz. Littlepod false flax

NL., from Gr. *mikros*, little, + *karpos*, fruit.
Pale yellow; late-April–June
Roadsides, prairie, waste ground
East half, primarily; intr. from Europe

Slender rough-pubescent annual. Stems erect, unbranched or with a few sharply ascending branches, up to about 7 dm tall. Basal leaves spatulate or narrowly oblanceolate, 2–8 cm long, with rounded or acute apexes, often gone by the later stages of flowering; upper leaves 1–10 cm long. Sepals narrowly elliptic to oblong, 2–2.5 mm long, with branched hairs and membranous margins. Petals pale yellow, often drying white, 3–4 mm long. Silicle 5–6 mm long (excluding the style) and 3–4 mm wide at maturity. Seeds orange-brown, often more or less oblong in outline but quite variable in shape, wingless, minutely roughened (use high magnification), mostly 1–1.25 mm long.

CAPSÈLLA Medic. Shepherd's purse

NL., from L. *capsa*, box or satchel, + L. diminutive suffix *-ella*, in reference to the fruits.

Erect, branched annual or biennial herbs. Leaves mostly basal, those on the stem much reduced. Flowers white (in ours) or pink, borne in terminal racemes. Petals obovate. Filaments slender. Short stamens each with a pair of minute glands. Pistil flattened. Silicle strongly flattened at right angles to the septum, obcordate to obtriangular, several- to many-seeded.

One species in Kansas.

C. búrsa-pastòris (L.) Medic. Shepherd's purse

NL., from L. *bursa*, purse (from Gr. *byrsa*, hide, skin, or leather), + *pastor, pastoris*, shepherd, the fruits of this species resembling small pouches.

White; late March–June (occasionally on into summer and fall)

Roadsides, fields, gardens, waste places

East two-thirds and a few northwest counties; intr. from Eurasia

157. *Capsella bursa-pastoris*

Winter annual. Stems erect, simple or branched above and/or from the base, up to about 6 dm tall. Basal leaves petioled, spatulate, oblanceolate, or elliptic in outline, 2–10 (–14) cm long, with margins occasionally entire but usually coarsely dentate or serrate or sharply pinnately lobed; stem leaves sessile, oblanceolate to elliptic or narrowly lanceolate, much smaller than the basal leaves, with auriculate bases and entire or coarsely to shallowly serrate margins. Stems and leaves with simple and branched hairs. Flowers white. Sepals broadly elliptic to oblong, 1.5–2 mm long, often tinged with purple, with a white membranous margin. Petals obovate, 1.5–2.5 mm long. Silicles triangular-obcordate, 4–8 mm long, truncate or shallowly notched at the apex and tipped by the very short style. Seeds orange-brown, oblong, wingless, about 1 mm long.

This plant is reputedly rich in vitamins C and K and in sodium and sulphur. The leaves may be used fresh in green salads, or the entire plant may be cooked in a small amount of boiling salted water. The seeds and young pods may be used as seasoning in soups and stems.

In southern Europe and southwest Asia, shepherd's purse has been prized since ancient times for its medicinal value. Dioscorides and Pliny (1st century) and Galen (2nd century) recommended the use of the seeds as a purgative, aphrodisiac, and abortifacient, and as a cure for sciatica, internal abscess, biliousness, and scant menstruation. The herbals of the Middle Ages advised the use of the whole plant for various other medicinal purposes, including aches and pains of the stomach and bowels, enlargement of the spleen, jaundice, gallstones, kidney stones, dysentery, scrofula, bladder troubles, gout, tumors, and abscess of the ear, throat, and neck, and as a hemostatic for stopping the flow of blood in cases of external and internal hemorrhage. However, at the close of the Middle Ages, and as the practice of medicine became more scientific, shepherd's purse was neglected and finally banished from the pharmacopoeias; but the people kept on using it as a home remedy. Then, during World War I when official hemostatics were running low in Germany, at a time they were most needed, search was made by physicians for suitable substitutes among the native flora. As a result, shepherd's purse was restored to its ancient place among preferred remedies for hemorrhage.

158. *Cardamine parviflora*

CARDÁMINE L. Bitter cress

NL., from Gr. *Kardamon*, a name used by Dioscorides for some cress.

Annual, biennial, or perennial herbs. Leaves simple and entire or pinnately lobed, or pinnately compound. Flowers white (in ours), borne in terminal and axillary racemes or panicles. Petals obovate or spoon-shaped, rarely absent. Filaments slender. Short stamens each with a semicircular gland. Pistil cylindric. Siliques slightly flattened, several- to many-seeded.

Kansas has 3 species, but 2 of these (*C. bulbosa* (Schreb.) B. S. P. and *C. pensylvanica* Muhl.) are rarely collected and are omitted here.

C. parviflòra L. Small-flowered bitter cress

NL., from L. *parvus*, little, + *flos, floris*, flower, i.e., having small flowers.

White; April–May

Open woods and prairies, often in moist places

East fifth, more common southward

Annuals or biennials with fibrous roots. Stems erect, unbranched or branched above and/or below, up to about 2.5 dm tall. Leaves deeply pinnately lobed to pinnately compound, mostly 2–4 cm long; lobes or leaflets of basal leaves varying from oblanceolate to obovate, nearly round, or "mitten-shaped"; lobes or leaflets of the stem leaves narrowly oblanceolate to linear. Flowers borne on spreading pedicels. Sepals oblong to linear, sometimes tinged with purple, mostly 1–2 mm long. Petals white, spatulate, 2–3.5 mm long. Fruits slightly curved or straight, erect, many-seeded, mostly 1.7–2.5 cm long at maturity. Seeds orange-brown, oblong, about 1 mm long, lacking a membranous wing or margin.

This species also occurs in the Old World. The American plants are segregated as *C. parviflora* L. var *arenicola* (Britt.) Schultz.

MUSTARD FAMILY

133

159. *Dentaria laciniata*

160. *Dentaria laciniata,* underground parts

CARDĀRIA Desv.

NL., from Gr. *kardia*, heart, from the heart-shaped fruit of the first-named species of this genus.

Perennial, branched herbs. Basal leaves petiolate, the margins sinuate, toothed, or pinnatifid, early deciduous; stem leaves sessile, toothed, with sagittate lobes clasping the stem. Flowers white, borne in terminal panicles. Petals with an obovate blade narrowed to a basal "claw." Filaments slender. Short stamens each with a pair of minute glands, and a single gland subtending each pair of long stamens. Pistil flattened. Silicle indehiscent, flattened somewhat at right angles to the septum, obcordate, 1- to 2-seeded.

One species in Kansas.

C. dràba (L.) Desv.   Hoary cress

Named after the genus *Draba*, also of the Mustard Family, in which early botanists included this species.
White; April–June
Roadsides, waste ground, prairie ravines
East two-fifths and Lane County; intr. from Europe

Perennials arising from rhizomes. Stems erect, up to 6 dm tall. Leaf blades oblanceolate below to elliptic or oblong above, 1.5–12 cm long and 0.5–4 cm wide, gradually reduced toward the inflorescence. Stems and under surface of leaves with simple hairs. Flowers white, crowded in the raceme. Sepals oblong to ovate, 1.25–2 mm long, with a white membranous margin. Petals 2.5–3.5 mm long. Silicle round or somewhat obcordate in outline, about 4 mm long at maturity (excluding the style), tipped by the persistent style and prominent stigma. Seeds orange-brown, ovate to elliptic, flattened, wingless, about 1.5 mm long.

DENTĀRIA L.   Toothwort

NL., from L. *dens*, tooth, in reference to the shape of the segments of the rhizome.

Low, erect perennials arising from rhizomes. Leaves 3, whorled, palmately divided, petioled. Flowers white to pale blue or pale pink, borne in a terminal raceme. Petals obovate to spoon-shaped. Filaments slender. Short stamens each subtended by a semicircular gland. Pistil cylindric. Siliques somewhat flattened, many-seeded.

One species in Kansas.

D. laciniàta Muhl.   Cutleaf toothwort

NL., from L. *lacinia*, fringe, in reference to the narrow segments of the leaves.
White to pale pink; April
Rich, moist woods
East fourth

Perennial from a deep-seated, fusiform tuberous rhizome. Stems single, unbranched, mostly 1.5–3 dm tall. Leaf blades mostly 4–6.5 mm long, divided nearly to the base, the lobes coarsely toothed and often with 1–4 smaller lateral lobes. Raceme mostly of 6–15 flowers. Sepals mostly oblong, 5–7 mm long, with a rounded or occasionally acute apex and membranous margin. Petals white to pale pink, 1–1.7 cm long. Siliques 2.5–4 cm long and 2–2.5 mm wide at maturity, tapering gradually into the slender beak. Seeds orange-brown, oblong, slightly flattened, about 2.5 mm long.

This delicate plant is one of the earliest to bloom in the woods of the eastern part of the state, along with the bloodroot, Dutchman's breeches, and golden corydalis.

DESCURĀINIA Webb & Berth.   Tansy mustard

NL., in honor of François Déscourain (1658–1740), French apothecary and botanist.

Erect annual or biennial herbs. Leaves 1, 2, or 3 times compounded. Flowers yellow, borne in terminal racemes. Petals obovate or spoon-shaped. Filaments slender. Glands extremely minute or absent. Pistil cylindric. Siliques more or less round in cross section, several- to many-seeded.

Two species in Kansas.

Key to Species:
1. Siliques club-shaped, mostly 6–11 mm long and 1–1.5 mm wide at maturity, seeds in 2 rows in each cell; sepals oblong or broadly elliptic, 1.25–1.5 mm long; inflorescence usually glandular .... *D. pinnata*
1. Siliques narrowly linear, mostly 1.5–3 cm long and 0.5–1 mm wide at maturity, with 1 row of seeds in each cell; sepals narrowly oblong to linear, (1.5–) 2–2.5 mm long; inflorescence not glandular .... *D. sophia*

D. pinnàta (Walt.) Britt.   Tansy-mustard

L. *pinnatus*, feathered, winged, from *pinna*, feather, wing, in reference to the dissected leaves.
Yellow; mid-April–early June
Roadsides, fields, waste ground, disturbed prairie
Throughout

Erect annual. Stems unbranched, or branched above and/or below, up to about 7 dm tall. Leaves pinnately compound, 1.5–10 cm long (including the petiole) below, gradually becoming smaller and less dissected toward the inflorescence; leaflets of the lower leaves pinnately lobed or dissected nearly or completely to the midrib. Stems with both branched and glandular hairs; leaves with branched hairs. Pedicels erect to ascending in flower, ascending to widely spreading in fruit. Sepals oblong or broadly elliptic, somewhat petal-like, 1.25–1.5 mm long, with membranous margins. Petals spoon-shaped, 1–1.5 mm long. Siliques club-shaped, ascending to erect, mostly 6–11 mm long and 1–1.5 mm wide at maturity, with 2 rows of seeds in each cell. Seeds orange-brown, more or less ellipsoid, wingless, about 1 mm long, with a minutely honeycombed surface (use high magnification).

161. *Descurainia pinnata*

D. sòphia (L.) Webb   Tansy-mustard

An old generic name.
Yellow; late-April–June
Dry, often sandy soil of flood plains, roadsides, waste ground, and disturbed prairie
Western three-fourths; intr. from Europe

Erect annual. Stems unbranched, or branched above and/or below, up to about 8 dm tall. Leaves pinnately compound, 2.5–11 cm long (including the petiole) below, gradually becoming smaller toward the inflorescence; leaflets lobed nearly or completely to the midrib. Stems glabrous or with branched hairs; leaves with branched hairs. Pedicels erect to ascending in flower, ascending to widely spreading in fruit. Sepals narrowly oblong to linear, green or slightly petal-like, (1.5–) 2–2.5 mm long, with very narrow membranous margins. Petals spoon-shaped, mostly 2–2.25 mm long. Siliques narrowly linear, straight, or slightly curved, mostly ascending and 1.5–3 cm long and 0.5–1 mm wide at maturity, with 1 row of seeds in each cell. Seeds orange-brown, ellipsoid, wingless, 1–1.2 mm long, the surface lustrous, only very faintly honeycombed, if at all (use high magnification).

162. *Descurainia sophia*

DRÁBA L.   Whitlow grass

NL., from the ancient name of a kind of cress, from Gr. *drabe*, sharp, burning—in reference to the pungent taste of the fruits and herbage.

Low annual, biennial, or perennial herbs. Leaves entire or toothed. Flowers yellow or white, borne in terminal racemes or panicles. Petals "clawed," the apexes rounded, emarginate, or 2-cleft; or petals linear or absent altogether. Filaments slender. Glands various. Pistil ovoid. Silicles flattened, mostly elliptic, several- to many-seeded.

Three species in Kansas.

Key to Species:

1. Petals 1.5–3 mm long, or absent; silicles mostly 3–4.5 mm long and about 1 mm wide; base of the inflorescence seldom more than 1 cm above the uppermost leaves .... *D. brachycarpa*
1. Petals 3–7 mm long (rarely absent); silicles 10–15 mm long and 1.5–3 mm wide; inflorescence borne atop a leafless peduncle, the lowest fruits usually 3–7 cm above the uppermost leaves 2
     2. Pedicels and axis of the inflorescence glabrous, often wiry .... *D. reptans*
     2. Pedicels and axis of the inflorescence hairy .... *D. cuneifolia*

D. brachycárpa Nutt.

NL., from Gr. *brachys*, short, + *karpos*, fruit, i.e., short-fruited.
Flowers white or greenish, inconspicuous; April
Rocky, open places in prairies; open woods; roadsides
East two-fifths, especially in the southeast eighth

Small annual or winter annual with a basal rosette of leaves. Stems erect, 3–18 cm tall, usually branched above and/or below. Basal leaves petiolate, the blades ovate to nearly round or obovate, 3–12 mm long, the margins entire or with a few teeth; stem leaves smaller, gradually becoming sessile and elliptic to oblong toward the inflorescence. Herbage and sometimes the sepals with branched hairs (use magnification). Flowers white or greenish, inconspicuous, borne in terminal racemes; pedicels ascending in flower, ascending or spreading widely in fruit. Sepals lanceolate to oblong or linear, (0.5–)

163. *Draba brachycarpa*

MUSTARD FAMILY

135

164. *Draba cuneifolia*

0.8–1 mm long, often purple, with a rounded apex and membranous margin. Petals spoon-shaped, 1.5–3 mm long, or absent. Silicles elliptic, mostly 3–4.5 mm long and about 1 mm wide at maturity. Seeds orange-brown, ellipsoid, flattened, wingless, about 0.8 mm long.

The apetalous blossoms of this species are cleistogamous—that is, they pollinate themselves before the buds open (from Gr. *kleistos*, closed, + *gamos*, marriage).

D. cuneifòlia Nutt.

NL., from L. *cuneus*, wedge, + *folium*, leaf, in reference to the leaves, which, although not
    wedge-shaped in their entirety, do have cuneate bases.
White; late March–mid-May
Dry, rocky or sandy soil, usually of prairies or roadsides
East third, primarily; more common southward

Annual or winter annual with a basal rosette of leaves. Stems erect, 0.4–1.8 dm tall, unbranched, or branched below and the branches ascending. Basal leaves sessile or with a very short petiole, the blades spatulate to oblanceolate or obovate, 0.6–3.5 cm long, and 0.3–1.6 cm wide, the margins entire or with 2–8 widely spaced teeth; stem leaves smaller, oblanceolate or obovate to rhombate or elliptic, gradually becoming sessile toward the inflorescence, the margins entire or with a few teeth. Herbage and sepals with branched hairs, sometimes with scattered simple hairs also (use high magnification). Flowers white; racemes elevated some distance above the leaves on a leafless peduncle; pedicels ascending in flower, ascending to spreading widely in fruit. Sepals lanceolate to elliptic or oblong, mostly 1.5–2.2 mm long, with a rounded apex, membranous margin, and simple and/or branched hairs. Petals spatulate, mostly 3.5–4 mm long, the apexes emarginate or 2-lobed. Silicles elliptic, 10–15 mm long and 2–3 mm wide at maturity, covered with simple appressed hairs. Seeds orange-brown, ellipsoid, flattened, wingless, about 1 mm long.

D. réptans (Lam.) Fern.

NL., from L. *repto, repo*, to creep, perhaps from the low habit of the plant.
White or greenish; April–May
Dry, rocky or sandy soil, primarily of prairies
East half and Meade County

Annual or winter annual with a basal rosette of leaves. Stems erect, 3–16 cm tall, usually branched above and/or below. Basal leaves sessile or with a very short petiole, the blades elliptic to oblanceolate or obovate, mostly 0.5–2 cm long and 0.2–1 cm wide, the margins usually entire, occasionally with a few teeth; stem leaves about the same size as or only slightly smaller than the basal leaves, ovate to elliptic or lanceolate, gradually becoming sessile toward the inflorescence, the margins usually entire, occasionally with a few teeth. Leaves and stem (excluding the peduncles) with both simple and branched hairs (use magnification). Flowers white or greenish; racemes elevated some distance above the leaves on a wiry, leafless, glabrous peduncle; pedicels ascending in flower, ascending to spreading widely in fruit. Sepals ovate to oblong or linear, mostly 1.5–2 mm long, often tinged with purple, with a rounded apex, membranous margin, and a few simple hairs. Petals "clawed," mostly 3–4 mm long, with an obovate blade and emarginate apex; or entirely absent. Silicles elliptic or linear, 10–15 mm long and 1.5–2.5 mm wide at maturity, glabrous or with appressed hairs. Seeds orange-brown, ellipsoid, flattened, wingless, about 0.8 mm long.

ERÝSIMUM L.   **Wallflower**

L. *erysimum*, from Gr. *erysimon*, ancient name of the hedge mustard, *Erysimum officinale*.

Erect annual, biennial, or perennial herbs. Leaves sessile, entire, toothed, or pinnately lobed. Flowers yellow to orange, borne in terminal racemes. Petals spoon-shaped or obovoid and gradually narrowed to the "claw." Filaments flattened, petal-like, and about the same width as the claws of the petals. Each short stamen with an annular or semicircular gland; each pair of long stamens subtended by a gland. Pistil linear. Silique more or less 4-angled, pubescent, many-seeded.

Two species in Kansas.

Key to Species:
1. Petals mostly 1–2 cm long; siliques about 2 mm wide at maturity, tipped by a prominent
    stigma; plants of western and north-central Kansas .... *E. asperum*
1. Petals mostly 7–8 mm long; siliques about 1 mm wide at maturity, the stigma incon-
    spicuous; plants primarily of eastern Kansas .... *E. repandum*

E. ásperum (Nutt.) DC.　　　　Plains erysimum, western wallflower

L. *asper, asperum*, rough, in reference to a character of the pods.
Yellow; late-April–June
Dry soil of prairies, pastures, roadsides
West third, and north half of the central third

Biennial. Stems erect, up to about 4 dm tall, longitudinally ridged, usually unbranched below the inflorescence. Leaves narrowly oblanceolate or linear below to linear or narrowly lanceolate above, 1.5–10 cm long and 0.2–1.5 cm wide; apex acute; margins entire or occasionally with a few widely spaced, outward-pointing teeth. Stems and leaves gray-green, densely covered with appressed, 2-pronged hairs (use magnification). Flowers yellow; inflorescence usually branched. Sepals oblong or linear, 8–10 mm long, with a rounded apex, pouchlike base, membranous margins, and appressed hairs. Petals mostly 1–2 cm long. Siliques 4-angled, longitudinally ridged, covered with hairs of the same type as found on the foliage, beaked by the style and prominent stigma, mostly 3.5–10 cm long and about 2 mm wide at maturity, ascending or widely spreading. Seeds orange, oblong, only slightly flattened if at all, 1.5–2 mm long, wingless or with a very narrow, crescent-shaped wing at the end opposite the point of attachment to the ovary wall.

165. *Erysimum asperum*

E. repándum L.

NL., from L. *repandus*, bent backward, turned up, undulate, from *pandus*, bent, crooked, curved,
　　　apparently in reference to the coarsely lobed or toothed leaf margins.
Yellow; April–June
Disturbed soil of fields, pastures, roadsides, and waste ground
East half and Meade and Barber counties; intr. from Europe

Annual. Stems erect, up to about 4 dm tall, unbranched or with ascending branches, longitudinally ridged but not as conspicuously so as the preceding species. Basal leaves (which are often gone at flowering time) entire or pinnately lobed with the sharp tips of the lobes pointing toward the base of the leaf; main stem leaves linear or very narrowly elliptic, mostly 2–8 cm long and 2–8 mm wide; apex acute; margins coarsely serrate or dentate or occasionally entire. Stems and leaves light green, with appressed, 2- and 3-pronged hairs (use magnification). Flowers yellow; inflorescence branched or unbranched. Sepals oblong or linear, 3–6 mm long, with a rounded or acute apex, membranous margins, and appressed hairs. Petals mostly 7–8 mm long. Siliques 4-angled, often slightly constricted between the seeds, covered with hairs of the sort found on the foliage, tipped by the inconspicuous stigma, mostly 4–8 cm long and about 1 mm wide at maturity, spreading widely or curving upward. Seeds orange-brown, oblong, slightly flattened, mostly 1.2–1.5 mm long, usually wingless but occasionally with a very narrow, crescent-shaped wing at the end opposite the point of attachment to the ovary wall.

166. *Erysimum asperum*

IODÁNTHUS T. & G.　　　　Purple rocket

NL., from Gr. *iodes*, violet-colored, + *anthos*, flower.

Erect perennials with pale violet flowers borne in terminal racemes. Petals triangular or obovate above and narrowed to a claw below. Filaments flattened and somewhat petaloid. Each short stamen subtended by an annular gland. Pistil cylindric. Siliques linear, round or nearly so in cross section, many-seeded.

One species in Kansas.

I. pinnatífidus (Michx.) Steud.　　　　Purple rocket

NL., having pinnately lobed leaves, from L. *pinnatus*, feathered, from L. *pinna, penna*, feather.
　　　In spite of the name, our plants usually have leaves with serrate, rather than pinnatifid,
　　　margins. Occasionally, 1 or 2 of the lowest leaves (which are usually gone at flowering
　　　time) will have 2 small, pinnate lobes at the base of the blade, but the rest of the margin
　　　will be serrate.
Pale lavender; May–June (occasionally blooming again in late summer or fall)
Moist woods in alluvial soil
East fifth, primarily; more common southward

Stems 1 or several from a short rootstock, unbranched or with a few branches above, up to 1 m tall. Leaf blades elliptic to lanceolate, 3–23 cm long and 1–6 cm wide; lower leaves tapering to a petiolelike base which has 2 auriculate lobes clasping the stem; upper leaves tending to have shorter "petioles" and/or to lack the clasping bases; margins wavy or serrate. Flowers pale lavender; inflorescence simple or branched. Sepals oblong

167. *Erysimum repandum*

MUSTARD FAMILY

137

to nearly linear, 3–6 mm long, with a membranous margin and resembling the petals in color. Petals 8–12 mm long. Siliques 2–4 cm long at maturity. Seeds oblong, wingless, 1–1.5 mm long.

168. *Iodanthus pinnatifidus*

169. *Lepidium oblongum*

LEPÍDIUM L.   Peppergrass

L. *lepidium*, ancient name of a plant, from Gr. *lepidion*, diminutive of *lepis*, scale, in reference to the fruit.

Annual, biennial, or perennial herbs. Leaves entire to toothed or pinnately lobed. Flowers small and inconspicuous, white or greenish, borne in crowded terminal racemes. Petals linear, spoon-shaped, or entirely absent. Stamens 6, or sometimes only 4 or 2; usually 2 in the species included here. Filaments slender. Two glands at the base of each stamen. Pistil flattened. Silicles flattened at right angles to the septum, ovate to circular or obovate with an apical notch, often winged, 2-seeded.

Five species in Kansas. The 3 most common are included in the following key.

Key to Species:
1. Petals present, 1–2 mm long .... *L. virginicum*
1. Petals absent or, if present, linear and much shorter than the sepals 2
   2. Upper stem leaves laciniate to pinnatifid; seeds wingless .... *L. oblongum*
   2. Upper stem leaves serrate to entire; seeds very narrowly winged (use magnification) ....
       .... *L. densiflorum*

L. densiflòrum Schrad.   Peppergrass

NL., from L. *densus*, thick, + *flos, floris*, flower, in reference to the crowded inflorescence.
Flowers greenish, inconspicuous; May–June, occasionally again in the fall
Dry, sandy or rocky soil of fields, roadsides, and waste ground
Throughout; intr. from Europe

Erect, bushy-branched annual up to about 5 dm tall. Leaves petioled below, gradually becoming sessile toward the inflorescence; basal leaves oblanceolate in outline, 1–11 cm long, the margins serrate to pinnately lobed and then the lobes usually with several teeth, often gone at flowering time; main stem leaves narrowly oblanceolate to linear, mostly 1.3–4 cm long, with serrate or entire margins, deciduous. Flowers greenish. Sepals oblong or broadly elliptic, often tinged with purple, about 0.8 mm long, with a rounded apex and membranous margin. Petals absent or linear and much shorter than the sepals. Silicles 2.5–3.5 mm long, with a narrow wing on the distal end. Seeds orange, flattened, ovate or shaped like the letter "D" in outline, narrowly winged (use magnification), mostly 1.5–1.8 mm long.

L. oblóngum Small

L. *oblongus*, longer than broad, the specific reference in this case uncertain, perhaps to the shape of the stem leaves, which sometimes are oblong.
Flowers greenish, inconspicuous; mid-April–June, occasionally again in the fall
Dry, sandy or rocky soil of roadsides, pastures, and waste ground
Central three-fourths

Annual. Stems much branched from the base, ascending or sprawling, up to 2 dm high. Leaves petioled below, becoming sessile toward the inflorescence; basal leaves petioled, mostly 2–4 cm long (including the petiole), oblanceolate or spatulate in outline, pinnately lobed, the lobes usually with 2–4 sharp teeth; stem leaves sessile, mostly 0.5–2 cm long, variable in outline, the lower ones pinnately lobed or merely lacerated, the upper ones sharply toothed. Flowers greenish. Sepals oblong to lanceolate, often tinged with purple, about 0.8 mm long, with an acute or rounded apex and membranous margin. Petals absent, or linear and much shorter than the sepals. Silicles 2.5–3.5 mm long, with a narrow wing at the distal end. Seeds orange, flattened, ovate or shaped like the letter "D" in outline, wingless, mostly about 1.5 mm long.

L. virgínicum L.   Poor man's pepper

"Of Virginia."
Flowers greenish-white, inconspicuous; late April–June, sometimes again in late summer and fall.
Prairie, roadsides, pastures, waste ground
East half, primarily

Erect annuals or biennials. Stems usually branched above, occasionally from the base, up to 5 dm tall. Leaves petioled below, gradually becoming sessile toward the inflorescence, 1.5–5 cm long; basal leaves usually pinnately lobed, the lobes sharply toothed or occasionally rounded; stem leaves oblanceolate below to very narrowly oblanceolate or linear above, the margins serrate or entire. Flowers greenish-white. Sepals oblong, often tinged with purple, about 1 mm long, with a rounded apex and membranous margin. Petals spoon-shaped, 1–2 mm long. Silicles about 3 mm long, tending to be concavo-convex with a sharp keel, but a distinct wing scarcely evident. Seeds orange, flattened, ovate or shaped like the letter "D" in outline, narrowly winged (use magnification), about 1.5 mm long.

LESQUERÉLLA S. Wats.   **Bladderpod**

Named after Leo Lesquereux, Swiss-born American botanist (1806–1889), assistant to Agassiz at Harvard and writer on fossil flora and North American mosses.

Annual, biennial, or perennial herbs with stellate hairs. Leaves entire or toothed. Flowers yellow, borne in terminal racemes. Petals obovate to spoon-shaped. Glands 2 at base of short stamens. Ovary globose or nearly so, often elevated on a short stalk (stipe). Silicle ellipsoid, obovoid, or globose, beaked by the slender style, few- to several-seeded. Seeds wingless.

Six species reported for Kansas. The 3 most widely distributed species are included in the following key. *L. ludoviciana* (Nutt.) S. Wats. is known from Cheyenne, Sherman, Logan, and Ellis counties only. It is a perennial with aboveground parts resembling our annuals *L. gracilis* and *L. gordonii*, but has a woody caudex somewhat like that in *L. ovalifolia*. It differs from all our species in having fruiting pedicels that arch downward and silicles covered with stellate hairs.

Key to Species (17):
1. Perennials from a woody caudex .... *L. ovalifolia*
1. Annuals 2
    2. Fruits pear-shaped, usually 5–7 mm long; fruiting pedicels ascending or spreading; plants primarily of the southeast one-eighth of Kansas .... *L. gracilis*
    2. Fruits spherical or nearly so, 3–5 mm long; fruiting pedicels usually becoming sigmoid; plants of southern and western Kansas .... *L. gordonii*

170. *Lepidium virginicum*

L. gòrdonii (Gray) Wats.

Named after Alexander Gordon, a Scottish-American botanist who botanized in present Wyoming Colorado, New Mexico, and Nebraska (1843–1848) and who first collected this plant.
Yellow; late April–mid-June
Sandy prairie
Southern Kansas, primarily, from Seward to Kiowa and Barber counties; also Trego and Ellis counties

Annuals. Stems erect or ascending, usually branched from the base, 6–24 cm tall. Leaves petioled below, sessile above; blades elliptic, linear, or spatulate, mostly 0.5–3 cm long and 1–5 (–8) mm wide; apex acute or somewhat rounded; base attenuate; margins entire or occasionally with a few teeth. Herbage gray-green due to the stellate pubescence. Sepals lanceolate, 3.5–4.5 mm long. Petals 4–7 mm long. Silicle globose, glabrous, 3–5 mm long (excluding the style); fruiting pedicels eventually developing a sigmoid curve. Seeds orange-brown, flattened, oval to nearly round in outline with a notch in 1 side, mostly 1.5–1.8 mm long.

171. *Lesquerella gordonii*

L. grácilis (Hook.) S. Wats.   **Slender bladderpod**

L. gracilis, slender.
Yellow; April–May (–June)
Prairie, usually in rocky soil
Southeast eighth, primarily

Annuals. Stems erect or ascending, usually branched from the base, 6–25 cm tall. Leaves petioled below, sessile above; blades elliptic to oblanceolate or spatulate (occasionally lyrate) below; narrowly oblanceolate to linear above, mostly 0.5–5 cm long and 1–7 mm wide; apex acute or rounded; base attenuate; margins entire or with 2–4 (–6) widely spaced teeth. Herbage light- or gray-green, sparsely to densely covered with minute stellate hairs. Sepals lanceolate to oblong with an acute apex, 3–5 mm long. Petals 5–9 mm long. Silicles more or less pear-shaped, glabrous, 5–7 (–10) mm long (excluding the style); fruiting pedicels ascending or spreading. Seeds orange-brown, flattened, oval to nearly round in outline, 2–2.5 mm long.

Our plants are recognized as ssp. *nuttallii* (Torr. & Gray) Rollins & Shaw.

172. *Lesquerella gracilis*

MUSTARD FAMILY

139

L. ovalifòlia Rydb.

Oval-leaf bladderpod

NL., from L. *ovum*, egg, hence egg-shaped or oval, + *folium*, leaf.
Yellow; mid-April–May
Rocky prairie hillsides and escarpments
West half

Perennial from a branched woody caudex. Stems erect, mostly 0.7–2 dm tall. Leaves petioled below, sessile above; blades of lower leaves nearly round in outline to broadly elliptic, obovate, or spatulate, 0.5–2 cm long and 0.3–1 cm wide; those of the upper leaves mostly narrowly spatulate or linear, 0.5–2 cm long and 1–3 mm wide; apex acute or rounded; base attenuate; margins entire. Herbage gray-green, densely covered with parasol-like stellate hairs. Sepals lanceolate or oblong, 5–6.5 mm long, somewhat saccate at the base, with an acute apex and membranous margin. Petals 9–14 mm long. Silicles globose, glabrous, 4–7 mm long (excluding the style); fruiting pedicels usually ascending. Seeds orange-brown, flattened, nearly round in outline with a notch in 1 side, sharply keeled around the circumference, mostly 2–2.5 mm long.

173. *Lesquerella ovalifolia*

NASTÚRTIUM R. Br.

Watercress

Name used by Theophrastus (died 285 B.C.), and Dioscorides and Pliny (1st century), from L. *nasus*, nose, + *torquere*, *tortum*, to twist—indicating our reaction to the pungent taste.

Ours glabrous, aquatic perennial herbs. Leaves pinnately compound. Flowers white, borne in terminal racemes. Petals obovate, "clawed." Filaments slightly dilated. Short stamens each with a pair of horseshoe-shaped glands. Pistil cylindric; stigma 2-lobed. Siliques linear, nearly round in cross section, many-seeded. Seeds in 2 rows in each cell, wingless.

One species in Kansas.

N. officinàle R. Br.

True watercress

NL. *officinalis*, pertaining to a workshop, from L. *officina*, workshop—in reference to the fact that the plant was kept at apothecary shops for medicinal uses.
White; April–November
In and around springs, spring-fed streams, and ditches
Scattered throughout; intr. from Europe

174. *Nasturtium officinale*

Smooth, succulent, pungent, floating or creeping perennial. Stems branched, rooting freely at the nodes. Although the leaves appear to be compound and petiolate, the base of the "petiole" is often auriculate, as is the base of the blade in many other representatives of the Mustard Family. Leaflets 3–11, elliptic to ovate, obovate, obcordate, or round in outline, with entire or wavy margins; the terminal leaflet 0.6–3 (–4.5) cm long and 0.5–2.5 cm wide, larger than the lateral leaflets. Sepals lance-oblong, 2–2.5 mm long, the outer 2 pouchlike at the base. Petals white, often veined with purple, 3–4.5 mm long. Filaments often purple. Siliques straight or slightly curved, 1–2 cm long (including the short beak). Seeds orange-brown, broadly elliptic to oblong or nearly round in outline, slightly flattened, about 1 mm long, with reticulate ornamentation.

The anthers of the 4 long stamens reach the height of the stigma and apparently effect self-pollination. There is, however, a chance of cross-pollination by bees, flies, and beetles that work the flowers for nectar and pollen.

Inspecting the habit of growth of the watercress, one sees that young stems in the water produce nodal adventitious roots extending into the water only, but that when grown old the stems sink to the bottom and form new roots anchoring them to the soil, in which condition they overwinter. In the spring they produce from axillary buds new foliaceous and flowering shoots with water roots at the nodes.

The early New Englanders used watercress as they and their ancestors had been accustomed to do in the homeland, for a salad, potherb, and garnish, and also medicinally for its antiscorbutic potency. Chemical analysis of the fresh herbage and seeds discloses a glucoside containing sulphur and nitrogen, which is soon broken down by its associate enzyme into a pungent volatile oil and grape sugar, and other compounds, including a large percentage of vitamins A and C and a smaller amount of D. These things taken internally might be expected to stimulate physiological processes; still we read with astonishment the extravagant claims for the therapeutic value of watercress that sprang up in the Middle Ages and have been repeated in popular herbals even to the present time. As an example, we give a somewhat abridged translation from the herbal of Hieronymus Bock (Strasbourg, 1577):

"All cresses taken internally kill worms, rectify the intestines, kidneys, and bladder, open the liver and spleen, heal and purify internal wounds, taken with olive oil eliminate poison. Cresses boiled in honey and wine, drunk before breakfast and evenings, dissolve obstinate phlegm and relieve coughing and difficult breathing. Seeds of cress taken with wine expell the afterbirth. Cresses boiled in goat's milk and drunk relieve pain in the chest. The juice of cresses held in the mouth clears out excreta from decayed teeth. Cresses when crushed and spread over the affected parts, together with the juice, remove the poisonous scab and scale or mange and the supperation of carbuncles. A similar application relieves the pain of lumbago and sciatica, reduces tumors, extracts arrow and thorn, and removes dandruff. The crushed seeds of cress held in the mouth dispel a condition of tongue-tie; put into the nose excite sneezing and arouse from a state of lethargy. The seeds of cress boiled in vinegar and applied dispel goiter and other lumps behind the ears. The seeds are in every way stronger and more potent than the herbage and for this reason are seldom to be taken internally except for a condition of lassitude and sluggishness. The seeds compounded with figs and laid on plaster-fashion restore lost sense of hearing."

Less exuberant and more conservative modern herbals are content with the statement that the fresh plant and its expressed juice are valued as a stimulant, stomachic, diaphoretic, alleviative of catarrh and rheumatism, and blood-purifier.

RORÍPPA Scop.   **Yellow cress**

NL., from L. *ros*, *roris*, moisture, dew, water, + *ripa*, bank—some species growing in wet places.

Annual, biennial, or perennial herbs. Leaves simple or compound. Flowers white or yellow, borne in terminal racemes. Petals spoon-shaped, or absent. Filaments slender. Short stamens each subtended by a reniform or nearly annular gland; long stamens alternating with conic glands. Pistil cylindrical. Silique ovoid to cylindric, sometimes sickle-shaped, nearly round in cross section, many-seeded.

Five species in Kansas. *R. curvipes* Greene and *R. truncata* (Jeps.) Stuckey are known from a few collections only and are not included here.

Key to Species (18):
1. Petals absent .... *R. sessiliflora*
1. Petals present 2
  2. Petals shorter than or equal to the sepals .... *R. palustris*
  2. Petals longer than the sepals 3
    3. Petals usually shorter than 2.5 (3) mm; style in fruit usually shorter than 0.8 (1.2) mm; anthers globose, 1–1.5 times as long as wide, notched at the apex .... *R. palustris*
    3. Petals longer than 2.5 (2) mm; style in fruit usually longer than 0.8 mm; anthers elongate, 1.5–3 times as long as wide, apiculate .... *R. sinuata*

R. palústris (L.) Bess.

L. *palustris*, marshy, boggy—in reference to the moist habitats preferred by this species.
Yellow or greenish-yellow, inconspicuous; mid-May–August
Moist places of flood plains, lake margins, mud flats, and ditches
Primarily north and east of a line from Jewell County to Cherokee County, with a few collections from farther west

Annuals, occasionally biennials or perennials. Stems erect, usually single and branched above (occasionally much branched below), 3–10 dm tall. Leaves simple, sessile or very short-petiolate; blades oblong to oblanceolate in outline, the lower or proximal one-half to two-thirds deeply pinnately lobed or cleft, the upper or distal one-half to one-third and the lobes of the lower portion crenate or coarsely and somewhat irregularly toothed, mostly 3–15 cm long and 0.5–7 cm wide; base auricled and clasping the stem. Flowers yellow or greenish-yellow, inconspicuous. Sepals ovate to oblong, mostly 1.25–2 mm long. Petals oblong to broadly spatulate, usually shorter than or about the same length as the sepals. Siliques mostly ellipsoid to oblong, short-cylindric, or obpyriform, mostly (3–) 5–6 mm long; pedicels ascending, spreading or recurved, 2–6 mm long. Seeds orange-brown or tan, more or less heart-shaped, minutely papillose, about 0.5 mm long.

Our plants have been called *R. islandica* (Oeder) Borbas, but a recent study by B. Jonsell (19) indicates that *R. islandica* is a European species distinct from our own, and the proper scientific name for our species is the one used here. The species is quite variable—or polymorphic (from Gr. *polys*, many, + *morphe*, form, i.e., occurring with a marked degree of variation in form)—and has been subdivided into 4 subspecies and 11 varieties. Our plants are recognized by R. L. Stuckey, the most recent monographer of the genus, as *R. palustris* (L.) Bess. ssp. *glabra* (Schulz) Stuckey var. *fernaldiana* (Butt. & Abbe) Stuckey.

R. sessiliflòra (Nutt. *ex* Torr. & Gray) A. S. Hitchc.   Sessile-flowered yellow cress

NL., from L. *sessilis*, sitting, + *flos, floris*, flower, i.e., sessile-flowered.
Flowers greenish, inconspicuous; May–September
Flood plains, margins of lakes and ponds, sand bars, seepy ravines, wet ditches
Eastern third, primarily, with a few collections from farther west

Glabrous annual (or biennial?). Stems erect, simple to much branched from the base, (1–) 2–5 dm tall. Leaves simple; blades of the lower leaves mostly ovate to elliptic, oblanceolate, or spatulate in outline, 1.5–12 cm long and 0.5–3.5 cm wide; apex rounded or acute; base long-attenuate, giving the appearance of a petiole, auricled or not auricled; lower leaves sometimes pinnately lobed below the middle of the expanded portion, otherwise the margins irregularly toothed to crenate, slightly wavy, or entire. Flowers greenish, inconspicuous. Sepals ovate to lanceolate or subulate, often tinged with purple, 1.5–2 mm long. Petals absent. Siliques cylindric, straight or somewhat curved, mostly 5–10 mm long; pedicels ascending or spreading, 0.5–1.5 mm long. Seeds light tan, heart-shaped, minutely pitted (use magnification), about 0.4–0.5 mm long.

R. sinuàta (Nutt. *ex* Torr. & Gray) A. S. Hitchc.   Spreading yellow cress

175. *Rorippa sinuata*

NL., from L. *sinuatus*, past participle of *sinuare*, to wind or to bend—in reference to the sinuous leaf margins.
Yellow; late April–June (occasionally into late summer or early fall)
Disturbed soil of flood plains, lake margins, fields, pastures, roadsides, and waste ground
Scattered throughout

Perennials arising from creeping rhizomes. Stems erect to decumbent, much branched, mostly 1.5–4 dm tall or long. Leaves simple, sessile; blades oblanceolate to spatulate in outline, often narrowed toward the base and resembling a petiole, mostly 2–10 cm long and 0.5–3.5 cm wide, base auricled or not auricled; margins coarsely toothed to pinnately lobed or divided nearly to the midrib. Flowers yellow. Sepals ovate to lanceolate or oblong, pouchlike at the base, 3–4 mm long, with a membranous margin. Petals oblong to oblanceolate or narrowly spatulate, 3–5 mm long. Siliques cylindric, slightly to strongly curved toward the axis of the inflorescence, 8–12 mm long (excluding the style); pedicels spreading widely, often with a sigmoid curve, 8–12 mm long. Seeds light tan, more or less heart-shaped although rather angular, minutely papillose, about 1 mm long.

This is the most commonly encountered of our species of *Rorippa*.

SELÈNIA Nutt.

NL., from Gr. *selene*, the moon, with allusion to the genus *Lunaria* (from L. *luna*, the moon), the fruits of the former genus resembling those of the latter.

Low annuals. Leaves once or twice pinnatifid. Flowers yellow, borne in leafy terminal racemes. Petals oblanceolate. Glands present between the petals and stamens, and each of the short stamens subtended by a horseshoe-shaped gland as well. Pistil cylindric. Fruit flattened, elliptic to oblong or somewhat crescent-shaped, terminated by the long slender style, few-seeded.

One species in Kansas.

S. aùrea Nutt.

L. *aureus*, golden, from *aurum*, gold, in reference to the yellow flowers.
Yellow; April–mid-May
Pastures, roadsides, rarely in open woods
Southeast eighth

Stems erect or ascending, usually branched from the base, 0.5–1.8 dm tall. Leaves once-pinnate, the lower (larger) ones 2–7 cm long and 0.5–1.2 cm wide, the lobes linear to long-triangular or occasionally ovate, often with 1 or more coarse teeth. Sepals yellow, petal-like, narrowly lanceolate to oblong or oblanceolate, somewhat pouchlike at the base, 4.5–5.5 mm long. Petals 8–11 mm long. Silicles as described for the genus, up to 2 cm long (excluding the style). Seeds orange, flattened, nearly round in outline, broadly winged, about 3 mm long.

SISÝMBRIUM L.

NL., from Gr. *sisumbrion*, an ancient name for some plant in the Mustard Family, possibly the watercress, but the exact identity is not known.

Annual herbs. Leaves entire to pinnately lobed. Flowers yellow, borne in terminal racemes. Petals obovate or spoon-shaped. Filaments slender. Short stamens usually subtended by ring-shaped glands. Ovary cylindric. Silique linear or sword-shaped, round or somewhat 4-angled in cross section, several- to many-seeded.

Two species in Kansas.

Key to Species (3):

1. Fruits awl-shaped, closely appressed to the rachis; pedicels erect, 2–3 mm long; fruits usually 1–1.5 cm long .... *C. officinale*
1. Fruits linear, widely spreading; pedicels divergent, 5–10 mm long; fruits 5–10 cm long .... *S. altissimum*

S. altíssimum L.   Tumble mustard

L. *altissimus*, tallest, from *altus*, tall or high.
Pale yellow; May–June
Fields, roadsides, waste ground, and other disturbed habitats
Eastern four-fifths; intr. from Europe

176. *Sisymbrium altissimum*

Stems erect, usually simple below and branched above, up to about 1 m tall. Lower leaves oblanceolate or spatulate in outline, mostly 0.4–2 dm long and (1–) 2–4 cm wide, pinnately lobed or divided nearly to the midrib, the margins of the segments coarsely toothed; upper leaves smaller, very finely dissected, the segments linear, about 0.5 mm wide. Lower leaves and lower portions of the stem sparsely covered with long, simple, spreading hairs. Flowers pale yellow, often drying cream-colored. Sepals subulate, somewhat pouchlike at the base, 3.5–5 mm long, with membranous margins. Petals spatulate, 6.5–8 mm long. Siliques linear, round in cross section, 5–10 mm long, spreading widely. Seeds brown, oblong, sometimes rather angular, wingless, about 1 mm long.

S. officinàle (L.) Scop.   Hedge mustard

ML. *officinalis*, pertaining to a workshop, from L. *officina*, workshop—in reference to the fact that the plant was kept at apothecary shops for medicinal uses.
Pale yellow; May–June
Pastures, roadsides, creek banks
East fifth and Saline County

177. *Sisymbrium officinale*

Stems erect, eventually becoming much branched, up to about 1 m tall. Leaves oblanceolate to spatulate in outline, the lower ones mostly 4–10 (–12) cm long and 1.5–6 cm wide, pinnately lobed or divided nearly or completely to the midrib, the margins of the lobes coarsely toothed. Leaves or bracts within the inflorescence much smaller, narrowly elliptic to linear, with hastate bases. Herbage and fruits glabrous or sparsely covered with simple hairs. Flowers pale yellow. Sepals oblanceolate to oblong, 1.5–2 mm long. Petals spatulate, about 3 mm long. Siliques awl-shaped, closely appressed to the axis of the inflorescence, 1–2 cm long. Seeds orange-brown, elliptic to triangular in outline, wingless, 1–1.5 mm long.

STANLÈYA Nutt.   Stanleya

NL., in honor of Lord Edward Stanley (1775–1851), a British ornithologist and one-time president of the Linnaean Society.

Annual, biennial, or perennial herbs. Leaves simple, entire to deeply pinnately lobed or cleft, petiolate to sessile and clasping the stem. Flowers quite showy, yellow to white, borne in large terminal racemes. Petals clawed, the expanded portion narrow. Stamens much longer than the petals; anthers often coiling after dehiscence. Silique linear, borne on a long stipe, round or somewhat flattened in cross section, several-seeded.

According to C. L. Hitchcock *et al.* in *Vascular Plants of the Pacific Northwest*, Part 2 (20), the species of *Stanleya* are believed to require selenium for proper growth, and therefore their occurrence is considered an indication of the presence of selenium in the soil.

S. pinnàta (Pursh) Britton   Desert prince's plume

L. *pinnatus*, feathered—in reference to the divisions of the leaf transverse to the midrib, like the pinnae of a feather transverse to the shaft, or possibly in reference to the plumelike appearance of the inflorescence.
Yellow; mid-May–June
Dry, rocky prairie
Northwest sixth and Hamilton and Edwards counties

Herbaceous, leafy perennial, commonly with several ascending stems 0.5–1 m tall arising from a woody base. Leaves glaucous and somewhat succulent or thickened, the lower ones pinnately lobed or cleft, the upper ones entire; blades lanceolate to elliptic, oblanceolate, or obovate in outline, those of the main leaves 5–12 cm long and 3.5–5 cm wide. Flowers yellow, borne on spreading pedicels in greatly elongating racemes. Sepals yellow, petal-like, linear, 0.7–1.7 cm long. Petals 0.9–1.2 (–1.7) cm long. Stamens protruding beyond the petals, nearly equal in length; anthers coiling as they dry. Glands 6, 1 encircling the base of each stamen. Pistil long-stipitate. Pods compressed-cylindric, somewhat constricted between the seeds, spreading or somewhat recurved at maturity, 2–4 cm long. Seeds orange-brown, narrowly elliptic to oblong or somewhat angular, 2–3 mm long.

Looking at a mummified herbarium specimen of this plant, one could never imagine how distinguished it is in the plains flora with its long plumes waving in the wind, a sight we fear Lord Stanley never saw.

See plate *46*.

178. *Stanleya pinnata*

THLÁSPI L. Pennycress

Gr. *thlaspi*, ancient name of a kind of cress.

Annual or perennial herbs. Basal leaves entire; cauline leaves with bases sagittate and clasping the stem. Flowers (in ours) white, borne in terminal racemes. Petals spoon-shaped to obovate. Filaments slender. Short stamens each subtended by a pair of semicircular glands. Pistil ellipsoid to obovoid, somewhat flattened. Silicles circular to obovate or obcordate, flattened at right angles to the septum, winged at the margin, notched at the apex, usually (6-) 8- or more-seeded. Seeds wingless.

Two species in Kansas.

Key to Species:
1. Foliage bright green; mature fruit 10–18 mm long, the distal notch deeper than wide; seeds concentrically ridged, about 2 mm long .... *T. arvense*
1. Foliage glaucous and light green; mature fruit 4–6 mm long, the distal notch wider than deep; seeds not ridged, about 1 mm long .... *T. perfoliatum*

T. arvénse L. Field pennycress

179. *Thlaspi arvense*

NL., from *arvum*, cultivated field, plowed land, this species often being found in such habitats.
White; April–May
Fields and waste places
East half, primarily, with a few scattered collections from the northwest fourth; intr. from Europe

Smooth, erect annual. Stems branched or unbranched below the inflorescence, up to 5 dm tall. Lower leaves petioled, spatulate to oblanceolate, 3–10 cm long (including the petiole), with entire or coarsely toothed margins; upper leaves sessile, elliptic or oblong-elliptic, 2–6 cm long, with coarsely toothed margins and sagittate bases clasping the stem. Flowers white. Sepals oblong to elliptic, 1.5–2.5 mm long, with membranous margins. Petals spatulate, 2–4 mm long. Silicles very broadly elliptic to round in outline, 10–18 mm long, 6- to 12-seeded, margin 2–4 mm wide near the middle; apical notch very narrow, 2–3 mm deep; pedicels ascending to spreading widely. Seeds dark red-brown (sometimes iridescent under illuminated magnification), broadly elliptic, slightly flattened, about 2 mm long, the surfaces ornamented with concentric ridges reminiscent of a fingerprint.

The young leaves of this species, when cut into small pieces, can be used as a tasty onionlike addition to green salads. The seeds may be used as a mustard substitute.

T. perfoliàtum L.

NL. *perfoliatus*, pierced through the leaf, from L. *per-*, through, + *foliatus*, -leaved, this species having leaves which completely clasp the stem.
White; April–early May
Roadsides and waste ground
East fourth; intr. from Europe

Smooth, erect annuals. Stems wiry, usually branched, up to 3 dm tall. Basal leaves petioled, spoon-shaped to oblanceolate, 1–3.5 cm long (including the petiole), with entire or finely toothed margins; upper leaves sessile, oblong to lanceolate, 1–3 cm long, with entire or shallowly toothed margins and sagittate bases clasping the stem. Flowers white, minute. Sepals ovate to oblong, often tinged with purple, 0.5–1 mm long, with membranous margins. Petals 2–3 mm long. Silicles cordate, 4–6 mm long, mostly 6-seeded; margin 0.5–1 mm wide near the middle; apical notch V-shaped, 1 mm or less deep; pedicels spreading widely. Seeds orange, elliptic to round in outline, slightly flattened, about 1 mm long.

CAPPARIDÀCEAE

Caper Family

180. *Cleome serrulata*

Named after the genus *Capparis*, the classical Latin name for a low prickly shrub of the Mediterranean region whose pickled buds are called capers.

Herbs, shrubs, trees, or vines (ours all herbs). Leaves alternate, simple or palmately compound with 3–5 leaflets, or sometimes with both compound and simple leaves on the same plant; stipules present as minute glands or spines, or absent. Flowers mostly bisexual, regular or irregular, borne singly or several together in the leaf axils or in terminal racemes. Sepals 4–8, distinct or more or less fused. Petals 4–8, usually 4, distinctly clawed, equal in size or 2 larger than the others. Nectiferous gland often present between petals and stamens. Stamens 6 to many. Pistil 1; ovary superior, 1-celled or 2-celled, with many ovules. Fruit an elongate capsule dehiscent by 2 valves. The ovary is often elevated above the insertion of the other parts of the flower by a thin extension of the receptacle in the form of a stalk, or stipe, and in that case the ovary and the fruits are said to be stipitate.

The family contains about 46 genera and 700 species, mostly of the Old World tropics. Cultivated members of the family include capers (*Capparis spinosa* L.) and the giant spiderflower (*Cleome spinosa* L.).

Four genera are native to Kansas.

Key to Genera:
1. Flowers pink (rarely white), petals (in ours) about 15 mm long .... *Cleome*
1. Flowers not pink; petals 10 mm long or shorter (rarely as long as 14 mm) 2
    2. Flowers yellow; plants not sticky; fruit 2-seeded, rhombic with rounded angles .... *Cleomella*
    2. Flowers white or cream-colored; plants sticky; fruit many-seeded, elongate 3
        3. Petals with a shallow notch at the apex, about 8 mm long; stamens 8 to many, conspicuously longer than the petals .... *Polanisia*
        3. Petals deeply and irregularly toothed, the larger 2 about 3–4 mm long; stamens 6–9, not conspicuous .... *Cristatella*

CLEÒME L.

Spiderflower

The derivation of the genus name is unknown, but Octavius Horatius, Roman physician of the 4th century, used it for a plant in the Mustard Family. Linnaeus later transferred it to its present use.

Erect annual or perennial herbs. Leaves palmately compound with varying numbers of leaflets (ours with 3), sometimes with spinose stipules. Flowers pink, purplish, or white, borne in terminal racemes. Sepals 4, separate or fused to one another. Petals 4, equal or nearly so in size. Nectar gland inconspicuous. Stamens usually 6, much longer than the petals. Pistil 1-celled, many-ovuled, and stipitate. Fruit an elongate capsule constricted slightly between the seeds.

One species in Kansas.

C. serrulàta Pursh

Bee spiderflower

NL., from L. *serrula*, little saw, in reference to the fine denticulations sometimes present on the margins of the leaves. The first part of the common name is appropriate because the flowers are noted for their yield of honey, and the species is sometimes cultivated for that purpose. "Spiderflower" seems to come from a resemblance between the spreading slender stamens and sprawling spider legs.

Pink, rarely white; July–early September
Sandy or rocky prairie, waste places
West half, primarily, with some collections from the northeast sixth

Robust annual. Stems erect, naked below and with ascending branches above, up to about 1 m tall. Leaflets 3, narrowly elliptic or lanceolate, those of the lower leaves mostly 3–5 cm long and 0.7–1.2 cm wide, becoming gradually smaller toward the inflorescence, glabrous or sparsely covered with kinky hairs; petioles 1–5 cm long; stipules minute. Flowers pink (rarely white), in dense terminal racemes which continue to elongate during the flowering season; a series of linear or lanceolate, leaflike bracts subtends the entire inflorescence, and, within the inflorescence, 1 bract subtends each flower. Sepals fused more than half their length, the calyx 3–4 mm long with 4 triangular lobes, often pink or purplish. Petals about 1.5 cm long, the expanded portion lanceolate or elliptic. Stamens with long green anthers which coil upon dehiscence. Capsule 3–6 cm long at maturity. Seeds brown, more or less round to ovate in outline, usually slightly flattened, 3–3.5 mm long, with a narrow wrinkled wing or excrescence of some type along the angles of the seed.

CLEOMÉLLA DC.   Cleomella

NL., from the genus *Cleome* + L. diminutive suffix *-ella*, little Cleome.

Erect annual or perennial herbs. Leaves palmately 3-foliate, the minute stipules threadlike. Flowers yellow, borne in terminal racemes. Sepals 4, minute, slightly fused at the base, eventually falling away. Petals 4, entire. Nectar gland conspicuous, 3-lobed. Stamens 6, equal in length, with anthers that coil tightly upon drying. Pistil 1-celled. Fruit a stipitate, several-seeded capsule which splits at maturity into 2 helmet-shaped segments.

One species in Kansas.

181. *Cleomella angustifolia*

C. angustifòlia DC.   Cleomella

NL., from L. *angustus*, narrow, + *folium*, leaf, in reference to the narrow leaflets of the tri-
foliolate leaf.
Yellow; July–September
Sandy soil of roadsides, prairies, flood plains, and salt flats
Central and south-central counties, mostly south of the Arkansas River

Erect annual. Stems naked below and with ascending branches above, up to about 1 m tall. Leaflets 3, linear-lanceolate to linear, mostly 1.5–3.5 cm long and 1–3 mm wide. Flowers yellow, in dense terminal racemes which continue to elongate during the flowering season; a series of linear leaflike bracts subtends the entire inflorescence and, within the inflorescence, 1 bract subtends each flower. Sepals lanceolate to triangular or ovate, about 1 mm long. Petals about 5 mm long, the expanded portion elliptic or lanceolate. Stamens with green anthers which coil tightly upon drying. Capsule 5–7 mm long and 5–8 mm wide. Seeds straw-colored or mottled brown, somewhat resembling a bird's head in shape, covered with minute tubercles or rough ridges, 2.5–3 mm long.

CRISTATÉLLA Nutt.   Cristatella

NL., from L. *cristatus*, crested, + L. diminutive suffix *-ella*, in reference to a small lobed appendage which arises from the receptacle.

Slight, glandular annuals. Leaves 3-foliolately compound, estipulate. Flowers white or yellowish, borne in terminal racemes. Sepals 4, fused at the base. Petals 4, clawed, the apex deeply and irregularly torn, the upper pair markedly longer than the other 2. Stamens 6–14. Pistil stipitate. The nectar gland is a tubular petaloid appendage, as long as the smaller petals, which arises between the posterior petals and the ovary. Capsule 1-celled, many-seeded, on a short stipe, opening longitudinally.

One species in Kansas.

182. *Cristatella jamesii*

C. jàmesii T. & G.   Cristatella

Named in honor of Edwin James (1797–1861), a surgeon-naturalist with Maj. Stephen H. Long's
1819 expedition to the Rocky Mountains and who was the first botanist to collect this
species.
Cream-colored; late June–early September
Sand dunes, sandy prairie
Southwest fourth and Harvey, Phillips, Rooks, Republic, and Cloud counties

Plants 0.8–3.5 dm tall. Leaflets linear, mostly 1–2.5 cm long; petioles 0.2–1 cm long below, becoming gradually shorter above. Sepals oblong or elliptic, glandular, about 2 mm long. Petals cream-colored, often purplish at the base, the expanded portion more or less obovate, the larger 2 mostly 3–4 mm long. Stamens 6–9. Capsule elongate, flat-tened, glandular, erect to ascending at maturity, 1–3 cm long. Seeds brown or tan, round in outline and somewhat flattened, about 1.8 mm long, covered with tubercles, some of which are opaque and pearlescent, others transparent.

POLANÍSIA Raf.   Clammy weed

NL., from Gr. *polys*, many, + *anisos*, unequal, because the many stamens are of various lengths.

Erect, glandular annuals. Leaves palmately 3-foliolate, estipulate. Flowers in ours white and purple. Sepals 4, slightly fused at the base. Petals 4, spoon-shaped, the apex 2-lobed or shallowly notched, 1 pair longer than the other. Nectar gland conspicuous, yellow or orange. Stamens 8–27, maturing at different times and of different lengths. Pistil 1-celled, many-ovuled, sessile or on a short stipe. Capsule elongate, dehiscing apically.

One species in Kansas.

P. dodecándra (L.) DC.   Clammyweed

NL., from Gr. *dodeka*, 12, + *andros*, male, the flowers often having 12 stamens. The common
 name, clammyweed, is quite appropriate since the herbage is covered with sticky glandular
 hairs.
Yellowish-white and purple; June–early October
Sandy or rocky prairie, flood plains, less frequently in rocky, open woods
Nearly throughout but more common in the western half

Stems erect, simple or branched, up to 2.5 dm tall (occasionally taller). Leaflets
ovate to lanceolate, elliptic, or oblanceolate, mostly 1.5–4 cm long, becoming much
smaller above; petioles mostly 1–4 (rarely up to 7) cm long. Stems, leaves, and fruits
covered with glandular hairs. Flowers in dense terminal racemes which continue to
elongate during the flowering season; on most plants, a series of ovate or elliptic, leaflike
bracts subtends the entire inflorescence and, within the inflorescence, 1 bract subtends
each flower. Sepals narrowly lanceolate, glandular, 3–4 mm long. Petals white, the 2
larger ones 4–12 (mostly 8–9) mm long, the expanded portion obcordate. Stamens 10–27
(often 12); filaments purple, threadlike, about the same length as to much longer than the
petals. Capsules flattened, narrowed below but lacking a true stipe, glandular, erect or
ascending at maturity, mostly 2–5 cm long and 0.5–1 cm wide. Seeds rust-brown or dark
brown, dull, round in outline, somewhat flattened, covered with minute, closely spaced
tubercles and sometimes with additional larger tubercles or rough ridges, 0.9–2.4 mm long.

183. *Polanisia dodecandra*

CRASSULÀCEAE   Stonecrop Family

Named after the genus *Crassula*, NL. diminutive of *crassus*, thick—in reference to the thickened
 leaves of these South African succulents.

Annual or perennial herbs or shrubs, mostly succulent. Leaves opposite, whorled,
or alternate, mostly simple, estipulate. Flowers bisexual, regular, frequently borne in
cymes. Sepals 4–5, distinct or slightly fused to one another at the base. Stamens 8–10,
hypogynous, or occasionally the petals fused basally and the stamens arising from the
corolla tube. Pistils 4–6, usually of the same number as the petals, entirely separate or
fused together at the base; each pistil with 1 style and stigma and subtended by a small
scalelike nectar gland. Fruit a follicle opening longitudinally and setting free few to
many small seeds.

A family of about 20 genera and 900 species, widely distributed, mostly in temperate and subtropical regions. Cultivated members of this family include species of *Sedum*,
Crassula, and *Kalanchoë*.

Kansas has 1 genus.

SÈDUM L.   Stonecrop

L. *sedum*, named used by Pliny for the houseleek, *Sempervirum tectorum*, which has rosettes of
 succulent leaves.

Succulent perennials, occasionally annuals. Leaves alternate, opposite, or whorled,
thick, often round in cross section. Flowers borne in a terminal inflorescence with 2–7
radiating, more or less horizontal, racemelike branches. Sepals 5. Petals 5, only slightly
(if at all) fused to one another at the base. Stamens 8–10, those situated directly in front
of a petal usually fused to the base of that petal. Pistils separate. Follicles with few to
many seeds.

Two species (possibly 3) in Kansas. *Sedum ternatum* Michx., a white-flowered
perennial, is reported from Cherokee County only and is not included here.

Key to Species:
1. Petals yellow; anthers yellow; follicles nearly perpendicular to the pedicel at maturity
 .... *S. nuttallianum*
1. Petals white to pale pink or purple; anthers dark red or purplish; follicles spreading
 somewhat but not perpendicular to the pedicel .... *S. pulchellum*

S. nuttalliànum Raf.   Yellow stonecrop

Named in honor of Thomas Nuttall (1786–1859), son of a Lancashire, England, printer, who
 became a pioneer in central United States and Pacific Coast botany, and professor of
 natural history and curator of gardens at Harvard.
Yellow; May–early July

Shallow soil over limestone or sandstone, in open oak woods
Chautauqua and Montgomery counties

Low, tufted, pale green annual. Stems erect or ascending, up to 10 cm tall. Leaves alternate, oblong, 3–8 mm long, deciduous. Flowers yellow. Sepals oblong or triangular, 2–3 mm long. Petals lanceolate, mostly 3–3.5 mm long. Anthers yellow. Follicles few-seeded, nearly perpendicular to the pedicel at maturity, about 3.5 mm long, tipped by a short style. Seeds oblong-ellipsoid, about 0.6 mm long.

184. *Sedum pulchellum*

S. *pulchéllum* Michx. — Rare stonecrop, rock moss

L. *pulchellus*, exquisite. "Stonecrop" is an old English name, probably suggested by the fact that in England, the commonest sedum, *Sedum acre* L., grows in the form of mats on stones and stone walls.
Pink to white; mid-May–late June
Rocky soil
Southeast eighth

Low winter annual with procumbent and erect stems up to 2.4 dm tall. Leaves alternate, linear-oblong (basal leaves in juvenile forms spatulate), 0.6–3 cm long, deciduous. Flowers pink to white. Sepals oblong or long-triangular, 2–4 mm long. Petals lanceolate to narrowly lanceolate, elliptic, or nearly linear, 4–8 mm long. Anthers dark red or purplish. Follicles few-seeded, spreading somewhat at maturity but not perpendicular to the pedicel, about 5 mm long, tipped with long acuminate styles. Seeds ellipsoid, about 0.8 mm long.

185. *Penthorum sedoides*

PENTHORÀCEAE — Penthorum Family

Erect herbaceous perennials. Leaves alternate, simple, estipulate. Flowers bisexual, regular, borne along the upper sides of the 2–4 scorpioid branches of cymes. Sepals 5, fused basally to form a short calyx tube. Petals absent. Stamens 10, perigynous. Pistils 5, rarely 6 or 8, half inferior, fused to one another nearly to the middle. Fruit an angular, 5-celled, many-seeded capsule with horned lobes which open circumscissilely.

A monogeneric family with 1 species in North America and 2 others in Japan and China.

PÉNTHORUM L. — Penthorum, ditch stonecrop

NL., from Gr. *pente*, 5, + *horos*, rule—5 or its multiple being the prevailing number of the floral parts.

Description of the genus that of the family. This genus is placed in the Crassulaceae by some authors, in the Saxifragaceae by others, or in its own family as is done here.

One species in Kansas.

P. *sedoìdes* L. — Virginia penthorum

NL., from the generic name *sedum*, + Gr. suffix *-oides*, similar to—the resemblance being in the aspect of flower and inflorescence.
Flowers light green or yellowish-green; July–early October
Moist ground and shallow water
East half, primarily

Perennial from slender rhizomes bearing fibrous roots along their length. Stems 1.5–8 dm tall. Leaves lanceolate to elliptic, short-petioled, often turning orange or reddish-brown on drying or with age, the main leaves (3–) 6–17 (–22) cm long and 1.5–4 cm wide; apex acute to acuminate; base attenuate; margins serrate. Flowers light green or yellowish-green, about 4 mm wide; branches of the inflorescence glandular-hairy. Calyx lobes triangular to ovate or oblong, 1–1.5 mm long. Seeds pale orange, ellipsoid, minutely spiny (use high magnification), about 0.5 mm long.

SAXIFRAGÀCEAE — Saxifrage Family

Herbs, shrubs, or small trees. Leaves mostly alternate, sometimes opposite or all basal, simple in ours, mostly estipulate. Flowers usually bisexual and regular, borne in cymes, racemes, or panicles. Sepals 4–5, often fused to one another, sometimes petaloid. Petals 4–5, borne on the receptacle or on a hypanthium, sometimes much reduced or entirely absent. Stamens of the same number as the petals and alternate with them, or twice as many as the petals. Pistils superior to inferior, often half-inferior; 1–5 in number; when 1, 1- to 5-celled, usually with as many styles and stigmas as cells. Fruit mostly a deeply lobed capsule with numerous seeds.

The family comprises about 80 genera and 1,200 species. Cultivated members of the family include the mock-orange (*Philadelphus* spp.), currants and gooseberries (*Ribes* spp.), and the ornamental species of *Hydrangea*, *Heuchera*, *Astilbe*, and *Saxifraga*.

The family is represented in Kansas by 5 native species in 4 genera, but only the most common herbaceous species is included here. *Ribes missouriense* Nutt., the wild gooseberry, is a woody shrub occurring in the eastern half of the state. *Ribes odoratum* Wendl., golden currant, also a shrub, is found throughout. *Hydrangea arborescens* L., a shrub up to 3 m tall, is known only from a small area of Cherokee County. *Saxifraga texana* Buckl., a small herbaceous perennial with a basal rosette of leaves and a congested scapose raceme of white flowers, is known from a small area of Chautauqua County only.

HEÙCHERA L. Alumroot

Named in honor of Professor Johann Heinrich Heucher (1677–1747), a German botanist and
custodian of the Botanic Garden in Wittenberg. The common name denotes the
astringent taste of the root.

Perennial herbs arising from a caudex or rhizome. Leaves all basal, palmately veined. Flowers bisexual, regular or somewhat irregular, borne in panicles. Calyx lobes 5, green or petal-like, diverging from the saucer-shaped or tubular hypanthium which is fused to the lower portion of the ovary. Petals 5, small (occasionally absent), borne on the hypanthium. Stamens 5. Pistil half-inferior; ovary 1-celled with 2 placentae; styles 2. Fruit a 2-beaked capsule opening between the beaks.

One species in Kansas.

H. hirsuticaùlis (Wheelock) Rydb. Alumroot

NL., from L. *hirsutus*, hairy, shaggy, + *caulis*, stem, i.e., having hirsute stems.
Flowers yellowish-green; late April–early June
Usually on rocky, wooded hillsides
East fifth

Plants up to about 1 m tall. Leaves long-petioled; blades cordate or nearly round in outline with a cordate base, mostly 2.5–12 cm long and 2.5–12 cm wide, with 7–11 shallow lobes, each lobe with several rounded sharp-pointed teeth. Flowers somewhat irregular, yellow-green (occasionally purple-tinged), 7–11 mm (including the exserted stamens) long; panicles elevated above the leaves on 1 or 2 scapes. Calyx and all herbage with glandular hairs, those of the petioles and lower peduncles long-spreading and conspicuous. Hypanthium short-tubular. Calyx lobes rounded, mostly 1.2–2.5 mm long. Petals spoon-shaped or narrowly oblanceolate, 2–3.5 mm long. Capsules about 10–12 mm long. Seeds black, narrowly ovoid, spiny, mostly 0.5–0.8 mm long.

186. *Heuchera hirsuticaulis*

187. *Heuchera hirsuticaulis* leaves

ROSÀCEAE Rose Family

Trees, shrubs, or perennial herbs. Leaves alternate, simple or compound, petiolate, usually with stipules. Flowers bisexual (unisexual in some species of *Fragaria*), regular, borne singly or in clusters. Sepals and petals usually 5 each, frequently fused basally to form a hypanthium. Stamens 5 to many, often 20, in 1 to several whorls of 5 stamens each, inserted on the rim of the hypanthium. Pistil(s) 1 to many, superior or inferior, simple or compound. Fruit an achene, follicle, pome, drupe, simple accessory fruit (as the strawberry), or aggregate accessory fruit (as the blackberry).

This large family of approximately 115 genera and 3,200 species includes many of our temperate-climate fruits such as the apple (*Malus*), pear (*Pyrus*), quince (*Cydonia*), raspberry, dewberry, blackberry (*Rubus* spp.), cherry, peach, plum, nectarine, apricot, almond (*Prunus* spp.), and strawberry (*Fragaria*), as well as numerous ornamentals such as *Spiraea*, *Cotoneaster*, firethorn (*Pyracantha*), hawthorn (*Crataegus*), flowering quince (*Chaenomeles*), and the various roses (*Rosa*).

Kansas has 13 genera and about 59 species. Only the most common herbaceous members of the family are included in the following key. For identification of trees and

188. *Agrimonia pubescens*

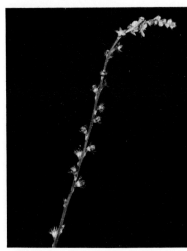

189. *Agrimonia pubescens* fruiting

shrubs in this and other families, I recommend either *Trees, Shrubs, and Woody Vines in Kansas* (1969) or *Woody Plants of the North Central Plains* (1973), both written by H. A. Stephens and published by the Regents Press of Kansas.

Key to Genera:
1. Leaves 3-foliolate or palmately once-compound 2
  2. Ovaries 2–10 .... *Gillenia*
  2. Ovaries many 3
    3. Styles threadlike, elongate, jointed above the middle (in our species); lower, middle, and upper leaves conspicuously different in shape .... *Geum*
    3. Styles short, often thickened, not jointed; leaves all alike or nearly so, except in size 4
      4. Flowers yellow (sometimes white in *P. arguta*); at least some leaves borne on the stem; receptacle dry in fruit .... *Potentilla*
      4. Flowers white; leaves all basal; receptacle red and fleshy in fruit .... *Fragaria*
1. Leaves pinnately compound .... *Agrimonia*

AGRIMÒNIA L.   Agrimony

L. *agrimonia*, name used by Pliny, Dioscorides, and others for a plant of this genus (*A. eupatoria*, so named by Linnaeus in honor of Mithridates Eupator, king of Pontus, who was said to be the first to use it for diseases of the liver).

Erect herbaceous perennials arising from short, stout rhizomes. Leaves compound, odd-pinnate, with small, paired or unpaired, toothed or entire, secondary leaflets interspersed with pairs of larger, toothed leaflets; stipules herbaceous, prominent, winglike, with lacerated margins. Flowers widely separated in slender racemes. Hypanthium hemispheric or obconic, fringed with hooked bristles. Stamens 5–15. Pistils 2, superior, simple. Fruits 2 achenes surrounded by the persistent, hardened, longitudinally grooved hypanthium.

Four species in Kansas. *A. gryposepala* Wallr., known from Marshall County, and *A. rostellata* Wallr., known from Douglas and Cherokee counties, are not included in the following key.

Key to Species:
1. Primary leaflets 9–11 (–17) per leaf; under surface of leaves, hypanthium, and axis of inflorescence glandular-dotted; stems densely pubescent with brownish hairs ...... *A. parviflora*
1. Primary leaflets 5–9 per leaf; under surface of leaves velvety; glands absent from all plant parts; stems usually only sparsely pubescent .... *A. pubescens*

A. parviflòra Ait.   Agrimony

NL., from L. *parvus*, little, + *flos, floris*, flower, i.e., small-flowered.
Yellow; July–early September
Moist open woods and thickets, moist prairies and prairie ravines, often in sandy soil
Scattered in the east half and in Comanche County

Stout plants up to about 1.3 m tall. Stems densely covered with brownish hairs. Leaves mostly 0.6–2 dm long; primary leaflets 9–11 (–17), elliptic to narrowly lanceolate, mostly 2.5–6 cm long and (0.5–) 1–2 cm wide, with an acute or short-acuminate apex and serrate margins; under surface of leaves glandular-dotted (use magnification), the veins brown-hairy. Hypanthium and axis of inflorescence glandular-dotted. Calyx lobes lanceolate to long-triangular, 1.5–2 mm long. Petals yellow (often fading in drying), oblanceolate to obovate, 2–3 mm long. Hypanthium (including bristles) in fruit 4–5 mm long.

A. pubéscens Wallr.   Soft agrimony

L. *pubescens*, grown downy or hairy—in reference to the stems and leaves.
Yellow; mid-July–early September
Open woods and thickets
East fifth and Pottawatomie County

Plants with slender, sparsely pubescent stems up to about 1.5 m tall, growing from short thick rhizomes that produce both fibrous and tuberous roots. The stem also develops adventitious roots as it ascends from the rhizome. Leaves mostly 0.8–2.2 dm long; primary leaflets 5–9, lanceolate to oblanceolate, 2.5–8 cm long and 1.3–4 cm wide, with crenate to serrate margins; under surface of leaves velvety. Hypanthium and axis of inflorescence lacking glands. Calyx lobes lanceolate to long-triangular, 1.25–2 mm long. Petals yellow (often fading in drying), obovate to oblanceolate (rarely oblong), 2–3 mm long. Hypanthium (including bristles) in fruit 4.5–6 mm long.

FRAGÀRIA L.   Strawberry

NL., from L. *fraga*, fragrant berry—the classical name Virgil (1st century B.C.), Ovid, and Pliny (1st century) used for the strawberry fruit, *fragrum* being the name of the plant; both names related to *fragrare*, to emit a fragrant odor—in reference to the fruit. "Strawberry" is from AS. *streawberige*, from *streaw*, straw or hay, + *berige*, berry—the straw apparently suggested by the runners covering the ground like thinly scattered straw. The name first appears in a Saxon list of plants of the 10th century. By 1265 the name is written "straberie," and in 1603 Ben Jonson writes "strawberry": "A pot of Strawberries gathered in the wood / To mingle with your cream."

Herbaceous perennials arising from a short, tough, erect rhizome bearing many coarse fibrous roots and spreading vegetatively by means of horizontal runners that root and form new plants which bloom and produce fruit the following summer. Leaves 3-foliolate, long-petioled, all basal. Flowers white, bisexual or both bisexual and unisexual, borne in few- to several-flowered terminal clusters. Hypanthium flattened. Sepals 5, alternating with 5 subtending sepal-like bracts. Petals 5. Stamens about 20. Pistils many, simple, 1-ovuled, each with a lateral style, borne over the surface of the convex receptacle. A ring of nectiferous tissue is situated between the inner circle of stamens and the base of the receptacle. Unisexual flowers lack functional pistils (in male flowers) or stamens (in female flowers), although rudimentary pistils are usually present. In bisexual flowers, the stigmas mature before the anthers of the same flower open to discharge pollen, so that cross-pollination may occur when insects—chiefly various species of bees, flies, and butterflies—go from older to younger flowers. In bisexual and female flowers, the receptacle becomes the enlarged, juicy, fragrant edible fruit, bearing small dry achenes on its surface. Enlargement of the receptacle takes place only when, after pollination, the egg in each ovule has been fertilized. If, as sometimes happens, the pistils over 1 side of the receptacle fail to receive pollen, that side fails to develop, while the other side enlarges and ripens.

190. *Fragaria virginiana*

One species in Kansas.

F. virginiàna Duchesne   Wild strawberry

"Of Virginia."
White; April–May
Roadsides, prairie, open woods
East third

Leaflets obovate to oblanceolate or broadly elliptic, 1.2–8 cm long and 1.4 cm wide, the apexes rounded, bases acute or oblique, and margins serrate, the serrations often incurved and bristle-tipped; whitened beneath with a bloom and coat of appressed hairs, green and sparsely hairy above; with short petiolules, or sometimes sessile; petioles with spreading hairs. Sepals lanceolate or lance-oblong, 4–7 mm long. Petals white, obovate, 6–10 mm long. Some plants with both bisexual and male flowers, other plants with female flowers only. Fruits ovoid, red, 1–1.5 cm thick, the achenes over its surface sunk in pits. Fruits on our plants ripen in mid-May to early June.

Our plants are all recognized as belonging to *A. virginiana* var. *illinoensis* (Prince) Gray.

The colonists in the Atlantic states found *F. virginiana* growing wild, as well as cultivated in Indian gardens. William Wood relates in his *New England's Prospect* (1634) that "there is, likewise, strawberries in abundance, verie large ones, some being two inches about; one may gather halfe a bushell in a forenoone." Also, Roger Williams tells us (1643) that "this berry is a wonder of all fruits growing naturally in these parts; it is of itself excellent, so that one of the chieftest doctors of England was wont to say that

God could have made, but never did, a better berry. . . . In some parts, where the natives have planted, I have many times seen as many as would fill a good ship within a few miles compass. The Indians bruise them in a mortar and mix them with meal and make Strawberry bread."

The Virginia strawberry was taken to England in the year 1629, and somewhat later to France. In 1712, Capt. Frezier of the French army took with him to France, from the province of Concepción, in Chile, some plants of the strawberry now known officially as *F. chiloensis* Duchesne, native to the west-coast mountain ranges of North and South America. This species has male and female flowers on separate individuals, or, less frequently, is polygamous as is *F. virginiana*. The Chilean plants were set out in a garden where plants of *F. virginiana*, *F. vesca* L., and *F. moschata* Duchesne (the latter 2 being native European species) were growing. All the Chilean plants brought by Frezier were pistillate and were incapable of producing fruit unless fertilized by pollen from 1 of the other neighboring species. This did happen, although not with satisfactory constancy. Thus the matter stood for more than 25 years, when a new kind of strawberry suddenly appeared among the seedlings of the Chilean, having polygamo-dioecious flowers like the Virginian and the large fruit of the Chilean, with a flavor reminiscent of the pineapple (*Ananus comosus* Merr.). The common name of this new berry, "pine," and its scientific name, *F. chiloensis* var. *ananassa* (Duchesne) Bailey, were suggested by the flavor of its fruits.

The question naturally arose as to how this new strawberry came into existence—was it a spontaneous variant or mutant of the Chilean, or was it a hybrid between the Chilean and one of the other companion strawberry species in the garden? Evidence—both morphological and cytological—indicates that *F. chiloensis* var. *ananassa*, which represents most of the forms cultivated today, is a hybrid between the Chilean and Virginian strawberries.

GÈUM L.   Avens

L. *geum*, the ancient Latin name for the common European Avens, *Geum urbanum* (L. *urbanus*, of or pertaining to a city or town—this species growing as a weed beside walls and hedges and among refuse heaps in towns, as well as in open woods). "Avens" is the English name of unknown signification.

Erect herbaceous perennials arising from a short, erect rhizome. Leaves lobed to pinnately deeply divided or compound, occurring in clusters at the base and alternately on the stem above; stipules fused to the petiole. Flowers solitary or only a few together at the end of the stem. Hypanthium shallow, cup- or top-shaped. Sepals 5, usually alternating with as many subtending bracts. Petals 5, obovate. Stamens many. Pistils many, 1-ovuled, covering a central elevation of the receptacle; style threadlike, strongly bent and jointed (near the distal end in our species). The portion of the style beyond the joint is deciduous; the portion below the joint becomes hardened, and the hooked tip aids in fruit dispersal by catching in animals' coats and people's clothing. Fruits achenes arranged in headlike clusters.

Three species reported for Kansas. *G. laciniatum* Murr. var. *trichocarpum* Fern. is extremely rare and is not included here.

While we never think of our species of *Geum* as being medicinal, they may be as much so as the European *G. urbanum* L., which Pliny in the 1st century cites as a drug plant, giving it a reputation that persisted in herbals of the Middle Ages and even, to some extent, to our own day. For instance, we read in one of the most renowned herbals, *Hortus sanitatis* (Garden of Health), printed in Mainz by Jacob Meydenbach (1491): "Where this herb is in the house, there the devil can do none of his works and flees from it, and therefore it is the most blessed of herbs, and if a man carries the herb about with him no venomous beast can harm him." For bodily afflictions, more credible powers are assigned to it, such as "to get the torpid liver going"; and when dried, pulverized, and drunk in wine it was reputed to be an antidote for poison and good for healing internal injuries. Wine in which the rhizome had been boiled was said to cleanse and improve all wounds, fistulas, and cancerous sores, and to expel filthy parasites. Later German herbals credit *G. urbanum* with being astringent, stimulant, and tonic, and valuable for treatment of diarrhea, dysentery, hemorrhoids, scrofula, gout, rheumatism, intermittent fever, dyspepsia, rickets, typhoid fever, liver complaint, and even for depression of the spirit and melancholy. Chemical analysis shows that the rhizomes contain tannin in relatively large amount, an aromatic, brownish-red ethereal oil, a variety of sugars, gum, and resin, a bitter principle, and salts of ammonia.

Key to Species (3):
1. Head of achenes sessile or nearly so; achenes hirsute toward the summit or completely glabrous; bractlets commonly present in the sinuses between the calyx lobes; flowers white .... *G. canadense*
1. Head of achenes conspicuously stipitate above the calyx; achenes sparsely hairy (use magnification); bractlets none; flowers yellow to cream-colored .... *G. vernum*

G. canadénse L.   White avens, red root

L. *-ensis, -ense*, suffix added to nouns of place to make adjectives—"Canadian," in this instance.
White; June–July
Woods and thickets
East half, primarily; a few records from farther west

Stems 1 to several, up to about 1 m tall. Basal leaves long-petioled, some nearly round to ovate with 3 shallow lobes, cordate bases, and crenate-serrate margins, others with 3 or 5 variously shaped (mostly ovate to obovate, entire to 3-lobed or 3-cleft) leaflets with serrate (occasionally crenate) margins, the terminal 3 leaflets 2.5–9 cm long and 2–6 cm wide, always much larger than the remaining 2 when 5 are present; stem leaves similar to the basal leaves, shorter-petioled, 3-lobed or with 3 oblanceolate to rhombate leaflets, becoming smaller toward the inflorescence; stipules prominent, toothed. Stems, petioles, under surface (and sometimes top surface) of leaves sparsely hairy. Sepals triangular to lance-acuminate, 4–6 mm long; bractlets in the sinuses about 1 mm long. Petals white, ovate to oblong, about the same length as the sepals. Numerous long silky hairs interspersed among the pistils. Achenes hairy, 2.5–4 mm long (excluding the style); achene head 1.5–2 cm long, sessile on the hypanthium.

G. vérnum (Raf.) T. & G.   Spring avens

L. *vernus*, of or belonging to spring, from *ver*, spring.
Yellow; mid-April–May (occasionally as late as July)
Moist woods and thickets
Eastern 2 columns of counties

191. *Geum canadense*

Stems usually several, erect or ascending, up to about 6 dm tall. Basal leaves long-petioled, some nearly round to ovate with 3 shallow lobes, cordate bases, and crenate-serrate margins, others with 3–9 highly lobed or dissected leaflets, the terminal leaflet (1.5–) 3–8 cm long and (1–) 2–6 cm wide, always larger than the other leaflets; stem leaves similar to the lobed and dissected basal leaves, shorter petioled, usually with only 3 leaflets; stipules prominent, toothed. Herbage sparsely hairy to nearly glabrous. Sepals triangular, 1.3–2 mm long, the sides straight or slightly convex; no bractlets present in the sinuses. Petals yellow (sometimes drying cream-colored or pale pink), obovate or short-oblong, about 1.2 mm long. No hairs present among the pistils. Achenes lacking hairs, 2.5–3 mm long (excluding the style); achene head about 0.8–1 cm long, raised above the hypanthium on a definite stipe.

GILLÉNIA Moench.   Indian physic

Dedicated to Arnold Gillen, an obscure German botanist and physician of the 17th century.

Erect herbaceous perennials arising from rhizomes. Leaves compound, 3-foliolate, nearly sessile. Flowers white or pink, scattered at or near the ends of the branches. Hypanthium narrowly campanulate. Stamens 10–20. Pistils 5, superior; styles threadlike. Fruits few-seeded follicles.

One species in Kansas.

192. *Gillenia stipulata*

G. stipulàta (Muhl.) Trelease   Indian physic

NL., stipulate, having stipules, from L. *stipula*, the diminutive of *stipes*, stalk or stem—the stipules in this species being leaflike and quite prominent.
White; June
Open, often rocky woods and thickets
Cherokee County only

Plants up to about 1 m tall. Leaves 4–10 cm long; under surface glandular-dotted and sparsely hairy; leaflets of lower stem leaves lanceolate, entire or pinnately lobed, with acuminate apexes and bases and double-serrate margins; leaflets of middle stem leaves narrowly lanceolate to narrowly elliptic, with double-serrate margins; stipules prominent, more or less ovate, with laciniate-toothed margins. Hypanthium 4–5 mm long. Calyx lobes triangular, with a short-acuminate apex and ciliate and/or glandular margins, 1–1.5 mm long. Petals white, narrowly oblanceolate, 8–12 mm long. Pistils 1- to 2-ovuled. Follicles 8–9 mm long (excluding the threadlike style), partially enclosed by the persistent hypanthium. Seeds reddish-brown, ellipsoid, longitudinally wrinkled, 3–4 mm long.

POTENTÍLLA L.   Cinquefoil

NL., from L. *potens*, potent, strong, the diminutive suffix *-illa*—that is, potent little plant—in reference to the medicinal value of some species. "Cinquefoil" is derived from Fr. *cinq*, from L. *quinque*, five, + OF. *foil*, from L. *folium*, leaf—the number of leaflets in each leaf in plants of this genus being five, as a rule.

Biennial or perennial herbs (rarely annuals) arising from short, erect rhizomes or rootstocks. Leaves pinnately or palmately compound. Flowers yellow or white, borne singly or in branched clusters. Hypanthium shallow to hemispheric, with a slightly nectiferous inner surface. Sepals 5, alternating with as many bracts. Stamens many. Pistils many, simple, 1-ovuled, borne on a dry, somewhat elevated receptacle; style usually long and threadlike and deciduous. Fruits achenes, usually enclosed by the persistent hypanthium and calyx.

193. *Gillenia stipulata* leaves

ROSE FAMILY

153

194. *Potentilla arguta*

195. *Potentilla norvegica*

196. *Potentilla recta*

Six species in Kansas. *P. paradoxa* Nutt. and *P. rivalis* Nutt. are known from a few widely scattered sites only and are not included in the following key.

Key to Species (3):
1. Flowers solitary on naked peduncles from the nodes of the stem .... *P. simplex*
1. Flowers few to many, in branched clusters 2
  2. Principal leaves below the inflorescence 3-foliolate .... *P. norvegica*
  2. Principal leaves below the inflorescence with 5 to several leaflets (those in the inflorescence smaller, often 3-foliolate or even simple) 3
    3. Leaves pinnately compound .... *P. arguta*
    3. Leaves palmately compound .... *P. recta*

P. argùta Pursh   Tall cinquefoil

L. *argutus*, clear, bright, from *arguo*, make clear—in allusion to the flowers.
White to pale yellow; June–July
Prairies, occasionally in open woodland
Scattered in the east third

Sturdy perennial herbs. Stems erect, usually single and unbranched below the inflorescence, up to about 1 m tall. Leaves pinnately compound, mostly basal, long-petioled, with 7–11 rounded leaflets with sharply serrate margins, the terminal leaflet the largest, 4–8 cm long and 2.5–4.5 cm wide; stem leaves shorter-petioled to sessile, with fewer leaflets. Herbage soft-hairy and glandular throughout. Flowers white to pale yellow, borne in congested, branched terminal clusters. Sepals lanceolate to ovate, 5–9 mm long, alternating with smaller, lanceolate bractlets. Petals obovate, 7–10 mm long. Styles spindle-shaped to narrowly crescent-shaped, nearly basal, deciduous. Achenes light brown, smooth, about 1 mm long.

P. norvégica L.   Norwegian cinquefoil

NL. *Norvegicus*, pertaining to Norway—this circumboreal species being native also to northern Europe.
Yellow; May–July
Prairie, open woods, riverbanks, fields, margins of ponds
East half, especially in the northern counties

Slender annuals or short-lived perennials. Stems erect, usually single, branched or unbranched below the inflorescence, mostly 2.5–6 dm tall. Main leaves 3-foliolate, petioled below and gradually becoming sessile above, the leaflets mostly oblanceolate or elliptic, 2–6 cm long and 1–2.5 cm wide, with rounded apexes, acuminate bases, and serrate margins. Herbage soft-hairy throughout. Flowers yellow, usually many, borne in open, branched terminal clusters. Sepals lanceolate to long-triangular, 4–5 mm long, alternating with lanceolate bractlets shorter or about the same length as the sepals. Petals obovate, 3–4 mm long. Styles broadest at the base, straight or slightly curved, deciduous. Achenes light brown, longitudinally ribbed, 1 mm long or shorter.

P. récta L.   Rough-fruited cinquefoil, sulfur cinquefoil

L. *rectus*, straight, upright—in reference to the erect stem.
Yellow; May–June (–July)
Prairie, roadside banks, waste ground, often in rocky or sandy soil
East third; intr. from Eurasia

Perennial from a short, erect rhizome with a cluster of thickened fibrous roots. Stems erect, slender, 1–3 per plant, mostly 3–7 dm tall. Leaves palmately compound, long-petioled below to sessile or nearly so above; leaflets 5, oblanceolate to narrowly oblanceolate, those of the main leaves mostly 2.5–7.5 cm long and 0.5–2.5 cm wide, with a rounded apex and serrate margins. Herbage sparsely hairy. Flowers yellow (often drying cream-colored), borne in rather loose, branched terminal clusters. Sepals lanceolate, 4–6 mm long, alternating with narrowly lanceolate bractlets about the same length. Petals obcordate, 7–10 mm long. Styles linear, straight, terminal or nearly so, deciduous. Achenes reddish-purple with lighter reticulations, about 1 mm long.

This species has been called *P. sulfurea* or *P. sulphurea* in earlier manuals.

P. símplex Michx.   Oldfield cinquefoil

L. *simplex*, simple or single—in reference to the flowers occurring singly, instead of in clusters, as is commonly the case in the genus.
Yellow; late April–May
Prairies, roadsides, occasionally in open woods
East fourth

Slender, supple perennial arising from a short rhizome. Stems ascending or decumbent, or eventually arching and rooting at the tips, mostly 1.5–3.5 dm long. Leaves long-petioled below to sessile or nearly so above; leaflets 5, obovate to oblanceolate or elliptic, those of the main leaves mostly (1.3–) 2–4.5 cm long and 0.8–2.2 cm wide, with a rounded apex and serrate margins. Herbage sparsely hairy, the under surface of leaves often with appressed silky hairs. Flowers yellow, borne singly in the axils of upper leaves. Sepals lanceolate to long-triangular with slightly convex sides, 2.5–3.5 mm long, alternating with narrowly lanceolate or elliptic bractlets about the same length. Petals obcordate, 4–5 mm long. Styles linear, straight, terminal or nearly so, deciduous. Achenes tan, wrinkled, about 1 mm long.

197. *Potentilla simplex*

LEGUMINOSAE   Legume Family

Herbs, shrubs, and trees. Leaves alternate, mostly pinnately or palmately compound, petiolate, mostly stipulate. Flowers bisexual, regular or irregular (see key to subfamilies below), inflorescence various. Sepals or calyx lobes 5. Petals or corolla lobes 5 (rarely absent or reduced to a single petal as in *Amorpha*). Stamens mostly 10, entirely free from one another, or all 10 fused by their filaments (monadelphous), or 9 fused and 1 remaining free (diadelphous). Pistil 1; ovary superior, 1-celled; style and stigma each 1. Fruit usually a legume (often called a pod), sometimes strongly contracted between adjacent seeds and then called a loment.

The legumes comprise one of the 3 largest families of flowering plants (the Composite Family is the largest; the Legume and Orchid families are the second largest). The legumes may be subdivided into 3 groups which are recognized as distinct families by some authors or as subfamilies by others. They are treated here as subfamilies. In either case, the characters which distinguish one group from another are the same and are outlined in the key below. The genera and species are arranged on the following pages according to the subfamily to which they belong. Names in parentheses in the key indicate alternate family or subfamily names encountered in various commonly used manuals.

The ancient Romans believed that the growing of plants of this family added to the fertility of the land. Pliny writes of the bean (*Vicia faba* L.) that it "fertilizes the ground in which it has been sown as well as any manure"; and Virgil advises in his *Georgics* (poems on husbandry): "Sow your golden grain on land where grew the bean, the slender vetch, or the fragile lupine."

Thus the matter stood, without explanation, until certain techniques of chemistry, microscopy, plant anatomy, and bacteriology were applied to the problem. There is space here only to summarize briefly the principal facts thus revealed.

Nitrogen is necessary for normal plant growth and development; it is especially important as a constituent of proteins. Although this element comprises nearly 80 percent by volume of the earth's atmosphere, this atmospheric nitrogen, or free nitrogen as it is called, cannot be utilized directly by most plants. The free nitrogen must somehow be combined with oxygen, hydrogen, or other elements to form water-soluble compounds, such as nitrates and ammonia, which plants are capable of absorbing through their roots. This process of incorporating atmospheric nitrogen into soluble nitrogenous compounds is referred to as "nitrogen fixation." Certain free-living soil bacteria, as well as some algae, are capable of fixing nitrogen, which they then utilize in their own physiological processes. When these small organisms die, their bodies decompose, and the nitrogen compounds become available to other organisms growing in the soil.

In the case of leguminous plants, certain soil bacteria enter the root hairs of the plant, penetrate into the cortex of the root, and stimulate the cortex cells to multiply and form small round lumps or nodules along the root. Within the nodules, the bacteria live and reproduce, deriving necessary carbohydrates, and perhaps other materials as well, from the tissues of the green plant. The plant benefits from the nitrogen compounds produced by the bacteria, part of these compounds being absorbed by the surrounding plant cells and translocated to other regions of the plant. Some of these compounds may be excreted into the soil where they are beneficial to other plants in the area. When the leguminous plant dies, or when it is plowed under as "green manure," the plant body decays and the nitrogenous compounds are released into the soil and become available to nonleguminous plants subsequently grown in that same place.

Of the approximately 12,000 species of legumes, about 10 percent have been tested for the ability to fix atmospheric nitrogen; of this 10 percent, about 90 percent appear to possess this ability. With only a few known exceptions, other kinds of flowering plants are apparently incapable of fixing nitrogen. The exceptions are primarily trees and shrubs and include members of the genera *Alnus* (alder), *Myrica* (such as *Myrica gale*, the bog myrtle), *Ceanothus*, *Shepherdia* (buffalo berry), *Hippophaë* (sea buckthorn), *Elaeagnus* (oleaster), *Casuarina* (shrubs and small trees native to Australia and New Zealand), *Artemisia* (western sage), *Opuntia* (prickly pear cactus), and *Chrysothamnus* (rabbitbrush).

Key to Subfamilies:
1. Flowers regular, stamens 10 or more, leaves twice-pinnately compound .... Mimosoideae (Mimosaceae), p. 156
1. Flowers somewhat irregular to distinctly bilateral, stamens 10 or fewer; leaves mostly pinnately or palmately compound, sometimes twice-pinnately compound, rarely simple 2
    2. Lateral petals (wings) covering the upper one (standard) in the bud, petals typically free from one another .... Caesalpinioideae (Caesalpiniaceae, Cassiaceae), p. 157
    2. Lateral petals (wings) enclosed by the upper one (standard) in the bud, the 2 innermost (lower) petals fused to one another at least basally to form the keel, flowers of the "sweet pea" type (except in *Amorpha*, where only the standard is present, and in *Petalostemon*, where the standard is accompanied by 4 petal-like sterile stamens) .... Lotoideae (Fabaceae, Papilionaceae, Papilionoideae), p. 160

MIMOSỖIDEAE   Mimosa Subfamily

Named after the genus *Mimosa*, with many species native to tropical and warm regions and noted for their sensitive leaves, whose responses to wind, touch, variations in light intensity, etc., suggested the NL. name (from Gr. *mimos*, mimic actor).

Herbs, shrubs, and trees. Leaves usually twice-pinnately compound, stipulate. Flowers bisexual, regular, very small, in our species borne in small pomponlike heads. Petals 3–6 (usually 5), free from one another or more or less united below, the corolla very short and inconspicuous. Stamens many, or of the same number as the petals, or twice as many, much longer than the corolla, the threadlike filaments providing most of the color of the inflorescence.

This group of legumes contains approximately 35 genera and 2,000 species, primarily of dry tropical or subtropical regions. Cultivated representatives include various species of *Acacia*, the sensitive plant (*Mimosa pudica* L.), and several species of *Albizzia*, small trees or tall shrubs known commonly as silk trees or mimosa trees.

Kansas has 5 genera and 8 species. Two of these, *Mimosa borealis* Gray (pink mimosa, cat's-claw mimosa) and *Prosopis juliflora* (Sw.) DC. var. *glandulosa* (Torr.) Cockll. (mesquite), are woody shrubs and are not included in the following key. *Acacia angustissima* (Mill.) Kuntz., a suffrutescent perennial similar to *Desmanthus illinoensis* (Michx.) MacM., is known from Montgomery, Chautauqua, Cowley, and Morton counties only and is also excluded.

Key to Genera:
1. Stems prickly; flowers pink or rose .... *Schrankia*
1. Stems without prickles; flowers white or greenish .... *Desmanthus*

DESMÁNTHUS Willd.   Bundleflower

NL., from Gr. *desme*, bundle, + Gr. *anthos*, flower—because the inflorescence is a head composed of many small flowers.

Shrubs or perennial herbs. Leaves twice-pinnately compound. Flowers white or greenish. Calyx campanulate, 5-toothed. Petals 5, separate or fused slightly. Stamens 5 or 10. Fruits oblong to linear, straight or curved, usually borne in clusters.

Three species in Kansas. *D. cooleyi* (Eat.) Trel. is known from Morton County only and is not included here.

Key to Species:
1. Stems erect; petals oblanceolate, about 1.5 mm long; fruits crescent-shaped, mostly 1.5–2.5 cm long and 4–5 mm wide; common throughout Kansas .... *D. illinoensis*
1. Stems ascending or decumbent; petals linear or very narrowly elliptic, about 3 mm long; fruits linear, straight or nearly so, mostly 3–6 cm long and 2–3 mm wide; central and southwest counties, not common .... *D. leptolobus*

D. illinoénsis (Michx.) MacM.   Illinois bundleflower

Although the species ranges from Ohio and Kentucky to Colorado and New Mexico, and from

South Dakota to Texas, the specimens by which Michaux, French botanist, established the species in 1803 were collected in Illinois Territory.
White; mostly mid-June–July (–August)
Prairies, flood plains, roadsides, waste ground
Throughout

Perennial herbs. Stems erect, somewhat woody, strongly angled, branched from the base, lacking prickles, mostly 0.3–1 m tall. Leaves 3.5–10 cm long (including the petiole); leaflets 1.5–2.5 mm long and about 0.75 mm wide, oblong or subulate, with an oblique base, apiculate apex, and ciliate or entire margins; stipules threadlike. Flowers white, the heads 0.9–2 cm wide. Calyx teeth about 0.5 mm long. Petals oblanceolate, about 1.5 mm long. Stamens 5. Fruits crescent-shaped, 2- to 5-seeded, 1–2.5 cm long and 4–5 mm wide. Seeds brown, flattened, ovate to variously 3-sided, 2.8–4.2 (–4.5) mm long.

D. leptolòbus T. & G.   Narrowpod bundleflower

NL., from Gr. *leptos*, narrow, + *lobos*, pod—the pods being linear and nearly straight
White; June–September
Sandy soil
Scattered in central and southwestern counties

Perennial herbs. Stems decumbent to ascending, much branched, up to about 6 dm tall, strongly ridged, the ridges somewhat roughened, at least in the upper parts of the plant. Leaves mostly 1.8–4 cm long (including the petiole); leaflets 1–3 mm long and 0.4–1 mm wide, similar to those of *D. illinoensis*; stipules bristlelike. Flowers white. Calyx teeth about 0.3 mm long. Petals linear or very narrowly elliptic, about 3 mm long. Stamens 5. Fruits linear, straight or nearly so, mostly 4- to 8-seeded, 3–6 cm long and 2–3 mm wide. Seeds brown, somewhat flattened, narrowly ellipsoid or oblong, (3–) 4.5–6 mm long.

SCHRÀNKIA Willd.   Sensitive brier

Named for F. von Paula Schrank (1747–1835), botanical and agricultural writer and director of the Botanical Garden in Munich.

Perennial herbs or shrubs, ours prickly. Leaves twice-pinnately compound, usually sensitive. Calyx much reduced. Corolla funnelform or cylindric, 5-lobed. Stamens 10 (occasionally 11–13). Fruits linear to oblong, strongly ribbed or 4-angled, usually prickly, separating along 4 lines.

Two species occur in Kansas. *S. occidentalis* (W. & S.) Standl. has been collected once, along a railroad in Stevens County, and may be merely a waif. It differs from *S. nuttallii* in having no lateral veins evident on the under surface of the leaflets.

S. nuttàllii (D. C.) Standl.   Sensitive brier, bashful brier

Named after Thomas Nuttall (1786–1859), plant explorer–naturalist and one-time professor of natural history and curator of the Botanic Gardens at Harvard.
Pink; June–July (–September)
Prairie
Throughout, except for some of the western counties

Perennial herbs, beset with rigid, curved prickles along the stem, petioles, and peduncles. Stems weak and sprawling, arched, or decumbent, up to about 1 m long. Leaves 5–11 cm long (including the petiole); leaflets 8–10, oblong, (3–) 3.5–7.5 mm long and 1–1.5 mm wide, with a sharply pointed apex, oblique base, and both midvein and lateral veins evident on the under surface; stipules subulate to acuminate. Flowers pink, the heads 1.5–3 cm wide. Calyx minute. Corolla about 2.5 mm long. Fruits linear, densely prickly, golden brown, 5–12 cm long and about 7 mm wide at maturity. Seeds brown, rather irregular and variable in shape but generally more or less ovate to oblong and slightly flattened, mostly 3–5 mm long, often parasitized by the larvae of some insect.

CAESALPINIOÌDEAE   Caesalpinia Subfamily

Annual, biennial, and perennial herbs, shrubs, trees, or vines. Leaves usually stipulate, pinnate or bipinnate in all but *Cercis* (redbud), whose leaves are simple though they may represent a single leaflet of an ancestral pinnate leaf. Flowers regular to somewhat irregular; bisexual, or unisexual and then monoecious or dioecious, or with bisexual and unisexual flowers on the same plant; the inflorescence type various. Sepals and petals usually 5, sometimes fewer, fused at the base to form a hypanthium, the petals conspicuous

198. *Desmanthus illinoensis*

199. *Schrankia nuttallii*

200. *Caesalpinia jamesii*

and colorful. Stamens usually 10, sometimes 5 or fewer, inserted on the hypanthium, usually free from one another.

This subfamily contains about 60 genera and 2,200 species, primarily of the Old World tropics. Cultivated members include redbud (*Cercis canadensis* L.), senna (*Cassia* spp.), *Poinciana*, and others.

Kansas has 6 genera and 8 species. *Cercis canadensis* L., *Gleditsia triacanthos* L. (honey locust), and *Gymnocladus dioica* (L.) K. Koch (Kentucky coffee tree) are trees and are not included in the following key, nor is *Hoffmanseggia glauca* (Ort.) Eifert, an herb reported from Morton County only.

Key to Genera:
1. Leaves pinnately once-compound .... *Cassia*
1. Leaves pinnately twice-compound .... *Caesalpinia*

CAESALPÍNIA L.

NL. *caesalpinia* from Andrea Caesalpino, Italian botanist.

Ours perennial herbs (trees, shrubs, or subshrubs in more southern latitudes). Leaves bipinnately compound; stipules small. Flowers yellow, irregular, bisexual, borne in terminal clusters. Calyx tube short, 5-toothed. Petals 5. Stamens 10. Pistil sessile or stipitate. Fruit ovate to crescent- or sickle-shaped, flattened, dehiscent; seeds few to several.

One species in Kansas.

C. jámesii (T. & G.) Fischer

Named after Edwin James (1797–1861), surgeon-naturalist who traveled with Maj. Stephen Long's expedition to the Rocky Mountains (1819–1820) and who first collected this species.
Yellow or cream-colored; late May–August
Dry prairies and dunes, usually in sandy soil
Scattered, primarily in the southwest fourth

Suffrutescent perennial with a fusiform woody taproot. Stems several, erect, 1.5–4 dm tall. Herbage minutely hairy and conspicuously dotted with black glands. Leaves 4–10 cm long (including the petiole); leaflets oblong, 3–7 mm long and 1.5–3.5 mm wide, with a rounded apex and acute base; stipules subulate to acuminate. Flowers borne in racemes 4–16 cm long (including the peduncle). Calyx 7–9 mm long, 1 lobe larger than the others and cupped around the stamens. Petals yellow or cream-colored, tinged with red near the base of the broadened portion, slightly longer than to slightly shorter than the sepals, unequal in size and shape, glandular-dotted. Anthers red. Fruits crescent-shaped, 2–2.5 cm long and 8–10 mm wide, covered with branched or tufted hairs and black glands. Seeds olive-green or brown, flattened, more or less ovate and 5–6.5 mm long.

CÁSSIA L.   Senna

Cassia is the name used by Dioscorides and Pliny (1st century) for the cassia-bark tree, a member of the Laurel Family, sometimes used as a substitute for cinnamon. Linnaeus (1753) appropriated the name for this genus of the Senna Family. "Senna" is from Arabic *sana*, for any plant of the genus *Cassia*.

Ours annual or perennial herbs (shrubs, trees, or woody vines farther south). Leaves even-pinnate, stipulate; petiole often with 1 or more glands. Flowers regular or irregular, bisexual, usually in axillary racemes, sometimes solitary or in terminal clusters. Hypanthium very short. Sepals slightly or quite different in size. Petals nearly equal or 1 larger than the others. Stamens 5 or 10, often differing in size, or some of them sterile. Fruits flattened (in ours), erect to pendant, indehiscent or dehiscent (sometimes violently so), few- to many-seeded.

Three species in Kansas.

Commercial senna, from the tropical *Cassia fistula* L., is used medicinally as a cathartic. According to John Kingsbury (21), "a number of species in the United States, including *C. fasciculata* Michx., possess the same or a similar active principle and if ingested in some quantity may cause distress and occasionally death in animals."

Key to Species (3):
1. Stamens partly imperfect, the upper 3 lacking normal anthers; pods normally more than 7 cm long, scarcely or tardily dehiscent; leaflets commonly more than 3 cm long; flowers in several axillary racemes .... *C. marilandica*
1. Anthers all perfect, but more or less unequal; pods less than 7 cm long, readily dehiscent, the valves then twisting spirally; leaflets commonly less than 2 cm long; flowers solitary

or in small clusters 2

   2. Stamens 10; longest anthers about 8 mm long, several times longer than the filaments; petals 1–2 cm long, nearly equal .... *G. fasciculata*

   2. Stamens 5; longest anthers 2–3 mm long, little longer than the filaments; longest petal 6–8 mm long, about twice as long as the other 4 .... *C. nictitans*

C. fasciculàta Michx.   Showy partridge pea

NL., from L. *fasciculus*, small bundle, bunch of flowers, or nosegay—applicable to the inflorescence or to a cluster of the flowering branches.

Yellow; July–October

Roadsides, disturbed prairie, and open woodlands

East half and southwest fourth

201. *Cassia fasciculata*

Erect annual. Stem branching freely from the base, 1.5–9 dm tall. Leaves 3–11 cm long; leaflets 18–20, oblong, (4–) 9–18 mm long and 2–4.5 mm wide, with a rounded mucronate apex and oblique base; the leaflets are sensitive to touch and fold snugly together when handled; petiolar gland just below the first pair of leaflets, conspicuous, sessile or on a short stalk. Stems, sepals, and fruits minutely hairy. Flowers showy, usually 2–4 together in the leaf axils, and commonly opening 1 at a time. Hypanthium scarcely evident. Sepals lanceolate to lance-acuminate, membranous except for an area along the midrib, 6–14 mm long. Petals yellow, 1–2 cm long, the lowest one larger than the others, which have a purple spot at the base. Stamens 10, eventually becoming purple, the longest ones about 8 mm long, all pollen-bearing and opening by terminal pores. Pistil silky, often arching away from the stamens. Fruits linear, flattened, straight or slightly curved, erect, mostly 9- to 15-seeded, 4–5.5 cm long and 5–6 mm wide, dehiscing along 2 sutures, the valves twisting spirally as they dry. Seeds dark brown, flattened, mostly rhomboidal, 3.5–4.5 mm long, with longitudinal rows of minute punctae (use magnification).

The flowers of this species are without nectar, but honeybees, bumblebees, and various other wild bees, coming to the flower for pollen, effect cross-pollination. The petiolar glands just below the first pair of leaflets supply nectar to several kinds of short-tongued bees and various other hymenoptera and to a variety of flies and ants. Our plant seems to get no return for this latter service. However, in the case of a tropical *Acacia* tree (Mimosa Family) which provides home for ants in its hollow stipular thorns, nectar from petiolar glands, as well as sausage-shaped food bodies on the tips of the leaflets, the ants put up a fight against any molestation of the tree.

C. marilándica L.   Wild senna

"Of Maryland."

Yellow; mid-July–August

Roadsides, rocky banks, and hillsides

East two-thirds

202. *Cassia marilandica*

Perennial with smooth green stems 1–2 m tall springing from a rootstock that bears thickened roots and reproduces by horizontal rhizomes. Leaves 1.2–3 dm long; leaflets 12–20, elliptic, (1.5–) 2.5–8 cm long and (0.5–) 1.5–3 cm wide, with an acute mucronate apex, and slightly oblique base; petiolar gland more or less ovoid, near the base of the petiole. Flowers borne in axillary and terminal panicles. Hypanthium very short. Sepals ovate, 5–7 mm long. Petals yellow (becoming white or cream-colored with brown veins when dry), obovate with a very narrow base to obtriangular, 8–11 mm long, the apex rounded or slightly emarginate. Stamens 10; the upper 3 erect, somewhat petal-like, and sterile; the middle 4 fertile and projecting forward; the lower 3 fertile but indehiscent, longer than the others, and projecting forward and downward; fertile stamens with apical pores. Pistil silky, with an upcurved style. Fruits dark brown, linear, flattened, straight or slightly curved, constricted and septate between the 10–25 seeds, pendulous, (4–) 6.5–8.5 cm long and 8–10 mm wide, tipped by the persistent curved style, scarcely or tardily dehiscent. Seeds slightly flattened, oblong or ovate, 4.5–5.5 mm long, gray, the central portion of each face dull, the rest of the seed coat lustrous or at least a darker gray.

The flowers produce no nectar, but bees, mostly bumblebees, come to them for pollen, landing on the pistil and lower 3 stamens. The bees bite open the 4 middle anthers, collect the pollen, and effect cross-pollination as they go from flower to flower. It is curious that 6 of the 10 stamens are nonfunctional, 3 by sterility, and 3, though fertile, by their posture and indehiscence. It would seem that the anthers of the 3 pollen-bearing lower stamens, by virtue of their proximity to the stigma, might effect self-pollination. However, since they do not automatically open and bees do not bite them open, their pollen goes unused. No pods develop when bees are excluded from the flowers.

The cathartic value of the leaves and pods, when gathered after the pods are ripe, is equal to that of the Alexandrian and East Indian sennas commonly imported for the drug trade.

C. níctitans L.   Sensitive pea

NL. *nictitans*, winking, from L. *nicto*, to wink, in reference to the opening and closing of the leaves.
Yellow; mid-August–early October
Wooded hillsides, usually in rocky or sandy soil
Southeast eighth

Erect annuals. Stems minutely hairy (use magnification), 1.5–4 dm tall, with branches ascending to spreading widely. Leaves 2–6 cm long; leaflets mostly 12–28, narrowly oblong, (4–) 7–16 mm long and 1–3 mm wide, with a rounded mucronate apex and oblique base, sensitive to touch; petiolar gland just below the 1st pair of leaflets, cup-shaped, on a short stalk. Flowers small, 1–3 together in the leaf axils, wilting by mid-day. Hypanthium scarcely evident. Sepals lance-acuminate, 3–5 mm long. Petals yellow, 3–8 mm long, the lowest one about twice as long as the other 4. Stamens 5; anthers pink, opening longitudinally, the longest ones 2–3 mm long. Pistil silky; style curving toward the stamens. Fruits linear, flattened, straight or slightly curved, erect, with minute appressed hairs, mostly 5- to 10-seeded, 2–3.5 cm long and about 4.5 mm wide, dehiscing along 2 sutures, the valves twisting slightly as they dry. Seeds black, shiny, minutely punctate, about 3.5 mm long.

The seeds of this plant are eaten by wild turkey and bobwhite quail.

203. *Cassia nyctitans*

LOTOÌDEAE   Bean Subfamily

Herbs, shrubs, trees, and woody or herbaceous vines. Leaves mostly alternate, compound (apparently simple in *Crotalaria*), stipulate. Flowers commonly bisexual, irregular, the inflorescence various but frequently a spike, raceme, or dense head. Sepals 5, more or less united at the base, sometimes appearing to be 4 or fewer through fusion of 2 or more. Especially characteristic of this group and of no other is the papilionaceous corolla ("butterfly-like," from L. *papilio*, butterfly). This consists of 5 unequal petals: The uppermost petal (see fig. 204) is designated the standard. The lateral 2 petals are similar to one another and are called wings. The 2 innermost and lowest petals are closely appressed, usually coherent or fused along their adjoining margins, and generally enclose the stamens and pistil; these petals comprise the keel of the corolla. The petals may taper gradually to their bases or may be abruptly narrowed to a very slender base (then said to be clawed). Sometimes 1 or more of the petals have either 1 or 2 basal lobes or auricles. Stamens usually 10, rarely fewer, usually monadelphous or diadelphous, sometimes entirely free from one another.

A subfamily of about 400 genera and 10,000 species. Many important food and fodder plants belong to this group: garden peas (*Pisum sativum* L.), lentils (*Lens culinaris* Medic.), chick peas or garbanzos (*Cicer arietinum* L.), peanuts (*Arachis hypogaea* L.), beans (*Phaseolus* spp.), clover (*Trifolium* spp.), alfalfa (*Medicago sativa* L.), soybeans (*Glycine max* Merr.), to cite a few examples. Ornamentals include *Wisteria*, lupine (*Lupinus* spp.), and sweet pea (*Lathyrus* spp.). Species of *Indigofera* are sources of indigo dye.

Numerous poisonous species belong to the Lotoideae, one of the most notorious being the rosary pea (*Abrus precatorius* L.). The seeds of this tropical woody vine are brilliant red with a contrasting black spot on one end and are sometimes incorporated into necklaces, swizzle sticks, and other trinkets sold to tourists in Mexico, Central America, and other tropical countries. The seeds attracted national attention several years ago when it was discovered that large numbers of novelty costume-jewelry pins made of the poisonous *Abrus* seeds had been imported and were being sold by a large mail-order department store company that was not aware of the potential danger. Rosary peas contain the powerful drug abrin, which is one of the most lethal poisons known to man— less than 1 seed, well chewed, is enough to kill an adult human.

Legumes, especially species of *Lespedeza*, *Psoralea*, and *Petalostemon*, are important constituents of prairie vegetation. Not only do they add to the fertility of the soil, but cattle and other grazers will selectively graze them for their high protein content.

The papilionaceous corollas characteristic of most members of this subfamily are adapted for pollination primarily by large bees (rarely by butterfles or birds) which have the strength necessary to creep into the flowers and force the petals apart. In general, the standard offers visual attraction, and the keel provides a landing place for prospective pollinators. The symmetric wings and/or color markings on the standard guide the insect toward the nectar or pollen which serves as a food attractant. The variation which can be seen in this overall plan is quite remarkable.

In pollen presentation, for example, the most primitive type requires that the upper edge of the keel be open, or else opens easily when depressed by a pollinator. The filaments and style are stiff and do not bend, so they protrude when a visitor lands and are concealed again when it leaves. This situation exists in *Trifolium* and *Astragalus*, among others. In *Lotus* and *Coronilla*, the keel is rostrate (beaklike) and the upper edge is closed except for the outermost tip. The distal ends of the filaments are dilated (swollen) and prevent the pollen, which is shed into the rostrum, from falling farther down into the keel. When the keel is pressed down, the ends of the filaments push the pollen mass out through the rostrum. In a third type of presentation, as in our species of *Lupinus*, the pollen is brushed out of the keel by the upper part of the style, which is sharply bent and densely hairy, as it grows in length.

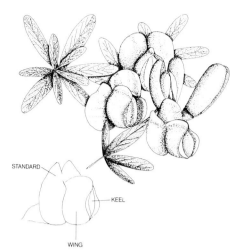

204. *Baptisia* flowers: an example of a papilionaceous corolla

A fourth, and most spectacular, form of pollen presentation is exhibited by those plants which have explosive blossoms. In these, the pistil and filament sheath are confined to the keel under pressure (which may come from 1 or more sources as described below). The weight of a pollinator causes the upper edge of the keel to rip open, allowing the pistil to rush out and spread a cloud of pollen. In *Medicago*, the necessary pressure is exerted by the pistil, which, after the explosion, lies against the standard. In *Genista* (represented in Kansas by ornamental species only), the pressure derives from the wings and keel, which, after the explosion, assume a downward position at right angles to their original one. In *Desmodium*, both movements occur. In some exotic species of *Cytisus* (again, known here only as cultivars), the stamens, which are alternately long and short, and the style curl up; when released, the 5 short stamens hit the pollinator in the abdomen and the 5 long ones (usually the style, too) hit it on the back. Explosive papilionaceous blossoms have pollen as their only food attractant and will discharge only once. Blossoms with nonexplosive pollen presentation may have nectar also and will function more than once.

In blossoms which offer nectar, the stamens are diadelphous, i.e., the upper stamen is free from the other 9, and 2 openings at its base offer access to the nectar at the base of the pistil. In addition, the bases of the petals clasp one another to form a sheath which can be entered from the front only, thus requiring a nectar gatherer to enter by the "front door" and pay his dues as a pollinator. The calyx tube may also be elongate, enhancing the protective effect of the petal bases. In *Trifolium pratense* L. (red clover), the bases of the petals are also fused to the base of the stamen sheath. In blossoms which offer pollen only, the stamens are monadelphous and the bases of the petals may be clawed, leaving spaces between them, there being no nectar to protect for specific pollinators.

There are a few radical deviations from the typical papilionaceous morphology and ecology. In *Petalostemon*, for example, the flowers are quite small, are short-tubular and nearly radial in symmetry, and have protruding, rather than concealed, stamens and pistils. The flowers are densely clustered into small terminal spikes or heads which serve as more effective visual attractants than would the tiny individual blossoms. Pollination is of the "mess-and-soil" type and occurs as insects crawl over the flowers gathering pollen. Flowers of *Vicia lathyroides* L. are self-fertilizing and exhibit a degeneration of parts, as do some of the flowers of *Amphicarpaea* (hog peanut). Often in *Lespedeza*, some of the flowers are cleistogamous and have no corolla at all. In *Clitoria*, the blossoms hang upside down, so that the standard provides a landing site for pollinators.

Kansas has 29 genera and about 111 species. Except for the woody *Amorpha canescens*, only the herbaceous species are included in the following key.

Key to Genera (3):

1. Flowers with 1 petal, purple with bright-orange anthers; leaves pinnately compound .... *Amorpha*
1. Not as above 2
　2. Leaves simple or 1-foliolate; a small annual with yellow flowers; fruit an inflated pod .... *Crotalaria*
　2. Leaves compound 3
　　3. Leaves evenly pinnate, or the terminal leaflet represented by a tendril .... *Vicia*
　　3. Leaves odd-pinnate, palmate, or 3-foliolate 4

205. *Amorpha canescens*

AMÓRPHA L.   Amorpha

NL., from Gr. *amorphos*, deformed—in reference to the incomplete corolla.

Shrubs or subshrubs. Leaves odd-pinnate. Flowers blue or purple (occasionally white), relatively small, borne in dense terminal racemes. Calyx tube obconic, the lobes equal in size, or the lowest 1 slightly larger. Standard more or less folded around the stamens; wings and keel absent. Stamens 10, protruding beyond the perianth; filaments united only at the base; anthers bright orange or yellow. Fruit oblong, often curved, somewhat flattened, indehiscent, 1- to 2-seeded.

Three species (all woody) in Kansas. Although a shrub, *A. canescens* L. is included here since it is often of rather small stature and might be mistaken for a herbaceous plant. *A. fruticosa* L. is easily recognized as a shrub, and *A. nana* Nutt. is known from Clark County only.

A. canéscens Pursh   Lead plant

L. *canescens*, becoming gray—as this plant does by its covering of fine hairs. The common name
    is suggested by the color of the foliage.
Purple; June–July
Prairie
East three-fourths

Erect or ascending subshrub, 0.3–1 m tall, with roots penetrating 2–5 m, depending upon the condition of the subsoil. Leaves mostly 3.5–10 cm long; leaflets 15–47, mostly lanceolate to elliptic or oblong, 8–14 mm long and (3–) 4–6 mm wide, with a rounded mucronate apex and slightly oblique base. Stems, under surface of leaves, and calyces densely woolly; upper surface of leaves darker green and not as woolly as the under side. Racemes 6–10 cm long. Calyx 3–4 mm long. Standard purple, 4–5 mm long. Fruit curved, woolly, glandular, about 4.5 mm long. Seeds pale orange-brown, elliptic with a slight beak on 1 end, 2–2.5 mm long.

The Indians report that they dried the leaves for tea and for pipe-smoking and laid small pieces of stem on a part afflicted with neuralgia and burned them there as a counterirritant.

Compare with *Petalostemon villosum*.

See plate 47.

AMPHICÁRPAEA Elliott ex Nutt.   Hog peanut

NL., from Gr. *amphi*, of both kinds, + *carpos*, fruit—plants of this genus having 2 kinds of
    flowers which produce 2 kinds of fruit.

Low, slender, twining perennial herbs. Leaves 3-foliolate, each leaflet stalked and subtended by a pair of minute, stipulelike structures called stipels. Flowers of 2 kinds: Those of the upper branches are complete, borne in axillary racemes, and produce flattened, somewhat curved pods which coil after dehiscence. Calyx slightly irregular, 4-lobed. Blade of the standard obovate, auricled, and narrowed to the base; blade of the wings oblong, auricled; keel petals narrowly obovate, nearly straight. Stamens 10, diadelphous. The flowers at the base of the plant are borne on threadlike, creeping branches, have only rudimentary petals (or none at all) and only a few stamens, and are self-pollinating. After pollination, the peduncles elongate and push the flowers into the soil where they produce fleshy, 1-seeded subterranean pods in the same manner that the fruits of the cultivated peanut plant are produced.

One species in Kansas.

206. *Amphicarpaea bracteata*

A. bracteàta (L.) Fern.   Hog peanut

NL., from L. *bractea*, small leaf—the flowers being subtended by small leaves, or bracts, as they are called.
Pale purple to white; August–September
Woods and thickets
East fifth

Stems slightly to densely hairy, 0.3–2 m long. Leaflets ovate to broadly ovate, occasionally rhombic, mostly 2–7 cm long and 1.5–7 cm wide; apex acute or short-acuminate; base rounded, that of the lateral leaflets usually oblique; both surfaces sparsely hairy. Racemes with 5–20 pale-purple to white flowers, each flower or sometimes a pair of flowers subtended by an obtriangular or fan-shaped bract. Calyx of aerial flowers 5–7 mm long. Standard 12–14 mm long, the margins recurved; wings and keel nearly straight, slightly shorter than the standard. Aerial fruits straight or slightly curved, 2- to 4-seeded, 1.5–3 cm long and about 7 mm wide. Seeds red-brown mottled with black, slightly flattened, round to somewhat 3-sided with rounded corners in face view, 3–3.5 mm long.

Compare with species of *Strophostyles* and *Apios*.

According to A. C. Martin (22), the seeds and leaves provide food for the ruffed grouse, ring-necked pheasant, and bobwhite quail, while meadow mice and white-footed mice eat the seeds.

The seeds are thought to have been an important food of the Indians of the central states, particularly in the Missouri Valley, who profited by the habit of field mice to store the seeds for winter use. It was the custom of the women to ferret out the mice's nests in autumn, sometimes acquiring several quarts of beans from a single nest.

ÀPIOS Fabricius   Groundnut

Gr. *apios*, pear—the underground tubers sometimes being pear-shaped.

Twining perennial herbs, with milky juice, arising from slender rhizomes. Some of the roots terminate with a starchy tuber, others enlarge in several places, thus producing a string of tubers. Leaves odd-pinnate with 5 or 7 stalked leaflets. Flowers borne in axillary racemes. Calyx tube with 4 very short lobes (sometimes scarcely evident) and 1 longer lobe. Blade of the standard round or obovate, becoming reflexed, auricled near the base; blade of the wings curved, more or less obovate to oblong, auricled, deflexed below the keel; keel strongly curved, blunt-tipped. Stamens 10, diadelphous. Pods linear, several-seeded, splitting into 2 twisted valves.

One species in Kansas.

A. americàna Medic.   Groundnut, potato bean

"American."
Brownish-purple or red-brown; August–September
Moist thickets
Known from scattered sites in the east half and from Meade County

Stems smooth or hairy, up to 2 m long. Leaflets ovate to lanceolate, 1.8–7 cm long and 1–4 cm wide; apex acute or short-acuminate, with the midvein extended into a tiny sharp point; base usually rounded or slightly cordate; both surfaces glabrous or very sparsely hairy. Racemes with 5 to many fragrant flowers; each node of the inflorescence subtended by 2 minute, ephemeral bractlets. Calyx 4–6 mm long, only the upper lobe prominent, the rest of the margin merely undulate. Standard white outside, red-brown inside, 9–13 mm long, the blade round or somewhat bean-shaped, apically notched; wings brownish-purple, the blade obovate; keel pale purple or greenish. Fruits 4.5–9.5 cm long and 5–6 mm wide. Seeds dark brown, plump, oblong, slightly flattened, 5–6 mm long.

The Indians, as well as the early settlers, found the cooked tubers palatable and nutritious, and the seeds were used by them as we use peas. According to early chronicles, the Pilgrims, during their first hard winters, "were enforced to live on groundnuts," which are more common eastward than in Kansas.

ASTRÀGALUS L.   Milkvetch, poison vetch, loco weed

Gr. *astragalos*, L. *astragalus*, ancient name of a plant in the Legume Family; also the name for vertebra or anklebone, which in ancient times was used as a form of dice. The name may have been applied because the sound of seeds rattling in the pods of some species resembles that of dice shaken in a thrower.

Annuals or herbaceous perennials with stout taproots or rhizomes and with or without aboveground stems. Leaves odd-pinnate with several to many leaflets. Flowers white, yellowish-white, pink or purple, borne in axillary racemes. Standard longer than the other petals, clawed, reflexed, the blade oblanceolate to nearly round; wings clawed, auricled, the apex broadly rounded or notched; keel petals rounded or acute, with a basal auricle. Stamens diadelphous. Fruits various in shape, beaked by the persistent style, usually dehiscent, 1- to many-seeded, subtended by remnants of the stamen tube and ruptured calyx.

A large and difficult genus. Sixteen species in Kansas. Only the 10 most commonly encountered are included in the following key. NOTE: Mature pods are usually necessary for positive identification. Also, measurements have been made from dried specimens, and the figures given for flower length may be somewhat shorter than those for fresh blossoms. To determine whether the hairs of the leaves are laterally or basally attached, prod the distal end of a hair (under magnification) gently with a needle or dissecting probe and note whether what appears to be the base of the hair is attached or is free and pivoting around some point farther up the hair.

Numerous species of *Astragalus* produce an alkaloidlike substance toxic to livestock, seriously deranging the functioning of the nervous system and resulting in loss of muscular control, and, in extreme cases, emaciation and death. This affliction was named *loco* (the Spanish word for "crazy") by stockmen, and plants which cause such severe poisoning are called locoweeds.

At least 24 species of *Astragalus* are obligate selenium accumulators or absorbers; these are used as selenium indicators. Others are facultative absorbers, and some are even inhibited by the presence of selenium in the soil. The absorbers, called poison vetches, are toxic to livestock to one degree or another, depending upon the soil in which they are growing and their stage of growth. The most troublesome of our species are *A. pectinatus* and *A. racemosus*, both obligate selenophiles.

The harmless species of the genus are called milkvetches.

Key to Species:
1. Hairs of the leaves laterally attached 2
  2. Robust plants up to 1.3 m tall, with 1 to few stems arising from rhizomes; fruits erect, completely 2-celled .... *A. canadensis*
  2. Low plants up to 6 dm tall, with several to many stems arising from a taproot; fruits variously oriented but always 1-celled 3
    3. Flowers mostly 8–11 mm long; calyx tube shallowly bell-shaped or short-cylindric, 2–4.5 mm long, the base not pouchlike on the upper side .... *A. lotiflorus*
    3. Flowers mostly 12–22 mm long; calyx tube cylindric, 6–9 mm long, the base pouchlike on the upper side .... *A. missouriensis*
1. Hairs of the leaves basally attached 4
  4. Calyx bell-shaped or short-cylindric 5
    5. Flowers 5–8 mm long, calyx 2–2.5 mm long, fruits ellipsoid to broadly ovoid; plants of the western half of Kansas .... *A. gracilis*
    5. Flowers 8–13 mm long, calyx 3–6 mm long; fruits crescent-shaped; plants primarily of extreme eastern Kansas .... *A. distortus*
  4. Calyx definitely tubular 6
    6. Fruits 1-celled, flowers white or yellowish 7
      7. Leaflets narrowly linear, similar to the rachis in appearance; fruit ovoid to ellipsoid, more or less round to elliptic in cross section .... *A. pectinatus*
      7. Leaflets oblong to narrowly elliptic or almost linear, but easily distinguished from the rachis; fruits ellipsoid to linear-oblong with a groove along the upper side, V- or heart-shaped in cross section .... *A. racemosus*
    6. Fruits 2-celled at least most of their length; flowers varying intensities and shades of purple or pink, rarely white or yellowish 8
      8. Herbage gray-green and silky; aboveground stems quite short, the stout peduncles appearing almost basal; fruits plump and crescent-shaped, often broadest near the base .... *A. mollissimus*
      8. Herbage more or less hairy, but not silky; aboveground stems well developed, the racemes obviously axillary; fruits not as above 9
        9. Flowers pink or pinkish-purple; stipules all free from one another ....
        .... *A. plattensis*
        9. Flowers purple to pink, white, cream, or greenish; at least the stipules at the lowest nodes fused and sheathing the stems .... *A. crassicarpus*

A. canadénsis L.   Canada milkvetch

L. *-ensis*, suffix added to nouns of place to make adjectives; *canadensis*—of Canada. "Milkvetch" is an old English name suggested by the belief that goats gave more milk when feeding on some plants of this species; but "vetch" is from OF. *Veche*, a corruption of L. *vicia*, name used by Pliny, Virgil, and others for some tendril-bearing plants of the Legume Family.

Greenish- or yellowish-white; late June–mid-August

Roadside banks, prairie hillsides, open woods and thickets
East three-fourths, primarily; more common east of the Flint Hills

207. *Astragalus canadensis*

Robust perennial herbs with horizontal or somewhat oblique rhizomes. Stems 1 or a few, erect to ascending or even prostrate, up to 1.3 m tall. Leaves 0.5–3.5 dm long; leaflets 13–35, thin, narrowly lanceolate to elliptic or oblong, mostly 1–4 cm long and 4–15 mm wide, on short stalks or petiolules; apex rounded; base rounded or acute; stipules membranous, triangular-acuminate. Stems and under surface of leaflets with straight, appressed, laterally-attached hairs, with 1 arm somewhat longer than the other. Flowers greenish- or yellowish-white, sometimes tinged with purple, 12–18 mm long; racemes many-flowered, (2.5–) 4–16 cm long, on stiffly upright peduncles. Calyx 6–8 mm long, tubular, somewhat oblique, the base pouchlike on the upper side, the teeth acuminate. Corolla slender, the blade of the standard only slightly reflexed. Fruits ascending or erect, short-cylindrical to ellipsoid, 2-celled, 10–15 (–17) mm long (excluding the beak) and about 5–6 mm wide, crowded on the mature fruiting spike. Seeds brownish-yellow, smooth, more or less obliquely heart-shaped to kidney-shaped, about 2 mm long.

A. crassicárpus Nutt.   Ground-plum milkvetch

NL., from L. *crassus*, thick + Gr. *karpos*, fruit—in reference to the plumlike pods.
Purple to pinkish, white, cream, or greenish; (late March–) mid-April–May
Rocky or sandy prairie hillsides and uplands
Throughout, more common eastward

208. *Astragalus crassicarpus*

Perennial herbs with a woody taproot which is sometimes quite enlarged and may have strong lateral roots. Stems several, mostly ascending or prostrate, 1–6 dm tall or long. Leaves 2–10 cm long; leaflets (11–) 15–27 (–33), narrowly elliptic or oblanceolate, less frequently obovate, 3–17 mm long and 1–6 mm wide; apex rounded or acute, often with a tiny sharp point; base rounded or wedge-shaped; lower surface hairier than upper surface; stipules somewhat membranous, lanceolate or lance-attenuate, free from one another. Leaves and stems with appressed or somewhat spreading, basally attached hairs. Flowers purple to pinkish, white, greenish, or cream-colored, 14–21 mm long; racemes mostly 5- to 25-flowered, on peduncles mostly 2–6 cm long. Calyx 7–9 mm long, tubular, with spreading or ascending, black or white hairs, the base of the calyx tube more or less pouchlike on the upper side, the teeth acuminate or long-acuminate. Blade of the standard moderately reflexed. Fruits initially plump and succulent, green or reddish or purplish, round to very broadly ellipsoid, ovoid, or obovoid, 2-celled, many seeded, 1.5–4 cm long, eventually resting on the ground, the thick walls becoming pithy, the fruit dehiscing only after a long period of drying and weathering. Seeds sooty-black, smooth or pitted, 2–4 mm long.

Most of our plants belong to *A. crassicarpus* var. *crassicarpus*, but var. *trichocalyx* (from Gr. *thrix, trichos*, hair, + *kalyx*, cup—because of the calyx which is more densely hairy and longer-haired than that of var. *crassicarpus*) is also known from Cherokee County. The plants are usually robust, with ascending stems, and racemes with 13–25 greenish or cream-colored flowers on stout peduncles 6–16 cm long. The fruits tend to be large (2.5–3.2 cm long).

The fruit of the ground-plum milkvetch is handsome and tempting in appearance and has been eaten raw, cooked, or made into spiced pickles. The Dakota Indians called the fruit *pe ta wote*, "food of the buffalo," while the Chippewas used it in compounding war medicines. Other groups, even after the original intent was long forgotten, put it into the water in which they soaked their seed corn before planting it.

See plate *48*.

A. distórtus T. & G.   Bent milkvetch

L. *distortus*, misshapen, dwarfish—probably in allusion to the bent pods.
Pinkish-purple to white; late April–May
Rocky or sandy prairie hillsides and uplands
Miami Co. south to Cherokee Co., Franklin south to Neosho, and Coffey, Woodson, and Sumner counties

209. *Astragalus distortus*

Short-lived perennial herbs with a woody taproot which often has strong lateral branches. Stems few to several, reclining or ascending, 1–2.3 dm tall or long. Leaves 4–10 cm long; leaflets 9–25, obovate to oblanceolate, wedge-shaped, or elliptic, 3–10 (–15) mm long and 1–6 mm wide; apex rounded, truncate, or notched; base acute or wedge-shaped; stipules herbaceous and free at the upper nodes, membranous and united around the stem at the lower nodes. Stems, lower leaflet surfaces, and calyx with sparse appressed hairs. Flowers purplish-pink (drying bluish) to pale lavender or white, 8–13 mm long;

racemes 5- to 21-flowered, the peduncles erect or ascending but becoming prostrate in fruit. Calyx 3–6 mm long, bell-shaped or short-tubular, somewhat oblique, the teeth acuminate. Blade of the standard slightly to moderately reflexed. Fruits spreading, crescent-shaped, 1-celled, many-seeded, 13–25 mm long and 3–7 mm wide, glabrous, dehiscent only after a period of weathering on the ground. Seeds brown, smooth or pitted, 1.6–2.6 mm long.

A. grácilis Nutt.   Slender milkvetch

L. *gracilis*, slender—applicable to stems and leaflets.
Light purple or lavender; May–mid-July
Dry, rocky or sandy prairie hillsides and uplands
West half, especially the northwest fourth

Slender, highly variable perennial which has in the past been divided into 2 species on the basis of growth habit and leaflet and fruit characteristics. Study of the species throughout its range has shown, however, that the various characters used to distinguish 2 taxa grade imperceptibly from one extreme to the other and occur in various combinations throughout the range, so that recognition of 2 species does not seem realistic. Stems erect, ascending, or nearly prostrate, 1.5–4 dm tall, much branched from the base in the lower forms. Leaves 2–9 cm long; leaflets 9–17, linear to narrowly oblong or wedge-shaped, 0.4–2.5 cm long and 1–2.5 (–4) mm wide; apex usually notched, sometimes truncate or rounded; stipules of the upper nodes papery, united around the stem. Herbage with sparsely scattered, appressed, basally attached hairs. Flowers light purple or lavender (often drying bluish), 5–8 mm long; racemes slender, loose, 12- to many-flowered, on long, naked peduncles. Calyx bell-shaped, 2–2.5 mm long, the teeth triangular or short-acuminate. Blade of the standard moderately to strongly reflexed. Fruits spreading or eventually drooping, ellipsoid to broadly ovoid, 5- to 9-seeded, 4.5–8 (–10) mm long including the beak and 3–5 mm wide, transversely wrinkled and with appressed hairs. Seeds brown or greenish, smooth or minutely pitted, 2.3–3.3 mm long.

210. *Astragalus gracilus*

A. lotiflòrus Hook.   Low milkvetch

L. *lotus*, Gr. *lotos*, name now applied to a genus of the Legume Family, + L. *flos, floris*, blossom, *floreus* of flowers—with flowers resembling those of *Lotus*.
Greenish-white, sometimes purplish; late April–May
Rocky or sandy prairie hillsides and uplands
West three-fourths and Woodson County; more common northward

Low, tufted, highly variable perennials with a taproot and eventually branching caudex. Stems few to many, 0.5–2 dm tall, the lower or outer ones prostrate or ascending, the upper or innermost erect. Leaves 2–11 cm long; leaflets mostly 7–17, elliptic or oblong (occasionally broadly elliptic or obovate on the lower leaves), 5–20 mm long and 1.5–7 mm wide; apex rounded or acute; base acute or rounded; lower surface more densely pubescent and usually lighter green than the upper surface; stipules lanceolate, somewhat membranous. Herbage with appressed silky hairs, those of the leaflets laterally (rather than basally) attached and with 1 arm longer than the other. Flowers of 2 kinds—normal flowers displayed conspicuously in dense, ovoid or round, 5- to 17-flowered heads on elongate peduncles, and reduced, inconspicuous, cleistogamous flowers borne 1–3 together on very short peduncles at the base of the plant, these usually found on different plants or at least not produced simultaneously by the same plant. Normal flowers greenish-white, sometimes tinged with purple or blue, 8–11 mm long, the standard moderately to strongly reflexed, with a bell-shaped or very short-tubular calyx with long-acuminate teeth. Cleistogamous (from Gr. *kleistos*, closed, + *gamos*, marriage—because the flowers fertilize themselves without ever opening) flowers white (drying yellowish), 4–7 mm long, the petals often shorter than the calyx teeth. Fruits from both kinds of flowers similar, ellipsoid or fusiform and often somewhat crescent-shaped, ours mostly 1.2–3 cm long (including the beak) and 5–8 mm wide, 1-celled, covered with straight, appressed hairs or rather kinky, spreading hairs. Seeds brownish-green, sometimes mottled with purple, obliquely heart-shaped, irregularly pitted, 1.5–2.5 mm long.

211. *Astragalus lotiflorus*

A. missouriénsis Nutt.   Missouri milkvetch

L. *-ensis*, suffix added to nouns of place to make adjectives, i.e., of Missouri.
Purplish-pink (rarely white or deep purple); late April–June
Prairie hillsides and uplands, on rocky or sandy soil
West three-fourths

Low, tufted perennial with a taproot, resembling *A. lotiflorus* in aspect. Stems several to many, (0.6–) 1–2 dm tall, the lower or outer ones prostrate or ascending, the upper or innermost erect. Leaves 4–11 cm long; leaflets mostly 11–17, elliptic to oblanceolate or less frequently obovate, 6–14 mm long and 2–6 mm wide; apex rounded or acute; base triangular to ovate or lanceolate, inconspicuous. Herbage silvery or gray-green from the dense covering of silky, appressed, laterally attached hairs. Flowers purplish-pink (drying bluish), rarely white or deep purple, 12–22 mm long; racemes mostly 5- to 12-flowered, eventually elevated above the leaves on stout, naked peduncles. Calyx 9–12 mm long, tubular, slightly oblique, the base pouchlike on the upper side, the teeth long-acuminate. Blade of the standard moderately to strongly reflexed. Fruits ascending or spreading, somewhat obliquely oblong-cylindrical or ellipsoid, 1-celled, many-seeded, 1.3–2.5 cm long (excluding the beak) and 7–10 mm wide, covered with appressed hairs, often persistent on the plant and becoming black with age. Seeds brown, pitted or wrinkled, 2–3 mm long.

212. *Astragalus missouriensis*

A. mollíssimus Torr.   Woolly loco

L. *mollissimus*, most soft, from *mollis*, soft—because of the covering of silky hairs. Spanish *loco*, foolish, insane—because of the behavior of animals that have too freely browsed this species.
Purple; late April–June
Dry, rocky or sandy prairie hillsides and uplands
West half

Gray-green, silky perennial herb 1–3.5 dm tall, with a woody taproot and very short (if any) aboveground stems. Leaves (0.5–) 0.9–2.2 dm long, ascending or arching, leaflets 15–27, broadly obovate to oblanceolate, elliptic, or infrequently lanceolate, 5–25 mm long and 2–15 mm wide; apex rounded or acute, sometimes mucronate; base rounded or acute; stipules free from one another, long-triangular, membranous but silky. Herbage densely covered with long, silky, basally attached hairs. Flowers purple (drying bluish), rarely yellow or white, 15–20 mm long; racemes 10- to 40-flowered, 4–10 cm long (longer in fruit), on long, stout, naked peduncles which become prostrate as the fruits mature. Calyx 9–13 mm long, tubular, slightly oblique, the base pouchlike on the upper side, the teeth long-acuminate. Blade of the standard slightly to moderately reflexed. Fruits plump and crescent-shaped, often broadest near the base, 2-celled most of the length, many-seeded, 1.4–2.5 cm long (including the beak) and 4–7 mm wide, usually glabrous.

213. *Astragalus mollissimus*

Our plants all belong to *A. mollissimus* var. *mollissimus*.

The plant is fair enough to the eye but is poisonous to all kinds of livestock, especially horses. Several poisonous substances—including an alkaloid, or similar compound, named locoine—have been discovered in the herbage, and some are apparently excreted into the nectar, causing serious destruction of bees when alfalfa and the sweet clovers have given out. Fortunately, cattle do not like it and ordinarily leave it alone. However, in early spring when few other plants are up, or during periods of drought when other forage has been exhausted, livestock may eat it.

A. pectinàtus Dougl. ex G. Don

L. *pectinatus*, comblike, from *pecten*, comb—in reference to the leaves.
Yellowish; mid-May–June
Dry prairie hillsides and uplands
North and central parts of the west third

Coarse, bushy, gray-green perennials with a thick woody taproot. Stems stout, ascending or reclining but with the tips ascending, 1.5–6 dm long. Leaves 4–10 cm long; leaflets mostly 7–15, linear, stiff, 1.3–7 cm long and 0.5–1 (–2) mm wide, curved toward the rachis; stipules membranous, united and sheathing the stem. Herbage with appressed, basally attached hairs. Flowers yellowish, 17–22 mm long; racemes 12- to many-flowered, eventually becoming quite loose. Calyx tubular, 8–10 mm long, the teeth short-acuminate or triangular. Corolla rather narrow, the standard moderately to strongly reflexed. Fruits eventually drooping, ovoid to ellipsoid, 1-celled, many-seeded, 1.1–2 (–2.5) cm long (including the beak) and about 6–8 mm wide, transversely wrinkled, the suture quite prominent. Seeds light- or pinkish-brown, smooth or minutely pitted, 3–3.6 mm long.

214. *Astragalus pectinatus*

Presence of this plant is a positive indication of selenium in the soil.

A. platténsis Nutt. ex T. & G.

L. *-ensis*, suffix added to nouns of place to make adjectives; *plattensis*—of the Platte region.
Pink or pinkish-purple; late April–May

Rocky or sandy prairie hillsides and uplands
Central third, primarily, and farther west

Low perennial herb. Stems several, reclining or ascending, 0.8–2.5 dm tall or long, often departing several inches below ground from the woody taproot. Leaves 3–10 cm long; leaflets mostly (9–) 15–23, obovate to oblanceolate or elliptic, 3–15 mm long and 1.5–5 mm wide; apex rounded or notched; base acute; upper surface glabrous; stipules membranous and united around the stem on the lower nodes (sometimes only on below-ground parts of the stem), free on the upper nodes and lanceolate- or ovate-acuminate. Herbage with long, spreading or somewhat appressed, straight or kinky, basally attached hairs. Flowers pink or pinkish-purple (drying bluish), 12–20 mm long; racemes initially dense and headlike, 6- to 15-flowered, on stout peduncles. Calyx 7–10 mm long, tubular, slightly oblique, the base pouchlike on the upper side, the teeth long-acuminate. Blade of the standard eventually becoming moderately to strongly reflexed. Fruits spreading or ascending, ovoid or nearly globose, 2-celled, many-seeded, 1–2 cm long (excluding the beak) and 1–1.3 cm wide, fleshy but becoming leathery or quite hard, sparsely hairy. Seeds pinkish-brown to sooty-black, smooth, 2.5–3.3 mm long.

215. *Astragalus racemosus*

A. racemòsus Pursh

L. *racemus*, bunch or cluster of grapes; *racemosus*, full of clusters.
White or yellowish; mid-May–June
Dry, rocky or sandy prairie hillsides and upland
West half and a few northern counties farther east

Perennial herbs, bushy in habit, with a gnarled rootstock and woody taproot. Stems several, erect or ascending, up to 6 dm tall. Leaves 0.4–1.5 dm long; leaflets mostly 9–29, oblong to narrowly elliptic or almost linear, 1–4 cm long and (1–) 2–9 mm wide; apex rounded or acute, base acute, margins rolled upward when dry; stipules membranous, those of the lower nodes united around the stem, those of the upper nodes free. Herbage light- or yellowish-green, with appressed, basally attached hairs. Flowers white or yellowish, sometimes tinged with pink or purple, 13–17 mm long, spreading or drooping; racemes many-flowered, 4–10 cm long, on stout peduncles. Calyx 5–8 mm long, tubular, slightly oblique, the base pouchlike on the upper side, the teeth acuminate. Blade of the standard moderately to strongly reflexed. Fruits drooping, leathery, 1-celled, 12- to many-seeded, ellipsoid to linear-oblong, with a longitudinal groove along the upper side that gives it a V- or heart-shape in cross section, 1–3 cm long and 3–8 mm wide, elevated above the calyx on a short stalk, glabrous or sparsely hairy. Seeds brown, smooth, 2.6–3.3 mm long.

Presence of this plant is a positive indication of selenium in the soil.

BAPTÌSIA Vent.   False indigo, wild indigo

NL., from Gr. *baptisis*, a dipping, as in dyeing, from *baptizein*, to dye—some species having been used as a source of an indigolike dye.

Herbaceous perennials. Stems erect or ascending, single at ground-level and dichotomously branched above. Leaves palmately 3-foliolate, stipulate. Herbage, flowers, and fruits often turning black with age or when dried. Flowers white, cream-colored, yellow, or blue, borne in stout racemes. Calyx 4- to 5-lobed and somewhat 2-lipped. Blade of the standard reniform to nearly circular with an apical notch, the sides flexed backward; wings more or less oblong, about the same length as and enfolding the keel. Stamens 10, free from one another. Pistil stipitate. Pods many-seeded, rather large, inflated, turning black with age, beaked with the persistent style, and elevated on a stalk or stipe beyond the base of the persistent calyx.

Three species in Kansas.

Regarding the common name of the genus, one should note that the indigo dye of commerce is not obtained from the *Baptisias* but from the shrubby *Indigofera tinctoria* L., a native of southern Asia, and *I. suffruticosa* Mill. of the West Indies, both formerly cultivated in the South and now growing wild there.

Key to Species:
1. Flowers blue or light purple, expanded portion of the standard 15–20 mm long; stems erect .... *B. australis*
1. Flowers white, cream-colored, or pale yellow 2
  2. Flowers white, expanded portion of the standard 9–12 (–14) mm long; stems erect, 1–2 m tall; herbage hairless, often glaucous; pistil and fruits hairless .... *B. leucantha*
  2. Flowers cream-colored or pale yellow, expanded portion of the standard 14–18 mm long; stems forming a low, bushy clump 3–8 dm tall; herbage, pistil, and fruits hairy .... *B. leucophaea*

BEAN SUBFAMILY

169

216. *Baptisia australis*

217. *Baptisia leucantha*

B. austrális (L.) R. Br.   Blue false indigo

L. *australis*, southern.
Blue; May–June
Rocky or sandy prairie
East two-thirds

Plants erect, up to 1.5 m tall, the branches ascending or spreading. Leaflets obovate to oblanceolate or elliptic, those of the main leaves 1.5–3.5 cm long and 4–11 (–20) mm wide; apex rounded or acute; base wedge-shaped. Herbage smooth. Flowers blue (rarely white), often fading to pink or turning black when dried; raceme erect. Calyx 9–12 mm long, 4-lobed, the upper lobe apically notched, margins of the lobes densely woolly. Expanded portion of the standard 15–20 mm long, that of the wing 18–22 mm long. Fruits erect, black, woody, often glaucous, ellipsoid, mostly 4–5 cm long (excluding the style), tapering gradually to the persistent style.

Our plants are *B. australis* var. *minor* (Lehm.) Fern.

See plate *49*.

B. leucántha T. & G.   Atlantic wild indigo

NL., from Gr. *leukos*, white + *anthos*, flower, i.e., white-flowered.
White; mid-May–July (–mid-August)
Prairie
East fourth

Plants erect, 1–2 m tall, the branches ascending. Leaflets oblanceolate to narrowly elliptic (occasionally obovate), those of the main leaves 2.5–5 cm long and 0.7–1.5 (–2.5) cm wide; apex rounded or acute; base wedge-shaped or acuminate. Herbage smooth, often glaucous. Flowers white; raceme erect. Herbage and/or flowers often turning black with age or upon drying. Calyx 8–10 mm long, 4-lobed, the inner surface of the lobes white-woolly, at least near the margins. Expanded portion of the standard 9–12 (–14) mm long; that of the wing 14–17 mm long. Fruits black, glaucous, woody, ellipsoid to nearly globose, 2–2.5 (–4) cm long (excluding the stipe and persistent style). Seeds 3.5–5 mm long, covered with resinous droplets.

See plate *50*.

B. leucophaèa Nutt.   Plains wild indigo

NL. *leucophaea*, light brown, from Gr. *leukos*, white, without color, + *phaios*, dun-colored, dull brown.
Cream-colored to pale yellow; mostly late April–mid-June
Prairie
East half

Stems ascending and spreading, forming a bushy clump 3–8 dm tall. Leaflets oblanceolate to elliptic, often narrowly so, those of the main leaves 3.5–6 cm long and 0.8–1.5 cm wide; apex rounded or acute; base wedge-shaped. Herbage hairy. Flowers cream-colored or pale yellow; raceme eventually becoming declined, with the flowers oriented along the upper side. Calyx 8–11 mm long, hairy, 4-lobed, the upper lobe apically notched. Expanded portion of the standard 14–18 mm long; that of the wing 16–22 mm long. Pistil densely hairy. Fruits black, woody, ellipsoid, 4–5 cm long, gradually tapering to the persistent style, the surface reticulately veined and covered with short hairs.

CORONÍLLA L.   Crown vetch

L. *coronilla*, dim. of *corona*, crown—in reference to the inflorescence.

Herbs or shrubs native to western Eurasia and northern Africa. Leaves odd-pinnate. Flowers yellow, pink, or white, borne in axillary umbels. Calyx 2-lipped, the lower lip 3-toothed, the narrow upper lip apically notched. Petals clawed, auricled. Blade of the standard round; that of the wings ovate-oblong; keel crescent-shaped. Stamens 10, diadelphous. Fruits round or 4-angled in cross section, transversely jointed.

One species in Kansas.

C. vària L.   Crown vetch

L. *varius*, variable.
Pink and white; late May–September
Roadside embankments, creek banks, rocky prairie ridges
Scattered in the east half; intr. from Europe

Bushy perennial with ascending or spreading stems up to 5 dm tall. Leaves mostly 4–12 cm long; leaflets 11–21, lanceolate to oblong or oblanceolate, (0.5–) 1–2 cm long; apex rounded or truncate, mucronate; base rounded or acute. Flowers pink and white, the globose umbels borne on peduncles 4–12 cm long. Calyx about 2.5 mm long. Petals 10–11 mm long. Fruits linear, round in cross section, 2–4 cm long, with 3–7 1-seeded segments. Seeds dark brown, oblong, slightly flattened, about 3–3.5 mm long and 1–1.2 mm wide.

Crown vetch is often planted on slopes to help control soil erosion.

CROTALARIA L.   Rattlebox

NL., from Gr. *krotalon*, rattle—in reference to the inflated pods with seeds loose after ripening.

218. *Crotalaria sagittalis*

Annual or perennial herbs (shrubs in the tropics). Leaves apparently simple (probably representing a single leaflet of an ancestral pinnate leaf) or 3-foliolate. Flowers usually yellow, borne in racemes. Calyx 5-lobed, somewhat 2-lipped. Blade of the standard more or less round to ovate, auricled; wings oblong, not auricled; keel-petals connivent on both margins, strongly angled on the lower side. Stamens 10, monadelphous, 5 long ones with nearly globose anthers alternating with 5 short ones with linear anthers. Fruits various in shape, 2- to many-seeded.

One species in Kansas.

Several species of *Crotalaria* are known to be toxic to livestock. *C. spectabilis* Roth., native to the Old World tropics but introduced in the southeastern United States, contains the alkaloid monocrotaline.

C. sagittàlis L.   Rattlebox, arrow crotalaria

NL., from *sagittalis*, of or pertaining to an arrow, from L. *sagitta*, arrow—in reference to the form of the stipules of the upper leaves.
Yellow; June–September
Prairie, open woods, often in sandy soil
East two-fifths, primarily

Slender, soft-hairy annual. Stems erect, 1–4 dm tall. Leaves apparently simple, elliptic to elliptic-lanceolate, mostly 1.5–6 cm long, and 0.6–1.6 cm wide; stipules on lower leaves none or quite minute, while on upper leaves they are nearly as large as the leaves and resemble narrow, down-pointing arrowheads adherent to the stem for most of the length of an internode. Flowers yellow, solitary or in 2- to 4-flowered racemes. Calyx 7–11 mm long, the lobes long-acuminate. Petals yellow, short-clawed, about the same length as or shorter than the calyx. Fruits ellipsoid, 2–3.5 cm long and 0.7–1.3 cm wide, bulging more at the stylar end and becoming leathery and nearly black when ripe. Seeds gray-brown, glossy to lustrous, obliquely cordate, about 2.5 mm long.

The herbage of this species, either fresh or dried in hay, and the seeds are suspected of causing livestock poisoning, but the poisonous principle, if one exists, has not been identified.

DÀLEA Juss.   Dalea

Named in honor of Samuel Dale, English botanist (1659–1739).

Annuals, herbaceous perennials, and shrubs, mostly inhabiting dry or desert places. Leaves mostly odd-pinnate with 3–11 leaflets, gland-dotted. Flowers small, in heads or spikes. Calyx 5-toothed. Corolla imperfectly papilionaceous; petals clawed, the standard inserted at the bottom of the calyx, the keel and wings fused with the slit tube of the 9–10 monadelphous stamens. Pistil 1- to 3-ovuled. Fruit indehiscent, 1-seeded, enclosed by the persistent calyx.

Six species in Kansas. *D. jamesii* (Torr.) T. & G., known from Morton County only, and *D. nana* Torr., from Finney and Clark counties, are both silky-haired, yellow-flowered perennials and are described briefly with *D. aurea*.

Key to Species:
1. Leaflets glabrous 2
   2. Plants annuals; leaflets 15–41; spikes densely crowded; flowers about 5 mm long ....
   .... *D. leporina*
   2. Plants perennials; leaflets 9–13; spikes loosely flowered; flowers 9–12 mm long ....
   .... *D. enneandra*
1. Leaflets with hairs 3
   3. Stems prostrate; flowers red to purple .... *D. lanata*
   3. Stems erect; flowers yellow .... *D. aurea*

219. *Dalea aurea*

220. *Dalea enneandra*

221. *Dalea enneandra*, close-up of flowers

D. aùrea Nutt. Silktop dalea, golden dalea

L. *aureus*, golden, from *aurum*, gold—in reference to the flower color.
Yellow; late June–early September
Prairie
West three-fourths

Perennial with a deep, strong taproot. Stems rather stout, usually several from a broad crown, unbranched above, up to 6.5 dm tall. Leaves mostly 1.5–3 cm long; leaflets 5–9 (mostly 5), elliptic to oblanceolate or obovate, 5–20 mm long and 2–8 mm wide, the apex acute or rounded, the base acute. Stems and leaves (at least the lower surface) with silky appressed hairs. Flowers yellow, crowded in terminal spikes 1–5 cm long and 1.5–2 cm wide, each flower subtended by a silky appressed bract with a spinose tip. Calyx 6–8 mm long, also silky, the lobes long-acuminate. Blade of the standard sagittate, about 4 mm long, short-clawed; wings oblong; keel oblong, about 7 mm long and closely enveloping the stamens and pistil. Seeds about 2 mm long, broadly ellipsoid, the end nearest the hilum truncate and with a stubby beak protruding at a right angle to the long axis of the seed.

Bees alighting on the wings and keel depress the keel below the anthers and stigma, permitting cross-pollination by pollen brought from other flowers and the picking up of fresh pollen. Self-pollination can occur within the keel, since the anthers and stigma mature at the same time and stand at the same level.

D. jamesii, known from Morton County, is usually less than 1 dm tall, and has thick, gnarled caudexes and 3-foliolate leaves. *D. nana*, in Finney and Clark counties, has slender stems 1–2 dm tall, 5- to 9-foliolate leaves, and in general has a looser, more open habit than *D. jamesii*. Both have yellow flowers which may turn brown or pink when dried, unlike those of *D. aurea* which remain yellow.

D. aurea resembles some species of *Petalostemon* but may be distinguished by its 9–10 stamens, there being only 5 fertile stamens in *Petalostemon*.

D. enneándra Nutt. Plume dalea

NL., from Gr. *ennea*, nine, + *aner*, *andros*, man, but when used as a suffix, as in this instance, male or masculine—in reference to the stamens or male members of the flower, the flowers of this species having 9 stamens instead of the 10 usual for the genus.
White; June–August (–September)
Dry, rocky or sandy prairie
West two-thirds; a few collections from farther east

Smooth perennial with a strong, deep taproot. Stems erect, slender, 0.4–1 m tall, unbranched and naked below but with several to many tenuous, well-foliated branches and branchlets above. Leaves 1–2 cm long; leaflets 5–11, narrowly elliptic to linear, 4–10 mm long; and 0.5–1.5 mm wide, black-dotted on the lower surface and each subtended by a reddish-brown gland on the somewhat leaflike rachis. Flowers white, borne in slender, loose-flowered terminal spikes, each flower subtended and enveloped by a broadly ovate to cordate, gland-dotted bract with membranous margins. Calyx 7–9 mm long, the lobes long-acuminate and conspicuously plumose. Standard long-clawed, 7–8 mm long, the blade more or less cordate; wings oblong, auricled, short-clawed, about 5 mm long; keel obliquely elliptic, 9–12 mm long. Seeds about 2.5 mm long, narrowly ovoid, the end nearest the hilum truncate and with a beak protruding at a right angle to the long axis of the seed.

D. lanàta Spreng. Woolly dalea

L. *lanatus*, woolly, from *lana*, wool, down on leaves or fruit.
Red to purple; July–early September
Sandy soil of prairie and flood plains
Southwest fourth and Cowley County

Perennial from a long woody root with orange bark. Stems prostrate, up to 5 dm long. Leaves 1.5–2.5 cm long; leaflets 9–13, oblanceolate to obovate or obtriangular, 5–10 mm long, and 1.5–6 mm wide, the apex rounded or truncate, the base acute, each subtended by a gland on the leaf rachis. Herbage densely woolly; fruits, calyx, flower bracts, and under surface of leaves also dotted with yellow glands (use magnification). Flowers red to purple, borne in elongate axillary spikes 3–9 cm long, each flower subtended by a persistent, obovate, gland-dotted, appressed bract with an abruptly acuminate apex. Calyx 5–6 mm long. Blade of the standard cordate and somewhat hoodlike, about 2 mm long; wings short-clawed, the blades auricled, oblong, about 3.5 mm long; keel oblong to elliptic, auricled, about 3.5 mm long; petals with a few scattered glandular dots.

Seeds gray- or orange-brown, ellipsoid, beaked, slightly flattened, about 2 mm long, often parasitized by some insect.

D. leporìna (Ait.) Bullock   Foxtail dalea

L. *leporinus*, of hares, from *lepus*, hare—probably in reference to the silky inflorescence which resembles a rabbit's foot
Pink to white; (July–) September
Sandy soil
Known from widely scattered sites across the state

Erect annual with an orange taproot which produces long, strong, nearly horizontal lateral branches as well as fine fibrous branchlets with many nodules harboring nitrogen-fixing bacteria. Stems 2–10 dm tall, with ascending branches. Leaves mostly 3–6 cm long; leaflets 15–41, elliptic to oblanceolate, 3–10 mm long and 1–3 mm wide, the apex rounded or somewhat truncate, the base acute, each subtended by an orange gland. Stems, under surface of leaves, bracts, calyx, fruit, and anthers gland-dotted. Flowers pink to white, about 5 mm long, borne in dense, silky, long-peduncled spikes 1.5–5.5 cm long; bracts subtending the flowers broadly obovate with a long-acuminate tip and membranous margins, falling away as the flowers bloom. Calyx silky, about 3.5 mm long. Blade of the standard about 2.5 mm long, "spade-shaped" with hastate basal lobes; blade of the wings oblong, auricled, about 2 mm long; keel about 1.5 mm long. Seeds gray-brown, ovoid with a short stubby beak, about 2 mm long.

DESMÒDIUM Desv.   Tickclover

NL. *desmodium*, like a chain, from Gr. *desmos*, bond or chain—in reference to the jointed pods. The common name calls attention to the fact that the leaves are 3-foliolate, as those of clover, and that the pods with their hooked hairs cling like a tick.

Perennial herbs (in North America) with a short rootstock and a strong, deep taproot. Leaves 3-foliolate, mostly alternate, each lateral leaflet subtended by a minute stipulelike bract called a stipel, the terminal leaflet by 2 stipels. Hooked hairs usually present on the fruits and rachis of the inflorescence, sometimes on other parts as well. Flowers small, white to pink or purple, borne in elongate, simple or branched racemes; within the raceme, flowers grouped in 2's or 3's and each group subtended by 2 small bracts. Calyx 5-lobed and more or less 2-lipped, the upper 2 lobes fused most or all their length, the lower 3 distinct. Standard narrowed at the base but not clawed, the blade oblong to nearly round; wings oblong, short-clawed; keel nearly straight, long-clawed. Stamens 10, diadelphous in most species. Fruit indehiscent, constricted between the seeds and usually separating into 1-seeded joints or segments covered with hooked hairs.

This is a difficult group to work with, and identification is greatly facilitated by having specimens with flowers and with fruits at least partially matured.

Twelve species known for Kansas. Four of these—*D. marilandicum* (L.) DC., *D. nudiflorum* (L.) DC., *D. pauciflorum* (Nutt.) DC., and *D. rotundifolium* (Michx.) DC. —are known only from a few sites in Cherokee County and are not included in the key.

The seeds of *Desmodium* species are eaten by bobwhite quail.

Key to Species (3):

1. Stamens monadelphous; calyx scarcely 2-lipped, the lobes less than half as long as the calyx tube; fruit long-stipitate, the stipe much longer than the persistent remains of the stamens; joints of the fruit triangular or semi-obovate, straight or slightly concave on the glabrous dorsal margin; leaves crowded at the top of an unbranched stem .. *D. glutinosum*
1. Stamens diadelphous; calyx distinctly 2-lipped, the lobes more than half as long as the calyx tube; stipe of the fruit shorter than the remains of the stamens; joints of the fruits triangular, rhombic, or rounded, the dorsal margin pubescent with hooked hairs, never concave 2
  2. Terminal leaflet linear to narrowly lanceolate, even in well-grown leaves not more than 15 mm wide, or not more than a fourth as wide as long 3
    3. Leaflets thick, strongly reticulate; fruit joints 2 (–3), rounded on both margins ......
    .... *D. sessilifolium*
    3. Leaflets thin, not strongly reticulate; fruit joints 3–5, triangular or somewhat rhombic in outline .... *D. paniculatum*
  2. Terminal leaflet lanceolate to ovate or broadly elliptic, in well-grown leaves regularly more than 15 mm wide, or more than a fourth as wide as long 4
    4. Leaves pubescent beneath with hooked hairs (visible with 10X lens) 5
      5. Joints of the fruit slightly curved or angled on one margin, broadly angled on the other, 7–13 mm long, averaging 9 mm; plant usually bearing several divergent racemes; leaves thin, not prominently reticulate when fresh .... *D. canescens*
      5. Joints of the fruit rounded on both margins, 4.5–7 mm long, averaging 5.5 mm;

bearing a single raceme which is terminal and usually unbranched; leaves firm, strongly reticulate .... *D. illinoense*
4. Leaves glabrous or pubescent beneath, but without hooked hairs 6
  6. Lower margin of the joints of the fruit gradually curved, nearly semicircular in outline 7
    7. Flowers 8–13 mm long; joints of the fruit 3–5; leaflets lanceolate to lance-oblong, the terminal leaflet 5–10 cm long .... *D. canadense*
    7. Flowers 3–4 mm long; joints of the fruit 1–3, usually 2; leaflets ovate to elliptic or short-oblong, the terminal leaflet 1–2.5 cm long .... *D. ciliare*
  6. Lower margin of the joints of the fruit abruptly curved or angled near the middle, the joint triangular or somewhat rhombic 8
    8. Flowers 5–8 mm long; apex of leaflets rounded or acute, never acuminate; stipules linear-subulate or threadlike, usually early deciduous .... *D. paniculatum*
    8. Flowers 8–10 mm long; apex of leaflets usually acuminate, sometimes acute; stipules lanceolate, 1–2 mm wide, persistent, often longer than 1 cm ....
.... *D. cuspidatum*

D. canadénse (L.) DC.   Canada tickclover, tick trefoil

L. *-ensis, -ense,* suffix added to nouns of place to make adjectives—"Canadian," in this instance.
Purple; August–early September
Prairies, roadsides, open woods
Scattered in the east half

    Stems erect, hairy, 1 to few, up to 2 m tall. Leaves 6–14 cm long (including petiole); stipules deciduous, narrowly lanceolate to linear-lanceolate with a long-acuminate apex, up to 1 (–1.5) mm wide at the base; leaflets lanceolate to lance-oblong, occasionally elliptic, more or less hairy but lacking hooked hairs (use magnification), the terminal leaflet 5–10 cm long and (1.5–) 2–3.5 cm wide, the apex acute to slightly rounded, the base rounded to broadly acute. Flowers purple, 10–12 (–15) mm long, the racemes terminal, branched. Calyx 2-lipped, 5–7 mm long, the lobes longer than the calyx tube. Stamens diadelphous. Joints of the fruits 1–5, 5–7 mm long, slightly convex on 1 margin, quite rounded on the other. Seeds brown, broadly ellipsoid to narrowly ovoid, about 3–4 mm long.

D. canéscens (L.) DC.   Tick trefoil, beggar's lice

L. *canescens,* becoming gray, hoary.
Pink; August–early September
Open woods (especially alluvial woods), roadsides, prairie
Widely scattered in the east half

    Stems erect, hairy, 1 to several, up to 1.5 m tall. Leaves mostly 6–16 cm long (including the petiole); stipules hairy, ovate or triangular with an acuminate apex, mostly 3–4.5 mm wide; leaflets lanceolate to ovate, at least the lower surface roughened with hooked hairs (use magnification), the terminal leaflet 3–9 cm long and 1.5–4.5 cm wide, the apex acute to rounded, often mucronate, the base rounded or nearly truncate, often oblique on the lateral leaflets. Flowers pink, 8–10 mm long, the racemes branched. Calyx 2-lipped, 4–6 mm long, the lobes longer than the calyx tube. Stamens diadelphous. Joints of the fruits 2–6, 6–9 (–11) mm long, slightly convex on 1 margin, broadly angled on the other. Seeds brown, bean-shaped, about 3.5–4 mm long (only a few seen).

D. ciliàre (Willd.) DC.

NL., ciliate, having hairlike processes along the margin, from L. *cilium,* eyelid.
Pink; mid-August–early September
Sandy, open woods
Woodson, Wilson, Neosho, and Montgomery counties

    Stems erect, hairy or becoming smooth with age, up to 1 m tall. Leaves 2–3.5 cm long (including the petiole); stipules subulate to threadlike; leaflets ovate to elliptic or short-oblong, softly hairy, the terminal leaflet 1–2.5 cm long and 8–12 mm wide, the apex rounded or acute, the base truncate to rounded or broadly acute. Flowers pink, 3–4 mm long, the delicate racemes terminal, branched. Calyx 2-lipped, 2–3 mm long, the lobes as long as or longer than the calyx tube. Stamens diadelphous. Joints of the fruits 2 (–3), 4–6 mm long, slightly convex on 1 margin, quite rounded on the other.
    Compare with *Lespedeza capitata* and *L. stuveii.*

D. cuspidàtum (Willd.) Loud.   Tick trefoil, beggar's lice

L. *cuspidatus,* pointed—in reference to the leaflets.
Pink or lavender; late July–early September

Wooded hillsides and creek banks, occasionally roadsides
East third, especially northward

Stems erect, smooth or with hooked hairs, up to 1.5 m tall. Leaves 9–22 cm long (including the long petiole); stipules lanceolate with an acuminate apex, 1–2 mm wide and up to 12 mm long, usually persistent; leaflets ovate to lanceolate, glabrous or pubescent but lacking hooked hairs, the terminal leaflet 5–12 cm long and (2.5–) 4–7 cm wide, the apex usually acuminate, sometimes acute, the base rounded to broadly acute. Flowers pink or lavender, 8–10 mm long, the racemes terminal, branched. Calyx 2-lipped, 4–5 mm long. Stamens diadelphous. Joints of the fruits (3–) 4–7, 7–11 mm long, convex or nearly straight on the upper margin, angled on the other.

D. glutinòsum (Schindl.) Wood.   Tick trefoil, beggar's lice

L. *glutinosus*, sticky—in reference to the hooked hairs on the fruits and rachis of the inflorescence.
Pink; late June–early September
Woods
East fourth, primarily

Stems erect, unbranched below the inflorescence, up to 4 dm tall (excluding the inflorescence). Leaves (1–) 1.4–3 dm long (including the petiole), crowded into a single verticel, or whorl, at the top of the stem; leaflets sparsely hairy, commonly ovate to deltoid with an abruptly acuminate apex and truncate, rounded, or broadly acute base, the terminal leaflet 8–13 cm long and 5–12 cm wide. Flowers pink, rarely white, 6–8 mm long, the raceme simple or branched, elevated above the whorl of leaves on a naked peduncle. Calyx slightly irregular but not 2-lipped, 1.5–2.5 mm long, the lobes or teeth much shorter than the calyx tube. Stamens monadelphous. Joints of the fruit 2–3, 8–10 mm long, the upper margin glabrous, straight or concave, the lower one broadly U-shaped, usually obliquely so. Seeds brown, flattened, about 7 mm long, with a thin seed coat.

D. illinoénse Gray   Illinois tickclover, tick trefoil

"Of Illinois."
Purplish; June–July
Prairie
East half

Stems stout, erect, hairy, unbranched, up to 2 m tall. Leaves mostly 7–15 cm long (including petiole); stipules persistent, lanceolate to ovate with a long-acuminate apex, 10–20 mm long and 2–6 mm wide; leaflets leathery, conspicuously net-veined, pubescent with hooked hairs (use magnification), drying yellowish-green, lanceolate to ovate, the apex acute or rounded, often mucronate, the base rounded or truncate, the terminal leaflet 3–8 cm long and 1.5–5 cm wide. Flowers purplish, 7–10 mm long, the racemes terminal, the rachis thick, usually unbranched. Calyx 2-lipped, 4–6 mm long, the lobes as long as or longer than the calyx tube. Stamens diadelphous. Joints of the fruits 3–6, 5–7 mm long, rounded on both margins. Seeds plump, ovoid or short-oblong, about 3–3.5 mm long.

Bumblebees are the most frequent insect visitors to this plant, although other bees and butterflies are also attracted by the nectar secreted at the base of the stamen tube.

D. paniculàtum (L.) DC.   Tick trefoil

NL. *paniculatum*, having panicles, from L. *paniculus*, tuft.
Purplish; July–early September
Rocky or sandy, wooded hillsides, occasionally roadsides or prairie
East fifth, a few records from farther west

Stems erect, sparsely to quite hairy, up to 1.2 m tall, usually branched above. Leaves 2.5–15 cm long (including petiole); stipules subulate to threadlike, early deciduous to semipersistent; leaflets ovate to lanceolate, narrowly elliptic, or narrowly lance-oblong, sparsely hairy to softly tomentose on the lower surface but lacking hooked hairs (use magnification), the terminal leaflet 2–10 cm long and 1–4 (–5) cm wide, the apex acute or rounded, the base rounded or acute. Flowers purplish, 5–8 mm long, the racemes terminal, branched, often leafy. Calyx 2-lipped, 3–4 mm long, the lobes as long as or longer than the calyx tube. Stamens diadelphous. Joints of the fruits 3–5, 6–11 mm long, slightly angled or convex on 1 margin and angled (rarely rounded) on the other, more or less triangular to obliquely rhombic in outline. Seeds brown or tan, 3.5–4 mm long.

222. *Desmodium cuspidatum*

223. *Desmodium glutinosum*

224. *Desmodium glutinosum* fruits

BEAN SUBFAMILY

175

225. *Desmodium illinoense*

226. *Desmodium illinoense* fruits

227. *Desmodium paniculatum*

This is an extremely variable species with regard to pubescence; leaf shape, size, and firmness; and petiole length, and has been divided into as many as 5 species by some authorities.

D. sessilifòlium (Torr.) T. & G.   Sessile tickclover

NL., from L. *sessilis*, pertaining to sitting—the leaf blade in this species sitting close to the stem instead of standing away from it on an elongate petiole.
Purple or white; July–September
Prairie, open woods, roadsides, and railroad embankments, usually on rocky or sandy soil
East half south of the Kansas River

Stems erect, hairy, unbranched below the inflorescence, up to 1.3 m tall. Leaves sessile or nearly so; leaflets firm, conspicuously net-veined, hairy but lacking hooked hairs (use magnification), linear-oblong or narrowly elliptic, the apex acute or somewhat rounded, the base rounded or acute, the terminal leaflet 1.5–6.5 cm long and 0.5–1.2 cm wide. Flowers purple or white, 3–5 mm long, the racemes terminal and branched. Calyx 2-lipped, 2–3 mm long, the lobes about as long as the calyx tube. Stamens diadelphous. Joints of the fruits 2–3, 4–6 mm long, rounded on both margins. Seeds light tan, narrowly ovoid, slightly flattened, about 2.5–3.5 mm long.

Both long- and short-tongued bees gather nectar from this species, apparently effecting cross-pollination.

GLYCYRRHĪZA L.   Licorice

L. *glycyrrhiza*, from Gr. *glykyrrhiza*, licorice-root, from *glykys*, sweet, + *rhiza*, root. Licorice is from L. *liquiritia*, a corruption of *glycyrrhiza*.

Perennial herbs. Leaves odd-pinnate, gland-dotted. Flowers white, yellow, or blue, borne in dense axillary racemes. Calyx 5-lobed, 2-lipped, the upper 2 lobes fused part of their length. Petals usually acute, the standard tapering to the base but not clawed, the wings and keel shorter than the standard, oblong, clawed at base. Stamens 10, diadelphous; anthers alternately large and small. Fruits indehiscent, 1-celled, few-seeded, inflated and spiny in our 1 species.

G. lepidòta Pursh   Wild licorice, American licorice

NL., from Gr. *lepidotos*, scaly, from *lepis*, scale—the leaves being minutely scaly when young.
Yellowish-white; May–June
Prairies, pastures, flood plains, waste places, along railroads
Throughout, but less common in the east fourth

Erect perennial with a long, branched, sweet root which reaches a depth of 2.5–4 m. Stems 1 to several, little branched (if at all), 0.5–1 m tall. Main leaves 6–14 (–17) cm long; leaflets (11–) 15–19, lanceolate to elliptic, 1.5–4.5 cm long and 4–15 mm wide, the apex acute and mucronate, the base acute. Herbage minutely gland-dotted (use magnification). Flowers yellowish-white, 9–12 mm long, subtended by lanceolate bracts, borne in racemes on stout axillary peduncles. Calyx 5–7 mm long. Standard lanceolate, strongly keeled at the base, about 12 mm long; wings closely enveloping the keel, the blades oblong, about 6 mm long, with a shallow basal lobe; blade of the keel petals oblong-elliptic with a shallow basal lobe. Pistil bristly. Fruits ellipsoid, 1–2 cm long, densely covered with hooked bristles which cling to the fur of animals, thus effecting seed dispersal. Seeds gray-brown, dull, short-reniform, plump, 2.5–3 mm long.

The licorice of commerce and pharmacy is *G. glabra* (L. *glaber*, smooth, without hairs or scales, in reference to the smooth capsules), a shrub native to the Mediterranean region. The bitter-sweet, yellowish substance extracted from the root of this species in hot water gelatinizes on cooling and has been employed medicinally as a mild purgative and to soothe inflamed mucous membranes of the throat and air passages, just as it was used, according to Egyptian papyri, thousands of years before the Christian Era. It may be presumed that our native species possesses in some degree, at least, the curative virtues of the official Old World species, since the Teton Dakota Indians used a decoction of the root as a febrifuge, chewed the root and held it in the mouth for toothache, and steeped the leaves and applied them to the ears for earache.

LESPEDÈZA Michx.   Bush clover, lespedeza

A misspelling of the name "Cespedes," after V. M. de Cespedes, Spanish governor of Florida (1784–1790) who gave aid to André Michaux in his botanical exploration in North America (1785–1796).

Mostly annual or perennial herbs. Leaves 3-foliolate; leaflets entire, stalked, and lacking stipels. Flowers small, yellowish-white to purple, borne singly or in clusters. Calyx lobes 5, oblong to linear-subulate or acuminate. Standard nearly round to oblong or obovate, erect or spreading, short-clawed; wings oblong, clawed, auricled; keel more or less obovate. Stamens 10, diadelphous. Fruits flattened, membranous, indehiscent, 1-seeded, resembling a single joint of a loment (a jointed pod, as occurs in the closely related genus *Desmodium*).

In our perennial species there are 2 kinds of flowers—those with a corolla and those without. In white-flowered species, the petaliferous flowers are fertile and far outnumber the apetalous ones. In purple-flowered species, the apetalous flowers, which are cleistogamous, are quite abundant and often tend to mature fruits and seeds more frequently than the petaliferous ones. The apetalous flowers are typically segregated into short-stalked axillary clusters but may be intermingled with the petaliferous flowers.

Ten species in Kansas. The 6 most commonly encountered species are included in the following key.

Ingestion of moldy *Lespedeza* hay has been known to cause death by hemorrhaging in cattle. For more information, see discussion of sweetclover poisoning under *Melilotus officinalis*.

228. *Desmodium sessilifolium*

Key to Species:
1. Plants low annuals, much branched from the base; stipules conspicuous, membranous, lanceolate to ovate, persistent .... *L. stipulacea*
1. Plants perennials of varying habits; stipules rather inconspicuous, hairlike or awl-shaped, persistent or deciduous 2
  2. Flowers yellowish-white, with or without a purple spot 3
    3. Leaflets mostly 2–4 cm long, conspicuously silky-pubescent; flowers borne in globose heads; sepals completely separate, 8–10 (–12) mm long .... *L. capitata*
    3. Leaflets mostly 0.5–2.5 cm long, not conspicuously silky; flowers borne in axillary clusters of 1–3; sepals united basally, the upper 2 for more than half their length, 2–4.5 mm long .... *L. cuneata*
  2. Flowers purple to pink 4
    4. Rather delicate plants; leaflets thin, elliptic (often broadly so), 0.5–2 cm wide; flowers 7–8 mm long, the clusters on long threadlike peduncles .... *L. violacea*
    4. Stout plants with tough, densely leafy stems; leaflets 2–11 mm wide; flowers 4–7 mm long, in short-stalked clusters 5
      5. Leaflets narrowly elliptic to narrowly oblanceolate, oblong or linear oblong, more than 3 times as long as wide .... *L. virginica*
      5. Leaflets oblong to elliptic or obovate, less than 3 times as long as wide .... *L. stuevei*

229. *Desmodium sessilifolium* fruits

L. capitàta Michx.   Roundhead lespedeza

L. capitatus, headlike or globose—the flowers occurring in headlike clusters. The Pawnee Indians called the plant *parus-as*, rabbit foot, for the terminal cluster of softly hairy flower heads.
White or yellowish; August–September
Prairie, prairie roadsides, less frequently in open oak woods, usually in sandy or rocky soil
East two-thirds

Erect perennial herb. Stems rigid, densely leafy, simple or branched above, up to about 1.2 dm tall. Leaves short-petioled or nearly sessile; stipules awl-like, persistent; leaflets elliptic to oblong or oblanceolate, prominently pinnately veined, usually conspicuously covered with appressed silky hairs, mostly 2–4 cm long and 5–12 mm wide, the apex acute or rounded, mucronate, the base rounded or cuneate. Flowers mostly petaliferous and fertile, crowded in globose heads which are axillary or borne in clusters at the tops of the stems. Calyx 8–10 (–12) mm long, the sepals separate, long-acuminate, approximately equal in length. Petals white or yellowish, the standard with a purple spot and about the same length as the calyx; wings longer than the keel. Fruit 4–5 mm long. Seeds brown to tan or green, shiny, ellipsoid, slightly flattened, 2.5–3 mm long.

L. hirta (L.) Horn., known from Cherokee County, somewhat resembles *L. capitata* but differs in having leaves which are broadly elliptic to obovate and lack a covering of silky hairs and in having looser, more cylindrical flower clusters.

230. *Glycyrrhiza lepidota*

L. cunèata (Dumont) G. Don.   Chinese bush clover

L. cuneata, wedge-shaped—in reference to the leaves.
Yellowish-white with purple or pink; late August–early October
Rocky prairie, roadsides
East third, especially southward; intr. from eastern Asia

Erect, somewhat shrubby perennial. Stems clustered, up to about 1 m tall, with numerous ascending branches, the branches and main stems densely leafy. Leaves short-petioled; stipules threadlike or awl-shaped, deciduous; leaflets oblanceolate or wedge-

BEAN SUBFAMILY

177

shaped, 5–25 mm long and 1.5–6 mm wide, the lower surface with appressed silky hairs (but not conspicuous as in *L. capitata*), the apex rounded or truncate, conspicuously mucronate, the base wedge-shaped. Flowers yellowish-white marked with purple or pink, borne in clusters of 1–3 in upper leaf axils, apetalous flowers scattered among the petaliferous ones. Sepals 2–4.5 mm long, long-acuminate, fused basally except the upper 2 which are fused more than half their length. Standard 5–7 mm long; wings about the same length as or shorter than the keel. Fruit 3–5 mm long, longer than the calyx. Seeds tan or greenish, mottled with brown, shiny, ellipsoid to ovoid, slightly flattened, 1.5–2.5 mm long.

This plant is valuable for erosion control and soil improvement, often surviving on poor soils where other legumes will not, and provides food and cover for wildlife.

231. *Lespedeza capitata*

L. stipulàcea Maxim.   Korean lespedeza

NL., from L. *stipula*, small stalk. Botanically the term *stipule* refers to the usually small, paired, leaflike structures which occur at the base of the petiole of some plants.
Pink or purple; August–September
Roadsides, waste ground, disturbed prairies
East third; intr. from eastern Asia

Low annual. Stems erect or reclining, much branched from the base, up to 4 dm tall or long; hairs on the stem appressed toward the apex of the plants. Leaves short-petioled; stipules conspicuous, lanceolate to ovate, membranous, conspicuously nerved, persistent; leaflets obcordate to obovate, (6–) 10–15 mm long and 4–10 mm wide, the apex rounded or notched, the base acute, margins and lower midvein ciliate, especially when young. Petaliferous flowers pink or purple, 7–8 mm long, borne in leafy racemes. Calyx 2–2.5 mm long, the lobes rounded or acute, shorter than the calyx tube. Fruit 2.5–3 mm long. Seeds brown or nearly black, shiny, ellipsoid, slightly flattened, about 2.5 mm long.

Common lespedeza, *L. striata* (Thunb.) H. & A., is recorded from at least 1 location in the state. It is an annual closely resembling *L. stipulacea* but has stem pubescence appressed toward the base of the plant and lacks conspicuous ciliate hairs on the margins and lower midveins of young leaves.

Korean lespedeza was introduced into the United States in 1919 for use as pasturage, hay, and cover and is especially important in the northern states where the short growing season does not permit common lespedeza to set seed.

232. *Lespedeza stipulacea*

L. stuèvei Nutt.

Named in 1818 for its discoverer, Dr. W. Stueve of Bremen.
Purple; September
Sandy prairie, open oak woods
Southeastern fourth, especially southward

Perennial herbs. Stems erect, 1 to several, simple or branched above, up to 8 dm tall, the upper portions densely leafy. Leaves short-petioled; stipules hairlike or awl-shaped, persistent; leaflets oblong to elliptic or obovate, mostly 0.6–2 cm long and 4–11 mm wide, the apex rounded, acute, or truncate, mucronate, the base rounded or acute, both surfaces but especially the lower one covered with silky appressed hairs. Petaliferous flowers purple, 4–6 mm long, mostly located in short-stalked clusters at the tips of the stems. Calyx about 3 mm long, the sepals united about half their length. Wings about the same length as the keel. Apetalous flowers abundant in short-stalked axillary clusters, producing many fruits. Fruits quite hairy, 5–7 mm long. Seeds greenish, sometimes mottled with purple, shiny, narrowly ovoid, slightly flattened, 2.5–3 mm long.

L. violàcea (L.) Pers.

L. *violaceus*, violet colored.
Purple; July–September
Wooded hillsides, roadsides, prairie, waste ground, usually in rocky soil
East fourth

Rather delicate perennial herbs. Stems erect to ascending, branched, 2.5–6 dm tall, glabrous or with closely appressed or ascending hairs. Leaves petioled; stipules hairlike, persistent; leaflets thin, elliptic, often broadly so, rarely obovate, mucronate, minutely hairy on the lower surface, mostly 0.8–3.5 cm long and 0.5–2 cm wide, the apex and base rounded. Petaliferous flowers purple, 7–8 mm long, borne in 4- to 6-flowered racemes on threadlike peduncles. Sepals 3–4 mm long, fused basally, the upper 2 fused at least half their length. Wings about 1 mm shorter than the keel. Fruits 5–6 mm long. Seeds ellipsoid, slightly flattened, about 3 mm long (only a few seen).

233. *Lespedeza violacea*

L. repens (L.) Bart. and *L. procumbens* Michx. are known from scattered sites in the east third of Kansas. In the field they may be distinguished from *L. violacea* on the basis of their prostrate or procumbent habit. *L. procumbens* has stems and petioles with spreading hairs, while *L. repens* has stems and petioles glabrous or with appressed or ascending hairs.

L. virgínica (L.) Britt.   Slender lespedeza

"Virginian."
Purple or pink; late July–early October
Prairie, open woods, creek banks, roadsides, often on rocky soil
Southeast sixth and Douglas and Franklin counties

Perennial herb. Stems several, erect, simple or branched above, densely leafy, up to about 1 m tall. Leaves petioled; stipules hairlike, persistent; leaflets narrowly elliptic to narrowly oblanceolate, oblong, or linear, especially the lower surface with appressed silky hairs, 1–3 cm long and 2–6 cm wide, the apex rounded, mucronate, the base rounded or acute. Petaliferous flowers purple or pink, 5–7 mm long, borne in short-stalked axillary clusters on the upper parts of the stem and branches. Sepals acuminate, 3–4 mm long, united about half their length, except the upper 2 which are united well over half their length. Fruit 4–6 mm long. Seeds greenish or tan, shiny, ellipsoid, slightly flattened, about 2.5–3 mm long.

234. *Lespedeza virginica*

LÒTUS L.   Trefoil, birdsfoot trefoil, bastard indigo

According to Fernald (9), an Ancient Greek plant name used in many senses.

Herbs or suffrutescent plants. Leaves odd-pinnate with (1–) 3–5 leaflets. Flowers small, yellow, white, or red, borne singly or in clusters. Calyx lobes 5, elongate. Petals clawed; standard obovate; keel petals usually united along both margins, distinctly beaked. Stamens 10, diadelphous, all or some of the filaments enlarged at the tip. Fruit several-seeded, oblong to linear, round in cross section (in ours).

Two species in Kansas. *L. corniculatus* L., an introduced perennial with 5-foliolate leaves and yellow or orange-red flowers in clusters of 3–7 is known from a few scattered counties only. It has been shown to occasionally produce toxic amounts of a cyanogenic substance.

L. purshiànus (Benth.) Clem. & Clem.   Prairie trefoil

Named after Frederick Pursh (1774–1820), a botanist active in North America in the early 1800s, author of *Flora americae septentrionalis*, the first work to treat North American botany on a continental scale and the first to include significant numbers of species from west of the Mississippi River, many of which had been collected by the Lewis and Clark Expedition (1804–1806) and by Archibald Menzies, sailing with Capt. Vancouver along the Pacific Coast.
Yellowish-white, tinged with red; July–early September
Sandy soil of prairies, pastures, and roadsides
Central third, especially southward, and extreme eastern counties.

235. *Lotus purshianus*

Erect, bushy-branched annuals, up to 6 dm tall. Stems with long, spreading hairs, becoming glabrate with age. Leaves 3-foliolate, sessile or with short petioles; stipules reduced to small glands; leaflets elliptic or narrowly lanceolate, 8–22 mm long and 2–9 mm wide, more or less densely covered with long hairs, apex acute, often with a sharp rigid tip, base acute or rounded. Flowers borne singly or in pairs in the leaf axils. Calyx (4–) 6–8 mm long, the lobes longer than the calyx tube. Standard 6–7 mm long, tinged with red; wings oblong, auricled; keel crescent-shaped, about as long as the standard. Fruit linear, 2–3.5 cm long, mostly 4- to 9-seeded, the 2 segments twisting longitudinally after dehiscence. Seeds mottled, short-oblong, plump, about 2.5–3 mm long and 2 mm wide.

LUPÌNUS L.   Lupine

L. lupinus, Ancient Latin name for plants of this genus, from *lupus*, wolf—apparently from an old belief that it destroys the soil.

Annual or biennial herbs. Leaves palmately compound with 3–10 leaflets. Flowers white, yellow, pink, or blue, showy, borne in terminal racemes or spikes. Calyx 2-lipped, the lower lip or lobe longer than the upper. Blade of the standard nearly round, the sides strongly bent backward; wings united at the tip and enclosing the crescent-shaped keel except for its slender curved beak. Stamens monadelphous; 5 anthers linear, alternating with 5 much smaller, round anthers. Fruits oblong, flattened, 2- to several-seeded, opening along 2 sutures.

Many, but not all, species of *Lupinus* are toxic to livestock to some degree and under certain range conditions.

L. pusíllus Pursh

Rusty lupine

L. *pusillus*, very small, insignificant—here used in reference to the low habit of growth.
Blue or purple, or tinged with pink or almost white; May–June (–mid-July)
Sandy prairie
West fifth, especially southward, and Norton, Graham, and Rooks counties

236. *Lupinus pusillus*

Erect winter annual which produces a low rosette of leaves the first fall, then blooms the next spring. Flowering plants much branched from the base, 0.8–1.6 dm tall, with conspicuous spreading hairs. Leaves with 5 or 7 leaflets (3 on the lowest leaves) and long petioles which are broadened and somewhat membranous at the base and united with the stipules; leaflets elliptic or oblanceolate, often narrowly so, (1–) 1.5–3.5 cm long and 3–7 mm wide, often folding along the midvein, apex acute or rounded, base acute or wedge-shaped. Racemes 2–7 cm long, short-peduncled or sessile. Flowers blue, purple, tinged with pink, or almost white, 9–12 mm long, subtended by inconspicuous membranous bracts. Calyx long-hairy, the 2 lobes or lips longer than the calyx tube. Fruits tough, long-hairy, 2-seeded, about 2 cm long excluding the slender, curved, hairless style which may eventually be broken off. Seeds tan or brown, mottled with darker brown, nearly round to more or less obliquely ovate, flattened, about 6 mm long.

Compare with *Psoralea cuspidata* and *Psoralea esculenta*.

MEDICÁGO L.

Medic

NL., from L. *medica*, for a kind of clover (alfalfa) introduced from ancient Media into Greece.

Annuals, herbaceous perennials, or rarely shrubs. Stems angled or square. Leaves 3-foliolate. Flowers small, blue, violet, or yellow, borne in headlike or somewhat elongate racemes. Calyx 5-lobed, the lobes similar but unequal in length. Standard obovate to oblong, longer than the wings; keel blunt, shorter than the wings. Stamens 10, diadelphous. Fruits 1- to several-seeded, usually indehiscent, straight or coiled, sometimes spiny.

Three species known for Kansas. *M. minima* (L.) L. is introduced and quite rare and therefore not included here. *M. sativa* L., cultivated alfalfa, is common as an escape along roadsides, in old fields, and on waste ground.

237. *Medicago sativa*

M. lupulìna L.

Black medic, nonesuch

NL. *lupulinus*, little hop—in reference to the hopslike clusters of little pods.
Yellow; May–September
Roadsides, old fields, waste ground, disturbed prairie
Scattered throughout, more common eastward; intr. from Europe

Annual. Stems ascending or reclining, branching, up to 4 dm long. Leaflets cuneiform to obovate or rhombic, mostly 0.5–1.5 cm long and 0.5–1 (–1.5) cm wide; apex rounded or sometimes notched, base acute, sometimes broadly so; margins finely toothed; stipules fused to the petiole 2–3 mm, those of the lower leaves winglike and deeply toothed, those of the upper leaves lanceolate, long-acuminate, and entire. Herbage and fruits finely hairy, sometimes with glandular hairs also. Flowers yellow, 1.5–2 mm long, in crowded, ovoid, headlike racemes 4–7 mm long (up to 15 mm in fruit). Calyx lobes long-acuminate, with green midribs. Petals deciduous after flowering. Fruits 1-seeded, brown or black, thin-walled, more or less reniform and obscurely coiled, about 2–2.5 mm long, not enclosed by the perianth, the surface strongly veined. Seeds greenish-yellow, smooth, broadly ellipsoid, about 1.5 mm long.

This European species, now widely distributed in North America, adapts itself to various conditions of climate and soil and has become a valuable addition to pasturage in the higher mountain ranges. It was, indeed, in recognition of its good behavior and forage value that the people of Britain called it "nonesuch." According to Prior's *Popular Names of British Plants* (23), this is the species recognized in Ireland as the true shamrock.

Compare with *Trifoloium campestre* and *T. dubium*.

238. *Medicago lupulina*

MELILÒTUS Mill.

Sweetclover

NL., from Gr. *melilotus*, for a kind of clover, from Gr. *meli*, honey, + *lotos*, the name of several kinds of plants.

Annual or biennial herbs. Leaves 3-foliolate, the terminal leaflet stalked, the margins coarsely to finely toothed. Flowers small, white or yellow, borne in slender elongate racemes. Calyx lobes more or less alike, separate; standard usually longer than the others; wings and keel coherent. Stamens 10, diadelphous. Fruit ovoid, wrinkled, somewhat flattened or nearly globose, usually 1-seeded and indehiscent.

Two species in Kansas.

Key to Species:
1. Flowers white .... *M. alba*
1. Flowers yellow .... *M. officinalis*

M. álba Desr.   White sweetclover

L. *albus*, white.
White; June–November
Roadsides, railroad embankments, fields, waste places
Throughout, more common eastward; intr. from Eurasia

Biennial, with a short rootstock and strong, deep taproot. Stems slender, erect, branching, up to 3 m tall. Leaflets broadly oblanceolate to elliptic or oblong, those of the main leaves mostly 1–3.5 cm long and 4–15 mm wide, apex rounded, base acute or wedge-shaped, margins coarsely serrate, sometimes only finely serrate or entire on uppermost leaves; stipules bristlelike. Flowers white, 3–4 mm long, in loose, slender racemes 4–12 cm long and about 3 mm wide. Calyx lobes long-acuminate, the sinuses between them rounded. Fruits 1-seeded, brown, papery, reticulate, about 2.5–3 mm long. Seeds yellowish, smooth, ellipsoid, about 2 mm or slightly longer.

The following structural details of the flower are coordinated in a manner to insure cross-pollination: There are 10 stamens, the upper 1 free, the other 9 united into a tube which encloses the ovary and lower part of the slender style. Both stamens and pistil are enveloped by the keel; the 2 wings are slightly fused to the keel, and a fingerlike projection from each wing hooks over the keel somewhat back from the middle. Nectar is secreted basally inside the staminal tube and can be reached by the proboscises of insects through the gap in the tube provided by the free stamen. Thus when the wings of the corolla are pressed down by a bee, for instance, the keel also is brought down enough to bring the stigma and anthers into contact with the bee's body, and cross-pollination may then take place as the bee visits other sweetclover plants. When insects are excluded by covering the flowers with muslin, no seeds are produced.

A native of Eurasia as far eastward as Tibet, this sweetclover was first recorded in this country in Virginia in 1739, and its cultivation was begun here in Alabama in 1856. It was early cultivated as a honey plant, but now it is highly valued also for hay, forage, and soil renovation. The amount of honey obtained from sweetclovers varies with the locality, the time of summer, and the weather, the yield being greater westward from the Mississippi than eastward, greater early in summer than later. Only about a third as much honey is yielded in a cool, wet summer as in a warm, dry one.

The herbage of sweetclover develops a delightful vanillalike fragrance on drying, because of an increase in its coumarin content, making it desirable for scenting clothing and linen. It is recorded that when these plants became disseminated in this country, the Indians hung them in their abodes for the pleasure of their fragrance.

M. officinàlis (L.) Lam.   Yellow sweetclover

NL., from L. *officina*, manufacturing laboratory, such as one where drugs are prepared for use.
   As a plant name *officinalis* or *officinale* signifies that the plant is or formerly was kept for medicinal use by pharmacists.
Yellow; May–September
Roadsides, waste ground, disturbed prairie
Throughout, more common eastward; intr. from Eurasia

A biennial very similar to *M. alba* but not so tall and with lower, more spreading branches and a smaller root system. Flowers yellow (sometimes drying cream-colored), 5–6 mm long, in loose racemes (2.5–) 4–10 cm long and about 1.5 cm wide. Leaves and fruits as in *M. alba*.

This is the species that was recommended by Hippocrates (died 4th century B.C.), Theophrastus (died c. 285 B.C.), and Dioscorides and Pliny (1st century) as an astringent and mollifying agent in the treatment of inflamed and swollen parts, and for taking internally for stomach and intestinal ulcers. Down through the Middle Ages it was employed in the preparation of salves, plasters, and poultices for application to swollen joints and glands, boils, and other abscesses, and decoctions of the herbage were applied to wounds and taken internally for chronic bronchial catarrh. Then, Gerard's herbal (13) informs us in good Elizabethan English that "melilote boiled in sweet wine untile it be soft, if you adde thereto the yolke of a rosted egge, the meale of Linseed, the roots of

239. *Melilotus alba*

240. *Melilotus officinalis*

Marsh Mallowes and hogs greeace stamped together, and used as a pultice or cataplasma, plaisterwise, doth asswage and soften all manner of swellings."

Sweetclover poisoning in cattle, and to a lesser extent in sheep and horses, is caused by the ingestion of molded sweetclover hay. The toxic substance, dicoumarin, is produced when certain fungi act on coumarin, a compound normally present in sweetclover, and causes the animal to die of internal (and sometimes external) hemorrhages. The substance acts as an anticoagulant and has been used in human medicine for this purpose. A similar bleeding syndrome in cattle may also be caused by eating moldy *Lespedeza* hay.

241. *Oxytropis lambertii*

OXÝTROPIS DC. Crazyweed

NL., from Gr. *oxys*, sharp, + *tropis*, keel—in reference to the sharp beak of the keel characteristic of the flowers in this genus.

Perennial herbs, mostly with leaves and scapes arising at ground level from a several-crowned taproot. Leaves odd-pinnate with many leaflets. Flowers yellowish-white to bright pink or purple, borne in spikes, quite similar to those of *Astragalus* except for the keel and the nearly equal divisions of the 5-lobed calyx. Wing-petals usually 2-lobed or notched at the apex; tip of the keel abruptly narrowed into a slender, straight or curved appendage called the beak. Stamens 10, diadelphous. Fruits ovoid to somewhat cylindric, longitudinally grooved on the upper side, partially or incompletely 2-celled by the intrusion of a partition from the upper suture.

Two species in Kansas, both producing symptoms of loco poisoning as discussed for *Astragalus*.

Key to Species:
1. Each hair attached to plant precisely at what appears to be the base of the hair; flowers usually whitish; plants of the northwest sixth of Kansas .... *O. sericea*
1. Point of attachment of each hair actually a short distance above what appears to be the base of the hair; flowers usually purplish-pink; plants of the west two-thirds of Kansas
.... *O. lambertii*

O. lambértii Pursh Lambert crazyweed

In honor of Aylmer Bourke Lambert, English botanist.
Various shades of purplish-pink to nearly white; late April–June
Prairie hillsides and uplands, usually on rocky or sandy soil
West two-thirds

Herbaceous perennial, up to 5 dm tall, with a strong, deep taproot and a crown with branches so short that the plant seems stemless. Leaves 0.5–1.5 (–2) dm long, mostly erect or ascending; leaflets (7–) 11–19, oblong to linear-lanceolate or linear, mostly 1–3.5 cm long and 2–7 mm wide, apex acute, base acute. Herbage and fruits silky, the hairs attached at a point near the base of each hair, with a short section protruding backward from the point of attachment. (This character is most easily seen—under magnification, of course—by gently prodding the distal end of the hair and noting whether what appears to be the base of the hair is attached or is free and pivoting around a point a short distance—about 0.25 mm—up from the base.) Flowers various shades of purplish-pink (drying purplish-blue) to nearly white, 16–23 mm long; spikes 0.5–1.2 dm long, elevated above the leaves on leafless stalks. Calyx tubular, 8–10 mm long, the teeth short-acuminate or triangular with concave sides. Standard somewhat reflexed; wings enveloping and longer than the keel. Fruits short-cylindric, tapering gradually into a slender beak and persistent style, 2–2.5 cm long (including style), and 5–6 mm wide, subtended by the persistent calyx, opening along 1 side.

All parts of this plant are toxic and addictive to horses, cattle, sheep, and goats, causing a dazed condition, lack of muscular control, and sometimes death. Crazyweed is relatively unpalatable, and livestock usually leave it alone except when forced by hunger to eat it. Once they have acquired a liking for it, they will seek it out, even when better pasturage is available.

O. serícea Nutt. Crazyweed

NL., from Gr. *serikos*, silken.
Whitish with purple keels; May–mid-June
Prairie hillsides and uplands, usually on sandy or rocky soil
Northwest sixth

Perennial herbs, up to 5 dm tall, quite similar to *O. lambertii*, with a branched woody caudex. Leaves 0.4–2 dm long, mostly erect or ascending; leaflets 11–25, elliptic to oblong, mostly 1.3–3 cm long and 2–7 mm wide, apex acute, base acute. Herbage and fruits silky, but the hairs differing from those of *O. lambertii* in that each is attached precisely at its base. Flowers whitish (rarely purple) with purple keels, 14–20 mm long; spikes 0.5–1 dm long, usually elevated above the leaves. Calyx tubular, 9–12 mm long, the teeth acuminate. Fruits as in *O. lambertii*.

242. *Petalostemon candidum*

PETALOSTEMON Michx.   Prairie clover

NL., from Gr. *petalon*, thin sheet or leaf of metal—in botany applied to the petals which are usually thin, + *stemon*, thread—in botany the stamen, because of the threadlike filament. The genus name (often pronounced *Petalóstemon*) calls attention to the union of stamens and petals—a very unusual feature in the Legume Family.

Perennial herbs, ours with a woody rootstock and long, stout roots. Leaves odd-pinnate, gland-dotted; stipules inconspicuous, bristlelike, often deciduous. Flowers small, white or yellowish to pink or rose-purple, borne in dense terminal spikes or heads. Calyx usually strongly ribbed, the 5 lobes similar in size and shape. Corolla not papilionaceous, consisting of a single petal, the standard, with a threadlike claw and obovate or round blade. Fertile stamens 5, protruding conspicuously, fused toward the base into a sheath open along the upper side, alternating with 4 petal-like staminodes (sterile stamens) with spreading, colored blades. Fruit obliquely obovate to semicircular, somewhat flattened, thin-walled, 1-seeded.

Seven species in Kansas. *P. compactum* (Spreng.) Swezey is known only from Morton, Stevens, and Grant counties. It is our only species of prairie clover with pale yellow flowers; it has silky spikes as long as 15 cm.

Key to Species:
1. Principal leaves with 11 or more leaflets; plants densely gray-hairy (less frequently only sparsely hairy) .... *P. villosum*
1. Principal leaves with 5–7 (–9) leaflets; plants glabrous to sparsely hairy, rarely more densely covered with long, soft hairs 2
  2. Flowers bright violet to pink (rarely white) 3
    3. Bracts evenly hairy, usually deciduous, by anthesis; calyx strongly 10-ribbed, densely covered with long, spreading hairs up to 1 mm long .... *P. tenuifolium*
    3. Bracts with a transverse band of appressed hairs just below the base of the narrowed tip, persistent; calyx not as above .... *P. purpureum*
  2. Flowers white 4
    4. Spikes globose to ovoid, not over 1.5 cm long; bracts shorter than the calyces, scarcely discernible, even in bud .... *P. multiflorum*
    4. Spikes becoming cylindric in fruit, usually longer than 1.5 cm; bracts as long as or longer than the calyces, at least in bud 5
      5. Spikes loosening in fruit so the calyces are not immediately contiguous, stem surface of rachis visible; calyx usually minutely hairy, its ribs drawn out into flanges or fine wings .... *P. occidentale*
      5. Spikes not loosening in fruit, the surface of the rachis entirely occupied by the contiguous flowers; calyx (except teeth) glabrous, the ribs usually appearing as rounded ridges, less frequently somewhat flangelike .... *P. candidum*

P. cándidum (Willd.) Michx.   White prairie clover

L. *candidus*, of a dazzling white—in reference to the corolla.
White; mid-June–July
Prairie, occasionally in rocky, open woods
East half, with a few records from farther west

Glabrous perennials arising from a woody rootstock. Stems 1 to several, erect, branched above, 3–7 dm tall. Main leaves 3–5 cm long; leaflets 7 on main leaves, 3–5 on upper leaves, narrowly oblanceolate to narrowly elliptic or linear, 6–30 mm long and (1–) 2–6 mm wide, often folded along the midrib, the apex sharp-pointed, the lower surface minutely gland-dotted (use magnification). Flowers white, about 6 mm long. Spikes (1–), 2–6 cm long and 7–9 mm wide (measured from calyx tip to calyx tip), densely flowered, the rachis (when the flowers are removed) completely covered by attachment scars; bracts gland-dotted, ciliate along the margins but otherwise glabrous, deciduous after anthesis. Calyx 3–4 mm long, glabrous, with rounded or flangelike ribs, and a ring of amber glands near the top of the calyx tube (use magnification). Seeds brown or buff, asymmetrically reniform, 1.5–2 mm long.

In the central counties where the range of this species overlaps with that of *P. occidentale*, one may find plants intermediate in some respects between the 2 species.

243. *Petalostemon multiflorum*

244. *Petalostemon occidentale*

245. *Petalostemon purpureum*

P. multiflòrum Nutt.   Round-headed prairie clover

NL., from L. *multus*, many, + *flos, floris*, flower—in reference to the numerous flower heads.
White; mid-July–early September
Rocky prairie and roadside banks, usually on limestone soil
Kiowa County and east third, except for extreme northeast and southeast (i.e., primarily in the
    Flint Hills)

Glabrous perennial arising from a woody rootstock. Stems several to many, erect, branched, up to about 6 dm tall, usually naked below at flowering time. Leaves mostly 1.5–2.5 cm long; leaflets 5–9, oblong or narrowly elliptic to linear, often folded along the midrib, 5–14 mm long and 0.75–2 mm wide, the lower surface conspicuously gland-dotted. Flowers white, about 7 mm long. Spikes globose, 0.8–1.5 cm long; bracts inconspicuous, deciduous. Calyx about 3 mm long, with a ring of small amber glands at the top of the calyx tube, glabrous except for minute hairs on the teeth. Seeds brown or buff, asymmetrically reniform, about 2 mm long.

P. occidentàle (Gray) Fern.   Western prairie clover

L. *occidentalis*, of the west—this species occurring primarily in the western portions of the
    plains states.
White; late June–mid-September
Dry, rocky or sandy hillsides and banks
West half, primarily

Glabrous perennials from a woody rootstock. Stems several, erect to ascending or even prostrate, much branched (in this respect resembling *P. multiflorum*), 3.5–8 dm tall. Leaves mostly 1.5–4 cm long; leaflets 5–7, oblanceolate below to linear above, mostly 0.5–2 cm long and 1–3 (–5) mm wide, apex usually rounded, lower surface gland-dotted. Flowers white, about 5–6 mm long. Spikes mostly cylindric, 1–4 cm long and 6–8 mm wide (measured from calyx tip to calyx tip), eventually becoming more loosely flowered than in *P. candidum*, the attachment scars on the rachis (when the flowers are removed) separated by intervening areas of stem; bracts gland-dotted, minutely ciliate along the margins but otherwise glabrous, deciduous after anthesis. It is not uncommon to find some smaller, globose spikes on a plant, in addition to the more prevalent cylindrical ones. Calyx about 3.5 mm long, minutely hairy to nearly glabrous, with flangelike or winglike ribs, and a ring of amber glands near the top of the calyx tube. Seeds brown or buff, asymmetrically reniform, sometimes with 1 end somewhat truncated, about 2–2.5 mm long. Considered by some authorities to be merely a variety of *P. candidum*, and then called *P. c.* var. *oligophyllum* (NL., from L. *oligos*, small, thin, + *phyllon*, leaf—the leaves being much smaller than those of the more eastern plants assigned to *P. candidum* var. *candidum*).

P. purpùreum (Vent.) Rydb.   Purple prairie clover

L. *purpureus*, for purple of various shades, including violet.
Reddish-purple; mid-June–July (rarely as late at mid-August or September)
Prairie, usually on rocky or sandy soil
Throughout

Stems several from the woody rootstock, usually erect but occasionally ascending or even prostrate, 2–6 dm tall. Leaves mostly 1–3.5 cm long; leaflets (3–) 5, linear, 5–17 mm long and 0.5–1.5 mm wide, the margins rolled toward the upper surface, the lower surface gland-dotted. Stems and leaves usually glabrous or only sparsely hairy. Flowers reddish-purple (rarely white), about 6 mm long. Spikes 1–4.5 cm long and 7–10 (–12) mm wide (measured from calyx tip to calyx tip); bracts persistent after anthesis, with a transverse band of appressed hairs, the dark tip abruptly acuminate. Calyx about 3.5 mm long, densely covered with appressed silky hairs imparting a silver appearance to the spike. Seeds buff-colored, broadly elliptic to short-oblong, usually truncated at 1 end and with a short, stubby beak at that end, 1.5–2 mm long.

Plants from the southwestern and extreme western counties tend to be smaller in stature and to have the spikes borne on longer peduncles (about a fourth of the height of the plant) than those from the east; they are placed by some authorities in a separate species, *P. arenicola* Wemple.

See plate *51*.

P. tenuifòlium A. Gray

NL., from L. *tenuis*, thin, + *folium*, leaf, in reference to the narrow leaflets.
Reddish-purple; late May–mid-July
Dry, rocky prairie
Southwest sixth and Wallace and Gove counties

Similar in size and aspect to *P. purpureum*. Spikes 1.5–7 cm long, becoming looser after anthesis than in *P. purpureum*; bracts gradually acuminate, falling soon after anthesis, evenly hairy except for the dark, glabrous tip. Calyx about 4.5 mm long, densely covered with spreading or ascending hairs up to 1 mm long.

P. villòsum Nutt.   Silky prairie clover

L. *villosus*, hairy, shaggy.
Lavender; July–August
Sandy prairie, frequently on stabilized dunes
Mostly south and west of the Arkansas River

Gray-green, soft-hairy perennials with a woody rootstock and rhizomes. Stems several, erect to ascending or prostrate, 1.5–6 dm tall. Leaves mostly 1.5–3.5 (–4.5) cm long; leaflets (9–) 13 (–15), elliptic or oblong, 5–14 mm long and 1–4 mm wide, the lower surface gland-dotted. Flowers pale lavender with yellow or orange stamens, about 7 mm long. Spikes (1.5–) 2.5–11 cm long and 7–9 mm wide (measured from calyx tip to calyx tip); bracts silky, inconspicuous, deciduous even before anthesis. Calyx about 4 mm long, densely covered with long, spreading hairs which become tawny-colored with age. Seeds buff-colored, narrowly ovoid with a short stubby beak, about 2 mm or slightly longer.

Compare with *Amorpha canescens*.

246. *Petalostemon tenuifolium*

PSORÀLEA L.   Scurfpea

NL., from Gr. *psoraleos*, scabby—in reference to the many superficial glands.

Perennial herbs. Leaves palmately compound with 3 or 5 leaflets or pinnately 3-foliolate, petiolate, stipulate. Stems, leaves, and especially the calyx and fruit conspicuously gland-dotted (except in *P. esculenta* which has inconspicuous glands or none at all). Flowers small, white or blue-purplish, borne in spikes or racemes. Calyx 5-lobed, the lowest lobe longest. Standard usually clawed, the blade obovate to round, somewhat reflexed or spreading; wings clawed, auricled, about as long as the standard, longer than the keel. Stamens usually 10 (rarely 9), diadelphous. Fruit a small, papery, indehiscent, 1-seeded pod.

Ten species in Kansas. For *P. hypogaea*, see comments under *P. cuspidata*.

247. *Petalostemon villosum*

Key to Species (24):
1. Leaves pinnately 3-foliolate, the stalk of the terminal leaflet longer than those of the lateral leaflets .... *P. psoralioides*
1. Leaves palmately 3- or 5-foliolate, the leaflets all sessile or with stalks of equal length 2
  2. Calyx 2–3 mm long; fruit exposed 3
     3. Fruit longer than broad; flowers lavender to violet; glandular dots on leaves not variable in size 4
        4. Leaflets linear, about 2 mm wide; pedicels slender, about 8 mm in length ....
        .... *P. linearifolia*
        4. Leaflets obovate to oblong, 3–8 mm wide; pedicels 2–4 mm long .... *P. tenuifolia*
     3. Fruit almost globose; flowers whitish except for purple tip on keel; glandular dots on leaves variable in size .... *P. lanceolata*
  2. Calyx (at least lower tooth) 5 mm or more in length; fruit partially or completely hidden by the enlarging and persistent calyx 5
     5. Flowers at anthesis in well-separated whorls; calyx about 6–7 mm long 6
        6. Lower calyx tooth subulate, equaling or longer than the corolla; plant usually conspicuously silvery, but upper leaf surfaces often greenish .... *P. argophylla*
        6. Lower calyx tooth broadly lanceolate, shorter than corolla; plant greenish to gray with stiff appressed hairs, but seldom silky or silvery .... *P. digitata*
     5. Flowers at anthesis crowded in dense spikes; calyx usually about 10 mm long 7
        7. Hairs spreading .... *P. esculenta*
        7. Hairs appressed .... *P. cuspidata*

P. argophýlla Pursh   Silverleaf scurfpea

NL., from Gr. *argos*, shining, + *phyllon*, leaf—because of the silvery hairs over the herbage.
Purplish; June–early August
Prairie hillsides and upland, often on rocky or sandy soil
North half and south-central counties

Erect, silver-haired perennials from a dark, woody root. Stems simple below with several open, spreading or ascending branches above, 2.5–6 dm tall. Upper leaves with 3 leaflets, the lower with 5; leaflets elliptic to oblanceolate or obovate, 1.5–4 cm long and 0.6–1.7 cm wide, the apex rounded or acute, usually with the midvein extended into a small sharp point, the base rounded or acute, both surfaces (but especially the lower) with appressed silky hairs which may nearly obscure the dark glandular dots. Flowers purple (usually drying brown), sessile, borne in few-flowered whorls which become well separated within the spikes. Calyx 7–11 mm long, densely silky, the lobes acuminate. Petals usually shorter than the longest calyx lobe.

Like *P. tenuiflora*, this plant scatters its seed in the tumbleweed manner. The plants have little nutritive value for livestock and have been suspected of causing severe poisoning in a child who ate a quantity of the seeds.

248. *Psoralea cuspidata*

P. cuspidàta Pursh

L. *cuspidatus*, pointed—probably in reference to the calyx lobes.
Blue or purplish; May–June
Rocky prairie uplands and slopes
West half, primarily

Perennials from a long, woody root or rootstock with papery bark. Stems erect to ascending or decumbent, 2.5–6 dm tall or long, with appressed hairs. Upper leaves with 3 leaflets, the lower with 5; leaflets oblanceolate to obovate, 1.5–4 cm long and 0.5–2 (–2.5) cm wide, the middle leaflet longest, the apex acute or rounded, usually with a small sharp point, the base acute, the lower surface with appressed hairs. Flowers blue or purple, (12–) 15–18 mm long, subtended by long-acuminate bracts, borne in racemes 3.5–6 (–8) cm long and about 3.5 cm wide, the stalk of the flower cluster longer than the leaves. Calyx dark purple, pouchlike on the upper side, 8–12 mm long, with lanceolate-acuminate lobes, becoming less conspicuously hairy and more conspicuously gland-dotted with age. Fruits thin-walled and papery, 6–8 mm long, terminated by the persistent style, subtended by fragments of the stamen tube, and enclosed in the persistent, much-enlarged calyx. Seeds gray-brown, sometimes mottled, shiny, very broadly elliptic to nearly round, slightly flattened, about 4–5 mm long and 3–3.5 mm wide.

Compare with *Lupinus pusillus* and *Psoralea esculenta*.

P. hypogaea Nutt. is known from Morton, Finney, Hamilton, and Cheyenne counties. It resembles *P. cuspidata* but differs in being much smaller in stature, in having the leaves and flowers crowded on a short stem at or very near ground level, and in having a woody, turnip-shaped storage root about 5 cm below ground.

249. *Psoralea digitata*

P. digitàta Nutt. ex T. & G.

L. *digitatus*, having fingers (i.e., digitate)—in reference to the palmately compound leaves.
Blue or purplish; June–early July
Sandy prairie, especially in sand-hill areas
South of a line drawn from Stanton Co. to Ellsworth and Harvey counties

Slender, open-branched perennial with a long, woody root or rootstock with papery bark. Stems erect, 3–8 dm tall. Herbage with appressed hairs. Uppermost leaves with 3 leaflets, lower ones with 5 or 7; leaflets linear to very narrowly oblong or oblanceolate, mostly 1.5–4.5 cm long and 3–7 mm wide, the middle leaflet longest, apex rounded or acute, usually with a small sharp point, base acute or attenuate. Flowers blue or purple, 8–10 mm long, borne in few-flowered whorls which become separated within the racemes. Calyx broadly bell-shaped, densely silky, 6–7 mm long at flowering time, expanding as the fruits mature, the acuminate lobes eventually becoming widely separated by rounded sinuses. Fruits thin-walled and papery, 6–8 mm long, terminated by the persistent style, subtended by fragments of the stamen tube, nearly as long as the persistent, enlarged calyx. Seeds gray-brown, shiny, very broadly elliptic, slightly flattened, about 4–4.5 mm long and 3 mm wide.

250. *Psoralea esculenta*

P. esculénta Pursh   Prairie turnip, Indian potato

L. *esculentus*, edible—in reference to the root.
Blue or purplish; May–early June
Prairie, usually in rocky, limestone soil
East half and northwest fourth

Perennials with top-shaped or oblong storage roots 1.5–5 cm thick. Stems erect, 1–4 dm tall, usually single and unbranched except for the long, stout axillary peduncles which may give the impression of lateral branches. Stems, petioles, and peduncles with long, spreading hairs; herbage and fruits not gland-dotted. Leaflets 5 (occasionally 3 on the lowest leaves), elliptic to oblanceolate, mostly 2–5 cm long and 6–16 mm wide, the middle 1 usually longest; apex rounded or acute, sometimes with a small sharp point; base attenuate or acute; lower surface with long appressed hairs. Flowers blue or purple (drying brown), 1.5–2 cm long, borne in dense spikes 4–6 cm long and about 2.5 cm wide (to 8 cm long in fruit) on long, spreading or ascending peduncles. Calyx bell-shaped, 13–15 mm long at flowering time, expanding as the fruits mature, the base pouchlike on 1 side, the lobes sword-shaped or acuminate. Fruits thin and papery, about 2 cm long, the style persisting as a long, straight beak which protrudes conspicuously beyond the calyx. Seeds gray or gray brown, sometimes mottled, nearly round to broadly elliptic or narrowly ovate, slightly flattened, 4–6 mm long and 3–4.5 mm wide.

In the days when the Indians roamed the prairie the tuberous roots of this species furnished a significant part of their food. Since the tops break off and are dispersed by the wind after the fruits ripen in July and August, leaving no sign for the location of the roots, the women and children were abroad with long sharp sticks, digging the roots before the tops were gone. They then peeled the tubers, braided them in a strand by their tapering extensions, and hung them up to dry for winter use. If the peeled tubers were to be used immediately they were boiled or roasted in the ashes of a dying fire. Nuttall, the naturalist, who explored westward to the Pacific in 1834, said of these tubers: "The taste is rather insipid but not disagreeable either raw or boiled." Fernald (9) describes the raw roots as having a "somewhat starchy and sweetish turniplike taste." The cooked roots, in my experience, have a distinctly leguminous flavor somewhat resembling raw peanuts.

251. *Psoralea linearifolia*

P. lanceolàta Pursh

L. *lanceolatus*, spearlike—in reference to the leaves, some of which are oblanceolate in outline.
White or purplish; mid-May–early July
Dry, sandy prairie
West half, primarily, especially in the southwest

Perennials from woody rhizomes or rootstocks. Stems erect, mostly 2–4 dm tall, with several ascending branches. Herbage and fruits densely gland-dotted. Leaflets 3, oblanceolate below to narrowly oblong or linear above, 1.5–4.5 cm long and 0.5–8 mm wide, apex acute or rounded on the lower leaves, with a small sharp point; base attenuate; lower surface with a few appressed hairs (use magnification). Flowers white or purplish, 5–7 mm long, borne in long-peduncled racemes 1.5–3 cm long. Calyx bell-shaped, about 2 mm long, the triangular lobes or teeth shorter than the calyx tube. Fruit more or less globose, about 5 mm long, with a short beak, densely gland-dotted, much longer than the persistent calyx. Seeds reddish-brown, round in outline, about 4 mm long.

P. linearifòlia T. & G.

NL., from L. *linearis*, linear, + *folium*, leaf—in reference to the narrow leaflets.
Blue or purplish; mid-June–August
Prairie, usually on rocky, limestone (caliche) soils
Scattered in the west third, primarily southward

252. *Psoralea psoralioides*

Slender, open-branched plants with woody rhizomes or rootstocks. Stems erect, mostly 3–7 dm tall. Herbage and fruits densely gland-dotted. Leaflets 3 on lower leaves, sometimes only 2 or 1 on upper leaves, very narrowly oblanceolate or linear, 1–5 cm long and 1–4.5 mm wide, the lower surface sparsely hairy. Flowers light blue or purplish, 8–10 mm long, borne in loose racemes 2–8 cm long. Calyx bell-shaped, hairy, about 4–5 mm long at flowering time, enlarging as the fruits mature, the lobes triangular or somewhat rounded. Fruits ovate to obovate, somewhat flattened, 7–8 mm long including the beak. Seeds dark brown or gray-brown, sometimes mottled, broadly elliptic or bean-shaped, slightly flattened, about 5 mm long.

P. psoralioìdes (Walt.) Cory   Sampson's snakeroot

NL., from the genus *Psoralea* + the suffix *-oides*, like, i.e., resembling *Psoralea*, this plant having been placed first by botanist Thomas Walter in the genus *Trifolium*.
Purplish; mid-May–mid-June
Prairie or open woods, often in sandy soil
Extreme southeast

Slender, erect perennials with a fibrous-rooted underground stem terminated by a spindle-shaped storage root. Stems simple, or branched at ground level, 3–6 dm tall. Leaflets 3, elliptic to linear-lanceolate, mostly 3–8.5 cm long and 6–13 (–17) mm wide (those of the lowermost leaves sometimes much smaller), each attached to the petiole by a small stalk of its own; apex and base rounded. Stems and leaves sparsely hairy. Flowers 5–6 mm long, purplish (drying tan), borne in dense axillary spikes 1–3.5 (–6) cm long and 1–1.5 cm wide which gradually elongate and become more open as the fruits mature; peduncles much longer than the subtending leaves. Calyx 2.5–4 mm long, the lobes short-acuminate, with long silky hairs, especially along the margins. Fruits broadly and obliquely heart-shaped, flattened, transversely wrinkled, 3.5–5 mm long. Seeds brown, about 3 mm long.

253. *Psoralea tenuiflora*

P. tenuiflòra Pursh   Scurfpea, wild alfalfa

NL., from L. *tenuis*, thin, slight, few, + *flos, floris*, flower—the flowers being both small and
    relatively few per raceme.
Purplish; late June–mid-August
Prairie
Throughout

Erect perennials from a woody rootstock, apparently reproducing vegetatively by means of rhizomes. Stems erect, 1 to many per plant, simple below, with several spreading or ascending branches above, 0.3–1.3 m tall. Leaflets 3 on most leaves (5 on the lower ones, which tend to be early deciduous), oblanceolate to elliptic, 0.5–3.7 cm long and 2–8 mm wide, apex rounded or acute, usually mucronate, base acute. Herbage with appressed hairs and black glandular dots. Flowers purplish, 4–7 mm long, borne in axillary racemes on peduncles longer than the subtending leaves. Calyx 2–3 mm long, densely glandular. Fruits obovate or oblong, with a triangular beak, slightly flattened, densely gland-dotted, 6–7 mm long. Seeds gray-green to orange-brown, sometimes purple-spotted, ovate to oblong or nearly round, slightly flattened, mostly 4–5.5 mm long.

Plants from the eastern half of the state tend to be larger in all respects and to have more crowded racemes than plants from the western half of the state; they are separated by some authors as a distinct species, *P. floribunda* Nutt., but I will include them here as a variety, *P. tenuiflora* var. *floribunda* (Nutt.) Rydb. The plants of the western half of the state then fall into var. *tenuiflora*. The eastern plants have calyces with spreading, sometimes kinky hairs and acute lobes, whereas those of the western plants have appressed hairs and short-acuminate lobes. In some of the central counties where the 2 varieties overlap, and even as far east as Chase County, some plants have measurements and calyx characters intermediate between those of the eastern and western extremes.

254. *Psoralea tenuiflora* var. *floribunda*

Observations of insects associated with this species of *Psoralea* were made by A. K. Tatschl (25). The plants are pollinated by several species of short-tongued mining and burrowing bees (Family Andrenidae) and by a species of sweat bee (Family Halictidae). They are parasitized by several insects, including a moth, *Schinia jaguarina*, whose larvae eat the seeds; a weevil, *Apion ablitum*, whose larvae eat the inner parts of the flower buds and then pupate within the dried bud; and by a species of *Diabrotica*, a chrysomelid beetle which eats the flowers.

SOPHÒRA L.   Sophora

NL., from Arabic *sufayra*, yellow—the color of the dry immature flower buds of some species from
    which a yellow dye is obtained.

Herbaceous perennials similar to *Astragalus* (ours) or shrubs or small trees. Leaves odd-pinnate with many leaflets. Flowers white to yellow, purple, or pink, borne in racemes. Calyx 5-toothed. Standard rounded, tapered to the base; wings straight, clawed; keel nearly straight, clawed, auricled, with a beaklike projection in our single species. Stamens 10, united at least part of their length. Fruit leathery, indehiscent, round in cross section, stipitate, more or less constricted between the seeds.

One species in Kansas.

S. nuttallìàna Turner   Silky sophora, Nuttall's sophora

Named after Thomas Nuttall (1787–1859), explorer-naturalist and one-time professor of natural
    history and curator of the Botanic Gardens at Harvard, and the first botanist to collect
    this plant.
White or yellowish, purple basally; May–mid-June
Prairie, usually on sandy soil
West half and a few counties as far east as Saline County

255. *Sophora nuttalliana*

Silky, gray-green herbaceous perennials forming colonies by means of creeping rhizomes. Stems erect, branched from the base, mostly 1–4 dm tall. Leaflets 7–25, elliptic to oblanceolate, 4–13 mm long and 1.5–8 mm wide, often folded along the midrib, the apex acute to rounded, truncate, or emarginate, the base acute. Flowers white, in loose racemes 2.5–9 cm long. Calyx 5–8 mm long, pouchlike on 1 side at the base, the lobes triangular to short-acuminate, shorter than the calyx tube. Standard 12–15 mm long, strongly recurved. Stamens united almost half their length on the lower side, grading to only basally united on the upper side of the stamen tube. Fruit 1- to 4-seeded, usually beaked, mostly 2–4 cm long.

Occasionally, individuals of this species are attacked by a rust fungus. The clusters of aeciospores, which rupture the epidermis of the leaves, appear as small dark spots which superficially resemble the glandular dots found in genera such as *Psoralea, Dalea, Petalostemon,* and *Glycyrrhiza.*

STROPHOSTÝLES Ell.   Wild bean vine

NL., from Gr. *strophe,* twining or twisting, and *stylos,* pillar—in reference to the style (so named because this projection from the ovary is often pillarlike), which in this instance curves to conform with the strongly recurved keel enveloping it.

256. *Strophostyles helvola*

Annual or perennial, prostrate or climbing herbs. Leaves 3-foliolate, the leaflets stalked (stipellate). Flowers white to pink-purple, borne in short, long-peduncled racemes or heads. Calyx 4-lobed, the lowest lobe longest. Blade of the standard round or broadly ovate, auricled; blade of the wings curved-oblong; keel broadest below the middle, the distal half quite narrow and curved upward, but not coiled. Stamens 10, diadelphous. Upper side of the style hairy near the stigma. Fruits linear, straight or slightly curved, nearly round or slightly flattened, several-seeded, the valves twisted after dehiscence.

Two species in Kansas.

Galactia volubilis (L.) Britt., milk pea, is sometimes confused with species of *Strophostyles.* Milk pea is a twining perennial legume with 3-foliolate leaves, the leaflets nearly round to very broadly elliptic or broadly ovate and mostly 1–4 cm long. The small purplish flowers with straight keels are borne in short-stalked, few-flowered axillary racemes, each flower subtended by a pair of minute bracts.

See also *Amphicarpaea bracteata.*

Key to Species:
1. Leaflets rhombic to ovate or ovate-oblong, either entire or more or less 3-lobed, 1.5–5 cm wide; flowers 9–12 mm long, the inflorescences on stout stalks (2–) 8–22 cm long; calyx glabrous or nearly so; fruits 6–9.5 cm long; seeds 6–8 mm long .... *S. helvola*
1. Leaflets lanceolate to elliptic or linear, never lobed, 0.2–1.6 (–2) cm wide; flowers 6–8.5 mm long, the inflorescences on threadlike stalks (1–) 4–8 cm long; calyx hairy; fruits 2.5–4 cm long; seeds 2.5–4 mm long .... *S. leiosperma*

S. hélvola (L.) Ell.   Trailing wild bean

L. *helvolus,* light yellow—reference unclear, probably to the flowers.
Greenish-purple or pinkish-yellow; mid-July–September
Prairie ravines, open woods, creek banks, roadsides, usually in sandy soil
East half and some south-central counties

Annual, loosely twining, usually prostrate vine. Stems branched at the base, more or less hairy, up to 2.5 m long. Leaflets highly variable in shape, ranging from rhombic to ovate or ovate-oblong, the margins quite entire to deeply 3-lobed, 2–7 cm long and 1.5–5 cm wide, both surfaces or just the lower surface with sparsely scattered, appressed hairs; apex acute or slightly rounded, mucronate; base rounded or acute. Flowers greenish-purple to pinkish-yellow, 9–12 mm long, borne in few-flowered heads on stout stalks (2–) 8–22 cm long. Bracts subtending the individual flowers acute, at least as long as the calyx tube. Calyx glabrous or with only a few hairs. Fruits 6–9.5 cm long. Seeds brown, scurfy, oblong, somewhat 4-angled in cross section, mostly 6–8 mm long and 3.5–4 mm thick.

Of the 10 stamens, 9 are united in a sleeve around the style. The tenth stamen, by standing free, gives access to nectar secreted at the base of the pistil. Pollen is discharged over the hairy style, and when a honeybee or bumblebee presses the keel down to get at the nectar, the style springs up and deposits pollen from its hairy surface upon the bee's thorax. This pollen the bee transports to the stigma of other flowers, at the same time receiving increments of pollen from their styles—and so on from flower to flower.

S. leiospérma (T. & G.) Piper   Wild bean vine

NL., from Gr. *lius,* smooth to the touch, and *sperma,* seed, i.e., smooth-seeded.
Pink or pale purple; July–mid-October
Sandy prairie, occasionally in alluvial thickets or woods
Scattered nearly throughout except in the Flint Hills

Annual, loosely twining, usually prostrate vine. Stems branched at the base, more or less hairy. Leaflets lanceolate to elliptic or linear (rarely ovate), 1.5–6 cm long and

0.2–1.6 (–2) cm wide, both surfaces with appressed hairs; apex rounded to acute, mucronate; base rounded; margins entire. Flowers pink or pale purple, 6–8.5 mm long, borne in few-flowered heads on slender threadlike stalks (1–) 4–8 cm long. Bracts subtending the individual flowers acute, shorter than to as long as the calyx tube. Calyx quite hairy. Fruits 2.5–4 cm long. Seeds gray or brown mottled with black and/or white, shiny or with a dull scurfy coat that rubs off easily, short-oblong, 2.5–4 mm long and 2–3 mm thick.

257. *Strophostyles leiosperma*

258. *Stylosanthes biflora*

STYLOSANTHES Sw.   Pencil flower

NL., from Gr. *stylos*, pillar, column, + *anthos*, flower, from the stalklike calyx tube.

Herbaceous perennials. Leaves 3-foliolate. Flowers small, usually yellow. Calyx tube 4- to 5-lobed, the lowest lobe the largest. Standard obcordate to broadly obovate or round, longer than the other petals, narrowed at the base; wings oblong to obovate, with 2 small auricles at 1 side of the base of the blade; keel curved upward, nearly as long as the standard. Stamens monadelphous, 5 of the anthers oblong and about 0.5 mm long alternating with 5 smaller, globose ones. Fruit an indehiscent, 1- to 2-jointed pod (loment), the lower segment, when 2 are present, usually sterile, stalklike, and persistent.

One species in Kansas.

S. biflòra (L.) BSP.   Pencil flower

NL., from L. *bis*, 2, + *flos, floris*, flower—i.e., 2-flowered.
Orange-yellow; late June–July (–October)
Prairie, roadsides, open woods, sandy or rocky soil
Southeast eighth and Rice and McPherson counties

Perennial with long, woody roots. Stems several, erect to spreading or ascending, more or less branched, 1.5–3.5 dm tall, minutely hairy or sparsely bristly. Leaves petiolate; stipules fused with the base of the petiole and united to form a cylinder around the stem, the free portions linear or subulate and with bristly margins; leaflets elliptic or oblanceolate to narrowly elliptic (the lowermost sometimes obovate), mostly 1–3 cm long and 2–7 mm wide, apex acute, spiny, base acute, margins sparsely bristly, lower surface with conspicuous veins. Flowers orange-yellow, usually drying cream-colored or purplish, 8–10 mm long, borne singly or in few-flowered spikes at the ends of branches. Calyx about 4 mm long, the lower and lateral lobes with acute apexes and as long as or longer than the calyx tube, the upper 2 lobes rounded and united nearly to the apex. Fruits hard, D-shaped, slightly flattened, minutely hairy, 3.5–5 mm long, terminated by the short, curved or coiled style. Seeds tan, slightly flattened, about 2.5 mm long and 1.5 mm wide, triangular, with the longest side convex.

TEPHRÒSIA Pers.   Tephrosia, hoary pea

NL., from Gr. *tephros*, ash-colored, hoary—a characteristic feature of the genus.

Hairy herbaceous perennials, more or less woody at the base with stems rising in a clump from a branched rootstock and a strong, deep root. Leaves odd-pinnate with 5 to many leaflets. Flowers yellow, cream-color, pink, or purple, borne in upright racemes. Calyx 5-lobed, the 2 upper lobes slightly shorter than the lower 3. Petals all clawed; blade of the standard round or broadly ovate; wings slightly coherent with the incurved keel. Stamens 10, diadelphous. Fruit linear to linear-oblong, several-seeded, dehiscent.

One species in Kansas.

This is a large and widespread genus with about 250 species distributed in warm regions around the world. Many of these produce small amounts of rotenone and related compounds and have been used by primitive groups as fish poisons for trapping fish. The technique most frequently used is to pulverize or chop the root or bark (sometimes the leaves) and to throw this material into a pool or netted-off section of a stream. The poison stuns the fish, by acting on the respiratory system, and the fish float to the surface where they are easily caught.

More recently, various species of *Tephrosia*, including *T. virginiana*, have been investigated as a source of rotenone for insecticides. It has also been learned that certain fresh-water snails known to be hosts of the organism responsible for schistosomiasis or bilharzia in man, and essential for its complete life cycle, are sensitive to the fish-poisoning species of *Tephrosia*. In some tests, these natural poisons were more effective than copper sulfate in controlling snails. Rotenoid compounds have advantages over some synthetic pesticides in being of low toxicity to warm-blooded animals (except pigs) and in decomposing quickly on exposure to light or heat, thereby failing to accumulate in soil, water supplies, or food chains.

T. virginiàna (L.) Pers.   Virginia tephrosia, goat's-rue

"Virginian."
Cream-colored and purple or pink; late May–early July
Prairies and open woods, usually on rocky or sandy soil
East half, excluding the Flint Hills and northernmost counties

Stems several, unbranched, 2–4 dm tall, sparsely to densely hairy. Leaflets 13–23, broadly to narrowly elliptic or oblanceolate, 1–3 cm long and 3–10 mm wide, smooth or with sparse appressed hairs above, densely silky and gray below; apex mucronate, that of lateral leaflets acute or rounded, that of terminal leaflets usually broadly rounded or truncate; base acute. Flowers borne in short, dense racemes 3–6 cm long. Calyx bell-shaped, silky, 7–8 mm long, the lobes triangular- or lanceolate-acuminate, longer than the calyx tube. Standard cream-colored, 15–18 mm long, spreading or somewhat reflexed, silky on the back; wings pink or purplish, rather obliquely obovate, shorter than the standard; keel broadly crescent-shaped, cream-colored, pink, or purplish, enveloped by the wings. Fruits linear-oblong, flattened, silky, erect or spreading, straight or arching downward slightly, 2.5–5.5 cm long and 3.5–6 mm wide, 3- to 8-seeded, the valves twisting longitudinally after dihescence. Seeds brown or greenish, flattened, nearly round to short-oblong, 3–4 mm long and about 3 mm wide, with a white membrane flaking loose from the surface.

TRIFÒLIUM L.   Clover

Name used by Pliny for clover, from L. tri-, combining form of tris, tres, 3, + folium, leaf—in reference to the compound leaves with 3 leaflets, characteristic of these plants.

Annual, biennial, or perennial herbs. Leaves 3-foliolate (rarely with more leaflets), the leaflets finely toothed; stipules fused to the petioles. Flowers yellow, white, pink, rose-purple, or red, borne in sessile or peduncled heads or in short headlike spikes. Calyx campanulate with 5 linear or awl-shaped teeth. Petals separate or more or less united into a tube, usually withering and persistent after anthesis; standard ovate to oblong or obovate, folded about the wings in bud but reflexed in the open flower; wings more or less hooked over the keel and somewhat adherent to it. Stamens 10, diadelphous. Fruit short, membranous, 1- to 6-seeded, indehiscent, often enclosed by the persistent calyx and corolla.

Eight species in Kansas. Only the 6 most common are included here.

Looking at a flower, we see nothing of the pistil and stamens, but when we press down on the wing-keel combination, we uncover the style and stamens in a huddle in front of the passage to the nectar found at the bottom of the corolla tube. A bee gathering nectar would press down the keel and touch the exposed stigma and anthers with its thorax, thereby pollinating the stigma with pollen previously picked up, while getting more pollen to be carried to still other flowers. Darwin's experiments have shown that without the aid of insects, principally bees, no seeds or at most a negligible number are produced.

Key to Species (26):
1. Flowers yellow 2
  2. Flower heads 15–40 flowered; standard prominently veined (at least when dried); petioles mostly longer than leaflets .... *T. campestre*
  2. Flower heads 5–10 flowered; standard not prominently veined; petioles mostly shorter than the leaflets .... *T. dubium*
1. Flowers not yellow 3
  3. Flowers rose to purplish-pink, subsessile (pedicels not exceeding 1 mm), frequently not recurved in fruit; stems hairy .... *T. pratense*
  3. Flowers white or pale pink, distinctly stalked (pedicels exceeding 2 mm in length), usually recurved in fruit; stems hairy or glabrous 4
    4. Calyx lobes narrowly lanceolate or bristlelike, more than 2 times longer than the tube; stems hairy or glabrous .... *T. reflexum*
    4. Calyx lobes triangular or lanceolate, up to 2 times as long as the tube; stems glabrous 5
      5. Stems prostrate, rooting at nodes, heads on long scapes arising at ground level; flowers usually whitish or pale pink; calyx lobes 0.8–1.1 times as long as the tube .... *T. repens*
      5. Stems ascending or erect; heads on short peduncles at apex of stem; flowers usually pinkish; calyx lobes 1.2–2.0 times as long as the tube .... *T. hybridum*

T. campéstre Schreb.   **Large hop clover**

L. campestre, of fields.

Yellow; May–mid-July (–September)
Roadsides, pastures, and rocky, open woods
East two-fifths, more common southeastward; intr. from Europe

Annual or winter annual. Stems ascending or reclining, round in cross section, hairy, up to 4 dm long. Leaflets obovate to oblanceolate or elliptic, glabrous, 5–15 mm long and 3–8 mm wide, the terminal leaflet on a longer stalk (petiolule) than the lateral ones; apex rounded or notched; base acute; margins finely serrate; petioles mostly longer than the leaflets; stipules lanceolate, somewhat oblique, fused to petiole about half their length. Flowers yellow (drying light brown), 3.5–4.5 mm long; heads long-peduncled, mostly 6–15 mm long and 6–10 mm wide, with 20–40 flowers. Calyx lobes long-acuminate or bristlelike. Blade of the standard 2–4 mm wide, prominently veined (at least when dried). Fruit 1-seeded. Seeds yellowish or brownish, ellipsoid, slightly flattened, smooth, about 1 mm long.

Compare with *Medicago lupulina*.

259. *Trifolium campestre*

T. dùbium Sibth.   Little hop clover, suckling clover

L. *dubius*, doubtful.
Yellow; May–June (–September)
Rocky or sandy pasture or open woods
Southeast 6 counties and Anderson County; intr. from Europe

Annual or winter annual similar to *T. campestre*, but with smaller flowers in 3- to 12-flowered heads. Petioles usually shorter than the leaflets. Calyx lobes linear or long acuminate. Standard 3–4 mm long, the blade 1–2 mm wide, not prominently veined. Seeds broadly ellipsoid, somewhat flattened, about 1.2–1.5 mm long.

T. hỳbridum L.   Alsike clover

L. *hybrida*, *hibrida*, mongrel.
White or pinkish; May–October
Roadsides, waste ground, lawns
Scattered in the east third; intr. from Europe

Smooth perennial herbs, 3–8 dm tall, much resembling *T. repens* but with erect or ascending stems not rooting at the nodes. Leaflets obovate or broadly elliptic, 1.5–3.5 cm long and 0.7–2.5 cm wide; apex rounded or slightly notched; base broadly acute or short-acuminate; margins finely serrate; stipules membranous with 3 veins. Flowers white or pinkish; 5–8 mm long; heads globose, many-flowered, about 2 cm across in ours, though some strains produce much larger heads. Fruits 2- to 3-seeded. Seeds light green to greenish-black, heart-shaped, about 1.5 mm long.

260. *Trifolium pratense*

T. praténse L.   Red clover

L. *pratensis* or *pratense*, growing in a meadow.
Rose to purplish pink; April–November
Roadsides, fields, lawns, waste ground, prairie
East two-fifths, primarily; scattered farther west; intr. from Europe

Biennial or short-lived perennials with a deep taproot and a rootstock that gives off short stems with only long-petioled leaves, and longer ascending or erect stems, up to 8 dm tall, with short-petioled leaves having elliptic, membranous, conspicuously veined stipules. Leaflets mostly ovate to broadly elliptic, 2–5 cm long and 1–3.5 cm wide, soft-hairy, marked above with a light-green crescent or inverted V which fades on drying; apex rounded (occasionally notched); base broadly acute; margins finely serrate, at least on youngest leaflets, the older leaflets often quite entire. Flowers rose to purplish-pink (rarely white), 12–18 mm long; heads many-flowered, ovoid to nearly globose, sessile or short-stalked, (1.5–) 2–2.5 cm long and 2–2.5 cm wide. Calyx more or less hairy, 9–10 mm long, the tube membranous and prominently veined, the lobes green, long-acuminate. Standard longer than the wings and keel, the blade oblong, not reflexed, enveloping the other petals. Seeds green tinged with purple, smooth, ellipsoid with a slight protuberance above the hilum, 2 mm long.

Pollination is effected by bumblebees especially, which have enough weight and strength to push down the keel-wing combination, and a proboscis long enough to reach the nectar at the bottom of the corolla-filament tube, which the shorter proboscis of the honeybee and of most wild bees is unable to do. With a covering of netting Darwin protected 100 flower heads of red clover from insects' visits and found that no seeds whatever were produced, while 100 unprotected heads freely visited by bumblebees produced 2,720 seeds.

Attesting the high economic value of red clover, a *U.S. Farmers' Bulletin* asserts that it is "the cornerstone of agriculture in the north-central states," while a voice from Europe declares that "red clover has contributed even more to the progress of agriculture than the potato itself." For, as Gerard says in his herbal (13), "Oxen and other cattell do feed on the herb, as also calves and young lambs."

Herbals of the Middle Ages also speak of the curative value of red clover. The learned Abbess Hildegard, for instance, in her *Physika* (12th century) recommends it for "Verdunkelung der Augen," "clouding of the eyes"—apparently meaning cataract. The list of human afflictions the decoctions of red clover were claimed to cure strains credulity, but we find admissible the statement in an old herbal that the administration of red clover "makes for easy breathing, a happy mood and good appetite," for we recall how heartening the mere sight and smell of blooming clover can be. The inflorescences apparently do contain some sedative compound and have been used to make an effective cough syrup.

261. *Trifolium reflexum*

T. refléxum L.   Buffalo clover

NL., bent or turned back, from L. *flexus*, bend.
Pink and white; May–August
Prairie, open woods, usually on acid soil
Primarily in the east 3 columns of counties south of the Kansas River

Annual. Stems ascending, branched from the base, glabrous to more or less hairy, up to 5 dm tall. Leaflets obovate to oblanceolate or broadly elliptic, 1–2.5 (–3.5) cm long and 0.8–1.5 cm wide; apex rounded or occasionally notched; base acute or short-acuminate; margins finely serrate; stipules large and conspicuous, leaflike, lanceolate to ovate. Flowers 8–11 mm long, in globose, many-flowered heads 2.5–4 cm across. Calyx 7–9 mm long, with a short, membranous, strongly veined tube and linear or long-acuminate lobes. Standard reddish (occasionally white), 5–6 mm wide; wings and keel pink or white; entire corolla turning brown with age. Fruit 2- to 4-seeded. Seeds yellowish, about 1–1.2 mm long and 1 mm wide.

262. *Trifolium repens*

T. répens L.   White clover

L. *repens*, creeping.
White or pinkish; May–October
Lawns, roadsides, fields, pastures, waste ground
East half, primarily; intr. from Europe

Mat-forming perennial 0.5–2 (–3) dm tall. Stems glabrous, prostrate, rooting at the nodes. Leaves held erect on petioles 3–20 cm long; leaflets ovate to nearly round, obovate, or obcordate, 1–2.5 cm long and (0.5–) 1–2 cm wide; apex rounded or notched; base acute; margins finely toothed. Flowers white or pinkish, 8–9 mm long; heads globose, about 2 cm across, the stalk or peduncle much longer than the leaves. Calyx 5–6 mm long, the lobes acuminate. Fruit 2- to 3-seeded. Seeds yellow, smooth, heart-shaped, 1 mm or slightly longer.

This is perhaps the most widely distributed of all legume species, occurring throughout the temperate and subtropical regions of the world. It is one of the best clovers for pastures, being hardy, nutritious, and well liked by livestock. In addition, it is valuable in erosion control, soil improvement, and as a honey plant.

Cross-pollination, accomplished primarily by honey bees, is necessary for seed set. In an experiment similar to that conducted with red clover, Darwin found that 20 heads on plants covered with netting to prevent insect pollination produced only a single, aborted seed, while 20 unprotected heads yielded 2,290 seeds.

VÍCIA L.   Vetch

Classical Latin name for a vetch.

Annual or perennial, trailing or climbing herbs. Leaves even-pinnate with a terminal, usually-branched, coiling tendril. Flowers yellow, white, purple, or blue, borne in axillary racemes. Calyx lobes 5, alike or dissimilar. Standard apically notched, narrowing to a broad claw which overlaps the wings, its blade obovate to nearly round; wings oblong or narrowly obovate, adherent to and usually longer than the keel. Stamens 10, diadelphous. Pistil sessile or on a short stalk (stipe); style slender with a tuft or ring of hairs at the summit. Fruit flattened or round in cross section, 2- to many-seeded, dehiscent.

Five species in Kansas. Our 2 most common species are included in the following key.

The vetches are eminently useful plants, the herbage providing important constituents of hay, and the seeds, pods, and leaflets affording foods for many kinds of birds. One species, *V. faba* L., the broad bean or Windsor bean, is eaten by man, and all species are valuable for renovating and holding the soil.

Some species of *Vicia* have been connected with poisoning in livestock—the toxic principle being a cyanogenetic glycoside, sometimes contained in the seeds, which causes liver lesions and photosensitization. In addition, among human beings, rare individuals develop a type of acute hepatitis called "favism" from eating raw or partially cooked seeds or inhaling the pollen of *V. faba*. For more details regarding these problems, see Kingsbury (21).

Kansas has 3 species in a closely related genus, *Lathyrus*, not included here because the plants are so infrequently found. This genus differs from *Vicia* primarily in having pistils of the flowers with a flattened style with a line of hairs along the inner surface, rather than a round style with a ring of hairs at the end.

263. *Vicia villosa*

Key to Species:
1. Stipules with but a single hastate basal lobe; stems and peduncles with spreading hairs; calyx strongly pouchlike on 1 side, the pedicel appearing to be attached laterally ...... *V. villosa*
1. Stipules with 2 or more teeth; stems and peduncles with appressed hairs; calyx not as above, the pedicel attached basally or nearly so .... *V. americana*

V. americàna Muhl.   Stiffleaf vetch, American vetch

"American."
Bluish-purple; May–June
Roadsides and prairie, usually on rocky or sandy soil
Scattered throughout except the southeast sixth

Herbaceous perennial. Stems trailing on the ground or climbing on other vegetation. Leaves 2–7 cm long (including petiole and tendril); leaflets 8–18, elliptic to linear or linear-oblong, 8–33 mm long and 1–5 mm wide; stipules coarsely dentate. Flowers bluish-purple (drying blue), 15–20 mm long, in loose, 3- to 9-flowered axillary racemes on slender peduncles. Calyx tube oblique but only slightly pouchlike at the base, the teeth broadly triangular on the lower side grading toward acuminate above, the longest lobe 1–4 mm long; pedicel attaching basally. Corolla rather broad, blade of the standard obovate, strongly reflexed, about 12–14 mm wide. Fruits 2.5–3.5 cm long, about 7–8 mm wide.

Two varieties in Kansas. Most of our plants belong to var. *minor*, which has thick leaflets narrowly oblong or elliptic to linear, 1–3 cm long and 1–4 mm wide. Variety *americana*, which occurs in Doniphan, Atchison, Leavenworth, Douglas, and Rooks counties, has thin leaflets which are broadly oblong or elliptic, 1–4 cm long and 4–14 mm wide.

V. villòsa Roth.   Hairy vetch

L. *villosus*, hairy, shaggy—descriptive of the stems, leaves, and peduncles of this plant.
Bluish-purple; May–September
Fields, roadsides, waste ground
Scattered, primarily in the east half; intr. from Eurasia

Slender annual or biennial vine, sprawling and forming mats on the ground, or climbing by tendrils when suitable support is available. Leaves 6–12 cm long (including petiole and tendril); leaflets 10–24, oblong to elliptic, mostly 1.5–3 cm long and 3–7 mm wide, opposite or alternate along the rachis; apex acute or rounded, with a tiny, sharp point; base rounded or wedge-shaped; stipules lanceolate to ovate, each with a single hastate basal lobe. Herbage with long, spreading hairs. Flowers bluish-purple (drying blue), 9–16 mm long, in loose, long-peduncled, 10- to 30-flowered axillary racemes on stout peduncles. Calyx tube oblique and pouchlike on 1 side at the base, the lobes or teeth long-acuminate on the lower side grading toward short-acuminate above, the longest lobe 2–5 mm long; pedicel appearing to attach laterally. Corolla rather slender, blade of the standard oblong, about 7 mm wide, the tip slightly reflexed. Fruits oblong, flattened, 2–2.5 cm long and 0.7–1 cm wide, mostly 3- to 5-seeded. Seeds dark brown, globose, about 4 mm in diameter.

V. dasycarpa Ten. is an introduced annual, quite similar to *V. villosa*, and is known from scattered sites in the east third. It differs from the above species in having the hairs of the racemes short and appressed, or incurved, rather than long and spreading and in having the longest calyx lobe only 1–2 (–2.5) mm long.

GERANIÀCEAE

Geranium Family

Annual, biennial, or perennial herbs. Leaves alternate or opposite and simple, usually round to kidney-shaped in outline, frequently palmately lobed and veined, petioled and stipulate. Flowers bisexual, regular, borne in clusters, rarely singly. Sepals 5, free from one another. Petals 5, free from one another, deciduous. Nectariferous glands usually present, alternating with the petals. Stamens 10, rarely 5. Pistil 1; ovary superior, 5-lobed and 5-celled, each cell 2-ovuled; styles 5, united below with a slender extension of the ovary axis to form a structure called the stylar column, then branching out again above; stigmas 5. Fruit a septicidal capsule with a prominent beak formed by the stylar column, separating at maturity into 5, 1-seeded sections. The unique structure of the pistil in this family allows an equally unique method of seed dispersal. As the fruits and seeds mature and the fruit wall dries, the 5 sections of the ovary separate at the base; the elastic styles separate from the inner column of the beak, except at the tip, and coil outward and upward rather explosively, flinging the seeds away from the plant in the process.

The family includes about 11 genera and 850 species of widespread distribution. The cultivated geranium (*Pelargonium zonale* Ait.), crane's bill (*Geranium* spp.), and stork's bill (*Erodium* spp.) belong to the Geraniaceae.

Kansas has 4 species in 2 genera. *Erodium cicutarium* (L.) L'Her., stork's bill, introduced from Europe, is reported from 7 central and south-central counties. It is similar in size and growth habit to *Geranium carolinianum* but has pinnately compound leaves with finely pinnately divided leaflets. The small pink flowers have only 5 fertile stamens, and the styles twist spirally, rather than coiling upward, as the sharp-pointed, fusiform segments of the fruit separate.

GERÀNIUM L.

Geranium

NL., from Gr. *geranion*, diminutive of *geranos*, crane—in reference to the beaked fruit.

Ours annual, biennial, or perennial herbs. Leaves opposite or mostly basal in ours, palmately lobed or divided. Flowers borne singly or in clusters of 2 to many, frequently in pairs on axillary peduncles. Stamens 10 (rarely 5), 5 long ones with nectiferous basal glands alternating with 5 short ones. Mature pistil segments separating along the inner suture, the recurved stylar portions of the beak recurved and coiled, but not spirally so, the inner surface smooth.

Three species in Kansas. *G. pusillum* L. is known from 5 counties in the eastern third of the state. See comments under *G. carolinianum*.

The old English and European herbals recommend the extracts of various species of *Geranium* for a variety of afflictions, such as nosebleed, hemorrhoids, wounds, inflamed eyes, sore mouth and throat, diarrhea, diseases of the kidney and bladder, and even cancerous sores. Of an English geranium "found neare to common high waies, desart places, untilled grounds, and especially upon mud walls almost every where," Gerard says, "The herbe and roots dried, beaten into a most fine pouder, and given halfe a spoonful fasting, and the like quantitie to bedwards in red wine, or old claret, for the space of one and twenty daies together, cure miraculously ruptures or burstings, as my selfe have often proved, whereby I have gotten crownes and credit . . . and a decoction of the herbe made in wine, prevaileth mightily in healing inward wounds, as my selfe have likewise proved." Various groups of American Indians used the spotted geranium, *G. maculatum*, for sundry similar purposes; more information is given with that species description.

Key to Species:

1. Erect perennials with branched, knobby rhizomes; petals 12–18 mm long; plants of rich woods in Cherokee County and our extreme northeastern counties .... *G. maculatum*
1. Erect to ascending or prostrate annuals or biennials with slender taproots; petals 4–6 mm long; plants of waste ground, fields, and other disturbed areas, primarily of the east half .... *G. carolinianum*

G. *caroliniànum* L.

Carolina geranium

"Of Carolina."
Pink to nearly white; April–August
Waste ground, fields, roadsides, open or disturbed areas of prairies and woods
East half, primarily

264. *Geranium carolinianum*

Common annual or biennial weed with a slender taproot. Stems erect to ascending or nearly prostrate, much branched, 1.5–4 (–6) dm tall. Main leaves mostly 2–5 cm wide, deeply palmately divided nearly to the petiole into 5–9 main segments, each of which is further divided into several toothed, oblong or linear lobes. Herbage and fruits more or less hairy. Flowers pink to nearly white in several dense terminal clusters of 4–12 subtended by reduced leaves. Sepals ovate to lanceolate, 4.5–6 mm long at anthesis (including the tail-like extension of the midrib), enlarging somewhat in fruit, with ciliate membranous margins. Petals oblanceolate, 4–6 mm long, the apex rounded or slightly notched. Stamens with membranous filaments widened toward the base. Fruits (including stylar column) 15–18 mm long, coarsely hairy. Seeds orange-brown or tan, finely reticulate, short-oblong, plump, about 1.7–2 mm long.

The blooming period is so much prolonged by the flowers of a cluster opening only 1 or 2 at a time that the ripe, purplish-black, pink-tipped fruits of the earliest-blooming flowers—each enhanced by a ruff of persistent sepals—provide a handsome setting for flowers of later appearance. By midsummer, the leaves also exchange their green for more colorful oranges, pinks, and purples.

G. pusillum L. (from L. *pusillus*, very small), introduced from Europe, resembles *G. carolinianum*. It is known from Atchison, Douglas, Coffey, Chautauqua, and Cowley counties and seems to be establishing itself in our area. In general aspect, it is a smaller, finer plant than the Carolina geranium and has red-violet flowers 3–4 mm long at anthesis. The sepals have acute or rounded tips lacking the tail-like projections characteristic of the Carolina geranium.

G. maculàtum L. — Spotted geranium

NL., *maculatus*, spotted, from L. *macula*, spot or stain, in reference to the green-spotted rhizome.

Rose-purple; mid-April–mid-May (–June)

Moist, rich, wooded hills and bluffs

Extreme northeast counties and Cherokee County

265. *Geranium maculatum*

Perennial, 2.5–4 dm tall, from a brown, green-spotted, branched, knobby rhizome, bearing a sparse cluster of long-petioled basal leaves and an erect stem from the apex of the youngest rhizome segments. Leaves deeply palmately divided, the main ones 8–14 cm wide, the 5–7 segments coarsely toothed or cut. Herbage more or less hairy. The flowering stem bears a single pair of short-petioled leaves, above which are usually 2–3 erect branchlets, each with a cluster of lavender or rose-purple flowers subtended by a sessile bract resembling but smaller than the other leaves. Sepals more or less elliptic, (7–) 8–12 mm long (including the long tail-like extension of the midrib), with minute appressed hairs and long, spreading hairs, some of which are gland-tipped (use magnification), the margins ciliate and/or membranous. Petals obovate, entire, 1.2–1.8 cm long. Stamens 10, the filaments united basally; individual pollen grains large enough to be seen with the unaided eye.

Chemical analysis shows this species to contain 2 astringents—tannic and gallic acids. The rhizome was listed in the *U.S. Pharmacopoeia* from 1820–1916 as an astringent and was described in the 23rd edition of the *Dispensatory of the United States* as being effective in the treatment of sore throat and mouth ulcers. Chippewa and Ottawa tribes boiled the entire plant and drank the tea to cure diarrhea. The Meskwakis steeped the rhizomes in water to prepare a rinse for inflamed gums. The Cherokees boiled the rhizome together with wild grape, and the resultant concoction was used as a mouthwash for children with thrush. The Pillager Ojibwas also used a wild geranium to treat mouth ulcers. The Indians of Great Manitoulin Island in Lake Huron applied the dried, powdered rhizome to bleeding wounds to promote coagulation. Clinical experiments with rabbits have since shown that the tannins contained in the plant do increase blood clotting.

OXALIDÀCEAE — Oxalis Family

Herbs or shrubs, rarely trees. Leaves alternate or all basal, ours 3-foliolately compound, petioled. Flowers bisexual, regular, borne in clusters. Sepals 5. Petals 5, sometimes united at the base. Stamens 10–15, basally united (monadelphous). Pistil 1; ovary superior, 5-celled with 2 to many ovules per cell; styles and stigmas 5. Fruit a capsule splitting open longitudinally along the back of each cell.

The family includes approximately 7 genera and 1,000 species, mostly pantropical in distribution.

Kansas has 1 genus.

OXÁLIS L.

Wood sorrel

L. and Gr. *oxalis*, a kind of rumex.

Herbaceous perennials with a sour juice. Leaves alternate or all basal. Flowers purplish-pink or yellow (rarely white). Leaves 3-foliolate (absent on fall-blooming plants of our *O. violacea*), with or without stipules, the leaflets obcordate, often reflexed, folded upward along the midrib. Petals slightly fused at the base. Sepals persistent in fruit. Stamens 10, 5 short ones alternating with 5 longer ones, the filaments widened and united toward the base. Capsule 5-angled.

Four species in Kansas. *O. corniculata* L., a widely distributed tropical weed, is introduced and known from a few localities. It resembles *O. dillenii* but has creeping stems which root at the nodes and brown seeds lacking white transverse ridges.

Key to Species:

1. Flowers purplish-pink (rarely white); tips of sepals bright orange; leaves and flowers arising directly from a scaly, bulblike underground part .... *O. violacea*
1. Flowers yellow; sepals entirely green; leaves and flowers borne on aboveground stems 2
    2. Plants with rhizomes; stipules lacking; stem and petioles with blunt, septate hairs (use high magnification); seeds entirely orange-brown, lacking white ridges .... *O. stricta*
    2. Plants lacking rhizomes, though decumbent stems may root at the nodes; stipules present; stem and petioles with pointed, nonseptate hairs (use high magnification); seeds orange-brown with whitish transverse ridges .... *O. dillenii*

266. *Oxalis stricta*

O. díllenii Jacq.

Yellow wood sorrel, oxalis

Named after Johann Jakob Dillen (1684–1747), a German-born professor of botany at the University of Oxford, England.
Yellow; late April–June, again late August–September
Fields, roadsides, lawns, waste ground, prairie, open woods
East two-thirds

Herbaceous perennials. Stems erect to reclining, branched, up to 4 dm tall, the prostrate stems sometimes rooting at the nodes, but true underground rhizomes never present. Leaves long-petioled, borne on the stem; leaflets yellow-green, obcordate, mostly 5–15 mm long and 5–15 mm wide. Stems and petioles with appressed, nonseptate hairs (use high magnification). Flowers yellow, 6–12 mm long, in 2- to several-flowered axillary umbels. Sepals lanceolate or oblong, 4–5 mm long, with appressed hairs. Capsules erect, linear, angled, hairy, 1–2 cm long, the pedicels spreading or descending. Seeds ovate to broadly elliptic, acute at the narrow end, orange-brown with whitish transverse ridges, about 1.3 mm long, with a glossy outer coat that peels away as the seeds mature.

O. strícta L.

Yellow wood sorrel, oxalis

L. *strictus*, erect, upright, probably in reference to the erect fruiting pedicels.
Yellow; late-May–early October
Open woods, creek banks, roadsides, prairie, usually in moister soils than *O. dillenii*
East two-fifths, primarily

Herbaceous perennials with underground rhizomes. Aboveground stems erect or prostrate, usually unbranched or with only a few branches, up to 5 dm tall. Leaves long-petioled, borne on the stem; leaflets yellow-green, obcordate, those of the main leaves mostly 1–2.5 cm long and 1.5–3 cm wide (conspicuously larger than in *O. dillenii*). Stems and petioles with spreading, septate hairs, in addition to appressed, nonseptate hairs (use high magnification). Flowers yellow, 6–10 mm long, in 2- to many-flowered axillary clusters which may be simple or branched. Sepals lanceolate or oblong, 3–4 mm long, glabrous or with a few appressed hairs. Capsules erect or spreading, linear, angled, glabrous (occasionally sparsely hairy toward the base), mostly 8–15 mm long, the pedicels spreading or ascending. Seeds as in *O. dillenii* but only rarely with white markings on the transverse ridges.

O. violàcea L.

Violet wood sorrel, oxalis

L. *violaceous*, violet in color.
Purplish pink, rarely white; mid-April–May (–June), again August–mid-October
Prairie, open woods, roadsides, usually on rocky or sandy soil
East two-fifths and Ellsworth and Barber counties

Herbaceous perennials with leaves and flowers arising directly from a scaly bulblike structure which produces offsets. Leaves long-petioled, the leaflets broadly obcordate, light gray-green, glaucous, somewhat succulent, 5–12 (–20) mm long and 1–3 cm wide. Flowers purplish-pink (rarely white), borne above the leaves in 1 or 2, 2- to 4-(8-)flowered clusters 5–18 cm tall. Sepals elliptic, 4–6 (–8) mm long, the tips bright orange. Petals 1–2 cm long, broadly rounded. Styles branches hairy. Capsules ovoid, 4–5 mm long. Seeds orange, broadly ellipsoid, somewhat pointed at each end, rugose, ornamented with minute white papillae (use high magnification), about 1.5 mm long.

Both the leaves and flowers respond sensitively to light and darkness, closing at night or on cloudy days and opening on return of sufficient daylight.

Some groups of Indians thought to increase the swiftness of horses by dosing their feed with the dried, powdered bulbs, and Indian and pioneer children chewed the sour leaves and flowers, which contain calcium oxalates and oxalic acid. In lieu of gooseberries and rhubarb, the pioneers used *Oxalis* leaves to make tart pies.

See plate 52.

267. *Oxalis violacea*

LINÀCEAE Flax Family

Mostly annual or perennial herbs, sometimes shrubs. Leaves alternate, rarely opposite, narrow, entire, sessile, with or without stipules. Flowers bisexual, regular, usually in terminal racemes or panicles. Sepals 5, distinct or fused basally, the calyx persisting in some species, deciduous in others. Petals 5, distinct, soon deciduous. Stamens 5, alternating with the petals, the filaments somewhat flattened and monadelphous below. Nectiferous glands or staminodes sometimes present. Pistil 1; ovary superior, 3- to 5-celled, sometimes more or less completely divided into twice as many; style 3–5, distinct or more or less united; stigmas terminal and somewhat enlarged (capitate) or extended a short distance down the inner faces of the style branches. Fruit usually a septicidal capsule.

The family consists of about 9 genera and 200 species distributed primarily throughout the temperate regions of the Northern and Southern hemispheres. The cultivated flax plant (*Linum usitatissimum* L.), source of linen fiber and linseed oil, belongs to this family.

One genus in Kansas.

LÌNUM L. Flax

L. *linum*, flax plant.

Annual or perennial herbs. Leaves alternate or opposite, the lower ones frequently falling early. Herbage often pale green or gray-green. Flowers blue or yellow (pale, dull orange in 1 of ours), as described for the family, the corolla quickly deciduous. Capsule ovoid to globose or somewhat flattened from above, completely or partly 10-celled.

Six species in Kansas. *L. hudsonioides* Planch. is known only from Sedgwick County. It differs from *L. rigidum* in having sepals without glandular teeth, and from *L. sulcatum* in having the styles fused more than half their length and the fruit separating into 5, rather than 10, segments. *L. medium* (Planch.) Britt., recorded from Cherokee and Crawford counties, is also similar to *L. sulcatum* but differs in having wider leaves (mostly 2–4 mm wide) with a rounded or acute apex, only the inner sepals with stalked glands along the margins, and smaller (mostly 1.6–2.1 mm long), depressed-globose fruits. The blue-flowered cultivated flax was grown in eastern Kansas, primarily east of the Flint Hills and south of the Kansas River, by early settlers. Within this area, 63,478 bushels of flaxseed were produced in 1873 and 2,173,800 bushels in 1890. It has been recorded as escaped in numerous counties in the eastern half; but most records are from at least 25 or 35 years ago, and it probably has not maintained itself in the wild.

Key to Species (64):
1. Flowers blue .... *L. pratense*
1. Flowers yellow to pale, dull orange 2
    2. Styles separate or nearly so; fruit ultimately separating into 10 1-seeded segments ......
    .... *L. sulcatum*
    2. Styles united to above the middle; fruit separating into 5 2-seeded segments ...... *L. rigidum*

L. praténse (Nort.) Small Meadow flax

L. *pratensis, pratense*, growing on or found in meadows, from *pratum*, meadow.

Blue; May–early August
Prairie, roadsides
Scattered in the west three-fifths, more common in extreme south-central counties

Delicate, erect annual, 0.5–3 (–4) dm tall, usually branched above and/or from the base. Leaves alternate, linear or very narrowly elliptic, mostly 5–15 mm long and 0.5–1.5 mm wide; stipular glands absent. Flowers blue, solitary in upper leaf axils or in leafy, few-flowered, branched clusters. Sepals ovate to broadly lanceolate or elliptic, 3–4.5 mm long, with membranous margins. Petals (5–) 8–12 mm long. Styles separate, the stigmas enlarged. Capsules straw-colored, broadly ovoid, 5–6 (–7) mm long, separating into 10 1-seeded segments, the pedicels spreading or drooping at maturity. Seeds reddish-brown, shiny, ovate, flattened, about 3.5–4 mm long.

L. usitatissimum, cultivated flax, is a taller, more slender plant which has the inner sepals with ciliate margins, the stigmas slender, and fruiting pedicels erect or ascending at maturity.

In times past, Indians of western America used the strong bast fibers of the stems of *L. pratense* for making string and cord needed in constructing baskets, mats, fishnets, and snowshoes.

268. *Linum pratense*

L. rígidum Pursh   Stiffstem flax, yellow flax

L. rigidus, stiff, hard, inflexible—probably in reference to the rigid branches and stems.
Yellow to pale, dull orange; (April–) mid-May–August
Dry, rocky or sandy prairie
West three-fifths

Smooth annual. Leaves alternate, linear, sharp-pointed, the larger ones mostly 0.8–3.5 cm long and 0.8–2 mm wide, the upper ones sometimes with glandular teeth; stipules usually represented by dark sessile glands. Flowers in branched terminal clusters. Sepals lanceolate, 4.5–7 mm long (sometimes to 10 mm in var. *berlandieri*), eventually deciduous, all 5 glandular-toothed, bristle-tipped, and with a prominent winged midrib. Petals broadly obovate, 8–18 mm long, hairy inside at the base. Stamen tube very short, the filaments slender. Styles united to within 0.5–1.1 mm of the stigmas. Capsules straw-colored, ellipsoid to ovoid, persistent, mostly 4–4.6 mm long, separating into 5 2-seeded segments. Seeds reddish-brown, narrowly ovoid-elliptic, flattened, about 2.5–3 mm long.

There is evidence that plants of this species are poisonous to sheep, but they grow too sparsely in our area to cause concern.

269. *Linum rigidum* var. *rigidum*

The species is represented in Kansas by 3 varieties which intergrade somewhat and which tend, in Kansas, to have the same range, though var. *berlandieri* is more common in the west fifth and extreme southern counties of the west half. Individuals of var. *rigidum* are mostly 3–4 dm tall, with relatively long, ascending branches above and open inflorescences with yellow flowers; while those of var. *berlandieri* are mostly 0.5–1.5 dm tall, rather densely and compactly branched from the base, and have yellow to pale dull-orange petals with a brick-red area below the middle. Plants of var. *compactum* have yellow flowers and tend to be intermediate between our other 2 varieties in height, amount of branching, and degree of compactness. The following characters are more reliable than growth habit for distinguishing among the varieties.

Key to Varieties (65):
1. Styles less than 5 mm long 2
  2. Stipular glands present .... *L. rigidum* var. *rigidum*
  2. Stipular glands absent .... *L. rigidum* var. *compactum*
1. Styles more than 5 mm long 3
  3. Fruit thin-walled (dark seeds evident through the wall), ellipsoid, the base rounded; petals yellow .... *L. rigidum* var. *rigidum*
  3. Fruit thick-walled, opaque, broadly ovoid, tapering abruptly at the base; petals yellow to dull orange, brick red below the middle .... *L. rigidum* var. *berlandieri*

L. sulcàtum Ridd.   Grooved flax, yellow flax

L. sulcatus, furrowed, plowed—in reference to the longitudinally grooved branches.
Yellow; June–mid-September
Prairie, usually on sandy or rocky soil
East half, primarily, with a few records scattered in the west half.

Smooth, slender, erect annuals 3.5–7 dm tall, the stems simple below with several ascending branches above. Leaves mostly alternate (the lowermost sometimes opposite), linear to narrowly lanceolate, 10–25 mm long and 1–2.5 mm wide, the apex sharply pointed; stipules present as dark glands. Flowers yellow, in branched terminal clusters.

270. *Linum sulcatum*

FLAX FAMILY

199

Sepals lanceolate, 3.5–5 mm long (up to 7 mm in fruit), with an acuminate apex, prominent midrib, and all 5 sepals with glandular teeth along the margins. Petals obovate, 5–10 mm long; their hold on the receptacle is so fragile that they are apt to drop away when an attempt is made to pick the flowers. Stamen tube very short (0.2–1.8 mm long). Styles united at the base only. Capsules straw-colored, ovoid or more or less globose, rounded or acute at the apex, 2.5–3 mm long, separating into 5 pairs of 1-seeded segments, subtended by the persistent calyx. Seeds reddish-brown, shiny, elliptic, flattened, about 2 mm long.

271. *Tribulus terrestris*

ZYGOPHYLLÀCEAE — Caltrop Family

After the genus *Zygophyllum*, NL., from Gr. *zygon*, pair, + *phyllon*, leaf—the leaf in that genus having a single pair of leaflets. "Caltrop" comes from L. *calx*, heel, + ML. *trappa*, trap, and is the English name for the military implement (*tribulus*) described under the genus *Tribulus* below.

Mostly shrubs, herbaceous perennials, or annuals. Leaves mostly opposite, pinnately compound with an unpaired terminal leaflet, stipulate. Flowers bisexual, regular, borne singly in the leaf axils. Sepals usually 5 and free from one another. Petals of the same number as the sepals, also free from one another. Stamens twice as many as the petals. Pistil 1; ovary superior, mostly 5- to 10-celled; style and stigma 1. Fruit a capsule, splitting at maturity into segments or *mericarps* as many as the cells.

The family contains about 27 genera and 200 species widely distributed in tropical and temperate regions and is especially well represented in the drier areas around the Mediterranean Sea.

Kansas has 2 species representing 2 genera. *Kallstroemia parviflora* Nort., a prostrate to decumbent annual resembling the following species, is recorded from a number of counties, primarily in the west half, but is rarely encountered. It differs from *Tribulus terrestris* in having larger flowers (which are orange) and leaves and fruits with a central apical beak, separating into 10 mericarps and lacking spines.

TRÌBULUS L. — Tribulus

L. *tribulus*, Gr. *tribolos*, from *tri-*, 3, + *bolos*, prong—the Latin name for a caltrop, a 4-pronged, iron implement of war for scattering over the ground, 3 prongs serving as feet, the fourth upthrust to impede a cavalry charge.

Prostrate to ascending annual or perennial herbs with numerous branches radiating from a central taproot. Leaves even-pinnate with 6–14 oblong to elliptic or ovate leaflets. Flowers yellow, small. Sepals 5, deciduous. Stamens 10, in 2 whorls of 5 each, the outermost opposite the petals and fused to the base of them. Nectar glands present between the stamens. Pistil 5-celled, each cell with 3–5 ovules. Fruit hard and spiny, splitting into 5 indehiscent mericarps, each of which is partitioned internally into 3–5 1-seeded compartments.

One species in Kansas.

T. terréstris L. — Puncture vine, bull-head

L. *terrestris*, belonging to the earth or ground, from *terra*, earth, ground, soil—in allusion to the habit of the plant to sprawl over the ground.
Yellow; mid-June–mid-October
Roadsides, railroad embankments, waste ground, and other disturbed areas, especially in sandy soil
Throughout, more common westward; intr. from the Old World

Prostrate annuals, often forming dense mats. Leaves mostly 1–5 cm long; leaflets 4–13 mm long and 1–5 mm wide, mostly oblong to elliptic or lanceolate, usually oblique at the base; stipules small and inconspicuous. Herbage with long, white, appressed or spreading hairs. Flowers yellow, borne singly in the leaf axils. Sepals lanceolate or elliptic, 2.5–4 (–5) mm long. Petals more or less obovate with an acute base, about 3–5 mm long. Fruit 5-angled, mostly 9–12 mm wide at maturity (excluding the large spines, which are 4–7 mm long). Each mericarp has 2 prominent spines and resembles the head of a horned animal, thus accounting for the common names "bull-head" and "goat-head," as well as providing well-known hazards for bicycle riders and barefoot children. The seeds are shaped like teardrops truncated at the broader end, about 3–4 mm long, and are reported to be viable for up to 8 years.

According to Kingsbury (21), ingestion of the plant, which contains steroidal saponins, causes hepatogenic photosensitivity in livestock. The problem may be complicated by high levels of nitrate in the plant or selenium in the soil (and hence in other plants eaten by the livestock).

POLYGALÀCEAE Milkwort Family

272. *Polygala alba*

Annual or perennial herbs or, in the tropics, shrubs or small trees. Leaves opposite, alternate, or whorled, simple, mostly entire, sessile, mostly estipulate or with stipular glands. Flowers bisexual, irregular, borne singly or in terminal or axillary spikes or racemes. Sepals 5, distinct, the 3 outer ones usually small and herbaceous, the 2 inner ones, called the wings, often enlarged, colored, and petal-like. Petals 3 or 5, the 2 lateral ones free, or sometimes united at the base with the lower one, which is keel-shaped and so called the keel and is often tipped with a crest or fringe. Stamens basically 10, but usually only 8 and monadelphous, united below into a sheath open on its upper side, and more or less united with the petals; anthers 1-celled, opening by a terminal pore or crack. Pistil 1; ovary superior, 2-celled with 1 ovule in each cell; style enlarged at the top, undivided or 2- to 4-lobed; stigma 1, terminal or recessed between the lobes of the style. Fruit usually a 2-seeded capsule which opens loculicidally.

A widely distributed family of about 10 genera and 700 species. Species of *Polygala* and a few other genera are sometimes cultivated as ornamentals.

One genus in Kansas.

POLÝGALA L. Milkwort, polygala

Classical Latin name for the herb milkwort, from Gr. *polys*, much, + *gala*, milk—because, as Pliny reported, "taken in drink, it increases the milk in nursing women."

Annual or perennial herbs. Leaves alternate, opposite, or whorled. Flowers small, in racemes, as described for the family, usually with 3 petals, the style bent and the stigma 2-lobed. Capsule small, somewhat flattened, with or without a wing or margin. Seeds usually hairy and with a 2-lobed aril.

The lowest petal, designated the keel, encloses the pistil and stamens similarly to the keel of the Lotoidae of the Legume Family. The bent style is enlarged upward, and terminates with a spoon-shaped projection, behind which is the sticky stigma; the anthers are in a position to shed pollen into this "spoon." An insect, usually a bumblebee or butterfly, coming for nectar secreted at the base of the ovary, gets pollen from the "spoon" on its proboscis and deposits it on the sticky stigma of the next flower visited, thus effecting cross pollination.

Several species of *Polygala* have been used medicinally. *P. senega* L., known in Kansas from Cherokee County only but more common eastward, furnishes the dried root officially known in pharmacy as senega. Before the settling of the Europeans in America, the Seneca Indians were employing the chewed root, placed on the wound, as an antidote against the bite of poisonous snakes. In addition, Ottawas and Chippewas used a tea prepared from the root as an abortifacient, and the Nishinam made a tea from the entire plant to treat diarrhea. White settlers and European physicians eventually came to use the plant also for diseases of the respiratory tract. W. C. Stevens (28) reminisced about "hive syrup," a compound having senega as 1 of its principal ingredients and which was, during his childhood, a popular remedy for the croup. This species was listed in the 23rd edition of the *Dispensatory of the United States* and in the *U.S. Pharmacopoeia* from 1820 to 1936 and its potency attributed there to saponins contained in the root. As with many medicines, LARGE DOSES MAY BE POISONOUS.

Five species in Kansas. *P. senega*, mentioned above, is not included in the following key. It is a white-flowered perennial, 1–5 dm tall, with clustered stems and linear-lanceolate to lanceolate or elliptic leaves 0.4–2.5 cm wide.

Key to Species:
1. Perennials with a woody crown and several to many stems; flowers white .... *P. alba*
1. Slender annuals with a weak taproot and 1 stem; flowers greenish, white, pink, or purplish .. 2
  2. Corolla 6–10 mm long, nearly 3 times as long as the wings, the flowers appearing tubular .... *P. incarnata*
  2. Corolla less than 5 mm long, approximately equaling or shorter than the wings, the flower not appearing tubular 3
    3. Leaves whorled; flowers greenish or white, in tapering racemes .... *P. verticillata*

3. Leaves alternate; flowers pink or purplish and white, in dense headlike or short-cylindric racemes .... *P. sanguinea*

273. *Polygala incarnata*

274. *Polygala verticillata*

P. álba Nutt.   White polygala

L. *albus*, white.
White; mid-May–August
Dry, rocky or sandy prairie hillsides
West half and to Washington and Clay counties in the north

Smooth, erect perennial with a woody crown and deep, slender taproot. Stems several to many, 2–3.5 dm tall, angled in cross section. Leaves alternate, narrowly oblanceolate to linear, 0.5–3 cm long and 0.5–2 mm wide, the margins revolute. Flowers white, in slender terminal racemes. Wings elliptic, about the same length as the 3 mm corolla. Capsule elliptic or oblong-elliptic, smooth, 1- to 2-seeded, 2.5–3 mm long. Seeds black, narrowly lanceolate in outline, about 2.3–2.5 mm long, covered with appressed silky hairs, the lobes of the aril overlapping the body of the seed most or all their length.

The dried roots of this plant, like those of *P. senega*, have the pharmaceutical name "senega" and possess the same medicinal virtues as that species. The Sioux Indians used the root, scraped and boiled in water, to treat earache.

P. incarnàta L.   Pink polygala

L. *incarnatus*, made flesh, i.e., flesh-colored—in reference to the pink flowers.
Pink or purplish; June–October
Grassland, pastures, prairies, usually sandy soils
East half

Smooth, slender, erect annual, whitened with a waxy "bloom" that can be rubbed off. Stem simple or once or twice dichotomously branched, longitudinally grooved, mostly (1–) 2–4 dm tall. Leaves alternate, linear or awl-shaped, distantly spaced, erect or ascending along the stem, mostly 1–1.8 cm long and 1–2 mm wide, often deciduous by fruiting time. Flowers pink or purplish, appearing tubular, in dense cylindrical racemes. Wings lanceolate to linear-oblong or oblanceolate, green with petaloid margins, much shorter than the 6–10 mm corolla. Capsule ovate, 2-seeded, 2.5–3.5 mm long. Seeds black, short-oblong, covered with appressed or slightly spreading hairs, about 1.8–2 mm long excluding the 2-lobed aril which is attached to an abruptly constricted beak and which for most of its length does not overlap the body of the seed.

P. sanguínea L.   Purple polygala

NL., from L. *sanguineus*, blood-red.
Pink or purplish and white; mid-June–mid-October
Moist, sandy, prairie, occasionally in open woods or calcareous soil
South and east of a line drawn from Barber to Harvey and Wyandotte counties, especially in the southeast tenth

Smooth, slender, erect annual with a weak taproot. Stem branched or not, angled in cross section or slightly winged, mostly 1–3.5 dm tall. Leaves alternate, linear or narrowly elliptic, erect or ascending along the stem, mostly 1–3 cm long and 1.5–4.5 mm wide. Flowers pink or purplish and white, in dense headlike or short-cylindric racemes. Wings ovate, longer than and concealing the 1.8–2 mm corolla. Capsule more or less obovate, 2-seeded, 2–3.5 mm long, with a flat, sterile basal portion. Seeds black, ovoid, about 1.5 mm long, covered with spreading hairs, the 2 lobes of the aril attached to the orange beak of the seed and overlapping the body of the seed most of their length.

See plate 53.

P. verticillàta L.   Whorled polygala

NL. *verticillatus*, from L. *verticillus*, for the flywheel or whirl of the ancient, hand-operated spinning-spindle—the several leaves which radiate from a node in this species suggesting the spokes of a wheel.
White; June–October
Prairie or open woods, usually on sandy or rocky soil
East four-fifths

Delicate inconspicuous annual with a weak taproot. Stem usually branched, (0.5–) 1–2 dm tall. Leaves in whorls of (2–) 4–5, narrowly elliptic to linear, mostly 0.8–3 cm long and 0.5–3 mm wide. Flowers white or greenish (sometimes tinged with purple), in small tapering racemes. Wings ovate to broadly elliptic or obovate, longer than or about the same length as the 1.5 mm corolla. Capsule broadly elliptic, 2-seeded, about 1.5 mm long.

Seeds black, narrowly lanceolate in outline to ellipsoid, about 1.5 mm long, covered with appressed hairs, the 2 lobes of the aril overlapping the body of the seed most of their length.

EUPHORBIÁCEAE   Spurge Family

Herbs (ours) or shrubs exuding a bitterly pungent, milky secretion when injured. Leaves alternate, opposite, or whorled, simple, petioled or sessile, usually stipulate. Flowers regular, unisexual, the male and female flowers occurring on the same plant (monoecious) or on different plants (dioecious), the flowers sometimes enclosed in an involucre that resembles a calyx. Calyx usually present and of 5 distinct or slightly united sepals. Corolla usually absent, when present consisting of 5 distinct petals. Male flowers with 1–8 (to many) stamens, the filaments separate or united. Female flowers with a single pistil; ovary superior, 3-celled, rarely 1-celled, with 1–2 ovules in each cavity, the style often branched. Fruit usually a capsule which splits into 3 1- to 2-seeded segments which then open along the inner surface. Seeds often with a prominent outgrowth, known as a caruncle, above the hilum.

A large family containing approximately 283 genera and 7,300 species widespread in distribution. Many members of the family are known to be poisonous. Economically important Euphorbs include the true rubber tree (*Hevea* spp.), the castor bean or castor oil plant (*Ricinus communis* L.), and several species of *Manihot*, the source of cassava, manioc, and tapioca. The poinsettia (*Euphorbia pulcherrima* Willd.), crown-of-thorns (*Euphorbia milii* Ch. des Moulins), and various species of *Croton* are used as ornamentals.

Kansas has 8 genera. *Phyllanthus caroliniensis* Walt. is known from the extreme southeast only. It is a delicate, erect, monoecious annual, about 2–4 dm tall, with obovate or oblanceolate leaves about 5–20 mm long, minute flowers in small axillary clusters, and 2-seeded capsules 1.5–2 mm wide.

Key to Genera (3):
1. Juice milky; perianth none; male flowers of a single stamen, female flowers of a single pistil, neither with a perianth; 1 female and few to many male flowers arising within a cup-shaped involucre simulating a bisexual flower with a stipitate pistil .... *Euphorbia*
1. Juice not milky; perianth present in male or female flowers or both; flowers clearly unisexual, the male and female differing in position or size or character of the bracts 2
   2. Hairs (use magnification) branched or stellate .... *Croton*
   2. Hairs simple or lacking 3
      3. Foliage armed with stinging hairs .... *Tragia*
      3. Foliage without stinging hairs 4
         4. Flowers all or partly in elongate terminal spikes; leaves serrate 5
            5. Leaf teeth glandular; flowers subtended by saucer-shaped glands; ours stout perennials .... *Stillingia*
            5. Leaf teeth not glandular; flowers subtended by palmately cleft bracts; ours inconspicuous annuals .... *Acalypha*
         4. Flowers in axillary spikes or racemes; leaves serrate or entire 6
            6. Leaves almost always toothed; flowers without petals .... *Acalypha*
            6. Leaves entire; flowers with petals .... *Argythamnia*

ACALÝPHA L.   Copperleaf, three-seeded mercury

NL., from Gr. *akalephe*, nettle—the herbage of these plants resembling that in the Nettle Family.

Inconspicuous annuals (ours), less frequently perennial herbs, rarely subshrubs. Leaves alternate, petioled, stipulate. Flowers small and inconspicuous, lacking petals, borne in terminal or axillary spikes, male and female flowers on the same plant (ours), either in the same or different spikes, or on different plants. Male flowers with 4 sepals (use high magnification) and usually 4–8 basally united stamens, each flower subtended by a minute bract. Female flowers with 3–5 sepals and a globular, 3-celled pistil, enclosed by a lacerated, sometimes leaflike bract. Capsule somewhat 3-lobed with 1 seed per cell (only 1 developing in *A. monococca*), subtended and sometimes enveloped by the persistent bracts. Seeds ovoid, variously roughened or pitted, with a dry white excrescence (caruncle) near the pointed end, usually orange to dark brown but sometimes becoming grayish or pearlescent as an outer seed coat loosens.

Four species in Kansas.

The seeds of *Acalypha* species are known to comprise a small portion of the diet of the mourning dove, American pipit, Botteri sparrow, and swamp sparrow.

275. *Acalypha rhomboidea*

Key to Species:

1. Male and female flowers on separate spikes (but still on the same plant); mature capsules with green spinelike protuberances, not concealed by the female bracts .... *A. ostryaefolia*
1. Flowering spike with female flowers at the base and male flowers at the apex; mature capsules sparsely hairy but lacking green protuberances, concealed by the female bracts 2
   2. Fruit 1-seeded; leaves narrowly lanceolate to linear or linear-oblong .... *A. monococca*
   2. Fruit 3-seeded; leaves broader 3
      3. Leaves broadly ovate to lanceolate, but usually more or less rhombic, the margins coarsely serrate or serrate-crenate; female bracts cut into (5–) 7–9 (–11) coarse sharp lobes, with long sharp hairs and glandular hairs .... *A. rhomboidea*
      3. Leaves mostly lanceolate or elliptic, occasionally somewhat rhombic, the margins entire to shallowly crenate or finely serrate; female bracts with (8–) 10–14 (–16) fine sharp lobes or teeth, with long sharp hairs but lacking glandular hairs .... *A. virginica*

A. monocócca (Engelm.) L. Mill.

NL., from Gr. *monos*, single, sole, + *kokkos*, pit—in reference to the fact that only 1 cell of each capsule matures a seed.
Flowers minute, inconspicuous, greenish; mid-July–mid-October
Prairie hillsides, open woods, roadcuts, frequently in shallow, rocky soil
East half, primarily in the east two-fifths

Erect, branched annuals, mostly 1.5–4 dm tall. Stems with soft, incurved hairs (use magnification). Leaf blades narrowly lanceolate to linear-oblong or linear, those of the main leaves mostly 2–5.5 cm long and 4–12 mm wide, with 3 prominent veins arising at the base, the lower surface and, to a lesser degree, the upper sparsely covered with appressed hairs; apex acute or slightly rounded; base acute or rounded; margins entire or very shallowly serrate; petioles of the main leaves mostly 2–7 mm long. Flowers minute, greenish, borne in axillary spikes with female flowers at the base and male flowers at the top of the same spike (rarely a few spikes of male flowers only). Bracts of the female flowers mostly with 9–13 short lobes or teeth, sparsely covered with short appressed hairs and with longer, sharp, marginal cilia. Capsules 1-seeded, minutely hairy, about 2–2.5 mm long, concealed by the female bracts. Seeds orange-brown, ovoid to broadly ovoid, about 1.5–2 mm long, with longitudinal rows of minute tubercles.

A. ostryaefòlia Riddell   Hop-hornbeam copperleaf

NL., from *Ostrya*, the name of the hop-hornbeam genus, + L. *folium*, leaf, because of the similar leaves.
Flowers minute, inconspicuous, greenish; July–mid-October
Wooded creek banks and flood plains, prairie ravines, roadsides, fields, often on rocky soil
East half, primarily, especially in the east third

Erect, dark-green annuals, mostly 1.5–7 dm tall, usually branching throughout. Stems minutely hairy. Leaf blades thin, ovate to lanceolate or broadly elliptic, those of the main leaves 4–10 cm long and 1.5–6 cm wide, with 3 or 5 prominent veins arising at the base, glabrous or with scattered appressed hairs and marginal cilia; apex acuminate or acute; base cordate; margins serrate; petioles of the main leaves mostly 2–9 cm long. Flowers minute, greenish, the male flowers in dense axillary spikes, the female flowers in loose terminal spikes. Bracts of the female flowers deeply cut into 9–19 narrow, attenuate or subulate lobes, nearly glabrous or sparsely covered with short, spreading hairs and a few glandular hairs. Capsules 3-seeded, about 2–3 mm long, not concealed by the female bracts, the upper half covered with soft, green, spinelike projections. Seeds orange-brown or gray, broadly ovoid, tuberculate, about 2 mm long.

A. rhomboìdea Raf.

NL., from Gr. *rhomboides*, from *rhombus*, a parallelogram with unequal adjacent sides, + *oides*, -oid, like, having the shape of—in reference to the more or less rhombic leaves of this species.
Flowers minute, inconspicuous, greenish; late July–October
Moist, rich soil of wooded creek banks and bluffs, prairie ravines, and occasionally lake margins
East half, primarily; west to Comanche County in the south

Erect annuals, mostly 1.5–6 dm tall, eventually branching at least below. Stems with soft, incurved hairs (use magnification) and sometimes longer, spreading hairs also. Leaf blades thin, broadly ovate to lanceolate but usually more or less rhombic, those of the main leaves 3–10 cm long and 1.5–5 cm wide, with 3 prominent veins arising at the base, sparsely covered with appressed hairs (use magnification) and, when dried, the upper surface covered with white dots (probably calcium oxalate crystals in epidermal cells) visible with a 10X hand lens; apex acuminate to acute or rounded; base acute or rounded; margins coarsely serrate or serrate-crenate; petioles of the main leaves mostly 1.5–7 cm long. Flowers minute, greenish, borne in small axillary spikes with female flowers at the base and male flowers at the top of the same spike. Bracts of the female flowers mostly cut into 5–11 coarse, sharp lobes, sparsely covered with long sharp hairs and long glandular hairs, especially along the margins, sometimes becoming copper-colored with age. Capsules 3-seeded, about 2.5 mm long, concealed by the female bracts, hairy. Seeds orange to dark brown, ovoid, 1.5–2 mm long, with longitudinal rows of minute pits.

276. *Acalypha virginica*

A. virgínica L.   Virginia copperleaf

"Virginian."
Flowers minute, inconspicuous, greenish; (July–) mid-August–mid-October
Usually in moist, rocky or sandy soil of woods or creek banks, less frequently in prairie or along roadsides
East half

Erect annuals, mostly 2–7 dm tall, usually branching below, often becoming copper-colored. Stems with soft, incurved or sometimes spreading hairs. Leaf blades thin, mostly lanceolate or elliptic, occasionally somewhat rhombic, those of the main leaves with 3 prominent veins arising at the base, 3–7 cm long and 1–3 cm wide (those of the secondary axillary branches and new terminal growth usually smaller), with scattered appressed hairs when young but becoming glabrate with age; apex acute or rounded; base wedge-shaped to acute or rounded; margins entire to shallowly crenate or finely serrate; petioles of the main leaves 0.5–3.5 cm long. Flowers minute, greenish, borne in small axillary spikes with female flowers at the base and male flowers at the top of the same spike. Bracts of the female flowers mostly with 10–14 fine, sharp lobes or teeth, densely covered with long, straight hairs and marginal cilia but lacking glandular hairs. Capsules 3-seeded, hairy, about 2 mm long, concealed by the persistent bracts. Seeds orange-brown, or dark brown, ovoid, about 1.5 mm long, with longitudinal rows of minute pits.

ARGYTHÁMNIA P. Br.   Wild mercury

NL., from *argyos*, silver, + *thamnos*, a bush or shrub.

Perennial herbs. Leaves alternate, usually prominently 3-veined from the base. Herbage often with malpighiaceous hairs—that is, straight, appressed, sharp-pointed hairs attached at some more or less central point rather than at 1 end (named after Marcello Malpighi, 17th century scientist who described such hairs). Flowers small, in axillary racemes, usually monoecious (sometimes dioecious), with a few female flowers at the base of the raceme, the rest male. Male flowers with 5 sepals, 5 petals, 5 nectar glands, 7–10 fertile stamens in 2 whorls with the filaments united most of their length to form an androphore, and with or without sterile stamens. Female flowers with 5 sepals which enlarge in fruit, 5 petals well-developed or not so (sometimes absent), 5 nectar glands opposite the sepals and inserted on a nectariferous disk below the sessile, more or less globose, 3-celled pistil which has 3 forked styles, the stigmas rounded or else flattened and expanded. Fruit a 3-lobed capsule splitting into 3 1-seeded segments around a persistent central column. Seeds lacking a caruncle.

Two species in Kansas. *A. humilis* (Engelm. & Gray) Muell. Arg. is known from a few counties in the southwest sixth. The plants have rather weak, much-branched, trailing or spreading stems up to about 5 dm tall, elliptic or oblanceolate leaves 1–5 cm long, and flowers in congested racemes shorter than the subtending leaves at anthesis.

A. mercurialìna (Nutt.) Muell. Arg.

NL., from the genus *Mercurialis*, also in the Spurge Family.
Flowers small, inconspicuous, greenish or purplish; June–early August
Rocky, prairie hillsides and uplands, usually in calcareous soil
Scattered in the south half, also in Ellis County

Stems several, stout, unbranched, 1.5–5 dm tall. Leaves elliptic to lanceolate or oblanceolate, entire, sessile, sometimes purplish, mostly (1–) 2–7 cm long and 1–3 cm wide; apex rounded or acute; base acute or somewhat rounded. Herbage with malpighian hairs. Racemes at anthesis much longer than the subtending leaves, though the staminate portion usually falls away by the time the fruits mature. Male flowers: sepals lanceolate or oblong-lanceolate, 2–3 mm long; petals purplish, the blades lanceolate or oblanceolate, narrowed basally to a short claw, about 3–3.5 mm long, alternating with 5 fingerlike glands; stamens usually 8. Female flowers: sepals lanceolate, about 4.5 mm long; petals usually absent; glands 5, fingerlike, pistil hairy, the stigmas flattened and expanded. Capsule deeply 3-lobed, hard, often purplish, 6–7 mm long and 9–10 mm wide, subtended by the persistent sepals. Seeds brown or whitish, nearly globose, minutely roughened, about 4–5 mm long.

CRÒTON L.   Croton, doveweed

NL., from Gr. *kroton*, tick, the ancient name of the castor oil plant, *Ricinus communis*—because the seed of that plant resembles a tick, as do the seeds of this genus.

Herbs, shrubs, or trees, ours all strongly scented annuals with white or rusty, stellate hairs. Leaves alternate, petioled, with inconspicuous, often early-deciduous stipules. Flowers small, inconspicuous, in terminal spikes or racemes, male and female flowers on the same or different plants. Male flowers: sepals (4–) 5 (–6), more or less united at the base; petals absent or of same number as sepals; nectar glands absent or present opposite the sepals; stamens 5 to many, separate. Female flowers: sepals 4–9, more or less united at the base; petals absent in ours; a 4- or 5-lobed nectariferous disk usually present below pistil; pistil 3-celled (2-celled in *C. monanthogynus*), with 2–3 styles, each 1, 2, or 3 times forked. Capsules 3-seeded (1-seeded in *C. monanthogynus*). Seeds lustrous, more or less ovoid in outline, lenticular in cross section but more compressed on the ventral (inner) side, with a prominent knoblike caruncle at the narrow end above the hilum.

Five species in Kansas. *C. lindheimerianus* Scheele is known from Kiowa, Barber, and Harper counties only. It would key out below with *C. capitatus* but differs from that species in being less woolly and more profusely branched (especially from the base), in having only 6 stigmas, and in having the mature capsules merely subtended but not concealed by the calyx.

The closely related genus *Crotonopsis* Michx. is represented in southeastern Kansas by 1 species, *C. elliptica* Willd., a wiry annual herb with opposite leaves, peltate rather than stellate hairs, and indehiscent, 1-seeded fruits.

Some species of *Croton*, especially *C. tiglium* L., are noted for the production of croton oil in their seeds—a purgative so potent that 1 drop of it has the potency of 1 or more ounces of castor oil and 10 drops are enough to kill a dog. It has been used as a blistering agent and as a stimulant for the scalp in cases of falling hair, but extreme caution must be taken to dilute it properly to avoid excessive blistering. Its use is NOT recommended.

C. texensis and *C. capitatus* are thought to be poisonous to livestock, and even honey made from *Croton* nectar may be toxic to humans.

In spite of its poisonous properties, the seeds of *Croton* species are eaten by various birds and a few small mammals. The seeds account for a significant portion of the summer and/or fall diet of the eastern white-winged dove (25–50 percent), ground dove (10–25 percent), and mourning dove (10–25 percent), as well as contributing lesser amounts to the diets of the western white-winged dove, bobwhite quail, redwing blackbird, cowbird, Sprague's pipit, prairie chicken, cardinal, American pipit, and others. The prairie pocket mouse and Pacific pocket mouse also eat the seeds (4).

Key to Species:

1. Leaves toothed, with 1 or 2 glands at the summit of the petiole .... *C. glandulosus*
1. Leaves entire or merely undulate, not glandular 2
   2. Male and female flowers on separate plants .... *C. texensis*
   2. Male and female flowers on same plant 3
      3. Styles 2, each forked once, the stigmas therefore 4; ovary 2-celled, capsule 1-seeded
          .... *C. monanthogynus*
      3. Styles 3, each forked 2–3 times, the stigmas therefore 12–24; ovary 3-celled, capsule normally 3-seeded .... *C. capitatus*

C. capitàtus Michx.   Woolly croton

L. *capitatus*, headlike, from *caput*, head—in reference to the close cluster of flowers at the ends of the branches.
Flowers inconspicuous, greenish; (late May–) August–mid-October
Prairie hillsides, open woods, usually in rocky or sandy soil
East two-thirds

Plants gray-green or light green, densely hairy, mostly 2–10 dm tall, usually branched above. Leaf blades lanceolate to oblong, less frequently elliptic, those of the upper leaves mostly (0.6–) 1–2 cm wide (lower leaves usually gone by flowering time); apex and base rounded or acute; margins entire; petioles 1–3 cm long. Male and female flowers in the same spike, male above female. Male flowers: calyx deeply 5-lobed, about 2 mm long, the lobes triangular, densely hairy on the outside; petals 5, oblanceolate, membranous, with fimbriate margins, shorter than the sepals, alternating with minute glands; stamens 7–12, separate. Female flowers: calyx deeply 6- to 9-lobed, about 5 mm long at anthesis, the lobes linear or oblong; petals absent; styles 3, each 2 or 3 times dichotomously branched so that there are usually 10 or 12 stigmas. Capsule globose, woolly, 6–9 mm long, 3-seeded, concealed by the persistent calyx. Seeds tan, brown, or mottled, round in outline, mostly 4–5 mm long.

C. glandulòsus L.   Glandular croton

L. *glandulosus*, glandular, possessed of glands—in reference to the glands at the base of the leaf blade.
Flowers inconspicuous, greenish; mid-June–mid-October
Prairie, roadsides, open woods, especially in sandy soil
East two-thirds, especially southward

Plants light yellow-green, mostly 1–5.5 cm tall, branched above or from the base, hairy but not woolly. Leaf blades mostly ovate to oblong, occasionally lanceolate, mostly 1–4.5 cm long and 0.5–1.5 (–3) cm wide; apex rounded or nearly acute; base rounded, with 2 obconic glands (use magnification) on the lower surface; margins coarsely serrate; petioles 0.2–2 (–3) cm long. Male and female flowers in the same spike, male above the female. Male flowers: sepals (4–) 5, lanceolate, about 1.5 mm long, hairy; petals oblanceolate, membranous with ciliate margins, about the same length as the sepals; stamens 7–13, separate. Female flowers: sepals 5, oblanceolate, about 3 mm long, hairy; petals absent, disk present under pistil; styles 3, each 2-forked almost to the base, stigmas 6. Capsules nearly globose or short-oblong in outline, finely hairy, 4–5 mm long, 3-seeded, subtended but not concealed by the persistent, enlarged sepals. Seeds dark brown or mottled, oblong-ovate, minutely pitted, mostly 3.5–4 mm long.

277. *Croton glandulosus*

C. monanthógynus Michx.   Single-fruited croton

NL., from Gr. *monos*, 1, + *anthos*, flower, + *gyne*, woman—the female flowers often occurring singly.
Flowers inconspicuous, greenish; mid-June–mid-October
Prairie hillsides and uplands, overgrazed pastures, roadsides, occasionally in open woods, frequently in rocky, calcareous soil
East two-thirds, primarily

Plants gray-green or light yellow-green, mostly 1–3.5 dm tall, branched above or, less frequently, from the base, the branches ascending or widely spreading. Leaf blades broadly ovate to lanceolate, oblong, or elliptic, those of the upper leaves mostly 1–3.5 cm long and 0.5–2 cm wide (lower leaves usually gone by flowering time); apex rounded or acute, with a tiny sharp point; base rounded or acute; margins entire; petioles 0.5–2.5 cm long. Male and female flowers in the same spike, male above the 1 or 2 female flowers. Male flowers: calyx deeply 4-(5-) lobed, about 1–1.5 mm long, the lobes triangular or acute, hairy on the outside; petals elliptic or oblanceolate, membranous, with ciliate margins, shorter than the sepals, alternating with glands; stamens 4–7, separate. Female flowers: sepals 4–5, elliptic or linear, irregular, narrower than in male flowers, hairy; petals absent; glands and disk present at base of pistil, staminodes sometimes present; styles 2, each 2-forked almost to the base, stigmas 4. Capsules ovoid, finely hairy, 4–5 mm long, 1-seeded, subtended but not concealed by the persistent calyx. Seeds gray, brown, or mottled, broadly ovate to nearly round in outline, minutely pitted, mostly 3–3.5 mm long.

278. *Croton texensis,* female plant

C. texénsis (Klotzsch) Muell. Arg.   Texas croton

"Of Texas."
Flowers inconspicuous, greenish; late May–early October
Sandy soil of prairies, flood plains, and roadsides
West three-fifths

Plants gray-green or light yellow-green, mostly 2–8 dm tall, usually branched above. Leaf blades mostly narrowly lanceolate to oblong or linear-oblong, less frequently narrowly elliptic or almost linear, those of the upper leaves mostly 2–6 (–8) cm long and 4–15 (–20) mm wide (lower leaves usually gone at flowering time); apex and base rounded or acute; margins entire; petioles 0.5–2.5 (–3.5) cm long. Male and female flowers in racemes on separate plants. Male flowers: calyx deeply 5-lobed, about 2–2.5 mm long, the lobes triangular, densely hairy on the outside; petals absent; glands visible with 10X hand lens; stamens 8–12, separate. Female flowers: calyx as in male flowers; petals absent; nectariferous disk present at base of pistil; styles 3, each 2 or 3 times dichotomously branched so there are at least 12 stigmas. Capsule globose, warty, hairy, mostly 5–7 mm

279. *Croton texensis,* male plant

SPURGE FAMILY

207

long, 3-seeded, subtended by but not concealed by the persistent calyx, the central columella and 3 radiating septa of the ovary usually remaining for a while after the seeds have fallen. Seeds brown or mottled, nearly round to very broadly ovoid in outline, mostly 3.5–4 mm long.

The powdered leaves have been burned as a natural insecticide.

EUPHÒRBIA L.   Spurge, euphorbia

NL. and L., from Gr. *euphorbion*, a plant named after Euphorbos, a celebrated Greek physician (1st century B.C.). "Spurge" comes from OF. *espurge*, for any plant of the genus *Euphorbia*, from L. *expurgo*, to purge or cleanse, in reference to the purgative effects of these plants.

Annual (rarely biennial) or perennial herbs with a milky, irritant, more or less poisonous juice. Leaves alternate, opposite, or whorled, sessile or short-petioled; stipules present or absent, sometimes minute or glandlike. Flower small, monoecious. Male flowers consisting of a single stipitate stamen, usually arranged in clusters of 1 to several subtended by a small bract. Female flowers composed of a rudimentary, 3-lobed calyx and single stipitate, globose, 3-lobed pistil with 3 cells and 3 2-lobed styles. One female flower borne with several clusters of male flowers in a cuplike involucre called a cyathium which bears on its rim 1–5 nectar glands, each with or without a white or colored wing or appendage, alternating with small leaflike lobes which represent the tips of the bracts which were united to form the involucre. This entire structure is sometimes mistaken for a single flower. These small inflorescences may be borne singly or many together in a branched terminal cluster subtended by showy, white or colored, leaf- or petal-like bracts as in *E. marginata* (snow-on-the-mountain) or *E. pulcherrima* (poinsettia). In the latter 2 cases, several cyathia subtended by the conspicuous petal-like bracts are often mistakenly taken for a single flower. Capsule with 3 rounded lobes and 3 cells, 1 seed per cell, the stipe lengthening and, in some species, arching over the rim of the cyathium. Seeds with or without a caruncle. NOTE: Mature fruits and seeds are quite helpful and sometimes absolutely necessary for positive identification of *Euphorbia* species.

This is the largest genus of the family, containing approximately 1,600 species found in subtropical to warm-temperate regions around the world. The species exhibit various habits of growth, from tall and rigidly upright to creeping flat upon the ground and forming mats, a form of common occurrence on the shoulders of graded or paved roads.

Nineteen species in Kansas. The 14 most common species are included in the following key.

Many species of spurge are known to have caused or thought to have caused poisoning in humans or livestock. The acrid, toxic principle causes severe irritations and lesions in the mouth and gastrointestinal tract, rarely death. Photosensitization in lambs has been attributed to *E. maculata*, and some Australian species contain cyanogenetic compounds. The toxic properties of the spurges were probably responsible also for the use of various species for medicinal purposes.

The seeds of *Euphorbia* species are eaten by various birds including the mourning dove, ground dove, greater prairie chicken, American pipit, horned lark, several quail species, chipping sparrow, and others. The Merriam kangaroo rat eats both the seeds and plants, and antelope and mountain sheep are known to graze on the plants.

Key to Species:
1. Glands of cyathia completely lacking petal-like appendages 2
   2. Glands 4 or 5, oblong or elliptic, flat or convex .... *E. spathulata*
   2. Glands usually 1 or 2, cup-shaped with an inflexed margin 3
      3. Leaves of 2 distinctly different shapes—those of the stem entire and linear to narrowly lanceolate, all or mostly alternate, hairy below primarily along veins, glabrous above; those subtending the cyathia usually lobed or panduriform, green, typically red basally; plants primarily of the east third of Kansas .... *E. cyathophora*
      3. Leaves relatively uniform in shape throughout—those of the stem all opposite, hairy on both surfaces; those subtending the cyathia green and typically mottled with red spots, usually cream-colored basally; throughout Kansas .... *E. dentata*
1. Glands of cyathia with petal-like appendages which are sometimes conspicuous but in most species quite minute 4
   4. Leaves alternate, opposite, or whorled, the bases symmetrical; stipules absent or minute and glandlike 5
      5. Stem leaves all opposite, petioled, linear to narrowly lanceolate .... *E. hexagona*
      5. Stem leaves alternate, sessile to very short-petioled, oval to oblong 6
         6. Plants annual, scattered nearly throughout Kansas; leaves and bracts subtending the cyathia with broad, white (rarely pink), petal-like margins; glands 4 ....

.... *E. marginata*

  6. Plants perennial from a deep subterranean rootstock, occurring primarily in the east third of Kansas; leaves and bracts subtending the cyathia reduced in size and without petal-like margins; glands 5 .... *E. corollata*

4. Leaves all strictly opposite, the bases usually strongly oblique; stipules relatively well developed though never very prominent 7

  7. Capsules variously hairy 8

    8. Seeds with narrow, sharp, transverse ridges; capsule primarily hairy on the angles and at the apex .... *E. prostrata*

    8. Seeds not as above; capsule uniformly hairy 9

      9. Styles forked about one-third their length; seeds with faint transverse ridges or wrinkles, plants of the east two-thirds of Kansas .... *E. maculata*

      9. Styles club-shaped, entire or slightly notched at the apex but not forked; seeds with a whitish, irregularly thickened outer coat over a brown inner coat; plants primarily of the west four-fifths of Kansas .... *E. strictospora*

  7. Capsules glabrous 10

    10. Plants perennial with a gnarled woody taproot, primarily of the southwest sixth of Kansas .... *E. fendleri*

    10. Plants annual 11

      11. Plants typically rooting at the lower nodes; stipules united into a relatively prominent white scale .... *E. serpens*

      11. Plants not rooting at the nodes; stipules usually not united 12

        12. Tips of stems and lower surface of leaf bases variously hairy; leaves often with a red blotch on the upper surface .... *E. nutans*

        12. Plants completely glabrous throughout 13

          13. Leaves entire; seeds varying from roughened and distinctly angled in western Kansas to smoothly and scarcely if at all angled in the east .... *E. missurica*

          13. Leaves minutely serrate or entire; seeds sharply 4-angled in cross section, the faces with broad, rounded transverse ridges .. *E. glyptosperma*

280. *Euphorbia corollata*

E. corollàta L.   Flowering spurge

L. *corollatus*, possessed of a corolla, from *corolla*, little crown—in reference to the small petal-like appendages of the glands.
White; July–mid-September (–early October)
Prairie, roadsides, pasture, open woods
East third, primarily

Erect, smooth or hairy, perennial herbs with a dark, stout rootstock. Stems 0.3–1 m tall, usually branched only within the inflorescence. Stem leaves alternate, sessile or with a very short petiole, usually crowded, mostly oblong or elliptic, less frequently somewhat lanceolate or oblanceolate, 1.5–6.5 cm long and 2–18 mm wide, gradually decreasing in size upward, at least the lower ones eventually deciduous; apex usually rounded, occasionally slightly notched or acute; base rounded or acute; margins entire; stipules minute, glandlike. Cyathia borne in a terminal, umbel-like compound inflorescence, each cyathium with 5 greenish-yellow, rounded or transversely oblong or reniform glands, each with a white petal-like appendage 0.5–4.5 mm long. Male flowers 20–30 per cyathium, in clusters of 4–6. Styles thick, forked about half their length. Mature capsules smooth, 3–4 mm long, opening explosively. Seeds white or gray, smooth or with shallow depressions, ovate to nearly globose, about 2–3 mm long, lacking a prominent caruncle.

Close observation shows that within a given cyathium, the flowers are proterogynous—that is, the stigmas of the female flower are in condition to receive pollen before the male flowers are ready to deliver it. At this stage the stigmas appear above the rim of the cup where insects taking nectar from the nectar glands would be apt to touch them with pollen brought on their mouth parts from the anthers of older flowers. After pollination the ovary emerges and is brought to a lower level by the elongation and deflection of its pedicel. In the meantime the filaments of the staminate flowers elongate and the anthers open above the rim of the cyathium.

The insects most significant for the pollination of Euphorbias are flies and short-tongued bees, and above all, hover flies, those wondrous, diaphanous little creatures that for a while hold themselves suspended at a fixed point in the air, then suddenly dart sidewise to another point, and so on. But they do from time to time alight on flowers for refueling with pollen and nectar, their mouth parts being marvelously designed for this purpose. When feeding on Euphorbias they scrape off pollen and ingest it through a mouth opening, and they lick up the films of nectar excreted over the surface of the nectar glands. Of course they go from flower to flower with some pollen adhering to their mouth parts, and the stigmas when emerging near a nectar gland would quite certainly receive from the flies some of this pollen.

In early medical practice in this country the bark of the root of this plant was sometimes used as an emetic and cathartic, but its potency too often amounted to violence. At the present time it is listed among our native poisonous plants. Cattle avoid it in pastures but accept it dried in hay, sometimes with serious consequences. If, when the plant is cut or bruised, its oozing white latex gets on the skin some degree of dermatitis may result. Yet this danger is so slight that florists are accustomed to gather the blooming stems from the wild for making bouquets.

This species has caused poisoning of livestock in Indiana (21), but the flowers, fruits, and leaves are eaten by the wild turkey (4).

281. *Euphorbia cyathophora*

E. cyathophòra Murray   Painted euphorbia

NL., from L. *cyathus*, Gr. *kyathos*, cup or ladle, + -*phore*, from Gr. *phoreus*, bearer—possibly in reference to the cyathia.
Flowers inconspicuous, greenish; July–October
Woodlands, shaded roadsides, creek and river banks
East third, primarily

Erect annual, glabrous or nearly so, with orange taproots and coarse lateral roots. Stems eventually branched, up to 1 m tall. Leaves mostly alternate and petioled below, opposite and shorter-petioled or sessile above, highly variable in shape even on a single plant, the blades mostly 3.5–10 cm long and 0.2–6 cm wide, lanceolate to elliptic, oblanceolate, linear, or the broader ones more or less fiddle-shaped, the uppermost typically with a red base; margins entire, toothed, or lobed; stipules absent, or present as minute glands. Cyathia borne in small terminal clusters, each cyathium with a fringed rim and 1 or 2 greenish, oblong, deeply cupped glands lacking petal-like appendages. Male flowers 30–50 per cyathium. Styles thick, forked over half their length. Mature capsules smooth, 3–4 mm long and 6–8 mm wide. Seeds dark brown or rusty brown, broadly ovoid, truncate or slightly concave at the broader end, mostly 2.5–3 mm long, covered with rough tubercles, lacking a prominent caruncle.

282. *Euphorbia dentata*

E. dentàta Michx.   Toothed spurge

L. *dentatus*, toothed, in reference to the leaf margins.
Flowers green, inconspicuous; late May–mid-October
Roadsides, fields, overgrazed pastures, creek and river banks, waste areas
Throughout

Erect, rough-hairy annual with a taproot and coarse lateral roots. Stems often branched from the base, 1.5–5 dm tall. Leaves opposite, petioled, the blades broadly ovate or somewhat rhombic to lanceolate, elliptic, lance-linear, oblanceolate, or rarely obovate, those of the main leaves mostly 2–8 cm long and 0.4–4 cm wide; apex acute or acuminate; base wedge-shaped or attenuate; margins shallowly to coarsely serrate; stipules minute, glandlike. Figs. 282 and 283 illustrate some of the variation in leaf shape and size which may be expected in this species. Cyathia borne in small terminal clusters, the subtending leaves or bracts usually pale green or whitish at the base, each cyathium with a fringed rim and (1–) 2 (–5) greenish, broadly obconical or somewhat 2-lobed glands lacking petal-like appendages. Male flowers 25–40 per cyathium, in clusters of 5–8. Styles thick, forked about half their length. Mature capsules smooth or somewhat rough-hairy. Seeds whitish to brown or black, very broadly ovoid, somewhat angled, about 2.5–3 mm long, with rather distantly spaced tubercles and a flat caruncle.

This plant is poisonous to livestock, but the flowers, fruits, and leaves are eaten by the wild turkey (4).

283. *Euphorbia dentata*

E. féndleri T. & G.

Named in honor of Augustus Fendler (1813–1883), who first collected this species.
Flowers minute, greenish; May–late August
Dry, rocky or sandy soil of prairie hillsides and uplands
Southwest sixth, primarily

Low, glabrous, cespitose perennial, often light green in color, with a gnarled woody taproot. Stems several to many, prostrate to ascending or erect, branched, (2–) 5–12 cm long. Leaves opposite, short-petioled, the blades nearly round to ovate, cordate, deltoid, lanceolate or broadly elliptic, 3–7 mm long and 2–8 mm wide; apex acute or rounded; base rounded, truncate, or cordate, usually oblique; margins entire; stipules linear, about 1 mm long. Cyathia borne 1 or 2 together, terminal or axillary, each cyathium with 4 greenish or reddish, oblong glands lacking petal-like appendages. Male flowers 13–25 per cyathium, in clusters of 2–5. Styles thick, forked about half their length.

Mature capsules smooth, about 1.5–2.5 mm long, usually 1- or 2-seeded. Seeds whitish or tan, ovate-oblong, somewhat 4-angled, smooth or with a few faint wrinkles, lacking a prominent caruncle.

E. glyptospérma Engelm.

NL., from Gr. *glyptos*, carved or engraved, + *sperma*, seed.
Flowers minute, greenish or white; June–mid-October
Dry, sandy or rocky soil of roadsides, waste ground, pastures, disturbed prairie, and river banks
West four-fifths and a few northeast counties

Glabrous, prostrate annual. Stems branched, mostly 0.5–3.3 dm long. Leaves opposite, short-petioled, the blades more or less oblong, often slightly curved, 3–13 mm long and 1–5 mm wide; apex rounded; base rounded, oblique; margins very minutely serrate (use magnification), at least at the apex; stipules membranous, cut into threadlike segments. Cyathia borne singly in the leaf axils or forks of the stem, each cyathium with 4 reddish, oblong or elliptic glands with white petal-like appendages shorter than the gland is wide. Male flowers 2–7 per cyathium. Styles forked about one-third their length, the stigmas enlarged. Mature capsules glabrous, about 1.2–1.8 mm long. Seeds whitish or light brown, ovate or ovate-oblong, about 0.7–1.2 mm long, sharply 4-angled in cross section, the faces with broad, rounded transverse ridges, lacking a prominent caruncle.

E. geyeri Engelm., a prostrate annual, is known from scattered counties, primarily in the southwest and central parts of the state, and is sometimes confused with *E. glyptosperma*. The former, however, has entire leaves and smooth seeds. It differs from *E. missurica* in having capsules 2 mm or less in length and glands with appendages less than 0.5 mm long.

E. hexágona Nutt.

L. *hexagona*, six-angled. According to Fernald (9) this inappropriate name was published as of Nuttall by Sprengel to whom Nuttall trustingly sent material, Nuttall himself slightly later formally describing it as *E. heterantha* (diverse-flowered) because of the separation of the sexes in the cyathia.
Flowers minute, greenish; late July–early October
Dry, sandy soils of river banks, roadsides, and prairie
West two-thirds plus a few counties farther east

Delicate, yellowish-green annual. Stems erect, 2–5 dm tall, with several to many slender branches. Leaves opposite, short-petioled, the blades mostly narrowly elliptic or linear (sometimes threadlike on the secondary branches), less frequently more broadly elliptic or narrowly oblanceolate, 1–5.5 cm long and 0.5–1.0 (–1.3) mm wide; apex pointed; base acute or wedge-shaped; margins entire; stipules minute and glandlike or absent. Cyathia borne singly in the upper forks and leaf axils, each cyathium with 5 greenish, oblong or kidney-shaped glands with erect, yellowish-green, triangular-ovate appendages about 0.7–1 mm long. Male flowers 20–40 per cyathium. Styles forked half or more of their length, the stigmas enlarged. Mature capsules glabrous, erect or strongly reflexed, about 3–5 mm long. Seeds ovoid, tubercled, white or gray all over or black or brown with lighter tubercles, about (2.5–) 3 mm long, lacking a prominent caruncle.

E. maculàta L.   Eyebane

L. *maculatus*, spot, spotted—in reference to the purplish spot on the leaf.
Flowers minute, greenish; July–October
Dry, usually sandy, disturbed soils of roadsides, waste ground, prairie, and creek and river banks
East two-thirds

Low annual. Stems branched, mostly prostrate, some ascending, more or less hairy, 0.5–4.5 dm long. Leaves opposite, short-petioled, the blades mostly oblong, less frequently elliptic or somewhat lanceolate, often curved slightly, 4–15 mm long and 2–6 mm wide, the upper surface glabrous, often with a reddish blotch along the midvein, the lower surface hairy, green or reddish; apex acute or rounded; base rounded or truncate, oblique; margins serrate (sometimes only minutely so); stipules membranous, long-acuminate, the tip often lacerated. Cyathia borne singly at the nodes of short, congested lateral branches, each cyathium with 4 reddish, oblong glands with minute, white or red, petal-like appendages up to about 0.6 mm long but often much shorter. Male flowers 2–5 per cyathium. Styles forked about one-third their length, the stigmas somewhat enlarged. Mature capsules hairy, about 1.5 mm long. Seeds whitish or brown, angled, about (0.8–) 1 (–1.2) mm long, with faint transverse ridges or wrinkles, lacking a prominent caruncle.

E. lata Engelm. is a low perennial known primarily from the southwest tenth of Kansas. The leaves are ovate to long-triangular, sometimes somewhat curved, but the revolute margins make the leaves look almost linear. *E. humistrata* Engelm. ex Gray, a prostrate annual, occurs rarely in eastern Kansas. It differs from *E. maculata* primarily in having stems occasionally rooting at the nodes, leaves round to short-oblong or ovate, and seeds which are nearly smooth to granular or roughened.

E. marginàta Pursh   Snow-on-the-mountain

L. *marginatus*, with a margin or border, from *margino*, to enclose with a border—in reference to the white-margined leaves and bracts.
Flowers white or pinkish, small; July–October
Prairie, roadsides, flood plains, more abundant in disturbed areas
Nearly throughout, less common in the extreme east and southeast

284. *Euphorbia marginata*

Stout, erect annual. Stems 0.3–1 m tall, unbranched below the inflorescence, sparsely covered with long, spreading or somewhat kinky hairs. Leaves sessile or short-petioled, those of the main stem alternate, mostly ovate to lanceolate or elliptic, glabrous, 2–8 cm long and 1.5–3.5 cm wide, with a rounded, acute, or short-acuminate, mucronate apex, rounded or acute base, and entire margins; leaves of the upper branches and inflorescence opposite or whorled, lanceolate to oblong, elliptic, or oblanceolate, smaller than the lower leaves; stipules small, falling early. Cyathia borne in umbel-like terminal clusters subtended by conspicuous bracts with wide white or pink margins, each cyathium with 4 greenish, oblong or nearly circular glands with prominent white or pink, petal-like appendages. Male flowers 60 or more per cyathium, in clusters of 12–25. Styles forked about two-thirds their length. Mature capsules about 4–6 mm long, covered with long, soft hairs. Seeds white to light tan, broadly ovoid to nearly globose, prominently tubercled, about 3–4 mm long, lacking a prominent caruncle.

E. marginata is native to the Middle West, but it is frequently cultivated as an ornamental, so it is well to know that the latex which flows copiously from it when wounded contains a poisonous substance, euphorbon, little understood, but known to produce inflammation of the skin similar to that caused by poison ivy. According to Kingsbury (21), this caustic juice has been used for branding cattle in preference to a hot iron. The plant is poisonous to livestock, though they usually do not eat it, and honey made from the nectar is apparently poisonous to humans and to young bees (29).

E. missùrica Raf.

"Of Missouri."
Flowers small, white or pink; June–early October
River banks, prairie pastures, in sandy soil
Nearly throughout

285. *Euphorbia missurica*

Glabrous annuals. Stems erect, ascending, or sometimes decumbent, branched at the base and above, mostly 1–9 dm tall. Leaves opposite, short-petioled, the blades elliptic to oblong, narrowly wedge-shaped, or linear, 0.6–2.5 cm long and 1.5–7 (–10) mm wide; apex rounded, truncate, mucronate or notched; base wedge-shaped to acute or rounded, sometimes slightly oblique; margins entire; stipules linear or triangular-acuminate, free or partly united, deciduous. Cyathia borne singly in the upper forks or in clusters of 2–3, each cyathium with 4 greenish-yellow, elliptic glands with white or pink, narrowly ovate to nearly round, petal-like appendages 0.5–2.5 mm long. Male flowers 10–76 per cyathium. Styles forked half or more of their length. Mature capsules glabrous, about 2–3 mm long. Seeds broadly ovoid and round or slightly 3-angled in cross section with a whitish, evenly or unevenly thickened outer coat to triangular-ovoid and strongly 3-angled in cross section with a heavier white, irregularly thickened outer coat. Seeds whitish, lacking an obvious caruncle.

This is a variable species which has been divided by some authors into 2 varieties—an eastern variety (var. *missurica*) with thin wiry branches, rather distinct internodes, narrow leaves (1–2.5 (–4) mm wide) and distinctly angled, roughened seeds 1.2–1.5 mm long, and, in Kansas, a western variety (var. *intermedia* (Engelm.) Wheeler) which has stouter stems, shorter internodes, wider leaves ((1.5–) 3–5 mm wide) and seeds about 1.8 mm long, with rounded angles or none at all and smooth, evenly whitened or somewhat blotchy (but not rough) seeds. The ranges of the 2 varieties overlap in central Kansas, and the distinguishing characters tend to intergrade. The seed characters—shape, size, and seed coat—seem to be the most reliable means of separating the varieties, if one wishes to do so.

E. nùtans Lag.   Spurge

NL., from L. *nuto, nutatus*, to nod or droop—possibly in reference to the gynophore, which is strongly reflexed at maturity.

Flowers small, greenish; July–October
Roadsides, waste ground, pastures, flood plains
East two-thirds

Annuals. Stems erect to decumbent, branched, up to 7.5 dm tall. Leaves opposite, short-petioled, the blades lanceolate to oblong, usually oblique or even somewhat curved, mostly (1–) 1.5–3.5 cm long and 3–15 mm wide, the upper surface often blotched with red, lower surface with 3 prominent veins arising from the base, often reddish; apex rounded or acute; base rounded or somewhat cordate, usually oblique; margins serrate, at least on the distal half of the blade; stipules triangular, quite small. Cyathia borne in small terminal clusters, each cyathium with 4 greenish, round or elliptic glands with a white or pinkish appendage usually about 0.5 mm long. Male flowers 5–28 per cyathium. Styles forked more than half their length. Mature capsules glabrous, about 1.5–2 mm long. Seeds black or dark brown, ovoid or oblong-ovoid, angled, wrinkled, about 1–1.5 mm long, lacking a prominent caruncle.

E. prostràta Ait.

L. *prostatus*, down flat—descriptive of the growth habit of this (as well as many other) species of *Euphorbia*.
Flowers minute, greenish; late June–September
Disturbed soil of roadsides, fields, waste ground, weedy prairie, creek banks, and open woods
Primarily in the north half and the central three-fifths of the south half

286. *Euphorbia nutans*

Low, hairy annual herb much resembling *E. maculata*, *E. glyptosperma*, and *E. strictispora*. Stems mostly prostrate, branched, 0.5–4 dm long. Leaves opposite, short-petioled, the blades mostly ovate to broadly elliptic or short-oblong, hairy at least on the lower surface, 3–15 mm long and 2–8 mm wide; apex rounded; base rounded or slightly cordate, oblique; margins minutely serrate, at least near the apex; stipules minute. Cyathia borne singly in the leaf axils, each cyathium with 4 reddish, elliptic or oblong glands with minute pinkish appendages. Male flowers 3–6 per cyathium. Styles forked nearly to the base, the stigmas purplish, enlarged. Mature capsules about 1–1.5 mm long, hairy on the angles and at the apex. Seeds gray or whitish, ovate-oblong, angled, about 1 mm long, the faces with rather sharp transverse ridges, lacking a prominent caruncle.

E. sérpens H. B. K.   Spurge

L. *serpens*, snake, from serpo, to creep—in reference to the growth habit of this species.
Flowers minute, greenish; July–October
Roadsides, waste ground, disturbed areas in prairie and open woodlands
Nearly throughout, uncommon in the northwest sixth

287. *Euphorbia spathulata*

Glabrous, prostrate annuals. Stems branched, up to 4 dm long, rooting at the lower nodes. Leaves opposite, short-petioled, the blades nearly round to ovate, broadly elliptic, or short-oblong, 2–10 mm long and 2–7 mm wide; apex rounded, sometimes shallowly notched; base truncate to rounded or somewhat cordate, often oblique; margins entire; stipules minute, membranous, those of opposing leaves united (rarely free). Cyathia borne singly in the leaf axils or in the forks of the stem, each cyathium with 4 yellowish-green or reddish, oblong glands with a white or pink appendage up to 1 mm long (sometimes quite minute, rarely absent). Male flowers 3–12 per cyathium, in clusters of 1–3. Styles forked about half their length, the stigmas somewhat dilated. Mature capsules glabrous, 1–1.5 mm long, usually 1- to 3-seeded. Seeds whitish, smooth, ovoid, somewhat angled, about 1 mm or slightly longer, lacking a prominent caruncle.

E. spathulàta Lam.

NL., from L. *spatha*, a broad flat instrument, from Gr. *spathe*, a broad blade—probably in reference to the leaves, possibly to the leaflike bracts subtending each flower.
Flowers minute, greenish; mid-May–early July (–September)
Prairie, usually in dry, rocky, calcareous soil, less frequently in sandy or moist soil
East three-fifths and Meade and Morton counties

Glabrous, erect annual (rarely biennial), usually unbranched below the inflorescence, 2–4.5 dm tall. Leaves sessile, those of the main stem alternate, spatulate or oblanceolate, mostly (0.7–) 1–3.5 (–4.5) cm long and 0.4–1 cm wide, with a rounded or acute apex, tapered, rounded, or slightly clasping base, and finely serrate margins; leaves of the upper branches and inflorescence opposite, oblong to lanceolate or ovate, 3–15 mm long, with a rounded or acute apex, clasping base, and finely serrate margins; stipules absent. Cyathia borne singly in the forks of the inflorescence, each cyathium with 4 greenish, oblong or elliptic glands lacking petal-like appendages. Male flowers

5–8 per cyathium, in clusters of 1–2. Styles slender, forked about half their length. Mature capsules papillose, about 2–3 mm long. Seeds brown, ovoid, finely reticulate, about 1.5–2 mm long, with a yellowish caruncle.

E. cyparissias L., an introduced European perennial, is known from northeastern Kansas and a few other widely scattered counties. It has 1 to many erect stems 1–3 dm tall, mostly linear or filiform leaves, cyathia subtended by conspicuous, ovate to cordate or deltoid bracts and borne in terminal umbel-like clusters, and capsules with short styles. *E. podperae* Croiz., also a perennial, is known from 4 Flint Hills counties. It has flower clusters and bracts resembling those of *E. cyparissias* but is a more robust plant 4–9 dm tall with broader, linear-oblong leaves, and exserted styles which are longer than the immature capsules.

E. strictóspora Engelm.

NL., from L. *strictus*, p. part. of *stringo*, to draw tight, + Gr. *spora*, seed—possibly in reference to the congested lateral branches on which the flowers and fruits are borne.
Flowers minute, greenish; late June–mid-October
Dry, rocky, gravelly, or sandy soil of roadsides, waste ground, fields, and disturbed prairie
West four-fifths, primarily; less common toward the southeast

Hairy, prostrate annual. Stems branched, 0.5–4.5 dm long. Leaves opposite, short-petioled, the blades mostly oblong, sometimes ovate-oblong or nearly round, 5–13 mm long and 1–6 mm wide; apex rounded or acute; base rounded, oblique; margins serrate, at least on the upper half of the leaf; stipules membranous, triangular, lacerated, adjacent pairs united or free. Cyathia borne singly in the leaf axils on crowded, leafy lateral branches, each cyathium with 4 reddish, elliptic to oblong glands with white appendages shorter than the gland is wide. Male flowers 3–9 per cyathium. Styles club-shaped, entire or slightly notched at the apex. Mature capsules hairy, about 1.5–2 (–2.3) mm long. Seeds with a whitish, irregularly thickened outer coat over a brown inner coat, ovate to oblong-ovate, sharply 4-angled in cross section, 1–1.5 mm long, lacking a prominent caruncle.

STILLÍNGIA Garden

Named in honor of Dr. Benjamin Stillingfleet (1702–1771), English naturalist.

Smooth perennial herbs or shrubs with milky sap. Leaves alternate, short-petioled, glandular-toothed, with stipules much reduced and glandular. Flowers small, without petals, monoecious, the male and female mixed in terminal spikelike clusters. At each node of the inflorescence is a minute bract flanked by 2 larger glands. Male flowers with an asymmetrical, cuplike, 2-lobed calyx and 2 stamens, borne singly or in clusters at the upper nodes. Female flowers with a 3-lobed calyx and a nearly round, 3-celled, 3-ovuled pistil with 3 unbranched styles. The lower portion of the ovary eventually becomes thickened and hard. Capsules shallowly 3-lobed, the upper part dehiscing both loculicidally and septicidally and separating from the lower, persistent, hardened part. Seeds with a conspicuous caruncle.

One species in Kansas.

S. sylvática L.   Queen's delight

NL., of the woods, from L. *sylva*, forest, woods—sometimes used in the sense of "growing wild," as opposed to being cultivated.
Flowers yellowish-green; mid-May–July
Loose sandy soil of sand dunes, prairie, and flood plains
South-central sixth and Morton and Montgomery counties

Perennial herbs arising from a woody rhizome with thick fibrous roots. Stems erect, about 0.3–1 m tall, often producing a whorl of branches beneath the first-formed flower cluster. Leaf blades mostly elliptic to lance-elliptic or oblanceolate (the lowermost sometimes obovate), 3–9 cm long and 0.7–3 cm wide; apex acute or slightly acuminate, occasionally rounded on the lower leaves; base acute or wedge-shaped; margins crenate to finely serrate, the teeth alternating with minute deciduous glands (use magnification). Spikes terminal, many-flowered, 3–13 cm long. Capsules about 12–15 mm long. Seeds brown, lustrous, broadly ovate, slightly flattened, about 8 mm long and 6 mm wide, with a cream-colored caruncle.

The bobwhite quail eats the seeds of this plant.

The roots, harvested in late summer or early fall, have had various medicinal uses (30). They have been used to treat constipation, syphilis, menstrual irregularity, pulmonary diseases, and liver ailments, and some Indian women drank an infusion of the boiled roots after childbirth.

TRAGIA L.
Tragia, noseburn

NL., from L. *tragus*, goat. The genus name was given in honor of Hieronymus Bock (1498–1554), also known as Tragus (German *Bock* meaning "goat").

Perennial herbs, sometimes woody at the base, with 1 to many stems arising from a woody crown and taproot, often with stiff stinging hairs. Leaves alternate, petioled or sessile, stipulate. Flowers small, green, lacking petals, with 1–2 female flowers situated below several to many male flowers in the same raceme, each flower subtended by a small bract. Male flowers with 3–6 (–7) calyx lobes, 2–4 (–10) stamens, the filaments united at least at the base, and a rudimentary ovary; borne on slender pedicels which abscise near the middle. Female flowers with 3–6 (–7) calyx lobes and a somewhat compressed, globose pistil covered with stinging hairs, with 3 styles united at the base. Capsule 3-lobed, separating explosively into 3 1-seeded segments and a persistent central column. Seeds dark brown, sometimes mottled, globose, lacking a caruncle.

Two species in Kansas.

Key to Species:
1. Leaf blades mostly ovate to lanceolate or triangular-lanceolate, 1.5–6 cm long and 7–30 mm wide. Calyx lobes of female flowers longer than the pistil at anthesis. Male flowers 14–75 per raceme .... *T. betonicifolia*
1. Leaf blades mostly narrowly oblong or lance-oblong (occasionally ovate), 1–3.5 cm long and 2–12 mm wide. Calyx lobes of female flowers shorter than the ovary at anthesis. Male flowers 2–30 per raceme .... *T. ramosa*

T. betonicifòlia Nutt.

NL., from the genus *Betonica* (now included in the genus *Stachys*), betony, in the Mint Family, + L. *folium*, leaf—the leaves of this species resembling those of *Stachys* (formerly *Betonica*) *officinalis* (L.) Trev.
Flowers minute, greenish; June–mid-August
Rocky or sandy prairie, usually on hillsides or banks, less frequently along roadsides or on waste ground
East half, primarily

Stems several to many, erect to ascending or the outer ones decumbent, 1.5–4 (–5.5) dm tall. Leaf blades mostly ovate to lanceolate or triangular-lanceolate, 1.5–6 cm long and 7–30 mm wide; apex acute; base cordate or sometimes truncate; margins serrate. Stems, petioles, lower leaf surfaces, inflorescence, and fruits with stinging hairs. Racemes 1–3 (–6) cm long. Male flowers 14–75 per raceme, about 1–2 mm long, mostly with 3–4 calyx lobes and 3–4 stamens with thickened filaments. Female flowers 1 or 2 per raceme, about 2–3 mm long at anthesis (longer in fruit), with the 6 calyx lobes longer than the pistil. Capsule 4–5 mm long and 7–10 mm wide. Seeds 3–4 mm long.

T. ramòsa Torr.

L. *ramosus*, full of boughs, branching.
Minute, greenish; mid-June–mid-September
Dry, rocky prairie hillsides and banks
Central three-fifths, primarily, especially in the north-central counties

Stems 1 to many, erect to ascending or the outer ones decumbent, (0.8–) 1–3 dm tall. Leaf blades mostly narrowly oblong or lance-oblong, occasionally ovate, 1–3.5 cm long and 2–12 mm wide; apex acute; base acute to broadly rounded, truncate, or somewhat cordate; margins serrate. Stems, petioles, lower leaf surfaces, inflorescence, and fruits with stinging hairs. Racemes 0.5–1.5 cm long. Male flowers 2–30 per raceme, about 1–2 mm long, mostly with 3–4 calyx lobes and 3–6 stamens with thickened filaments. Female flowers 1 or 2 per raceme, about 0.8–2.5 mm long at anthesis (to 3 mm long in fruit), with the 6 calyx lobes shorter than the ovary. Capsule 3–4 mm long and 6–8 mm wide. Seeds 2.5–3.5 mm long.

288. *Tragia betonicifolia*

289. *Tragia ramosa*

CALLITRICHÀCEAE
Water Starwort Family

Annual or perennial, terrestrial or aquatic herbs. Leaves opposite, simple, entire, petioled or sessile, estipulate. Flowers unisexual, either monoecious or dioecious, regular, borne singly in the leaf axils. Sepals and petals absent. Male flowers consisting of a single stamen. Female flowers consisting of a single pistil; ovary superior, 2-lobed, 2-

celled; styles 2, threadlike; stigmas 2. Fruit 2-lobed, splitting at maturity into 4 achene-like sections.

The family contains a single genus, *Callitriche*.

CALLÍTRICHE L.   Water starwort, water chickweed

NL., from Gr. *callos*, beautiful, + *thrix*, hair.

Delicate plants with limp stems. Flower and fruit characters as described for the family.

Two species in Kansas. *C. terrestris* Raf. is known from a few eastern localities only. It is a much smaller plant than *C. heterophylla* and grows on damp soil rather than in the water or stranded on mud.

C. heterophýlla Pursh

NL., from Gr. *heteros*, other, different, + *phyllon*, leaf—because of the variation in leaf types on a single plant or on different plants.
Flowers minute, green; mid-March–July
In water or stranded on mud of lake margins, farm ponds, swampy areas
Scattered in the east three-fifths

Stems 0.2–2 dm long, branched from the base or unbranched. Leaves 4–12 mm long and 0.5–5 mm wide, the submersed leaves usually linear, the floating leaves (or all the leaves of stranded plants) narrowly oblanceolate to spatulate or obovate, those at the tip of the stem often crowded into a rosette. Flowers subtended by 2 minute, white inflated bracteoles. Styles 1–6 mm long. Fruit 0.5–1.3 mm long and about as wide as long or slightly narrower.

Ducks eat the plants.

BALSAMINÀCEAE   Balsam Family

Herbaceous annuals or suffrutescent undershrubs (ours all rather succulent annuals). Leaves alternate, opposite, or whorled, simple, petiolate, estipulate. Flowers bisexual, irregular, borne singly or in few-flowered clusters. Sepals 3–5, sometimes petaloid, the posterior one prolonged backward into a tubular nectiferous spur. Petals 4–5, distinct, or fused and appearing to be fewer in number. Stamens 5, fused to one another or at least coherent around the pistil. Ovary superior, 5-celled; stigmas 1–5, sessile or nearly so. Fruit a fleshy capsule which opens explosively at maturity.

The family includes 4 genera and about 500 species, most of which occur in the tropics of Africa and Asia. A few species of *Impatiens* are cultivated as ornamentals.

Kansas has but 2 species, these both in the genus *Impatiens*.

IMPÁTIENS L.   Balsam, jewelweed, touch-me-not

NL., from L. *in-*, not, + *patiens*, patient (*in-* before *b* or *p* becomes *im-*)—because the ripe seed pods snap open elastically and scatter the seeds when touched.

Smooth, somewhat succulent annuals. Leaves alternate or whorled, usually coarsely toothed. The flowers, borne on long, drooping peduncles, may be of several types. The conspicuous flowers which attract one's attention are *chasmogamous* (from Gr. *kasmos*, open, + *gamos*, marriage)—that is, they enlarge in the normal manner and open before pollination occurs.

Through the years, there has been considerable controversy among botanists regarding the structure of the perianth in the genus *Impatiens*. Outermost are 2 lanceolate green structures readily identified as sepals. Next are 4 distinct petals or, at least, petal-like structures, 1 of which is in the form of an open sack that is abruptly constricted at the base to produce a slender, bent spur or nectar receptacle. The question that has arisen is whether all of these colored structures are indeed petals, or whether some of them are sepals. Because the 2 obvious sepals sometimes bear spurs, this character is considered by several authorities (31) of the genus to belong to members of the sepaline whorl and, therefore, the spurred, saclike, petaloid member of the perianth is apparently also a sepal, though disguised as a petal. The petal-like upper member of the perianth sometimes bears 2 spurs and apparently represents 2 sepals united. This would make a total of 5 sepals: 2 greenish and 3 petal-like. There then remain the 2 lateral petaloid members of the perianth, each of which is 2-lobed and which never bear spurs and which are thought to represent 4 petals fused in 2 pairs, the fifth petal probably having been lost through evolution.

The structure of these flowers is highly modified to increase the possibility of cross-pollination. The 2 broad, lateral petals extend downward and outward and form a vestibule before the entrance to the saclike posterior sepal. Five short stamens project downward from their insertion on the receptacle and hold their anthers, which are united into a cylinder, close together over the stigma in a position to be touched by bees and hummingbirds intent on sucking nectar from the spurlike nectar receptacle. From the inner face of each stamen filament, a broad scale projects inward and all 5 scales are usually fused to one another above the stigma, thus preventing self-pollination. The anthers soon drop away, exposing the stigma to cross-pollination by pollen brought from younger flowers by hummingbirds, bumblebees, and other long-tongued bees.

When the ovary has become a ripe capsule, the partitions between the 5 cells break down, leaving intact the central column of united placentae to which the numerous seeds were attached. Now the capsule is set to spring apart into 5 divisons or "valves" whenever a touch sets it off, in such a way that the 5 valves, while remaining attached to the tip of the column, coil snappily upward, throwing out and scattering seeds.

In addition to chasmogamous flowers, many species of *Impatiens* also produce small, inconspicuous *cleistogamous* flowers (from Gr. *kleistos*, closed, + *gamos*) which pollinate themselves and never open. These flowers are about 3 mm long when pollination occurs. Sepals and usually petals are present, though they never expand, and the spur may be merely rudimentary. The anthers have only 2 pollen sacs each instead of 4, are not actually fused to one another, and have only a few pollen grains; the scales which protect the stigmas of cleistogamous flowers from self-pollination are absent. Fewer ovules, and therefore fewer seeds, are produced by the pistils.

Intermediate forms, called *pseudocleistogamous* flowers, may sometimes be found. These apparently are produced when drought, excessive heat, or other unfavorable conditions cause buds which would normally produce chasmogamous flowers to develop imperfectly. These flowers are self-pollinating (though the exact means by which this occurs is not clear) and fruits are produced.

The young shoots of our species may be eaten as a green vegetable, though it is recommended that the cooking water be changed once.

The nectar of the flowers is collected by ruby-throated hummingbirds, and the seeds of our species are eaten by the white-footed mouse, ruffed grouse, ring-necked pheasant, great prairie chicken, and bobwhite quail (22).

Key to Species (3):
1. Flowers orange (rarely lemon yellow or white) with many reddish-brown spots; the saccate sepal conic, at its mouth about half as wide as long; spur curved forward close to the sepal, about 6–9 mm long .... *I. biflora*
1. Flowers usually pale yellow (rarely white) with a few reddish-brown spots; saccate sepal broadly rounded at its summit, two-thirds to fully as wide as long; spur deflexed at a right angle to the sepal, about 4–6 mm long .... *I. pallida*

I. pállida Nutt.   Pale touch-me-not

L. *pallidus*, pale—in reference to the flowers.
Pale yellow; late June–early September
Moist woods and thickets
East fifth

Erect, branched annuals up to about 2.5 m tall, much like *I. biflora*. Leaf blades thin, 2.5–13 cm long and 2–6.5 cm wide. Flowers pale yellow (rarely white), with or without some reddish-brown spots, the normal (i.e., chasmogamous) flowers 2–3 (–3.5) cm long, the spur about 4–6 mm long, bent at an angle of about 45°–90° relative to the sepal, never coiled. Seeds black, oblong to ovate, slightly flattened, abruptly constricted at one end into a short blunt beak, about 5–6.5 mm long, more or less covered with sharp wrinkles or netlike ridges.
See plate 54.

I. biflòra Walt.   Spotted touch-me-not

NL., from L. *bi-*, *bis-*, 2, + *flos*, *floris*, flower—probably because the flowers are often borne in pairs, but possibly in reference to the 2 kinds of flowers produced.
Orange; late June–mid-September
Moist woods, creek banks, and seepy areas
East fifth, especially in the north, and a few collections from farther west

290. *Impatiens pallida*

291. *Impatiens pallida*, close-up of flowers

BALSAM FAMILY

217

Erect, branched annuals up to about 1.5 m tall. Leaf blades thin, ovate or somewhat rhombic to lanceolate or elliptic, mostly 2–9 (–12) cm long and 1.5–5.5 cm wide; apex mucronate; base acute or acuminate, often oblique and with 2–5 glandular projections; margins crenate or serrate, the teeth mucronate. Flowers orange (rarely lemon yellow or white) with many reddish-brown spots, the normal (i.e., chasmogamous) flowers about 2–3 cm long, the spur about 6–9 mm long, strongly recurved against the base of that sepal, sometimes coiled once. Smaller cleistogamous flowers may also be present. Capsules somewhat club-shaped with an acute apex, more tapered toward the base, about 2 cm long. Seeds dark brown, oblong to ovate, slightly flattened, abruptly constricted at one end into a short beak, about 4–5 mm long, covered with netlike ridges.

This plant has had various medicinal uses through the years. Potawatomi Indians used the juice of the leaves to treat the itch caused by poison ivy, while the Omahas and the Blackfoot Indians used the crushed leaves and stem for curing rashes and eczema.

MALVÀCEAE   Mallow Family

Herbs and shrubs, trees in the tropics. Leaves alternate, simple, petioled, stipulate, mostly palmately veined. Herbage frequently with stellate hairs. Flowers mostly bisexual (including all of ours), regular, frequently subtended by an involucre of separate or united bracts. Sepals 5, usually united at the base, the calyx bell-shaped. Petals 5, basally united and fused to the androecium. Stamens many, monadelphous and forming a tube around the pistil; anthers 1-celled; pollen grains large enough to be viewed easily with a hand lens, often spiny. Pistil superior, 2- to many-celled; styles and stigmas as many as the cells, mostly free above and united below. Fruit a loculicidal capsule or a schizocarp separating into as many discoid, usually 1-seeded segments (mericarps) as there are cells.

This is a large family, comprising approximately 82 genera and 1,500 species, of world-wide distribution but especially well represented in the American tropics. Cultivated members of this family include the cotton plant (*Gossypium* spp.), hollyhock (*Althea rosea* L.), rose-of-sharon (*Hibiscus syriacus* L.), and okra (*Hibiscus esculentus* L.).

Eight genera have species native or naturalized in Kansas, and the cultivated hollyhock is sometimes found as an escape. *Malvella leprosa* (Ort.) Krap. is known from Meade County only and is not included here.

Key to Genera:
1. Calyx subtended by a whorl of 3 or more bracts 2
  2. Bracts 6 or more (often 12) .... *Hibiscus*
  2. Bracts 3 3
    3. Petals truncate or rounded .... *Callirhoë*
    3. Petals obcordate or notched 4
      4. Calyx usually angled or winged in bud; corolla yellow or orange; stamen tube with filaments departing at apex only; stigmas apical, noticeably enlarged .... *Malvastrum*
      4. Calyx not angled or winged; corolla white to pale pink or lavender in ours; stamen tube with filaments departing along much of its length; stigmas lateral, apex of the style branches not enlarged .... *Malva*
1. Calyx not subtended by a whorl of bracts 5
  5. Petals white to pink or purplish red; stamen tube with filaments departing along much of its length; stigmas lateral; leaves divided quite or nearly to the base .... *Callirhoë*
  5. Petals yellow to orange or brick-red (rarely white in *Sida spinosa*); stamen tube with filaments departing at the apex only; stigmas apical, noticeably enlarged; leaves various 6
    6. Flowers in ours usually salmon-colored, 12–20 mm long; fruit a more or less hemispheric schizocarp, the mericarps beakless .... *Sphaeralcea*
    6. Flowers in ours usually yellow (rarely white in *Sida spinosa*), 8–10 mm long; fruit not as above 7
      7. Leaves broadly heart-shaped, 5–15 cm wide; fruit in ours a fluted, short-cylindric, several-celled capsule, each cell tipped by a persistent divergent awn and opening along the top and sides nearly to the base .... *Abutilon*
      7. Leaves mostly ovate to lanceolate or oblong, 3 cm wide or narrower; fruit a 5-celled schizocarp, the mericarps in ours 2-beaked, opening between the beaks .... *Sida*

ABÙTILON Mill.   Indian mallow

NL., from Arabic *aubutilun*, the name given by Avicenna, celebrated Mohammedan physician (980–1037).

Annual or perennial herbs, suffrutescent or woody in some areas. Leaf blades often heart-shaped. Flowers usually orange or yellow, borne singly or in clusters, not subtended by a whorl of bracts. Pistil usually more than 10-celled, with 2–9 ovules per cell; stigmas apical, headlike. Fruit short-cylindric or ovoid, the cells opening along the top and sides nearly to the base.

One species in Kansas.

292. *Callirhoë alceoides*

A. theophrásti Medic.   Velvetleaf, piemarker, Chingma abutilon

Named after Theophrastis (370–285 b.c.), a Greek scientist who is sometimes called the Father of
    Botany. The common name "piemarker" comes from the fruit which, with its radiating
    carpels, resembles that implement. Chinese *chi'ng ma* means "China jute."
Yellow; May–October
Roadsides, fields, pastures, waste places
East two-fifths, with a few collections from farther west; intr. from southern Asia

Sturdy, unbranched annuals, up to 2 m tall, the herbage yellowish-green, covered with soft, velvety stellate hairs. Leaf blades broadly heart-shaped, palmately veined, (3–) 6–18 cm long and (3–) 5–15 cm wide, long-petioled; apex acuminate; base cordate; margins crenate or dentate; stipules early deciduous. Flowers yellow or orangish-yellow, in clusters of 1–3 in the leaf axils, on peduncles shorter than the petiole, sometimes also in few-flowered terminal clusters. Calyx bell-shaped at anthesis, 6–9 mm long (persistent and becoming larger and more bowl-shaped in fruit), the lobes triangular-acuminate to somewhat ovate-acuminate or oblong. Corolla lobes broadly obtriangular, 8–10 mm long. Stamen tube with filaments departing at the apex. Fruit a short-cylindric, several-celled, fluted capsule, 1.5–2 cm long with each lobe tipped by a persistent divergent awn. Seeds brown, obliquely heart-shaped, flattened, sparsely hairy, about 3 mm long and 3 mm wide.

In its native land, this plant has been of economic importance since man began to make use of thread, cordage, and woven fabric, the abundant, strong fibers of the stem providing material for these uses.

CALLÍRHOË Nutt.   Poppy mallow

L. *callirhoë*, name of an ocean nymph in Grecian mythology, from Gr. *kallirrhoos*, beautiful-
    flowing. The dieresis over the *e* indicates that letter should be pronounced separately
    from the *o*, thus giving the name 4 syllables.

Annual or perennial herbs, the latter often with a woody, much-thickened taproot. Leaf blades mostly palmately lobed or divided. Flowers bisexual, showy, white to pink or purplish-red, either with or without a subtending whorl of 3 small bracts, borne singly on axillary peduncles. Apex of corolla lobes rounded or truncate, usually finely and somewhat irregularly toothed. Stamen tube with filaments departing along much of its length. Pistil with 10 to many cells, 1 ovule per cell; stigmas lateral, the style branches threadlike. Fruit a flattened, discoid schizocarp.

Four species in Kansas.

Key to Species:
1. Flowers closely subtended by 3 narrow bracts .... *C. involucrata*
1. Flowers not subtended by bracts 2
    2. Flowers pale pink to white, 12–23 mm long; plants primarily of east two-thirds of
       Kansas .... *C. alcaeoides*
    2. Flowers purplish-red or rose, 20–30 mm long; Cherokee County only .... *C. digitata*

C. alcaeoìdes (Michx.) Gray   Light poppy mallow

NL., from the L. *alcea*, for the European mallow, *Malva alcea*, + the Gr. suffix *-oides*, similar
    to—in reference to the style branches and leaves.
Pink or white; April–mid-June
Prairie, frequently on rocky soil
East two-thirds and Meade County

Slender, erect to ascending perennials with a thickened, often turnip-shaped tap-root and usually 3–7 unbranched flowering stems 1.5–4.5 dm tall. Basal leaves with blades triangular or cordate, entire with crenate margins to palmately lobed or deeply cleft, (2–) 4–9 cm long and 2.5–9 cm wide, petioles 4–15 cm long; stem leaves shorter-petioled, the blades smaller, triangular to ovate, rhombic, or oblanceolate in general outline, usually more deeply cleft and with narrower segments than the basal leaves; stipules ovate to lanceolate or lance-oblong, sometimes oblique, 3–8 mm long. Herbage sparsely covered with a mixture of straight, stellate, and/or 2-pronged hairs. Flowers borne singly in the leaf axils or in several-flowered terminal clusters, pink or white (sometimes drying lavender or becoming salmon-colored with age). Calyx 7–11 mm long at anthesis, the lobes triangular to triangular-acuminate or narrowly triangular, woolly on the inside at the tip, about the same length as or slightly longer than the calyx tube; subtending bracts absent. Corolla 12–23 mm long, the lobes wedge-shaped with a truncate apex. Fruit about 8–9 mm wide, backs of the mericarps bumpy or wrinkled, sparsely hairy, the lateral faces reticulate.

293. *Callirhoë digitata*

294. *Callirhoë digitata* leaves

295. *Callirhoë involucrata*

C. digitàta Nutt. Finger poppy mallow, fringed poppy mallow

L. *digitatus*, having fingers or toes, from L. *digitus*, finger, toe—in allusion to the fingerlike divisions of the leaves.
Purplish-red or rose; (April–) mid-June–early August
Rocky soil of prairie or open woods
Cherokee County

Slender, erect perennials with a thickened taproot, conspicuous early in the flowering season because of its height (up to 1 m) relative to the surrounding plants. Stems 1 to few, unbranched below the inflorescence. Leaves few, mostly basal, the blades 5- to 7-cleft or parted, 5–12 cm long and 6–20 cm wide, the segments linear to threadlike, 0.5–4 mm wide; petioles mostly 6–20 (–30) cm long; stipules usually deciduous by flowering time. Herbage glabrous or very sparsely hairy. Flowers purplish-red or rose, borne on pedicels 2–17 cm long in branched terminal clusters. Calyx 7–11 mm long at anthesis, the lobes triangular-acuminate to lance-attenuate, woolly on the inside, about the same length as the calyx tube; subtending bracts absent. Corolla 2–3 cm long, the lobes wedge-shaped with a truncate apex. Fruit about 8–11 mm wide, backs of the mericarps bumpy or wrinkled, glabrous, the lateral faces reticulate.

C. leiocarpa Martin is an annual with light pink to purplish-red flowers which has been collected in Elk County. It resembles *C. digitata* somewhat but is a more compact plant (up to 4.5 dm tall) and has a slender taproot. The fruits are encircled by a white papery collar at the base of the beaks.

C. involucràta (T. & G.) Gray Low poppy mallow

NL., having an involucre, from L. *involucrum*, wrapper, covering—in reference to the bracts subtending the calyx.
Purplish-red; late April–mid-October
Prairie, roadsides, pasture, waste ground, often in disturbed soil
Nearly throughout except for the extreme northeast, east, and southeast; more common westward

More or less rough-hairy, procumbent to ascending perennials with a deep, thickened taproot. Stems few to several, 1.5–6 dm long. Leaf blades round to cordate in general outline, palmately cleft, lobed, and toothed, 3–8 cm wide, the lower surface more or less densely covered with stellate hairs, the upper surface with straight hairs; petioles mostly 2.5–15 cm long; stipules ovate or cordate, oblique, 4–13 mm long. The earliest or lowermost leaves are usually cleft halfway or slightly more to the base and have the main lobes relatively broad (5–10 mm wide at the base of the sinus) with rounded teeth or subdivisions; the later leaves are progressively more deeply cleft with narrower (0.5–2.5 mm wide), sharply pointed subdivisions and teeth. Flowers mostly borne singly in the leaf axils on pedicels 2.5–12 cm long. Calyx broadly bell-shaped, 1.5–2 cm long at anthesis, with prominent whitish veins, the calyx tube with long, spreading hairs, the calyx lobes long-triangular to linear-oblong or lance-attenuate with an acute apex and stellate hairs, longer than the calyx tube; subtending bracts linear to linear-lanceolate or oblanceolate, 7–15 mm long. Corolla 2–3.5 cm long, the lobes obovate to obtriangular, purplish-red with a white base. Fruit mostly 9–12 mm wide, backs of the mericarps bumpy or wrinkled, more or less hairy, the lateral faces reticulate. Seeds brown, lustrous, very broadly elliptic to nearly round in outline, with a notch at the hilum, 2.5–3 mm long.

Though low-lying, slender, and somewhat vinelike, the plant is made conspicuous by the rich coloring of its flowers and is a striking feature in the landscape when it occurs in patches and colonies, as frequently is the case in western Kansas where it is noted for its resistance to drought. It is easily established in a wildflower garden when seeds are sown where the plants are to remain undisturbed.

The Indian names for this plant meant "medicine" or "smoke-treatment–medicine." The Indians boiled the root and drank the decoction for intestinal pains, and they burned dried roots and inhaled the smoke for head and bronchial colds.

HIBÍSCUS L. Rose mallow

NL., from Gr. *hibiskos*, marsh mallow.

Herbaceous annuals or perennials or shrubs. Flowers large and showy, each lasting but 1 day, subtended by a whorl of linear bracts (usually 12). Corolla lobes usually obovate. Stamen tube with filaments departing along most of its length. Pistil with 5 cells, the style with 5 radiating branches, each with a conspicuous, apical, headlike or peltate stigma. Fruit a loculicidal capsule, nearly round to ovoid or somewhat angular, enclosed by the persistent, enlarged calyx. Seeds several to many in each cell.

Three species in Kansas.

Several of the hardy North American species, including our own, were brought into cultivation in the United States and Europe during the 19th century. The various species hybridize rather easily, and numerous hybrids have been produced artificially in order to improve upon the color, shape, and life of the individual blossoms. Both our perennial species readily withstand transplanting and will thrive in good garden soil with moderate amounts of water. They may also be grown from seed and will flower the first season if sown early (32).

Key to Species:

1. Annuals up to 5 dm tall; leaves divided nearly to the base; petals usually 2–2.5 cm long
.... *H. trionum*
1. Herbaceous perennials 1–2 m tall; leaves merely serrate, crenate, or lobed; petals 6 cm or longer 2
   2. Leaves glabrous or nearly so, most of them hastate, glabrous, the margins serrate; plants primarily of the east fourth of Kansas .... *H. militaris*
   2. Leaves densely covered with soft, velvety hairs, the blades lanceolate to ovate, the margins crenate or serrate; known from Bourbon and Cherokee counties only ....
.... *H. lasiocarpos*

296. *Hibiscus lasiocarpos*

H. lasiocárpos Cav.　　　　　　　　Woolly rose mallow

NL., from Gr. *lasios*, hairy or woolly, + *karpos*, fruit, i.e., hairy-fruited.
White or pale pink with a red or purplish center; July–August
Swampy areas
Bourbon and Cherokee counties

Erect herbaceous perennial, 1–2 m tall, with large fleshy roots. Leaf blades lanceolate to ovate, 6–16 cm long and 3–9 cm wide, densely covered with soft stellate hairs; apex acuminate; base rounded or slightly cordate; margins serrate or crenate, at least above the middle. Flowers white or pale pink (often drying yellowish) with a red or purplish center, borne singly in the leaf axils or also in few-flowered terminal clusters, subtended by a whorl of linear-acuminate bracts 2–3 cm long. Calyx soft-hairy, (2–) 2.5–3.5 cm long at anthesis, the lobes broadly to narrowly triangular-acuminate. Corolla lobes obovate, 7.5–10 cm long. Capsule short-cylindric, hairy.

297. *Hibiscus militaris*

H. militáris Cav.　　　　　　　　Halberdleaf rose mallow

L. *militaris*, pertaining to a soldier, from L. *miles, militis*, soldier—because the leaves frequently are shaped like the head of a halberd spear.
Pink or white, with a purple center; July–October
Lake margins, swampy areas, wet roadside ditches
East fourth, primarily, with scattered records as far west as Sedgwick and Barber counties

Glabrous, erect herbaceous perennial, 1–2 m tall, with large fleshy roots. Leaf blades mostly triangular-hastate, 3–12 cm long and 2–12 cm wide; margins serrate. Flowers pink or white (drying yellowish), with a purple center, borne singly in the leaf axils, subtended closely by a whorl of linear-acuminate bracts 2–3 cm long. Calyx glabrous or nearly so, 2.5–4 cm long at anthesis, the lobes triangular with an acute or short-acuminate apex. Corolla lobes obovate, 6–8 cm long. Capsule glabrous or nearly so, about 3 cm long. Seeds 3–4 mm long, densely covered with short, rust-colored hairs.

298. *Hibiscus trionum*

H. triònum L.　　　　　　　　Flower-of-an-hour

NL., from Gr. *tri-*, 3, + *onyx*, claw—because of the narrow, often clawlike divisions of the leaf.
The flowers open in the morning but close around 9 o'clock unless the sky is overcast.
"Good night at noone" was an old English name for this plant.
Yellow or cream-colored with purple center; June–October
Waste ground, roadsides, fields, pastures
East three-fourths, primarily; more common eastward; intr. from Europe

Erect, sparsely hairy annuals, up to 6 dm tall. Leaf blades 3-lobed or 3- or 5-parted (the lowermost 2 or 3 sometimes nearly entire), 2–5 (–7) cm long and 3–6 cm wide, the segments mostly obovate to oblanceolate or cuneate-oblong in general outline and toothed or pinnately lobed. Flowers light yellow or cream-colored, with a dark red or purplish center, borne singly in the leaf axils or also in few-flowered terminal clusters, subtended

MALLOW FAMILY
221

by a whorl of ciliate, linear bracts 7–9 mm long. Calyx membranous with prominent dark veins, 10–15 mm long at anthesis. Corolla lobes obovate, 2–2.5 cm long. Anthers and pollen bright orange-yellow. Capsule ovoid, hairy, becoming black, 1–1.5 cm long. Seeds brown, obliquely heart-shaped, papillose, about 2 mm long and 2 mm wide.

See plate *55*.

299. *Malva neglecta*

MÁLVA L.   **Mallow**

L. *malva*, mallow; related to Gr. *malakos*, soft—in allusion to the mucilage in roots, stems, leaves, and fruits of plants in the genus *Malva*.

Annual, biennial, or perennial herbs. Leaves round or reniform, toothed, lobed, or divided. Flowers borne singly or in axillary clusters, subtended by a whorl of 3 bracts. Apex of corolla lobes often deeply notched. Stamen tube with filaments departing along much of its length. Pistils with 10–20 cells, 1 ovule per cell; stigmas lateral, the style branches threadlike. Fruit a flattened schizocarp, the segments (mericarps) beakless, indehiscent.

Four species occur in Kansas. *M. parviflora* L., *M. rotundifolia* L., and *M. sylvestris* L. are introduced from Eurasia but rarely are encountered.

M. neglécta Wallr.   **Common mallow**

L. *neglectus*, neglected, slighted—because of its weedy character and inconspicuous flower.
White to pale lilac; May–October
Disturbed soil of fields, roadsides, waste ground, river banks, and prairie
East two-thirds, a few records from farther west; intr. from Eurasia

Prostrate or ascending biennials with a thick taproot. Stems usually branched near the base, 1.5–10 dm long. Leaf blades round or reniform, very shallowly lobed, 1.5–7 cm wide; base cordate; margins crenate; stipules ovate or lanceolate. Herbage more or less densely covered with stellate hairs and some simple hairs. Flowers white to pale lilac, in fascicles of 1–3 in the leaf axils. Calyx 4–7 mm long, the lobes ovate-acuminate, the subtending bracts linear or subulate, 3–5 mm long. Corolla lobes obcordate, apically notched, mostly 8–10 mm long. Mature fruit crenate in outline, 5–7 mm across, short-hairy, the mericarps rounded on the back and not wrinkled, the lateral faces smooth. Seeds brown, in side view nearly round, notched at the hilum, but essentially wedge-shaped (as the mericarps are), about 1.5–1.8 mm long.

The flowers are fitted for cross-pollination by their proterandry: the anthers open and expose their pollen, then curve downward before the styles spread apart above the anthers and expose the stigmatic surfaces to contact with insects seeking pollen or nectar secreted by the ringlike nectar gland at the base of the staminal tube. Insect visitors are honeybees, wild bees, butterflies, and beetles—the bees gathering nectar and pollen, the butterflies sipping nectar, the beetles eating pollen, and all of them effecting cross-pollination as they go from younger flowers with freshly opened anthers to older flowers with outspread stigmas. It is to be noted, however, that as a final act the styles entwine with the anthers, and if cross-pollination has not already happened, self-pollination might then occur.

Although this mallow is rated as a weed, it has both food and medicinal value. The Greeks, Romans, and Egyptians long ago used the young shoots and their fruits or "cheeses" for greens and salad. Also, because of their mucilage content, the leaves were boiled to make hot fomentations for relief from pain and poultices for sores; and an infusion of the leaves was drunk to alleviate coughs and colds.

MALVÁSTRUM Gray

NL., from L. *malva*, mallow, + -*astrum*, a diminutive suffix which has a derogatory implication—hence, "wild mallow," or "resembling but not equalling the ornamental mallows."

Annual or perennial herbs or subshrubs. Leaf blades linear-lanceolate to triangular-ovate, unlobed or obscurely 3-lobed, the margins toothed or crenate. Flowers terminal or axillary, subtended by 3 small bracts, opening for a short time in late morning or early afternoon. Calyx usually angled or winged in bud with 5 obscure nectaries inside at the base of the calyx tube. Corolla yellow or orange-yellow, the lobes obovate, shallowly notched and slightly oblique at the apex. Stamen tube up to 3 (–6) mm long, the filaments all departing at the apex or within the upper 1 mm or so. Pistil with 5–16 cells, 1 ovule per cell; stigmas apical, capitate. Fruit a flattened, discoid schizocarp.

One species in Kansas.

M. híspidum (Pursh) Hochr.

L. *hispidus*, rough, shaggy, hairy, bristly—in reference to the rough pubescence of the herbage.
Yellow; July–mid-October
Roadsides, prairie, open woods, frequently on shallow soil over limestone
East third, primarily; scattered in the west two-thirds

Erect, rough-hairy annual, 1.5–5 dm tall, with spreading or ascending branches, somewhat resembling *Sida spinosa*. Leaf blades oblong to lance-oblong or elliptic, 1.5–4 cm long and 4–12 mm wide; apex acute or minutely mucronate; base acute or rounded; margins serrate or dentate; petioles 5–11 mm long; stipules linear or threadlike. Herbage with 4- to 6-armed stellate hairs and some simple hairs. Flowers borne singly or in few-flowered clusters, opening for a short time in late morning or early afternoon. Calyx angled or winged in bud, 3–7 mm long with triangular lobes at anthesis, becoming much larger with broadly cordate-acuminate lobes when in fruit. Corolla yellow, 2.8–4.5 mm long. Ovary 5- (rarely 6-) celled, each cell with 1 ascending ovule. Fruit 5.5–7.5 mm wide, the mericarps rounded on the back, hairy, dehiscent, arranged in a starlike whorl. Seeds reddish-brown, about 2.5 mm long, short-reniform or nearly round in outline, with a notch at the hilum.

The flowers are self-pollinating, the style branches recurving and pushing the stigmas against the anthers by anthesis. In some instances, the blossoms are cleistogamous, i.e., they pollinate themselves without ever opening (33).

300. *Sida spinosa*

SÍDA L.

NL., from a name used by Theophrastus for some plant in the Mallow Family.

Annual or perennial herbs, sometimes suffrutescent. Flowers mostly small, white to yellow, orange, or reddish-purple, single or clustered, axillary or terminal, not subtended by bractlets in our species. Calyx often winged or angled in bud. Stamen tube with filaments departing at the top only. Pistil with 5–15 cells, 1 ovule per cell; stigmas apical, headlike. Fruit a schizocarp enclosed by the persistent, enlarged calyx, the segments (mericarps) dehiscent or indehiscent.

One species in Kansas.

S. spinòsa L.   Prickly mallow

L. *spinosus*, thorny—in reference to the rigid, spinelike protuberances which subtend the petioles of some plants.
Yellow; July–October
Disturbed soil of fields, roadsides, pastures, creek banks, and waste ground
East half and Sheridan County; intr. from the tropics

Erect, minutely and softly hairy annuals. Stems tough and fibrous, branched, up to about 7 dm tall. Leaf blades mostly ovate to lanceolate or oblong, (1–) 2–5.5 cm long and 0.7–3 cm wide; apex rounded or acute; base truncate or rounded, occasionally cordate; margins serrate, the sides of the teeth sometimes rounded; petioles 0.5–3 cm long, on some plants subtended by a rigid prickle; stipules linear or threadlike, deciduous. Herbage often tinged with purple. Flowers yellow (rarely white), in 1- to few-flowered axillary and terminal clusters, on pedicels shorter than to as long as the subtending petiole, opening mornings for a short time. Calyx 5–7 mm long at anthesis, the lobes triangular or triangular-acuminate. Corolla about 8–9 mm long. Fruit ovoid, about the same length as the persistent calyx, splitting into 5 2-beaked, reticulate mericarps which open between the beaks. Seeds brown, 2–2.5 mm long, shaped rather like a longitudinal quarter of an apple.

SPHAERÁLCEA St.-Hil.   Globe mallow, false mallow

NL., from Gr. *sphaira*, a sphere, + *alkea*, ancient name of the European mallow, *Malva alcea*—in allusion to the fruits, which may be hemispheric to short-conic, and to the resemblance of the flowers in this genus to those of the European mallow.

Mostly perennial herbs, less frequently annuals, biennials, or suffrutescent perennials. Leaves merely toothed to deeply 3- or 5-parted or dissected. Flowers yellowish to orange or brick-red (usually salmon-colored in ours), in clusters, usually subtended by 3 deciduous bracts but not in *S. coccinea*. Stamen tube with anthers only at the tip. Pistil usually with 10–15 cells, 1 ovule per cell; stigmas apical, only slightly enlarged. Fruit a more or less hemispheric schizocarp, the mericarps often remaining attached to the central column after maturity.

Three species in Kansas. *S. angustifolia* (Cav.) D. Don. occurs in Hamilton and

301. *Sphaeralcea coccinea*

Kearny counties only and may be as tall as 1.5 m. It has salmon-colored flowers (in our area) and linear- or oblong-lanceolate leaves with toothed margins. A southern species, *S. fendleri* Gray, is recorded from Pawnee, Barber, and Sedgwick counties but apparently is only occasionally adventive in Kansas.

S. coccínea (Pursh) Rydb.　　　　　　　　　　　　Scarlet globe mallow

L. *coccineus*, scarlet.
Salmon-colored; late April–mid-July
Sandy or rocky soil of prairie and roadsides
West two-thirds

Erect, gray-green or light-green perennial, spreading by means of creeping woody roots. Stems several, erect to ascending, 1–3 dm tall. Leaf blades 1–4 cm long and 1–4 cm wide, deeply divided into 3 or 5 segments, the segments more or less obovate to oblanceolate or wedge-shaped in general outline, entire or with a few teeth or lobes; stipules small, deciduous. Herbage more or less densely covered with stellate hairs. Flowers usually salmon-colored in Kansas (varying to brilliant orange or nearly red in some parts of the United States), borne singly in the upper leaf axils and in terminal racemes, not subtended by small bracts. Calyx 4–8 mm long at anthesis, the lobes mostly triangular to lance-acuminate. Corolla 12–20 mm long, the lobes obcordate. Fruit about 6–8 mm wide, subtended by the persistent and slightly enlarged calyx, the mericarps wrinkled and more or less hairy on the back, reticulate on the sides except for the beaklike apex.

The Blackfoot Indians chewed the plant and applied the mucilaginous paste to scalds, burns, and other sores (34).

HYPERICÀCEAE　　　　　　　　　　　　　　St. John's-wort Family

Herbs, shrubs, or some small trees in the tropics. Leaves opposite or whorled, simple, sessile, estipulate, often with transparent or black dot. Flowers bisexual, regular, borne in terminal clusters. Sepals 4–5, often united basally. Petals 4–5, often narrowed to a petiolelike base (claw). Stamens many, distinct or the filaments somewhat fused at the base, sometimes clustered into 3 or 5 groups. Pistil 1; ovary superior, 3- to 5-celled with parietal placentae; styles 3–5, separate or fused basally. Fruit a many-seeded capsule.

A family of about 50 genera and 1,000 species, mostly of the tropics. Some of the woody members of the family (*Hypericum* spp. and *Ascyrum hypericoides* L.) are cultivated as ornamentals.

Two genera in Kansas. *Ascyrum hypericoides* L., our only representative of that genus, is known from Chautauqua and Cherokee counties only. It is a low, woody perennial somewhat resembling our species of *Hypericum*, but the few flowers have 4 sepals, 4 petals, and 2 styles.

HYPÉRICUM L.　　　　　　　　　　　　　　　　St. John's-wort

NL., from Gr. *hyperikon*, the name of obscure origin used by Dioscorides and Pliny (1st century)
　　for plants supposed by commentators on Pliny to be *Hypericum perforatum* or *H. crispum*
　　of Linnaeus.

Glabrous herbs (ours) or shrubs. Leaves mostly opposite, entire, usually with translucent and/or black glandular dots. Flowers yellow in ours. Sepals 5. Petals 5, the apex oblique. Stamens few to many, often clustered or united into 3 or 5 groups. Styles separate or fused part or all of their length. Capsules mostly 1-celled, sometimes partially or completely 3- or 5-celled, tipped by the persistent style(s). Seeds cylindric, the surface reticulate or honeycomblike.

Seven species in Kansas. *H. pyramidatum* Ait., known from Doniphan County only, is a conspicuous perennial up to 1.5 m tall with flowers 4–5 cm across and ovoid, 5-celled fruits 1.5–2 cm long. *L. majus* (Gray) Britt., recorded from Ellsworth, Reno, and Harvey counties, resembles *H. mutilum* somewhat, but differs from that species in being somewhat taller, having the branches of the inflorescence strongly appressed, rather than widely spreading, and in having lanceolate sepals 5–6.5 mm long and ovoid capsules 5.5–7.5 mm long.

Several species of St. John's-wort have been used as dye plants.

Key to Species:
1. Style 1, persistent as a single straight beak on the capsule and eventually split by its dehiscence; stigma minute, never capitate .... *H. sphaerocarpum*

1. Styles 3, separate to the base, often divergent; stigmas capitate **2**
   2. Petals conspicuous, longer than the sepals; stamens many, clustered into 3 or 5 groups; capsule completely 3-celled, covered with amber-colored resin tubes or pustules **3**
      3. Branches finely ridged; petals with black dots along the margin only .... *H. perforatum*
      3. Branches not ridged; petals with black dots and line throughout .... *H. punctatum*
   2. Petals inconspicuous, shorter than the sepals, stamens 10 (–20), evenly spaced; capsule 1-celled, lacking resin tubes or pustules **4**
      4. Leaves narrowly elliptic to linear or awl-shaped, 0.5–2 mm wide, strongly ascending or nearly erect; seeds brownish-orange, 0.8–1.2 mm long, the surface honeycomblike
      .... *H. drummondii*
      4. Leaves mostly lanceolate to elliptic, 3–14 mm wide, spreading; seeds light tan, about 0.6–0.7 mm long, the surface not honeycomblike .... *H. mutilum*

302. *Hypericum perforatum*

H. drummóndii (Grev. & Hook) T. & G.   **Nits-and-lice**

Named for its discoverer Thomas Drummond (1780?–1835), a Scottish naturalist who, under the patronage of Sir William Jackson Hooker, collected plant and animal specimens in North America in 1825–29 and 1830–34.
Yellow; June–October
Prairie, wooded areas, pastures
Southeast eighth

Slender, erect annual. Stem 1–8 dm tall, eventually with several to many finely ridged ascending branches. Leaves ascending or nearly erect, narrowly elliptic to linear or awl-shaped, with many translucent glandular dots, mostly 1–2 cm long and 0.5–2 mm wide; apex usually attenuate. Sepals narrowly lanceolate or awl-shaped, basally united, 4–5 mm long. Petals yellow, 2–3 mm long, lasting only 1 morning. Stamens 10–20, evenly spaced. Styles 3, united slightly at the base; stigmas headlike. Capsule ovoid, 1-celled, about 5 mm long (excluding the persistent styles), lacking resin pustules or tubes, subtended by the old flower parts. Seeds brownish-orange, oblong-cylindric, 0.8–1.2 mm long, with a honeycomblike surface.

H. mùtilum L.

NL., from L. *mutilus*, maimed, cut off, shortened, the Linnaean type specimen being merely a fragment of a plant.
Yellow; July–September
Moist soil of prairie ravines, open woods, creek banks, and lake margins
Scattered in the east two-fifths

Slender, erect annuals or short-lived perennials. Stems 1 to few, finely ridged, 1–8 dm tall, eventually widely branched above. Leaves mostly lanceolate to elliptic, 1–3.5 cm long and 3–14 mm wide with many translucent glandular dots (use magnification), and 3 or 5 veins arising at the base; apex cordate. Sepals elliptic to linear-oblong with an acute apex, slightly united at the base, 3–4.5 mm long. Petals yellow, 1–2 mm long. Stamens 10, evenly spaced. Styles 3, united slightly at the base; stigmas headlike. Capsule ellipsoid, 1-celled, 3–5 mm long (excluding the persistent styles), lacking resin pustules or tubes (use magnification), subtended by the old flower parts. Seeds light tan, oblong-cylindric, about 0.6–0.7 mm long.

H. perforàtum L.   **Common St. John's-wort**

L. *perforatus*, pierced through—in reference to the many translucent spots in the leaves.
Yellow; June–October
Prairie, roadsides, moist ravines
East three-fifths, primarily; intr. from Europe

Erect herbaceous perennials, spreading by means of rhizomes. Stems 1 to several, 4–8 dm tall, with many short, ascending, decussate, finely ridged, leafy branches. Leaves oblong to linear- or lance-oblong or elliptic, with many transparent dots, 5–20 (–30) mm long and 1.5–5 (–9) mm wide, those of the branches smaller than those of the main stem; apex rounded, less frequently acute; base rounded or acute. Sepals linear to narrowly lanceolate, *not* black-dotted, 9–13 mm long, the apex oblique and black-dotted. Stamens many, clustered into 3–5 groups, each anther with a black glandular dot. Styles 3, completely separate; stigmas headlike. Capsule ovoid, 3-celled, 5–8 mm long (excluding the persistent styles), usually with many resin pustules (use magnification), subtended by the old flower parts. Seeds dark brown, oblong-cylindric, about 1 mm long, with a netlike or honeycomblike surface (use high magnification).

The anthers and stigmas of this species mature at the same time, and a large variety of insects coming for nectar and pollen insure both self- and cross-pollination.

303. *Hypericum punctatum*

304. *Hypericum sphaerocarpum*

There seems to be a final promotion of self-pollination when the petals on withering become erect and bunch the anthers around the stigmas.

Plants of this species were anciently reputed to have medicinal value. An infusion of the flowering tops in water was administered as a nervine in cases of hysteria and somnambulism, as a diuretic, and as an astringent for internal inflammation. Gerard tells of putting triturated leaves, flowers, and seeds in a glass of olive oil, which he then set in the sun for several weeks; then he strained the oil and put in a fresh trituration for standing in the sun as before, finally obtaining "a most pretious remedie for deep wounds and those that are thorow the body, for the sinues that are pricked, or any wound made with a venomed weapon." Occult influences also were attributed to these plants. The dark spots on the petals were supposed to represent drops of St. John's blood, and if a plant was hung in the window on the anniversary of St. John's birth (reputed to be June 24), the home was guarded against "ghosts, devils, imps, and thunderbolts." After its introduction to North America, the Menominee Indians adopted it as a medicinal plant and combined it with black raspberry roots in a tea designed to treat tuberculosis.

Common St. John's-wort was the first plant recorded as a cause of photosensitization in animals. Horses, cattle, or sheep which have eaten the flowering plants become hypersensitive to sunlight and develop a dermatitis of unpigmented areas of skin not protected by a dense coat of hair. The dermatitis usually progresses into puffiness, running sores, and, eventually, death of the outer tissues in affected areas. If the animals are not exposed to strong sunlight, the symptoms do not appear. The toxic substance is thought to be a photodynamic compound called hypericin, but the mechanism of its action is not understood. For a more detailed description of photosensitivity, see Kingsbury (21).

H. perforatum has long been used as a source of dye. The flowers, with alum as a mordant, yield a yellow dye. The leaves, also with alum, give a red dye.

H. punctàtum Lam.   Spotted St. John's-wort

NL., *punctatum*, punctate, dotted over, from L. *punctum*, point, small round dot—in allusion to the black-dotted leaves and petals.
Yellow; June–August
Prairie, open woods, roadsides, often in moist soils
East fourth

Erect herbaceous perennials. Stems sparingly branched or unbranched below the inflorescence, black-dotted, not ridged, mostly 5–7 dm tall. Leaves ovate to lanceolate, elliptic, or oblanceolate, conspicuously gland-dotted, (2–) 2.5–4.5 (–5) cm long and 0.7–2 cm wide; apex rounded; base at least somewhat cordate and clasping the stem. Sepals lanceolate with an acute or acuminate apex, basally united, 3–5 mm long, with numerous black or amber resinous dots and lines. Petals oblong or oblanceolate, 5–7 mm long, the apex oblique, the many black lines and dots not restricted to the apex. Stamens many, clustered into 3–5 groups, each anther with a black glandular dot. Styles 3, separate; stigmas headlike. Capsule ovoid, 3-celled, 4–5 (–6) mm long (excluding the persistent styles), covered with amber-colored resin tubes (use magnification), subtended by the old flower parts. Seeds orangish-brown, oblong-cylindric, about 0.8 mm long, with a honeycomblike surface (use high magnification).

H. sphaerocárpum Michx.

NL., from Gr. *sphaira*, sphere, ball, + *karpos*, fruit—the fruits being more or less spherical.
Yellow; May–October
Prairie, roadsides, open oak woods, often on shallow, rocky soil
East fourth

Erect, herbaceous or suffrutescent perennials spreading by means of rhizomes. Stems several, 2–5 dm tall, unbranched or with few to several finely ridged, ascending branches terminated by flower clusters. Leaves narrowly elliptic or oblong to narrowly lanceolate or oblanceolate, with numerous transparent glandular dots (1.2–) 2–5.5 cm long and (2–) 4–12 mm wide; apex and base rounded or acute. Sepals broadly ovate to lanceolate with an acute or rather abruptly acuminate apex, basally united, not black-dotted, deciduous, 7–9 mm long. Stamens many, evenly spaced. Style 1; stigma minute, not headlike. Capsule spherical to ovoid, 1-celled, 4–5 mm long (excluding the persistent style), lacking resin pustules (use magnification), subtended by the old calyx and stamens, but not the corolla. Seeds black, oblong-cylindric to ellipsoid, sometimes slightly curved, 2.5–3 mm long, with a honeycomblike surface (use high magnification).

CISTÀCEAE   Rockrose Family

Herbs or low shrubs. Leaves alternate, opposite, or whorled, simple, sessile, stipulate or estipulate. Flowers bisexual, regular, borne singly or in clusters. Sepals 5, in 2 series, the outer 2 smaller than the inner 3. Petals 5, 3, or none. Stamens few to many, arising beneath the ovary from an elongated projection of the receptacle. Pistil 1; ovary superior, 1-celled or imperfectly 3-celled; style 1 or lacking; stigmas 3–5 or 1. Fruit a tough loculicidal capsule, frequently enclosed by the persistent calyx.

The family comprises about 8 genera and 175 species. A few species of *Cistus* and *Helianthemum* are cultivated as garden plants.

Kansas has 2 genera. *Helianthemum bicknellii* Fern., though rarely encountered, is recorded from 13 counties in the east two-fifths of Kansas. It is an erect herbaceous perennial with wiry stems 2–6 cm tall and narrowly elliptic or oblanceolate leaves. The earliest-blooming flowers have 5 conspicuous yellow petals which soon fall; the later-blooming flowers have no petals at all.

LÉCHEA L.   Pinweed

Named after Johan Leche (1704–1764), Swedish botanist.

Herbaceous or suffrutescent perennials, mostly with overwintering basal shoots. Leaves alternate (occasionally opposite or whorled), estipulate. Flowers small and inconspicuous, mostly in terminal clusters. Outer 2 sepals linear to lanceolate and leaflike, inner 3 broader, dry and membranous. Petals 3, reddish or brownish, shorter than the sepals. Stamens mostly 5–15. Pistil incompletely 3-celled with 2 ovules per cell; style lacking; stigmas 3, plumose. Capsule 1- to 6-seeded, dehiscing along 3 lines.

Two species in Kansas. *L. mucronata* Raf. (*L. villosa* Ell. of some authors) is known from sandy areas in Rice, McPherson, Reno, Harvey, and Cherokee counties. It is a more robust, more conspicuously hairy plant up to 9 dm tall with leaves 1–7 mm wide.

L. tenuifòlia Michx.

NL., from L. *tenuis*, thin, + *folium*, leaf, i.e., thin-leaved.
Flowers minute, dark red; June–October
Sandy soil of prairie, open woods, pastures
East fourth, primarily; a few collections as far west as Stafford County

Herbaceous perennials with a slender taproot and a small, woody crown. Flowering stems several, slender, mostly erect, much branched above, 1–2.5 dm tall, subtended by several short, leafy, prostrate or ascending, nonflowering basal shoots. Leaves of the main stems linear, 3–15 mm long and about 1 mm wide, sharply pointed, soon falling. Herbage, especially stems and calyx, with mostly appressed hairs. Outer sepals 2–3 mm long; inner sepals lanceolate to ovate, keeled along the midvein, as long as or shorter than the outer ones. Capsules ovoid, 2- to 3-(5-)seeded, 1.4–1.7 mm long. Seeds reddish-brown, narrowly to broadly ovate in outline, nearly flat on 1 side and strongly convex on the other, 0.7–1 mm long.

VIOLÀCEAE   Violet Family

Ours all herbs, but some vines, shrubs, and small trees in the tropics. Leaves alternate, frequently clustered in a basal rosette, simple, mostly petioled but sessile or nearly so in *Hybanthus*, stipulate. Flowers bisexual, irregular, mostly solitary. Sepals 5, free from one another or slightly united at the base. Petals 5, the lowest petal swollen at the base and prolonged into a spur. Stamens 5, the anthers close together or coherent into a sheath around the ovary, the 2 lowest with nectar glands at the base, the filaments prolonged beyond the anthers into colored, membranous appendages. Pistil 1; ovary superior, 1-celled but with 3 parietal placentae; style 1, much enlarged toward the stigma and bent downward at the tip. Fruit a many-seeded capsule splitting into 3 segments, each bearing a line of seeds down the middle. As each segment dries, it slowly folds together lengthwise, and the resultant pressure on the seeds throws them several feet through the air. Most species produce reduced cleistogamous flowers in addition to the showy flowers described above.

A family of about 15 genera and 800 species. Numerous species of violets and pansies (*Viola* spp.) are cultivated as ornamentals.

Two genera in Kansas.

Key to Genera (3):

1. Sepals with earlike basal lobes projecting backward toward the peduncle; stamens free from one another, though the anthers are somewhat connivent, the 2 lower ones spurred .... *Viola*
1. Sepals without such lobes; stamens united into a sheath surrounding the pistil, none of them spurred .... *Hybanthus*

HYBÁNTHUS Jacq.   Green violet

NL., from Gr. *hybos*, hump-backed, + *anthos*, flower.

Erect, leafy-stemmed herbaceous perennials. Leaves entire, sessile or nearly so. Flowers greenish-white, small, few, borne singly on drooping axillary pedicels. Sepals lacking basal auricles (earlike lobes). Petals about the same length, the lowermost pouch-like at the base. Stamens united into a sheath around the pistil, none of them spurred. Style hooked at the tip.

Two species in Kansas.

Key to Species:

1. Plants 4–8 (–10) dm tall; leaves oblanceolate or elliptic, 6.5–15.5 cm long and 1.5–5.5 cm wide; extreme east and southeast Kansas .... *H. concolor*
1. Plants 1–3 dm tall; leaves linear to narrowly elliptic or oblanceolate, 1–5.5 cm long and 1–8 mm wide; capsules globose or ovoid, 4–6 mm long; plants from farther west .... *H. verticillatus*

H. cóncolor (T. F. Forst) Spreng.

L. *concolor*, colored uniformly.
Greenish-white; late April–May
Rocky, wooded hillsides
Extreme east and southeast

Stems 1 to several, unbranched, 4–8 dm tall, more or less hairy, arising from a rhizome with a mass of fibrous roots. Leaves petioled below, sessile above, the blades oblanceolate or elliptic 6.5–15.5 cm long and 1.5–5.5 cm wide, rather thin, minutely and sparsely hairy; apex abruptly acuminate; base wedge-shaped or acuminate; stipules narrowly elliptic to linear, 4–20 (–30) mm long and 0.5–4 mm wide. Flowers greenish-white, 3.5–5 mm long. Calyx lobes linear or awl-shaped. Capsules green, oblong to ellipsoid, 1.5–2 cm long. Seeds cream-colored, smooth, nearly spherical to broadly ovoid, 4–5 mm long.

H. verticillàtus (Ort.) Baill.

L. *verticillatus*, whorled—the prominent stipules much resembling the leaves and giving the appearance of a whorled leaf arrangement at each node.
White; mid-May–June
Prairie pastures, grasslands
Logan, Finney, and Meade counties east to Pottawatomie, Wilson, and Montgomery counties; more common southward

Plants with severals stems, 1–3 dm tall. Leaves sessile, linear to narrowly elliptic or narrowly oblanceolate in ours, 1–5.5 cm long and 1–8 mm wide; stipules resembling the leaves, 4–30 mm long and 0.5–2 mm wide. Flowers white, sometimes tinged with purple, 3–5 mm long. Calyx lobes lanceolate. Capsules straw-colored at maturity, globose to broadly ovoid, slightly 3-angled, 4–6 mm long. Seeds eventually becoming glossy black, ovate to nearly round in outline, slightly flattened, 2–2.5 mm long.

VIÒLA L.   Violet

The classical name for the violet.

Annual, biennial, or perennial herbs. Leaves all basal, or alternate on an above-ground stem, mostly long-petioled and cordate or triangular in outline, entire or deeply palmately divided. Flowers yellow or various shades of purple or blue to white, borne singly on long, apically reflexed peduncles which are naked except for a pair of small bracts near the middle. Sepals separate, with earlike basal lobes or "auricles" projecting backward toward the peduncle. Petals definitely unequal in length, the lateral 2 and sometimes the lowermost with a cluster of hairs at the corolla throat and therefore said to be "bearded"; all the petals usually with dark-purple lines, probably nectar guides, radiating from the corolla throat. Anthers converging around the style but not actually united, shedding pollen inwardly, the 2 lowermost stamens with spurlike nectar glands projecting backward into the prominently spurred nectar receptacle of the lowest petal. Seeds usually ovoid, smooth, usually with a membranous caruncle near the acute end.

Twelve species in Kansas.

This large genus (about 450 species) exhibits great diversity in vegetative characters, but—except for the cultivated pansy, whose large, flat corolla might distract attention from the basic violet characters—the flowers differ only slightly from species to species.

The reproductive behavior of violets is quite interesting. The scalelike tips of the stamen filaments project beyond the anthers and overlap snugly around the style, thus preventing the dry pollen from dropping onto the stigma and effecting self-pollination. How then does pollen ever reach the stigma? Christian Konrad Sprengel (35), a major contributor to the field of pollination ecology, demonstrated the answer. Holding a flower above the level of his eyes, to keep it in its natural pose while looking at it from below, he lifted the end of the style with a splinter, causing the scalelike tips of the anthers to be separated and allowing the imprisoned pollen to fall out. From this he concluded that a bee or butterfly standing on the lowest petal, facing the throat of the corolla, and reaching with its proboscis into the nectar receptacle would certainly lift the style and cause pollen to fall out over the proboscis and eventually to be transferred to the stigmas of other flowers.

The showy flowers described above are produced mostly during the spring, and in spite of their complex mechanism for cross-pollination frequently bear no seeds at all. During the summer, the plants produce inconspicuous, reduced flowers consisting of 5 sepals, 2 rudimentary petals, 2 functional stamens, and a pistil. These flowers are cleistogamous (from Gr. *kleistos*, closed, + *gamos*, sexual union)—that is, they never open. The pollen germinates while in the anther, and the pollen tubes grow into the ovary cavity, delivering sperm into the ovules, whose eggs are then fertilized, frequently producing more seeds by self-pollination than the showy flowers do by cross-pollination. Only *V. pedata* lacks cleistogamous flowers.

Hybrids between various of our species are occasionally found.

NOTE: For all perennial violets, but especially for our stemless blue violets with uncut leaves, it is best to collect plants with "mature" leaves, in May or later. The earliest leaves are usually smaller and differ somewhat in shape from the later leaves which are used for descriptions and keys.

Key to Species:
1. Plants annual .... *V. rafinesquii*
1. Plants perennial 2
  2. Leaves borne on erect, aboveground stems; flowers yellow in ours 3
    3. Leaf blades ovate to lanceolate or elliptic, nearly entire, borne all along the stem; plants primarily of the northwest fourth .... *V. nuttallii*
    3. Leaf blades heart-shaped, usually serrate or crenate, mostly borne some distance above the base of the stem; plants primarily of the east third .... *V. pubescens*
  2. Leaves arising directly from an underground rhizome; flowers various shades of purple or blue (rarely white) in ours 4
    4. Largest leaves cordate or reniform, the margins merely serrate or crenate 5
      5. Leaf blades *typically* moderately to densely hairy on 1 or both surfaces (occasionally nearly glabrate); sepals ciliate below the middle .... *V. sororia*
      5. Leaf blades typically quite glabrous 6
        6. Leaf blades with a somewhat attenuate apex which bears fewer, more widely spaced teeth than the rest of the margin; sepals not ciliate .... *V. missouriensis*
        6. Leaf blade uniformly toothed to the acute apex; sepal ciliate below the middle or not so .... *V. pratincola*
    4. Largest leaves deeply toothed, lobes cut, or incised, or if merely crenate or serrate then the blades lanceolate or oblong-lanceolate 7
      7. Leaves lanceolate to oblong-lanceolate, longer than wide, the base cordate or sagittate, the margin *typically* more coarsely and deeply toothed toward the base; Allen and Bourbon counties south .... *V. sagittata*
      7. Leaves deeply palmately lobed, cut, incised, or dissected 8
        8. Petals all the same color, arranged in a partially closed fashion somewhat like the blossoms of a sweet pea; anthers not protruding conspicuously ...... *V. pedatifida*
        8. Petals all the same color or the upper 2 dark purple and the lower 3 lavender, arranged in an open, rather flat manner somewhat like those of a cultivated pansy; anthers orange, protruding conspicuously .... *V. pedata*

V. missouriénsis Greene — Missouri violet

"Of Missouri."
Light blue or purple; mid-April–mid-May
Mostly along moist, wooded creek banks, occasionally along roadsides or in prairie
East half, primarily; scattered records from the west half

Herbaceous perennials, mostly 10–23 (–28) cm tall, with flowers and leaves arising directly from a thick rhizome. Blades of the main leaves triangular or less frequently cordate, 2.5–5 (–10) cm long and 3–6 (–10) cm wide; apex acuminate; base truncate to cordate or broadly sagittate; marginal teeth usually more widely spaced toward the apex. Herbage entirely glabrous. Flowering peduncles about as long as the petioles. Sepals lanceolate to lance-linear or linear-oblong, 6–8 mm long, the margins not ciliate. Petals light blue or purple, 14–21 mm long, the lateral 2 bearded. Cleistogamous flowers produced during the summer. Capsules greenish or straw-colored, sometimes purple-spotted, 10–12 mm long. Seeds about 2 mm long.

V. nuttállii Pursh

Named after Thomas Nuttall (1786–1859), English-American naturalist.
Yellow; late April–May
Prairie hillsides, usually in rocky, calcareous soil
Northwest fourth, primarily

Glabrous herbaceous perennials, 0.8–3 dm tall, with leaves and flowers borne on erect, aboveground stems, the leaves borne all along the stem. Leaf blades ovate to lanceolate or elliptic, 3–7 cm long and 1.5–2.5 mm wide, the margins entire, with a few widely spaced teeth, or merely uneven; stipules lanceolate to lance-attenuate and membranous at the lower nodes, becoming oblong or elliptic and leaflike at the upper nodes, 3–15 mm long. Flowering peduncles longer than the petiole of the sudtending leaf. Sepals lance-attenuate to awl-shaped, 6–7 mm long. Petals yellow, 12–15 mm long. Capsules ovoid to short-elliptic, 6–8 mm long. Seeds reddish-brown or straw-colored, about 2.5 mm long.

V. pedàta L. — Bird's-foot violet

L. *pedatus*, shaped like a foot—the deeply dissected leaves resembling a bird's foot (a characteristic not limited to this species, however).
Entirely lavender, or 3 petals lavender and 2 dark purple; March–May
Hillsides in prairies or open woods, often in sandy, acid soil
Southeast eighth plus McPherson, Douglas, and Franklin counties

Herbaceous perennials, 0.9–2 dm tall, with a short, erect rhizome, the plants resembling the smaller-flowered *V. pedatifida*. Leaves all basal, the blades of the main leaves 2–4 cm long, deeply palmately divided into linear or narrowly elliptic segments, these sometimes further divided or with a few teeth; blades of the earliest or lowermost leaves smaller and less deeply divided, the segments broader. Flowers fragrant, the peduncles as tall as or taller than the leaves. Sepals linear to linear-obong or awl-shaped, 9–13 (–15) mm long. Petals assuming a wide-open, flattened position reminiscent of the cultivated pansy, all lavender in var. *lineariloba* DC., or the upper 2 dark purple and the lower 3 lavender in var. *pedata*, beardless, the lowermost 16–27 mm long. Anthers and their terminal appendages orange, protruding conspicuously from the corolla throat. Capsules green, smooth, 8–12 mm long (excluding the style, which may persist awhile). Seeds about 1.5 mm long, ovoid, the broad end truncate.

This is our only violet which does not produce cleistogamous flowers.

According to Jacob Bigelow (36), "The violets are generally mucilaginous plants, and employed as demulcents in catarrh and strangury. Some of them . . . contain *emetin* in their substance. The *Viola pedata*, a native species retained in the Pharmacopoeia [from 1820–1870], is considered a useful expectorant and lubricating medicine in pulmonary complaints, and is given in syrup or decoction." The roots and aboveground parts have been used as a mild laxative and to induce vomiting (30).

See plates *56* and *57*.

V. pedatífida G. Don — Prairie violet

NL., from L. *pedatus*, from *pes, pedis*, foot, + *fidus* (from *findere*, to cleave), cleft—the leaves being cleft like a bird's foot (a characteristic not limited to this species).
Purple; mid-April–May (rarely again in the fall)
Prairie, rarely in open woods
East two-fifths, especially the east fourth

305. *Viola pedata* var. *pedata*

Herbaceous perennials, mostly 0.5–2 dm tall, with flowers and leaves arising directly from a short, diagonal or nearly vertical rhizome, resembling the larger-flowered *V. pedata*. Blades of the main leaves 1.5–5.5 cm long, and 1.7–7.5 cm wide, deeply palmately divided into linear, narrowly elliptic, or linear-oblong segments, these sometimes further divided or with a few teeth; blades of the earliest or lowermost leaves smaller and/or less deeply divided. Flowering peduncles usually as tall as or taller than the leaves. Sepals lanceolate to lance-oblong or linear-lanceolate, 5–10 mm long. Petals purple, the lower 3 bearded, the lowermost 10–22 mm long. Anthers not protruding conspicuously. Cleistogamous flowers produced during the summer. Capsules greenish or straw-colored, smooth, ellipsoid, 10–13 mm long. Seeds yellowish or tan, about 2 mm long.

306. *Viola pedatifida*

V. pratíncola Greene   Common blue violet, meadow violet

NL., from L. *pratum*, meadow + *incola*, inhabitant.
Blue or purplish; April–May (rarely again in the fall)
Prairie, roadsides, open woods
East three-fourths, primarily; a few records from the west fourth

Herbaceous perennials, mostly 7–27 cm tall, with flowers and leaves arising directly from a thick rhizome. Blades of the main leaves cordate to somewhat triangular, 3.5–7 cm long and 3.5–8 cm wide; apex acuminate; base sagittate or cordate; marginal teeth evenly spaced or more distantly spaced toward the apex. Herbage entirely glabrous. Earliest flowers on peduncles reaching above the leaves, later flowers on peduncles shorter than to only as long as the petioles. Sepals broadly lanceolate or lance-acuminate to lanceolate or lance-oblong, 6–8 mm long, the auricles and lower margin ciliate or glabrous. Petals blue or purplish, 13–20 mm long, the lateral 2 bearded, the lowermost sometimes so. Cleistogamous flowers produced during the spring and summer. Capsules greenish or straw-colored, 9–12 mm long. Seeds black, or gold mottled with black, about 2 mm long.

307. *Viola pratincola*

Our common blue violet is listed in earlier manuals under the name *V. papilionacea* Pursh. The specimen upon which that name was based, however, proved to be an individual of an already-named species, *V. sororia*. Although many manuals adopted the use of the former name, it now appears that there really is no such plant (37), and our plants have been going by the wrong name for many years. In addition, our common blue violet is very closely related to 2 other of our species—*V. sororia* and *V. missouriensis*—and the 3 are, for several reasons, difficult for the inexperienced to distinguish. First, the differences among them are few and slight and have to do primarily with hairiness of the plants and leaf size and shape. Second, leaf shapes and sizes usually vary on a given plant between the earliest leaves to those produced in May or later, and leaves produced in late summer or fall may be different yet. Third, the immediate habitat of a plant may influence the vegetative growth produced. For example, a small short plant with few leaves, growing in a bright, dry, exposed site, may produce many large leaves when transplanted to a moist shady location. Fourth, it is not uncommon to find plants which are intermediate in several characters between any 2 or these 3 species. These intermediates may be hybrids, or it simply may be that there are not actually 3 separate species but only 1 which exhibits a great deal of variation. Before we can understand this group of violets as well as we would like, it may be necessary for some ambitious person to undertake a thorough study of the effect of habitat differences upon vegetative growth and the progression of leaf shapes through the season. More information about fruits and seeds and controlled hybridization experiments might also be helpful.

See plate *58*.

V. pubéscens Ait.

308. *Viola pubescens*

NL., *pubescens*, downy, hairy, from L. *pubes*, *pubis*, the down or hair of adulthood and maturity.
Yellow; April–May
Moist, rich soil of alluvial woods
East third, primarily

Herbaceous perennials, 0.7–3.5 dm tall, with leaves and flowers borne on erect, aboveground stems. Except for a few long-petioled basal leaves, most of the leaves are borne quite some distance above the base of the stem. Leaf blades cordate, 2–6 cm long and 3.5–7.5 cm wide, the margins usually serrate or crenate; stipules ovate or lanceolate, oblique, rarely awl-shaped, (4–) 7–17 mm long. Herbage usually nearly glabrous in ours, occasionally sparsely to moderately hairy. Flowers borne on axillary peduncles longer than the petiole of the subtending leaf. Sepals narrowly lanceolate to lance-oblong or awl-shaped, 6–8 mm long. Petals yellow, 10–15 mm long. Capsules ellipsoid, sometimes broader toward the base, 7–10 mm long, glabrous or woolly. Seeds straw-colored, truncate at the broad end, 2–2.5 mm long.

Most of our plants are recognized as *V. pubescens* var. *eriocarpa* (Schwein.) Russell. The more typically hairy plants, var. *pubescens*, tend to occur in the northeastern states and Great Lakes area.

A tea made from the roots of this plant was used by the Ojibwas to treat sore throats and by the Potawatomi Indians as a heart medicine. Whites used an extract of the roots as an emetic and alterative (38).

See plate *59*.

309. *Viola rafinesquii*

V. rafinésquii Greene   Field pansy, Johnny-jump-up

Named after Samuel Constantine Rafinesque-Schmaltz (1783–1840), Sicilian botanist who came to this country in 1814, wrote floras of various regions, a monograph on roses in North America, and *Medicinal Flora of the U.S.A.*
Pale purple; March–June
Prairie, open woods, often in disturbed habitats such as roadsides, pastures, waste grounds, or in poor soil
East three-fifths, primarily; a few records from the northwest

Small branching annual, 4.5–27 cm tall. Lower leaves distinctly petioled, the blades broadly ovate to round or somewhat reniform, 5–12 mm long and 5–13 mm wide, the margins with a few teeth; upper leaves gradually becoming less distinctly petioled, the blades broadly elliptic to oblanceolate or spatulate, 7–25 mm long and 3–15 mm wide, margins with 1 or 2 pairs of teeth or the uppermost entire; stipules conspicuous, deeply palmately or pectinately divided. Capsules straw-colored or greenish at maturity, 4–6 mm long. Seeds smooth, oblong or ellipsoid, about 1.2–1.5 mm long.

The vernacular name "Johnny-jump-up" is well bestowed upon this small plant that is up from its seedbed in the early morning of spring and is blooming before April is past. The name "field pansy" is also a good one, for this species is not far from being a replica in miniature of the European *Viola tricolor* (from L. *tri*, 3 + *color*, color, in reference to the 3 colors discernible in the flowers—yellow, blue, and red—the last 2 appearing in combination as purple). The pansy of our gardens is the product of selection by plant breeders of variants of *V. tricolor* that have obtained twice the number of chromosomes normal to the wild violets.

310. *Viola rafinesquii* fruiting

V. sagittàta Ait.

NL., from L. *sagitta*, arrow, in reference to the shape of the leaves.
Purple or blue; April–May
Prairie hay meadows, rarely in open woods
Southeast 6 counties, from Allen and Bourbon counties south

Herbaceous perennials, mostly 5–17 cm tall, with flowers and leaves arising directly from a short, more or less horizontal rhizome. Blades of the main leaves narrowly cordate or sagittate or sometimes oblong, 2–5.5 cm long and 1.5–2.5 cm wide, apex rounded or acute, base cordate or sagittate, margins serrate or crenate, often more deeply toothed or incised toward the base; blades of the earliest leaves usually smaller, the base often truncate. Flowering peduncles as tall as or taller than the leaves. Sepals lanceolate to linear-lanceolate or linear-oblong, 6–10 mm long. Petals purple or blue, all bearded, the lowermost 13–20 mm long. Anthers protruding conspicuously. Cleistogamous flowers produced during the summer. Capsules greenish or straw-colored, smooth, ellipsoid, about 11 mm long.

This species frequently hybridizes with other species in nature.

311. *Viola sagittata*

V. soròria Willd.   Sister violet, downy blue violet

NL., from L. *sororius*, sisterly, from *soror*, sister—suggested by the likeness between this and other species in size and form and their convergence by intergrading variants.
Lavender to dark purple; April–July
Woods, usually in mostly rich soil, occasionally in prairie
East fourth

Herbaceous perennials, mostly 5–23 cm tall, with flowers and leaves arising directly from a thick rhizome. Blades of the main leaves usually cordate, sometimes verging on triangular, 3–8 (–9) cm long and 4–7 (–10) cm wide; apex acute or acuminate; base cordate or sagittate; margins evenly toothed. Petioles and lower leaf surface moderately hairy to very sparsely hairy or nearly glabrous and the petioles merely ciliate. Flowering peduncles mostly reaching above the leaves. Sepals lanceolate to lance-oblong, 6–8 mm long, the margins ciliate below the middle and on the auricles. Petals various shades of lavender to reddish-purple or dark purple, 16–20 mm long, the lateral 2 bearded. Cleistogamous flowers produced during the summer. Capsules purplish or purple-spotted, about 7–10 mm long. Seeds cream mottled with rust or sometimes very dark, about 2 mm long.

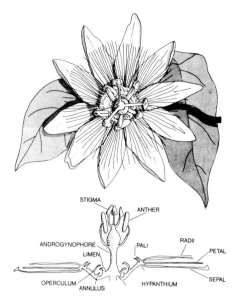

312. Flower structure in the passionflower

PASSIFLORÀCEAE   Passionflower Family

Ours herbaceous perennial vines climbing by means of coiling axillary tendrils. Leaves alternate, simple or compound, lobed or unlobed, petiolate, stipulate. Flowers bisexual, regular, usually axillary and borne in pairs. Sepals 4–5, free from one another or fused basally, often petal-like. Petals 4–5 or absent, sometimes smaller than the sepals, free from one another or fused slightly at the base; sepals and petals both arising from a fleshy hypanthium. Pistil 1, often elevated on a small stalk (gynophore); ovary superior, 1-celled; styles 3–5; stigmas 3–5, capitate, clavate, or discoid. Stamens 4–5, the filaments united around and sometimes fused to the gynophore, the compound stalk thus formed being called an androgynophore. Arising from the hypanthium are several additional series of parts which serve as nectar guides and nectar guards; these vary in exact structure among the various genera and species of the family and are discussed in more detail under the following genus. Fruit a capsular berry.

A family of approximately 12 genera and 60 species, primarily of the American tropics. More than 25 species of *Passiflora* are cultivated for their edible fruits and their attractive flowers and foliage. Some tropical genera are poisonous and have been used to stupefy fish or to treat poison darts. Others have been used medicinally.

One genus in Kansas.

PASSIFLÒRA L.   Passionflower

NL., from L. *passio*, suffering, in the religious sense referring to the sufferings of Christ, + *flos*, *floris*, flower. The Spanish physician N. Monardes suggested the name *Flos passiones* (1593), affirming that one could see in this flower symbols of the scourging and crucifixion of Christ—a concept discussed in more detail below.

Mostly herbaceous or woody perennial vines, usually climbing by tendrils. Leaves simple, usually palmately lobed, the petioles often glandular. Sepals 5, petal-like (in ours). Petals 5. Between the perianth and stamens there arise several series of parts which collectively are referred to as the corona (from L. *corona*, crown). In precise detail, the parts of the corona differ considerably among the various species of *Passiflora*, but in general their nature is as shown in fig. 312 and as described below. Outermost are 1 or 2 series of narrow or even threadlike structures, the radii (also referred to in some manuals as the outer corona), which are petal-like in texture and sometimes transversely colored in different shades. These tend to be widely spreading or even reflexed and serve as nectar guides. Next come 1 or more series of usually smaller, shorter, filamentous structures, the pali (or inner corona). These tend to be erect or, as the blossom is opening, inclined toward the androgynophore and thereby serve to protect the nectar below from dilution by rainwater or dew. Below the pali is a usually delicate and much-folded membrane, the operculum, which is also inclined toward the androgynophore. Internal anatomy indicates that the radii, pali, and operculum all evolved as outgrowths from the inner surfaces of the sepals and petals. Below the operculum is a ringlike, more or less well developed outgrowth from the inner surface of the hypanthium called the annulus. In the trough or groove below the annulus (or below the operculum, if an annulus is not evident) and at the base of the androgynophore, glandular epidermal cells produce nectar to attract pollinators. The innermost part of the corona is the limen, a more or less well developed, ringlike or shelflike structure which projects outward from the base of the androgynophore and which, along with the operculum (and the annulus, if it is well-enough developed), serves to protect the nectar, apparently either from dilution or from nonpollinating nectar thieves. Stamens and pistil as described for the family; anthers and stigmas quite large. Fruit berrylike, fleshy, indehiscent, often edible.

The unusual structure of the blossoms has given rise to an extensive collection of folklore. One interpretation of the flower parts was given by the Italian J. B. Ferrari in his book *On the Culture of Flowers* (39): "This flower is a miracle for all future time; God, in His love, has therein with His own hand symbolized the sufferings of Christ; calyx divisions are tipped with thorns typifying the crown of thorns; the innocence of the Saviour is manifest in the whiteness of the petals; the thready crown of the nectary calls

313. *Passiflora incarnata*

to mind His rent garments; the column in the middle of the flower represents the reed, and the pistil on the end of the column is the sponge filled with vinegar; the 3 styles are the 3 nails; the 5 anthers are the 5 wounds; the 3-lobed leaves are the spear; the tendrils are the scourge; only the cross is lacking because gentle and merciful nature would not permit representation of the ultimate agony."

Another interpretation (40) suggests that "the bud of the flower represents the Eucharist; the sepals and the petals, the Apostles of Christ minus Peter and Judas; the corona, the halo or the crown of thorns; the 5 stamens, the hammers or wounds; the 3 styles, the nails; and the 3 bracts on the peduncle, the Trinity. In addition, the half-opened flower is said to represent the star of the East, and the long roots descending into the earth symbolize the triumph over hell and signify Christ's rebirth. Some species reportedly have 5 red spots on the corolla segments, the former being referred to as blood from the 5 wounds. The lower leaf surfaces of some species have round spots interpreted by some believers as representing the 30 pieces of silver for which Christ was betrayed."

Contrary to Ferrari's assertion that the cross is not represented, the stems of some *Passiflora* species are somewhat X- or cross-shaped in cross section.

Concerning the flower, Christian Konrad Sprengel, in *The Discovered Secret of Nature in the Structure and Fertilization of Flowers* (35) portrays it as a miracle in itself, unclouded by symbolism, bearing evidence of a rational ordering of nature manifest in various ways: its size and color making it conspicuous to insects at a distance; the radiating filamentous nectar-guides by which an insect is directed straightway to the nectar located at the bottom of the hypanthium in a ring-form trough, and well protected from rain by an overhanging rim of the trough, and above this, all the way round, by a canopy of parallel filaments with spaces narrow enough to exclude rain, while permitting the entrance of a bumblebee's proboscis; then, in a freshly opened flower, the recurving of the filaments so that the anthers hang down where their pollen would be brushed off on the bee's back; later the styles bending down lower than the anthers, where the stigmas would receive pollen from the back of a bee coming from younger flowers, and so securing cross-pollination; the bee, too, a miracle matching the flower.

The carpenter bee has been found to be the main pollinator of *P. edulis* Sims., and some passionflowers are pollinated by hummingbirds or other nectar-garthering birds.

Two species in Kansas.

Key to Species:

1. Leaves deeply 3-lobed, the lobes acuminate, minutely toothed; flowers purple or pink, about 4–6 cm wide; fruits yellow, up to 5 cm long .... *P. incarnata*
1. Leaves shallowly 3-lobed, the lobes broadly rounded, not toothed; flowers greenish-yellow, 2–2.5 cm wide; fruits purple, about 1.5 cm long .... *P. lutea*

P. incarnàta L.   Maypop passionflower

L. *incarnatus*, made flesh, flesh-colored—in reference to the corona.
White with a purple or pink corona; July–October
Roadside, thickets, old fields, rocky slopes
Cherokee, Crawford, Bourbon, Montgomery, Woodson, Cowley, and Wyandotte counties

Stems up to 8 m long. Leaf blades deeply 3-lobed, (4–) 6–11 (–15) cm long and 6–14 cm wide, the margins finely toothed; petioles 2–5 (–8) cm long, with 2 conspicuous glands near the apex or at least above the middle; stipules bristlelike, inconspicuous. Flowers 5.5–8.5 cm wide, borne singly on long pedicels in the leaf axils, subtended by 3 small leaflike bracts, each bract with 2 conspicuous glands. Hypanthium short-campanulate. Sepals lance-oblong, greenish, white or pale lavender (often drying bluish), somewhat keeled, the apex hoodlike and with a short awn, 2–3.5 cm long (excluding the awn). Petals more or less oblong, white or pale lavender, about the same length as the sepals. Radii and pali purple or pink. Stamens 5, the filaments united around the gynophore below, free above. Style branches and stigmas 3. Fruits yellow, ovoid or nearly globose, about 5 cm wide, edible. Seeds reddish-brown, obovate, somewhat flattened, deeply pitted, 6–7 mm long.

This plant has had various medicinal uses. An extract of the plant acts as a depressant on the motor nerves of the spinal cord, and in the late 1800s it was used to relieve insomnia, to soothe nerves, and to treat epilepsy. The dried flowering tops were listed officially in the *National Formulary* from 1916 to 1936 and have been used in preparations to treat hemorrhoids, burns, and skin eruptions. The plant is reputed to be an aphrodisiac, and in Bermuda it is used as a perfume base. The Houma Indians used the pulverized roots in water as a systemic tonic, and other Indian groups used the plants to induce vomiting or to treat swellings or sore eyes.

P. lùtea L. Yellow passionflower

L. *luteus*, golden-yellow, the 3 series of corona threads being yellowish.
White to greenish-yellow; mid-June–August
Rocky, wooded creek banks and bluffs
Montgomery, Labette, Cherokee, and Crawford counties

Stems up to 3 m long. Leaf blades more or less bean-shaped in general outline, with 3 shallow (rarely lobed halfway to the leaf base in ours) rounded lobes, (2–) 3–8 cm long and 4–12 cm wide, the margins not toothed; petioles (1–) 2–4.5 cm long, glandless. Flowers 1–2 cm wide, usually in pairs in the leaf axils, on slender pedicels lacking bracts. Hypanthium very short, bowl-shaped. Sepals pale green, linear-oblong to lance-oblong, 0.5–1 cm long. Petals white to greenish-yellow, linear about 0.5 cm long. Radii greenish-white. Pali white tinged with pink at the base. Stamens 5, the filaments united around the gynophore below, free above. Style branches and stigmas 3. Fruits purple, ovoid to nearly globose, about 1 cm wide. Seeds pinkish-brown, more or less round to obovate, somewhat flattened, about 4.5 mm long, the base acute, the apex contracted into a small point, the surface ornamented with 6–9 transverse ridges, these with smaller vertical ridges.

LOASÀCEAE Loasa Family

From the genus *Loasa*, a South American name.

Annual, biennial, or perennial, erect, spreading, or climbing herbs, beset with barbed, sticky, or stinging hairs (none of ours climbing or with stinging hairs as exhibited in the South American *Loasa*). Leaves alternate or opposite (ours alternate), sinuate-margined, lobed, or pinnatifid, sessile or petiolate, estipulate. Flowers bisexual, regular, solitary or clustered. Sepals 4–5, persistent. Petals 5, deciduous, sometimes appearing to be 10 in number because of 5 petaloid staminodia. Stamens many, free from one another or fused together in clusters opposite the petals, the outer ones often imperfectly developed, sometimes petal-like. Pistil 1; ovary inferior or almost completely so, 1-celled with 1–5 placentae; style and stigma 1. Fruit a loculicidal capsule in ours.

The Loasa Family includes approximately 15 genera and 250 species, primarily of western South America.

One genus in Kansas.

MENTZÈLIA L. Mentzelia

Named after Christian Mentzel (1622–1701), court physician to the elector of Brandenburg and writer on botanical subjects.

Annual or perennial herbs with barber hairs. Leaves entire or lobed, sessile or petiolate. Flowers borne singly in the leaf axils or several in terminal clusters. Perianth as described for the family. Stamens few to many, the filaments threadlike or petaloid; petaloid staminodes present or lacking. Ovary with 3–5 placentae and few to many ovules; style(s) threadlike, stigmas 3, minute. Capsules dehiscing at the apex.

Four species in Kansas. *M. albescens* (Gill. & Arn.) Griseb. is known from Cherokee County only, where it has been collected from abandoned mining sites, and may not be naturalized in our area.

Key to Species:
1. Petals 5, cream-colored to yellow or orangish, 8–11 mm long; seeds not winged ....
.... *M. oligosperma*
1. Petals apparently 10, white to cream-colored, 2–8 cm long; seeds winged 2
   2. Petals 4–8 cm long; stamens 3–5 cm long .... *M. decapetala*
   2. Petals 2–4 cm long; stamens 0.4–2.5 cm long .... *M. nuda*

M. decapétala (Pursh) Urban & Gilg. Tenpetal mentzelia

NL., from Gr. *deka*, 10, + *petalon*, thin sheet or foil of metal—in botany applied to the petals. One of the vernacular names is "candleflower," in allusion to the daytime appearance of the closed flowers. The names "sand lily" and "chalk rose" are often used locally, but the plant belongs to the Loasa Family, which is far removed in all its internal and external characters from the Lily Family, and in respect to the flower is more like the cactus than the rose.
Cream-colored; July–October
Dry, rocky prairie hillsides and banks, frequently on seleniferous, calcareous soil
West two-fifths

314. *Mentzelia decapetala*

Stout, erect, rough-hairy herbaceous perennials with a strong, deep-set, spindle-shaped, fleshy taproot. Stem eventually branched above, 0.3–1 m tall. Leaves petioled below, sessile above, lanceolate or oblanceolate to narrowly elliptic or almost linear in general outline, with pinnately toothed or incised margins, mostly 4–12 cm long and (0.6–) 1.5–2.5 cm wide. Herbage densely covered with retrorsely barbed hairs. Calyx lobes lanceolate or long triangular, with an attenuate apex, 2–4 cm long (measured from the point of departure from the ovary). Petals apparently 10, cream-colored (drying brown), narrowly oblanceolate or spatulate, with an acuminate apex, 4–8 cm long. Stamens many, 3–5 cm long. Capsules narrowly conical to ellipsoid, 2.5–4.5 cm long, crowned by the persistent calyx lobes. Seeds light tan, ovate to broadly elliptic, flattened, minutely papillose, 3–4 mm long, the margin winged.

The fragrant flowers open about 5 P.M. and remain open until dawn, longer when the sky is overcast. The plants do not transplant readily but may be started in the garden from seed if sown where the plants are to remain.

M. nùda (Pursh) **T. & G.**

L. *nudus*, bare, naked.
White or cream-colored; July–October
Dry, sandy, gravelly or rocky soil of prairie, pastures, roadsides, or flood plains
South of a line drawn from Cheyenne County to Trego and Sedgwick counties; more common
    westward and southwestward

Erect, rough-hairy herbaceous perennial with a strong, deep, spindle-shaped taproot. Stem eventually branched above, 0.3–1 m tall. Leaves petioled and oblanceolate below to sessile and linear-oblong or lanceolate above, with coarsely serrate or dentate margins, mostly 2–9 cm long and 0.5–2 cm wide. Herbage with stout, retrorsely barbed hairs interspersed with slender hairs which have an apical whorl of retrorse barbs (use high magnification). Calyx lobes lanceolate to long-triangular, with an attenuate apex, 1.7–3 cm long (measured from the point of departure from the ovary). Petals apparently 10, white or cream-colored (usually drying brown), oblanceolate or spatulate with an acuminate apex, 2–4 cm long. Stamens many, 0.4–2.5 cm long. Capsules cylindrical or somewhat conical, 1.5–3 cm long, crowned by the persistent calyx lobes. Seeds light tan, more or less ovate, flattened, minutely papillose, 4–5 mm long, the margin with a broad membranous wing.

M. oligospérma Nutt.   Stickleaf mentzelia

NL., from Gr. *oligos*, few, + *sperma*, seed, i.e., few-seeded—the capsule having about 3 seeds.
Cream-colored to yellow or orangish; late June–early October
Dry, rocky, usually calcareous soil of prairie hillsides and ravines, roadside banks, and waste areas
Scattered nearly throughout but more common in the north-central and eastern counties

Erect, rough-hairy herbaceous perennial with an enlarged, spindle-shaped taproot. Stem openly branched, 0.3–1 m tall. Leaves sessile, mostly lanceolate to ovate or rhombic, 1–7 cm long and 0.7–3.5 cm wide, coarsely toothed, sometimes with 2 or 4 more prominent teeth or lobes toward the leaf base. Herbage densely covered with retrorsely barbed hairs. Calyx lobes awl-shaped or linear-attenuate, 4–6 mm long (measured from the point of departure from the ovary). Petals 5, cream-colored to yellow or orangish (drying light tan), obovate to oblanceolate with a rounded or acute, mucronate apex, 8–11 mm long. Stamens 25–35, 6–10 mm long. Capsules narrowly cylindrical or slightly conical, usually slightly curved, 7–13 mm long. Seeds gray, oblong or elliptic in outline, concave on 1 side, convex on the other, about 3.5–4.5 mm long, lacking a marginal wing.

The leaves, brittle stems, and fruits, with their dense covering of barbed hairs, catch on clothing or the fur of animals, thus dispersing the seeds. This is all to the good of the species but is no small nuisance to victims, for the leaves become flattened against clothing and the hooks so incorporated with its fabric as to necessitate removal by scraping with a knife; and in the case of sheep the herbage sometimes gets so enmeshed in the fleece as to lower its market value.

CACTÀCEAE   Cactus Family

Named after the Linnaean genus *Cactus*, a name superseded as a genus name, though popularly
    any plant in the family is a cactus.

Ours fleshy perennials. Stems continuous and ball-shaped or cylindric, or of flattened or cylindric sections called "joints"; the surface smooth, with prominent nipplelike projections (tubercles) or longitudinally ridged, spiny, covered with a thick, waxy epidermal layer which retards water loss. Leaves alternate, simple, cylindric or narrowly conical, fleshy, small and inconspicuous and soon falling, or entirely absent. The regular leaves and the spines, which are actually modified leaves, arise from round or elongated areas called areoles, which are complex buds. The surface of each areole is brown or blackish, hard, and rough, and is sometimes covered with white, brown, or blackish wool. At each areole, there may be several types of spines. There are large, rigid spines often subtended by much smaller, shorter bristles called glochids (the "ch" pronounced like "k"). The larger spines may be straight, curved, or hooked, with a smooth or barbed surface. The glochids are minutely barbed and, in spite of their small size, may cause more discomfort than the large spines to someone handling the cactus since they readily detach from the plant and become embedded in fingers and clothing. Flowers large and showy though sometimes greenish, bisexual, regular, usually borne singly although there may be several on a plant. Sepals and petals numerous, fused basally to form a hypanthium above the ovary, the inner sepals usually petal-like and grading into the petals. Stamens many, spirally arranged and arising from the hypanthium. Pistil 1; ovary usually inferior, 1-celled, with many ovules on several parietal placentae; style 1; stigmas as many as the placentae, radiating. Fruit a large berry; either fleshy or dry at maturity, with several to many seeds imbedded in its pulp.

A large family of about 100 genera and between 1,200 and 1,800 species, almost entirely of the New World. Numerous species of cacti are grown as ornamentals, and the fruits of some are edible.

Three genera native to Kansas.

The cacti exhibit several adaptations to life in an arid climate. Most of each plant is composed of water-storage tissue which is covered with a thickened, waterproof skin that retards, very effectively, loss of water by evaporation into the air. In addition to vertical anchoring roots, most cacti have shallow horizontal systems of long, slender, fleshy roots which quickly absorb rainwater, even from a light shower. The spines serve to protect the plants and fruit from browsing herbivores, and, in some species, at least some of the spines are deflected downward and also direct the flow of dew drops or rainwater to the base of the plant where it may be absorbed by the roots rather than remaining on the plant to evaporate later. As cacti shrink during long periods of drought, the spines of some kinds intermesh and provide an additional outer covering or barrier to water loss or evaporation by trapping a layer of humid air against the surface of the cactus and deflecting the hot, drying wind.

Key to Genera:

1. Stems composed of series of cylindrical or flattened joints, not ribbed; large spines subtended by small white glochids (barbed bristles); young stems with a slender, cylindrical or narrowly conical, fleshy leaf at each areole .... *Opuntia*
1. Stems continuous, globose to ovoid or short-cylindrical in ours, prominently tubercled or longitudinally ridged; glochids absent, though the large spines may be subtended by woolly hairs; leaves absent 2
   2. Areoles borne on prominent tubercles; flowers produced on the season's new growth at the apex of the stem .... *Coryphantha*
   2. Areoles borne on longitudinal ridges; flowers nearly always produced on areas 1 to several years old, clearly below the apex of the stem or branch .... *Echinocereus*

CORYPHANTHA (Engelm.) Lem. Nipple cactus

NL., from Gr. *koryphe*, head, top, + *anthos*, flower—because the flowers develop near the apex of the stem.

Plants cylindric or somewhat globose, the stems simple or branched, sometimes forming mounds of several hundred plants or branches. Surface with conspicuous nipplelike projections (tubercles), the tubercles separate from one another. Leaves essentially absent. Spines smooth. Areoles with 1–10 central spines which are straight, curved, hooked, or twisted, and which more or less gradually grade into the 5–40 radial spines. Flowers borne near the apex of the stem. Hypanthium funnelform. Fruit green or red, lacking spines and glochids.

Two species in Kansas. Because of a great deal of disagreement among botanists regarding the definition of a genus in the cactus family, plus the ever-changing state of our knowledge regarding the biology of this group of plants, one may find our 2 species of nipple cactus included in other manuals under various genera, including *Cactus*, *Mammillaria*, *Neomammillaria*, or *Neobesseya*.

Key to Species:

1. Flowers pink to purplish-red; areoles with usually 3–6 central spines; mature fruit green; seeds brown, reticulate .... *C. vivipara*
1. Flowers pale yellowish or straw-colored, the petals tinged with pink or dull red below; areoles lacking central spines; mature fruit red; seeds black, punctate .... *C. missouriensis*

315. *Coryphantha vivipara*

C. missouriénsis (Sweet) Britt. & Rose

NL., "of Missouri."
Pale yellowish or straw-colored; April–May (occasionally, blooming again in the fall)
Dry, rocky prairie hillsides and uplands, frequently in calcareous soil
Scattered from Norton, Trego, and Ford counties east to Wabaunsee, Wilson, and Montgomery counties

Aboveground part of stem 2–6 cm tall (including the spines). This species produces vegetative basal offshoots more freely than does *C. vivipara*, and it is not unusual to find a cluster with as many as 50–60 plants or branches. Areoles woolly, lacking central spines but usually with 10–16 radial spines. Flowers 2–4.5 cm long, the petals pale yellowish to straw-colored, often tinged beneath with pink or dull red, the sepal margins ciliate but the hypanthium not spiny. Mature fruits red, about 1–1.5 cm long. Seeds black, broadly ovoid to globose, sometimes 3-angled toward the hilum, 1.8–2.25 mm long, the surface punctate (use magnification).

C. vivípara (Nutt.) Britt. & Rose

L. *viviparus*, bringing forth its young alive, from L. *vivus*, alive, + *parere*, to bear or bring forth—this species vegetatively producing clusters of offspring around its base.
Pink to purplish-red; May–August
Dry, rocky or sandy prairie hillsides and uplands
West two-thirds; more common westward

Aboveground part of the stem 2–9 cm tall (including the spines), often with several vegetative offshoots. Areoles woolly, usually with 3–6 straight, reddish central spines spreading at various angles and 9–16 smaller, white radial spines. Flowers pink to purplish-red, 3–4 cm long, the sepal margins ciliate but the hypanthium not spiny. Mature fruit green. Seeds reddish-brown, narrowly D-shaped but narrower toward the hilum, 1.2–1.8 mm long, the reticulations of the surface rather thick so that the surface almost appears punctate.
See plate *60.*

ECHINOCÉREUS Engelm.   Hedgehog cactus

NL., from Gr. combining form *echino*, from *echinos*, hedgehog or sea urchin, + the name of the closely related genus *Cereus* (L. *cereus*, wax candle, from L. *cera*, wax)—plants of the genus *Echinocereus* having cylindric stems with longtiudinal spiny ridges, remindful of a hedgehog.

Plants cylindric, usually simple in ours, sometimes with 1 or more globose branches arising at or below ground level (often appearing to be a separate plant), with 5–14 longitudinal ribs. Leaves essentially absent. Areoles with 1–2 (–16) smooth, awl-like or somewhat flattened central spines and 3–32 radial spines which in ours are shorter and smaller than the central spine and radiate more or less at right angles to it; there are no glochids, but the young areoles are woolly. Flowers borne on old growth. Hypanthium funnelform or conical. Fruit globose or ellipsoid, with deciduous, spiny areoles.
One species in Kansas.

E. viridiflòrus Engelm.

NL., from L. *viridis*, green, + *flos, floris*, flower.
Flowers yellowish-green; May–September
Sandy, gravelly, or rocky soil of prairies slopes and uplands
West column of counties

Stems ovoid, with 10–14 ribs, the aboveground part 2–7.5 cm tall, often found with clusters of 3–4 vegetative offshoots. Areoles elliptic, usually with 1 central spine and 8–16 radial spines. Flowers yellowish-green, about 1.8–2.5 (–3.5) cm long, the hypanthium spiny. Fruit green at maturity, 6–9 mm long.

OPÚNTIA Mill.   Prickly pear, cholla

NL., from L. *opuntia*, a name used by Pliny for some plant whose identity we do not know now.

Plants with a series of cylindrical or flattened and padlike joints without ribs, the areoles borne on elongate tubercles in the treelike chollas (pronounced "choy yas") or on a flat stem surface in the prickly pears. Leaves cylindrical or narrowly conical, succulent, soon falling. Areoles usually with 1–10 spines. Flowers arising within spiny areoles. Hypanthium short, deciduous. Fruit fleshy or dry at maturity, often spiny.

The opuntias, though the most widely distributed group of cacti in the New World, are not well understood botanically. They display an amazing amount of variability in spininess; shape, color, and size of the fruits; and flower and seed size. In some cases, the variations seem to be genetically determined; in others they seem to be caused by environmental differences. For example, cacti growing in shaded areas tend to produce narrower and thinner joints, have fewer and shorter spines and larger flowers and fruit, and to be less highly colored (41). These variations have prompted botanists to erect many new species based on plants which were merely environmental variants or growth forms. This in turn led to a great deal of confusion regarding the correct names to be used in any geographical region, including Kansas. The occurrence of hybrids between various species has further clouded the issue.

Five species occur in Kansas, only 3 of which are very common. *Opuntia imbricata* (Nutt.) Haw., our only treelike cactus, is known from scattered sites in the west half and is sometimes planted near homes as an ornamental. *O. phaeacantha* Engelm., a prickly pear, is known from Morton County. It differs from the common *O. macrorhiza* in having at least some of the spines at each areole flattened at the base. There are some cactus colonies in western Kansas which appear to consist largely of hybrids between *O. macrorhiza* and *O. phaeacantha*, and in our eastern counties some specimens of *O. macrorhiza* exhibit some of the characters of a more eastern species, *O. compressa*. The group needs more study by cactus specialists before it can be thoroughly understood, and the beginning botanist should feel neither surprised nor discouraged should he find a prickly pear which does not fit nicely into any one of the following descriptions.

The prickly pears have been and are still used for food more frequently than any other cacti. According to A. C. Martin *et al.* (22), the stems, fruits, and seeds form an appreciable part of the diet of at least 44 kinds of birds and mammals. Ranchers sometimes singe the spines and glochids from the plants to provide forage for livestock. Wood rats may even use a large clump of *Opuntia* as a naturally fortified den site.

In the past, Indians of the Southwest gathered the fruits of fleshy-fruited prickly pears and dried them for use during winter months. The seeds were ground and also used. Today, prickly pear jelly, made from the fruits, is a regional delicacy offered by hostesses in New Mexico and Arizona. The fruits may also be used in preserves and conserves, and the joints, called *nopales* in Mexico, may be sliced or chopped and used in scrambled eggs, relishes, stew, beans, casseroles, and other dishes. The following recipe for prickly pear jelly is borrowed from H. A. Stephens (41): Remove glochids from fruits. Put fruits in kettle, cover them with water, crush, and cook them until tender. Pour off juice and save it. Again cover the fruits with water, cook, and reserve juice. A third cooking may be necessary. To each 2½ cups of juice, add ¾ cup water, ¼ cup lemon juice, 5 cups sugar, and 1 package pectin. Follow directions on the pectin box.

It is not uncommon to find a clammy white fuzz on the joints of prickly pear-type *Opuntias*. The fuzz is a protective covering manufactured by a scale insect, *Dactylopius coccus* Costa, which feeds on the cactus juices. The mature female insects are the source of a reddish-pink or magenta dye called cochineal produced by the natives of Mexico. Seventy thousand insect bodies are required to make 1 pound of the dye.

There are a few records of medicinal uses of *Opuntia* species. The Blackfoot Indians removed warts and moles by lacerating these growths and rubbing glochids (or cochineal "fuzz"?) into them. This tribe also used the peeled stem joints to dress wounds. Nineteenth-century physicians used baked joints for this same purpose and also treated pulmonary complaints with a drink made from boiled juice.

Key to Species:
1. Fruit at maturity usually red or reddish-purple (less frequently green or salmon-colored), fleshy and more or less juicy, usually with glochids but lacking spines; throughout Kansas .... *O. macrorhiza*
1. Fruit initially green or reddish but becoming tan and dry as the seeds mature, with both glochids and spines 2
   2. Plants of small stature, the joints short-cylindric or almost ovoid, 25–38 mm long; joints produced by this season's growth readily detached after flowering; scattered in the west half .... *O. fragilis*
   2. Plants larger, the joints obovate or nearly round, definitely flattened, 50–125 mm long, none readily detached; west fourth .... *O. polyacantha*

316. *Opuntia macrorhiza*

O. frágilis Nutt.   **Little prickly pear, fragile prickly pear**

L. *fragilis*, easily broken, brittle—in reference to the ease with which some of the joints detach
    from the plant.
Flowers green or yellowish; June–July
Sandy or gravelly soil of prairie hillsides and uplands and pastures
Scattered in the west half

A low, mat-forming plant, up to 1.5 dm tall, of much smaller stature than our
other prickly pears. Stems with 1 or more short-cylindric to obovoid or somewhat padlike
joints 2–4.5 cm long and 1–3 cm wide, appearing almost cylindrical in small plants.
Areoles woolly when young, later with a few short, yellowish glochids and (1–) 4–7 (–10)
smooth, dark-brown to straw-colored, needlelike spines. Flowers rarely produced. Petals
yellow or greenish 1.5–2 cm long. Fruits ovoid, dry when mature, 1.2–2.5 cm long, with
both glochids and spines. Seeds light tan, more or less circular in outline, about 7 mm
long, the margin pinkish but scarcely thinner than the body of the seed.

In late summer and fall, the season's new joints have matured and will detach
easily from the plant when brushed by a passing animal or person. If caught in fur or
clothing, they may be carried some distance before dropping or being thrown to the
ground, where they may root and start a new plant.

O. macrorhìza Engelm.   **Common prickly pear**

NL., from Gr. *makros*, long, large, + *rhiza*, root—the main roots of this species often being
    enlarged and tuberous.
Yellow, occasionally dull reddish-yellow or copper-colored; May–October
Dry soil of prairies, pastures, and roadsides
Nearly throughout; more common westward

Plants usually low and spreading, branched, each branch composed of a series of
1–3 (–5) flattened, padlike joints, each of which is obovate to broadly elliptic or nearly
round in outline, (5–) 8–12 cm long and 5–10 cm wide. Where prostrate joints rest on
the ground, they may take root. In this manner, a single plant may eventually produce
a clump 3–7 m across. Areoles with a dense tuft of reddish-brown glochids and, the upper
ones at least, with white or gray, needlelike spines 0.5–3.5 cm long, many of which are
angled downward. (Plants growing in moist, shaded situations may lack needlelike
spines.) Petals yellow, often red at the base, occasionally entirely dull reddish-yellow or
copper-colored, wedge-shaped or narrowly obcordate, 2.3–4 cm long, the apex notched or
truncate and greenish or salmon-colored, obovoid to obconic or club-shaped, fleshy at
maturity, with glochids but lacking spines, 2.5–5 cm long, the apex with a circular,
concave scar where the sepals and petals were attached. Seeds light tan or bone white,
more or less circular in outline, flattened, 3–6.5 mm long, with a thick marginal wing
0.5–1 mm wide.

Our common prickly pear is included in some older manuals as *O. humifusa* or
O. compressa.

O. polyacántha Haw.   **Plains prickly pear**

NL., from Gr. *poly-*, many, + *akantha*, thorn—in reference, obviously, to the numerous spines.
Yellow or purplish-pink; June–September
Sandy or gravelly soil of prairie and roadsides
West fourth

Plants usually low and spreading, forming clumps, composed of 1 or more flattened,
padlike joints which are obovate or nearly round in outline, 3–12 cm long and (2.5–)
5–12 cm wide. Areoles with a tuft of grayish-brown or golden glochids and 6–16 smooth,
reddish, yellowish, or, with age, gray or white, needlelike spines 0.5–4 cm long, some
spreading, some angled downward, and occasionally some curly or tangled and threadlike.
Petals yellow, sometimes tinged below with red or pink, or entirely purplish-pink, about
2–3 cm long, with apex truncate or rounded. Fruits dry at maturity, 1.8–2.5 (–3.5) cm
long, with both spines and glochids, the apex with a shallow, concave scar where the
sepals and petals were attached. Seeds light tan or whitish, flattened, more or less circular
in outline but somewhat irregular, 6–7.5 mm long, with an uneven marginal wing
1–2.5 mm wide.

LYTHRÀCEAE   **Loosestrife Family**

Mostly herbs, sometimes shrubs, and in the tropics sometimes trees. Leaves mostly opposite, sometimes whorled or alternate, simple, sessile or petioled, stipules minute or absent. Flowers bisexual, usually regular, solitary or in clusters in the axils. Sepals 4–6, united with the petals and stamen filaments to form a tubular or bell-shaped hypanthium. Calyx teeth often alternating with an equal number of accessory teeth or "intersepalar appendages." Petals usually 4–6, appearing crumpled, inserted on or just below the rim of the hypanthium, or petals entirely absent. Stamens few to many, frequently twice as many as the petals, inserted on the hypanthium at various levels. Pistil 1; ovary superior, sessile or on a short stalk, 2- to 6-celled; style and stigma 1, the latter usually discoid or capitate. Fruit a dehiscent capsule enveloped by the persistent hypanthium.

The family includes approximately 25 genera and 550 species, most abundant in tropical America. The crepe myrtle (*Lagerstroemia indica* L.), purple loosestrife (*Lythrum salicaria* L.), and several species of *Cuphea* are among the cultivated members of the Lythraceae.

Five genera in Kansas. *Didiplis diandra* (Nutt.) Wood, water-purslane, an aquatic, is recorded from Cherokee, Saline, and Jackson counties only and is not included below.

Rhexia mariana L., in the Melastomaceae, a family closely related to the Lythraceae, is known from Crawford and Cherokee counties. It is an erect, branched perennial, 2–6 dm tall, with opposite, lanceolate to elliptic, serrate leaves; conspicuous, spreading, glandular hairs; purplish-pink petals 1.2–2 cm long; conspicuous, curved, yellow anthers 5–7 mm long; and globose capsules concealed by an urn-shaped hypanthium.

Key to Genera:
1. Hypanthium bell-shaped to globose or hemispheric, about as broad as long; calyx teeth and petals (when present) usually 4 or 5 2
  2. Capsule rupturing irregularly; appendages of the hypanthium subulate (awl-shaped); flowers in small axillary clusters of 3–5 .... *Ammania*
  2. Capsule separating into 2–4 valves; appendages of the hypanthium broadly triangular; flowers usually borne singly in the leaf axils .... *Rotala*
1. Hypanthium tubular or obconical, its length greatly exceeding its diameter; calyx teeth and petals usually 6 3
  3. Flowers irregular; hypanthium pouchlike on 1 side .... *Cuphea*
  3. Flowers regular; hypanthium not as above .... *Lythrum*

AMMÁNNIA L.   Ammannia

Named after Johann Ammann (1699–1741), professor of natural history in St. Petersburg, Russia.

Glabrous annuals. Stems often 4-angled. Leaves opposite, sessile, entire. Flowers small and inconspicuous, regular, borne singly or in axillary clusters of 3–5. Calyx lobes 4, minute. Petals small, deciduous, or absent. Stamens 4–8. Pistil 2- to 4-celled. Capsule incompletely 2- to 4-celled, dehiscing irregularly.

Two species in Kansas.

Key to Species:
1. Peduncles longer than the pedicels, mostly 3–5 mm long .... *A. auriculata*
1. Peduncles very short, no longer than the pedicels .... *A. coccinea*

A. auriculàta Willd.

NL., from L. *auricula*, dim. of *auris*, ear, i.e. "little-eared"—in reference to the earlike lobes at the base of each leaf.
Purplish-red to white; July–October
Wet soil of creek banks, pond and lake margins, ditches, marshy areas
Scattered nearly throughout except for the extreme north and west

Erect, branched annuals 0.7–5 dm tall, much resembling *A. coccinea* but less common. Leaves mostly linear, 1.5–9 cm long and 2–11 mm wide; apex attenuate; base auricled, clasping the stem. Flower clusters on peduncles 3–5 mm long. Hypanthium (including the tiny toothlike sepals) 2–3 mm long at anthesis, enlarging in fruit. Petals purplish-red, 1–1.5 mm long, falling early. Capsule globose, membranous, 2–4 mm long, enveloped by the persistent, enlarged calyx. Seeds reddish-brown, ovate or 3-angled in outline, concavo-convex, about 0.4 mm long.

A. coccínea Rottb.   **Purple ammannia**

L. *coccineus*, scarlet red.
Lavender-pink; late June–October
Wet soil of creek banks, pond and lake margins, ditches, marshy areas
Nearly throughout but less common westward

317. *Ammania coccinea*

318. *Cuphea petiolata*

Erect, branched annuals with a tuft of fibrous roots. Stems 1–5 dm tall, 4-angled. Leaves linear or narrowly elliptic 0.5–7.5 cm long and 2–12 mm wide; apex acute or attenuate; base auricled, clasping the stem. Flower clusters sessile or nearly so, the peduncles less than 3 mm long. Hypanthium at anthesis narrowly bell-shaped, 4-angled, and 4-creased, 2–3 mm long (including the tiny toothlike sepals), enlarging in fruit. Petals lavender or pink, 2–3 mm long, falling early. Capsule globose, membranous, 2–4 mm long, enveloped by the persistent, enlarged calyx. Seeds reddish or yellowish-brown, ovate or 3-angled in outline, concavo-convex, about 0.4 mm long.

The anthers face inward and can pollinate the stigma, but cross-pollination can also take place through the visits of bees, flies, and butterflies that are known to visit the flowers. As the capsules ripen, the very thin, fragile walls might too soon crack open and spill the immature seeds but for the fact that the hypanthium tightly covers and protects it.

CŮPHEA P. Br.   Cuphea

NL., from Gr. *kyphos*, bent, humpbacked, in reference to the gibbous or spurred calyx exhibited by members of this genus.

Annual or perennial herbs or shrubs, often clammy, with glandular hairs. Leaves opposite, entire, petioled in ours. Flowers irregular, borne singly or in pairs in the leaf axils, or in terminal spikes or racemes. Hypanthium tubular, with 6 prominent ribs and 6 calyx teeth, the teeth alternating with 6 small appendages, the tube pouchlike or spurred at the base on the upper side. Petals 6, dissimilar. Stamens 11–12, departing from the hypanthium at different heights. Pistil 1- or 2-celled, with a curved gland situated at the base near the spur of the hypanthium; style slender, stigma 2-lobed. Capsule ovoid, or ellipsoid, few-seeded, opening along 1 side, enclosed by the persistent calyx.

One species in Kansas.

C. viscosíssima Jacq.   Clammy cuphea

NL., from L. *viscosus*, sticky, + the diminutive suffix -*issimus*.
Magenta; June–October
Open woods, occasionally in prairie or pastures
East fifth, mostly south of the Kansas River, and Coffey County

Erect annual, 2–6 dm tall, with numerous ascending branches. Leaf blades lanceolate or elliptic, 1–4 cm long and 0.5–1.5 cm wide. Stems, petioles, and hypanthiums with viscid, usually purple hairs. Flowers borne singly in the leaf axils. Hypanthium 8–13 mm long (including the calyx teeth). Petals magenta, 2–5 mm long, the upper ones larger than the lower ones. Capsules ellipsoid, membranous, 6–8 mm long. Seeds brown, very broadly ovate or nearly round in outline, flattened, biconvex, slightly notched at 1 or both ends, 2–2.5 mm long.

LÝTHRUM L.   Loosestrife

NL., from Gr. *lythron*, clotted blood—because of the flower color of *L. salicaria*, the European species which Dioscorides called *Lythrum*. "Loosestrife" amounts to a translation of the genus name *Lysimachia*. According to Pliny, the Eurasian *Lythrum salicaria* (which he called *Lysimachia*, a name which clung to it through the Middle Ages) was fastened to the yoke of teams of oxen which had gotten out of hand. The magic of the plant loosened and exorcised the evil spirit that caused their strife, and the animals settled down and pulled together in unison again.

Annual or perennial herbs or shrubs. Stems 4-angled. Leaves opposite, alternate, or whorled, sessile, entire. Flowers regular, usually axillary, often of 2 or 3 kinds with regard to stamen and style length. Hypanthium cylindric, 8- to 12-ribbed, with 4–7 teeth alternating with minute appendages. Petals 4–6 (rarely absent). Stamens 4–12, arising low on the hypanthium. Pistil 2-celled; style threadlike. Capsule cylindrical, membranous, many-seeded, opening irregularly (or opening septicidally), enclosed by the persistent hypanthium.

Two species occur in Kansas. Both were formerly treated as a single species under the name *L. alatum*. That name, however, apparently belongs to a plant which occurs only in southern Georgia but which has not been found in recent years, and the correct names of our plants are those used below.

Charles Darwin made a prolonged study of the common European *L. salicaria* and found among the different individual plants 3 forms of flowers: those with (1) a long style, a whorl of medium-length stamens, and a whorl of short stamens, (2) a medium-length style, a whorl of long stamens, and a whorl of short stamens, (3) a short style, a whorl of long stamens, and a whorl of medium-length stamens. Thus in any one flower the stigma would not touch the anthers and become self-pollinated, while bees and butterflies in sipping nectar from the flowers of many different plants would get pollen on their proboscises, heads, and bodies at the right levels to pollinate the stigmas of all 3 lengths of styles. In ours, there are 2 lengths of styles and correspondingly 2 lengths of stamens involved in the game of cross-pollination.

We never hear our native species of *Lythrum* spoken of as medicinal plants, yet *L. salicaria* has long been recommended for its healing properties. Among the old herbalists Dioscorides (1st century) is the earliest authority quoted for this species, thus: "The herb is tart and strong in taste, of an astringent and refrigerant nature, good for stanching both outward and inward bleeding; sap extracted from the leaves and drunk stops blood-spitting and dysentery, and sour wine in which the leaves have been boiled when taken internally will have the same effect; and if the plant is set afire it gives off a pungent vapor and smoke that drives away serpents; and flies cannot stay in a room where this smoke is." Whatever medicinal virtues the European species possessed might be found in our own (28).

319. *Lythrum dacotanum*

Key to Species:

1. Leaves of inflorescence typically linear-lanceolate; appendages of hypanthium much longer than the teeth; pollen green .... *L. dacotanum*
1. Leaves of inflorescence typically linear-oblong; appendages of hypanthium about as long as the teeth; pollen yellowish or white .... *L. californicum*

L. califórnicum T. & G.

"Californian."
Purplish-pink; June–early September
Moist soil of prairie ravines, seeps, lake and pond margins, and ditches
East two-thirds, primarily, especially the central third

Glabrous, erect, branched perennials, usually 3–7 dm tall, with stems arising from slender, fibrous-rooted rhizomes. Leaves alternate, lanceolate below to elliptic or linear-oblong above, 0.5–4.5 cm long and 1–7 mm wide. Flowers borne singly in the leaf axils toward the ends of the stems and branches. Hypanthium tube narrowly obconical, 12-ribbed or -winged, 5–7 mm long, with 6 triangular-acuminate calyx teeth alternating with 6 acute, fleshy appendages about the same length as the teeth. Petals bright purplish-pink, obovate, 5–7 mm long. Stamens usually 6, the pollen yellow or white. Seeds buff-colored or orangish-brown, about 0.5–0.7 long.

See plate *61*.

L. dacotànum Nieuw.

NL., probably from Gr. *daketon,* a biting animal, possibly in reference to the subulate appendages which alternate with the sepals.
Purplish-pink; May–August
Moist soil of prairie ravines, seeps, lake and pond margins, and ditches
Our records are from the east half and from the central counties of the west half, probably nearly throughout

Glabrous, erect, branched perennials, 3–8 dm tall, with stems arising from a slender, branched, fibrous-rooted rhizome. Leaves alternate, lanceolate below to linear-lanceolate above, 0.5–5 cm long and 2–15 mm wide. Flowers borne singly in leaf axils toward the ends of the stems and branches. Hypanthium tube narrowly obconical, 12-ribbed or -winged, 5–6 mm long, with 6 triangular-acuminate calyx teeth alternating with 6 longer, attenuate appendages. Petals bright purplish-pink, obovate, 5–7 mm long. Stamens usually 6, the pollen green. Seeds reddish-brown, about 0.5–0.7 mm long.

ROTÁLA L.

NL., an incorrect diminutive of L. *rota,* wheel, i.e., "little wheel," in reference to the whorled leaves of the first-named species.

Glabrous, annual or perennial herbs. Leaves opposite or whorled, sessile, entire. Flowers small and inconspicuous, regular, borne singly in the leaf axils or in terminal racemes. Hypanthium bell-shaped or globose. Calyx lobes 3–6, sometimes alternating with intersepalar appendages. Petals 3–6, minute, deciduous, or entirely absent. Stamens 1–6. Pistil 2- to 4-celled. Fruit a 2- to 4-valved, septicidal capsule with transverse striations.

One species in Kansas.

LOOSESTRIFE FAMILY

R. ramòsior (L.) Koehne   **Toothcup**

L. *ramosus*, full of branchs, branching.
White or pink; July–October
Wet soil of lake and pond margins and ditches, occasionally in low wet woods or seepy ravines
Scattered in the east half

Erect to ascending or prostrate, branched herbs, 1–3 dm tall, sometimes rooting at the lower nodes. Leaves opposite, oblanceolate or elliptic, 1–4 cm long and 2–10 mm wide, narrowed to a petiolelike base, the apex rounded or acute. Flowers borne singly in the leaf axils. Hypanthium 2–4 mm long. Calyx teeth 4, 1 mm long, alternating with 4 triangular appendages about the same length. Petals 4, white or pink, 1 mm long, soon falling. Capsule membranous, 2–4 mm long, enclosed by the persistent hypanthium. Seeds brown, nearly round to ovate or somewhat 3-angled in outline, plano-convex, 0.3–0.5 mm long.

ONAGRĂCEAE   Evening Primrose Family

Herbaceous annuals, biennials, or perennials, sometimes woody near the base, rarely shrubs or trees. Leaves alternate or opposite, sessile or petioled, simple, stipules absent or falling early. Flowers bisexual, usually regular, borne singly or in spikes or racemes. Sepals and petals usually 4, rarely 2, 3, 5, or 6. Stamens of the same number or twice as many as the corolla lobes. Pistil 1; ovary inferior, mostly 4-celled; style 1, capitate to peltate or notched, or with radiating branches. Hypanthium united with the inferior ovary and sometimes extending beyond it. Fruit nutlike and indehiscent or a loculicidal capsule. Seeds with or without a tuft of hairs, called a coma, at one end.

NOTE: All measurements of sepals, petals, and hypanthium in this family are made from the point of departure from the ovary. Characteristics of the mature fruits and seeds, along with those of the flowers, are used in various of the keys, and one will do well to collect specimens with both flowers and fruits when possible.

The family comprises about 20 genera and 650 species, all North Temperate and chiefly American. Numerous species within the family are cultivated as ornamentals, but otherwise the family is of little economic importance.

Seven genera native to Kansas.

Key to Genera:
1. Sepals, petals, and stamens 2 each; leaves opposite .... *Circaea*
1. Sepals and petals 4 or 5, stamens as many as or twice as many as the petals 2
  2. Sepals persistent in fruit; hypanthium not prolonged beyond the ovary .... *Ludwigia*
  2. Sepals deciduous in fruit; hypanthium obviously prolonged beyond the ovary 3
    3. Stigma entire 4
      4. Stigma club-shaped; seeds with a tuft of hairs at 1 end .... *Epilobium*
      4. Stigma more or less peltate; seeds without a tuft of hairs .... *Calylophus*
    3. Stigma deeply 4-lobed or -branched 5
      5. Flowers regular; fruit a many-seeded, dehiscent capsule .... *Oenothera*
      5. Flowers regular or irregular; fruit hard, indehiscent 6
        6. Flowers borne in the axils of upper leaves, these not much smaller than the lower leaves, OR the plants lacking an aboveground stem .... *Oenothera*
        6. Flowers borne in essentially naked inflorescences with leafless branches, the plants definitely with elongate, aboveground stems 7
          7. Hypanthium funnelform or tubular, but never threadlike; each filament with a small scale at the base (except in *G. parviflora*); leaves abruptly becoming smaller in the inflorescence .... *Gaura*
          7. Hypanthium threadlike; filaments without scales; leaves gradually becoming smaller in the inflorescence .... *Stenosiphon*

CALÝLOPHUS Spach.

NL., from Gr. *kalyx*, cup, cover, outer envelope of a flower, + *lophos*, mane, cress, comb, tuft, ridge; probably in reference to the keeled sepals of the type species, *C. serrulatus*.

Annual or perennial herbs, sometimes woody at the base. Leaves alternate, sessile or petioled; stipules absent. Flowers regular, yellow, borne singly in the axils of upper leaves, opening only during the dim light of early morning or dusk or on cloudy days. Sepals 4, deciduous in fruit. Petals 4. Stamens 8. Pistil 4-celled, each cell with 2 rows of ovules. Stigma discoid or somewhat globose. Hypanthium well developed above the ovary. Fruit a cylindric, many-seeded capsule dehiscing longitudinally and loculicidally. Seeds without a coma.

Some authorities combine this genus with *Oenothera*.

Three species in Kansas. *C. drummondianus* Spach. is known from Seward and Meade counties only. It strongly resembles *C. serrulatus* but tends to have larger flowers (mostly 9–21 mm long) and has the stigma usually exserted at least to the end of the outer anthers or positioned outside of the circle of anthers. Unlike *C. serrulatus*, *C. drummondianus* must be cross-pollinated.

Key to Species:
1. Sepals with prominent raised or keeled midribs; flower buds sharply 4-angled ...... *C. serrulatus*
1. Sepals lacking prominent midribs; flower buds round in cross section .... *C. hartwegii*

C. hartwégii (Benth.) Raven

Named after Karl Theodor Hartweg (1812-1871), native of Karlruhe, Germany, who collected
      for the Royal Horticultural Society of London in Mexico, central and western equatorial
      America, and Jamaica (1836–42), and in California (1845–47).
Yellow; late April–August (–September)
Dry, rocky and/or sandy soil of prairie hillsides and uplands
West two-fifths and Barber County

320. *Calylophus hartwegii* ssp. *fendleri*

Bushy perennials arising from a branched woody caudex. Stems several, erect, to ascending or decumbent, 0.7–2 (–4) dm tall, sometimes woody near the base. Leaves sessile, mostly linear to very narrowly oblanceolate, 8–36 mm long and 1–3 mm wide, the apex acute, the margins entire or with a few shallow teeth. Herbage and fruits (in our area) usually with minute appressed hairs. Unopened flower buds round in cross section. Hypanthium 3.8–6.8 cm long, funnelform above with a long slender tube below. Sepals reflexed but with spreading or ascending tips, 0.8–1.8 cm long, lacking a prominent keel or midrib. Corolla yellow (drying cream-colored or pinkish), 1–2.3 cm long, the lobes broadly ovate to nearly round or slightly rhombic. Stigma discoid, at least slightly 4-lobed, eventually protruding beyond the anthers, requiring cross-pollination. Capsules 1.5–3 cm long and 2.5–4 mm wide, 4-angled in cross section. Seeds dark brown, smooth, irregular and angular in shape, about 2 mm long.

Most of ours plants are placed by authorities into var. or subspecies *lavandulifolius* (from *Lavandula*, the genus to which lavender belongs, + *folium*, leaf). Glabrous plants with larger leaves and flowers are known from Barber, Meade, and Morton counties and are placed in subspecies *fendleri* (*Oenothera fendleri* of some manuals). Plants with long, spreading hairs and leaves with truncate or clasping bases occur in Barber, Clark, and Morton counties, along with the more common subspecies *lavendulifolius*, and belong to subspecies *pubescens*.

321. *Calylophus serrulatus*

C. serrulàtus (Nutt.) Raven   **Toothleaf evening primrose**

NL., from L. *serrula*, little saw—in reference to the fine teeth of the leaf margin.
Yellow; mid-May–August
Prairie
Nearly throughout; more common westward

Erect perennials, 1.3–5.5 dm tall, with a long, tough woody taproot, the stems often branched from the base and more or less woody below. Upper portions of the stem, lower surface of younger leaves, hypanthiums, and capsules more or less densely covered with appressed hairs. Leaves sessile or very gradually narrowed to a petiolelike base, mostly very narrowly elliptic to linear, folded along the midrib, 1.5–7 cm long and 2–7 mm wide, the apex acute, apiculate, the margins finely serrate. Unopened flower buds sharply 4-angled. Hypanthium funnelform, 3–20 mm long. Sepals reflexed, 3.5–6 mm long, strongly keeled along the midrib. Corolla yellow (drying white or cream-colored), 4–15 mm long, the lobes broadly obovate to obcordate. Stigma more or less discoid and somewhat 4-angled with undulate margins, not elevated beyond the anthers, frequently self-pollinated. Capsules 1.5–4 cm long and about 1.5 mm wide, 4-angled in cross section, superficially resembling a portion of the stem. Seeds brown, smooth, angular, about 1–1.3 mm long.

While this species appears in most counties in the state, it is so drought-resistant that its frequency increases westward until, besides the grasses, it may be the species most frequently seen in some areas. The folded posture of the leaves is an adaptation to a dry habitat, presenting the edges of the leaves, rather than their broad surfaces, to the highest intensity of sunlight and heat around midday, thus reducing the amount of water lost through the leaf surfaces.

CIRCAÈA L. — Enchanter's nightshade

NL., from Circe, name of the enchantress in Homer's *Odyssey*—in reference to the holdfast character of the fruits.

Perennial herbs. Leaves opposite, petioled. Flowers small, white, few, in terminal racemes. Sepals and petals 2 each, fused to form a short, tubular hypanthium above the ovary. Stamens 2. Pistil with 1–2 1-ovuled cells; stigma 2-lobed or -cleft. Fruit small, indehiscent, covered with hooked bristles.

One species in Kansas.

C. quadrisulcàta (Maxim.) Franch. & Sav. var. *canadensis* (L.) Hara

NL., from L. *quadri-*, four, + *sulcatus*, grooved, furrowed, in reference to the grooves on the fruits.
White; mid-June–July
Moist, rich, woodland soil
East third, especially in the east fourth

Delicate perennials 2–5 (–10) dm tall. Leaf blades thin, ovate to lanceolate or, occasionally, deltoid, 3–12 cm long and 2.5–7 cm wide, the apex acute or acuminate, the base truncate to broadly rounded or cordate, the margins shallowly toothed; petioles mostly 0.5–5 cm long. Racemes 15- to 30-flowered, (2–) 4–9 cm long (excluding the peduncle), simple or with a few branches. Calyx lobes ovate- or lanceolate-acuminate, 2–2.5 mm long. Corolla lobes broadly obcordate, about 1.25–2.25 mm long. Fruit 2–4 mm long, longitudinally grooved, densely covered with hooked bristles, 2-celled and 2-seeded.

EPILÒBIUM L. — Willow herb

NL., from Gr. *epi-*, on, upon, + *lobos*, lobe or pod—the perianth seeming to be based on the inferior ovary.

Annual or perennial herbs (rarely subshrubs) growing in moist or wet habitats. Leaves alternate or opposite, toothed; stipules absent. Flowers regular, borne singly in the axils of upper leaves. Calyx 4-lobed, deciduous. Corolla 4-lobed, white, pink, or purplish. Hypanthium tubular. Stamens 8. Pistil 4-celled; stigma club-shaped. Fruit a linear, many-seeded capsule opening along 4 lines. Seeds with a tuft of hairs (coma) at 1 end.

Two species in Kansas. *E. leptophyllum* Raf. is known from a few scattered counties only and differs from the following more common species in having leaf margins entire (or nearly so) and revolute and in having stems round in cross sections, lacking parallel lines of minute hairs extending downward from the base of each leaf.

E. coloràtum Biehl. — Purpleleaf willow-weed

L. *coloratus*, colored or variegated—in reference to the red veins of the leaves.
Pink or white; June–mid-September
Wet soil
North half, primarily

Erect, branched perennial herbs, up to 1 m tall, with a short rhizome and fibrous roots. Stems with 2 parallel lines of minute hairs extending downward from each leaf base (use magnification). Leaves opposite, short-petioled, the blades narrowly lanceolate to lance-linear or narrow elliptic, pinnately veined, those of the main leaves mostly 3–10 cm long and 0.5–2.5 cm wide, the apex acute or attenuate, the base rounded on the larger leaves, acute or wedge-shaped on the smaller leaves, the margins serrate. Calyx 2.5–4.5 mm long, the lobes lanceolate to oblong-elliptic. Corolla pink or white, slightly longer than the calyx, the lobes oblong, deeply notched at the apex. Hypanthium extended a short distance only above the ovary. Capsule 3–5 cm long and about 1 mm wide. Seed body lance-ellipsoid, 1–1.5 mm long, coma about 8 mm long.

GAÙRA L. — Gaura

NL., from Gr. *gauros*, stately—suggested by the tall growth of the type species, *G. biennis*.

Annual, biennial, or perennial herbs. Leaves alternate, sessile. Flowers usually somewhat irregular, subtended by much-reduced, early-deciduous bracts, borne in long terminal spikes, racemes, or panicles, usually opening about sunrise or sunset and lasting about 12–16 hours. Hypanthium tubular. Sepals 4, reflexed. Petals 4, clawed, the blades elliptic to rhombic or oblanceolate, widely spreading and positioned in the upper half of the flower. Stamens 8, each filament bent downward somewhat and usually with a basal scale. Pistil 4-celled (sometimes 1-celled if the partitions fail to develop), 1 ovule

322. *Circaea quadrisulcata*

323. *Circaea quadrisulcata*, close-up of flowers

324. *Circaea quadrisulcata* fruiting

EVENING PRIMROSE FAMILY
246

(rarely 2) per cell; style bent downward with the stamens; stigma 4-lobed. Fruit a hard, indehiscent, nutlike, 1- to 4-seeded capsule.

Four species in Kansas. Except for *G. parviflora*, all our species are cross-pollinated by moths. In accordance with this, *G. coccinea*, *G. longiflora*, and *G. villosa* tend to have relatively showy, irregular flowers, characteristics frequently associated with pollination by insects, birds, or other animals. In addition, their stamens have minute basal scales which project inward and effectively close the throat of the corolla to short-tongued, nonpollinating nectar thieves. In contrast, *G. parviflora*, which is self-pollinating, has quite small, regular flowers and lacks staminal scales.

Key to Species:

1. Fruit with a slender stipe 2–5 (–10) mm long; plants of southwestern Kansas .... *G. villosa*
1. Fruit sessile or nearly so 2
  2. Fruits more or less abruptly constricted to a thick, cylindrical base nearly half the length of the body; low perennial 1–6 dm tall .... *G. coccinea*
  2. Fruits not as above; annuals or biennials up to 4 m tall 3
    3. Sepals 7–11 mm long; anthers 2.5–4.5 mm long, the filaments with basal scales; fruit lanceolate to rhombic or ovate in side view, narrowly winged on the angles; east half of Kansas, especially the east third .... *G. longiflora*
    3. Sepals 3.5–4.5 mm long; anthers 0.5–1 mm long, the filaments lacking basal scales; fruit fusiform or club-shaped, 4-angled but not winged; throughout Kansas except for the extreme southeast .... *G. parviflora*

325. *Epilobium coloratum*

G. coccínea Nutt. ex Pursh   Scarlet gaura

L. *coccineus*, scarlet-colored.
White to pinkish or orangish-red; May–mid-July
Prairies, roadsides, and railroad right-of-ways, usually in sandy soil
West three-fifths and Riley and Geary counties

Low herbaceous perennials arising from a branched woody caudex. Stems several to many, erect or ascending, branched, 1–6 dm tall. Herbage gray-green, more or less densely covered with minute appressed hairs. Leaves linear to narrowly elliptic, 0.7–6.5 cm long and 1–15 mm wide, the apex acute, the base acute or rounded, the margins entire or with a few shallow teeth. Racemes 1.5–23 cm long. Hypanthium narrowly tubular, mostly 5–10 mm long. Sepals subulate, 5–8 mm long. Petals white to pinkish or orangish-red, 3–7 mm long. Anthers red, 3–5 mm long, the filaments with basal scales. Fruit sessile or nearly so, 4–8 mm long, the upper half pyramidal, the lower half constricted and cylindrical.

G. longiflòra Spach

NL., from L. *longus*, long, + *flos*, *floris*, flower—i.e., "long-flowered."
White or becoming pinkish; (July–) August–October
Prairie, roadsides, railroad right-of-ways, fields, waste ground, occasionally in open woods or thickets
East half, especially the east third

326. *Gaura coccinea*

Robust, branched winter annuals, 0.5–4 m tall, with a fleshy taproot. Herbage covered with soft, minute, nonglandular appressed hairs, sometimes with glandular hairs also. Leaves narrowly elliptic, 1.5–15 cm long and 0.2–2.7 cm wide, the apex usually acute, the base acute or wedge-shaped, the margins entire or remotely and shallowly toothed. Inflorescence much branched, the flowers opening near sunset. Hypanthium narrowly tubular, mostly 7–10 mm long. Sepals subulate, 7–11 mm long. Petals white or becoming pinkish, clawed, 6–9 mm long. Anthers reddish or purplish, 2.5–4.5 mm long, the filaments with basal scales. Fruits lanceolate to rhombic or ovate in side view, narrowly winged on the angles, mostly 6–8 mm long at maturity, the pedicel 1 mm or shorter.

G. parviflòra Dougl.   Small-flowered gaura

NL., from L. *parvus*, little, puny, + *flos*, *floris*, flower.
White to pink; late May–early September
Prairie, roadsides, flood plains
Throughout except for the extreme southeast

Tall, erect annuals or biennials with a taproot. Stems usually unbranched below the inflorescence, eventually becoming quite robust, 0.3–2 m tall, with long, spreading silky hairs intermixed with shorter ones. Leaves mostly ovate to lanceolate or narrowly elliptic, covered with minute velvety hairs, 3–11 cm long and 0.7–4 cm wide, the apex acute to acuminate or attenuate, the base acute or acuminate, the margins entire or remotely and shallowly toothed. Inflorescence eventually with few to many branches,

327. *Gaura longiflora*

EVENING PRIMROSE FAMILY

247

328. *Gaura villosa*

329. *Ludwigia alternifolia*

the bracts falling early. Hypanthium narrowly tubular, mostly 2–4 mm long. Sepals narrowly oblong to subulate, 3.5–4.5 mm long. Petals white or becoming pinkish, clawed, 4–7 mm long. Anthers reddish, about 0.5–1 mm long, shedding pollen directly onto the stigma, the filaments lacking basal scales. Fruit fusiform or somewhat club-shaped, 4-angled, 5–11 mm long.

According to Krochmal (30), a poultice made from the crushed plant, harvested while in bloom, has been used to treat muscular pains and arthritis.

G. villòsa Torr.   Hairy gaura

L. *villosus*, hairy.
White to pink; late May–mid-September
Prairie, roadsides, flood plains, usually in sandy soil
South half of the southwest fourth

Erect perennials with a deep, woody rootstock. Stems 6–18 dm tall, frequently with a whorl of branches just below the inflorescence, more or less hairy below, glabrous within the inflorescence. Leaves oblanceolate to narrowly elliptic or linear, more or less densely covered with long, soft appressed hairs, 2–8 cm long and 2–20 mm wide, the apex acute, mucronate, the base acute or rounded, the margins entire or shallowly and remotely toothed. Inflorescence eventually much branched. Hypanthium short-tubular, 2–4 mm long. Sepals subulate or linear, 4–7 mm long. Petals white to pink, mostly 6–8 mm long. Anthers reddish, 2.5–6 mm long, the filaments with basal scales. Fruit body lanceolate to oblong in outline, winged on the angles, 6–10 mm long, on a slender stipe 2–5 mm long.

LUDWÍGIA L.   False loosestrife

Honoring C. G. Ludwig (1709–1773), botanical explorer and writer, professor of medicine in Leipzig.

Annual or perennial herbs of wet places, creeping in mud or floating, the submersed portions often spongy. Leaves alternate or opposite, with minute stipules. Flowers regular, yellow in ours, borne singly in the axils of upper leaves. Sepals 4–5 (rarely 6), persistent in fruit. Petals 4–5 (rarely 6) or absent. Stamens as many as or twice as many as the sepals. Stigmas globose or hemispherical. Hypanthium not extended beyond the ovary, often 4-angled or -winged. Fruit a 4-celled, many-seeded capsule dehiscing longitudinally or by a terminal pore. Seeds without a coma.

Five species in Kansas. *L. glandulosa* Walt., known from Cherokee County only, is an erect, much-branched herb up to 1 m tall, with petal-less flowers and small, cylindrical fruits 2–8 mm long and 1.5–2 mm thick. *L. polycarpa* Short & Peter, known from a few scattered sites in the east half, is also an erect, branched plant up to about 1 m tall but has minute greenish petals (sometimes none at all) and bell-shaped capsules 4–7 mm long and 3–5 mm thick.

Key to Species:
1. Leaves opposite; corolla absent .... *L. palustris*
1. Leaves alternate; corolla present (although readily falling) 2
    2. Erect plants with rather brittle stems; sepals and petals 4 each; stamens as many as the sepals, in 1 series; fruits more or less globose .... *L. alternifolia*
    2. Creeping or floating plants, the stems not brittle; sepals and petals 5 each; stamens twice as many as the sepals, in 2 series; fruits long-cylindric .... *L. peploides*

L. alternifòlia L.   Seedbox

NL., from L. *alternatus*, alternate, + *folium*, leaf—the leaves occurring singly at different heights, not opposite at the same node.
Yellow; June–September
Usually in moist or wet soil of marshes, pond and lake margins, ditches and seepy ravines or low areas in prairies
East half

Erect, minutely hairy perennial herbs with a cluster of thickened roots. Stems branched above, up to 1 m (–2 m) tall, somewhat woody and rather brittle. Leaves alternate, the lower ones usually gone by flowering time, the upper ones sessile or very short-petioled, lanceolate to narrow elliptic or almost linear, 2–9 cm long and 0.2–1.5 cm wide, the apex acute, the base wedge-shaped, the margins entire. Sepals 4, lanceolate to ovate-acuminate, 5–11 mm long, often tinged dark red, eventually deciduous. Petals 4, obovate, yellow, about as long as the sepals, usually falling easily. Stamens as many as the sepals, in 1 series. Capsules more or less globose but square above and rounded below,

4-angled, 5–7 mm long and 5–6 mm wide, dehiscing through a terminal pore. Seeds tan, oblong-ellipsoid, about 0.5–0.8 mm long.

See plate *62*.

L. palústris (L.) Ell.   Water purslane

L. *paluster, palustris,* marshy, swampy—descriptive of the habitat of this species.
Flowers green; June–September
Pond margins, marshes, wet ditches, stream banks
East half, primarily; a few records from farther west

330. *Ludwigia peploides*

Glabrous plants with rather flaccid stems up to 5 dm long, either floating or creeping and rooting at the nodes, producing erect stems 0.4–3 dm tall. Leaves opposite, mostly 1.5–3.5 (–4.5) cm long and 0.3–1.5 cm wide (including the petiolelike base), the blade elliptic to lanceolate or ovate, abruptly constricted into a tapering, winged, petiole-like base, the apex acute or short-acuminate, the margins entire. Sepals 4, ovate-acuminate, about 1.5 mm long. Petals absent. Capsules sessile, (2.5–) 4–5 mm long (including the persistent sepals) with 4 prominent veins (corresponding to the midveins of the sepals) alternating with 4 thinner areas which rupture to release the seeds. Seeds tan, more or less oblong, about 0.5 mm long.

L. peploídes (H. B. K.) Raven   Floating water-primrose

NL., from *Peplis,* a genus in the Loosestrife Family + the suffix *-oides,* like, similar to.
Yellow; late June–September
Mud or shallow water at margins of ponds, lakes, and streams
East three-fifths

Smooth, nearly glabrous, perennial herbs with stems 3 to many dm long floating or prostrate on the mud and rooting at the nodes. Leaves opposite, the blades mostly 1–10 cm long and 0.5–2.5 cm wide, broadly elliptic to obovate or oblanceolate with a rounded apex or lanceolate to narrowly elliptic with an acute apex, the base always tapered more or less gradually to a distinct petiole. Flowers borne on elongate pedicels, the hypanthium subtended by 2 minute bracts. Sepals 5, lance-oblong to very narrowly triangular, 5–10 mm long (to 12 mm in fruit). Petals 5, yellow, 8–17 mm long, obovate with a rounded or very shortly acuminate apex. Stamens twice as many as the sepals, in 2 series. Capsules cylindrical, (1–) 2.5–4 cm long, on pedicels 1–6 cm long, rather woody and dehiscing rather irregularly after a period of weathering. Seeds about 1–1.3 mm long, arranged in a single vertical row in each cell and partially imbedded in the inner fruit wall.

OENOTHÉRA L.   Evening primrose

L. *oenothera,* a name used by Pliny for a plant (identity unknown to us) reputed to produce sleep when its juice was drunk in wine.

Annual, biennial, or perennial herbs, sometimes woody at the base. Leaves alternate, sessile or petioled; stipules absent. Flowers regular, yellow to white, pink, or rose, borne singly in the leaf axils or in terminal clusters, opening in the dim light of early morning or evening or on cloudy days. Hypanthium well developed beyond the ovary, usually narrowly funnelform. Sepals 4, usually lance-linear to linear-attenuate, reflexed, often united at the tips, deciduous in fruit. Petals 4. Stamens 8, the anthers yellow, linear, pivoting freely about their point of attachment to the filament. Pistil 4-celled; style slender; stigma deeply 4-lobed or -branched. Fruit a capsule, either indehiscent or longitudinally dehiscent. Seeds without a coma.

Seventeen species in Kansas. *O. engelmannii* (Small) Munz and *O. nuttallii* Sweet are known only from Morton and Graham counties, respectively, and are not included here.

Key to Species:
1. Fruit club-shaped to broadly ellipsoid, the basal part sterile and narrowed, the distal part thicker, fertile, and ribbed or winged; seeds clustered, stalked, neither in definite rows nor angled 2
   2. Perennial with leaves lanceolate to narrowly elliptic, oblong, or oblanceolate, shallowly toothed to pinnatifid, 2–9 cm long and 3–28 mm wide; petals white, rose, or purplish, 2–5 cm long .... *O. speciosa*
   2. Slender annual with threadlike leaves 1–2.5 cm long and less than 1 mm wide; petals yellow, 3–6 mm long .... *O. linifolium*
1. Fruit body ovoid to cylindric or prismatic, sometimes winged; seeds usually sessile and in definite rows 3
   3. Seeds angled, in several rows in each cell; fruit ovoid, sharply 4-angled or -winged; low,

bushy perennials with petals white to pink with red spots or stripes; west third, primarily .... *O. canescens*
3. Seeds not angled (or if seeds angled, then fruits cylindric, and plants up to 2 m tall and little branched, if at all); arranged in 1 or 2 rows in each cell; flowers white, pink, or yellow 4
   4. Fruit broadly winged its *entire length*; seeds in 1 or 2 rows in each cell and with minute corky tubercles (use magnification) 5
      5. Petals 15–22 (–35) mm long; hypanthium 25–45 mm long; fruit body 3–4 mm thick, the wings 2–3.5 mm wide; northwest fourth and Hodgeman County ....
.... *O. fremontii*
      5. Petals 30–65 mm long; hypanthium 50–120 mm long; fruit body 7–10 mm thick, the wings 1.5–2.5 cm wide; east three-fourths, more common eastward .. *O. macrocarpa*
   4. Fruit either not broadly winged at all OR broadly winged only on the distal one-half to two-thirds of its length; seeds usually in 2 rows in each cell, lacking minute corky tubercles 6
      6. Fruit body ovoid with 4 winged angles in the distal part at least; low, acaulescent plants with yellow petals 1–2.5 cm long, hypanthium 4–8 cm long, and leaves usually deeply pinnatifid .... *O. triloba*
      6. Fruit body cylindrical to slightly enlarged upward or lance-ovoid, not winged at all or winged in the basal part only 7
         7. Seeds sharply angled (use magnification), horizontal; fruits gradually narrowed toward the tip 8
            8. Petals 2.5–4 (–6) cm long; sepals 2.5–3.5 (–5) cm long; style slightly longer than the petals; primarily south and/or west of the Arkansas River .... *O. hookeri*
            8. Petals 0.8–2.5 cm long; sepals 0.7–2.5 cm long; style 3–17 mm long 9
               9. Plants densely grayish, strigose or hirsute with few to no gland-tipped hairs evident in the inflorescence; leaves rather thick .... *O. strigosa*
               9. Plants more green, not gray-hairy, some with gland-tipped hairs present in the inflorescence; leaves thinner .... *O. biennis*
         7. Seeds not sharply angled, aligned vertically in the cells 10
           10. Flowers white (often becoming pink with age or upon drying), the mature buds nodding 11
              11. Seeds smooth, in 1 row in each cell; gray-green perennial with slender rhizomes .... *O. pallida*
              11. Seeds in 2 rows in each cell 12
                 12. Annual or winter annual with a taproot; throat of hypanthium lacking conspicuous white hairs; petals 18–42 mm long; fruit 2–4 cm long; seeds pitted .... *O. albicaulis*
                 12. Perennial with slender rhizomes 2–10 cm below the soil surface; throat of hypanthium with long, conspicuous white hairs; petals 9–13 mm long; fruit 0.8–2 cm long; seeds minutely tubercled ....
.... *O. coronopifolia*
           10. Flowers yellow (often drying cream-colored or pinkish), the mature buds erect 13
              13. Flowers in terminal spikes, the upper leaves reduced to bracts; stems erect; leaves entire or remotely and shallowly toothed, never pinnatifid
.... *O. rhombipetala*
              13. Flowers in upper leaf axils, not in definite spikes; stems solitary and erect to several and decumbent; main leaves of the stem(s) usually pinnatifid (or at least deeply toothed) 14
                 14. Sepals 1.3–3 cm long; petals 2–3.6 cm long; primarily in the south half of the west three-fifths .... *O. grandis*
                 14. Sepals 6–11 mm long; petals 6–11 (–18) mm long; nearly throughout but more common eastward .... *O. laciniata*

O. albicaùlis Pursh   White-stemmed evening primrose

NL., from L. *albus*, white, + *caulis*, stem.
White; late April–June
Sandy soil of disturbed habitats such as roadsides, pastures, and flood plains
West half, especially the west fourth

    Low annuals or winter annuals with a taproot. Stems 0.8–3 dm tall, branched from the base, the branches erect to ascending or decumbent. Herbage densely covered with minute appressed hairs intermixed with much larger spreading hairs (use magnification). Basal leaves spatulate or oblanceolate, entire to toothed or pinnatifid, 2–8.5 cm long and (0.5–) 1–2.5 cm wide, the upper leaves oblanceolate to elliptic or lanceolate in general outline but mostly pinnatifid, mostly 2–7 cm long and 0.5–2.5 cm wide. Flowers with a faint or unpleasant odor, borne singly in the upper leaf axis, the buds nodding, opening near sunset. Hypanthium 2–3 cm long. Sepals 1.5–2.5 cm long, the tips united. Petals white (becoming pink), obcordate, 1.8–4.2 cm long. Stigma branches linear. Capsules cylindric, 2–4 cm long, superficially resembling a section of the stem. Seeds sessile, ascending, in 2 rows in each cell, golden brown, more or less obovoid or broadly ellipsoid with a short beak at the apex, pitted, about 1 mm long.

O. biénnis L.   Common evening primrose

L. *biennis*, lasting for 2 years, from *biennium*, space of 2 years—indicating the lifetime of an
    individual of this species. "Evening primrose" was first applied to this species—the one
    used by Linnaeus in establishing the genus—because the flowers open in the evening and
    have the light-yellow color of the English primrose or cowslip, *Primula veris*.
Yellow; August–mid-October
Disturbed habitats such as fields, roadsides, pastures, and weedy flood plains
Throughout, more common eastward

Erect biennials quite similar to *O. strigosa* but greener in color and bearing some glandular hairs in the inflorescence and having thinner leaves than that species. Stems 0.5–2 m tall. Leaves spreading or ascending, 8–17 cm long and 2–5 cm wide, mostly lanceolate to narrowly elliptic or nearly linear, the margins nearly entire to shallowly and remotely dentate. Flowers borne in terminal spikes, the buds erect, opening near sunset. Hypanthium 2–4 cm long, usually bearing some glandular hairs. Sepals (0.7–) 1–2 cm long, the tips free or united at anthesis. Petals yellow (often drying cream-colored or pinkish), usually obcordate, 1–2 cm long. Style about as long as the stamens, the stigma lobes oblong. Capsules cylindric, 2–4 cm long, thicker near the base, dehiscent. Seeds sessile, horizontal, in 2 rows in each cell, brown, angular, 1–2 mm long.

O. canéscens Torr. & Frem.   Spotted evening primrose

L. *canescens*, gray, hoary.
White to pink with red spots or stripes; June–mid-September
Sandy prairie
West third, primarily

Low, bushy perennials with stems erect to decumbent, 1–2 dm tall, arising from slender rhizomes. Herbage more or less densely covered with stout appressed hairs (use magnification). Leaves sessile or very short-petioled, mostly lanceolate to linear-lanceolate, 7–20 mm long and 1.5–4 mm wide, the margins entire or with a few shallow teeth. Flowers sessile and solitary in the upper leaf axils, the buds erect, opening near sunset. Hypanthium 5–8 mm long. Sepals red-spotted, 8–10 mm long, at least some of them united. Petals pink (less frequently white) with red spots or stripes, obovate, 10–14 mm long. Pollen cohering in cobwebby masses. Style protruding just beyond the anthers, the stigma with 4 elongate lobes. Capsules ovoid, sharply 4-angled or winged, 8–9 mm long (including the beak), covered with appressed hairs. Seeds sessile, in several rows in each cell, light brown, angular, keeled or minutely winged on the angles, about 1–1.5 mm long.

O. coronopifòlia T. & G.

NL., from L. *coronopus*, Gr. *koronopous*, crowfoot.
White; late May–August
Sandy soil of roadsides, flood plains, pastures, and prairie
West fifth

Low perennials with stems branched from the base, 1–3.5 dm tall, arising from slender rhizomes 2–10 cm below the soil surface. Herbage covered with inconspicuous appressed hairs intermixed with larger spreading hairs. Leaves usually pinnatifid (less frequently merely toothed or nearly entire), 1–3.7 cm long and 4–13 mm wide. Flowers borne singly in the upper leaf axils, the buds nodding. Hypanthium 12–25 mm long, often reddish, with numerous long white hairs at the throat. Sepals 10–14 mm long, usually separating by anthesis. Petals white (drying pinkish), obcordate, 9–13 mm long. Stigma lobes oblong. Capsules oblong to ellipsoid, 8–20 mm long. Seeds sessile, ascending, in 2 rows in each cell, light brown, more or less obovoid, minutely tubercled, about 2 mm long.

O. fremóntii S. Wats.   Fremont evening primrose

Named after Capt. John C. Fremont.
Yellow; May–mid-August
Exposed sites in dry, rocky, calcareous prairie
Northwest fourth and Hodgeman County

Low, silvery herbaceous perennials, 0.5–3.5 dm tall, with 1 to several stems arising from a thickened woody root and caudex, the latter often branching below ground level.

331. *Oenothera fremontii*

332. *Oenothera grandis*

333. *Oenothera laciniata*

Leaves spatulate to very narrowly oblanceolate or elliptic (occasionally nearly linear), 4–11 cm long and 3–11 mm wide, entire or with a few remote, shallow teeth, the bases very gradually narrowed to a petiolelike base. Flowers borne singly in the upper leaf axils, the buds erect. Hypanthium 2.5–4.5 cm long. Sepals 2–2.5 cm long, the tips united. Petals yellow, obcordate to broadly obovate, 1.5–3.5 cm long. Pollen cohering in cobwebby masses. Style protruding beyond the anthers, the stigma with 4 elongate lobes. Capsules 1.5–3.3 cm long, indehiscent, the body itself square in cross section and 3–4 mm thick, the wings 2–3.5 mm wide. Seeds sessile, 8–10 per cell, in definite rows, brown, 2.5–3 mm long, with minute corky tubercles (use magnification).

This is one of our rarest plants, its total known distribution being limited to northwest and north-central Kansas and Franklin and Webster counties, Nebraska. Within the area just described, however, it is relatively common.

O. grándis (Britt.) Smyth

L. *grandis*, large, magnificent, noble—in reference to the flowers.
Yellow; May–early August (sometimes again in the fall)
Sandy soil of disturbed habitats such as roadsides, fields, flood plains, and pastures
Primarily in the south half of the west three-fifths

Erect, sparsely hairy annuals much resembling *O. laciniata* (and, in fact, once considered a variety of that species) but with larger flowers. Stems simple or with several branches at the base and/or above, 1–7.5 dm tall, the epidermis often exfoliating with age. Leaves as for *O. laciniata*, 2–9 cm long and 0.3–3.2 cm wide. Flowers borne singly in the leaf axils, the mature buds erect, opening at sunset. Hypanthium 2.8–5 cm long. Sepals 1.3–3 cm long. Petals yellow (sometimes drying cream-colored or pinkish), obcordate, 2–3.6 cm long. Stigma branches linear. Capsules cylindric, sparsely hairy, 2.5–4.5 cm long. Seeds sessile, ascending, golden brown, somewhat flattened and/or angled, pitted, about 1–1.5 mm long.

O. hoòkeri T. & G.

Named in honor of Sir William Jackson Hooker (1841–1865), author of *Flora boreali-americana* (1833–1840) and numerous other botanical works.
Pale yellow; late July–mid-October
Sandy soil of roadsides, pastures, flood plains
Mostly south and/or west of the Arkansas River

Erect biennials or short-lived perennials, up to 2 m tall, much resembling *O. biennis* and *O. strigosa* but with much larger flowers. Leaves mostly lanceolate to elliptic, 4–15 cm long and 0.8–3 cm wide, the margins nearly entire to remotely and shallowly dentate or serrate. Flowers borne in terminal spikes, the buds erect, opening near sunset. Hypanthium 2.5–5.5 cm long. Sepals 2.5–3.5 (–5) cm long. Petals pale yellow (often drying pinkish), obcordate, 2.5–4 cm long. Style slightly longer than the petals, the stigma branches linear. Capsules cylindric, thicker near the base, dehiscent, 2–3.5 cm long, hairy. Seeds sessile, horizontal, in 2 rows in each cell, brown, angular, 1–1.5 mm long.

O. laciniàta Hill   Cutleaf evening primrose

NL., divided into small parts, from L. *lacinia*, small piece or part—in reference to the lobes or divisions of the leaves.
Yellow; mid-April–July
Fields, prairies, roadsides, waste ground, frequently in sandy soil
Nearly throughout, more common eastward

Erect annuals or biennials with a slender taproot. Stems simple or branched, 0.7–3 dm tall. Herbage nearly glabrous to sparsely hairy. Basal leaves spatulate to oblanceolate, (2–) 4–16 cm long and 0.4–2 cm wide, the margins entire or with 1–3 pairs of sharp teeth or lobes below the middle; main stem leaves oblanceolate to oblong or lanceolate in general outline, usually pinnatifid, 2–6 cm long and 0.5–1.5 (–2) cm wide. Flowers borne singly in the leaf axils, the mature buds erect, opening near sunset. Hypanthium 17–30 mm long. Sepals 6–11 mm long, free or some united at anthesis. Petals yellow (sometimes drying pinkish, broadly obovate or obcordate, 6–11 mm long. Capsules cylindric, sparsely hairy, 2.4–4 cm long. Seeds sessile, ascending, in 2 rows in each cell, golden brown, more or less ovoid and somewhat angled, pitted, about 1.5 mm long.

O. linifòlia Nutt.   Flax-leaved evening primrose

NL., from *Linum*, the genus to which flax belongs, from L. *linum*, flax plant, + *folium*, leaf—in reference to the narrow leaves of this species which resemble those of various species of *Linum*.

Yellow; May–early July
Prairie, occasionally in open woods
Southeast eighth and Sumner County

Slender, erect annuals, 1.4–4.5 dm tall, with ascending branches. Herbage glabrous below, minutely hairy and/or glandular in the inflorescence. Leaves mostly threadlike, 1–2.5 cm long and less than 1 mm wide. Flowers borne in long-peduncled, few- to several-flowered terminal spikes 1–10 (–15) cm long, opening near sunrise, the buds erect. Hypanthium about 2 mm long. Sepals 1.5–2 mm long, the tips free or at least some of them united. Petals yellow (drying cream-colored or pink), obcordate, 3–6 mm long. Style about as long as the stamens, the stigma merely 4-lobed. Capsules 5–8 mm long and 2.5–3 mm thick, broadly club-shaped or ellipsoid, 4-angled or very narrowly winged, minutely hairy. Seeds clustered, stalked, not in definite rows, brown, ellipsoid, somewhat angled, about 1 mm long.

334. *Oenothera linifolia*

O. macrocárpa Nutt.   Ozark sundrops, Missouri evening primrose

NL., from Gr. *makros*, long, + *karpos*, fruit, in reference to the very large capsules.
Yellow; May–June (–mid-July)
Prairie slopes and uplands and roadcuts, usually on shallow, rocky, calcareous soil
East three-fourths, more common eastward

Low, silvery, drought-resistant perennials, 0.7–6.5 dm tall, with few to several decumbent or erect stems arising from a deep, thick root. Herbage in ours usually covered with white appressed hairs. Leaves narrowly lanceolate to elliptic, oblanceolate, or spatulate, 4–12 cm long and 0.5–3.5 cm wide, very gradually narrowed to petiolelike bases, the margins entire or rarely with a few barely discernible teeth. Flowers quite showy, borne singly in the leaf axils, the buds erect, opening in the evening. Hypanthium 5–12 cm long. Sepals 2–6 cm long, sometimes red-spotted, the tips united. Petals yellow (sometimes drying reddish), broadly obovate to slightly obcordate, 3–6.5 cm long, the apex apiculate. Stamen filaments somewhat flattened, enlarged toward the base, 2.5–3.5 cm long, the pollen cohering in cobwebby masses. Style protruding beyond the anthers, the stigma with 4 linear branches. Mature capsules mostly 5–8 cm long, the body square in cross section, 7–10 mm thick, tough but not woody, the wings 1.5–2.5 cm wide, giving the fruit an ovate or nearly round, apically notched outline in side view.

335. *Oenothera macrocarpa*

Most of our plants belong to var. *macrocarpa*. Entirely glabrous plants, in Kansas known from Meade, Clark, and Barber counties, are placed in var. *oklahomensis*. Densely hairy plants with ovate to broadly lanceolate leaves are known from Kiowa County and belong to var. *incana*.

O. macrocarpa is included in some manuals as *O. missouriensis* Sims.

Pollination is effected by night-flying hawkmoths, and possibly also by humming-birds, as in the case of *O. biennis* and *O. speciosa*.

See plate *63*.

O. pállida Lindl.

L. *pallidus*, ashen, pale.
White; late May–August (–September)
Dry, sandy soil of dunes, sand hills, flood plains, and roadsides
West half

Erect, gray-green perennials, 1–4 dm tall, with several ascending or sprawling branches, producing slender rhizomes. Epidermis of the stem more or less hairy, usually white and exfoliating. Leaves short-petioled, quite variable in shape, ovate or nearly round to lanceolate, elliptic, oblong, or linear, 1–5 (–7.5) cm long and 3–13 mm wide, the margins nearly entire to shallowly toothed or more or less pinnatifid. Flowers fragrant, borne singly in the upper leaf axils, the buds nodding, opening near sunset. Hypanthium 2–4 cm long. Sepals 2–2.5 cm long, often marked with red, usually united at the tips. Petals white (often drying pinkish), obovate, 1.3–2.5 cm long. Stigma branches linear. Capsules narrowly cylindric, 3–4 cm long, superficially resembling a piece of the stem. Seeds sessile, ascending, in 1 row in each cell, blackish, more or less ellipsoid, smooth, about 2 mm long.

336. *Oenothera macrocarpa* fruiting

O. rhombipétala Nutt. ex T. & G.   Rhombic evening primrose

NL., from L. *rhombus*, Gr. *rhombos*, mathematical figure with 4 sides whose opposite sides and angles are equal, + L. *petalum* or Gr. *petalon*, for a thin sheet or foil of metal, in botany applied to the thin leaves of the corolla, i.e., the petals—the petals of this species being rhombic or diamond-shaped.

Yellow; July–August (–mid-October)

Sandy soil of dunes, prairie, flood plains, pastures, and roadsides

Central three-fifths, primarily

Erect, soft-hairy biennials with a single stem 0.6–1.3 m tall. Epidermis of the stem thin and papery, tending to exfoliate with age. Leaves short-petioled or sessile, the blades narrowly oblanceolate to elliptic or linear-lanceolate below to linear-lanceolate or lanceolate above, 1.5–9 cm long and 3–17 mm wide, becoming gradually smaller upward, the margins entire or remotely and shallowly toothed. Flowers borne in many-flowered terminal spikes, the buds erect, opening near sunset. Hypanthium 2.5–3.5 cm long. Sepals 1.2–3 cm long, often tinged with red, usually united. Petals yellow, rhombic, 1.8–2.7 cm long. Stigma branches linear. Capsules cylindric, curved, tapering slightly toward the apex, 1.3–2.4 cm long. Seeds sessile, ascending, in 2 rows in each cell, more or less ellipsoid but somewhat angular, minutely pitted (see magnification), about 1–1.4 mm long.

337. *Oenothera pallida* ssp. *latifolia*

O. speciòsa Nutt.   Showy evening primrose

L. *speciosus*, glorious.

White (rarely rose); mid-May–early August

Roadsides, waste ground, disturbed areas in prairie

East half, primarily; scattered records as far west as Sheridan County

Herbaceous perennials, 2–7 dm tall, with several ascending or decumbent stems arising from a slender rhizome. Herbage covered with minute appressed hairs. Leaves 2–9 cm long and 0.3–2.8 cm wide, lanceolate to narrowly elliptic, oblong, or narrowly oblanceolate in general outline, gradually tapered to the base, the margins shallowly toothed to more or less deeply pinnatifid. Flowers borne in few-flowered, open spikes in the upper leaf axils, the buds nodding, opening in the evening (or on cloudy days) and remaining open through the morning. Hypanthium 1–2 cm long. Sepals 1.8–3.5 cm long, reflexing as the flower opens while more or less cohering at the tips. Petals white (rarely rose) with a yellow base, the white fading to pink, obovate to obcordate, 2–5 cm long. Pollen cohering in cobwebby masses. Style protruding beyond the anthers, the stigma lobes linear. Capsules 1.3–2 cm long, somewhat club-shaped, the fertile portion 3–5 mm thick, with 4 stout ribs alternating with 4 wings less than 1 mm wide. Seeds clustered, stalked, not in definite rows, brown, asymmetrically obovoid, about 1 mm long.

See plate 64.

338. *Oenothera rhombipetala*

O. strigòsa (Rydb.) Mack. & Bush

NL. *strigosus*, having sharp, straight, stiff appressed hairs.

Yellow; late June–September

Disturbed habitats of roadsides, fields, weedy flood plains, and pastures

Nearly throughout, more common eastward

Erect, strigose, gray-green biennials, 0.5–2 m tall, quite similar to *O. biennis*. Leaves usually ascending or held nearly erect, the blades mostly oblanceolate to lanceolate, narrowly elliptic or almost linear, 3.5–14 cm long and 0.8–2.6 cm wide, becoming gradually smaller upward and into the inflorescence, the margins remotely dentate, either flat or undulate. Flowers borne in terminal spikes with few or no glandular hairs, the buds erect, opening near sunset. Hypanthium 2–4 cm long. Sepals 1–2 cm long. Petals yellow (often drying pinkish), 0.8–2.3 cm long. Style about as long as the stamens, the stigma lobes oblong. Capsules 2–4 cm long, cylindric, thicker near the base, dehiscent, hairy. Seeds sessile, horizontal, in 2 rows in each cell, brown, angular, 1–2 mm long.

O. trilòba Nutt.   Three-lobe evening primrose

NL., from L. *tri-*, combining form of *tris* or *tres*, 3, + L. *lobus*, lobe—the petals sometimes being slightly 3-lobed.

Yellow; April–mid-May

Roadsides. pastures, prairie, usually in rocky, calcareous soil

East three-fourths, more common to the south and east

Low, acaulescent, winter annuals, biennials, or short-lived perennials, 0.6–2.2 dm tall, with a taproot. Leaves few to many in a rosette, 2.5–18 cm long and 0.4–5 cm wide, oblanceolate in general outline, remotely serrate to mostly deeply pinnatifid. Flowers sessile, the buds erect, opening near sunset. Hypanthium 3.8–8 cm long. Sepals 0.7–1.8 cm long, the tips usually united. Petals yellow (usually drying cream-colored or purplish), broadly obovate, sometimes appearing slightly 3-lobed, 1–2.5 cm long. Stamen filaments

339. *Oenothera speciosa*

flattened; pollen cohering in cobwebby masses. Style about as long as the stamens, the stigma lobes linear. Capsules 1.3–2.5 cm long, indehiscent, obovoid in general outline, hard, with 4 broad, reticulate-veined wings, often occurring in short, crowded, pine-cone-like spikes at the top of the root crown. Seeds sessile, in 2 rows in each cell, dark brown, about 2 mm long.

STENOSÌPHON Spach. Stenosiphon

NL., from Gr. *stenos*, narrow, + *siphon*, tube—the hypanthium tube extended above the ovary being extremely narrow.

A genus composed only of the following species and with the characters of this species.

S. linifòlius (Nutt.) Britt. Flax-leaved stenosiphon

NL., from L. *linum*, flax plant, + *folium*, leaf—calling attention to the similarity of the leaves to those of the flax plant, *Linum usitatissimum*.
White; late May–mid-October
Dry, rocky prairie hillsides, usually in calcareous soil
Nearly throughout, but more common westward

Slender, erect, biennial or short-lived perennial herb with a deep, woody taproot. Stems simple or with a few ascending branches, woody near the base, up to about 2 m tall. Herbage glaucous and glabrous below, minutely glandular-hairy within the inflorescence. Leaves alternate, sessile, estipulate, the blades lanceolate to lance-linear, 1.5–6 cm long and 4–15 mm wide, the apex acute or attenuate, the base slightly rounded, the margins entire. Flowers regular, borne in slender terminal spikes, each flower subtended by a linear-attenuate bract 3–5 mm long (somewhat longer in fruit). Hypanthium threadlike, 7–11 mm long. Sepals 4, subulate, 4–6 mm long, deciduous. Petals 4, white, clawed, 4–7 mm long, the blades broadly ovate to triangular or somewhat rhombic. Stamens 8. Style elongate, threadlike; stigma deeply 4-lobed. Fruit a minutely hairy, indehiscent, 1-seeded capsule 3–3.5 mm long, ovoid with an acuminate apex, triangular in cross section, 6- (8-)ribbed, transversely wrinkled between the ribs.

As summer advances, and especially during a dry period, transpiration is often reduced by casting off the leaves, photosynthesis still being carried on by chloroplasts abounding in the superficial tissues of the stem, while flowering and fruiting continue their course.

Only this 1 species of *Stenosiphon* is known to exist, and its range is restricted to a strip of territory extending from Nebraska down through Kansas, Colorado, Oklahoma, and Texas and into Mexico. The inflorescence and its individual flowers are much like those of *Gaura*, but *Stenosiphon* has a narrower, almost threadlike hypanthium tube extending beyond the top of the ovary. In comparison with that of *Gaura*, the limited range of *Stenosiphon* and its single species would indicate a more recent origin, possibly by mutation or cross-fertilization among the *Gauras*.

340. *Oenothera triloba*

341. *Stenosiphon linifolius*

HALORAGÀCEAE Water Milfoil Family

Ours all herbaceous, perennial, aquatic plants. Leaves alternate, opposite, or in whorls of 3–6, simple, when submersed pinnately parted into hairlike divisions, mostly estipulate. Flowers usually unisexual and monoecious, or unisexual and bisexual flowers on the same plant, regular, quite small, borne singly or in axillary clusters or terminal spikes. Calyx lobes 2–4 or absent. Petals 2–4 and deciduous or entirely absent. Stamens 4 or 8. Pistil 1; ovary inferior, 1- to 4-celled; styles of the same number as the cells; stigmas often featherlike. Fruit nutlike, sometimes separating into 4 1-seeded, indehiscent segments called mericarps.

The family includes about 8 genera and 100 species and is cosmopolitan in distribution.

One genus in Kansas.

MYRIOPHÝLLUM L. Water milfoil

NL., from Gr. *myrios*, numberless, + *phyllon*, leaf, in reference to the numerous, hairlike divisions of the submersed leaves.

Perennial, aquatic herbs with usually whorled leaves with fine, hairlike, pinnate divisions. Flowers minute, in the axils of upper leaves or in terminal spikes. Calyx lobes or teeth 4. Petals 4, or absent. Stamens 4 or 8. Pistil 4-lobed and 4-celled, almost appearing to be 4 separate pistils; stigmas 4. Fruit 4-lobed.

Three species in Kansas.

Key to Species:

1. Leaves of the middle and lower parts of the stem alternate to nearly opposite or irregularly scattered, less than 1 cm apart; mericarps with 2 prominently tubercled vertical ridges
.... *M. pinnatum*
1. Leaves of the middle and lower parts of the stem definitely whorled 2
    2. Leaves more than 1 cm apart; bracts of the inflorescence shorter than the flowers, ovate to oblong, entire or finely toothed; stamens 8; mericarps rounded on the back, smooth or finely tubercled .... *M. spicatum*
    2. Leaves less than 1 cm apart; bracts of the inflorescence much longer than the flowers, lanceolate to oblong or oblanceolate, spiny-toothed or lobed; stamens 4; mericarps with 2 tubercled vertical ridges on the back .... *M. heterophyllum*

M. heterophýllum Michx.

NL., from Gr. *heteros*, different, + *phyllon*, leaf—the submersed leaves differing in shape from the emergent leaves.
Flowers minute, greenish; June–August
Shallow water of small ponds, lake margins, and small streams
Scattered in the east half, primarily

Plants growing submersed. Stems usually branched. Leaves definitely in whorls of 4–6, those on the middle and lower parts of the stem usually much less than 1 cm apart; blades of 2 different kinds—those of the submersed leaves 2–5 cm long with 7–10 pairs of limp, threadlike divisions, the emergent leaves growing near the surface 0.7–3 cm long and merely toothed to pinnately lobed. Flowers minute, greenish, borne in whorls of 4–6 in terminal spikes, bisexual or with the lower flowers female and the upper male, each subtended by a firm lanceolate to elliptic or oblanceolate bract, much longer than the flowers, with toothed or narrowly pinnately lobed margins, and 1 or more minute membranous bractlets or bractioles. Calyx with 4 triangular-acuminate lobes. Petals 4. Stamens 4. Stigma under magnification neither conspicuously papillose nor recurved. Mature fruit 1.5–2 mm long, broadly ovoid, each mericarp with 2 vertical rows of hard tubercles.

M. pinnàtum (Walt.) BSP.

NL., pinnate, from L. *pinnatus*, feathered, from *pinna*, *penna*, feather—in reference to the leaf divisions.
Flowers minute, greenish; July–September
Shallow water or mud of lake and pond margins, occasionally in marshy areas or temporary pools
Scattered in the east four-fifths

Plants submersed in shallow water or rooting in mud. Stems often much branched. Leaves alternate, nearly opposite, or irregularly scattered, less than 1 cm apart; blades of 2 different kinds—those of the submersed leaves mostly 1.5–3 cm long with about 5 pairs of limp, hairlike divisions, the emergent leaves near the surface of the water 0.7–2 cm long and merely toothed or with a few pinnate lobes. Flowers minute (1.5–2 mm long), greenish or purplish, borne singly in the axils of the emergent leaves, bisexual or with the lower flowers female and the upper male. Calyx lobes 4. Petals 4. Stamens 4. Stigma under magnification papillose and erect or somewhat recurved. Mature fruit 1.5–2 mm long, broadly ovoid, each mericarp with 2 vertical rows of hard tubercles.

M. spicàtum L.

L. *spicatus*, pointed, from L. *spika*, point, ear of corn.
Flowers minute, greenish or purplish; mid-May–September
Shallow water of ponds and lake margins
Known from widely scattered sites across the state

Plants usually growing submersed. Stems branched or not, purplish when living but often becoming pale green or whitish when dry. Leaves definitely in whorls of 3–4, those on the middle and lower parts of the stem 1 cm or more apart; blades 1.5–3 cm long, with 6–11 pairs of limp, threadlike divisions. Flowers minute, greenish, borne in terminal spikes with female flowers in widely spaced whorls below and male flowers in densely crowded whorls at the tip of each spike, each flower subtended by a small, ovate to oblong, entire or finely toothed bract shorter than the flowers and 1 or more minute bractlets or bractioles. Male flowers 2.5–3.5 mm long at anthesis, each consisting of 4 minute sepals, 4 purplish petals, 8 stamens, and a rudimentary pistil. Female flowers about 1–1.5 mm long at anthesis, with 4 triangular calyx teeth, 4 keeled petals about 0.5 mm long, and 4 prominent (under magnification), papillose, eventually recurved stigmas. Mature fruit 2–3 mm long, nearly globose, the mericarps rounded on the back, smooth or finely tubercled.

UMBELLÍFERAE   Parsley Family, Carrot Family

The typical inflorescence in this family is the umbel, an indeterminate cluster with the pedicels of the flowers all radiating from a common point like the rays of an umbrella. The term "umbel" comes from L. *umbella*, meaning "parasol" or "sunshade." This is the Linnaean name for the family and is often retained in spite of its not having the typical *-aceae* family ending. Other names used for this family include Ammiaceae and Apiaceae, from the genera *Ammi* and *Apium*, respectively. The *International Code of Botanical Nomenclature* permits the use of the long-used "Umbelliferae," but the only correct alternate is "Apiaceae."

Annual, biennial, or perennial herbs, usually with alternate, compound or decompound leaves basally sheathing the hollow stems. Flowers small, usually bisexual, sometimes unisexual and bisexual flowers on the same plant, regular, borne in simple or compound umbels, or in heads in a few genera. When the umbels are compounded, the ultimate subdivisions are called umbellets. At the base of a simple or a compound umbel there may be small leaflike or bristlelike bracts which are referred to, as a group, as the involucre. Umbellets may also be subtended by similar small bracts which then compose the involucel. The stalks of the umbels are called peduncles, those of the umbellets are called rays, and those of the individual flowers are pedicels. Calyx of 5 toothlike sepals or absent. Petals 5, usually with an abruptly constricted, inflexed tip. Stamens 5. Pistil 1; ovary inferior, 2-celled; styles 2, usually with an enlargement called the stylopodium at the base; stigmas 2, minute. Fruit a schizocarp (from Gr. *schizein*, to split, + *karpos*, fruit) which at maturity splits into 2 indehiscent, 1-seeded segments called mericarps (Gr. *meris*, part, + *karpos*), each usually suspended from a slender stalk or carpophore (Gr. *karpos* + *pherein*, to bear). The surface of the fruit frequently has longitudinal ridges, ribs, or wings, and sometimes microscopic oil tubes in the intervals between these ribs and on the inner surfaces of the mericarps.

Identification of members of this family is usually simplified by having both flowers and mature fruits.

A very large family of about 200 genera and 3,000 species, world-wide in distribution but primarily of the Northern Hemisphere. Many of our table vegetables and seasoning plants are represented in this family: celery (*Apium graveolens* L.), parsley (*Petroselinum hortense* Hoffm.), carrot (*Daucus carota* L.), parsnip (*Pastinaca sativa* L.), dill (*Anethum graveolens* L.), caraway (*Carum carvi* L.), and chervil (*Anthriscus cerefolium* Hoffm.), for example. Other members such as water hemlock and poison hemlock contain poisonous resins or alkaloids, and extreme caution should be exercised in identifying any wild members of this family for eating purposes.

Kansas has 31 genera, counting escaped cultivars such as celery and parsnip. Only the most common representatives are included here.

Key to Genera (3):
1. Inflorescence neither a true umbel nor a compound umbel 2
  2. Inflorescence a dense, globose or ovoid head; each flower subtended by a spine-tipped bractlet .... *Eryngium*
  2. Inflorescence not as above; individual flowers not subtended by bractlets, though each umbellet may have a whorl of bractlets 3
    3. Cauline leaves once deeply palmately divided .... *Sanicula*
    3. Cauline leaves pinnately dissected .... *Torilis*
1. Inflorescence a true umbel or a compound umbel 4
  4. Fruit and ovary hairy, bristly, spiny, or tuberculate (use 10X hand lens) 5
    5. Plants with mature fruits 6
      6. Fruit roughly tuberculate .... *Spermolepis*
      6. Fruit hairy, bristly, or spiny 7
        7. Fruit narrow and elongate, the bristles inconspicuous, mostly appressed, neither hooked nor barbed .... *Osmorhiza*
        7. Fruit not as above 8
          8. Fruit with simple hairs (not hooked or barbed) .... *Lomatium*
          8. Fruit spiny or bristly, the spines or bristles hooked or barbed 9

BÉRULA Hoffm.   Water parsnip

L. *berula*, ancient Latin name of an herb.

Aquatic herbaceous perennials. Leaves alternate, once pinnately compounded. Flowers white; in compound umbels, the umbels and umbellets subtended by narrow bracts. Calyx teeth minute, subulate. Stylopodium conical. Fruit broadly ovate to round, flattened laterally, the outer wall thickened, corky, and with many oil tubes.

Only 1 species in the Western Hemisphere.

342. *Berula erecta*

B. erécta (Huds.) Cov.   Stalky berula

L. *erectus*, upright.
White; June–October
Streams, marshes, springs, often in sandy soil
Scattered primarily in the west three-fifths

Plants glabrous, erect or somewhat reclining, branched, 2–8 dm tall, with a cluster of fibrous roots, spreading vegetatively by means of rhizomes. Leaf blade 6–22 cm long and 4–16 cm wide, varying from lanceolate to oblong or ovate in outline, the margins finely serrate to pinnatifid or lacerated, sometimes nearly twice compounded. Umbels with 6–15 rays of different lengths, the bracts of the involucre linear, entire to toothed or occasionally lacerated and resembling the leaflets, 4–12 (–20) mm long; umbellets subtended by 4–8 linear or lanceolate bracts 1–5 mm long, the flower pedicels 2–5 mm long. Petals white, 0.5–1 mm long. Fruits 1.5–2 mm long.

CHAEROPHÝLLUM L.   Chervil

NL., from Gr. *chairephyllon*, chervil, from Gr. *chairein*, to rejoice, + *phyllon*, leaf—in allusion to the graceful fernlike leaves which in some species have an agreeable odor also.

Annual or biennial herbs with taproots or tubers. Leaves alternate, twice pinnately compounded. Flowers white in ours, minute, in compound umbels. Involucre usually absent, but umbellets subtended by numerous, conspicuous leaflike bractlets. Calyx teeth not evident. Stylopodium conical. Fruit linear or narrowly oblong to lanceolate, usually narrowed toward the apex and either rounded or narrowed at the base, flattened laterally, with small oil tubes alternating with the prominent ribs.

343. *Chaerophyllum procumbens*

Two species in Kansas.

Key to Species:

1. Stems rather weak, often branched from the base, the lowest branches often prostrate; finest subdivisions of the leaflets ovate to oblong, 1–2 mm wide; fruits 1–6 per umbellet, the pedicels threadlike, not enlarged upward .... *C. procumbens*
1. Stems more rigid, the branches all ascending; finest subdivisions of the leaflets linear to oblong or elliptic, 0.5–1.5 mm wide; fruits 3–20 per umbellet, the pedicels rather thick, often enlarged upward .... *C. tainturieri*

C. procúmbens (L.) Crantz.   Spreading chervil

L. *procumbere*, to bend or lean forward—in reference to the posture of the leaf rosettes and
    lower branches.
Flowers minute, white; late March–June
Moist soil of open woods, thickets, roadsides, and waste ground, occasionally in prairie
East two-fifths and Ellsworth County

Winter annuals with taproots, the seeds germinating in the fall and producing a rosette of prostrate, ternately compound leaves with pinnatifid leaflets that remain green all winter. Stems in spring glabrous or sparsely hairy, 0.7–4 dm tall, the lower branches prostrate, the upper ones ascending. Main leaves 2–6 cm long (excluding the petiole) and 1.5–6.5 cm wide; leaflets pinnately lobed or cut nearly to the midrib, 1–3 cm long and 1–2.5 cm wide, the finest subdivisions ovate to oblong, 1–2 mm wide. Petals white, 0.5–1 mm long. Fruits 5–7 mm long, 1–6 per umbellet, the pedicels threadlike, not enlarged upward.

As small and as few as the flowers are, they are visited for their nectar by various kinds of bees, flies, and bugs which bring about cross-pollination, the more certainly because the flowers are proterandrous, the anthers discharging their pollen before the stigmas are ready to receive it.

C. tainturièri Hook.

Named after L. F. Tainturier des Essarts who sent plants of Louisiana to Sir William Hooker from 1824–1836.
Flowers minute, white; mid-May–June
Roadsides, prairie pastures, rocky limestone soils
Eastern two-fifths

Erect annuals much resembling *C. procumbens* but generally taller, stiffer, more erect, and more conspicuously hairy than that species. Stems sparsely to moderately hairy, 1–5 dm tall, the branches ascending. Main leaves 2–8 cm long (excluding the petiole) and 1.5–5.5 cm wide; leaflets pinnately cut nearly to the midrib, 7–20 mm long and 5–15 mm wide, the finest subdivisions usually linear, less frequently oblong or elliptic, 0.5–1.5 mm wide. Petals white, 0.5–1 mm long. Fruits 5–6 mm long, broadest below the middle, the pedicels 3–20 per umbellet, tending to be enlarged upward and thicker when the fruits are mature than in *C. procumbens*.

344. *Chaerophyllum tainturieri*

CICÙTA L. Water hemlock

The classical Latin name for this and allied species.

Perennial, quite poisonous herbs with clustered, tuberous storage roots. Leaves alternate, 1, 2, or 3 times pinnately or ternately compounded; lateral veins from the mid-vein traceable to a sinus between teeth, whence a branch continues to the tooth. Flowers white, in compound umbels usually lacking an involucre, though small bractlets may subtend the umbellets. Calyx teeth well developed. Stylopodium flattened. Fruit nearly round to ovate or elliptic, flattened laterally, the prominent ribs rounded and corky.
One species in Kansas.

C. maculàta L. Water hemlock, spotted cowbane

L. *maculatus*, spotted, from *macula*, spot or blemish—the stems being more or less purple-spotted.
White; June–September
Marshy swamps, riverbanks, usually rocky limestone soils
East two-thirds, a few records farther west

345. *Cicuta maculata*

Stout, erect, glabrous perennials, 0.6–2 m tall, with a cluster of fascicled tuberous roots. Lowest portion of the stem enlarged, chambered. Leaves odd-pinnate, the lower ones 2 or 3 times compounded, 8–30 cm long (excluding the petiole) and 8–25 cm wide, the uppermost sometimes only 3-foliolate; leaflets narrowly elliptic to lance-linear, 7–12 cm long and 1–3 cm wide, apex usually attenuate, base rounded or acute, often oblique, margins serrate. Umbels with about 18–27 rays which are unequal or nearly equal in length and subtended by 1 or 2 linear bracts; umbellets 10- to many-flowered, subtended by a whorl of linear or awl-shaped bractlets. Calyx teeth ovate- or triangular-acuminate, about 0.5 mm long. Petals white, 1–1.5 mm long, clawed, the blade obovate or round. Fruits broadly ovate to nearly round, 2.5–4 mm long and 2–3 mm wide.

This plant is poisonous to humans and to all classes of livestock. The toxic principle is known as cicutoxin, which acts directly on the central nervous system, and is associated with the yellow, oily liquid present in the underground parts and, to a lesser extent, the lower portion of the stem (21). Early spring growth is toxic, as are even old, dead roots. A piece the size of a walnut, if ingested, is enough to kill a cow. According to Hardin and Arena (42), children have been poisoned by using whistles and peashooters made from the hollow stems. Usually, however, poisoning in humans comes about when someone mistakes the tuberous roots for those of an edible plant such as wild parsnip (*Pastinaca sativa*) or jerusalem artichoke (*Helianthus tuberosus*).

It goes without saying that we should learn to recognize this most poisonous of all our flowering plants. The fascicled tuberous roots and the chambered basal portion of the stem, both of which exude a yellowish, oily substance when cut, are useful recognition features.

It is stated in some publications that this is the plant used by the Athenians for the execution of Socrates, but after reading in Plato of the symptoms said to be recorded by Phaedo, a friend of Socrates who stayed by him to the end, we may be sure that it was not *Cicuta* but *Conium* that was used, causing a painless and gradual paralysis of the heart, and the muscles used in breathing. We quote Plato's *Phaedo* under *Conium maculatum*. In contrast to the effect of *Conium*, the symptoms in a fatal case of *Cicuta* poisoning are nausea, stomach pain, diarrhea, difficult breathing, rapid and weak pulse, convulsions, and finally death.

The North American Indians knew the lethal power of this plant, and sometimes their old men who felt themselves outmoded and humiliated by their waning skills committed suicide with a brew of *Cicuta*. Indian children, too, have been known to seek the same means of release from the chagrin of a reprimand.

CONÌUM L.   Poison hemlock

NL., from Gr. *koneion*, hemlock.

Fatally poisonous biennials. Leaves alternate, 2 or 3 times pinnately compounded. Flowers white, in compound umbels with both involucres and involucels of slender bracts. Calyx teeth not evident. Stylopodium short-conic, winged. Fruit broadly ovoid, somewhat flattened at the sides, the ribs prominent, wavy.

One species in Kansas.

C. maculàtum L.   Poison hemlock

L. *maculatus*, spotted, from *macula*, spot or blemish—in reference to the dark-purple spots at the base of the stem and the leaf rachis.
Flowers white; May–June
Moist soil of roadsides, flood plains, waste ground
East three-fourths and north part of west fourth; intr. from Eurasia

Robust, erect biennials, mostly 0.6–2 (–3) m tall, with a stout, hollow, purple-spotted stem; delicate, glabrous, fernlike leaves; and a thick taproot. Leaf blades 1.5–3 dm long and 0.5–3 dm wide; leaflets ovate to lanceolate, the smallest subdivisions toothed or pinnately lobed. Umbels with about 8–17 unequal rays, umbellets mostly 9- to 13-flowered; bracts of the involucre lanceolate, those of the involucel ovate, both with white, petal-like margins. Petals white, 1–1.5 mm long, clawed, the blade apex notched. Fruits 2–4 mm long.

All parts of this plant are poisonous to humans and to livestock, the fruits and leaves more so than the roots, and the former becoming more potent as the fruits become nearly ripe. The plants contain at least 5 closely related alkaloids, including coniine, N-methyl coniine, conhydrine, lamba-coniceine, and pseudoconhydrine. The alkaloid which predominates in the plant varies with the stage of growth and geographic area (21).

Undoubtedly it was a decoction of the seeds, and possibly of the leaves, of this plant that was used by the Athenians for putting Socrates to death. The physiological effect of the decoction we learn from Plato's *Phaedo*. We quote from Jowett's translation Phaedo's answer to a question by Echecrates about the last hours of Socrates: [Speaking to the servant who brought the cup] "Socrates said: You, my good friend, who are experienced in these matters, shall give me directions how I am to proceed. The man answered: You have only to walk about until your legs are heavy, and then to lie down, and the poison will act. At the same time he handed the cup to Socrates, who in the easiest and gentlest manner, without the least fear or change of color or feature, looked at the man with all his eyes. . . . Then holding the cup to his lips, quite readily and cheerfully he drank of the poison . . . and he walked about until, as he said, his legs began to fail, and then he lay on his back, according to the directions, and the man who gave him the poison now and then looked at his feet and legs; and after a while he pressed his foot hard and asked him if he could feel; and he said no; and then his leg, and so upwards and upwards and showed us that he was cold and stiff. And he felt them himself, and said: When the poison reaches the heart, that will be the end. He was beginning to grow cold about the groin, when he uncovered his face, for he had covered himself up, and said (they were his last words)—he said: Crito, I owe a cock to Asclepius; will you remember to pay the debt? The debt shall be paid, said Crito; is there anything else? There was no answer to this question; but in a minute or two a movement was heard, and the attendants uncovered him; his eyes were set, and Crito closed his eyes and mouth. Such was the end, Echecrates, of our friend, whom I may truly call the wisest, and justest, and best of all the men whom I have ever known."

CRYPTOTAÈNIA DC.   Honeywart, wild chervil

According to M. L. Fernald (9), NL., from Gr. *kryptos*, hidden or secret, + *tania*, a fillet, referring to the concealed oil tubes on the fruits.

346. *Conium maculatum*

347. *Cryptotaenia canadensis*

Smooth, perennial herbs. Leaves alternate, with 3 leaflets, the lateral pair oblique and sometimes deeply lobed. Flowers white, in compound umbels, the rays unequal in length, the umbellets few-flowered, usually lacking involucres and involucels. Calyx teeth quite minute, or absent. Stylopodium narrowly conical. Fruit linear-oblong, slightly flattened laterally, with low, rounded ribs.

One species in Kansas.

C. canadénsis (L.) DC. — Wild chervil

L. *-ensis*, suffix added to nouns of place to make adjectives—"Canadian," in this instance.
Flowers minute, white; May–September
Moist woods
East fourth

Slender, erect plants, 4–8 dm tall, with ascending branches and a cluster of somewhat thickened roots. Leaves long-petioled below, becoming shorter-petioled upward, the blades 8–14 cm long and 11–14 cm wide; leaflets lanceolate to ovate or rhombic, 4–15 cm long, the apex acute or acuminate, the base usually abruptly narrowed to a petiolelike base, the margins double-serrate, the lateral leaflets oblique and entire or more or less deeply 2-lobed. Umbels with 2–4 ascending or spreading rays. Petals white, obcordate, about 0.5 mm long. Fruits dark brown, prominently ribbed, 5–9 mm long.

This plant is cultivated in Japan, and the roots, stems, and leaves are cooked and eaten.

CYMÓPTERIS Raf.

NL., from Gr. *kyma*, wave, + *pteron*, wing—the wings of the fruit often being undulate or wavy.

Low, perennial herbs with fleshy, deep-seated taproots. Aboveground stems very short, arising at or shortly above ground level from a vertical subterranean stem. Leaves arising near the base of the plant, mostly pinnately or twice pinnately compound, the leaflets entire or variously toothed or lobed. Flowers white, yellow, or purple, in compound umbels with or without involucels but usually lacking an involucre. Calyx teeth minute or absent. Stylopodium lacking. Fruit ovoid to oblong, flattened dorsally, the lateral and usually 1 or more of the dorsal ribs broadly winged.

Two species in Kansas.

Key to Species:
1. Bracts of the involucels entirely herbaceous .... *C. acaulis*
1. Bracts of the involucels with broad, papery or petal-like margins .... *C. montanus*

348. *Cymopteris acaulis*

C. acaùlis (Pursh) Raf.

NL., from the Gr. prefix *a-*, without, + *kaulos*, stem—this species having no conspicuous aboveground stem.
White; April–May
Dry, rocky and/or sandy prairie hillsides, often on chalky outcrops
West half

Glabrous, acaulescent plants, 1–2 dm tall. Leaves once or twice compounded, the blades 2–5 cm long and 1–4 cm wide; leaflets lanceolate to ovate in general outline, several times pinnately lobed or toothed, mostly 7–20 mm long and 4–10 mm wide. Umbels compact, the rays short and thick; umbellets subtended by a whorl of entirely herbaceous bracts. Petals white, about 1 mm long. Fruits ovoid to short-oblong, 4–6 mm long and 4–6 mm wide, the wings quite conspicuous, often purple-tinged.

C. montànus (Nutt.) T. & G.

L. *montanus*, pertaining to mountains, from *mons*, *montis*, mountain.
White or purplish; April–May
Dry prairie hillsides, often on rocky or sandy soils
West half and Clay County

Pale green, acaulescent plants, 1–2 dm tall. Leaves somewhat succulent, once or twice compounded, the blades 3.5–7 cm long and 3–4 cm wide; leaflets lanceolate to ovate or rhombic in general outline, several times pinnately lobed or toothed, (5–) 10–20 mm long and 5–10 mm wide, roughened along the margins and veins. Umbels compact, the rays short and thick; umbellets (and occasionally the umbels also) subtended by a whorl of bracts with broad, papery or petal-like margins. Petals white or purplish, about 1 mm long. Fruits ovoid to nearly round, mostly 7–12 mm long and 7–12 mm wide, the wings quite conspicuous, often purple-tinged.

DAÙCUS L.   Carrot

L., from Gr. *daukos*, for a kind of carrot or parsnip. The name was applied to several species of the Carrot Family by Theophrastus (died c. 285 B.C.), and Dioscorides (1st century), who distinguished between the wild and the cultivated forms. Since the Middle Ages the name has been used only for the carrot. "Carrot" is from L. *carota*, from Gr. *karoton*, name used for the carrot by Athenaeus (2nd and 3rd centuries).

Erect, more or less hairy annual or biennial herbs with taproots. Leaves alternate, pinnately decompound, the leaflets linear, lanceolate, or oblong. Flowers white or yellowish (rarely pink), the outer ones often larger and more irregular than the central ones, in compound umbels with rays that curve inward after the flowers have matured. Involucre and involucels present or absent. Calyx teeth minute or absent. Fruit oblong to ovoid, flattened dorsally, the primary ribs filiform and bristly, the secondary ribs winged and armed with a single row of barbed or glochidiate prickles.

Two species in Kansas. *D. pusillus* Michx., an annual of smaller stature than *D. carota* and with umbels lacking the reddish central flowers usually found with that species, is reported from Douglas and Cherokee counties only. The fruits of *D. carota* are broadest at the middle, and each wing bears 12 or more rounded prickles; the fruits of *D. pusillus* are broadest below the middle, and each wing bears 1–8 flat prickles.

349. *Daucus carota*

D. caròta L.   Wild carrot, Queen Anne's lace

L. *carota*, old generic name for the carrot.
White; May–October
Roadsides, pastures, creek banks, and other disturbed habitats
East third, primarily, with scattered collections as far west as Sheridan County; intr. from Eurasia

Biennial plants with a strong taproot, occasionally blooming here the first year, but usually producing only a rosette of pinnately decompound, fernlike leaves, followed the second year by an erect, hairy flowering stem 0.3–1 m tall. Stems nearly glabrous to conspicuously and coarsely hairy, eventually branching. Leaves much dissected, the blades usually triangular in general outline, 7–17 cm long and 4–11 cm wide; leaflets highly variable in aspect, pinnately lobed or cut nearly to the midrib, the smallest subdivisions usually 0.5–2 mm wide and sharply pointed; petioles winged, clasping the stem. Umbels long-peduncled, subtended by a whorl of conspicuous, pinnately lobed bracts with linear or threadlike subdivisions; rays many, the outer ones longer than the inner and tending to curve inward, especially after fertilization; umbellets many-flowered, subtended by a whorl of linear, entire or few-toothed bractlets. Calyx teeth not evident. Petals white (those of the center flower usually pink to red or maroon), 0.5–2 mm long, clawed, those of the marginal flowers usually larger and more irregular than the rest. Stylopodium compressed; fruits bristly, 3–4 mm long. When the fruits are ripe, the incurved peduncles usually spread out again, giving the fruits better opportunity to hook onto passing animals.

The domesticated carrot has been derived from the wild carrot by giving seedlings the advantage of cultivation in a deep, fertile soil and by selection of the seeds from the more promising individuals generation after generation. However, it does not take long for the domesticated carrot to revert to the wild condition, also known as Queen Anne's lace, after it escapes from cultivation. This process, of course, does not take place in any single plant, but comes about as seedlings of the cultivated carrot, which still possesses the genetic potential of the wild carrot, germinate and mature without anyone weeding out the individuals whose roots lack the size and flavor desirable of garden plants.

This species, even in the wild state, has been put to numerous uses. An extract from the root has been the source of a yellow coloring for butter, and oil from the seeds is included in some perfumes. The pink or red flowers from the center of each umbel reputedly have aphrodisiac properties. An infusion of the entire plant, mordanted with alum and cream of tartar, produces a yellow dye. Medicinal applications have been many, and the seeds were official in the *U.S. Pharmacopoeia* from 1820 to 1882 as a diuretic, stimulant, and menstrual excitant. The root has also been used to treat jaundice and threadworm. The Mohegan Indians steeped the blossoms to prepare a treatment for diabetes, and the Crow Indians used the plant for ceremonial purposes and in some medications.

In some parts of North America, the fruits are eaten, to a small extent, by the ruffed grouse, ring-necked pheasant, cotton rat, pine mouse, and the Townsend mole (34).

ERÝNGIUM L.   Eryngo

NL., from Gr. *eryngion*, name used by Hippocrates, Father of Medicine (c. 460–359 B.C.), for one of his medicinal plants, evidently the species now known as *Eryngium campestre*.

350. *Eryngium yuccifolium*

Annuals, biennials, or perennial herbs with stout taproots or rootstocks with fibrous roots and prostrate or erect aboveground stems. Leaves alternate, entire or pinnately or palmately lobed. Flowers white or purple, sessile, in dense many-flowered heads rather than in a typical umbel, each flower subtended by a small bractlet, the heads and branches of the inflorescence subtended by a whorl of leaflike bracts. Sepals free from one another, conspicuous. Stylopodium lacking. Fruit globose or obovoid, only slightly compressed, covered with tubercles, papillae, or scales.

Three species in Kansas. *E. prostratum* DC., a small, prostrate perennial with heads up to 9 mm long, is known from 1 site in Cherokee County only.

Key to Species:
1. Plants pale green; leaves linear, entire, parallel-veined, bristly-margined; heads globose, lacking a whorl of bracts at the apex .... *E. yuccifolium*
1. Plants eventually becoming purplish, at least in the inflorescence; leaves oblanceolate to ovate or round in general outline, palmately lobed, the lobes pinnatifid and spiny-margined; heads ovoid, topped by an apical whorl of bracts .... *E. leavenworthii*

E. leavenwórthii T. & G.   Leavenworth eryngo

Named after its discoverer, Dr. Melines Conkling Leavenworth (1796–1862), physician, botanist, assistant surgeon in the U.S. Army, and nephew of Gen. Henry Leavenworth (1783–1834) for whom Leavenworth, Kansas, was named.
Purple; August–October
Prairie hillsides, roadsides, pastures, frequently on dry, rocky, shallow soil over limestone
East two-fifths

Erect annuals with a taproot. Stems branched above, 0.5–1 m tall. Main leaves sessile or very short-petioled, broadly obovate to round in general outline, palmately lobed, mostly 2.5–5 cm long, the lobes pinnately lobed or toothed, spiny-margined. Inflorescence branched, the heads purplish, ovoid or oblong, on short peduncles, subtended by and terminated by whorls of leaflike bracts. Each individual flower subtended by a linear bract about 1 cm long, with 3–7 spiny, attenuate lobes near the apex. Sepals 4–7 mm long, oblong, with 3 or 5 spiny, attenuate lobes at the apex. Petals magenta, 3–3.5 mm long, with 4 longitudinal wings and 2 basal auricles. Stamens blue, the filaments threadlike, protruding conspicuously beyond the corolla. Fruits oblong, 2–4 mm long, densely covered with white, inflated, papillose, club-shaped scales.

The plants may be propagated successfully by sowing the seeds, as soon as they are ripe, although they will not germinate until the following spring.

For use in winter bouquets, hang the plants upside down in a dimly lighted place until the stems are dry and will no longer droop.

See plate *65.*

E. yuccifòlium Michx.   Button snakeroot, rattlesnake master

NL., from the genus *Yucca*, + L. *folium*, leaf—signifying a leaf like that of a yucca plant.
White or greenish; June–October
Prairie, usually in rocky soil
East fourth and Butler County

Rigidly erect, herbaceous perennials with a cluster of thickened fibrous roots. Stems usually unbranched below the inflorescence, 0.3–1.5 dm tall. Leaves sessile, linear-attenuate, parallel-veined, bristly-margined, mostly 1–5 (–8) dm long and 7–25 mm wide, becoming gradually smaller above. Inflorescence branched, the heads greenish, globose, 1.5–3 cm long, subtended by a whorl of inconspicuous green bracts. Each individual flower subtended by a lance-attenuate bractlet about 8–10 mm long. Sepals rigid, ovate-acuminate, about 3 mm long. Petals white, about 2 mm long. Stamen filaments threadlike, eventually protruding conspicuously beyond the corolla and bracts. Fruits 4–8 mm long, covered with flat, stramineous scales.

As to the medicinal reputation of the *Eryngium* since the time of Hippocrates, Gerard's herbal, published in Engand in 1597, says, speaking of *Eryngium maritimum*: "The roots if eaten are good for those that be liver sick; and they ease cramps, convulsions and the falling sickness. If conditioned or preserved with sugar, they are exceeding good to be given to old and aged people that are consumed and withered with age, and who want natural moisture." But by 1847, R. Egglesfeld Griffith, M.D., could say of it in his *Medical Botany*, "In fact, no article of the Materia Medica was at one time in greater repute than this, nor [is there] any one that has more completely been forgotten."

At the same time he says that *E. yuccifolium* "has been used with some success as a diaphoretic and expectorant. In large doses it proves emetic. The root, which is the part that is officinal, is pungent, bitter and aromatic." According to Weiner (34), the Creeks used the plant as a diuretic, and both the Creeks and Koasati used it to induce vomiting prior to a hunt or important ceremonies. The Natchez Indians of Mississippi inserted wads of the chewed stem and leaves into the nostrils to stop nosebleed. White settlers in North America adopted various uses of the plants, and physicians later accepted it for use as a stimulant, diuretic, and expectorant. Krochmal and Krochmal (30) list 11 uses for the roots, including the treatment of liver ailments, rattlesnake bite, rheumatism, respiratory ailments, and urinary tract problems. C. F. Millspaugh, in his classic work *American Medicinal Plants, an Illustrated and Descriptive Guide to the American Plants Used as Homeopathic Remedies* (43), listed 18 additional uses for it, including treatment of impotency and of exhaustion from sexual depletion.

LOMÀTIUM Raf.   Lomatium

351. *Lomatium foeniculaceum*

NL., from Gr. *lomation*, small border—in allusion to the wings of the fruits.

Low perennial herbs with a thickened taproot and lacking aboveground stems. Leaves basal, pinnately or ternate-pinnately decompound. Flowers yellow, white, or pinkish, in compound umbels, the rays usually unequal in length, the umbellets several- to many-flowered, involucre small or absent, involucels present. Calyx teeth minute or absent. Stylopodium lacking. Fruit oblong to nearly round, dorsally flattened, lateral ribs with conspicuous flat wings, the others slender.

Two species in Kansas. *L. orientale* Coulter & Rose, with white flowers, is reported from several central and western counties but is rarely encountered.

L. foeniculàceum (Nutt.) T. & G.   Carrotleaf lomatium

NL., from the genus *Foeniculum*, fennel, + L. suffix *-aceus*, having the nature of—from the resemblance between these two plants.
Greenish-yellow; March–May
Prairie hillsides, rocky limestone soils
East three-fifths

Perennial, acaulescent herbs, (0.3–) 1–4 dm tall, with a strong, deep, somewhat swollen taproot. Leaf blades ovate in general outline, pinnately decompound, the leaflets finely divided into linear segments; petioles much expanded at the base. Umbels with 5 to many unequal rays, lacking an involucre; umbellets subtended by an involucel of united, woolly, membranous bracts. Petals yellow, 1–1.5 mm long, obcordate. Mature fruits broadly elliptic, winged, 5–8 mm long and 3.5–6 mm wide, glabrous to more or less pubescent.

This species has been divided into 2 species, subspecies, or varieties by some authorities, primarily according to the degree of hairiness of the ovaries and mature fruits. Within any single colony, one finds plants with ovaries varying from glabrous to more or less hairy, though the ratio of one to the other will differ. If one wishes to attach names to these variants, plants with ovaries and fruits glabrous (rarely pubescent) may be assigned to var. *daucifolium*; those with ovaries moderately to densely pubescent and mature fruits pubescent to nearly glabrous, to var. *foeniculaceum*.

OSMORHÌZA Raf.   Sweet Cicely

NL., from Gr. *osme*, smell or odor, + *rhiza*, root.

Erect, perennial herbs with thick fascicled roots and branched aboveground stems. Leaves ternate or ternate-pinnate, the leaflets lanceolate to ovate or nearly round, the margins serrate or pinnately lobed. Flowers white or greenish in ours, in compound umbels, rays few, unequal in length, involucre usually lacking, involucels present or absent, rays and pedicels ascending to spreading or widely divergent. Calyx teeth absent. Stylopodium conical. Fruit linear to oblong; cylindric to club-shaped; obtuse, tapering, beaked, or constricted at the apex; rounded or caudate at the base; compressed laterally; the filiform ribs acute.

Two species in Kansas, *O. claytonia* (Michx.) Clarke is known only from Doniphan, Leavenworth, and Wyandotte counties. This species lacks the anise scent of *O. longistylis* and has styles never longer than 1.5 mm.

O. longistỳlis (Torr.) DC.   Anise root

NL., from L. *longus*, long, + Gr. *stylos*, pillar or column, used in botany for the style of the pistil, i.e., "long-styled."
White; April–July
Moist soil of wooded hillsides and creekbanks
East third

352. *Osmorhiza longistylis*

353. *Osmorhiza longistylis* fruits

354. *Polytaenia nuttallii*

266

Stems 3–7 dm tall, branched above. Leaves long-petioled below, nearly sessile above, the leaflets thin, ovate to lanceolate, elliptic, or rhombic, serrate or double-serrate and often pinnately lobed or incised toward the base as well, (2.5–) 5–9 cm long and 1.5–6 cm wide. Umbels with 3–8 rays, subtended by a few narrowly elliptic or linear-attenuate bracts; umbellets several-flowered, bractlets of the involucel elliptic, attenuate, becoming sharply reflexed. Petals white, broadly elliptic to obovate, usually appearing apically notched. Mature fruits dark, oblong-fusiform, often curved slightly, the apex beaked by the persistent styles, the base tail-like, the ribs with appressed white bristles, the entire fruit 1.5–2 cm long.

The aromatic roots have sometimes been used as the source of an oil similar to anise (*Pimpinella anisum* L.).

Densely hairy individuals are segregated by some authors into var. *villicaulis* Fern., the less glabrous plants then being placed in var. *longistylis*.

See plate *66*.

POLYTAÈNIA DC.

NL., from Gr. *polys*, many, + *taina*, thin strip of ribbon—because the many crowded oil tubes in the fruit give that effect in cross section.

Perennial herbs with spindle-shaped taproots and stout, erect, branched, above-ground stems. Leaves pinnately dissected (bipinnate or ternate-pinnate), the leaflets deeply incised or merely lobed or crenate. Flowers yellow, in compound umbels with small, several- to many-flowered umbellets, lacking involucres but with involucels of small, linear or threadlike bracts. Calyx teeth ovate or triangular with acute or acuminate apex. Stylopodium absent. Fruit broadly ovate to nearly round or obovate, dorsally flattened, the dorsal ribs inconspicuous, the lateral ribs with a thick corky wing.

One species in Kansas.

P. nuttállii DC.   Prairie parsley

In honor of Thomas Nuttall (1786–1859), professor of natural history at Harvard.
Yellow; mid-April–June (–July)
Prairie, often in rocky limestone soil
East third and Edwards County

Stem erect, minutely hairy, 5–10 dm tall, eventually branching above. Leaf blades twice pinnately compounded or ternate-pinnate, triangular in general outline, mostly 6–18 cm long and 7–15 cm wide, becoming smaller and shorter-petioled toward the inflorescence; leaflets 2–4 cm long, ovate to lanceolate, pinnately lobed and/or toothed, the lobes reaching most or all the way to the midrib; petiole bases enlarged, sheathing the stem. Umbels with 10–20 stout rays somewhat unequal in length; umbellets many-flowered, crowded, subtended by a whorl of linear bractlets. Calyx teeth 0.3–0.5 mm long. Petals yellow or greenish, the blade 1–1.8 mm long, oblong, the apex abruptly constricted and folded inward, making the blade appear apically notched, the short claw minutely spurred (use magnification). Mature fruits oblong to broadly elliptic or nearly round in outline, 5–8 mm long, with a broad, thick corky wing.

PTILÍMNIUM Raf.   Mock bishop's-weed

NL., from Gr. *ptilon*, feather, wing, leaf, + *limnion*, diminutive of *lime*, marsh, lake, pool—apparently in allusion to the finely divided leaves and to the habitat.

Slender, erect annuals with fibrous roots. Leaves alternate, twice-pinnately compound with threadlike leaflets, or in some species (not ours) the leaves represented merely by the petiole. Flowers white, in compound umbels with involucre and involucels. Calyx teeth triangular or lanceolate. Stylopodium conical. Fruit ovoid to nearly round, flattened laterally, the lateral ribs small or with corky wings.

Two species in Kansas. *P. capillaceum* (Michx.) Raf. is known only from Crawford County. It differs from the following species in having styles 0.2–0.5 mm long, broadly triangular calyx lobes, 3-cleft involucral bracts, and usually 3 leaf segments at each joint of the leaf rachis.

P. nuttállii (DC.) Britt.

Named for its discoverer, Thomas Nuttall (1786–1859), professor of natural history at Harvard.
White; mid-June–mid-August
Moist prairie, occasionally on waste ground or in idle fields
Cherokee, Labette, Crawford, Neosho, and Bourbon counties

Plants eventually branched above, 3.5–6 dm tall. Leaves up to 9 cm long and

4 cm wide, the subdivisions less than 0.5 mm wide, usually just 2 at each joint of the rachis. Umbels with 20–30 rays unequal in length; umbellets 20- to 30-flowered; bracts of the involucre and involucel threadlike and entire. Calyx lobes lance-attenuate or ovate-attenuate, about 0.5 mm long, persistent on the fruit. Petals white, 0.8–1.5 mm long, the apex abruptly constricted and inflexed and therefore appearing notched. Mature fruits 1.5–2 mm long, the ribs rounded and somewhat corky.

SANÍCULA L. Sanicle, black snakeroot

NL. and ML., from L. *sanare*, to heal—the herbals of the Middle Ages praising the healing power of the European species, with unjustified extravagance.

Biennial or perennial herbs. Leaves alternate, long-petioled at the base and gradually becoming shorter-petioled above, the blades simple in ours but palmately divided into 3–5 (–7) segments. Flowers greenish-white or -yellow, bisexual and male flowers on the same plant, in compound umbels with unequal rays and conspicuous leaf-like involucres. Sepals conspicuous, fused to one another at the base. Stylopodium absent. Fruit ovoid or oblong to nearly round, slightly flattened laterally, covered with hooked bristles.

Two species in Kansas.

Key to Species:

1. Petals white; calyx lobes awl-shaped; male flowers usually 3 per umbellet .... *S. canadensis*
1. Petals greenish-yellow; calyx lobes ovate to lanceolate; male flowers 12–25 per umbellet
.... *S. gregaria*

355. *Polytaenia nuttallii*

S. canadénsis L.

L. *-ensis*, suffix added to nouns of place to make adjectives, "Canadian," in this instance.
White, inconspicuous; June–August
Rich, moist soil of wooded hillsides and creek banks
East two-thirds, primarily; a few records from the west third

Smooth, erect biennials, 3–8 dm tall, with a short rhizome bearing fibrous roots. Stem dichotomously branched above. Leaf blades 3–11 cm long and 6–14 cm wide, usually deeply 3-parted, the middle segment elliptic to oblanceolate, the lateral 2 more or less ovate to lanceolate, at least those of the lower leaves often 2-parted nearly to the base, the margins coarsely serrate or double-serrate, often lacerated near the apex as well. Umbels with 2–3 rays; umbellets crowded, 4- to 6-flowered, with 2–3 male flowers and 2–3 bisexual flowers, subtended by a whorl of minute bractlets. Calyx lobes awl-shaped, rigid, about 0.75 mm long. Petals white, shorter than the sepals. Styles inconspicuous, shorter than the calyx. Mature fruits usually 3 per umbellet, globose to ovoid, 3–6 mm long.

356. *Ptilimnium nuttallii*

S. gregària Bickn.

L. *gregarius*, pertaining to a flock or herd, common.
Greenish-yellow, inconspicuous; mid-May–June
Moist, rich soil of wooded hillsides, bluffs, and creek banks
East fourth, primarily

Smooth, erect perennials, 4–8 dm tall, much resembling *S. canadensis*. Flowers greenish-yellow, inconspicuous, the umbellets 15- to 30-flowered, the male flowers 12–25 per umbellet, either in separate umbellets or intermixed with bisexual flowers. Calyx lobes ovate to lanceolate, about 1 mm long. Styles eventually protruding beyond the calyx.

Earlier manuals erroneously referred to our plants as *S. marilandica* L., a more northern and eastern species which does not actually occur in Kansas.

SPERMÓLEPIS Raf. Scale-seed

NL., from Gr. *sperma*, seed, + *lepis*, scale—alluding to the bristly fruits.

Smooth, delicate, erect annuals with slender taproots. Leaves alternate, ternately decompound, with linear or threadlike leaflets. Flowers white, in compound umbels with a few rays, the umbellets 2- to 6-flowered, with involucels of a few small linear bracts but lacking involucres. Calyx teeth absent. Stylopodium low-conical. Fruit ovoid, somewhat flattened laterally, inconspicuously ribbed, smooth to tubercled or spiny.

Three species in Kansas. *S. divaricata* (Walt.) Raf. and *S. echinata* (DC.) Heller are uncommon and are not treated in detail here. *S. echinata* is often branched from the base and has fruits covered with hooked bristles. *S. divaricata* resembles our more common

357. *Sanicula canadensis*

PARSLEY FAMILY

267

358. *Sanicula canadensis* fruits

359. *Sanicula gregaria*

360. *Taenidia integerrima*

S. inermis but has spiny fruits, and the rays of the umbels are about equal in length, widely spreading in fruit.

S. inérmis (Nutt.) Math. & Const.

L. *inermis*, unarmed, without weapons.
White, inconspicuous; June
Prairie, pastures
East two-thirds and Scott County

Plants slender, dichotomously branched above, 2–5.5 dm tall. Leaves finely dissected, 1.5–4.5 cm long, the ultimate subdivisions threadlike. Umbels with 5–11 erect, unequal rays. Petals white, to about 0.5 mm long. Mature fruits round to ovate in outline, tubercled 1.5–2 mm long.

TAENÍDIA Drude   Yellow pimpernel

NL., from Gr. *tainidion*, a little band—in reference to the small, scarcely prominent ribs of the fruits.

Smooth, glaucous, erect, perennial herbs with taproots. Leaves alternate, 1- to 3-times pinnately or ternately compound. Flowers yellow, in delicate compound umbels lacking involucres and involucels. Calyx teeth and stylopodium absent. Fruit elliptic to broadly ovate-oblong, flattened laterally, with low ribs.

A genus containing but a single species.

T. integèrrima (L.) Drude

NL., from L. *integer*, whole, entire, + the suffix -*imus*, -*ima*, indicating a condition or state—in reference to the margins of the leaflets.
Yellow; May–mid-June
Wooded hillsides, bluffs, and creek banks, usually in moist, rocky soil
East fifth

Slender plants 3–5 dm tall. Leaves long-petioled below, becoming shorter-petioled above; leaflets ovate to lanceolate, elliptic, or oblanceolate, entire or deeply 2- or 3-lobed or -parted, 1.5–5 cm long and 0.7–2 cm wide, the apex rounded or acute, the base often oblique. Umbels with 7–14 slender rays unequal in length; umbellets with 15–20 flowers on threadlike pedicels. Petals yellow, about 1 mm long, the apex abruptly constricted and inflexed and therefore appearing notched. Mature fruits 3–4 mm long.

THÁSPIUM Nutt.   Meadow parsnip

NL., a play upon *Thapsia*, the name of a related genus, so called from the peninsula of Thapsos.

Erect, branched, perennial herbs. Leaves alternate, simple below to ternately compound above with toothed leaflets. Flowers yellow, greenish, or purple, in compound umbels lacking involucres. Calyx teeth ovate or obovate. Stylopodium absent. Fruit oblong or ovoid, slightly flattened or nearly round in cross section, at least some of the ribs prominently winged.

One species in Kansas.

T. barbinòde (Michx.) Nutt.

NL., from L. *barba*, beard, + *nodus*, knot, swelling, joint of the stem (node)—in reference to the minutely hairy nodes of this species.
Pale yellow or cream-colored; mid-May–June
Wooded hills, bluffs, and creek banks, usually on moist, rocky limestone soil
East fifth, primarily in the east column of counties

Smooth, erect plants, 5–10 dm tall, with a short, erect rhizome bearing thick fibrous roots. Upper nodes of the stem minutely hairy. Leaves twice ternately compound, at least above, ovate or triangular in general outline, mostly 5–12 cm long and 3.5–15 cm wide, long-petioled below, becoming smaller and quite short-petioled above; leaflets ovate to lanceolate 2.5–5 cm long, coarsely serrate and often with 1–4 deep basal lobes as well. Umbels mostly 3–5 cm wide (to 8 cm in fruit), with 10–15 rays nearly equal in length; umbellets many-flowered, subtended by a few linear-attenuate bractlets. Calyx teeth ovate, about 0.5 mm long. Petals pale yellow or cream-colored, about 1 mm long, the blade oblong or broadly elliptic, the apex abruptly constricted and inflexed, giving the blade a notched appearance. Mature fruits broadly elliptic to oblong in side view, scarcely flattened, 4–5.5 mm long, most of the ribs with wings 1–1.5 mm wide.

TÒRILIS Adans. Hedge parsley

NL., possibly suggested by L. *torulus*, tuft of hairs—in reference to the thin bristles covering the fruits.

Hairy, erect or decumbent, branched annuals with slender taproots. Leaves alternate, 1 or more times pinnately compound. Flowers white, in compound umbels with involucels of linear bracts, involucres present or lacking. Calyx teeth evident or not so. Stylopodium conical. Fruit ovoid to narrowly oblong, flattened laterally, covered with hooked bristles or merely with tubercles.

One species in Kansas.

T. arvénsis (Huds.) Link.

L. *arvensis*, pertaining to a cultivated field, from L. *arvum*, cultivated field, plowed land.
White; mid-June–July
Disturbed habitats such as roadsides, river banks, pastures, and waste ground, and open areas in prairies
East half, primarily; intr. from the Old World

Slender, erect, rough-hairy, grayish-green annuals, 1.5–8 dm tall, with ascending branches. Leaves mostly once-pinnate (the lowermost sometimes twice-pinnate), the blades triangular in general outline, 3.5–12 cm long and 4–10 cm wide; leaflets ovate to lanceolate, pinnatifid at the base to merely sharply incised or toothed at the apex, 1.5–5.5 cm long. Umbels long-peduncled, with 2–10 unequal rays, subtended by 1 threadlike bract; umbellets few-flowered, the bractlets linear-attenuate and inconspicuous. Petals white, about 0.5–1.5 mm long, those toward the periphery of each umbellet usually 2-lobed and larger than those toward the center. Mature fruits 3–5 mm long, densely covered with barbed bristles which readily catch in clothing.

ZÍZIA W. D. J. Koch Golden alexanders

In honor of J. B. Ziz, German botanist.

Erect, branched perennial herbs with a cluster of thickened roots. Leaves alternate, simple or compound below to 1- to 3-times compound above. Flowers yellow, in compound umbels, with involucels of small linear bracts but lacking involucres, the central rays usually shortest and the center flower or fruit of each umbellet sessile. Calyx teeth evident. Stylopodium absent or much flattened. Fruits glabrous, ovate to oblong, flattened laterally, with threadlike or narrowly winged ribs.

Two species in Kansas. *Z. aptera* (Gray) Fern. is known from Cherokee County only. It differs from the following species in having entire, cordate basal leaves and crenate, rather than serrate margins, at least on the lower leaves.

Z. aùrea (L.) W. D. J. Koch Golden zizia

L. *aureus*, golden.
Yellow; late April–early July
Wooded hillsides and creek banks, less frequently in prairie or along roadsides, often in moist, rich, rocky soil
East third and Ottawa County

Smooth, erect, branching perennials, 3–7 dm tall, arising from a rhizome producing both thickened and fine fibrous roots. Leaves once or twice ternately compound, long-petioled below to sessile above; leaflets ovate to lanceolate, 3.5–7 cm long, with serrate margins, often with 1 or 2 deep lateral lobes, the lateral leaflets often oblique. Umbels with 10–20 rays subequal in length; umbellets several- to many-flowered, crowded, subtended by a few inconspicuous, awl-shaped or attenuate bractlets 1–3 mm long. Petals yellow, 1–1.5 mm long, the apex abruptly constricted and curved inward and appearing apically notched. Mature fruits oblong, 3–4 mm long, with narrowly winged ribs.

Though the individual flowers are insignificant in size, the multitude of them in a compound umbel makes a showy inflorescence attractive to an astonishing variety of insects, including bees, flies, butterflies, and beetles. Cross-pollination by these agents is insured by the fact that the flowers are proterandrous, the stamens projecting beyond the perianth and discharging pollen before the stigmas of the same flower are mature.
Compare with *Thaspium barbinode*.

361. *Thaspium barbinode*

362. *Thaspium barbinode* fruits

363. *Torilis arvensis*

PARSLEY FAMILY

269

364. *Zizia aurea*

365. *Zizia aurea*, close-up of flowers

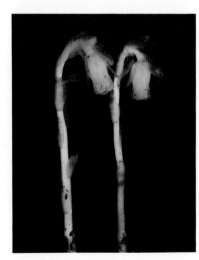

366. *Monotropa uniflora*

ERICÀCEAE   Heath Family

Trees or shrubs, less frequently vines or fleshy parasitic or saprophytic herbs. Leaves mostly alternate, simple (reduced to scales or bracts in *Monotropa*), petiolate or sessile, estipulate. Flowers bisexual, regular, borne singly or in various types of terminal or axillary clusters. Sepals 4–7, free or fused, usually persistent. Petals 4–7, usually fused together at least part of their length. Stamens as many as or twice as many as the petals or corolla lobes. Pistil 1; ovary superior or inferior, mostly 5-celled; style 1; stigma 1, sometimes with rays or lobes of the same number as the cells of the ovary. Fruit a capsule or a berry.

A widely distributed family of about 75 genera and 2,000 species. Cultivated members of the family include the blueberry (*Vaccinium* spp.), cranberry (*Vaccinium macrocarpon* Ait.), *Rhododendron* spp., heather (*Calluna vulgaris* Hull), heath (*Erica* spp.), trailing arbutus (*Epigaea repens* L.), and wintergreen (*Gaultheria procumbens* L.).

Kansas has 2 genera. The *Vacciniums*, being woody, are not included here.

MONÓTROPA L.   Indian pipe

NL., from Gr. *monos*, one, + *tropos*, turn—the top of the flowering stem being turned to one side.

Low, fleshy, parasitic or saprophytic herbs with matted fibrous roots. Herbage and flowers pinkish or white in ours, apparently without chlorophyll, turning black upon drying. Leaves reduced to sessile scales or bracts. Flowers 1 to several, nodding. Sepals 2–5, free from one another, eventually deciduous. Petals 4–5, free from one another, the corolla urn-shaped or narrowly bell-shaped in ours. Stamens 8 or 10, the filaments subulate, hairy, the anthers dehiscing through 2 slits at the top. Ovary superior, 4- to 5-celled, with a short, thick style and a broad, peltate stigma. Fruit an erect, many-seeded, loculicidal capsule. Seeds minute.

Two species in Kansas. *M. hypopithys* L. is quite rare and is known from Douglas and Cherokee counties only. It differs from *M. uniflora* primarily in having few to many flowers 0.8–1 cm long borne in racemes.

This genus is sometimes placed in a separate family, the Monotropaceae.

M. uniflòra L.   Indian pipe

NL., from L. *unus*, one, + *flos*, *floris*, flower—having a single flower.
White to pale pink; late August–early October
Moist, rich humus of oak-hickory forests
Central counties of the east fifth and Neosho County; rather rare

Plants smooth, waxy white, eventually becoming pinkish and then black, mostly 0.8–2 dm tall, the stems unbranched, arising singly or in clusters of 2 to several. Leaves or bracts of the stems more or less oblong or elliptic, 0.5–2 cm long and 3–8 mm wide, smaller toward the base of the plant. Flowers colored as the rest of the plant, borne singly, terminal, nodding at anthesis but becoming erect as the fruits mature, 1.2–2 cm long. Sepals 2–4, resembling the upper bracts of the stem, soon falling away. Petals usually 5, oblong-spatulate, the apex truncate or very broadly rounded, the base somewhat pouch-like. Stamens usually 10. Style short, broad. Capsule broadly ellipsoid, about 1 cm long (excluding the persistent style and stigma), often persisting along with the old stems into the next flowering season. Seeds light brown or pale orange, rather membranous, elongate, about 0.5–1 mm long.

PRIMULÀCEAE   Primrose Family

After the genus *Primula*, whose type species is *Primula veris*, from L. *primulus*, the diminutive of *primus*, the first, + *veris*, of spring.

Annual or perennial herbs. Leaves all basal or on the stem above in pairs or whorls, usually simple, sessile or petiolate, estipulate. Flowers bisexual, regular, borne singly in the leaf axils or in various types of terminal or axillary clusters. Calyx usually with 5 lobes. Petals 5, rarely absent, fused together at least part of their length, rarely free from one another. Stamens as many as the petals or corolla lobes, inserted on the corolla tube or at its base. Pistil 1; ovary superior (half-inferior in *Samolus*), 1-celled; style 1;

stigma 1, capitate or truncate. Fruit a capsule splitting, at least apically, into 2–8 segments, or splitting horizontally around the circumference.

The family includes about 28 genera and more than 800 species especially abundant in North Temperate regions of the world. Cultivated members of the family include the primrose (*Primula* spp.), rock-jasmine (*Androsace* spp.), *Cyclamen* spp., especially *Cyclamen persicum* Mill, creeping jenny (*Lysimachia nummularia* L.), and others.

Kansas has 6 genera. *Centunculus minimus* L. is known from Chautauqua County only and is not included here.

Key to Genera (3):
1. Leaves in a basal rosette; inflorescence a scapose, bracted umbel 2
  2. Small annuals; corolla lobes spreading or ascending .... *Androsace*
  2. Perennial herbs up to 4.5 (–6) dm tall; corolla lobes sharply reflexed from the base ....
  .... *Dodecatheon*
1. Leaves all or mostly arranged along the length of the stem; inflorescence not as above 3
  3. Leaves alternate; flowers white .... *Samolus*
  3. Leaves usually opposite or whorled; flowers white, pink, red, blue, or yellow 4
    4. Flowers yellow or pale yellow; capsules splitting longitudinally .... *Lysimachia*
    4. Flowers white, pink, red, or blue; capsules splitting horizontally .... *Anagallis*

ANAGÁLLIS L.   Pimpernel

Gr. and L. *anagallis*, name used by the ancients for species of European pimpernel, 1 of which has been introduced into this country. Pliny speaks of a red-flowered and a blue-flowered kind, calling the former male and the latter female, although both are actually bisexual. "The juice of either plant," he says, "applied with honey, disperses films upon the eyes, bloodshot in those organs resulting from blows and argema." And then he remarks, as if with approval, "Some persons recommend those who gather it to prelude by saluting it before sunrise, and then, before uttering another word, to take care and extract the juice immediately."

Annual, rarely perennial, herbs with low, leafy, decumbent stems. Leaves opposite or whorled, entire, sessile or petiolate. Flowers solitary on slender pedicels in the leaf axils. Sepals 5, united slightly at the base. Corolla wheel-shaped, with only a short tube. Stamens 5, the filaments bearded, arising near the base of the corolla tube. Capsule membranous, many-seeded, dehiscing by splitting transversely around the circumference.

One species in Kansas.

A. arvénsis L.   Scarlet pimpernel

L. *arvensis*, pertaining to a cultivated field, from *arvum*, cultivated field, plowed land.
Red to salmon-colored; May–August
Fields, roadsides, waste ground, occasionally in prairies or open woods
Scattered in the east half; intr. from Eurasia

Weak, glabrous annuals, 1–4.5 dm tall, the stems 4-angled, eventually branching from the base. Leaves opposite, sessile, ovate to lanceolate or broadly elliptic, 6–12 mm long and 2–7 mm wide; apex usually acute; base rounded; margin entire. Flower pedicels slender, about as long as to much longer than the leaves, recurved in fruit. Calyx 3–4 mm long, the lobes lanceolate, with membranous margins. Corolla red to salmon-colored (rarely white or blue), 3–6 mm long, the lobes obovate. Capsules globose, membranous, about 4 mm long, subtended by the persistent calyx. Seeds short-conic and somewhat angular, 1–1.25 mm long, dark brown, covered with rust-colored reticulations.

ANDRÓSACE L.   Rock jasmine

NL., from L. *androsaces*, name used by Pliny for a plant or polyp impossible to identify from his description: "The androsace is a white plant, bitter, without leaves, and bearing arms surmounted with follicules, containing the seed."

Small, erect annuals with a basal rosette of sessile leaves. Flowers quite small, in terminal umbels subtended by a whorl of leaflike bracts, elevated on naked peduncles arising from the crown. Calyx 5-lobed, bell- or top-shaped. Corolla urn-shaped or funnelform, constricted at the throat, the tube shorter than the calyx. Stamens included within the corolla tube. Style very short; stigma headlike. Capsule enclosed by the calyx, opening at least partway down into 5 longitudinal segments.

One species in Kansas.

Though the flowers are insignificant, they attract bees, flies, and butterflies by proffering a film of nectar excreted over the surface of the ovary. These effect cross-pollination, but self-pollination can also take place, because the stigma and anthers are in proximity and mature at the same time.

A. occidentàlis Pursh

Western rockjasmine

L. *occidentalis*, western.
White; April–May
Dry soil of waste ground, roadsides, pastures, and other disturbed sites
East two-thirds, primarily

367. *Androsace occidentalis*

Minutely hairy annuals, 2–10 cm tall, often with reddish foliage. Leaves oblanceolate to elliptic, oblong, or nearly linear, 7–17 mm long and 1–4 mm wide; apex acute; margins with a few teeth near the apex or entire. Calyx bell-shaped, 1.5–2 mm long, the teeth or lobes triangular or linear-elliptic, green or reddish. Corolla white, inconspicuous, and soon ruptured by the enlarging fruit, about 2.5 mm long, the lobes oblong. Capsules ovoid or globose, about 3 mm long. Seeds dark brown, angular-conical, about 1 mm long, with a minutely honeycombed surface (best seen with high magnification and oblique light).

DODECÁTHEON L.

Shooting star

L. from Gr. *dodekatheos*, ancient name of a plant (probably *Primula vulgaris*), from *dodeka*, 12, + *theos*, god—so called by Pliny because these plants were believed to be under the care of the 12 greater gods. The common name was possibly suggested by the legend that fallen stars come up as wildflowers. Also, the form and pose of this flower are reminiscent of a comet streaking through space, the forward pointing anthers representing the head, and the sharply reflexed ribbonlike petals the backward-streaming tail.

Smooth, perennial herbs with fibrous roots. Leaves all basal. Flowers in terminal umbels on naked peduncles, the pedicels arching, subtended by a whorl of bracts. Calyx bell-shaped, deeply 5-lobed. Corolla with a shallow tube and 5 elongate, reflexed lobes. Stamens 5, the filaments very short and more or less united at the base, the anthers elongate and converging around the style. Capsules erect, many-seeded, dehiscing at the apex, subtended by the persistent calyx.

One species in Kansas.

368. *Dodecatheon meadia*

D. meàdia L.

Common shooting star

Named after Dr. Richard Mead (1673–1754), English physician.
White to pink or brilliant magenta; mid-April–May
Prairie, often in rocky limestone soil, occasionally in open woods
Southeast corner, north to Bourbon County, west to Greenwood, Elk, and Chautauqua counties

Plants 1–4.5 dm tall, with a basal rosette of broadly spatulate leaves 4–15 cm long and 1–3.5 cm wide. Flowers borne on a single, erect, naked peduncle, the pedicels eventually arching downward. Calyx 5–11 mm long, reflexed and concealed by the corolla at the time of flowering. Corolla tube white with dark markings at the throat, the lobes white to pink or brilliant magenta, 1.5–2.5 cm long. Anthers linear, bright yellow, 8–10 mm long, converging in the form of a narrow cone and dehiscing toward the inside around the style, which extends beyond the stamens and holds the capitate stigma outside the anther cone. Capsules very narrowly ovoid, reddish-brown, 11–15 mm long.

Small wild bees seem to be the most frequent visitors, and when they are clinging to the anther cone and sucking nectar from inside the corolla tube, pollen falls from the cone upon the under side of the body, as well as upon the stigma. This gives self-pollination a chance, while cross-pollination would ensue when the bees visit other flowers. Again, self-pollination has another chance when the anthers brush past the stigma as the corolla falls off. Thereafter the pedicels and the lobes of the calyx become erect. When the seeds ripen and the capsule splits open at the apex, the tiny seeds are shaken out and scattered by the wind.

See plate 67.

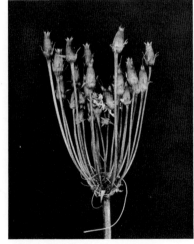

369. *Dodecatheon meadia* fruits

LYSIMÁCHIA L.

Loosestrife

L. *lysimachia*, the name of a plant which, according to Pliny, was discovered by Lysimachos, a high officer in the army of Alexander the Great. That plant, however, has been determined to be what is now called *Lythrum salicaria*. The name Lysimachos had a meaning in the Greek language, being compounded of *lysis*, loosing, + *mache*, battle, strife, and from this we get "loosestrife."

Perennial herbs with opposite or whorled, petiolate leaves. Flowers mostly yellow or yellow-orange, borne singly in the leaf axils or in terminal or axillary clusters. Sepals 5 (–6), free from one another or fused slightly at the base. Corolla bell- or wheel-shaped, 5- (or 6-) lobed, the corolla tube very short. Stamens 5, free from one another or monadelphous, sometimes alternating with small staminodes (sterile, reduced stamens). Capsules few- to many-seeded, globose or ovoid, 5-valved, the style persisting on 1 valve.

Three species in Kansas. *L. nummularia* L., moneywort or creeping jenny, is a creeping ornamental ground cover with yellow flowers and opposite round leaves and is occasionally found as an escape in moist woods. *L. hybrida* Michx. is recorded from Douglas and Miami counties only and differs from our common species in having leaves which are narrowly lanceolate to linear and taper gradually to a winged petiole.

370. *Lysimachia ciliata*

L. ciliàta L. — Fringed loosestrife

L. ciliatus, provided with cilia or a fringe of hairs like eyelashes, from *cilium*, eyelid—the petioles
   of this plant being fringed with hairs.
Yellow; May–July
Moist or wet soil of prairie swales and wooded areas
East half, primarily; more common eastward

Plants glabrous, up to 1 m tall, spreading by means of slender rhizomes. Stems somewhat 4-angled, usually producing short branches from the upper axils. Leaves opposite, the blades usually lanceolate, less frequently ovate or lance-oblong, (2.5–) 4–10 cm long and 1.5–4.5 cm wide, the apex acute or acuminate, the base rounded or short-acuminate; petioles conspicuously ciliate and united around the stem. Calyx 6–9 mm long, cleft nearly to the base, the lobes lanceolate to linear-attenuate. Corolla yellow, the lobes broadly ovate to round or obovate, 9–15 mm long, the apex erose. Stamens 5, free from one another, alternating with minute, linear or spoon-shaped staminodes. Capsules many-seeded, globose, beaked by the persistent style, about 6 mm long (excluding the style). Seeds brown with a reticulate seed coat, 2–2.5 mm long, with one face flat or slightly convex and more or less oblong to rhombic and 2 faces concave and forming a ridge at the hilum (point of attachment to the placenta).

SÀMOLUS L. — Water pimpernel

According to Fernald (9), an ancient name, probably of Celtic origin, said to refer to curative
   properties of this genus in diseases of cattle and swine.

Smooth, somewhat succulent perennial herbs. Leaves entire, in basal rosettes and alternate on the stem. Flowers small, white, on wiry pedicels in terminal clusters. Calyx bell-shaped, with 5 triangular lobes. Corolla white or pink, bell-shaped, the 5 lobes alternating with minute scalelike staminodes (sterile, reduced stamens). Stamens 5, on, and included within, the corolla tube. Ovary about two-thirds inferior, style short or none; stigma slightly enlarged. Capsule globose, 5-valved at the apex, many-seeded.

Two species in Kansas. *S. cuneatus* Small is known from Barber County only.

S. parviflòrus Raf.

NL., from L. *parvus*, small, + *flos, floris*, flower, i.e., "small-flowered."
White; June–August
Wet soil of springs, seeps, pond margins, creek banks
Southern counties

Plants mostly 2–4 dm tall, usually branched. Leaves spatulate to obovate, 2–7 cm long and 1–3 cm wide, the apex rounded, the base acuminate. Flowers borne on thread-like pedicels, each with a tiny bract, in airy racemes or panicles. Calyx 1.5–2 mm long, glabrous. Corolla white, 2–3 mm wide. Capsules 2–3 mm long.

GENTIANÀCEAE — Gentian Family

Smooth annual or perennial herbs with bitter juice. Leaves opposite, simple, entire, sessile or petiolate, estipulate. Flowers bisexual, regular, usually showy, borne singly or in various types of clusters. Sepals 4–5, fused to one another, persistent. Petals 4–5, fused to one another, the corolla usually tubular, salverform, bell-shaped, or wheel-shaped. Stamens of the same number as the corolla lobes, arising on the corolla tube. Pistil 1; ovary superior, 1-celled but with 2 parietal placentae, or sometimes 2-celled; style 1; stigma 1, sometimes 2-lobed. Fruit a many-seeded, septicidal capsule.

A family of about 80 genera and 900 species, widely distributed but most abundant in temperate regions. Cultivated members of the family include the gentians (*Gentiana* ssp.), centaury (*Centaurium scilloides* Druce), prairie gentian (*Eustoma grandiflorum* (Raf.) Shinners), and the water snowflake (*Nymphoides indicum* Kuntze).

371. *Eustoma grandiflorum*

The family is represented in Kansas by 4 genera. *Centaurium texense* (Griseb.) Fern. is known in Kansas from Wilson and Montgomery counties only. It is a small, much-branched, pink-flowered annual resembling our species of *Sabatia* but differing from them in having smaller flowers with keeled sepals; a narrow corolla tube much longer than the abruptly spreading, narrowly elliptic corolla lobes; and linear to narrowly elliptic leaves.

Key to Genera (3):

1. Corolla wheel-shaped, the corolla tube much shorter than the corolla lobes or calyx lobes .... *Sabatia*
1. Corolla tubular to funnelform or bell-shaped, the corolla tube as long as or longer than the corolla lobes or calyx lobes 2
    2. Flowers blue or white, the areas of the corolla tube between the lobes often plaited (folded like a fan); foliage not glaucous .... *Gentiana*
    2. Flowers light purple with a darker purple center, the areas of the corolla tube between the lobes not plaited; foliage glaucous .... *Eustoma*

EŪSTOMA Salisb.

NL., from Gr. *eu*, good, + *stoma*, mouth—the mouth of the corolla becoming wide open.

More or less glaucous, annual or short-lived perennial herbs with taproots. Stems erect or ascending. Leaves sessile, clasping the stem. Flowers showy, long-pedicelled, borne singly or in clusters. Calyx deeply cleft, the lobes keeled and long-attenuate. Corolla bell-shaped. Stamens 5–6, on the throat of the corolla; anthers oblong, versatile. Ovary 1-celled; style slender; stigma prominently 2-lipped. Capsule ellipsoid, 2-valved, many-seeded, beaked by the persistent style.

One species in Kansas.

E. grandiflòrum (Raf.) Shinners   Russell prairie-gentian

NL., from L. *grandis*, large, great, noble, sublime, + *flos, floris*, flower—in reference to the spectacular flowers. This species has been called *E. russellianum* (Hook.) Sweet by some manuals, the name being in honor of Alexander Russell (died 1786), English physician, botanical explorer, and writer.

Light purple with a darker center; July–September
Moist, sandy soil of low prairie, salt flats, and flood plains
Southwest fourth, primarily; with records as far east as Rice, Reno, and Harper counties and northward in Cheyenne and Logan counties

Erect, annual or short-lived perennials with light bluish-green, glaucous herbage. Stems 1 to few, branched above, mostly (2–) 3–5 dm tall. Leaves lanceolate to lance-oblong or elliptic, 1–6 cm long and 0.5–2.5 cm wide; apex acute, or rounded and mucronate; base rounded, clasping the stem. Flowers usually light purple with a dark purple center, rarely white in our area (forms with petals pink, yellow, or white tinged with purple are reported from Texas), usually several, on elongate pedicels in branched terminal clusters. Calyx 12 cm long. Corolla 3–4.5 cm long, the lobes obovate to oblanceolate or nearly oblong, much longer than the corolla tube, the apex rounded or abruptly short-acuminate, the margins usually finely erose. Capsule 1–2.5 cm long. Seeds about 0.5 mm long, the surface honeycombed (use high magnification).

As in the gentians, the flowers are adapted for cross-pollination by bees. When a flower first opens, the anthers discharge their pollen, but the stigma lobes are still immature. A bee coming to the flower at this time would pick up pollen but could not transfer it to the stigma of that flower. When the anthers are empty and shriveled, the style elongates and the stigma lobes expand and spread apart in readiness to receive enough pollen brought from younger flowers to fertilize the crowd of ovules in the 1-celled ovary. As in the case of *Sabatia*, this is a good example of proterandry, although the styles and stigmas of these related flowers behave in different ways.

GENTIĀNA L.   Gentian

Named after Gentius, Illyrian king, 2nd century B.C., reputed to have recommended the European yellow gentian (*G. lutea*) as a cure for the bubonic plague.

Usually glabrous, annual or perennial herbs. Leaves sessile or nearly so. Flowers showy, borne singly or in clusters. Calyx bell-shaped, obconic, or tubular. Corolla lobes usually much longer than the tube, often alternating with membranous, folded areas called "plaits." Ovary ellipsoid or cylindric, with a short, stout style or none; stigma 2-lobed; capsule ellipsoid, 2-valved, beaked by the persistent style and/or stigma.

Two species in Kansas.

Key to Species:
1. Flowers deep blue, the corolla lobes widely ascending to spreading; anthers free from one another .... *G. puberulenta*
1. Flowers yellowish- to greenish-white, the corolla lobes erect or slightly incurved; anthers united in a ring .... *G. alba*

G. álba Muhl.   White prairie gentian

L. *albus*, white, although the flowers of this species are yellowish or greenish.
Yellowish- to greenish-white; July–September
Sandy soil of open, wooded hillsides
Shawnee, Jefferson, Leavenworth, Douglas, and Cherokee counties, uncommon

372. *Gentiana puberulenta*

Stout, erect perennial. Stems 1 to few, erect, unbranched, glabrous, 5–8 dm tall. Leaves sessile, lanceolate, 3–10 cm long and 2–4 cm wide; apex acute to somewhat attenuate; base rounded or cordate and clasping the stem; margins not revolute. Flowers yellowish- to greenish-white, in a crowded, few- to several-flowered, terminal cluster closely subtended by a whorl of leaves. Calyx funnelform, 12–15 mm long, the lobes triangular-acute or -acuminate, eventually spreading or reflexed. Corolla obconical or narrowly bell-shaped, 2.5–4.5 cm long, the lobes ovate or ovate-acuminate, erect or slightly incurved, longer than the erose plaits. Anthers united in a ring. Capsules about 2.5 cm long (including the style), enclosed by the persistent calyx and corolla. Seeds tan, oblong to elliptic, flattened, 2–3 mm long, with a broad membranous wing.

G. puberulénta Pringle   Prairie gentian, downy gentian

NL., from L. *puberulus*, possessed of downy hairs.
Deep blue; August–October
Prairie
East third

Erect biennial or perennial with a vertical rootstock and thick lateral roots. Stems 1 to few, erect, usually unbranched, rather rigid, minutely hairy, 2.5–4.5 dm tall. Leaves sessile, stiff, lanceolate to lance-linear or lance-oblong, 2–5 cm long and 0.7–2 cm wide; apex usually acute or rounded; base rounded or somewhat clasping the stem; margin revolute and sometimes minutely ciliate. Flowers deep blue, borne in crowded, few-flowered terminal clusters. Calyx bell-shaped, 2–3.5 cm long, the lobes linear, rigid, alternating with membranous areas of the calyx tube. Corolla funnelform, 3–4.5 cm long, the lobes triangular or ovate-acuminate, ascending or spreading, longer than the fringed plaits. Anthers free from one another. Capsules about 3.5–4 cm long (including the persistent style), surrounded by the persistent calyx and corolla. Seeds light tan, more or less elliptic, flattened, 1.5–2 mm long, with a broad membranous wing.

The anthers discharge pollen before the stigmas in the same flower are ready to receive it, so that cross-pollination is accomplished by pollen brought from younger to older flowers. The flowers open 1 or 2 at a time in the morning, closing in the evening, the same flower repeating the process for several days. They remain closed in dull stormy weather, to the disappointment of bumblebees, their most frequent visitors, which must enter the flower bodily to get the nectar secreted at the base of the corolla.

The plants are not long-lived and, since little success attends transplanting them from the prairie to the garden, they should be propagated by seeds planted, as soon as they are ripe, where they are to grow. This species is one of the best for cut flowers, since both leaves and flowers keep their freshness for a long time.

The gentians have been employed in the practice of medicine from the pre-Christian era down to the present. Besides its use in bolstering patients prostrated by the plague, Dioscorides and Pliny in the 1st century recommended the yellow gentian (*G. lutea*) for treatment of poisonous bites and stings, for promoting the healing of wounds and ulcers, for convulsions, and for ailments of the stomach and liver. Concerning the popularity of the gentian in the folk medicine of the Middle Ages and in his own day, Hieronymous Bock wrote in 1551: "The most used root in Germany is the Gentian. The common man knows no better stomach medicine than the Gentian," and the clerical empiric, Sebastian Kneipp, wrote: "Whoever has a little garden should have in it, (1) some sage, (2) some wormwood, (3) some gentian. Then he has his dispensary right at hand."

So stood the reputation of the yellow gentian in Europe; and for England we have the report of the celebrated herbalist, John Parkinson, in his *Paradisus Terrestris*, written in the early 17th century: "The wonderful wholsomnesse of Gentian cannot bee easily knowne to us by reason our daintie tastes refuse to take thereof, for the bitternesse sake; but otherwise it would undoubtedly worke admirable cures, both for the liver,

373. *Sabatia angularis*

374. *Sabatia campestris*

stomacke and lunges. It is also a speciall counterpoison against any infection, as also against the violence of a mad dogges bite."

While the yellow gentian is the species above referred to, many other species, including our prairie gentian, contain the same active principle, a bitter glucoside, that makes the yellow gentian effective.

See plate *68*.

SABATIA Adans.   Sabatia, marsh pink, rose gentian

Named in honor of Liberatus Sabbati, 18th-century Italian botanist.

Erect, glabrous, annual or perennial herbs. Leaves linear to ovate, usually sessile. Flowers showy, in branched clusters. Calyx with 5–12 slender lobes. Corolla wheel-shaped, with 5–12 spreading lobes and a shorter corolla tube. Stamens alternating with corolla lobes, the anthers spirally twisted after anthesis. Style slender, 2-lobed or -cleft. Capsule ovoid to cylindric, eventually 2-valved, long enclosed by the persistent, wilted corolla.

Two species in Kansas.

Key to Species:

1. Branches of the inflorescence opposite; calyx tube bowl-shaped, 1.5–2 mm long, lacking green ribs or wings, the tips of the calyx lobes usually curled inward .... *E. angularis*
1. Branches of the inflorescence alternate; calyx tube bell-shaped, 3–6 mm long, membranous with 5 green ribs or wings, the tips of the calyx lobes always straight and ascending or spreading .... *S. campestris*

S. angularis (L.) Pursh   Rose pink

L. *angularis*, having angles—in reference to the prominently angled or winged stems.
Pink; July–August
Prairie and open woods
Crawford and Cherokee counties

Glabrous, erect annual resembling *S. campestris*, 3–6.5 dm tall, with thickened fibrous roots. Stems conspicuously angled or winged, usually unbranched below the inflorescence. Leaves cordate below to ovate or lanceolate above, 1–4 cm long and 0.5–3 cm wide, the apex usually acute or acuminate. Flowers on long pedicels in much-branched, many-flowered terminal clusters, the branches of the inflorescence opposite, the latter therefore appearing trichotomous. Calyx tube 1.5–2 mm long, bowl-shaped, lacking green ribs or wings; calyx lobes 5, green, awl-shaped to linear-attenuate, 0.5–1 cm long, the tips often curled inward. Corolla pink with a greenish or yellowish throat, 15–22 mm long, the lobes obovate to oblanceolate. Stamens bright yellow. Style deeply 2-cleft. Capsule broadly ellipsoid, thick-walled, about 9 mm long (excluding the style).

S. campestris Nutt.   Meadow gentian

L. *campestris*, of a level field, from *campus*, field.
Pink; June–August
Roadside ditches, prairies, moist limestone soil
Southeast eighth

Glabrous, erect annual 1–3.5 dm tall. Stems angled or narrowly winged, unbranched below the inflorescence. Leaves ovate with a rounded apex below to lanceolate with acute to attenuate apex above, 0.5–2.5 cm long and 3–13 mm wide. Flowers on long pedicels in openly branched terminal clusters, the branches of the inflorescence alternate and the prolongation of the axis usually somewhat deflexed to one side and the cluster appearing to be dichotomous. Calyx tube 3–6 mm long, bell-shaped, membranous, with 5 green ribs or narrow wings; calyx lobes 5, green, attenuate, 1–2 cm long, the tips always straight and spreading or ascending. Corolla pink (rarely white) with a yellow and white throat, 16–25 mm long, the lobes obovate or less frequently oblong or elliptic, the apex rounded or abruptly acuminate. Stamens bright yellow. Style deeply 2-cleft. Capsule broadly ellipsoid, thick-walled, 8–11 mm long (excluding the style). Seeds gray, minute (about 0.3 mm long), with a honeycombed surface (use high magnification).

In these flowers we find a remarkable exhibition of proterandry, where, as the term implies, the anthers deliver their pollen before the stigmas of the same flowers are ready to receive it. When the flowers open, the 5 anthers on short filaments stand forward where insect visitors would pick up pollen but could not transfer it to the stigmas of the flower because the single style is lying flat, with its 2 long and slender branches so twisted together that the stigmatic lines along their inner surfaces are not exposed. Later, as the stamens retract and wither, the style rises to an erect position while its branches untwist and stand far apart so that insects, especially both the long-tongued and the short-tongued bees, that have picked up pollen from younger flowers would be able to pollinate them.

Apart from this very interesting behavior we may say of these flowers that they are among the loveliest of the sweet-scented and long-lasting native flowers.

See plate *69*.

375. *Amsonia tabernaemontana*

APOCYNĀCEAE   Dogbane Family

Trees, shrubs, vines, or perennial herbs, mostly with a milky juice. Leaves alternate or opposite, simple, sessile or petiolate, usually estipulate. Flowers bisexual, regular, borne singly or in clusters. Sepals 5, rarely 4, fused together part of their length. Petals 5, fused together, the corolla bell-shaped, urn-shaped, funnelform or salverform. Stamens as many as the lobes of the corolla, arising from the corolla tube, the anthers converging around the stigma. Pistils superior or slightly inferior; ovaries 2, but usually more or less united into a single pistil, though in our native species they are united only at the tip, with a single style or none and a massive stigma. Fruit (in ours) a follicle; seed with or without a tuft of silky hairs (the coma).

A family of about 180 genera and 1,500 species, world-wide in distribution but most numerous in the tropics. Among the cultivated members of the family are the oleander (*Nerium oleander* L.), yellow oleander (*Thevetia peruviana* Schum.), periwinkle (*Vinca* spp.), frangipani (*Plumeria* spp.), and amsonia (*Amsonia tabernaemontana* Walt.). This family is closely related to the Milkweed Family, and some suggestion of the complex gynostegium characteristic of that family may be seen in the genus *Apocynum* which has convergent stamens with broad, flattened filaments subtended by nectar glands.

Two genera in Kansas.

Key to Genera:
1. Leaves opposite; seeds with a tuft of silky hairs; flowers with nectar glands .... *Apocynum*
1. Leaves alternate; seeds without hairs; flowers lacking nectar glands .... *Amsonia*

AMSŌNIA Walt.   Amsonia

Named after one Dr. Amson, an 18th-century physician of Gloucester County, Virginia.

Perennial, mostly erect herbs. Leaves alternate, petioled in ours, estipulate. Flowers blue, few to many, in terminal clusters. Calyx deeply 5-lobed. Corolla with a slender funnelform tube, 5 abruptly spreading lanceolate lobes, and a hairy throat. Anthers lance-triangular, with rounded basal lobes, converging around the stigma but free from one another, completely enclosed within the corolla tube. Ovaries 2, only slightly united, lacking nectaries. Fruits many-seeded, round in cross section, cylindric or constricted between the seeds. Seeds rather corky, lacking a tuft of silky hairs.

One (possibly 2) species in Kansas.

A. tabernaemontàna Walt.   Willow amsonia

In honor of Jacobus Theodorus Tabernaemontanus, German botanist and herbalist of the 16th century.
Light blue; mid-April–May
Moist, rocky soil of creek banks and woods
Bourbon and Cherokee counties

Erect, smooth plants with several stems 0.4–1.3 m tall, from knobby woody rhizomes with fibrous roots. Leaves short-petioled, the blades thin, mostly lanceolate or elliptic, 4–12 cm long and 1–3 cm wide, the apex acute to acuminate, the base acute or wedge-shaped, the margins ciliate. Calyx glabrous, 0.5–1 mm long, the lobes triangular to triangular-attenuate. Corolla light blue, the tube about 7 mm long, hairy on the outside below the lobes, the lobes lanceolate to lance-oblong, 6–8 mm long. Fruits erect, glabrous, 8–12 cm long and about 3 mm wide, not constricted at all between the seeds. Seeds dark brown, oblong, truncate at either end, variously pitted and wrinkled, 5–11 mm long.

This species is available from floricultural supply houses and is easily propagated by division of the rhizome or by seeds.

A second species, *A. illustris* Woods., is also reported from Woodson, Crawford, and Cherokee counties. It is usually distinguished from *A. tabernaemontana* by its thicker, shinier leaves, sparsely hairy calyx, and its fruits which tend to be constricted between the seeds and which assume a lax or pendulous position.

376. *Apocynum cannabinum*

APÓCYNUM L.

Dogbane, Indian hemp

NL., from Gr. *apokynon*, dogbane, a name given by Dioscorides, Greek medical writer of the 1st century, to some plant with milky juice.

Erect, branched perennial herbs with tough fibrous bark and horizontal roots capable of producing buds and forming new aboveground stems. Leaves opposite (rarely whorled), sessile or petioled, entire, with minute stipules. Flowers small, white, greenish, or pale pink, on short pedicels in terminal and axillary clusters. Sepals united at the base. Corolla bell- or urn-shaped in ours, with small toothlike appendages projecting inward from near the base of each petal. Stamen filaments short, with winglike lateral extensions; anthers lance-triangular, with pointed, sagittate basal lobes, united around the stigma and prolonged into a cone beyond it. Ovaries two, slightly inferior, united at the apex, the stigma ovoid or clublike, essentially sessile. Fruits 2 per blossom, long-cylindric, many-seeded, hanging downward at maturity (in ours). Seeds with a tuft of silky hairs (the *coma*).

Two or three species in Kansas. *A. medium* Greene is known from Douglas, Bourbon, and Cherokee counties. It differs from the following in having pink or pink-striped flowers with calyx lobes much shorter than the corolla tube, the corolla lobes spreading, and the uppermost flower clusters elevated above the uppermost leaves.

Stem fibers of *Apocynum* have been used for making thread, cord, ropes, bags, fishnets, and other items. One of the most interesting early artifacts of this type known is a drawstring bag knotted from *Apocynum* fibers and dated at about 5000 B.C. The bag was found in a cave in Utah along with other artifacts of a hunting and gathering culture.

Key to Species:

1. Plants glabrous (rarely with petioles, lower leaf surfaces, or inflorescences minutely hairy); at least the lowermost leaves with cordate bases; petioles 1–3 mm long .... *A. sibiricum*
1. Plants glabrous to conspicuously hairy; none of the leaves with cordate bases; petioles 2–6 mm long .... *A. cannabinum*

A. *cannábinum* L.

Indian hemp, hemp dogbane

L. *cannabinum*, hempen, of hemp, from *cannabis*, hemp plant, + -*inus*, a suffix indicating quality or condition—the best fibers of this plant, like those of the hemp, being used for making cordage, rope, etc. "Bane" from AS. *bana*, murderer.

White or greenish-white; mid-May–mid-August

Prairie, roadsides, creek banks, waste ground

Throughout

Plants glabrous to conspicuously soft-hairy, up to 1.5 m tall, with ascending branches. Leaf blades mostly lanceolate or elliptic, 3–10 cm long and 0.7–4 cm wide, the apex acute or rounded, mucronate, the base acute or rounded, the petioles 2–6 mm long. Uppermost flower clusters always over-topped by the uppermost leaves. Calyx 2–3 mm long, the lobes lanceolate to lance-attenuate, most of them as long as the corolla tube. Corolla greenish-white, mostly 3–4 mm long, the lobes erect or ascending. Follicles up to 16 cm long and about 3 mm wide. Seed body 4–5 mm long, the coma white, 2.5–3 cm long.

As the key indicates, glabrous plants with sessile or nearly sessile leaves, some of which have cordate bases, have been called *A. sibiricum* Jacq. by some authorities. However, in Kansas at least, the characters which supposedly separate these from *A. cannabinum* appear to overlap and intergrade extensively. Both species occur throughout the state, though plants tentatively identified as *A. sibiricum* tend to occur in moister habitats than *A. cannabinum* and to bloom earlier. The leaves of the latter range from ovate to lanceolate, elliptic, or oblong. A closer study of these plants throughout their ranges seems necessary before we can be certain whether our common *Apocynums* represent 1 species or 2.

A variety of uses for *Apocynum cannabinum* and other species is recorded in the literature. Medicinal uses by Indians and whites are too many to enumerate. The dried underground parts of *A. cannabinum* were listed officially in the *U.S. Pharmacopoeia* from 1831 to 1916 and in the *National Formulary* from 1916 to 1960 as a cardiac stimulant.

When mordanted with alum, chopped plants of *A. cannabinum* produce a dark tan dye with wool. Mordanted with copper, they will dye wool black.

Although ingestion of dogbane produces marked physiological effects, even causing the death of cattle when the dried leaves are experimentally administered in sufficiently large doses, there is no record of serious poisoning of livestock under natural conditions. The bitter latex sap of the plant apparently renders it extremely unpalatable to livestock.

ASCLEPIADÀCEAE　　　　　　　　　　　　　　Milkweed Family

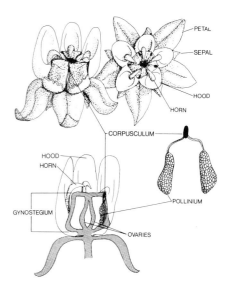

377. Flower structure in *Asclepias*

Perennial herbs, vines, or shrubs, rarely small trees, occasionally fleshy or cactuslike, usually with milky sap. Leaves opposite, whorled or rarely alternate, simple, mostly entire, sessile or petiolate, with minute stipules. Flowers bisexual, regular, usually borne in umbel-like clusters. The most distinctive feature of this family is the unique and rather complex structure of the flower (see fig. 377). Most Kansas members of this family belong to the genus *Asclepias*, to which the following description applies. The 5 sepals and 5 petals are typical in appearance and frequently are reflexed or curved downward with the petals obscuring the smaller sepals from view. Above the petals and usually elevated on a column is a whorl of 5 erect petaloid parts called hoods or nectar receptacles. The hoods vary a great deal in shape from one species to another, and this is indeed one of the major diagnostic characters used in identification. In general, however, the hoods or nectar receptacles tend to be shaped somewhat like small sugar scoops with the open or concave side toward the center of the flower. Arising from the base of each hood and also projecting toward the center of the flower in most species of *Asclepias* is a slender, curved structure called a horn. Tissue at the base of the horn secretes nectar which then accumulates in the bottom of the nectar receptacle and serves to attract insect pollinators. The 5 hoods are referred to collectively as the corona. Flowers of *Cynanchum* and *Matelia* have a corona but lack nectar horns. If one pulls the hoods aside, he sees a 5-angled headlike structure formed by the fusion and extreme modification of the 5 stamens and the pistil. This *gynostegium* (from Gr. *gyne*, woman, + *stege*, roof), as it is called, completely encloses the female part of the flower. The pistil is composed of 2 separate 1-celled superior ovaries and 2 styles which are apically united and share a single large 5-lobed stigma head. The stigma head is fused with stamen tissue to form the apex of the gynostegium; but the only regions of the stigma head which are receptive to pollen are the terminal tips of the 5 lobes, and these are enclosed within the gynostegium and not exposed to the outside. How then does pollination take place? Looking again at the outside of the gynostegium, one sees that there are 5 horny winglike projections, each with a vertical slit (pollinating crevice) that allows access to the inner chamber; behind each slit is a pollen-receptive lobe of the stigma head. Immediately above each pollinating crevice can be seen a tiny black seedlike corpusculum with a microscopic basal slit that tapers upward. By inserting a straw or needle into the pollinating crevice and gently moving the former upward, it is possible to tease the corpusculum loose from its normal position and to note that attached to it is a pair of threadlike connectives. Attached to the other end of each connective and originally tucked away in a small pocket on the outside of the gynostegium is a translucent mass of golden pollen. A corpusculum and its 2 connectives and pollen masses constitute a pollinium (pl. pollinia). In an intact blossom, each horny winglike projection with its pollinating crevice has a pollinium. An insect, crawling around on the flowers gathering nectar, may accidentally step into a pollinating crevice. In the process of removing his leg, he may pull it upward in such a manner as to catch it in the basal notch on the corpusculum. Larger insects, such as bumblebees and wasps, are strong enough to pull the entire pollinium out and to fly off with it still attached to a foot or leg. Weaker insects may remain trapped or escape only by leaving a leg behind. Landing on a second flower, the insect may chance to slip his foot with the attached pollinium into another pollinating crevice. The pollen masses are only weakly attached to the connectives, and, in addition, the edge of the pollinating crevice is rolled inward in such a manner as to allow the insect's foot and the attached pollinium to enter easily but to interfere with the removal of the latter. Thus, when the insect's foot is withdrawn, the pollen masses usually break loose, remaining within the pollinating crevice and up against the stigmatic tissue. Pollination has thus been effected. The fruit produced in this family is a spindle-shaped follicle or pair of follicles filled with numerous seeds, each with an elliptic, flattened body and a tuft of long silky hairs (the coma) derived from tissue of the placenta. Although similar in appearance to the tuft on the fruits of thistles, dandelions, and similar Composites, it is of quite different origin, the hairs of the pappus in the Compositae being an unusual form of calyx.

A large family of about 130 genera and 1,200 species, widely distributed in warmer regions of the earth but especially well represented in the tropics of South America. Ornamentals in the family include the wax plant (*Hoya carnosa* R. Br.), carrion flower (*Stapelia* spp.), Madagascar jasmine (*Stephanotis floribunda* Brongn.), and various species of milkweed (*Asclepias*). The rubber vine (*Cryptostegia grandiflora* R. Br.), a native of Africa, has been cultivated as a source of rubber as well as for ornament.

Kansas has 3 genera.

Key to Genera:

1. Stems not twining .... *Asclepias*
1. Stems twining 2
   2. Corolla white, the lobes erect at anthesis, the tips later spreading or recurved .. *Cynanchum*
   2. Corolla (in ours) brownish- or greenish-purple, the lobes spreading .... *Matelea*

ASCLÈPIAS L.   Milkweed, silkweed

Gr. *asklepias*, Latin *asclepias*, ancient name of some plant, from Asklepios, Greek god of medicine.

Mostly perennial herbs arising from stout roots or rhizomes. Stems erect, or decumbent, never twining. Leaves opposite or less frequently whorled or alternate, entire. Flowers borne in terminal and/or lateral, umbel-like clusters toward the tops of the plants. Calyx lobes divided nearly to the base. Corolla lobes elliptic or lanceolate, usually reflexed. Corona and gynostegium usually elevated on a column, but sessile in some species. Hoods or nectar receptacles varying from hooklike to clublike, usually with a horn or crest arising at the base and arching toward the gynostegium, the hoods either basally narrowed at the point of attachment to the column and then said to be stipitate, or not narrowed and then said to be sessile. Fruits and seeds as described for the family.

Twenty-two species in Kansas.

Key to Species (44):

1. Hoods narrowed to a stalklike base at the point of attachment to the column, the base not deeply saccate or pouchlike 2
  2. Horns gradually tapered and usually arching over the gynostegium, the latter about as long as broad or slightly longer or slightly shorter; column cylindric or conic, rarely obconic 3
    3. Hoods with a pair of marginal teeth 4
      4. Plants lacking milky sap; flowers red-orange to orange or yellow, 10–15 mm long .... *A. tuberosa*
      4. Plants with milky sap; flowers greenish-white or greenish tinged with purple or rose 5
        5. Leaves linear, 0.5–2.5 mm wide; flowers 6–8 mm long .... *A. verticillata*
        5. Leaves broadly ovate to lance-oblong, 3.5–7.5 cm wide; flowers 13–20 mm long .... *A. amplexicaulis*
    3. Hoods lacking marginal teeth 6
      6. Leaves opposite; flowers purplish-red (rarely white) .... *A. incarnata*
      6. Leaves densely spiraled; flowers white or greenish .... *A. pumila*
  2. Horns abruptly constricted near the tip and sharply inflexed toward or over the gynostegium, the latter usually broader than long; column broadly obconic or barely evident 7
    7. Hoods truncate or abruptly rounded at the tip, keeled on the back; horns fused half or more of their length with the hoods 8
      8. Plants persistently hairy; leaves definitely petioled, lanceolate or ovate below to oblong, nearly rectangular or obovate above; hoods broadly 2-lobed at the tip, the nectar horns united with the hoods about half their length .... *A. arenaria*
      8. Plants minutely hairy when young, but soon becoming nearly glabrous; leaves short-petioled to nearly sessile, mostly broadly obovate to short-oblong or nearly round; hoods truncate at the tip, the nectar horns united with the hoods their entire length .... *A. latifolia*
    7. Hoods gradually rounded to acuminate, not keeled; horns fused half or less their length with the hoods 9
      9. Hoods with sharply incised marginal lobes or teeth, usually broadly rounded on the back 10
        10. At least the middle leaves 4 per node; hoods delicately petal-like; corpusculum of each pollinium relatively small .... *A. quadrifolia*
        10. Leaves 2 per node; hoods fleshy and more or less inflated; corpusculum of each pollinium relatively massive 11
          11. Plants stout, usually hairy; leaves definitely petioled; flowers usually dull purplish-pink; common in the east three-fifths of Kansas .... *A. syriaca*
          11. Plants slender, glabrous; leaves sessile or nearly so; flowers yellowish-green tinged with purple; rare in the east fifth of Kansas .... *A. meadii*
      9. Hoods without sharply incised marginal lobes or teeth, usually broadly flattened on the back 12
        12. Hoods less than twice as long as the gynostegium, the tips very broadly rounded .... *A. sullivantii*
        12. Hoods more than twice as long as the gynostegium, the tips abruptly acute to acuminate 13
          13. Hoods 3–4 times as long as the gynostegium, spreading, relatively broad at the base, then abruptly constricted to a narrowly oblong tip; west three-fourths of Kansas .... *A. speciosa*

13. Hoods 2–3 times as long as the gynostegium, erect or only slightly spreading, lanceolate as viewed from the back; east 2 columns of counties and Coffey Co. .... *A. purpurascens*
1. Hoods very sessile, the basal attachment deeply saccate, the adnate horn (frequently absent) typically reduced to an inconspicuous and isolated crest or terminal appendage 14
14. Hoods sharply deflexed from the gynostegium, thence typically with ascending involute, conduplicate, or clublike tips 15
15. Hoods less than half as long as the corolla lobes; gynostegium about as long as broad, the wings very narrow and inconspicuous .... *A. viridis*
15. Hoods almost as long as the corolla lobes; gynostegium much broader than long, the wings very broad and conspicuous .... *A. asperula*
14. Hoods not deflexed from the gynostegium 16
16. Hoods with 2 external, longitudinal laminate flaps 17
17. Leaves rather strictly ascending; hoods about as long as the gynostegium, 3-toothed at the tip; wings along the pollinating crevices spurred toward the base .... *A. stenophylla*
17. Leaves widely spreading to somewhat reflexed; gynostegium protruding beyond the hoods, broadly emarginate to essentially entire; wings along the pollinating crevices not spurred .... *A. engelmanniana*
16. Hoods without such flaps 18
18. Leaves linear to linear-lanceolate; corona with a short but definite column; the gynostegium protruding prominently beyond the hoods .... *A. hirtella*
18. Leaves ovate to lanceolate, elliptic, oblong, or oblanceolate; corona sessile; gynostegium protruding but slightly beyond the hoods .... *A. viridiflora*

378. *Asclepias amplexicaulis*

A. amplexicaùlis Smith

Bluntleaf milkweed

NL., from L. *amplexus*, encircling, + *caulis*, stem—in reference to the leaf base.
Flowers greenish tinged with purple or rose; mid-May–June
Prairie, usually in sandy soil
East two-thirds

Erect herbaceous perennials with unbranched stems 3.5–7 dm tall. Leaves opposite, sessile, broadly ovate to lance-oblong, 5–11 cm long and 3.5–7.5 cm wide. Flowers greenish tinged with purple or rose, 13–20 mm long and 9–12 mm wide, borne in terminal clusters. Calyx lobes lanceolate, 2–4 mm long. Corolla lobes lance-oblong or elliptic, reflexed, 7–12 mm long. Column cylindric. Hoods stipitate, nearly cylindrical with a pair of marginal teeth, erect, 5–6 mm long; nectar horns tapered gradually and arching broadly over the gynostegium, the latter 2–3 mm long and 2–4 mm wide. Follicles erect, narrowly fusiform, smooth, 10–16 cm long and 1–2 cm thick. Seed body broadly elliptic, 6–9 mm long, the coma tawny, 4–6 cm long.

379. *Asclepias amplexicaulis* leaves

A. arenària Torr.

L. *arenarius*, of sand, sandy—descriptive of the habitat of this species.
Flowers greenish and cream-colored; mid-June–early August
Sandy soil of prairie, roadsides, riverbanks, and dune areas
West half, primarily

Erect, densely hairy perennial herbs with stout, usually unbranched stems up to 8 dm tall. Leaves opposite, petioled, the blades lanceolate or ovate below to oblong, nearly rectangular, or obovate above, 6–11 cm long and 3–9 cm wide, the apexes acute on lower leaves and rounded, truncate, or shallowly notched on the upper leaves, the bases acute or rounded on the lower leaves, rounded, truncate, or cordate on the upper leaves, the margins more or less undulate. Flowers greenish and cream-colored, 1–1.5 cm long and 1–1.3 cm wide, the clusters borne singly in the upper axils. Calyx lobes ovate-lanceolate, 5–7 mm long. Corolla lobes lanceolate, reflexed, 7–10 mm long. Column broadly obconic. Hoods stipitate, short-cylindric, keeled along the back, erect or somewhat ascending, broadly 2-lobed at the apex; nectar horns united with the hoods about half their length, abruptly constricted toward the tip and sharply incurved; gynostegium 2–3 mm long and about 4 mm wide. Follicles broadly spindle-shaped, finely hairy or nearly glabrous, 7–9 cm long and 1.5–2.5 cm wide. Seed body broadly elliptic, about 11 mm long, the coma tawny, about 2.5 cm long.

A. aspérula (Dcne.) Woods.

NL., from the genus *Asperula* in the Madder Family, the diminutive of *asper*, rough, harsh, i.e., somewhat or slightly rough or harsh.
Yellowish-green or pale green with dark purple; mid-May–June
Dry, rocky prairie uplands
Central third, primarily

380. *Asclepias hirtella*

381. *Asclepias hirtella* fruits

Herbaceous, more or less rough-hairy perennial with several decumbent stems 2–5.5 dm tall arising from a woody crown and a thick, tough main root giving off laterals at different levels. Leaves short-petioled, irregularly arranged, there being 1, 2, or 3 leaves at various nodes on the same stem, mostly lanceolate to linear-lanceolate, 5–11 cm long and 1–3 cm wide, the apex acute, frequently mucronate, the base acute or rounded. Flowers 10–20 mm long and 12–15 mm wide, yellowish-green or pale green, the hoods and/or other parts often tinged with purple, the clusters terminal. Calyx lobes lanceolate, 4–7 mm long. Corolla lobes 8–11 mm long, spreading, with ascending tips. Column not evident. Hoods quite sessile, fleshy, deflexed from the gynostegium, 5–7 mm long, the tips round, clublike, incurved; nectar horns absent; gynostegium about 2 mm long and 5 mm wide, the horny wings along the pollinating crevices very broad and conspicuous. Follicles narrowly spindle-shaped, smooth, minutely hairy, 4–13 cm long and 1–2.5 cm wide. Seed body 7–8 mm long, the coma pale tan, about 3 cm long.

Ingestion of approximately 1.2 percent of an animal's body weight of the green plant causes death (21).

A. engelmanniàna Woods.

Named after George Engelmann (1809–1884), a German-born physician and botanist who resided in St. Louis.
Pale green tinged with purple; late June–mid-September
Dry, rocky or sandy prairie
West two-thirds (especially the west half) and Anderson County

Erect herbaceous perennial with 1 or more stems 0.4–1.3 dm tall, simple or with a few branches. Leaves alternate, opposite, or irregularly arranged, sessile, widely spreading or reflexed, linear, 6–20 cm long and 1–3 mm wide, the apex attenuate, mucronate. Flowers pale green tinged with purple, 7–10 mm long and 5–9 mm wide, the clusters terminal and lateral at the upper nodes. Calyx lobes lanceolate to ovate-acuminate, 2–4 mm long. Corolla lobes reflexed, about 5 mm long. Column short and stout. Hoods sessile, 2.5–3 mm long, keeled along the back and with a pair of longitudinal laminate flaps, pouchlike at the base, the apex truncate or notched; nectar horns not evident; gynostegium 2–2.5 mm long and 3–3.5 mm wide, definitely protruding beyond the hoods; wings along the pollinating crevices not spurred. Follicles narrowly spindle-shaped, 8–12 cm long and about 1.5 cm wide. Seed body about 7 mm long, the coma pale tan, about 3 cm long.

This species may be confused easily with the more common *A. stenophylla*.

A. hirtélla (Pennell) Woods.

NL., from L. *hirtus*, hairy, rough, shaggy, + -*ella*, -*ellus*, a diminutive suffix—covered with short stiff hairs.
Pale green tinged with purple; June–early September
Prairie or occasionally in open woods, along roadsides, or on waste ground
East fifth

Stout, erect herbaceous perennial with 1 or more stems 3–13 dm tall, simple or with a few branches. Leaves alternate, opposite, or irregularly arranged, short-petioled, ascending, linear to linear-lanceolate, 5.5–15 cm long and acute or wedge-shaped, the margins revolute, rough. Flowers pale green tinged with purple, 6–10 mm long and 4–6 mm wide, the clusters terminal and lateral at the upper nodes. Calyx lobes lance-triangular, 1.5–3 mm long. Corolla lobes reflexed, 3–5 mm long. Column about as long as broad. Hoods sessile, 1.5–2 mm long, rounded on the back, pouchlike at the base, the apex rounded; nectar horns absent; gynostegium about 2 mm long and 2 mm wide, protruding prominently above the hoods. Follicles narrowly spindle-shaped, 6–15 cm long and 1–2 cm wide, covered with minute soft hairs. Seed body about 9 mm long, the coma pale tan, 3.5–4 cm long.

A. incarnàta L.   Swamp milkweed

L. *incarnatus*, made flesh, or incarnate—suggested by the color of the pleasantly scented flowers.
Purplish-red, rarely white; July–August
Wet places
Throughout

Erect perennial herbs with branched or unbranched stems from a stout, branching rootstock with a dense cluster of strong fibrous roots. Herbage glabrous or with fine minute hairs. Leaves opposite, petioled, the blades narrowly lanceolate to nearly linear, 6–14 cm long and 1–3.5 cm wide, the apex acute or attenuate, the bases acute to rounded or truncate. Flowers purplish-red (rarely white), 7–9 mm long and 4–7 mm wide, borne in lateral and terminal clusters. Calyx lobes linear-oblong, 1–2 mm long. Corolla lobes elliptic, reflexed, 3–6 mm long. Column cylindric. Hoods stipitate, scooplike, ascending, 1.5–3 mm long; nectar horns slender, longer than the hoods, tapered gradually and arching broadly over the gynostegium, the latter 2–3 mm long and 1.5–2 mm wide. Follicles erect, spindle-shaped, long-attenuate, smooth, glabrous or minutely hairy, 7–9 cm long and 8–12 mm wide. Seed body broadly elliptic, 7–10 mm long, the coma white, about 2 cm long.

See plate 70.

A. latifòlia (Torr.) Raf. Broadleaf milkweed

NL., from L. *latus*, broad, + *folium*, leaf.
Flowers greenish; July–August (–September)
Sandy or rocky prairie hillsides
West half, primarily

382. *Asclepias incarnata*

Stout, erect perennial herbs with stems 2–6 dm tall, usually unbranched. Herbage minutely hairy when young, becoming near-glabrous and glaucous with age. Leaves opposite, short-petioled, thick and somewhat leathery, mostly very broadly obovate to short-oblong or nearly round, 8–12 cm long and 6.5–10 cm long, the apex broadly rounded or more or less deeply notched, mucronate, the base cordate. Flowers greenish, sometimes tinged with purple, 10–15 mm long and 9–12 mm wide, the clusters terminal or lateral. Calyx lobes ovate-lanceolate, 3–4 mm long. Corolla lobes reflexed, 7–10 mm long. Column short and stout, broadly obconic. Hoods stipitate, short-cylindric, keeled along the back, truncate at the apex, erect or somewhat ascending, 3–4 mm long; nectar horn fused the length of the hood, the tip abruptly constricted and arching over the gynostegium, the latter 3–5 mm long and 3–4 mm wide. Follicles broadly spindle-shaped, smooth, nearly glabrous, 6–8 cm long and 1.5–3 cm wide. Seed body elliptic, about 7 mm long, the coma tawny, about 2 cm long.

This plant is toxic to livestock, with approximately 1 percent of an animal's body weight of the green plant causing death (21).

A. mèadii Torr.

Named for its discoverer, Samuel Barnum Mead (1799–1880).
Yellowish-green tinged with purple; late May–June
Ungrazed prairie
East fifth; rare

383. *Asclepias latifolia* fruiting

Slender, erect herbaceous perennials with 1 to few stems 2–6 dm tall. Herbage pale green, glabrous, glaucous. Leaves opposite, sessile, very narrowly lanceolate to ovate, 5–9 cm long and 1.5–4.5 cm long, the apex usually acute, the base acute to rounded or slightly cordate. Flowers greenish-yellow tinged with purple, 8–17 mm long and 7–14 mm wide, borne in a terminal cluster. Calyx lobes lance-triangular, 3–5 mm long. Corolla lobes reflexed, lance-oblong to elliptic, 6–9 mm long. Column short and stout. Hoods short-stipitate, erect, somewhat ovoid with 2 marginal teeth, rounded on the back, pouch-like toward the base, fleshy, 4–6 mm long; nectar horns shorter than the hoods, united basally with the hoods, constricted near the tip, curved inward, about as long as the gynostegium, the latter 2.5–3 mm long and 2–4 mm wide; corpusculum of each pollinium relatively massive. Follicles erect, narrowly fusiform, 8–10 cm long and 1 cm wide, glabrous. Seed body broadly elliptic, about 8 mm long, the coma white, about 4 cm long.

A. pùmila (Gray) Vail Plains milkweed

L. *pumilus*, dwarf.
White; July–August (–September)
Dry, rocky or sandy prairie hillsides and uplands
West three-fifths, primarily

Small, minutely and sparsely hairy perennial herbs with few to several slender stems 1.4–4 dm tall arising from a deep-seated rhizome. Leaves arranged in a dense spiral, essentially sessile, the blades narrowly linear or almost threadlike, 2–6 cm long and 0.5–1 mm wide, the apex sharply pointed, the margin revolute. Flowers white (occasionally tinged with pink or yellowish-green), borne in mostly terminal clusters. Calyx lobes lance-triangular to oblong-elliptic, 2–2.5 mm long. Corolla lobes elliptic, reflexed, 2–4 mm long. Column cylindric. Hoods stipitate, scooplike, erect, 1.5–2 mm long; nectar horns slender, almost twice as long as the hood, tapered gradually and arching over the gynostegium, the latter 1–3 mm long and 1.5–2 mm wide. Follicles erect, narrowly

384. *Asclepias pumila*

MILKWEED FAMILY

283

spindle-shaped, 4–8 cm long and 6 mm wide. Seed body broadly elliptic, 4–6 mm long, the coma white, about 2.5 cm long.

Ingestion of approximately 1–2 percent of an animal's body weight of the green plant causes death (21).

385. *Asclepias purpurascens*

A. purpuráscens L. — Purple milkweed

L. *purpurascens*, becoming purple.
Deep purplish-red; late May–early June
Rocky, open woods and thickets
East 2 columns of counties and Coffey County

Erect perennials with unbranched stems 4–9 dm tall. Flower stalks and lower leaf surface minutely soft-hairy. Leaves opposite, short-petioled, the blades usually lanceolate, less frequently ovate or elliptic, 7–23 cm long and 3–10 cm wide, the apex acuminate to acute or broadly rounded and mucronate, the base rounded to acute or short-acuminate, often oblique, the margins sometimes undulate. Flowers deep purplish-rose, 14–16 mm long and 7–10 mm wide, the clusters terminal and at the upper nodes. Calyx lobes very narrowly lanceolate or narrowly triangular, 3–4 mm long. Corolla lobes reflexed, 6–9 mm long. Column short and broad. Hoods short-stipitate, erect or ascending, lanceolate, 5–8 mm long, the margins inrolled and fused at the tip; nectar horns fused to the hoods about half their length, the tip abruptly constricted and arching toward the gynostegium, the latter 2.5-3 mm long and 2–4 mm wide. Follicles narrowly spindle-shaped, 10–16 cm long and 1–2 cm wide, minutely soft-hairy to nearly glabrous. Seed body elliptic, 5–6 mm long, the coma white, 3.5–4.5 cm long.

See plate *71*.

386. *Asclepias quadrifolia*

A. quadrifòlia Jacq. — Four-leaved milkweed

L. *quadri-*, 4, used in combination with other words, + *folium*, leaf—because the leaves along the middle of the stem are usually in whorls of 4.
Pink to nearly white; late April–June
Woods and thickets
Cherokee County

Graceful, erect perennial herb with 1 to few stems 2–7 dm tall arising from a fleshy rhizome. Herbage minutely soft-hairy. Leaves opposite below, 4 at a node above, petioled, the blades thin, lanceolate, (2.5–) 6–10 cm long and 3–3.5 cm wide, the apex acute to attenuate, the base acute or acuminate. Flowers pink to nearly white, 8–11 mm long and 6–9 mm wide, the clusters terminal or in the upper axils. Calyx lobes lanceolate, 1.5–3 mm long. Corolla lobes elliptic, reflexed, 5–6 mm long. Column short and stout. Hoods stipitate, ascending, 3–4 mm long, cupped at the base, 3-lobed, the middle lobe oblong-elliptic, rounded, the lateral lobes sharply pointed and perpendicular to the middle lobe; nectar horn fused with about half the length of the hood, rather broadly crescent-shaped, abruptly constricted near the tip, arching over the gynostegium, the latter 1.5–2 mm long and 1.5–2 mm wide; corpusculum of each pollinium relatively minute. Follicles erect, smooth, glabrous, narrowly spindle-shaped, 8–12 cm long and 4–8 mm thick. Seed body elliptic, about 7 mm long, the coma white or tawny, about 3.5 cm long.

387. *Asclepias quadrifolia* fruiting

A. speciòsa Torr. — Showy milkweed

L. *speciosa*, handsome—in reference to the flowers.
Purplish-rose; June–July
Moist soil of prairies, roadsides, waste ground, flood plains, and lake margins
West three-fourths

Stout, erect, unbranched herbaceous perennial 6–10 dm tall. Herbage, especially the lower leaf surface and the flower pedicels and peduncles, more or less densely woolly. Leaves opposite, short-petioled, cordate to ovate, broadly lanceolate, or (less frequently) oblong or elliptic, 7–14 cm long and 4–11 cm wide, the apex acute or rounded, mucronate, the base cordate to broadly rounded or nearly truncate, the margins sometimes undulate. Flowers purplish-rose, 18–27 mm long and 16–22 mm wide, the clusters borne laterally at the upper nodes. Calyx lobes lanceolate, 4–6 mm long. Corolla lobes reflexed, elliptic, 8–14 mm long. Column short and stout. Hoods short-stipitate, spreading, relatively broad at the base, then abruptly constricted to an elongate, narrowly oblong tip, 8–14 mm long; nectar horns united basally with the hoods, much shorter than the hoods, the tip abruptly constricted and incurved sharply toward the gynostegium, the latter 3–5 mm long and 2–4 mm wide. Follicles erect, spindle-shaped, more or less spiny, densely woolly, 9–12 cm long and 2–3 cm wide. Seed body elliptic, 6–9 mm long, the coma white, 3–4 cm long.

A. stenophýlla Gray — Narrow-leaved milkweed

NL., from Gr. *stenos*, narrow, + *phyllon*, leaf, i.e., "narrow-leaved."
Greenish-white; June–August
Prairie
Throughout, but more common eastward

Slender, erect, minutely hairy herbaceous perennial with 1 to several unbranched stems 3–8 (–10) dm tall arising from a vertical tuberous root that continues downward as a slender taproot. Leaves alternate, opposite, or irregularly arranged, sessile, usually ascending, linear, 6–20 cm long and 1–5 mm wide, the apex attenuate, mucronate. Flowers greenish-white, 7–9 mm long and 5–7 mm wide, the clusters terminal and lateral at the upper nodes. Calyx lobes lanceolate to long-triangular, 2–3 mm long. Corolla lobes eventually reflexed, 4–5 mm long. Column nearly obscured by the hoods. Hoods sessile, 2–4 mm long, keeled along the back and with a pair of longitudinal laminate flaps, the base pouchlike, the apex 3-toothed; nectar horns entirely fused with the hoods, evident only by the middle apical tooth on the hood; gynostegium 2–3 mm long and 1.5–3 mm wide, not protruding beyond the hoods; wings along the pollinating crevices spurred toward the base. Follicles narrowly spindle-shaped, 9–12 cm long and 7–8 mm wide, minutely hairy or nearly glabrous. Seed body 5–6 mm long, the coma pale tan, 3–3.5 cm long.

388. *Asclepias speciosa*

A. sullivántii Engelm. — Sullivant milkweed

Named after William Sterling Sullivant (1803–1873), authority on mosses.
Rose-purple; June–August
Moist soil of low prairies and roadsides
East half

Stout, erect, glabrous, glaucous perennial 5–9 dm tall, with a short, often-branched rhizome and a long, stout root. Leaves opposite, sessile or very short-petioled, broadly ovate to lanceolate, lance-oblong or elliptic, 8–14 cm long and 3–8.5 cm wide, the apex acute to broadly rounded or notched, mucronate, the base rounded to more or less cordate. Flowers rose-purple, 15–18 mm long and 10–12 mm wide, the clusters terminal and lateral. Calyx lobes lance-triangular, 3–6 mm long. Corolla lobes reflexed, lance-elliptic, 9–12 mm long. Column short and stout. Hoods short-stipitate, erect, very broadly rounded at the tip, 5–7 mm long; nectar horns united with the hood about half their length, more or less abruptly constricted toward the tip, about as long as the gynostegium and arching toward it, the latter 3–5 mm long and 3–5 mm wide. Follicles erect, broadly spindle-shaped, 8–10 cm long, more or less spiny. Seed body broadly elliptic, about 8 mm long, the coma white, about 4.5 cm long.

389. *Asclepias stenophylla*

A. syrìaca L. — Common milkweed

L. *syriacus*, Syrian—but the name is strangely misapplied, since this is a North American, not a Syrian species.
Dull purplish-pink to greenish-white; June–August
Prairie, pastures, roadsides, and waste ground
East three-fifths, primarily; a few records from northwest counties

Stout, erect perennial herbs with unbranched stems 0.6–2 m tall. Herbage, especially lower leaf surface, finely soft-hairy. Leaves opposite, short-petioled, the blades ovate to lanceolate, oblong, or elliptic, rarely oblanceolate, 1–2 (–3) dm long and 3.5–12 cm long, the apex acute, short-acuminate, or broadly rounded and merely mucronate, the base acute to broadly rounded, truncate, or cordate. Flowers fragrant, dull purplish-pink or sometimes greenish-white, 10–13 mm long and 7–10 mm wide, the clusters few to several at the upper nodes. Calyx lobes ovate-lanceolate, 3–4 mm long. Corolla lobes elliptic, reflexed, 7–10 mm long. Column very short and stout. Hoods short-stipitate, ascending, 3–6 mm long, fleshy, scooplike, with 2 triangular-acuminate lateral lobes, rounded on the back; nectar horns shorter than and united basally with the hoods, abruptly constricted near the tip, curved inward, the gynostegium 2.5–3 mm long and 2–5 mm wide; corpusculum of each pollinium relatively massive. Follicles erect, spindle-shaped, 7–12 cm long and 2–4 cm wide, the surface quite variable, smooth or with soft, long or short, spinelike protuberances, more or less densely hairy. Seed body elliptic, 6–8 mm long, the coma white, 3–4 cm long.

This species has been suspected of poisoning livestock (30).

The unopened flower buds and the very young fruits are edible when boiled for a few minutes in several changes of lightly salted water. They are quite tasty when served with butter or a cheese sauce.

390. *Asclepias sullivantii*

MILKWEED FAMILY

391. *Asclepias syriaca*

392. *Asclepias tuberosa*

393. *Asclepias tuberosa,* close-up of flowers

In Appalachia, the milky sap of the root has been used to treat warts and moles. A concoction of the root has also been used as a remedy for coughs, asthma, indigestion, constipation, and gonorrhea, and to induce vomiting and increase urine flow (30).

See plate 72.

A. tuberòsa L. Butterfly milkweed, pleurisy root

L. *tuber,* swelling or knob; *tuberosus,* full of swellings or knobs—in botany, possessed of tubers, the thick root suggesting the name.
Red-orange to orange or yellow; late May–early August (–September)
Prairie
East two-thirds

Erect herbaceous perennial with a deep, woody rootstock, lacking the milky juice typical of this genus. Herbage, especially the stems, covered with long, soft hairs. Stems few to several 3–8 dm tall, branching only in the inflorescence. Leaves alternate, short-petioled, crowded on the stem, the blades usually broadest near the base and tapering gradually to an acute or attenuate apex, mostly 4–8.5 cm long and 0.6–2 cm wide, the base rounded or truncate to cordate or somewhat sagittate, the margins revolute. Flowers red-orange, sometimes lighter orange or yellow, 10–15 mm long and 5–9 mm wide, borne in terminal clusters. Calyx lobes lanceolate to long-triangular, 2–3 mm long. Corolla lobes lanceolate to elliptic, reflexed, 5–8 mm long. Column cylindric. Hoods stipitate, more or less lanceolate and scooplike, with a pair of marginal teeth, erect, 4–6 mm long; nectar horns slender, arching gradually over the gynostegium, the latter 1.5–2.5 mm long and 1.5–3 mm wide. Follicles erect, narrowly spindle-shaped, smooth, 8–15 cm long and 1–1.5 cm wide, covered with minute, soft hairs. Seed body broadly oval, 5–7 mm long, the coma white, 3–4 cm long.

Flower color in *A. tuberosa* ranges from deep red-orange to a rich yellow, depending upon the amount of red pigment which is superimposed over the yellow carotenoid background pigments. Populations of this plant in central, northern, and eastern Missouri and adjacent portions of Illinois, Iowa, and Kansas tend to have deep red-orange flowers. As one travels outward in any direction from this central region, the flowers tend to have less and less red pigment and so grade toward orange and then yellow. Especially as one travels southwestward, in central Oklahoma and adjacent parts of Kansas and Texas, one may encounter large colonies with pure yellow flowers (74).

Various uses for this species have been recorded. Some western and Canadian Indian tribes boiled the young shoots and immature pods as potherbs. The Omahas and Poncas boiled the roots and drank the decoction for bronchial and pulmonary sickness. For healing wounds and old sores they applied chewed fresh roots or powdered dried roots. Vogel (45) lists numerous additional applications, drawn together from several sources. According to him, the Menominee pulverized the root and used it for cuts, wounds, and bruises, as well as mixing it with other roots for other remedies, and the Potawatomi apparently used it as a poison antidote. The Natchez boiled the roots and took a cupful at a time, believing the plant to be the best remedy for pneumonia and winter fever. The Catawba Indians made a decoction of the roots for dysentery. Whites also adopted this plant as a medicinal agent, and the dry root was listed officially in the *U.S. Pharmacopoeia,* 1820–1905, and the *National Formulary,* 1916–1936, as a diaphoretic and expectorant or, in larger doses, as an emetic and purgative.

See plate 73.

A. verticillàta L. Whorled milkweed

NL., from L. *verticillus,* for the small flywheel of stone or clay set on the old-time, hand-operated spinning spindle to add momentum and stability in twirling the spindle—the whorled leaves of this species suggesting the radiating spokes of a wheel.
Greenish-white; June–September
Prairie
Nearly throughout, but more common in the east three-fifths

Slender, erect perennial herbs with unbranched stems 2.5–11 dm tall arising from short vertical rhizomes producing fibrous roots. Herbage sparsely and minutely hairy, the stems with lines of minute hairs extending downward from each node (use magnification). Leaves whorled, usually 3 or 4 at a node, sessile or nearly so, linear, strongly ascending, 2.8–7.8 cm long and 0.5–2.5 mm wide, the apex sharply pointed, the margins revolute. Flowers greenish-white (occasionally tinged with purple), 6–8 mm long and 4–5 mm wide, borne in terminal and lateral clusters. Calyx lobes narrowly triangular, 1.5–2.5 mm long. Corolla lobes elliptic, reflexed, 3–4 mm long. Hoods stipitate, scooplike, with a pair of marginal teeth, 1.5–2 mm long; nectar horns slender, about twice as long as the hoods, tapered gradually and arching broadly over the gynostegium, the latter 1.5–2 mm long and 1.5–2 mm wide. Follicles erect, narrowly spindle-shaped, smooth, glabrous, 7–10 cm long and 5–8 mm wide. Seed body elliptic, about 5 mm long, the coma white, about 2.5 cm long.

This species is toxic to livestock, but more than 2 percent of the animal's weight must be eaten to cause death.

A. subverticillata (Gray) Vail, which strongly resembles *A. verticillata*, is reported from several far-western counties. The former differs from the latter in having leaves which are larger (2–13 cm long and 1–4 mm wide) and more lax and which bear, in some of the axils, dwarf branches with reduced, opposite leaves. It also is very poisonous to livestock, with ingestion of only 0.2 percent of an animal's body weight of the green plant required to produce death (21).

A. viridiflòra Raf.   Green milkweed

NL., from L. *viridis*, green, + *flos, floris*, flower.
Flowers pale green; June–early August
Prairie
Throughout

Herbaceous perennial with simple, ascending or reclining stems 2–10 dm tall. Leaves alternate, opposite, or irregularly arranged, short-petioled, ovate to lanceolate or elliptic, occasionally oblong or oblanceolate, 2.5–13 cm long and 1.2–3.5 cm wide, the apex usually short-acuminate, acute, or broadly rounded, mucronate, rarely notched, the base acute or rounded, the margins undulate. Flowers pale green, 10–12 mm long and 3–5 mm wide, the clusters numerous and usually lateral. Calyx lobes narrowly lanceolate to lance-attenuate, 3–5 mm long. Corolla lobes reflexed, 6–7 mm long. Column barely evident, if at all. Nectar hoods sessile, 3–5 mm long, oblong in back view, pouchlike at the base, the apex rounded; nectar horns absent; gynostegium 3–3.5 mm long and 2–3.5 mm wide, protruding but slightly beyond the hoods. Follicles narrowly spindle-shaped, 7–15 cm long and 1.5–2 cm thick, finely soft-hairy or nearly glabrous. Seed body 6–7 mm long, the coma pale tan, 3–5 cm long.

A. lanuginosa Nutt. resembles *A. viridiflora* somewhat and is known from several counties, primarily in central and northeastern Kansas. It is smaller in stature than *A. viridiflora*, has narrowly lanceolate leaves and terminal flower clusters only. The wings along the pollinating crevice are broadened to form a point below the middle of the gynostegium, while those of the more common *A. viridiflora* have a point above the middle.

See plate *74*.

A. víridis Walt.   Green antelopehorn

L. *viridis*, green, the flowers being greenish.
Greenish-white and reddish-purple; mid-May–July
Prairies and roadsides, often in rocky soil
East three-fifths, primarily

Low, light-green, sparsely and minutely hairy perennial herbs with a cluster of erect, ascending or reclining stems 2.5–5 dm tall arising from a woody crown, the main root thick and tough, giving off numerous horizontal laterals at different levels. Leaves irregularly arranged on the stem, short-petioled, the blades mostly lanceolate, lance-oblong or elliptic, 5.5–12 cm long and 2–6 cm wide, the apex acute or broadly rounded, frequently mucronate, the base wedge-shaped to acuminate, rounded, or slightly cordate. Flowers greenish-white with reddish-purple hoods, 13–18 mm long and 13–20 mm wide, the clusters terminal and lateral at the upper nodes. Calyx lobes lanceolate to narrowly triangular, 2–5 mm long. Corolla lobes 10–17 mm long, spreading, with ascending tips. Column not evident. Hoods quite sessile, fleshy, deflexed from the gynostegium, 4–6 mm long, the tips rounded, clublike, ascending; nectar horns absent; gynostegium 2.5–4 mm long and 3–4 mm wide, the horny wings along the pollinating crevices narrow and inconspicuous. Follicles broadly spindle-shaped to ovoid, minutely hairy or nearly glabrous, 6–13 cm long and 2–3 cm wide. Seed body elliptic, about 7 mm long, the coma white or pale tan, about 4 cm long.

CYNÁNCHUM L.

According to Fernald (9), an ancient name for some plant supposed to be poisonous to dogs, from Gr. *cyon*, dog, + *anchein*, to strangle.

Perennial vines with blossom, fruit, and seed characters resembling those of

394. *Asclepias verticillata*

395. *Asclepias viridiflora*

396. *Asclepias viridiflora*, close-up of flowers

MILKWEED FAMILY

287

397. *Asclepias viridis*

398. *Asclepias viridis* fruiting

399. *Cynanchum laeve*

Asclepias. Leaves opposite. Flowers in few- to many-flowered axillary umbel-like clusters. Corolla small, funnelform or bell-shaped, white to yellowish or yellow-green. Members of the corona flat, lacking nectar horns.

One species in Kansas.

C. laève (Michx.) Pers.   Sandvine

L. *laevis*, smooth—characterizing the whole plant.
White; late June–August
Thickets, creek banks, roadsides, fields, waste ground
East half, primarily

Slender, twining, herbaceous perennial vine with branching stems up to 2 m long. Leaves opposite, the blades triangular-lanceolate to deltoid, dark green with whitish veins, 3–12 cm long and 2–8 cm wide, the apex acuminate, the base usually sagittate with rounded lobes, occasionally merely cordate, the margins entire, the petioles 1–7.5 cm long. Flowers small, white, fragrant, borne in crowded umbel-like racemes 1.5–2.5 cm wide. Sepals lanceolate to narrowly lance-elliptic, 2–4 mm long, with membranous margins. Corolla long, the lobes oblong or lance-oblong, erect at anthesis but the tips eventually spreading or somewhat recurved. Column not evident. Members of the corona about as long as the corolla, more or less lanceolate, divided above into 2 long-attenuate segments. Follicles smooth, narrowly lanceolate in outline, somewhat angled in cross section, 8.5–14 cm long and about 3 cm wide. Seed body ovate, flattened, 7–9 mm long, with a marginal wing; coma white, 2–3.5 mm long.

Since these flowers lack functional nectar receptacles, the nectar collects in basal hollows of the stigmatic chambers, and bees and flies, visiting the flowers for nectar, incidentally pick up pollinia and transport them to the stigmatic chambers of other flowers.

Polygonum convolvulus and *P. scandens* are vining plants with herbage superficially resembling sandvine. The flower structure of the two genera is quite different, however.

MATÈLEA Aubl.   Anglepod

Derivation uncertain.

Ours herbaceous perennials with twining stems. Leaves opposite, more or less heart-shaped, petioled. Flowers few to many, in umbel-like axillary racemes in our species, basically resembling those described for *Asclepias*. Calyx 5-lobed, usually glandular on the inner surface. Corolla green to dark purple, wheel- or bell-shaped. Corona quite inconspicuous, usually merely cup- or ring-shaped, lacking nectar horns. Gynostegium broad and flattened. Follicles smooth, warty, or with spinelike protuberances. Seeds with a coma.

Two species in Kansas.

Key to Species:
1. Corolla lobes brownish-purple, oblong, 2–3 mm wide; corona fleshy, shorter than the gynostegium, with 10 rounded lobes; pedicels and calyx densely hairy with short glandular hairs and long nonglandular hairs .... *M. decipiens*
1. Corolla lobes brownish- to greenish-purple, linear-attenuate, 1–2 (–3 mm wide; corona not fleshy, at least as tall as the gynostegium, rimmed with acuminate teeth; pedicels and calyx glabrous or with a few cilia at the tips of the lobes .... *M. gonocarpa*

M. decípiens (Alex.) Woods.

L. *decipiens*, deceitful or puzzling, probably in reference to the taxonomic complexity of this genus.
Brownish-purple; late May–June
Woods and thickets
Cherokee and Neosho counties

Plants with stout twining stems. Herbage glandular-hairy. Leaf blades cordate, 5–15 cm long and 3–11 cm wide, both surfaces soft-hairy; petioles 3.5–8 cm long. Pedicels and calyx densely hairy with short glandular hairs and long nonglandular hairs, the calyx lobes lanceolate to lance-triangular or lance-attenuate, 3–7 mm long. Corolla lobes brownish-purple, oblong, 2–3 (–6) mm wide. Corona fleshy, shorter than the gynostegium, with 10 rounded lobes. Follicles spiny.

M. gonocárpa (Walt.) Shinners.   Anglepod milkvine

NL., from Gr. *gonia*, angle, + *karpos*, fruit—in reference to the 3–5, more or less definite, longitudinal angles of the pods.

Brownish- to greenish-purple; mid-June–mid-July
Woods and thickets
Cherokee County

Plants with wiry twining stems. Herbage hairy but not glandular. Leaf blades more or less cordate, 7.5–15 cm long and 4.5–10 cm wide, more or less hairy; petioles 2–10 cm long. Pedicels and calyx glabrous, the calyx lobes lance-attenuate to linear-attenuate, 3–5 mm long, occasionally with a few cilia at the tip. Corolla lobes brownish- to greenish-purple, linear- attenuate, 7–11 mm long and 1–2 (–3) mm wide. Corona not fleshy, at least as tall as the gynostegium, rimmed with acuminate teeth. Follicles smooth, glabrous, angled in cross section, 6–12 cm long and about 2–3 cm wide.

400. *Matelea gonocarpa*

CONVOLVULÀCEAE   Morning-glory Family

Annual or perennial, twining, trailing or rarely upright herbs or small trees, sometimes with milky juice. Leaves alternate, usually simple, entire, dentate, lobed, or dissected, absent or much reduced in the parasitic genus *Cuscuta*, sessile or petioled, stipulate. Flowers bisexual, regular, solitary or clustered in the leaf axils. Sepals 5, separate or fused together part of their length, persistent. Petals (4–) 5, fused together, the corolla tubular to funnelform, bell-shaped, or nearly wheel-shaped, more or less 5-angled or -lobed. Stamens (4–) 5, arising from the base of the corolla tube, or from between the corolla lobes in *Cuscuta*. Pistil 1; ovary superior, 2- to 3-celled with 2 ovules in each cavity; style and stigma usually 1 each (2 in *Evolvulus* and *Cuscuta*). A nectiferous disk is sometimes present around the base of the pistil. Fruit usually a globular capsule.

The family consists of about 50 genera and 1,600 species, widespread in temperate, subtropical, and tropical regions, especially in the tropics of Asia and the Americas. Economically important members of the family include the sweet potato (*Ipomoea batatas* (L.) Lam.), various noxious weeds such as bindweed (*Convolvulus* spp.) and dodder (*Cuscuta* spp.), and ornamentals such as the morning-glory (*Ipomoea purpurea* (L.) Roth.) and the wood rose (*Ipomoea tuberosa* L.).

The family is represented in Kansas by 4 genera.

401. *Matelea gonocarpa*

Key to Genera:
1. Plants yellowish-white to pale orange in color, lacking foliage leaves or chlorophyll, twining parasites on other plants .... *Cuscuta*
1. Green plants with normal leaves, not parasitic 2
　2. Styles 2, each 2-cleft; stigmas 4, threadlike; plants not twining or sprawling .... *Evolvulus*
　2. Style 1; stigmas 1 or 2; plants twining or sprawling (except in *Ipomoea leptophylla*) 3
　　3. Stigma or stigmas capitate or globose .... *Ipomoea*
　　3. Stigmas 2, threadlike or oblong .... *Convolvulus*

CONVÓLVULUS L.   Bindweed

NL., from L. *convolvere*, to entwine, *convolvo*, roll round—from the twining habit of plants of this genus.

Annual or perennial herbs, usually with twining or trailing stems. Leaves entire or lobed, sessile or petioled. Flowers white or pink, borne singly or in few-flowered axillary clusters. Sepals alike or nearly so, separate. Corolla funnelform or somewhat bell-shaped, 5-angled or very shallowly 5-lobed. Ovary 2-celled, or sometimes 4-celled by the intrusion of false partitions from the ovary wall; style 1; stigmas 2, threadlike, linear, or spindle-shaped. Capsules (1-) 2- to 4-celled, 2- to 4-seeded, separating into 2 or 4 segments.

Four species in Kansas. *C. pellitus* Ledeb., a double-flowered ornamental, is naturalized in some parts of the United States and is reported from scattered localities in eastern Kansas.

Key to Species:
1. Calyx enclosed by 2 large bracts; stigmas oblong or spindle-shaped, scarcely flattened if at all .... *C. sepium*
1. Calyx not enclosed by large bracts; stigmas linear or threadlike, flattened 2
　2. Plants gray-green, densely hairy; leaves with basal lobes coarsely toothed and/or lobed; sepals 6–12 mm long; southwest sixth .... *C. equitans*
　2. Plants darker green, glabrous or inconspicuously hairy; leaves with basal lobes entire; sepals 2–5 mm long; throughout .... *C. arvensis*

C. arvénsis L.

L. *arvensis*, inhabiting a field.
White or pinkish; May–October
Disturbed areas such as roadsides, pastures, cultivated and waste ground
Throughout; intr. from Europe or Eurasia

402. *Convolvulus arvensis*

Glabrous or more or less hairy perennial with slender, creeping rhizomes. Stems trailing or twining, up to 1 m long. Leaves hastate or sagittate, petioled, 1.5–5.5 cm long and 0.5–3.5 cm wide. Flowers open during early morning hours, borne singly on long axillary stalks with a pair of tiny bracts at the juncture of the pedicel and peduncle. Sepals obovate to oblong, 2–5 mm long, the apex truncate or notched, mucronate, the margins membranous. Corolla white or pinkish inside, entirely pinkish or with broad vertical pink stripes outside, shallowly 5-lobed, 1.5–2.5 cm long. Nectar disk present at base of ovary. Stigmas threadlike. Capsules broadly ovoid, mostly 7–10 mm long. Seeds dark brown, approximately quarter-spheroidal, minutely tubercled, about 4 mm long.

This European immigrant has become one of our worst weeds. Freely branching and rapidly growing, it is capable of invading large areas. While it is spreading over the surface, its slender horizontal rhizome is doing the same thing beneath the surface. When the plant is hoed out, pieces of rhizomes remaining in the soil give off upright shoots that soon cover the ground again. When the plant is plowed under, any portion of the rhizome can produce upright shoots. The new growths, of course, are nourished by food stored in the rhizomes, and if we keep the shoots cut away or ploughed under before the leaves have turned green and start photosynthesizing, the rhizome cannot replenish its lost food and will starve to death. This mechanical destruction of the plant is a long process, and a chemical method may well be employed where the use of the hoe is not feasible. This species and *C. sepium* are common in England and Ireland, and throughout Europe and Russian Asia, excepting in the extreme north. *C. sepium* is native to this country also, but *C. arvensis* is said to have been introduced in the 1870s through the importation of Turkey hard winter wheat.

See plate 75.

C. équitans Benth.

NL., from L. *equito, equitatus*, to ride on horseback.
White to pink; mid-May–August (–October)
Prairies and roadsides, in sandy and/or rocky soil
Southwest sixth

403. *Convolvulus equitans*

Gray-green, densely hairy perennials with a taproot, not creeping. Stems trailing or climbing, up to 2 m long. Leaves petioled, the blades quite variable in shape but more or less sagittate or hastate, the main portion of the leaf 1.5–4 cm long and 2–10 mm wide, the basal lobes usually coarsely toothed and/or lobed. Flowers borne singly on long axillary stalks with a pair of tiny bracts at the juncture of the pedicel and peduncle. Sepals ovate to lanceolate, 6–12 mm long, the apex usually acute, the margins membranous. Corolla white to pink, often with a red center, 10–16 mm long, 5-angled, the angles often sharply pointed. Stigmas linear, more or less flattened. Capsules broadly ovoid to nearly spherical, 9–13 mm long. Seeds dull black, rough, about 4.5–6 mm long.

C. sèpium L.

NL., from L. *sepes*, hedge—in allusion to the plant's habit of climbing over bushes and hedges.
White; May–early October
Disturbed areas such as roadsides, flood plains, fields, pastures, waste ground
Nearly throughout

Glabrous to more or less hairy perennials with creeping rhizomes. Stems trailing or twining, 2–3 m long. Leaves petioled, the blades hastate to sagittate or occasionally cordate, 3–10 cm long and 1.5–7.5 cm wide. Flowers borne on naked peduncles longer than the petiole of the subtending leaf. Calyx subtended and concealed by a pair of large ovate bracts. Sepals lanceolate, thin, 12–15 mm long. Corolla white, funnelform, 5-angled, 3–6 cm long. Stamen filaments broader and papillose toward the base. Nectar disk present around base of ovary. Stigmas oblong or spindle-shaped, scarcely flattened, if at all.

See plate 76.

404. *Convolvulus sepium*

CUSCÙTA L.

NL., from Arabic *kushuth, kashutha, kashuta*. "Dodder" is plural of Frisian *dodd*, bunch—in allusion to the bunches of flowers and threadlike stems entwining the stems of, or scrambling in a tangle over the tops of, host plants.

Yellowish or orange, twining, mostly parasitic, leafless and eventually rootless, annual or perennial herbs. Leaves represented only by minute scales. Flowers whitish, small and inconspicuous, in few- to many-flowered clusters. Calyx of 5 (rarely 4) sepals free or united. Corolla bell-shaped to short-tubular or urn-shaped, which withers after fertilization occurs, the margin 5- (rarely 3- or 4-) cleft. Each stamen usually subtended by a minute, scalelike, often fringed appendage, usually membranous, but fleshy in some species. Ovary 2-celled; styles 2 (rarely 1); stigmas 2, usually headlike. Fruit a 4-seeded, globular capsule, indehiscent or opening transversely near the base.

According to a recent monograph (46), there are about 150 species in the genus *Cuscuta*. This is a relatively difficult group to work with, but it seems worth including here, since the unusual coloration and structure of these parasitic plants frequently catch the eye and arouse the curiosity of the wildflower lover or field botanist. Of the 10 species in the state, the 4 most frequently collected are described here. *C. indecora* is illustrated as an example of a species which has a loose, trailing growth habit rather than a tight, ropelike habit such as that of *C. glomerata*.

Development of the dodder plant from independent seedling to parasite is an interesting sequence of events to follow. Upon germinating, the seedling appears as a mere thread, its caulicle, or primary stem, emerging first, while the radicle, or primary root, remains in contact with the endosperm until the nutrient substances are exhausted. Then the radicle emerges from the seed and penetrates the soil. As the seedling elongates, it begins to swing in a circle, thus increasing its chances of coming in contact with a suitable host. If such contact is made, the dodder twines around the host and puts forth rootlike, absorbing structures called "haustoria" which grow into the stem in such a way that the water-conducting, food-conducting, and storage tissues of the parasite become fused with the corresponding tissues of the host—a union as intimate and complete as that achieved between scion and stock after grafting or budding. After the first haustorial contacts are made, the true root of the parasite dies, severing contact with the soil.

A dye bath made from chopped dodder plants imparts a yellow color to wool mordanted with alum and cream of tartar (47).

See plate 77.

405. *Cuscuta indecora*

Key to Species:

1. Sepals free from one another, the calyx subtended by 1 or more bracts 2
   2. Flowers sessile, in dense, ropelike masses which obscure most of the *Cuscuta* stem; corolla tubular or urn-shaped, 5–6 mm long, 4-lobed .... *C. glomerata*
   2. Flowers on short pedicels, in loose, more or less rounded panicles which do not obscure the *Cuscuta* stem; corolla bell-shaped, 3–4 mm long, 5-lobed .... *C. cuspidata*
1. Sepals united, the calyx not subtended by bracts 3
   3. Corolla bell-shaped, the lobes 4, triangular, erect .... *C. polygonorum*
   3. Corolla urn-shaped, the lobes 5, acute or acuminate, reflexed .... *C. pentagona*

406. *Cuscuta glomerata*

C. cuspidàta Engelm.   Dodder

L. *cuspidatus*, pointed.
Whitish; August–September
Roadsides, flood plains, frequently in sandy habitats
Probably throughout

Stems evident among the short, usually rounded, branched flower clusters. Flowers mostly 3–4 mm long, on short pedicels. Sepals free from one another, shorter than the corolla tube, lanceolate to ovate or nearly round, appressed, subtended by 1 to few rounded bracts. Corolla bell-shaped with 5 acute, spreading lobes. Capsules globose, capped by the persistent corolla, opening irregularly. Seeds about 1.4 mm long.

Like *C. glomerata*, this species exhibits some preference for members of the Composite Family as hosts.

C. glomeràta Chois.   Knotted dodder

L. *glomeratus*, gathered into a ball or knot—in allusion to the dense masses of flowers.
Yellowish-white; late July–early October
Roadsides, thickets, creek banks
East two-thirds, primarily

Stems mostly obscured by the dense, ropelike masses of flowers, coiled tightly around the stem of the host plant. Flowers 5–6 mm long, sessile, subtended by acuminate, recurved bracts. Sepals free from one another, about as long as the corolla tube, lance-linear, the apex acute and spreading. Corolla tubular or urn-shaped with 4 usually acute, spreading or reflexed lobes. Capsules flask-shaped, capped by the persistent corolla, opening irregularly. Seeds about 1.7 mm long.

Although occurring on numerous species of woody and herbaceous plants, *C. glomerata* exhibits some preference for members of the Composite Family as hosts.

C. pentagòna Engelm.   **Dodder**

L. *pentagonus*, 5-angled, from L. *penta*, Gr. *pente*, 5, + *gonia*, angle.
Whitish; late June–early October
Flood plains, lake margins, roadsides, occasionally in prairie ravines
Nearly throughout

Stems evident among the short, loose, usually rounded, branched flower clusters. Flowers 1.5–4 mm long, on short pedicels. Sepals united, shorter than the corolla tube, the lobes broadly ovate, appressed, the calyx not subtended by bracts. Corolla urn-shaped, with 5 acute or acuminate, reflexed lobes. Capsules more or less globose to ovoid, opening irregularly. Seeds about 1 mm long.

Hosts of this species include *Solidago, Aster, Ambrosia, Artemisia, Euphorbia*, and others.

407. *Evolvulus nuttallianus*

C. polygonòrum Engelm.   **Dodder**

NL., of *Polygonum*, this species often parasitic on plants of that genus.
Whitish; June–October
Wet roadside ditches, creek and river banks, lake margins, sloughs
East half, primarily

Stems evident among the short, branched flower clusters. Flowers mostly 2–3 mm long, on short pedicels. Sepals united, as long as or longer than the corolla tube, the lobes triangular-ovate, often unequal, appressed, the calyx not subtended by bracts. Corolla bell-shaped, the 4 lobes triangular, erect. Capsules more or less globose, opening irregularly, about 2.5–3 mm long. Seeds about 1.5 mm long.

This species is found frequently on *Polygonum*, but also parasitizes *Impatiens, Penthorum, Lycopus*, and other genera.

EVÓLVULUS L.   **Evolvulus**

L. *evolvulus*, that which is rolled out, unrolled, from *e*, out, + *volvere*, to roll, in contrast to *Convolvulus*.

Low, erect or prostrate perennial herbs, sometimes woody near the base. Leaves entire, sessile or nearly so. Flowers blue to purplish or white, small, axillary, usually solitary. Sepals 5, separate. Corolla funnelform to wheel-shaped, the tube short. Ovary 2-celled; styles 2, each 2-cleft; stigmas 4, linear. Capsule 2-celled, 1- to 4-seeded.

One species in Kansas.

E. nuttalliànus R. & S.   **Nuttall evolvulus**

Named in honor of Thomas Nuttall (1786–1859), naturalist and professor of botany at Harvard.
Lavender to white; mid-May–early June
Rocky or sandy prairie hillsides and banks
West four-fifths, primarily; more common westward

Silvery-pubescent perennial with a taproot and short woody crown. Stems several to many, erect to ascending or some of them decumbent, 5–12 (–25) cm tall. Leaves elliptic, occasionally linear, densely hairy on both surfaces, 6–15 (–20) mm long and 1–6 mm wide, the apex acute or less frequently rounded. Sepals 4–5 mm long. Corolla lavender to nearly white, funnelform, 8–10 mm long. Capsules globose or nearly so, 4–5 mm long, the pedicels elongating slightly and curving downward. Seeds yellow to orangish-brown, smooth, quite variable in shape, about 2.5–3 mm long.

IPOMOÈA L.   **Morning-glory**

NL., from Gr. *ips, ipos*, for a kind of worm, + *homoios*, like—in reference to the twining stems.

Annual or perennial herbs, often twining or climbing, or woody trees, shrubs, or vines. Leaves simple (in all ours) or palmately compound, entire to toothed or lobed, sessile to petioled. Flowers large and showy, funnelform (in our common species) to salverform or bell-shaped, the margin entire to angled or 5-lobed. Ovary (1-) 2- to 3-(5-) celled; style 1; stigma 1 with 1–3 (–5) globose lobes. Capsules broadly ovoid to nearly globose, usually beaked by the persistent style base, 1- to 6-(10-)seeded, usually separating into 2 or more segments.

Eight species in Kansas. *I. coccinea* L. and *I. cristulata* Hallier, both escaped from cultivation, and *I. shumardiana* (Torr.) Shinners are but rarely or occasionally encountered and are not included below.

Key to Species:

1. Bushy perennial with a massive taproot, erect to decumbent stems, and linear-attenuate leaves .... *I. leptophylla*
1. Trailing or twining annuals or perennials with leaves more or less cordate, either entire or lobed 2
  2. Stigma entire or 2-lobed; ovary 2-celled; capsules 4-seeded 3
    3. Sepals rounded, glabrous; peduncles longer than the petioles; corolla 5–8 cm long .... *I. pandurata*
    3. Sepals acuminate, bristly-ciliate; peduncles short; corolla 1.5–2.3 cm long .... *I. lacunosa*
  2. Stigma 3- to 5-lobed; ovary 3- to 5-celled; capsules 6- to 10-seeded 4
    4. Sepals with short acute tips; leaf blades entire or occasionally 3-lobed .... *I. purpurea*
    4. Sepals with very narrow, elongate green tips much longer than the body; leaf blades 3- to 5-lobed or occasionally entire .... *I. hederacea*

408. *Ipomoea hederacea*

I. hederàcea Jacq.   Ivyleaf morning-glory

L. *hederaceus*, ivylike, from *hedera*, ivy—the resemblance to English ivy, *Hedera helix*, being in the lobed leaves.
Purple to blue or white; mid-June–October
Fields, roadsides, waste ground, and other disturbed areas
East half, primarily; scattered records from farther west; intr. from tropical America

More or less hairy, twining or trailing annual vine with a slender taproot. Stems 3–15 dm long, with retrorse hairs. Leaves simple, petioled, the blades cordate, usually deeply 3- or 5-lobed, but occasionally entire, 2.5–7 cm long and 2.5–9 cm wide. Flowers borne singly or in clusters of 3. Sepals about equal in length, 1.5–2.5 cm long, conspicuously hairy toward the base, lance-attenuate, the long-tapering tips much longer than the body. Stigma 3-lobed. Capsules about 10–13 mm long. Seeds dull black, about 4–5 mm long.

This is one of the most beautiful of morning-glories and a favorite in garden culture. Native to tropical America and possibly of the southern United States, it has escaped from gardens and become somewhat of a nuisance.

See plate *78*.

I. lacunòsa L.

L. *lacunosus*, full of holes, cavities, gaps—in reference to the open, netlike venation of the leaves.
White; (late June–) August–mid-October
Roadsides, fields, creek banks, waste ground, frequently in moist soil
East third, primarily

Sparsely to densely hairy, twining annual with a slender taproot. Stems 0.5–2.5 m long. Leaves simple, petioled, the blades cordate and entire to deltoid or more or less deeply 3-lobed, 1.5–10 cm long and 1.5–9 cm wide. Sepals oblong-elliptic, acuminate, 9–11 mm long, about equal in length, the margins bristly-ciliate. Corolla white, often tinged with purple, funnelform, sharply 5-lobed, 17–23 mm long. Stigma 2-lobed. Capsules broadly ovoid to nearly globose, 2-celled, 4-seeded, 9–10 mm long. Seeds reddish-brown, smooth, about 4–5 mm long.

409. *Ipomoea lacunosa*

I. leptophỳlla Torr.   Bush morning-glory, bigroot morning-glory

NL., from Gr. *leptos*, fine, narrow, + *phyllon*, leaf.
Pink or purplish; mid-June–early September
Sandy prairie, roadsides, disturbed ground
West two-thirds and Riley County

Glabrous, bushy perennials, up to about 1.3 m tall and 1.5 m across, with an enormously enlarged taproot. Stems branched, erect to ascending or decumbent. Leaves simple, short-petioled, the blades linear-attenuate, 2–8.5 cm long and 1–8 mm wide. Sepals broadly ovate to nearly round, 6–10 mm long, the innermost longer than the outer ones. Corolla pink or purplish, funnelform, 6–9 cm long. Stigma more or less 2-lobed. Capsules ovoid, 2-celled, 1.5–2 cm long (excluding the persistent style). Seeds brown, densely hairy, about 1 cm long.

The massive, spindle-shaped taproot becomes 15–20 cm in diameter and 1.3 m or more in length, then gives off several slender extensions that spread and descend to a depth of 3 m or more. In addition, many lateral branches are given off throughout the length of the massive part and radiate horizontally a distance of 3 m or more, branching and rebranching until the soil is fully occupied by absorbing rootlets. This voluminous system of roots enables the plant to withstand long drought periods without injury.

The 2 stigmas, elevated on a slender style above the anthers, are posed to receive

410. *Ipomoea leptophylla*

MORNING-GLORY FAMILY
293

pollen brought from other flowers by butterflies, moths, and long-tongued bees coming for nectar secreted around the base of the ovary.

The plant was used by the Pawnee Indians as a cardiac stimulant (48). See plate *79*.

411. *Ipomoea purpurea*

I. panduràta (L.) G. F. W. Mey.   Bigroot morning-glory, wild potato vine

NL., *panduratus*, from L. *pandura*, for a 3-stringed musical instrument with a high-vaulted back and long neck like a mandolin, said to have been invented by Pan, in Greek mythology the god of woods and shepherds. Although the instrument was not incurved on the sides like a fiddle, the term *panduratus* is applied to leaves constricted at the middle, or fiddle-shaped, as they sometimes are in this species.

White with a purple throat; June–mid-September
Roadsides, thickets, creek banks
East third

Glabrous or hairy, trailing or twining, herbaceous perennial vine with a long, vertical thickened root which may eventually weigh 10–30 pounds or more. Stems up to 4 m long. Leaves simple, petioled, the blades usually cordate, less frequently deltoid or with indented sides and almost 3-lobed, 2.5–9 cm long and 2.5–10 cm wide. Sepals broadly elliptic to oblong-elliptic, 15–20 mm long, about equal in length, the apex rounded, the the margins not ciliate. Corolla white with a purple throat, funnelform, shallowly 5-lobed, 5–8 cm long. Stigma 2-lobed.

I. purpùrea (L.) Roth.   Common morning-glory

L. *purpureus*, purple.
Bluish-purple to red or white; June–October
Fields, roadsides, waste ground
East three-fifths, primarily; intr. from tropical America

Twining or trailing annual vine similar to *I. hederacea*. Stems with retrorse hairs. Leaves simple, petioled, the blades cordate, usually entire but occasionally deeply 3-lobed, 3–7 cm long and 2.5–7 cm wide. Flowers borne in long-peduncled, umbel-like clusters. Sepals about equal in length, the innermost subulate, the outer ones lanceolate to lance-oblong with an acute apex, 1–1.5 cm long, conspicuously hairy toward the base. Corolla bluish-purple to red or white, funnelform, 4–6 cm long. Stigma 3-lobed. Capsules about 8–10 mm long. Seeds dark brown, minutely hairy, about 4–5 mm long.

According to Krochmal and Krochmal (30), the entire plant, including the seeds, has been used to prepare a laxative, and the seeds have been used as hallucinogens.

POLEMONIÀCEAE   Phlox Family

Annual, biennial, or perennial herbs, rarely vines, shrubs, or small trees. Leaves alternate or opposite, mostly simple, sessile or petiolate, estipulate. Flowers bisexual, usually regular, borne in clusters or rarely solitary in the leaf axils. Calyx and corolla both tubular below and 5-lobed above, the corolla trumpet-shaped or salverform to wheel- or bell-shaped. Stamens 5, arising from the corolla tube at more or less unequal heights. Pistil 1; ovary superior, 3-celled; style 1; stigmas 3 or 1 with 3 lobes. Fruit usually a many-seeded loculicidal capsule subtended by the persistent calyx.

A family of about 13 genera and 300 species, mostly of the western United States. Cultivated members of the family include species of *Phlox, Gilia, Cobaea,* and *Polemonium.*

Kansas has 4 genera.

Key to Genera (66):
1. Leaves pinnately compound, with distinct broad leaflets .... *Polemonium*
1. Leaves simple, but in some species cut into narrowly lanceolate to threadlike segments 2
  2. Lower or all leaves opposite .... *Phlox*
  2. Leaves alternate or basal 3
    3. Corolla appearing rotate (tube very short), or narrow funnelform with the tube
      flaring upward .... *Gilia*
    3. Corolla more or less salverform, with a nearly cylindrical tube longer than the lobes
      .... *Ipomopsis*

GÍLIA R. & P.   Gilia

Named in honor of Felipe Gil, Spanish astronomer and botanist of the last half of the 18th century.

Annual, biennial, or perennial herbs, sometimes woody at the base. Leaves alternate, entire to toothed or variously pinnatifid. Flowers blue or purplish, often with a yellow throat, borne singly, in pairs, or in bracted clusters. Calyx lobes separated by and bordered by dry transparent membranes. Corolla wheel-shaped to trumpet-shaped. Stamens even to moderately uneven in position. Capsule several- to many-seeded.

Three species in Kansas. *G. spicata* Nutt., with white flowers borne in spicate clusters, is known from Wallace, Scott, and Hamilton counties. *G. calcarea* Jones, with whitish to blue or purple flowers borne in open, much-branched clusters, is known from Hamilton County only. Both of these species have salverform, rather than wheel-shaped, corollas and lack the woody basal portion characteristic of *G. rigidula*.

G. rigidula Benth.   Needleleaf gilia

412. *Gilia rigidula*

NL., from L. *rigidus*, stiff, inflexible, $+$ *-ulus*, an adjective diminutive meaning "somewhat" or "rather."
Blue; late April–June
Rocky or sandy prairie hillsides, frequently in calcareous soil
South two-thirds of the west third

Low, glandular-hairy perennial, mostly 5–15 cm tall, branching and woody at the base. Leaves pinnatifid, with 3–5 linear needlelike subdivisions, 8–20 mm long. Flowers in few-flowered terminal clusters. Calyx bell-shaped, 6–8 mm long in flower (longer in fruit), the lobes with a green midvein, linear-attenuate. Corolla blue (rarely white) with a yellow center, wheel-shaped, the lobes ovate or broadly elliptic, 5–8 mm long, longer than the short corolla tube. Anthers yellow, protruding conspicuously beyond the corolla throat on threadlike filaments. Capsules ovoid, several- to many-seeded, 4–6 mm long, enclosed by the persistent, somewhat enlarged calyx. Seeds yellowish, irregularly angled, about 1.5 mm long.

Our plants were formerly called *G. acerosa* (Gray) Britt. but have now been placed in subspecies *acerosa* of *G. rigidula*.

IPOMÓPSIS Michx.

NL., from Gr. *ips, ipos*, a kind of worm, $+$ Gr. *Mopsos*, L. *Mopsos*, a celebrated soothsayer.

Annual, biennial, or perennial herbs or subshrubs. Leaves alternate, entire to pinnately lobed or dissected, with firm sharp tips. Flowers showy, purplish or red to pink or white, usually individually bracted, borne singly or in clusters. Calyx lobes sharp-tipped, partially united by dry thin membranes. Corolla regular or slightly irregular, salverform but lacking a definitely expanded throat. Stamens arising unevenly toward the top of the corolla tube, exserted. Stigmas 3.

Two species in Kansas. *I. laxiflora* (Coult.) V. Grant is known from Stanton and Stevens counties only. It resembles *I. longiflora* but is smaller and more delicate in all respects.

I. longiflòra (Torr.) V. Grant   White-flowered gilia

NL., from L. *longus*, long, $+$ *flos, floris*, flower—in reference to the long corolla tube.
White; August–October
Dry, sandy or rocky prairie
Southwest fourth, east to Kingman and Harper counties

Erect, branched annual, up to 6 dm tall, with a strong, deep taproot having many superficial, wide-spreading, finely branched laterals. Lower leaves 2.5–5 cm long, pinnately divided into linear or threadlike segments 0.5–1 mm wide; upper leaves undivided, linear. Flowers borne in branched terminal clusters. Calyx 5–7 mm long. Corolla white, the tube 2.2–4.5 cm long, the lobes 6.5–12 mm long, obovate to rhombic, with an abruptly acuminate apex. Capsules ellipsoid, 9–12 mm long. Seeds more or less elongate, somewhat angular, 2.5–4.5 mm long, minutely winged (see high magnification), sticky when wet.

The stamens are inserted on the corolla tube near the mouth, with 2 or 3 of the anthers exserted a short distance. They shed pollen on the proboscises of butterflies and night-flying moths and on the bills of hummingbirds, all of which are seeking nectar secreted around the base of the ovary. After the pollen is shed, the anthers reflex or fall away, and the stigmas spread apart to be available for cross-pollination.

POLEMÒNIUM L.   Polemonium, Jacob's ladder

NL., from L. *polemonia*, Gr. *polemonion*, ancient name of the Greek valerian (*Polemonium caeruleum*).

Mostly perennial herbs. Leaves alternate, pinnately compound. Flowers showy, blue to purple, borne singly or in clusters. Calyx wholly herbaceous, bell-shaped. Corolla short-funnelform to broadly bell-shaped. Stamens arising at about the same height near the top of the corolla tube, the filaments slender, declined toward 1 side of the corolla, and often hairy near the base. Style elongate; stigmas 3. Capsule ovoid, several- to many-seeded, enclosed by the persistent, enlarged calyx, separating into 3 segments.

One species in Kansas.

413. *Polemonium reptans*

P. réptans L.   Creeping polemonium

L. *reptans*, creeping—but this plant does not creep.
Pale blue; mid-April–May
Rich, moist, rocky, wooded hillsides
Extreme northeastern and extreme southeastern counties

Smooth perennial, 2–4 dm tall, with a short vertical rhizome which gives off strong fibrous roots and 1 or more rather fragile, ascending or spreading, branched stems. Lower leaves long-petioled, with 9–13 (–17) leaflets; upper leaves with fewer leaflets and gradually becoming sessile upward; leaflets ovate to lanceolate or elliptic, mostly 2–5 cm long and 0.4–2 cm wide. Flowers borne in terminal clusters. Calyx 5–7 mm long (5–12 mm in fruit), minutely hairy, the lobes triangular or acuminate. Corolla pale blue with a white center, 10–16 mm long. Stamens and style protruding slightly. Capsules several-seeded, about 5 mm long.

The flowers are proterandrous, the anthers opening before the 3 stigma lobes spread apart in readiness to receive pollen on their papillose inner surface. Nectar, secreted by an annular nectar gland at the base of the ovary, is sought after by bumblebees and honeybees, and pollen is taken by bees and some species of beetles, any of which may effect cross-pollination.

Plants are easily propagated by seeds sown in the fall. According to Steyermark (4), the roots have been used as a diuretic and as a remedy for kidney problems.

PHLÓX L.   Phlox

L. and Gr. for a kind of flower, from Gr. *phlox*, flame. Pliny mentions "the flame-colored flower, the name of which is phlox." Of another plant, he says it is "quite inodorous; which is the case also with the plant known by the Greek name of phlox." Finding the name to be of uncertain application in his day, Linnaeus gave it to this genus which he established from the type species *Phlox glaberrima*, a native of the southeastern United States, bearing pink or purple flowers.

Annual or perennial herbs or subshrubs. Leaves mostly opposite, sessile, entire. Flowers showy, often fragrant, solitary or in clusters. Calyx 5-ribbed, the lobe sharply pointed, often hairy on the inner surface, and very partially united by thin dry membranes. Corolla with a long, very slender, slightly curved tube and 5 abruptly spreading lobes. Stamens definitely arising at different heights. Style short or quite elongated; stigmas 3. Capsule ovoid, few- to several-seeded, subtended by the persistent calyx.

Four species native to Kansas. *P. andicola* Nutt. ex. Gray is known from Cheyenne County only. It is a low, white-flowered perennial with sharp-pointed needlelike leaves, somewhat resembling *P. oklahomensis*, but with the corolla lobes rounded rather than notched. The cultivated *P. paniculata* L. is rarely found as an escape.

Key to Species:
1. Low, sprawling plants 0.8–1.5 dm tall; flowers usually white, to corolla lobes notched at the apex .... *P. oklahomensis*
1. Erect plants 1.5–5 dm tall; flowers blue, purple, or brilliant pink (rarely white), the corolla lobes not notched 2
   2. Plants with sterile, basal, evergreen overwintering shoots; sepals barely awned; corolla usually light blue or lavender; plants mostly of woodlands .... *P. divaricata*
   2. Plants usually lacking sterile overwintering shoots; sepals definitely awned; corolla usually brilliant pink; plants mostly of prairies, though occasionally found in open woods .... *P. pilosa*

P. divaricàta L.   Sweet-william phlox

L. *divaricatus*, spreading—the habit of the basal overwintering shoots.
Blue or lavender; April–early June
Rich, moist soil of woodlands
East third

Soft-hairy, erect to decumbent perennial with 1 or more erect flowering stems 1.5–4 dm tall and several decumbent, sterile basal offshoots or stolons which persist through the winter months when the other stems have died and weathered off. Main leaves of the flowering stems lanceolate to elliptic or lance-oblong, 2.5–5 cm long and 0.5–2.5 cm wide, the apex acute or attenuate, the base cordate and clasping the stem or merely rounded. Leaves of the sterile offshoots lanceolate to elliptic or oblanceolate, 1.5–5.5 cm long and 0.3–2.3 cm wide, the apexes and bases usually acute. Flowers fragrant, borne in glandular-hairy terminal clusters. Calyx 5–9 mm long, the lobes long-attenuate. Corolla light blue or lavender, the slender tube 12–16 mm long, the lobes 9–13 mm long, usually obovate with a short-acuminate apex. Capsules 4–5 mm long. Seeds greenish-brown, broadly elliptic, somewhat flattened, minutely roughened, about 2.5 mm long.

Our plants are recognized as *P. divarica* subspecies *laphamii* (Wood) Wherry, which differs from the more eastern ssp. *divaricata* in being stronger-growing and longer-blooming, with flower color richer in blue, and petals without the notch at the apex that characterizes the species.

See plate *80.*

414. *Phlox divaricata*

P. oklahoménsis Wherry

Oklahoma phlox

"Of Oklahoma."
White to pale blue or pink; late March–mid-May
Prairie, usually in shallow soil over surfacing limestone
Butler, Cowley, Elk, and Chautauqua counties

Low, sprawling perennial 8–15 cm tall with stems somewhat woody at the base. Leaves narrowly lanceolate to linear-oblong or linear, 1–4 cm long and 1–4 mm wide, the margins, at least of the upper leaves, ciliate. Flowers delicately fragrant, borne in few- to several-flowered, pubescent terminal clusters. Calyx 5–10 mm long, the lobes subulate, hairy. Corolla white or, less frequently, pale blue or pink, the tube 8–12 mm long, the lobes 5–8 mm long, heart-shaped. Capsules about 4–5 mm long.

This is a rather rare plant, limited in its total distribution to a few counties in Kansas, northeastern Oklahoma, and Dallas County, Texas.

P. pilòsa L.

Prairie phlox

L. *pilosus*, downy—true of the herbage.
Brilliant pink to purplish, rarely white
Prairies, less frequently in open woods, usually in rocky limestone soil
East fourth

415. *Phlox oklahomensis*

Erect, more or less hairy perennials 2–5 dm tall with rhizomes, but usually lacking the sterile, basal overwintering shoots typical of *P. divaricata*. Main leaves mostly 3–9 cm long and 3–16 mm wide. Flowers borne in terminal clusters. Calyx 8–13 mm long, the lobes attenuate and rather rigid (said to be "awned"). Corolla usually brilliant purplish-pink, rarely white, the tube 10–16 mm long, the lobes (5–) 7–14 mm long, usually obovate with a rounded or short-acuminate apex. Capsules 5–7 mm long, with a minutely honey-combed surface.

Three subspecies of *P. pilosa* occur in Kansas. The plants occurring in Franklin and Miami counties northward bear only unbranched nonglandular hairs on the calyx and inflorescence branches and belong to ssp. *fulgida* Wherry. Most of the plants in the southeast eighth (Anderson and Linn counties south and southwestward to Elk, Chautauqua, and Cherokee counties) belong to ssp. *pilosa* (referred to as var. *virens* in some older manuals). Individuals of ssp. *ozarkana* Wherry are recorded from Montgomery and Cherokee counties. Both of the latter subspecies have calyces and inflorescence branches with glandular hairs, and those of ssp. *ozarkana* may be branched as well. Plants of ssp. *ozarkana* fall into the upper extremes of the size ranges given above and have upper leaves mostly linear to narrowly oblanceolate or linear-lanceolate and not noticeably broader at the base.

See plate *81.*

HYDROPHYLLÀCEAE

Waterleaf Family

Annual, biennial, or perennial herbs, rarely shrubs. Leaves mostly alternate, simple, linear to toothed or pinnately lobed or divided, more or less hairy, usually petioled, estipualte. Flowers bisexual, regular, solitary or borne in racemes which are sometimes 1-sided, or scorpioid, or in rounded clusters. Calyx of 5 partly united sepals, persistent in fruit. Corolla of 5 petals united below. Stamens 5, arising at or near the

416. *Phlox pilosa*

base of the corolla tube. Pistil 1; ovary superior or half-inferior, usually bristly-hairy, usually 1-celled but with 2 parietal placentae, the latter sometimes intruding into the ovary cavity in such a manner as to make the ovary appear 2-celled; styles 2, separate or more or less united; stigmas 2, capitate. Fruit usually a loculicidal capsule separating at maturity into 2 or 4 segments.

The family consists of approximately 18 genera and 300 species widely distributed on all continents except Australia. A few species are grown as ornamentals, but otherwise the family is of little economic importance.

Four genera in Kansas. *Nama stevensii* C. L. Hitchcock, a small, inconspicuous, grayish annual, usually much branched from the base and forming rounded clumps, is known from gypsiferous areas in Clark, Comanche, and Barber counties only. It has linear leaves 1–3 cm long and 1–3 mm wide and small, tubular, lavender flowers borne singly or in small clusters in the leaf axils.

417. *Ellisia nyctelea*

Key to Genera:

1. Flowers few, mostly borne singly; leaves opposite below (sometimes alternate above) .... *Ellisia*
1. Flowers several to many, in clusters; leaves all alternate 2
    2. Branches of the inflorescence repeatedly forked, not coiled, the cluster more or less headlike, not elongating after anthesis .... *Hydrophyllum*
    2. Branches of the inflorescence coiled at the tip, not forked, gradually straightening out and elongating after anthesis .... *Phacelia*

ELLÍSIA L.   Ellisia

Named in honor of John Ellis, English correspondent of Linnaeus, noted for his discovery of the animal nature of coral.

Small, delicate, branching annuals. Leaves opposite below, becoming alternate above, pinnately lobed. Flowers small, white to lavender, borne singly in or opposite the leaf axils, occasionally in a few-flowered terminal cluster. Corolla bell-shaped, about as long as or slightly longer than the calyx, the tube with 5 minute appendages. Stamen filaments included in the corolla tube. Ovary 4-ovuled; styles shorter than the corolla, united about half their length. Capsule 1-celled, separating into 2 segments, subtended by the enlarged, persistent calyx.

One species in North America.

E. nyctèlea L.   Ellisia

L. *Nycteleus*, Gr. *Nyktelios* (from Gr. *nyx, nyktos*, night), a surname of Bacchus because his mysteries were celebrated at night—the inference being that the flowers open in the night.
White or bluish; late April–mid-June
Moist, rich, shaded soil, usually in disturbed areas such as creek banks, roadside parks, waste ground
Throughout

Plants erect, more or less hairy, 7–27 cm tall, eventually becoming diffusely branched. Leaf blades 2–8.5 cm long and 1–5.5 cm wide, mostly oblanceolate to elliptic or lance-oblong in general outline, with 5–7 coarsely toothed, oblong, pinnate lobes reaching nearly to the midrib. Flowers minute, inconspicuous. Calyx at anthesis 2–5 mm long, the lobes lanceolate to triangular. Corolla white to bluish, 2–4 mm long. Fruiting pedicels arching downward, the capsules globose, sparsely hairy, 5–7 mm long, subtended by the much-enlarged star-shaped calyx. Seeds black, nearly globose, about 1.8–2.5 mm long, the surface minutely honeycombed (use magnification and side illumination).

HYDROPHÝLLUM L.   Waterleaf

NL., from Gr. *hydor*, water, + *phyllon*, leaf—the foliage of all species containing so much water as to make "washy" forage.

Mostly perennial herbs with rhizomes and somewhat fleshy fibrous roots. Leaves pinnately cleft. Flowers showy, white to purple, in compact, often headlike, branched clusters lacking a well-developed main axis. Calyx divided nearly to the base, sometimes with a small appendage in each sinus. Corolla bell-shaped to tubular, the tube with 5 pairs of longitudinal linear appendages forming nectiferous grooves. Filaments more or less bearded, protruding beyond the corolla tube, each flanked at the base by the nectiferous grooves mentioned above. Ovary 4-ovuled; style protruding beyond the corolla tube. Capsule globose, 1- to 3-seeded, separating into 2 segments.

Two species in Kansas. *H. appendiculatum* Michx., known from extreme northeast and east counties only, differs from the following species in having the upper leaves more or less circular in general outline and but shallowly palmately lobed.

H. virginiànum L.   Virginia waterleaf

"Virginian."
Lavender to white; May–June
Moist, rich, wooded hillsides and creek banks
East fourth, especially northward

Erect, branched herbaceous perennial 3–8 dm tall with a horizontal, scaly branching rhizome which gives off many strong, thickened roots along its length. Basal leaves long-petioled, the blades pinnately parted all or nearly all the way to the midrib, 4.5–10 cm long and 7–12 cm wide, the terminal segment triangular in general outline, deeply 3- or 5-lobed, the lateral segments obovate to oblanceolate, elliptic, or lanceolate, sometimes 2-lobed, the margins coarsely serrate; upper leaves much like the lower, but shorter-petioled and smaller. Calyx 3–6 mm long, the lobes linear, covered with minute appressed hairs intermingled with longer, conspicuous, spreading hairs or cilia. Corolla lavender to white, bell-shaped, 5-lobed, 8–10 mm long. Capsules 5–6 mm long. Seeds orangish-brown, 2.5–3.5 mm long, with relatively deep honeycomblike ornamentation.

The flowers are proterandrous—that is, the anthers discharge pollen before the stigmas of the same flower are in condition to receive it. Bees flying from plant to plant may effect cross-pollination by carrying pollen from younger to older flowers. Both bumblebees and honeybees are attracted to the nectar which is secreted around the base of the ovary and which rises in grooves formed by flaps of the corolla below the lobes—a neat little device that gives bees some depth of nectar from which to suck.

The young leaves of this and other species of *Hydrophyllum* can be used as salad greens or as cooked greens.

418. *Hydrophyllum virginianum*

PHACÈLIA Juss.   Phacelia

NL., from Gr. *phakelos*, bundle or fascicle—in reference to the flower clusters.

Usually hairy, annual, biennial, or perennial herbs. Leaves mostly alternate, entire or toothed to variously pinnately divided. Flowers blue or purplish to white, more or less conspicuous, in 1-sided, coiled clusters which elongate and straighten as they develop. Calyx divided nearly to the base, sometimes enlarging in fruit. Corolla wheel-shaped to bell-shaped or tubular, usually longer than the calyx, lobed to about the middle. Filaments equal in length, included in or protruding from the corolla tube, slender or dilated at the base, sometimes subtended by a pair of scales, or a gland bordered by parallel flaps between each filament pair. Styles 2, or 1 and then more or less deeply 2-forked or 2-cleft. Capsule ovoid to globose, with intruding placentae, 2- to many-seeded, separating into 2 segments.

Two species in Kansas. *Phacelia* is exclusively a New World genus of more than 100 species, most of them native to western North America.

419. *Hydrophyllum virginianum* leaf

Key to Species:

1. Delicate nonglandular plants; corolla bell-shaped, 8–12 mm across; southeast Kansas .... *P. hirsuta*
1. Robust glandular plants; corolla tubular, 3–5 mm across; south-central Kansas .. *P. integrifolia*

P. hirsùta Nutt.   Hirsute phacelia

L. *hirsutus*, rough with hairs, bristly—applicable to stems, leaves, calyx, and filaments.
Blue to lavender; mid-April–early June
Open woods, prairies, roadsides, usually in rocky, open places
Southeast corner, north and west to Bourbon, Allen, Wilson, and Montgomery counties

Erect, soft-hairy annuals or biennials. Stems eventually several to many times branched near the base, 1–3 dm tall. Basal leaves long-petioled, at least some of them divided to the midrib and essentially compound, often gone or withered by flowering time; upper leaves short-petioled or sessile, deeply pinnately lobed, 1–3.5 cm long and 0.6–2.5 cm wide, the lobes mostly oblong to linear-oblong or oblong-elliptic, acute or rounded. Inflorescences mostly 8- to 20-flowered, becoming quite loose and open. Calyx 4–7 mm long at anthesis, the lobes linear to linear-oblong or linear-elliptic. Corolla bell-shaped, with a short tube, 5–7 mm long, blue to lavender. Capsules globose, 3.5–4 mm long. Seeds rusty-brown, rather angular and variable in shape, about 1.5 mm long, with a reticulate ornamentation.

See plate *82*.

420. *Phacelia hirsuta*

P. integrifòlia Torr.

NL., from L. *integer*, whole, entire, + *folium*, leaf.
Lavender; July–mid-October

Prairie hillsides and banks, in gypsiferous soil
Kiowa, Comanche, Barber, and Harper counties

Robust, erect, glandular-hairy annuals or biennials 2–4 dm tall with a taproot. Main stem leaves petioled, lanceolate to lance-oblong, 2–12 cm long and 1.5–5.5 cm wide, the base cordate or truncate and frequently oblique, the margins irregularly crenate or toothed. Inflorescences many-flowered, crowded. Calyx about 3 mm long, the lobes elliptic. Corolla lavender, tubular, 4–6 mm long, the lobes much shorter than the tube. Capsules broadly ovoid, concealed by the persistent calyx, about 3.5 m long. Seeds black at maturity, oblong, 1.5–3 mm long, the abaxial surface deeply honeycombed, the adaxial surface with a prominent, sharp, funicular scar flanked by 2 deep grooves.

421. *Phacelia integrifolia*

BORAGINÀCEAE   Borage Family

Named after the genus *Borago,* not of our flora but represented in our gardens by the Mediterranean *Borago officinalis* L. *Borago,* first recorded in 13th-century herbals, is evidently derived from Spanish and Medieval Latin *borra* or *burra,* meaning "stiff hair," in reference to the hairs which often occur conspicuously on the stems and leaves.

Herbs, shrubs, or trees, rarely vines. Ours annual or perennial, bristly or rough-hairy herbs. Leaves alternate, simple, mostly sessile in ours, estipulate. Flowers bisexual, regular or (rarely) irregular, frequently arranged in clusters which, in bud, are circinately coiled, and which may be either helicoid (with all the flowers borne along 1 side) or scorpioid (with the flowers borne alternately on the right and the left sides). Sepals 5, separate or fused together part of their length. Petals 5, fused to one another, the corolla variously shaped, with 5 usually regular lobes. Stamens 5, inserted on the corolla tube. Pistil 1; ovary superior, 2- or 4-celled, deeply 4-lobed (except in *Heliotropium*); style 1, arising from the top of the ovary or emerging from down among the 4 lobes; stigmas usually 1 and headlike, sometimes 2 or 4. Fruit a 1- to 4-seeded nut or drupe, or 2 2-seeded or 4 1-seeded indehiscent mericarps, frequently very hard and bony, derived from the 4 lobes of the ovary, attached to a columnar or pyramidal extension of the receptacle called the gynobase.

A large family containing approximately 100 genera and 2,000 species. Cultivated members include Virginia bluebells (*Mertensia* spp.), forget-me-not (*Myosotis scorpioides* L.), comfrey (*Symphytum* spp.), lungwort (*Pulmonaria officinalis* L.), heliotrope (*Heliotropium arborescens* L.), and numerous others.

The family is represented in Kansas by native species in 7 genera and introduced species in 4 more. *Mertensia virginica* (L.) Pers. and *Echium vulgare* L., both with showy blue flowers, probably occur only as escapes from cultivation or are occasionally introduced from farther north or east. *Cynoglossum officinale* L., a European weed with dull red flowers 7–10 mm long, is recorded from scattered sites in the east third of the state. *Asperugo procumbens* L., also introduced from Europe, has been reported but once and apparently has not persisted in Kansas.

Key to Genera (3):
1. Ovary unlobed or shallowly 4-lobed; style terminal .... *Heliotropium*
1. Ovary deeply 4-lobed; style arising from among the lobes 2
  2. Mericarps attached by their base 3
    3. Style simple, stigma not lobed .... *Myosotis*
    3. Style 2-forked or stigma 2-lobed 4
      4. Corolla lobes rounded .... *Lithospermum*
      4. Corolla lobes acute or acuminate .... *Onosmodium*
  2. Mericarps attached laterally (the attachment area may be central, near the base, near the apex, or continuous over the whole length of the nutlet) 5
    5. Mericarps smooth or roughened, lacking prickles with retrorse barbs at the tip .... *Cryptantha*
    5. Mericarps conspicuously armed with prickles bearing retrorse barbs at the tip 6
      6. Pedicels erect or ascending in fruit; flower clusters bracted; mericarps attached all along the entire central angle; plants mostly annuals .... *Lappula*
      6. Pedicels recurved or deflexed in fruit; flower clusters naked or sparsely bracted; mericarps attached by an ovate or lanceolate area in the middle third of the inner face; plants mostly biennials or perennials .... *Hackelia*

CRYPTÀNTHA Lehm.

NL., from Gr. *kryptos,* hidden, + *anthos,* flowers.

Erect, branched, annual or perennial herbs, sometimes shrubs. Leaves narrow, entire. Flowers small, usually white, regular, borne in unilateral spikes or racemes which are coiled in the bud. Sepals but slightly united at the base, erect, linear or oblong, enclosing the mature fruits and falling with them. Corolla tubular, the lobes rounded and with small appendages which intrude into the corolla throat, sometimes with scales inside at the base of the tube also. Ovary deeply 4-lobed, 2- to 4-ovuled; style 1, arising among the ovary lobes; stigma headlike. Mericarps 1–4, more or less ovoid to 3-sided, rough or smooth, margined or marginless, but lacking bristles barbed at the tip, attached laterally by at least their lower half to the gynobase.

Three species in Kansas. *C. crassisepala* (T. & G.) Greene, an annual closely resembling *C. minima*, is known from a few scattered sites in the west half of the state. It is distinguished from the latter species primarily on the basis of its bractless, rather than bracted, flower clusters.

Key to Species:

1. Plants perennial; leaves 1–7 cm long and 2–10 mm wide, the upper surface covered with fine silky appressed hairs, the lower surface with coarser hairs arising from white siliceous disks; mericarps smooth .... *C. jamesii*
1. Plants annual; leaves 1–2.5 cm long and 1–5 mm wide, both surfaces covered with coarse, appressed or ascending hairs arising from white siliceous disks; mericarps tubercled .... *C. minima*

422. *Cryptantha jamesii*

C. jàmesii (Torr.) Payson

Named after Dr. Edwin James, compiler of records for Maj. Stephen H. Long's expedition to the Rocky Mountains (1819–1820), and first collector of this species.
White; mid-May–July (–early September)
Dry prairie hillsides and uplands, usually in sandy soil
Southwest fourth, primarily

Short-lived, pale green perennials with a woody taproot and branched caudex. Stems usually several, simple or branched at the base, 1.5–3 dm tall, covered with minute appressed hairs, and larger, spreading hairs. Leaves spatulate to narrowly oblanceolate or nearly linear, 1–7 cm long and 2–10 mm wide, the upper surface covered with silky appressed hairs, the lower surface also with coarser hairs arising from white siliceous disks. Flowers subtended by small bracts. Calyx 2–3 mm long, the lobes lanceolate. Corolla white, 5–6 mm long. Mericarps brownish mottled with white siliceous areas, lustrous, more or less triangular in cross section and definitely angled, about 2 mm long.

C. mínima Rydb.

L. *minimus*, least—in reference to the small stature of the plant.
White; May–June
Prairie hillsides and uplands, roadsides, waste ground, usually in dry, sandy soil
West half

423. *Cryptantha minima*

Rough-hairy, bristly annual. Stems branched, mostly decumbent or ascending, 0.5–2.5 dm tall, covered with minute appressed hairs and larger, spreading hairs. Leaves narrowly spatulate, 1–2.5 cm long and 1–5 mm wide, both surfaces covered with coarse, sharply pointed hairs arising from siliceous disks. Flowers subtended by small bracts. Calyx 2–2.5 mm long, the lobes nearly obscured by coarse hairs. Corolla white, 2.5–3 mm long. Mericarps more or less ovoid, not sharply angled, about 1.5 mm long, tubercled.

HACKÈLIA Opiz   Stickseed

In honor of Josef Hackel (1783–1869), Czech botanist.

Coarse biennial or perennial (rarely annual) herbs. Leaves broad. Flowers small, white or blue, in branched, one-sided terminal racemes. Sepals ovate to lanceolate or oblong, united but slightly at the base. Corolla short- to long-tubular, the lobes with small appendages that intrude into the throat. Stamens included within the corolla tube, the filaments slender, arising about halfway up the tube. Ovary deeply 4-lobed, 4-ovuled, the style arising among the lobes and scarcely (if at all) longer than they are; stigma headlike. Nutlets 4, ovoid, attached laterally to the pyramidal gynobase by an ovate or lanceolate area, the margins with narrow barbed prickles, the backs smooth or prickly, the fruiting pedicel recurved or deflexed.

One species in Kansas.

H. virginiàna (L.) I. M. Johnston

"Virginian."
White or pale blue; July–mid-August

Moist woods and thickets, rarely in prairie or waste areas
East half, primarily; a few records from farther west

Erect, rough-hairy, loosely branched annuals or biennials 3–9 dm tall. Leaves thin, lanceolate to elliptic, 4–22 cm long and 2.5–12 cm wide, the apex gradually acuminate, the margins entire. Flowers on short pedicels, subtended by small bracts, the clusters loose. Calyx 1–2 mm long, the lobes lanceolate. Corolla white or pale blue, 2–3 mm long, the tube about as long as the calyx. Mericarps covered with minute papillae (use high magnification) interspersed with conspicuous prickles with retrorsely barbed tips, 5–6 mm long.

HELIOTRÒPIUM L. Heliotrope

L. *heliotropium*, Gr. *heliotropion*, name used by ancient Greeks and Latins for the European heliotrope, from Gr. *helios*, sun, + *tropos*, a turning—Dioscorides and Pliny asserting that the flowers of their species turned toward the sun.

Annual or perennial herbs, some low shrubs in warmer climates. Leaves alternate (rarely opposite), sessile or petioled, entire. Flowers white to yellow or purple, usually in 1-sided, distinctly scorpioid clusters, with or without bracts. Sepals slightly united at the base, more or less unequal in length, persistent but not usually concealing the fruits. Corolla funnelform to salverform. Stamens included within the corolla tube, the anthers sessile or nearly so. Ovary 2- or 4-celled, scarcely if at all 4-lobed; style terminal, often very short or entirely absent, the stigma usually broadly to narrowly conical, receptive only in a band around the base. Fruit eventually separating into 2 hard, 2-celled, 2-seeded indehiscent mericarps or into 4 1-seeded indehiscent mericarps.

Four species in Kansas. *H. indicum* L., introduced from the American tropics via Asia, is recorded from Cherokee, Crawford, and Linn counties only. It is a coarse annual herb up to 1 m tall with broad petioled leaves; small blue or purplish flowers borne in 2 ranks in slender, elongating, scorpioid cymes; and sharply-ribbed mericarps.

Key to Species:
1. Perennial with a deep rhizome; plants succulent, glabrous, usually glaucous .. *H. curassavicum*
1. Annuals with taproots; plants not succulent, hairy 2
    2. Main leaves mostly 5–15 mm wide; corolla 15–22 mm wide, the lobes broader than long or but scarcely developed; mericarps 2; south half of the west two-thirds, and Cheyenne County .... *H. convolvulaceum*
    2. Main leaves 0.5–3 mm wide; corolla 5–6 mm wide, the lobes longer than broad; mericarps 4; southeast sixth .... *H. tenellum*

H. convolvulàceum A. Gray Bindweed heliotrope

Named after the genus *Convolvulus* in the Morning-glory Family, because the flowers of this species superficially resemble those of *C. arvensis*, the field bindweed.
White; July–mid-October
Sandy prairies and flood plains
South half of the west two-thirds, and Cheyenne County

Erect, gray-green annuals 0.8–3.5 dm tall, usually much branched from the base. Leaves mostly lanceolate to elliptic or lance-oblong, 1–2.5 cm long and 5–15 mm wide, the margins not rolled under. Herbage covered with coarse, appressed white hairs. Flowers fragrant, opening during early morning and evening, borne in the axils of the upper leaves or leaflike bracts or in the internodes. Calyx 5–7 mm long. Corolla white, 12–20 mm long, the narrow tube opening out into a broadly funnelform limb 15–22 mm across, the lobes definitely broader than long, if developed at all. Style elongate, the stigma more or less globose with an apical tuft of hairs. Mericarps 2, hairy, about 3 mm long.

H. curassàvicum L. Seaside heliotrope

"Of Curaçao," the main island of the Netherlands Antilles, off the northwest coast of Venezuela.
White; late May–September
Salt flats
Southwest fourth

Glabrous, glaucous, succulent perennials with a deep rhizome. Stems erect to decumbent, 1–4 dm tall. Leaves mostly spatulate or oblanceolate below to more or less elliptic or nearly linear above by autumn, 1–4.5 cm long and 2–25 mm wide. Inflorescences not bracted. Calyx 2.5–4 mm long, the lobes lance-linear. Corolla white, funnelform with 5 rounded lobes, 6–9 mm long. Style extremely short, the stigma conical, somewhat resembling a mushroom cap. Mericarps 2.5–3 mm long, glabrous, frequently remaining attached to one another.

424. *Heliotropium convolvulaceum*

425. *Heliotropium curassavicum*

H. tenéllum (Nutt.) Torr.　　　　　　　　Slender heliotrope

L. *tenellus*, slender—descriptive of the roots, stems, and leaves.
White; mid-June–early October
Prairie, pastures, usually in dry, rocky, shallow soil with surfacing limestone
Southeast sixth

Slender, gray-green, erect annual 1–3 dm tall with ascending branches. Leaves linear, acute, 1.3–4 cm long and 0.5–3 mm wide, the margins rolled under. Herbage covered with appressed white hairs. Flowers borne singly or a few together at the ends of the branches. Calyx 2–4 mm long, the lobes linear or subulate. Corolla white, funnel-form with elliptic or obovate lobes, 5–6 mm long and 5–6 mm across. Style short, stigma narrowly conical. Mericarps 4, minutely hairy, about 1.5 mm long, frequently remaining attached to one another.

NOTE: Compare with *Lithospermum arvense*.

LÁPPULA Moench　　　　　　　　　　　Stickseed

NL., from L. *lappa*, bur, + *-ula, -ulus*, an adjective diminutive suffix, i.e., "little bur"—the fruits in this genus usually armed with prickles.

Small, erect, rough-hairy, annual (rarely perennial) herbs. Leaves narrow, entire. Flowers small, regular, blue or white, in terminal, bracted, scorpioid racemes. Sepals united but slightly at the base. Corolla with a short tube and 5 ascending, rounded lobes, the throat closed by 5 scales. Stamens included within the corolla tube. Ovary deeply 4-lobed, 4-ovuled; style short, arising between the lobes; stigma slightly enlarged. Mericarps 4, quarter-ovoid, attached along their entire length to the conical or columnar gynobase which is much longer than broad, the outer faces bearing along their margins 1 or 2 rows of apically barbed prickles which sometimes are basally confluent to form a winglike or cuplike border, the actual faces minutely tubercled.

Three species in Kansas. *L. echinata* Gilib., introduced from Asia and the Mediterranean region, is known from a few scattered sites in the eastern third of the state. It differs from our 2 native species primarily in having the mericarps with marginal prickles in 2 rows.

Key to Species:
1. Mericarps with marginal prickles united most of their length to form a conspicuous, thickened, horseshoe-shaped structure .... *L. texana*
1. Mericarps with marginal prickles free all or most of their length .... *L. redowskii*

L. redówskii (Hornem.) Greene

Named after D. Redowsky, Russian botanist who published early in the 19th century.
Blue; May–June
Sunny, disturbed areas
West half, primarily, reaching eastward in the northwest fourth to Pottawatomie County; also recorded from Wilson and Wyandotte counties

Gray-green annuals 1–4.5 dm tall, branched from the base and/or above. Leaves spatulate below to narrowly oblong or elliptic above, 0.5–5 cm long and 5–8 mm wide, the apex rounded, the margins entire. Herbage covered with stiff, appressed or ascending hairs. Calyx 2–3 mm long at anthesis, the lobes linear to narrowly elliptic. Corolla blue, 3–3.5 mm long. Mericarps 2–3 mm long, the marginal prickles in a single row, distinct nearly to the base.

L. texàna (Scheele) Britt.

"Of Texas."
Blue; May–June
Prairie, pastures, roadsides, usually in rocky or sandy soil
West half

Gray-green annuals closely resembling *L. redowskii* in size, aspect, and most details. The distinguishing feature is the fruits, the mericarps having the marginal prickles united most of their length to form a conspicuous, thickened, horseshoe-shaped structure.

LITHOSPÉRMUM L.　　　　　　　　Gromwell, puccoon

NL., from Gr. *lithospermon*, the name of a European gromwell, from Gr. *lithos*, stone, + *sperma*, seed—the apparent seed (actually the fruit) being stony in appearance and hardness. "Gromwell," from OF. *gromil*, is the English vernacular name of *Lithospermum officinale*. "Puccoon" is the Omaha-Ponca Indian name for plants of this genus.

426. *Lappula redowskii*

427. *Lappula texana*

BORAGE FAMILY

303

428. *Lithospermum arvense*

429. *Lithospermum canescens*

Hairy, mostly annual or perennial herbs, often with thick roots which produce a red stain. Leaves in ours entire, usually narrow. Flowers yellow or yellow-orange to white in ours, borne singly in the leaf axils or crowded into a terminal leafy-bracted cluster. Sepals narrow, united slightly at the base, persistent. Corolla funnelform or salverform, the throat either hairy or with a small transverse fold or scale opposite each lobe or completely naked. Stamens included within the corolla tube. Ovary deeply 4-lobed, 4-ovuled; style threadlike, arising between the ovary lobes; stigmas 2, or 1 and 2-lobed. Nutlets 4, ovoid or angular, smooth or roughened, usually bony or stony, attached basally to a flat or broadly pyramidal receptacle.

Five species in Kansas. *L. latifolium* Michx., known from rich, moist woods in Miami, Douglas, Leavenworth, and Wyandotte counties only, is a perennial, 4–8 dm tall, with broad, thin, ovate or lance-ovate leaves and pale yellow flowers 5–7 mm long borne in the upper leaf axils.

Key to Species:

1. Annual or biennial; corolla white, about 5.5–8 mm long; plants of roadsides, waste places, and fields .... *L. arvense*
1. Perennial; corolla lemon-yellow to orange-yellow, 13–35 mm long; plants usually of prairie or prairie remnants 2
    2. Corolla lemon-yellow, the lobes erose and crinkly, the tube 13–30 mm long; mericarps pitted .... *L. incisum*
    2. Corolla orangish-yellow, the lobes entire, not crinkly, the tube 7–14 mm long; mericarps smooth 3
        3. Foliage soft-hairy; calyx 3–6 mm long at anthesis .... *L. canescens*
        3. Foliage rough-hairy; calyx 8–11 mm long at anthesis .... *L. carolinense*

L. arvénse L.   Corn gromwell

L. arvensis, pertaining to a cultivated field, from *arvum*, cultivated field, plowed land.
White; April–May
Roadsides, pastures, waste ground, and other disturbed habitats
East two-fifths, primarily; intr. from Europe

Hairy, light green, annual or biennial weed 1–5 dm tall, usually branched from the base. Leaves spatulate below to elliptic, oblong, or narrowly lanceolate above, 1–5 cm long and 1–10 mm wide, the apex rounded or acute. Flowers mostly borne singly in the leaf axils. Calyx 5–7 mm long. Corolla white, 5.5–8 mm long, the throat neither hairy nor appendaged. Mericarps light brown to gray, irregularly wrinkled and pitted, 2.5–3 mm long, subtended by but not concealed by the persistent calyx.

NOTE: Compare with *Myosotis verna* and *Heliotropium tenellum*.

L. carolinénse (Walt.) MacM.

"Of Carolina."
Orange-yellow; May–June
Sandy prairie
East half, primarily in the west half of that area

Erect, rough-hairy perennial with 1 to several stout stems 2–4.5 dm tall arising from a deep, strong root which yields a purple or reddish stain. Leaves mostly lanceolate to lance-linear, often folded along the midrib, 1–6 cm long and 2–11 mm wide, the apex and base acute or rounded. Flowers in terminal clusters. Calyx 8–11 mm long at anthesis. Corolla orange-yellow (fading to pale or brownish-yellow on drying), 16–25 mm long, the lobes rounded, entire, the tube hairy at the base within. Mericarps white, shiny, smooth, about 4 mm long.

See plate *83*.

L. canéscens (Michx.) Lehm.   Hoary gromwell, puccoon

L. canescens, growing white—the plant being whitened by a coating of short hairs.
Orange-yellow; mid-April–May
Prairie
East third

Erect, soft-hairy perennial with 1 to several stems 1.5–4 dm tall arising from a thickened root which yields a deep red or purple stain. Leaves narrowly lanceolate to lance-linear or narrowly elliptic, mostly 1–6.5 cm long and 4–15 mm wide, the apex rounded (occasionally acute), the base acute or rounded. Flowers sessile in terminal clusters. Calyx 3–6 mm long at anthesis. Corolla orange-yellow (fading to lemon-yellow on drying), 13–18 mm long, the lobes rounded, entire, the tube glabrous at the base. Mericarps white, shiny, smooth, about 2.2 mm long, seldom collected.

L. incìsum Lehm.   Narrowleaf gromwell, puccoon

L. *incisus*, cut into, in reference to the margins of the corolla lobes.
Yellow; April–May
Prairie
Throughout

Hairy, erect perennial with 1 or more stems 0.8–4 dm tall arising from a long, stout taproot. Basal leaves spatulate, often gone by flowering time; main stem leaves mostly linear to very narrowly elliptic or lance-linear, 1–7 cm long and 1–6 mm wide. Flowers of 2 kinds: those of early spring showy, chasmogamous, borne in crowded terminal clusters; those of late spring and early summer tiny and cleistogamous. Calyx of the showy flowers 9–10 mm long, the lobes linear-attenuate. Corolla yellow, salverform, 15–35 mm long, the throat appendaged, the lobes with crinkly margins. Mericarps about 4 mm long, white to buff, glossy, more or less ovoid with a basal collar, the adaxial side with a prominent keel and usually pitted also.

MYOSÒTIS L.   Forget-me-not

NL., from Gr. *myos*, of a mouse, + *ous*, ear—suggested by the small, soft leaves of some species.

Low, erect, mostly soft-hairy, annual, biennial, or perennial herbs. Leaves entire, narrow. Flowers small, regular, blue or white (rarely pink), borne in 1-sided, bractless or bracted racemes which straighten and elongate as they mature. Calyx lobes lanceolate or triangular. Corolla with a short tube and 5 rounded lobes, the throat with a small blunt arching appendage at the base of each corolla lobe. Stamens included within the corolla tube. Ovary deeply 4-lobed, 4-ovuled; style threadlike, arising between the lobes; stigma minute, not lobed. Nutlets 4, ovoid to ellipsoid, smooth and shiny, attached basally to the flat receptacle.

One species in Kansas.

M. vérna Nutt.

L. *vernus*, of spring.
White; late April–June
Prairie and open woods
East half, primarily

Erect, hairy annual. Stems 1–4 dm tall, simple or with few to several decumbent or ascending branches. Leaves spatulate below to mostly oblong above, 0.8–6 cm long and 1–11 mm wide, the apex rounded or acute, both surfaces covered with stiff, appressed hairs. Inflorescences terminal, only the lowermost flowers bracted. Calyx irregular, about 2.5 mm long, hairy, some of the hairs hooked. Corolla white, 2.5–3 mm long. Mericarps brownish at maturity, glossy, ovate, flattened, about 1.2–1.5 mm long, with a narrow wing around the margin, concealed by the persistent calyx.

Very young plants of this species are sometimes confused with the weedier, more common *Lithospermum arvense*. The latter species has larger flowers (corolla 5.5–8 mm long) and has the hairs on the base of the calyx stiff and appressed, neither spreading nor with hooked tips.

ONOSMÒDIUM Michx.   Marbleseed, false gromwell

NL., from the genus *Onosma* in this family. The ancient Greeks and Romans used the name for a plant which cannot be identified from what they say about it. Pliny's description is as follows: "The onosma has leaves some four fingers in length, lying upon the ground and indented like those of the anchusa; it has neither stem, blossom, nor seed." *Onosma* is from Gr. *onos*, ass or burro, + *osme*, odor, implying that the plant has the odor of that animal, a character not shared by our plants.

Coarse, erect, rough-hairy, leafy perennial herbs with taproots. Leaves entire, usually broad, strongly veined. Flowers regular, yellow or white to greenish- or grayish-white, borne in terminal, bracted, scorpioid spikes or racemes. Calyx persistent, with 5 narrow lobes. Corolla tubular or somewhat funnelform, with 5 erect, acute or acuminate lobes, the throat glabrous within and lacking appendages. Stamens included within the corolla tube, the anthers sagittate and sessile or nearly so. Ovary deeply 4-lobed, 4-ovuled; style threadlike, arising from among the lobes, protruding and long-persistent; stigma slightly enlarged, minutely 2-lobed. Nutlets 4, white to dingy-brown, smooth or with a few scattered pits, attached basally to the nearly flat receptacle.

One species in Kansas. *O. hispidissimum* Mack., included in earlier manuals, now is treated by most authorities as a variety of the following species.

430. *Lithospermum incisum*

431. *Myosotis verna*

O. mólle Michx.   **False gromwell**

L. *mollis*, soft.
Greenish- or yellowish-white; mid-May–early July
Dry, rocky prairie hillsides and banks, pastures, usually in calcareous soil
East four-fifths of the north half, east third of the south half

Stiff, upright, gray-green perennial. Stems rather stout, simple or branched, 3–8 dm tall. Leaves lanceolate to narrowly lanceolate or elliptic, 1–10 cm long and 0.4–2.5 cm wide, conspicuously veined. Herbage more or less densely covered with long, stiff, appressed, yellowish or grayish hairs. Calyx 9–11 mm long, the lobes linear-attenuate, erect, persistent but not concealing the fruits. Corolla greenish- or yellowish-white, 1–2 cm long, the lobes erect, covered on the outside with soft kinky hairs. Mericarps more or less broadly ovoid, extremely hard, 3.8–4.8 mm long.

VERBENÀCEAE   Verbena Family

Herbs, shrubs, or trees (ours all herbs), the stems or small twigs often 4-angled in cross section. Leaves usually opposite or whorled, usually simple, sessile or petiolate, estipulate. Flowers bisexual, mostly irregular (at least somewhat so), arranged in clusters. Sepals 5 (appearing to be 4 in *Phyla*), fused to one another part of their length, the calyx persistent. Corolla usually with 5 unequal lobes, sometimes slightly 2-lipped, tubular below and 4- to 5-lobed above. Stamens usually 4, inserted on the corolla tube. Pistil 1; ovary superior, 2- or 4-celled, and usually with as many lobes as cells; style 1, arising from the top of the ovary; stigma lobes usually 2. Fruit separating into 2 or 4 1-seeded mericarps.

The family includes approximately 76 genera and 3,375 species and subspecific taxa, primarily tropical in distribution but also well represented in temperate regions. Economically important members include *Tectona grandis*, the source of teak lumber, and numerous ornamentals such as the verbenas (*Verbena* spp.), *Lantana*, beauty-berry (*Callicarpa* spp.), and lemon verbena (*Aloysia triphylla* Britt.).

Kansas has 2 genera and 12 species.

Key to Genera:
1. Fruit separating into 4 1-seeded mericarps; corolla 5-lobed, usually only slightly irregular .... *Verbena*
1. Fruit separating into 2 1-seeded mericarps; corolla 4-lobed, usually plainly 2-lipped .... *Phyla*

PHÝLA Lour.   Frog fruit

NL., from Gr. *phule*, a clan or tribe—apparently in reference to the many-flowered inflorescences.

Perennial herbs with trailing or ascending stems, sometimes rooting at the nodes. Leaves opposite. Herbage with malpighiaceous hairs—i.e., straight, appressed, sharp-pointed hairs attached at some more or less central point rather than at 1 end (named after Marcello Malpighi, the 17th-century scientist who described such hairs). Flowers small, sessile, subtended by small bracts in dense, ovoid or globose to cylindric, long-peduncled axillary spikes. Calyx membranous, 2-keeled, or -winged and 2- or 4-toothed. Corolla slender, straight or incurved, irregular, more or less 2-lipped. Ovary 2-celled. Fruit small, dry, enclosed by and sometimes fused to the persistent calyx, eventually separating into 2 indehiscent mericarps.

Three species in Kansas. *P. incisa* Small, known from Barber County only, resembles *P. cuneifolia* somewhat but differs in having the floral bracts 2–3 mm long, with an acute appressed apex and peduncles 1.5–4 times the length of the subtending leaf.

Key to Species:
1. Leaf blades mostly widest at or below the middle, serrate from below the middle to the apex, at least some of them narrowed to a petiolelike base; internodes of prostrate stems not arching; spikes at first ovoid, 5–7 mm wide; flowers 3–4 mm long .... *P. lanceolata*
1. Leaf blades mostly widest above the middle and toothed only near the apex, not narrowed to a petiolelike base; internodes of prostrate stems arching; spikes at first globose, 8–12 mm wide; flowers 5–7 mm long .... *P. cuneifolia*

P. cuneifòlia (Torr.) Greene   Wedge-leaf frog fruit

NL., from L. *cuneus*, wedge, + *folium*, leaf.
Purplish-white; late May–September

Moist places along roadsides, in disturbed prairie, and along lake margins
West two-thirds, primarily; eastward to Shawnee County in the Kansas River valley

Low perennial with stems woody near the base, 2–7 dm long, prostrate, producing short, erect branches at the nodes, the internodes elongate and often arching. Leaves sessile, rather rigid, mostly narrowly lanceolate to nearly linear, 1–4.5 cm long and 3–20 mm wide, the apex acute, the base narrowly wedge-shaped or sometimes acute, the margins with 1–4 pairs of remote, sharp teeth above the middle only. Spikes 5–15 (–20) mm long and 8–12 mm wide, at first globose, elongating somewhat and becoming short-cylindric, the bracts often tinged with purple. Flowers purplish-white, 5–7 mm long, the subtending bracts about 5 mm long, the apex long-acuminate and eventually recurved.

432. *Phyla cuneifolia*

P. lanceolàta (Michx.) Greene   Northern frog fruit

L. *lanceolatus*, spearlike—in reference to the leaf shape.
Pale blue or purplish to white; late June–September
Moist soil of lake margins, creek beds, ditches
South and east of a line from Washington County to Gray and Meade counties

Low perennial with slender stems 2–6 dm long, ascending or prostrate and ascending at the ends, often rooting at the nodes. Leaves broadly lanceolate to elliptic, mostly 2–7 cm long and 1–3 cm wide, the apex acute, the base wedge-shaped or narrowed and petiolelike, the margins serrate to below the middle. Spikes 5–20 mm long and 5–7 mm wide, at first ovoid, elongating and becoming cylindric, the bracts often tinged with purple. Flowers pale blue or purple to white, 3–4 mm long, blooming in a circle successively from the base to the apex of the spike, presenting the anthers of the longer pair of stamens at the entrance of the open throat and above the stigma. Fruits broadly ovate to nearly circular, slightly flattened, 1–1.5 mm long, the apex usually apiculate.

Insect visitors to this plant include butterflies, long- and short-tongued bees, and various kinds of flies.

433. *Phyla lanceolata*

VERBÈNA L.   Verbena, vervain

L. *verbena*, classical name for branches of laurel, olive, myrtle, cypress, and other trees used in religious rites. L. *verbenaca*, ancient name for the European, medicinal *Verbena officinalis*, "Vervain," English equivalent of OF. *Verveine*, is now frequently applied to the upright verbenas in general.

Annual or perennial herbs with stems procumbent to ascending or erect. Leaves mostly opposite and toothed (less frequently lobed or incised). Flowers small, bracted, in simple or branched terminal spikes. Calyx usually tubular, 5-ribbed, 5-toothed, with 1 tooth usually shorter than the others. Corolla tubular, often curved, the rim 5-lobed and weakly 2-lipped. Stamens 4, didynamous, usually included within the corolla tube, the upper 2 sometimes without anthers; anthers with or without glandular appendages. Style usually short, with 1 smooth nonfunctional (or at least nonstigmatic) lobe and 1 larger, papillose stigmatic lobe. Ovary 4-lobed, 4-celled, 4-ovuled. Fruit mostly enclosed by the persistent calyx, separating at maturity into 4 dry, linear, 1-seeded indehiscent mericarps.

Nine species in Kansas. *V. ambrosifolia* Rydb., a low, decumbent plant resembling *V. bipinnatifida*, is known from Hamilton County only. It differs from the latter species primarily in having floral bracts shorter than the calyx and the calyx sometimes glandular. In addition, hybrids occurring naturally among several of our species may be found.

Key to Species:
1. Low, spreading, mostly prostrate or decumbent plants; flower clusters initially short and broad, gradually elongating as the season progresses 2
   2. Flowers small and inconspicuous, about 2–3 mm across, shorter than the prominent subtending bracts; fruits with the upper half reticulate, the lower half merely ribbed .... *V. bracteata*
   2. Flowers rather showy, 8–15 mm across, longer than the narrow subtending bracts; fruits punctate in vertical rows 3
      3. Flowers rose-purple or rose, 11–15 mm across, leaf blades mostly ovate to lanceolate in general outline, incised to pinnatifid, or some 3-lobed .... *V. canadensis*
      3. Flowers bluish-purple, 8–10 mm across; leaf blades mostly deltoid to broadly ovate in general outline, 2 or 3 times pinnately parted into narrow linear or oblong segments .... *V. bipinnatifida*
1. Taller, erect plants; flower clusters initially slender and linear 4
   4. Flowers white, 3–5 mm long, fruits faintly ribbed .... *V. urticifolia*
   4. Flowers blue to purple (rarely pink or white), 4–11 mm long; fruits not as above 5
      5. Leaves narrowly elliptic to narrowly oblanceolate, spatulate or nearly linear, 2–18 mm wide; fruits about 2.6 mm long, the upper half to two-thirds with a raised reticulate ornamentation, the lower half merely ribbed .... *V. simplex*

VERBENA FAMILY
307

434. *Verbena bipinnatifida*

435. *Verbena canadensis*

5. Leaves narrowly lanceolate to lanceolate or broadly elliptic, 0.8–11 cm wide 6

    6. Lower leaves petioled, often with a pair of sharply pointed, spreading basal lobes; calyx 2–3 mm long, rather sparsely and minutely hairy; corolla 4–5 mm long; fruits about 2 mm long, slightly rough but neither reticulate nor ribbed .... *V. hastata*

    6. Lower leaves sessile or nearly so, lacking basal lobes; calyx 4–5 mm long, densely and conspicuously hairy; corolla 7–11 mm long; fruits 2.5–3 mm long, the upper half or more with a raised reticulate ornamentation, the lower half or less merely ribbed .... *V. stricta*

V. bipinnatìfida Nutt.   Dakota verbena

NL., from L. *bi-*, 2 or twice, + *pinna*, feather, + *fidus*, cleft—indicating that the leaves are split into divisions like a feather and that these divisions are themselves split in a similar manner.
Bluish-purple; May–September
Prairie hillsides and banks, pastures, roadsides
Nearly throughout, but more common in the west three-fourths

    Low, more or less hairy perennial herb with branches prostrate to erect, 1–4 dm long, sometimes rooting at the lower nodes. Leaf blades mostly deltoid to broadly ovate in general outline, 1–4.5 cm long and 1–5 cm wide, 2 or 3 times pinnately parted into narrow linear or oblong segments. Flowers more or less showy, each subtended by a slender bract, the clusters at first short and headlike, the axis gradually elongating as the season progresses. Calyx 8.5–10 mm long, hairy but not glandular, the lobes slender and bristlelike, unequal in length. Corolla bluish-purple, 12–20 mm long, the expanded rim 8–10 mm across. Fruits about 2.5–3 mm long, constricted along the lines of cleavage, the mericarps pitted in vertical rows.

    See plate *84*.

V. bracteàta Lag. & Rodr.   Bigbract verbena

L. *bracteatus*, covered with thin plates of metal, from L. *bractea*, thin plate of precious metal; in botany applied to the usually thin, more or less scalelike leaves subtending flowers.
Purplish-blue; late May–early September
Prairie, pastures, roadsides, and waste ground
Throughout

    Low, more or less hairy perennial herb with branches prostrate or decumbent, 1–3.5 dm long. Leaf blades deltoid to broadly ovate or lanceolate (the uppermost sometimes oblanceolate) in general outline, 1.5–4.5 cm long and 1–3 cm wide, the margins toothed and pinnately incised or more or less 3-lobed. Flowers small and inconspicuous, each subtended and surpassed by a lance-linear bract, only a few open at a time, the clusters at first short and headlike, the axis gradually elongating as the season progresses. Calyx 3–4 mm long, the lobes short, coming together to enclose the maturing fruits. Corolla purplish-blue, 4–6 mm long, the expanded rim about 2–3 mm wide. Fruit about 2–2.2 mm long, constricted along the lines of cleavage, the mericarps with the upper half pitted, the lower half ribbed.

    The tiny flowers seem insignificant, yet they are visited by various butterflies and short-tongued bees.

V. canadénsis (L.) Britt.   Rose verbena

L. *-ensis*, suffix added to nouns of place to make adjectives—"Canadian," in this instance.
Rose-purple or rose; mid-April–early September
Prairie, pastures, open woods, or roadsides, usually on rocky hillsides and banks
East third, primarily

    Low, sparsely hairy perennial herb with stems prostrate or spreading, branched, (1–) 3–6 dm long, rooting at the lower nodes. Leaf blades mostly ovate to lanceolate in general outline, (1–) 3–9 cm long and 1–4.5 cm wide, the margins toothed, incised to pinnatifid, or some of the leaves 3-lobed. Flowers showy, the clusters at first broad and headlike, the axis gradually elongating as the season progresses, bearing new buds and flowers at the summit. Calyx 10–13 mm long, glandular-hairy, the lobes quite slender, unequal in length. Corolla usually rose to rose-purple (rarely white), 13–20 mm long, the expanded rim 11–15 mm across. Fruit about 2.6–3 mm long, constricted along the lines of cleavage, the mericarps pitted in vertical rows.

    Rose verbena is frequently transplanted into gardens, usually when in full bloom, and endures the ordeal remarkably well. However, even though it is a perennial, it is likely to be gone in 2 or 3 years under such treatment, and it is better to leave the plants to adorn the wayside, while gathering the seeds to plant in the garden.

    See plate *85*.

V. hastàta L.   Blue verbena, vervain

L. *hastatus,* armed with a spear, from *hasta,* spear—because of the shape of the leaves, some of
      the lower ones having divergent basal lobes giving the appearance of a halberd spear.
Bluish-purple; July–September
Moist soil of low prairie, ditches, and lake and pond margins
Nearly throughout, uncommon in the southwest sixth

436. *Verbena hastata*

Erect, rough-hairy perennial herb. Stems up to 1.5 m tall, branched above. Leaves (at least the lowermost) narrowed to a petiolelike base, widely spreading to ascending, the blades mostly lanceolate to narrowly lanceolate, the lowermost (and sometimes the upper ones as well) often with a pair of acute, spreading basal lobes, 2.5–18 cm long and 1–11 cm wide, the apex gradually acuminate, the margins coarsely serrate or double-serrate. Flowers small, borne in crowded, linear terminal spikes. Calyx 2–3 mm long, minutely hairy, the teeth triangular-acuminate. Corolla bluish-purple, 4–5 mm long, the expanded rim 2.5–4.5 mm across. Fruits about 2 mm long, scarcely if at all constricted along the lines of cleavage, the mericarps slightly rough but neither pitted nor prominently ribbed.

While the plant might be thought weedy in structure, its impressive fascicles of slender terminal flower spikes, blooming over a long period, would make it acceptable in a garden border of wild flowers. It is also of interest because, like the European vervain, *Verbena officinalis,* it has had a place in folk medicine. Concerning the latter species, the old herbals are forthright in praise; for example, Otto Brunfels' (1543): "Ironweed [*Eisenkraut,* German vernacular name for vervain] is esteemed by surgeons a perfect gem for the treatment of all wounds, be they fresh or putrid, and for all ulcers, wens, and hardened arteries. Ironweed-water drunk morning and evening opens the obstructed liver, expels tapeworm, clears the kidneys, reduces stone, relieves griping and heals internal ulcers," and so for many other ailments. Naturally, European immigrants to America used our native *V. hastata* as they had been accustomed to employ *V. officinalis.* However, we find Gerard rebelling, in his herbal, against extravagant credulity current through the Middle Ages to his own day: "Many odde old wives fables are written of Vervaine tending to witchcraft and sorcery, which you may reade elsewhere, for I am not willing to trouble your eares with reporting such trifles, as honest eares abhorre to heare."

437. *Verbena simplex*

Gerard is not implying, however, that the reputation of the vervain as a curative is all moonshine; indeed, chemical analysis shows its possession of various active principles, such as tannin and saponin and other glucosides.

Various American Indian tribes employed *V. hastata* as a medicinal. The Menominees used a decoction of the roots to treat cloudy urine. The Teton Dakotas made a tea from the leaves to cure stomachache. The dried aboveground parts were listed officially in the *National Formulary* from 1916 to 1926 as a diaphoretic and expectorant.

The small but numerous fruits of this and other verbenas are of undoubted value to a variety of seed-eating birds, like the many kinds of sparrows and others of the order of finches; but who would dream of the verbenas providing food for human beings? Yet it is on the record that Indians of California used to gather the seeds of *V. hastata* in quantity and after roasting them grind them into meal.

V. símplex Lehm.   Narrowleaf verbena, vervain

L. *simplex,* simple, uncompounded—referring to its few dispersed spikes, in contrast to the
      fascicled spikes of its allies, *V. hastata* and *V. stricta.*
Blue to purple; mid-May–August
Pastures, roadsides, prairie
East third, primarily

Slender, erect, sparsely hairy perennial, up to 0.5 m tall. Leaf blades narrowly elliptic to narrowly oblanceolate, spatulate, or nearly linear, 2.5–10 cm long and 2–18 mm wide, prominently veined, the margins serrate. Flowers in slender, solitary terminal spikes. Calyx 4–5 mm long, sparsely and minutely hairy, the teeth triangular-acuminate. Corolla blue to purple, 6–7 mm long, the expanded rim 4–6 mm across. Fruits about 2.6 mm long, the upper half to two-thirds with a raised reticulate ornamentation, the lower one-third to one-half merely ribbed.

Narrowleaf verbena is frequently associated with the more robust, less graceful *V. stricta* but is not as successful as the latter in establishing itself in run-down pastures. Comparing these 2 species further, we note that the taproot of *V. stricta* goes deeper, with fewer and coarser branches near the surface, to tap the soil moisture of deeper levels. The copious, fine, more superficial branches of the taproot of *V. simplex* adapt it to situations where water can be had in the upper soil layer.

VERBENA FAMILY

438. *Verbena stricta*

439. *Verbena urticifolia*

V. strícta Vent.   Hoary vervain

L. *strictus*, drawn tight or together—in reference to the stiffness and upright growth of the
    stems and leaves.
Blue or purple; mid-June–early September
Disturbed prairie, overgrazed pastures, roadsides, waste ground
Throughout

Stout, erect, more or less densely hairy perennial, up to 1.2 m tall. Leaves sessile or
nearly so, ascending or nearly erect, the blades lanceolate to broadly elliptic, 2–10 cm
long and 0.8–4.5 cm wide, the margins coarsely serrate to double serrate or lacerated.
Flowers in slender terminal spikes borne singly or several together. Calyx 4–5 mm long,
densely hairy, the teeth triangular-acuminate. Corolla blue or purple (occasionally white),
7–11 mm long and 4–6 mm across. Fruits 2.5–3 mm long, the upper half or more with a
raised reticulate ornamentation, the lower half merely ribbed.

This is the sturdiest and most widely distributed of all our verbenas. It, along with
the western ironweed, is one of the most frequent invaders of run-down and dried-out
pastures, where it forms societies of considerable size, then makes a fine mass-effect with
its abundant blue flowers and hoary foliage.

See plate *86.*

V. urticifòlia L.   White verbena, vervain

NL., from L. *urtica*, name of the stinging nettle genus, + *folium*, leaf—because of the re-
    semblance of the leaves of this verbena to those of some *Urticas*.
White; July–mid-August
Moist soil of creek banks, ditches, open woods, prairie sloughs
East two-thirds

Erect, rough-hairy biennial or perennial herb, up to about 1 m tall. Leaf blades
ovate to lanceolate, 2–17 cm long and 1–7 cm wide, the margins serrate. Flowers sessile,
subtended by minute bracts, widely spaced in slender, branched spikes. Calyx 1.5–2.5 mm
long. Corolla white, 3–5 mm long, the expanded rim 2–4 mm across. Fruit about 2 mm
long, widely spaced, completely filling the persistent calyx, the mericarps faintly ribbed.

LABIÃTAE (Lamiàceae)   Mint Family

Mostly herbaceous annuals or perennials, sometimes shrubs, trees, or vines. Stems
usually 4-angled in cross section; foliage frequently dotted with minute glands producing
aromatic oil. Leaves mostly opposite, occasionally whorled or all basal, simple, or compound, sessile or petiolate, estipulate. Flowers bisexual, usually irregular and 2-lipped
(regular or nearly so in *Mentha* and *Lycopus*), borne singly in the leaf axils, in axillary
whorls, or in terminal heads. Calyx usually 5-lobed or -toothed, sometimes 2-lipped.
Corolla of 5 fused petals, although sometimes obscurely so, obscurely to quite distinctly
2-lipped. Stamens 4, 2 long and 2 short, or only 2, inserted on the corolla tube. Pistil 1;
ovary superior, more or less deeply 4-lobed, the cells 2 or sometimes appearing to be 4;
style 1, either arising from down among the 4 lobes of the ovary or terminal when the
ovary lobes are partially united below; stigmas 2, minute. A nectiferous disk is sometimes
present beneath the ovary and may have an erect lobe appressed against the side of the
ovary and appearing to be a fifth lobe of the latter. Fruit ripening at maturity into 4
1-seeded mericarps, often referred to as nutlets or cocci, in the bottom of the persistent
calyx.

The Mint Family is quite large, containing about 180 genera and 3,500 species,
cosmopolitan in distribution but especially well represented in the Mediterranean region.
The family includes a number of plants important as culinary herbs, such as basil
(*Ocimum basilicum* L.), sweet marjoram (*Majorana hortensis* Moench), oregano or pot
marjoram (*Origanum vulgare* L.), thyme (*Thymus vulgaris* L.), sage (*Salvia officinalis* L.
and other species), savory (*Satureja* spp.), and rosemary (*Rosmarinus officinalis* L.). Essential oils are derived from sage, rosemary, mint (*Mentha* spp.), and lavender (*Lavandula*
spp.). Ornamentals, in addition to many of the aforementioned, include skullcap (*Scutellaria* spp.), *Coleus*, beebalm (*Monarda* spp.), dragonhead (*Dracocephalum* spp.), false
dragonhead (*Physostegia* spp.), catnip (*Nepeta cataria* L.), and many others. In addition,
extracts from horehound (*Marrubium vulgare* L.) are used in medicinal preparations
and candies.

The family is represented in Kansas by about 22 native and naturalized genera,
the most common of which are included here.

Key to Genera (3):

1. Calyx with a distinct cap or protuberance on the upper side of the tube (see plate *88*) .... *Scutellaria*
1. Calyx without a cap or protuberance on the tube 2
  2. Upper lip of the corolla very short and scarcely discernible, or cleft deeply and its lobes appearing laterally on the margins of the lower lip, the corolla therefore appearing 1-lipped .... *Teucrium*
  2. Upper lip of the corolla well developed and evident, entire or 2-lobed 3
    3. Stamens included within the corolla tube and not directly visible .... *Marrubium*
    3. Stamens exserted beyond the throat of the corolla tube, or even surpassing the corolla lips, and therefore directly visible 4
      4. Stamens 2 5
        5. Calyx distinctly 2-lipped, the lobes of the upper lip differing in size or shape or both from those of the lower lip (upper lip entire in 2 species of *Salvia*) 6
        6. Flowers in loose, few-flowered whorls in the axils of ordinary foliage leaves .... *Hedoma*
        6. Flowers in terminal racemes, spikes, or heads .... *Salvia*
        5. Calyx regular or very nearly so, its lobes essentially alike in size and shape 7
        7. Corolla very irregular, 15–50 mm long .... *Monarda*
        7. Corolla nearly regular, 5 mm long or shorter .... *Lycopus*
      4. Stamens 4 8
        8. Inflorescence essentially axillary in its general appearance, the whorls several to many, subtended by ordinary foliage leaves and separated by internodes of ordinary length, or the uppermost subtending leaves somewhat smaller and the upper internodes somewhat shorter (plants with axillary spikes or racemes not included here) 9
        9. Corolla almost equally 4- or 5-lobed 10
          10. Flowers 1–3 in each axil, 2–6 in each whorl .... *Isanthus*
          10. Flowers many in each axil .... *Mentha*
        9. Corolla very irregular, strongly 2-lipped, the upper lip concave and arched over the stamens 11
          11. Flowers distinctly pediceled, forming relatively loose clusters .... *Glecoma*
          11. Flowers sessile in the clusters 12
            12. Calyx lobes prolonged into short, stiff spines .... *Leonurus*
            12. Calyx lobes slenderly acuminate but not spiny .... *Lamium*
        8. Inflorescence generally terminal in its general appearance, the whorls solitary or few to many, subtended (all or chiefly) by bracteal leaves conspicuously different from the foliage leaves, or separated by much shorter internodes; whole inflorescences describable as a terminal spike, raceme, panicle, or head. (In some species, a very few whorls may also appear in the uppermost axils. Plants with lateral or axillary spikes or racemes are included here) 13
        13. Flowers single in the axil of each bracteal leaf, the whorls therefore with not more than 2 flowers .... *Perilla*
        13. Flowers 2 to many in the axils of each bracteal leaf, the whorls therefore with 4 or more flowers 14
          14. Stamens ascending under the upper corolla lip and not surpassing it 15
            15. Calyx distinctly 2-lipped and irregular .... *Prunella*
            15. Calyx nearly or quite regular, the lobes all alike or differing slightly in size only 16
              16. Lower whorls distinctly peduncled .... *Nepeta*
              16. Lower whorls sessile .... *Stachys*
          14. Stamens (at least some of them) protruding from the corolla 17
            17. Inflorescence a dense or loose raceme in which the component whorls are plainly discernible .... *Mentha*
            17. Not as above 18
              18. Inflorescence a number of terminal heads, often with secondary heads in some of the upper axils, never a spike or a raceme .... *Pycnanthemum*
              18. Inflorescence a dense spike, sometimes with 1 or 2 lower whorls separate; flowers sessile or nearly so .... *Agastache*

AGÁSTACHE Clayt. — Giant hyssop

NL., from Gr. *agastachys,* rich in grain, from *agan,* much, + *stachys,* spike, or ear of grain—suggested by the several upright, dense spikes of this species.

Perennial herbs with short, erect rhizomes. Leaves opposite, simple, toothed, mostly petioled. Flowers many, small, in dense, bracted whorls crowded into terminal, continuous or interrupted spikes. Calyx cylindric or somewhat obconic, nearly regular, slightly 2-lipped, usually 15-veined. Corolla conspicuously 2-lipped, the upper lip erect, 2-lobed, the lower lip spreading, 3-lobed, with the middle lobe crenate. Stamens 4, all fertile, protruding beyond the corolla. Ovary deeply 4-parted. Fruits smooth, ovoid. One species in Kansas.

440. *Agastache nepetoides*

A. nepetoìdes (L.) O. Ktze.   Giant hyssop

NL., from L. *nepeta*, the name of the genus to which the catnip belongs, + Gr. suffix *-oides*, like or resembling. "Hyssop" is the English vernacular name for *Hyssopus officinalis*, a European member of the Mint Family once used in medicine, now naturalized in this country.
Greenish-yellow; mid-August–September
Moist woodlands, especially in alluvial soils
East two-fifths, especially the east third

Stems erect, up to 1.5 m tall, with rigidly ascending branches above. Leaves thin, softly and minutely hairy, long-petioled below and becoming shorter-petioled upward, the blades ovate to lanceolate or (less frequently) elliptic, 1.5–16 cm long and 1–9 cm wide, the apex usually acuminate, the base rounded or slightly cordate, often oblique, the margins serrate, each tooth mucronate. Spikes continuous, 2–16 cm long and 1–1.5 cm across. Calyx obconic, 4–5 mm long, the lobes 1–1.5 mm long with convex margins. Corolla greenish-yellow, 7–8 mm long, curved outward. Mericarps brown, about 1.5–1.8 mm long, minutely hairy toward the apex, elliptic in outline, the inner face angled, the outer face convex and faintly veined.

GLECHÒMA L.   Ground ivy

NL., from *glechon*, the old Greek name for pennyroyal.

Creeping perennial herbs. Leaves petioled, the blades reniform or nearly round. Flowers blue, short-pediceled, borne 1–3 in the upper leaf axils. Calyx tubular, 15-veined, the lobes essentially alike but the upper slightly longer than the lower. Corolla 2-lipped, the upper lip erect, straight, notched at the apex; lower lip bent downward, broad, 3-lobed, the middle lobe as well as the throat bearing a fringe of papillae bordering the expanding throat. Stamens 4, inserted on the corolla tube in 2 pairs, the upper pair longer than the lower and both bearing anthers under the protection of the upper lip. One species in Kansas.

G. hederàcea L.   Ground ivy, gill-over-the-ground

NL., from the genus *Hedera*, ivy, i.e., "ivylike."
Light blue to purplish; mid-April–May
Roadsides, yards, stream banks, and cultivated ground
East fourth, primarily, but scattered records as far west as Scott County; intr. from Europe, sometimes cultivated

Stems slender, creeping, up to 5 dm long, rooting at the nodes and producing short, ascending branches. Leaf blades cordate, round with a cordate base, or reniform, 1–3.5 cm long and 0.8–4 cm wide, the margins crenate. Herbage sparsely hairy. Calyx 6–8 mm long, the lobes triangular-acuminate. Corolla light blue to purplish, purple-spotted, 10–15 (–20) mm long. Style inclined upward under the upper lip, eventually extending beyond it.

The 2 stigmatic lobes at first have their receptive surfaces pressed together. However, the lower lobe eventually bends backward, presenting its receptive surface for contact with incoming insects bearing pollen from younger flowers, and the upper lobe bends downward where insects may touch the receptive surface on backing out. Cross-pollination is effected by proterandry; that is, the anthers of this flower discharge pollen when the flower is young and before the receptive stigmatic surfaces spread apart—the time relationship of anther and of stigmatic behavior being so adjusted as to prevent self-pollination and ensure cross-pollination. Nectar glands subtending the ovary and enclosed by the corolla tube ensure the service of various species of bees, flies, and butterflies.

Since before the Middle Ages, at least, extracts of this plant have been used in folk medicine as a tonic, astringent, digestive, and diuretic. For colds of long standing, an infusion was made with 1 ounce of the herbage in 1 pint of boiling water, sweetened with honey, 1 wineglassful to be taken 3 or 4 times daily. In England, from early Saxon times down to the reign of Henry VIII, when the hop was introduced from the Netherlands, the leaves were steeped in hot beer to clarify it and improve its flavor.

This plant has been listed in earlier manuals as *Nepeta hederacea*.

HEDEÒMA Pers.   False pennyroyal

NL., from Gr. *hedys*, sweet, + *osma*, odor—the herbage and flowers smelling of oil of pennyroyal. The European pennyroyal is *Mentha pulegium* of Linnaeus but was known to the ancient Latins simply as *pulegium*, from L. *pulices*, fleas. This name was given to the plant because, as Pliny relates, "The blossom of it, fresh gathered, and burnt, kills fleas by its smell."

Annual or perennial herbs, sometimes woody at the base, mostly aromatic and hairy. Leaves opposite, simple, petioled or sessile, entire or toothed, often gland-dotted. Flowers small, on short pedicels in 1- to several-flowered axillary clusters. Calyx tubular, 13-nerved, more or less 2-lipped, bearded in the throat, the upper lip 3-toothed, the lower lip 2-cleft or -lobed. Corolla 2-lipped, the upper lip projecting forward, entire, notched, or 2-lobed; lower lip spreading, 3-cleft. Fertile stamens 2, with divergent anther sacs; sterile stamens 2 or none. Style 2-cleft, with inward-facing stigmatic surfaces, ascending with the stamens close under the upper lip. Mericarps 4, smooth, the surface usually faintly reticulate and with gelatinous projections which emerge from the areoli when the seed is wetted.

Three species in Kansas.

441. *Hedeoma drummondii*

Key to Species:
1. Perennial with a woody taproot and stems woody near the base; upper calyx teeth straight, connivent with the lower ones after the corolla has fallen; mericarps more or less ellipsoid, 1.2–1.5 mm long .... *H. drummondii*
1. Annuals with weak taproots, stems not woody; upper calyx teeth spreading or recurved, not connivent with the lower ones 2
    2. Leaves thin, narrowly lanceolate to elliptic, 3–10 mm wide, the margins usually with a few shallow teeth; mericarps 0.8–1 mm long, the outer face broadly elliptic to nearly circular; east fifth .... *H. pulegioides*
    2. Leaves leathery, narrowly elliptic to linear, 1–3 mm wide, the margins entire; mericarps about 1.2 mm long, the outer face elliptic; throughout except for the southwest .... *H. hispida*

H. drummóndii Benth.   Plains false pennyroyal

Named after James Drummond, English botanist, who in the second quarter of the 19th century explored the Hudson highlands, Texas, and Louisiana.
Purple; June–September
Rocky prairie hillsides and banks
West third, and north half of the central third

Perennials with a woody taproot. Stems several to many, erect to ascending or the lowermost decumbent, woody near the base, mostly 1–4 dm tall, covered with retrorsely curled hairs. Leaves sessile, elliptic to oblong, 7–17 mm long and 1–4 mm wide, the lower surface covered with antrorsely curled hairs. Calyx 5–7 mm long, covered with mostly antrorsely curled hairs, the tube constricted above the fruits, the lobes bristle-like, touching one another after the corolla falls. Corolla purple, 7–15 mm long, the upper lip strap-shaped, straight, concave. Mericarps brownish, about 1.2–1.5 mm long, more or less ellipsoid, the inner face slightly angled.

The Mescalero Apaches and Lipan Indians treated prolonged headaches by rubbing the aromatic twigs of this plant and inhaling the mintlike odor (45).

H. híspida Pursh   Rough false pennyroyal

L. *hispidus*, hairy, bristly—in allusion to the hairy calyx lobes and the bracts of the inflorescence.
Bluish-purple; May–mid-June
Prairie, roadsides, waste ground
Throughout except for the southwest eighth

Annual, 0.7–3 dm tall, with stems simple or branched near the base, covered with retrorsely curled hairs. Leaves sessile, narrowly elliptic to linear, 7–18 mm long and 1–3 mm wide, the apex rounded, the margins and sometimes the lower midvein ciliate, but the blade otherwise glabrous. Calyx 5–6 mm long, bearing coarse, spreading hairs on the veins, the tube constricted above the fruits, the lower 2 lobes bristlelike and ciliate, the upper 3 very narrowly triangular and recurved. Corolla bluish-purple, of 2 kinds, either protruding from the calyx tube and then 6–7 mm long, or scarcely longer than the calyx. Mericarps brownish, about 1.2 mm long, more or less ellipsoid, the inner face angled, the outer face convex.

Some of the Plains Indians used a tea prepared from the leaves of this species as a cold remedy and also as a flavoring agent to improve the appetite of the ill (45).

H. pulegióides (L.) Pers.   American false pennyroyal

Named after *Mentha pulegium*, the European or "true" pennyroyal. The *-oides* in the species name is a Gr. suffix meaning "similar to," the similarity being in their fragrant oils which are analogous in constitution and when administered medicinally produce similar physiological effects—namely, stimulant, diuretic, diaphoretic, emmenagogue, carminative, and rubefacient.

Bluish-purple; August–September
Rocky woods
East fifth and Pottawatomie County

Annuals 1.5–3 dm tall with spreading or ascending branches, the stems covered with retrorsely curled hairs. Leaves glabrous or nearly so, petioled, narrowly lanceolate to elliptic, 1–3 cm long and 3–10 mm wide, the apex rounded or acute, the margins entire or with a few shallow teeth. Calyx 4–6 mm long, constricted above the fruits, the lower 2 lobes bristlelike and ciliate, the upper 3 triangular and spreading or recurved. Corolla bluish-purple, 5–7 mm long. Mericarps black, 0.8–1 mm long, the outer face broadly elliptic or nearly circular, convex, the inner face slightly angled.

Several uses of this species by American Indians are recorded by Weiner (34) and Vogel (45). The Rappahannock Indians of Virginia prepared a tea from the leaves to treat menstrual cramps, while the Onondagas and Catawbas used it for headaches and colds, respectively. American settlers adopted some of the Indian uses for this plant and invented some of their own as well, including its utilization in the preparation of a chigger repellent and to induce abortions.

The dried leaves were official in the *U.S. Pharmacopoeia* from 1821 to 1916 as a stimulant, carminative, and emmenagogue. Pennyroyal oil was listed officially from 1916 to 1931.

ISÁNTHUS Michx. False pennyroyal

NL., from Gr. *isos*, equal, + *anthos*, flower, in reference to the nearly regular corolla.

Clammy-pubescent annual herbs. Leaves opposite, entire, short-petioled. Flowers blue, small, in 1- to 3-flowered axillary clusters. Calyx bell-shaped, 5-lobed, nearly regular, enlarging in fruit. Corolla irregular, the 4 upper lobes spreading or ascending, the lower lobe deflexed and usually longer than the others. Stamens 4, protruding slightly. Mericarps obovoid, attached laterally at the base, pubescent at the apex, the outer face prominently reticulate.

A genus consisting of 1 species.

I. brachiàtus (L.) B.S.P.

L. *brachiatus*, with arms or branches.
Blue; August–September
Shallow, rocky soil
East third

Plants much branched, 1.5–4 dm tall. Leaves mostly lanceolate to narrowly elliptic, 1.5–4 cm long and 3–15 mm wide, the apex acute, the base wedge-shaped. Calyx 3–4 mm long at anthesis, up to 8 mm long in fruit, the lobes lance-attenuate to triangular-attenuate. Corolla blue, 6–7 mm long. Mericarps about 2.5 mm long.

The branches of false pennyroyal often bear galls caused by the ovipositing of some insect.

LÀMIUM L. Deadnettle

NL., from Gr. *lamos*, throat—suggested by the wide-open throat of the corolla.

Annual, biennial, or perennial herbs. Leaves opposite, simple, petioled below and sessile above, often tinged with purple, rugose. Flowers purple or purplish-pink in ours, borne in whorls of 6–12 subtended by the upper leaves or leaflike bracts, the whorls sometimes crowded into terminal spikes. Calyx bell-shaped or tubular, with 5 nerves and 5 awl-shaped teeth. Corolla 2-lipped, the upper lip erect, ovate to oblong, hoodlike, narrowed at the base, the lower lip spreading, 3- or 4-lobed, the middle lobe spotted, merely apically notched to deeply divided and appearing as 2 lobes. Stamens 4, ascending under the upper lip, the shorter pair inserted on the upper side of the corolla tube, the anthers hairy. Style 2-cleft. Ovary 4-lobed. Mericarps smooth or tubercled.

Two species in Kansas.

Key to Species:
1. Flowers sessile in the axils of upper leaves separated by long internodes; calyx densely hairy; throughout Kansas .... *L. amplexicaule*
1. Flowers and the leaflike bracts crowded into terminal spikes; calyx sparsely hairy; east third .... *L. purpureum*

L. amplexicaùle L.   Henbit deadnettle

NL., from L. *amplexus*, encircling, + *caulis*, stem—in reference to the leaves. The common
   name implies that hens nibble the leaves, and that the leaves, although *very* faintly
   resembling those of a nettle, are inert because they lack stinging hairs.
Purple or purplish-pink; February–October
Lawns, roadsides, waste ground, cultivated ground
Throughout; intr. from Eurasia

Low, sparsely hairy biennial or winter annual, the stems 1–4 dm long, the lowest
branches bending to the ground and taking root. Lower leaves few, ovate to nearly
round, long-petioled, the upper leaves sessile, very broadly cordate or reniform, 7–15 mm
long, the margins with coarse, rounded teeth or lobes, the bases clasping the stem, the
internodes quite elongated. Flowers sessile, clustered in the axils of the upper leaves.
Calyx tubular, 5–7 mm long, densely hairy. Corolla purple or purplish-pink, 12–20 mm
long. Mericarps brown, spotted with white, obovate or oblong-obovate, about 2 mm long,
3-angled in cross section, the apex very broadly rounded.

The flowers are visited by bees seeking nectar from a gland at the base of the ovary.
Self-pollination is possible since the anthers and stigmas are close together and mature
at the same time. However, since the lower lobe of the stigma is reflexed downward
beyond the anthers, it is likely to be cross-pollinated by bees coming from other flowers.

442. *Lamium amplexicaule*

L. purpùreum L.   Deadnettle

NL., from L. *purpura*, purple—in reference to the flowers.
Purple or reddish-purple; April–mid-May
Roadsides, yards, creek banks
East third

Hairy, grayish-green annual. Stems 1 or few, erect to decumbent, 0.8–3 dm tall.
Leaves proper, all petioled, the blades mostly cordate, 1–4 cm long and 1.2–3.5 cm wide,
the lowermost smaller and often ovate, reniform, or nearly round. Flowers crowded into
a short, leafy-bracted terminal spike 2–7 cm long and 2.5–4 (–7) cm wide. Calyx narrowly
bell-shaped, hairy, 5–7 mm long, the tips of the lobes bristlelike. Corolla purple or
reddish-purple, 11–15 mm long. Mericarps brownish-gray, with or without white spots,
3-angled in cross section, oblong-obovate, about 2 mm long, the apex truncate.

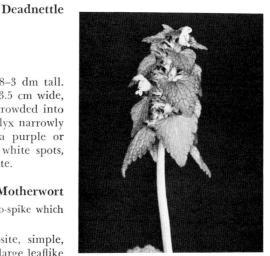

LEONÙRUS L.   Motherwort

NL., from Gr. *leon*, lion, + *oura*, tail—the series of axillary heads forming a pseudo-spike which
   in its immature state suggests the brush of a lion's tail.

Erect, aromatic annual, biennial, or perennial herbs. Leaves opposite, simple,
toothed or laciniate, petioled. Flowers white to pink, crowded in the axils of large leaflike
bracts, also subtended by smaller linear bracts, in long, interrupted terminal spikes. Calyx
narrowly bell-shaped, with 5 nearly equal, attenuate or awl-shaped teeth. Corolla 2-
lipped, the upper lip erect, somewhat hoodlike, entire, hairy, the lower lip spreading or
deflexed, 3-lobed, the middle lobe largest, narrowly oblong-obovate, entire, the lateral
ones oblong. Stamens 4, about equal, ascending under the upper lip of the corolla.
Ovary 4-lobed. Mericarps 3- or 4-angled, obpyramidal to club-shaped, the apex truncate
and hairy.

One species in Kansas.

443. *Lamium purpureum*

L. cardìaca L.   Common motherwort

NL., from Gr. *kardia*, heart—infusions of the plant having been used from the time of Theo-
   phrastus (died c. 285 B.C.) to allay palpitation of the heart due to a nervous condition,
   and as a simple tonic.
White to pink or purplish; June–August
Fields, roadsides, waste ground, thickets
Primarily in the north half of the east two-thirds; intr. from Asia, sometimes cultivated

Erect, soft-hairy perennials with rhizomes. Stems simple or with ascending
branches, 4–9 dm tall. Lower leaves palmately 5-lobed with toothed margins, 6–10 cm
long and 6–10 cm wide, the middle and upper leaves becoming gradually smaller and
narrower and 3-lobed or, the uppermost, merely toothed; upper surface rugose, the veins
prominent and hairy on the lower surface. Calyx 5-ribbed, 3–4 mm long, the lobes rigid,
spinelike, some of them recurved. Corolla white to pink or purplish, about 1 cm long,
the upper lip conspicuously hairy. Mericarps brown, sharply 3-angled, obpyramidal,
about 2 mm long.

Nectar is taken primarily by bumblebees and honeybees.

444. *Leonurus cardiaca*

The young plants may be eaten as a green vegetable, either raw or boiled.

An infusion of the dried leaves and flowering tops has been used in the treatment of rabies, heart palpitations, and delayed menses, and as a tonic and stimulant.

LÝCOPUS L.   Water horehound, bugleweed

NL., from Gr. *lyko-*, combining form of *lykos*, wolf, + *pous*, foot—the coarse teeth of the leaves apparently suggesting claws. "Bugleweed," from ML. *bugula*, originally applied to the European bugle, *Ajuga reptans*, also of the Mint Family.

445. *Lycopus americanus*

Perennial herbs, mostly with rhizomes. Leaves opposite, simple, with toothed or pinnatifid margins, sessile or petioled. Flowers small, white to pale pink or purple, in dense axillary whorls. Calyx bell-shaped to ovoid, with 4–5 ovate to subulate, erect or spreading teeth or lobes. Corolla nearly regular, 4-lobed, the upper lobe slightly larger than the others, entire or apically notched; throat of the corolla tube hairy. Fertile stamens 2, as long as the corolla or protruding slightly, sometimes accompanied by 2 minute staminodes. Mericarps 3-angled in cross section, widened toward the apex, with a corky ridge on the lateral angles and across the top, the rest of the surface more or less resinous.

Five species in Kansas. *L. asper* Green, *L. rubellus* Moench, and *L. uniflorus* Michx. are rarely encountered and are not included here.

Key to Species:

1. At least the lowermost leaves deeply cut or pinnatifid; calyx lobes lance-acuminate, longer than the mature fruits; apex of the mericarps broadly rounded .... *L. americanus*
1. All of the leaves more or less deeply toothed, none deeply cut nor pinnatifid; calyx lobes triangular or acute, shorter than the mature fruits; apex of the mericarps oblique, erose .... *L. virginicus*

L. americànus Muhl.   American bugleweed

"American."
White to pale purple; July–early September
Wet soil
Throughout

Plants 3–9 dm tall, tending to multiply by rhizomes and prostrate, rooting branches. Leaf blades minutely roughened, lanceolate or elliptic, mostly 1.5–8 cm long and 0.5–2 cm wide, the margins merely serrate to cleft or deeply pinnatifid. Calyx 2–2.5 mm long, the lobes lance-acuminate, about as long as the tube, protruding beyond the mature fruit, the midvein extended into an awnlike tip. Corolla white to pale purple, crimson-spotted, slightly longer than the calyx. Mericarps 1–1.5 mm long, the apex broadly rounded, occasionally somewhat uneven, the upper part of the inner face resin-dotted.

The flowers of American bugleweed are proterandrous; that is, the anthers of a flower open and liberate pollen before the 2 stigmatic branches of the style of the same flower have spread apart to receive pollen. The nectar secreted at the base of the ovary is sought after by honeybees, bumblebees, and other wild bees, as well as by some species of butterflies and flies, and cross-pollination can occur when insects go from younger flowers with protruding dehiscing anthers to older flowers with withered stamens but with stigmas spread apart and receptive to pollen.

L. virgìnicus   Virginia bugleweed

"Virginian."
White to pale purple; July–mid-September
Wet soil
East fifth, primarily

Erect perennials strongly resembling *L. americanus*. Leaves more or less deeply toothed but never deeply cut nor pinnatifid. Calyx about 1.5 mm long, the lobes triangular or acute, shorter than the mature fruit. Corolla definitely longer than the calyx. Mericarps about 1.5 mm long, the apex oblique and erose, the inner face resin-dotted (use magnification).

According to Henry Lyte, whose *A Niewe Herball or Historie of Plantes* was first published in 1578, the European *Lycopus europaeus* was called gipsy-wort "because the rogues and runagates which call themselves Egyptians do colour themselves black with this herbe." However that may be, the European species was used medicinally as an astringent and sedative, and our native *L. virginicus* was at one time officinal for the same purposes. Griffith's *Medical Botany* (50) has this to say of the latter species: "It acts like a mild narcotic and at the same time displays tonic powers. Those practitioners who have employed it are unanimous in declaring that it is an exceedingly valuable addition to the Materia Medica." Presumably, *L. americanus* might be found to possess similar virtues.

MARRŪBIUM L.   Horehound

L. name used for the horehound plant by the early writers on medical botany. The common name is derived from AS. *harhune*, from *har*, hoary, + *hune*, name of a plant—indicating the hoary felt of hairs over the plant.

Strongly aromatic perennial herbs. Leaves opposite, simple, petioled. Flowers small, white, sessile in dense, many-flowered axillary whorls. Calyx tubular, nearly regular, the throat hairy, the 5–10 teeth more or less spiny-tipped and spreading at maturity. Corolla distinctly 2-lipped, the upper lip erect, notched, the lower lip spreading, advanced, 3-lobed, the middle lobe largest; corolla tube swollen below the throat and closed above the swelling by the anthers and a fringe of hairs. Style short, arising between and at the base of the ovary lobes, stigma included within the corolla tube.

One species in Kansas.

M. vulgàre L.   Common horehound

L. *vulgaris*, common, from *vulgus*, the common people.
White; late May–mid-October
Disturbed habitats such as roadsides, pastures, yards, fields, and creek banks
East four-fifths, more common eastward; intr. from Eurasia, sometimes cultivated

Stems stout, branched, densely woolly, 3–7 dm tall. Leaf blades mostly very broadly ovate to broadly elliptic, more or less densely woolly, 3–5 cm long and 2–4.5 cm wide, the upper surface rugose, the lower surface prominently veined, the apex more or less rounded, the margins rather irregularly crenate or dentate. Calyx 4–5 mm long, the 10 veins extended into hooked spines. Corolla white, 5–6 mm long. Mericarps brown, about 2 mm long, 3-angled in cross section, somewhat flattened, the outer face convex, ovate to ovate-oblong, sometimes minutely roughened.

The stigmas and the anthers mature at about the same time, giving a chance of self-pollination, but cross-pollination is apt to be brought about by the action of bees, flies, and butterflies.

The nectar excreted in the corolla tube below the ovary is, in some localities, enough to be important in the honey crop, though the quality is not of the best.

Examining flower clusters of different ages, one sees that the ripe nutlets are contained in the persistent tubular calyx, and that the hooked tips of the 10 fine divisions of the calyx are fitted to play a part in the ultimate removal of the calyx and dispersal of the seeds.

Horehound was anciently employed by Egyptians and Europeans as a remedy for various ailments and even as magic to avert the malign influence of evil spirits or injury from dangerous animals. The herbals of the Middle Ages tell explicitly that it was used for liver complaint, migraine, phlegm, impetigo, indigestion, ordinary headache, earache, chronic catarrh, phthisis, and poisoning, as well as to overcome demoniacal possession; and Gerard's herbal (13) asserts that "it will tye the tongues of Houndes so that they shall not bark at you, if it be laid under the bottom of your feet."

The reputation of horehound survived such fictions of the Middle Ages, and the use of infusions and candy containing extracts of the herbage as a popular remedy for congestion in the respiratory system has continued into the present century. Though the species is no longer official in our pharmacopoeia its curative value was vouched for in early 20th-century medical literature. As an example we quote from W. Bogn's *Die Heilwerthe heimischer Pflanzen* (49): "The white Andorn [horehound] favorably affects the mucous membranes, resolves old, stubborn catarrhs, alleviates asthmatic troubles, and has found a place in the treatment of consumption. In larger doses it stimulates the vascular system, quickens excretion by the skin and kidneys, and acts as a laxative. It stimulates activity of the liver in jaundice, rectifies a run-down condition and thereby promotes the curing of anaemia."

What the horehound has to offer that might affect our physiological processes one way or another, chemical analysis shows to be: fat, wax, mucilage, resin, tannins, glucose, ethereal oil, and a bitter principle called marrubin.

MÉNTHA L.   Mint

L. *menta* or *mentha*, from Gr. *minthe*, name used by Theophrastus (died c. 285 B.C.) for a mint, presumably the peppermint—from a nymph of that name in Gr. mythology who was changed into a mint by Proserpine.

Aromatic perennial herbs, with ascending or erect stems, reproducing vegetatively by stolons or rhizomes. Leaves opposite, simple, toothed, sessile or petioled. Flowers small, white to pale blue or purple, in dense axillary whorls or in terminal spikes or heads. Calyx bell-shaped or tubular, 5-toothed. Corolla more or less 2-lipped, the upper lip entire or apically notched, the lower lip 3-lobed. Stamens 4, about equal in length, erect, included in or protruding beyond the corolla tube. Nutlets ovoid, smooth.

Two species reported for Kansas. The following is the only native species and the only one very commonly encountered.

M. arvénsis L.   **Wild mint**

NL., occurring in a cultivated field, from L. *arvum*, cultivated field or plowed land.
Pale blue to pale lavender; July–September
Moist or wet soil of creek banks, lake margins, prairie ravines, and low woods
East two-fifths, primarily; scattered records farther west

Erect perennial herb spreading by means of horizontal rhizomes. Stems 3–8 dm tall, eventually branching, sparsely covered with downward-curled hairs. Leaves petioled, the blades lanceolate to elliptic, 2–5 cm long and 0.5–3.5 cm wide, the apex and base acute or acuminate, the margins serrate, both surfaces minutely gland-dotted. Flowers borne in dense axillary whorls. Calyx tubular, 2.5–3 mm long, densely gland-dotted (use magnification), sparsely hairy. Corolla pale blue to pale lavender, 4–7 mm long. Mericarps tan, broadly ellipsoid to broadly ovoid, about 0.7–1 mm long, the basal region of attachment to the receptacle white, truncate.

Flowers of this and other species of *Mentha* are of 2 kinds, bisexual or female, both borne on the same plant. The bisexual flowers are proterandrous (i.e., the anthers mature and discharge pollen before the stigmas of the same flower are ready to receive it). Some time after the anthers have discharged their pollen, the style elongates far beyond the corolla and the stigma lobes spread apart, ready to receive pollen carried by bees, flies, hover flies, and butterflies.

The Chinese cultivate this species extensively and export the menthol obtained from the fragrant herbage. The volatile oil is distilled from numerous glands visible to the naked eye, though better seen with a hand lens.

Various American Indian tribes made use of wild mint, either as a beverage or for medicinal purposes. The Menominees combined wild mint with catnip and peppermint to prepare a tea or chest poultice to treat pneumonia, and several groups, including the Potawatomis and Pillagers, made a tea from the leaves and flowers to break fevers and/or to treat pleurisy. Other recorded uses of the plant include its employment as a carminative, to prevent vomiting, as an antispasmodic, and antirheumatic, to name a few.

MONÁRDA L.   Beebalm, horsemint, monarda

Named after Nicholas Monardes, 16th-century physician of Seville who wrote about medicinal and other useful plants of the New World.

Erect, aromatic annual or perennial herbs or shrubs. Leaves opposite, simple, sessile or petioled. Flowers conspicuous, attractive, crowded into dense headlike whorls subtended by leaflike bracts, the whorls terminating branches or arranged in interrupted spikes. Calyx tubular, 5-toothed, 13- to 15-nerved, the throat usually hairy. Corolla strongly 2-lipped, the upper lip narrow, straight or curved, the lower broader, spreading or curved downward, 3-lobed, the middle lobe the longest. Stamens 2, ascending under and usually longer than the upper corolla lip. Ovary deeply 4-parted, the style elongate and ascending under the upper lip. Mericarps smooth, oblong, mostly about 1.5–2 mm long.

Six species in Kansas. *M. bradburiana* Beck., known from Cherokee and Greenwood counties only, resembles *M. fistulosa* in having flowers in terminal heads but is shorter (3–6 dm) and has white to pale purple, purple-spotted flowers and leaves sessile or nearly so.

Key to Species:
1. Plants perennial; flowers borne in solitary terminal heads .... *M. fistulosa*
1. Plants mostly annuals; flowers borne in dense whorls arranged in interrupted spikes 2
    2. Calyx teeth triangular, more or less acuminate but not bristlelike; flowers pale yellow or whitish, purple-spotted; floral bracts yellowish or whitish, the tips acute but not bristlelike .... *M. punctata*
    2. Calyx teeth bristlelike; flowers white to pale pink or pale purple; floral bracts with bristlelike tips 3

3. Floral bracts frequently recurved, the upper surface densely covered with minute purple or whitish hairs .... M. citriodora
3. Floral bracts not recurved, the upper surface glabrous .... M. clinopodioides

M. citriodòra Cerv.   Lemon beebalm

NL., *citriodorus*, lemon-scented, from L. *citrus*, for the lemon tree, as well as the orange and lime tree, + *odoro*, give a fragrant smell—applicable to the whole plant.
White to pinkish, unspotted; late May–July
Rocky or sandy prairie, roadsides, occasionally in open woods
East three-fourths, primarily; more common southward

446. *Monarda citriodora*

Annuals with a taproot and many fibrous laterals. Stems simple or branched, 2–7 dm tall. Leaf blades oblong to lanceolate, 1.5–7.5 cm long and 0.6–2 cm wide, often folded along the midrib, the apex acute, the base tapering to a short petiole, the margins shallowly serrate, the lower surface minutely hairy, both surfaces gland-dotted. Flowers borne in interrupted spikes, the individual whorls 2–4 cm across, subtended by conspicuous leaflike bracts which are usually reflexed and have the upper surface densely covered with minute purple (less frequently whitish) hairs and are abruptly acuminate near the apex, ending in long, slender bristlelike tips. Calyx 13–17 mm long, the tube glabrous or finely hairy, the teeth subulate below, the tip quite prolonged and bristlelike, ciliate. Corolla white to pinkish, unspotted, 18–25 mm long, the upper lip strongly arched. Mericarps about 1.5–1.9 mm long.

M. clinopodioídes Gray   Ciliate beebalm

NL., like *Clinopodium*, the name of a genus in the Mint Family, from Gr. *kline*, couch or bed, + *pous*, *podos*, foot—because the apex of the corolla of *Clinopodium vulgare*, the type species, is turned out all around like the rim of the receptacle of a bed-caster. The similarity of this *Monarda* to that *Clinopodium* exists in its series of axillary flower clusters, the hairiness of the young stems, the erect bristlelike lobes of the calyx, and the hairy and ciliate character of its floral bracts.
White to purple; late May–early July
Sandy prairie
South-central sixth

447. *Monarda clinopodioides*

Annuals resembling *M. citriodora* but tending to be smaller in all respects. Leaf blades 1.5–5 cm long and 0.2–2 cm wide. Individual whorls of the inflorescence 1.5–3 cm across, the subtending bracts ascending or spreading, never reflexed, sometimes purple-tinged but glabrous except for the ciliate margins, the apex bristlelike as in *M. citriodora*. Calyx 12–14 mm long, the tube coarsely hairy, the bristlelike teeth ciliate, (3–) 4–6 mm long. Corolla white to purple, 17–20 (–25) mm long, the upper lip strongly arched.

M. pectinata Nutt., known from scattered sites in the west half of the state, will key out here with *M. clinopodioides* but differs from that species in having the calyx tube glabrous or nearly so, the lobes 2–3 (–4) mm long, and the floral bracts very minutely hairy (use magnification and side illumination) on the outer surface.

M. fistulòsa L.   Wild-bergamot beebalm

L. *fistulosus*, porous, full of holes, from L. *fistula*, pipe or tube—probably in reference to the clusters of persistent calyces after the corollas have fallen away.
Lilac to rose-purple; mid-June–mid-August
Prairie hillsides and banks, pastures, roadsides, occasionally in open woods, usually in rocky soil
East two-thirds

448. *Monarda fistulosa*

Perennial herbs with brittle, branched, clustered stems 5–9 dm tall arising from a branched rhizome. Leaves distinctly petioled, the blades lanceolate to narrowly deltoid, 2–6.5 cm long and 1–4.5 cm wide, the apex acute or attenuate, the base acute or truncate, the margins more or less coarsely and irregularly dentate or serrate, both surfaces gland-dotted and minutely soft-hairy, the lower surface much lighter green than the upper surface. Flowers borne in terminal heads 5–8 cm across. Calyx 7–10 mm long, the teeth triangular or acuminate with a long bristlelike tip. Corolla lilac to rose-purple, 19–25 mm long, the upper lip somewhat arched but essentially continuing straight out in line with the corolla tube, conspicuously hairy near the tip. Stamens protruding beyond the upper lip. Mericarps about 1.5 mm long.

M. fistulosa as treated here includes the plants separated in some earlier manuals as *M. mollis* L.

There is satisfaction in examining a flower critically enough to understand the character and workings of its several parts, especially where there are clever devices for making use of insects to effect cross-pollination. To make out some of the smaller struc-

449. *Monarda punctata*

tural details, a magnifier is often needed. In the young flower buds of *Monarda*, the corolla is just visible in the calyx throat, and the immature lips of the corolla are pressed tightly together, an apical flag of the lower lip covering the mouth of the tube while a tuft of hairs at the end of the upper lip may serve to keep out rain. When the corolla opens, the flap of the lower lip becomes quite evident, and the stamens and style which it had been covering are seen close to and projecting beyond the upper lip. At this stage the anthers open and expose the pollen, but the 2 stigma lobes have not yet spread apart to present their inner surfaces for pollination. As the flower ages, the anthers shrivel, the style elongates, and the stigma lobes spread out, ready to receive pollen brought from younger flowers. It will be found that the base of the corolla tube encloses the 4-lobed ovary and that the threadlike style starts out from a central point beween the lobes— each lobe containing a single ovule which is to become a seed. Nectar secreted around the base of the ovary is held in the corolla tube where bumblebees, honeybees, and butter- flies must probe with their proboscises to get it.

We admire this plant in many respects: It is so hardy, and so prolific by seeds and branching rhizomes that we can count on its presence year after year, even through periods of drought. We welcome it for the restful tone of its abundant, long-blooming flowers and the fragrance of herbage and inflorescence which persists after drying and can be used to scent clothes closets, bureau drawers, and pillows through many months. A refreshing tea can be made by boiling the dried leaves in water, with the addition of sugar and sliced lemon according to taste.

Various medicinal uses of this and other species of *Monarda* are on record. Several species, including *M. fistulosa* and *M. punctata*, are the source of thymol, a fragrant volatile oil which was listed officially in the *U.S. Pharmacopoeia* from 1882 to 1950 and has been in the *National Formulary* since 1950. It has been used to treat bacterial and fungal infections and infestations of worms, especially hookworms. Several Indian groups learned to extract the oil by boiling the dried plant in water, then used either the tea or the oil itself to relieve colds, fevers, sore throats, and bronchitis. According to Weiner (34), the Blackfoot Indians applied the boiled leaves to pimples to dry them up, while the Winnebagos used the extracted oil for the same purpose, a practice also followed by white settlers. Thymol combined with mercury or iodine has been employed in the preparation of surgical dressings in the form of ointments or powders.

M. punctàta L.   Western spotted beebalm

NL., from L. *punctus*, pricked in—in reference to the abundant glandular dots over both surfaces
    of the leaves. Our plants are recognized as belonging to subspecies *occidentalis*, a Latin
    word meaning "of the west," hence the common name.
Yellowish to whitish, purple-spotted; late May–July (–August)
Open places with sandy soil
Southwest fourth, east to Saline and Sumner counties

Gray-green, soft-hairy annuals, biennials, or perennials. Stems branched, 1.5–3.5 dm tall. Leafblades more or less narrowly lanceolate to elliptic, 1.5–5.5 cm long and 3–15 mm wide, often folded along the midrib, the apex rounded to acute, the base tapered to the short petiole, the margins shallowly serrate, both surfaces profusely gland- dotted, the lower surface and sometimes the upper surface minutely hairy. Flowers borne in interrupted spikes, the individual whorls (1.5–) 2.5–3 cm across, subtended by con- spicuous leaflike bracts which are pale yellow to whitish or rarely pinkish in color and taper to a sharp point. Calyx 5–6 mm long, the teeth triangular, corolla yellowish to whitish, purple-spotted, rarely purplish, 12–18 mm long, the upper lip strongly arched. Stamens protruding beyond the upper corolla lip. Mericarps about 1.2 mm long.

The leaves and flowering tops of this species were listed officially in the *U.S. Pharmacopoeia* from 1820 to 1882, as was Monarda oil. The plant is also a source of thymol, as discussed under *M. fistulosa*. *M. punctata* plants were used to induce sweating (primarily in the treatment of typhoid fever and rheumatism), to prevent vomiting, and to relieve flatulence. Extracted Monarda oil was taken internally for the same purposes or used externally as a stimulant, counterirritant, or as a blistering agent (45).

NÈPETA L.   Catnip

L., after the Etruscan town Nepete, in whose vicinity plants of this species grew wild.

Annual or perennial herbs. Leaves opposite, simple, petioled in ours. Flowers small, white to blue, crowded in terminal simple or branched spikes. Calyx tubular, 15- nerved, slightly irregular. Corolla tube dilated above into a wide-open throat, distinctly 2-lipped, the upper lip forming a roof over the stamens and style, the lower more or less deflexed, 3-lobed, the middle lobe the broader and slightly toothed. Stamens 4, 2 long and 2 short, the upper pair the longer.

One species in Kansas.

N. catária L.   Catnip

NL., *catarius*, pertaining to cats, from L. *catus*, tomcat. Also called catmint or catnep, according
   to Gerard's herbal "because cats are very much delighted herewith; for the smell of it is
   so pleasant unto them that they rub themselves upon it and wallow or tumble in it, and
   also feed upon the leaves very greedily."
White or pale purple, with red spots; late June–October
Disturbed habitats such as roadsides, creek banks, waste ground, old homesites
East half and northwest fourth, a few records from the southwest fourth; intr. from Eurasia

450. *Nepeta cataria*

Erect, soft-hairy perennial with a strong taproot, many laterals and fibrous rootlets, and a branching crown by which extensive colonies are sometimes formed. Stems erect, 0.6–1 m tall, with ascending branches. Leaf blades cordate to narrowly triangular, mostly 2–8 cm long and 1–4.5 cm wide, the base cordate or truncate, the margins coarsely serrate, the lower surface more densely hairy than the upper; petioles 1–4.5 cm long below, becoming shorter above. Bisexual and female flowers occurring on the same plant in terminal spikes 1.5–9 cm long and 1.5–3.5 cm wide at anthesis. Calyx 5–6 mm long, conspicuously hairy, the lobes narrowly triangular-attenuate. Corolla dull white or pale purple with red spots, 8–10 (–12) mm long. Mericarps brown, smooth, very broadly obovate to very short-oblong, somewhat flattened, about 1.5 mm long.

The slender style with its 2-lobed stigma extends to about the same level as the anthers, but self-pollination is prevented by the anthers' discharging their pollen before the stigma lobes in the same flower are spread apart to receive it. Later when the stigma is ready it must be pollinated by insects bringing pollen from flowers freshly opened— something sure to happen because nectar exuding from a gland subtending the 4-lobed ovary and rising a short distance in the corolla tube is repeatedly being sucked out by honeybees, bumblebees, and other wild bees, flies, hover flies, and butterflies.

People, too, have found uses for the catnip. Before the introduction of tea from China the British and Europeans were making a pleasant drink from its herbage. It was used medicinally for various distresses of body and mind, particularly as a carminative for relief from colicky pains, as a diaphoretic to induce perspiration in cases of colds and influenza, as a sedative for nervousness and hysteria, as an antispasmodic for convulsions; and a decoction sweetened with honey was employed to relieve coughing.

Griffith's *Medical Botany* (50), a standard work of its day, says this of catnip: "It is now seldom employed in regular medical practice, but is far more deserving of notice than many articles admitted into the officinal lists." According to Krochmal and Krochmal (30), Appalachian residents use catnip tea to treat colds, nervous conditions, stomach disorders, and hives, and smoke the dried leaves and stalks to relieve respiratory ailments. This species was listed officially in the *U.S. Pharmacopoeia* from 1842 to 1882 and in the *National Formulary* from 1916 to 1950.

PERÍLLA L.

Native East Indian name.

Annual herbs, often with purplish herbage. Leaves opposite, simple, long-petioled. Flowers small, white to purple, borne in loose, slender terminal and axillary racemes. Calyx bell-shaped, weakly 2-lipped at anthesis, the upper lip with 3 broadly triangular lobes, the lower with 2 deeper, narrowly triangular lobes. Corolla weakly 2-lipped, the lobes rounded. Stamens 4, straight, about as long as the corolla. Mericarps globose, the surface reticulate, enclosed by the enlarged, persistent, definitely 2-lipped calyx.

One species in Kansas.

P. frutéscens (L.) Britt.   Beefsteak plant

NL., *frutescens*, *fruticosus*, shrubby, bushy, from L. *frutex*, *fruticis*, shrub or bush.
White to purplish; September–mid-October
Moist soil of alluvial woods
East third, primarily, especially the east fourth; intr. from Asia

Erect, sparsely hairy plants with slender stems up to 1 m tall, superficially resembling several of our genera in the Nettle Family which tend to occur in the same type of habitat. Leaf blades broadly ovate, 2.5–10 cm long and 1.5–7 cm wide, the apex and base acuminate, the margins coarsely serrate or occasionally lacerate. Racemes 2–15 cm long. Calyx conspicuously hairy, 2–3 mm long at anthesis (up to 12 mm long and

451. *Physostegia virginiana*

definitely 2-lipped in fruit). Corolla white to purplish, 2.5–3.5 mm long. Mericarps brown, about 1.5 mm long.

At least one variety of this plant is grown as an ornamental for its bronze or purple foliage. The seeds are a source of oil which is or has been used in the manufacture of lacquer, artificial leather, paper umbrellas, printer's ink, and waterproof clothes (4).

PHYSOSTÉGIA Benth. — False dragonhead

NL., from Gr. *physa*, bladder, + *stege*, covering—in reference to the inflated condition of the mature calyx. The true dragonhead of Eurasia is *Dracocephalum* (NL., from Gr. *drakon*, dragon, + *kephale*, head)—the gaping mouth and throat of the corolla calling to mind representations of the fabled dragons of ancient lore.

Smooth, erect perennial herbs. Leaves opposite, simple, sessile (or the lower petioled), usually toothed. Flowers showy, in long, slender, simple or branched terminal racemes. Calyx regular, short-tubular or bell-shaped, 5-toothed, enlarging somewhat in fruit. Corolla 2-lipped, gradually widened upward, the tube much longer than the calyx; the upper lip straight, hoodlike, entire or slightly notched; lower lip about as long as the upper, spreading, 3-lobed. Stamens 4, ascending under the upper lip, the lower pair longer. Mericarps ovoid, smooth.

At least 1, possibly 3 species in Kansas. Most of our plants appear to fall into the following species.

P. virginiàna (L.) Benth. — Virginia lion's-heart

"Virginian."
Pink or purple-tinted; July–September
Low prairies
East third in the north, east fourth south of the Kansas River

Plants up to 1 m tall, the stems simple or giving off ascending branches above. Leaves mostly narrowly elliptic, narrowed to the base, 4–14 cm long and 1–3 cm wide, the margins sharply serrate. Herbage glabrous except for the minutely glandular-hairy calyx and inflorescence axis. Flowers on very short pedicels in racemes up to 6 dm long. Calyx 6–8 mm long, the teeth triangular to lance-acuminate, the tips somewhat recurved. Corolla 2.5–3 cm long. Mericarps about 3 mm long.

Our *Physostegias* have the virtue of being easily transplanted from the wild and multiplied by seeds and division of strong clumps. Having them near at hand in the garden, one can observe how their device for securing cross-pollination operates. One can clearly see that one pair of anthers stands a little in front of the other pair and that the style poses the 2-lobed stigma between them. Examining flowers of different ages, one sees that the anthers early break open and expose their pollen to contact with visitors—hummingbirds, butterflies, bumblebees, honeybees. Later the style bends downward, placing the stigma lobes, which at first were in contact but now spread apart, where the visitors carrying pollen from younger flowers will touch them. This very ancient device for securing cross-pollination by having the anthers and stigmas of a flower mature at different times is in use in many families of plants. Proterandry (anthers maturing first) is of frequent occurrence; proterogyny (stigmas maturing first) is less common but is the rule in the grass family and also occurs in *Verbascum*.

PRUNÉLLA L. — Selfheal

NL., from L. *pruna*, glowing coal, + the diminutive ending -*ella*—evidently alluding to the fact that in folk medicine decoctions of these plants were used for reducing internal and external inflammations.

Low, smooth perennial herbs. Leaves opposite, simple petioled. Flowers in 3's subtended by broad leaflike bracts and crowded into terminal spikes or heads. Calyx deeply 2-lipped, the upper lip with 3 broad, shallow, short-acuminate lobes, the lower lip with 2 lance-triangular lobes, the midveins of all 5 lobes prolonged as short awns. Corolla 2-lipped, the upper lip hooklike, entire, arched over the stamens, the lower lip deflexed, 3-lobed, the middle lobe largest and fringed. Stamens 4, in 2 pairs, 1 pair of anthers standing above and close to the other pair, the filaments 2-pronged, the inner prong bearing the anther. Ovary deeply 4-lobed. Mericarps 4, smooth.

One species in Kansas.

P. vulgàris L. — Selfheal, heal-all

L. *vulgaris*, common, from *vulgus*, the common people. As to the common names, we quote the Frenchman, Jean de la Ruelle (1474–1537), physician to Francis I, "No one who keeps *Prunella* (*P. vulgaris*) needs a surgeon."

Lavender and white; mid-May–mid-October
Low woods and thickets, roadsides in partial shade, yards, lake margins
East third, primarily, with a few scattered records farther west; intr. from Europe

Low perennial with a short rhizome. Stems erect to decumbent, eventually branching, 1.5–6 dm long. Leaf blades lanceolate, 2–10 cm long and 0.5–4 cm wide, the apex acute or rounded, the base acute or short-acuminate, the margins entire or very shallowly crenate or serrate. Flowers and subtending bracts crowded into short cylindric spikes 1–5 cm long and 1.5–2.5 cm wide, the bracts reniform, with an abruptly acuminate apex, conspicuously ciliate. Calyx 8–11 mm long, green or tinged with purple, the lower lip hirsute toward the base. Corolla lavender and white, 10–16 mm long. Mericarps brown, about 1.8–2 mm long, the outer face obovate, prominently veined.

The style, with 2 slender, vertically spreading stigmas, is positioned between the anthers and is ready for self-pollination as soon as the anthers open. However, nectar attracts honeybees, bumblebees, other long-tongued bees, and some kinds of butterflies, so that cross-pollination also is effected. After the corolla falls away, the lips of the corolla close together, covering the ovary as the fruit ripens.

Among the medicinal applications of this plant, we find the treatment of boils, gas, colic, sore throat, hemorrhages, and diarrhea.

452. *Prunella vulgaris*

PYCNANTHEMUM Michx.　　　　　　　　Mountain mint

NL., from Gr. *pyknos*, dense, + *anthemon*, flower—because of the many-flowered inflorescence.

Erect perennial herbs with rhizomes and a mintlike flavor. Leaves opposite, simple, entire or toothed, sessile or petioled. Flowers small, white to purple, crowded in dense, many-flowered terminal heads. Calyx tubular, regular or somewhat 2-lipped, 5-toothed or -lobed. Corolla 2-lipped, dotted with purple, the upper lip erect, entire or notched, the lower lip bent downward, 3-lobed. Stamens 4, protruding beyond the corolla in some flowers, not so in others, the lower 2 stamens longer than the upper 2. Mericarps glabrous, or the apex hairy.

Five species in Kansas. *P. virginianum* (L.) Durand & Jackson is known from a few northeastern and extreme southeastern counties only. It is distinguished from *P. tenuifolium* on the basis of the bracts subtending the flower heads, those of *P. virginianum* being oblanceolate with the tips acute or short-acuminate, thin and soft, without a prominent midvein. *P. albescens* Torr. & Gray is known from southeastern Cherokee County only. It has broadly lanceolate, serrate leaves with the under surface densely covered with minute white hairs and inflorescences more open and more noticeably branched than in our other species.

453. *Pycnanthemum pilosum*

Key to Species:
1. Leaves linear-attenuate, 1–3 mm wide, glabrous; heads 4–10 mm across, the outermost subtending bracts ciliate but otherwise glabrous or nearly so .... *P. tenuifolium*
1. Leaves lanceolate to narrowly elliptic, 4–15 mm wide, soft-hairy; heads (10–) 15–20 mm across, the outermost subtending bracts densely white-hairy .... *P. pilosum*

P. pilòsum Nutt.

L. *pilosus*, hairy.
White to pale purple; July–August
Moist, wooded banks and bluffs
Extreme northeast and southeast

Plants conspicuously soft-hairy, up to 1 m tall. Leaves lanceolate to narrowly elliptic, usually 2.5–7 cm long and 4–15 mm wide. Heads (10–) 15–20 mm across, the subtending bracts densely white-hairy. Corolla white to pale purple, 5–6 mm long.

P. tenuifòlium Schrad.　　　　　　　　Slender mountain mint

NL., from L. *tenuis*, thin, slight, few, + *folium*, leaf—in reference to the narrow leaves.
White to pale purple; June–July
Prairie, open woods, roadsides, frequently in rocky, sandy soil
East fourth, primarily, especially southward

Stems clustered, 0.4–1 m tall, branched above. Leaves sessile, glabrous, linear-attenuate, 1–3 cm long and 1–3 mm wide, dotted with fine glands filled with a fragrant volatile oil. Heads 4–10 mm across, closely subtended by many lance-acuminate to linear-attenuate bracts with the prominent midvein prolonged into a rather rigid tip, the margins densely ciliate but the bracts otherwise glabrous. Corolla white to pale purple,

454. *Pycnanthemum tenuifolium*

MINT FAMILY
323

455. *Pycnanthemum tenuifolium,* close-up of flowers

456. *Salvia pitcheri*

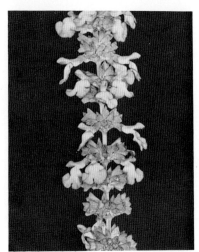

457. *Salvia pitcheri,* close-up of flowers

purple-spotted inside, 6–7 mm long. Mericarps dark brown, more or less oblong, about 1 mm long, smooth.

This species is listed in some earlier manuals as *P. flexuosum* (Walt.) BSP.

SÁLVIA L.   Salvia, sage

NL., from L. *salvare,* to heal, some species being medicinal.

Annual, biennial, or perennial herbs or shrubs, often aromatic. Leaves opposite or all basal, simple or compound, mostly petioled. Flowers mostly blue, red, or white, usually showy, in few- to several-flowered whorls arranged in interrupted spikes, racemes, panicles, or heads. Calyx 2-lipped, the upper lip entire or 3-toothed, the lower lip 2-toothed or -cleft. Corolla strongly 2-lipped, the upper lip flat or somewhat hoodlike, the lower lip spreading or drooping, 3-lobed, the middle lobe the largest, the tube elongate. Stamens 2, protruding beyond the corolla or ascending under the upper lip; filaments actually short, but appearing longer because of an elongated transverse connective which bears a 1-locular anther at the anterior end and a deformed 1-locular anther or none at all at the posterior end which protrudes backward into the throat of the corolla. Style apex 2-lobed. Nutlets smooth.

This is the largest genus in the Mint Family, with approximately 700 species distributed mostly in temperate and tropical regions around the world, especially in the tropics of South America.

Five species reported for Kansas. *S. lyrata* L., *S. nemorosa* L., and *S. pratensis* L. are rarely encountered and are not included here.

Key to Species:

1. Perennial, 0.6–1.5 m tall, with a short, thick rhizome; flowers 11–23 mm long; mericarps brown, resin-dotted, usually but 1 or 2 per flower .... *S. pitcheri*
1. Annual, 1–8 dm tall, with a taproot; flowers 7–9 mm long; nutlets light tan, not resin-dotted, usually 4 per flower .... *S. reflexa*

S. pítcheri Torr.   Pitcher's salvia

In honor of Dr. Z. Pitcher, U.S. Army surgeon and botanist.
Blue; late June–mid-October
Dry, rocky or sandy prairie hillsides and uplands, roadside banks
East three-fourths, primarily

Soft-hairy, erect perennial with stems 0.6–1.5 m tall arising from a short, thickened rhizome and stout, fibrous roots reaching down 2–2.5 m. Leaves oblanceolate below, usually becoming narrowly elliptic or linear above, tapering to a narrow base, 3–10 cm long and 4–20 mm wide, the apex acute or attenuate, the margins serrate, sometimes entire on the uppermost leaves. Flowers on short pedicels in branched terminal racemes. Calyx 5–8 mm long at anthesis, hairy between the veins as well as on them, the upper lip entire. Corolla blue, 11–23 mm long. Mericarps brown, ellipsoid, somewhat flattened, resin-dotted, 2–2.5 mm long, usually only 1 or 2 per flower maturing.

Study of the individual flowers reveals ingenious devices for ensuring cross-pollination: (1) The lobed, drooping, relatively large lower lip is overhung by the shorter, stiffer, concave upper lip, which protects the nectar from rain and shields the style and the 2 stamens close under it. (2) The lower lip provides a convenient place for honeybee and bumblebee visitors to stand while they push their heads into the throat of the corolla and thrust their sucking mouth parts to the bottom of the tube and draw away the nectar secreted by the nectar gland at the base of the ovary. Meanwhile the bees are incidentally operating an ingenious device concerned with cross-pollination. (3) The flowers are proterandrous; that is, the anthers discharge their pollen before the stigmas of that flower are ready to receive it. When the style of that flower has elongated and the stigmas spread apart, pollination must be done with pollen brought on the backs of bees from younger flowers whose anthers are freshly opened. How do the bees get this pollen on their backs? (4) The stamens are very unusual in that the connective between the 2 sacs of an anther, instead of being a narrow tissue, as is the rule, is elongated in the form of a bent rod, with its concave side forward, being so joined at a point below its middle to the short filament that it stands in a vertical position, much like a see-saw, its upper end bearing 1 anther sac filled with pollen and its lower end bearing the other sac, which is without pollen. (5) The 2 stamens of this kind stand in front of the corolla throat, and when a bee advances to get the nectar its head presses against the sterile posterior sac, shoving it inward and causing the long upper arm of the connective to sweep down and press the opened fertile anther sac against its back and deposit the pollen there. While this is going on, the immature style is out of the way, close under the upper lip. (6) The bee, now pollen-laden, flies away, and on coming to an older flower where the style with outspread stigma lobes has assumed a lower position in front of the corolla throat, brushes its back against the stigma and cross-pollinates it.

This interesting exhibit can be seen over a long period, for even when the blossoming stems are mown with the grass, new stems arise from the rootstock and bloom abundantly.

See plate *87*.

458. *Salvia reflexa*

S. refléxa Hornem. Lanceleaf salvia

L. *reflexus*, bent, or turned back—applicable to the upper branch of the 2-parted style.
Light blue; late May–early October
Disturbed habitats such as roadsides, fields, waste ground, bare areas in prairie, etc.
Throughout

Erect, branched, fragrant, minutely hairy annuals, 1–8 dm tall. Leaves petioled, mostly lanceolate to lance-oblong or narrowly oblong, 3–6.5 cm long and 4–15 mm wide, the apex usually rounded, the base gradually tapered to the petiole, the margins shallowly toothed. Flowers borne in loose, slender terminal racemes up to 10 cm long. Calyx 5–6.5 mm long, ciliate along the veins and margins, the upper lip entire. Corolla light blue, 7–9 mm long. Mericarps tan, smooth, 3-angled in cross section, the outer face elliptic, convex, 2–2.5 mm long, usually 4 per flower maturing.

SCUTELLÀRIA L. Skullcap

NL., from L. *scutulum*, little shield, from *scutum*, shield—the large crested calyx of the type species, *S. peregrina*, of eastern Europe calling to mind the ancient Greek and Roman crested helmet. "Skullcap," from the shape of the large fruiting calyx of *S. galericulata* (NL., from *L. galerum*, for a kind of priest's cap, or a helmetlike covering for the head, made of undressed skin, + a diminutive suffix).

Bitter but not aromatic annual or perennial herbs or subshrubs, often with rhizomes. Leaves opposite, simple, sessile or petioled. Flowers blue or purplish to pink or white, usually showy, borne singly in the leaf axils or in terminal or axillary racemes. Calyx strongly 2-lipped, both lips broadly rounded, entire, and about equal in length, the upper lip with a conspicuous transverse protuberance or crest. Corolla strongly 2-lipped, the upper lip entire or notched, the lower lip spreading, convex, apically notched, the corolla tube elongate, curved, ascending, and much longer than the calyx. Stamens 4, ascending under the upper lip, the anthers hairy, those of the lower stamens 1-celled, those of the upper stamens 2-celled.

Seven species in Kansas. *S. galericulata* L. and *S. brittonii* Porter are rare, known from Cheyenne County only, and are not included here.

Key to Species:
1. Flowers borne singly in the axils of middle or upper leaves 2
   2. Plants perennial from a woody caudex; Blue Hills region .... *S. resinosa*
   2. Plants perennial from creeping rhizomes forming beadlike tubers; east two-fifths of Kansas, primarily .... *S. parvula*
1. Flowers borne in terminal or axillary racemes 3
   3. Racemes mostly axillary; flowers 5–9 mm long; herbage glabrous .... *S. lateriflora*
   3. Racemes mostly terminal; flowers 15–25 mm long; herbage softly and conspicuously hairy .... *S. ovata*

S. lateriflòra L. Side-flowering skullcap

NL., from L. *lateris*, pertaining to the side, lateral, from *latus*, side, + *flos, floris*, flower—the flowers occurring on one side of or turned to one side of axillary or terminal racemes.
Blue to nearly white; mid-July–August (–early October)
Wet places
Nearly throughout, more common eastward

Glabrous perennial with slender, creeping rhizomes. Stems 0.3–1 m tall, eventually branching above. Leaves petioled, the blades thin, lanceolate to ovate, 2–8 cm long and 1–5 cm wide, the apex acute or slightly rounded, the base abruptly acuminate to truncate or slightly cordate, the margins coarsely serrate or crenate. Flowers blue to nearly white, 5–9 mm long, borne in many-flowered, 1-sided racemes. Mericarps yellowish, about 1.2–1.5 mm long, tubercled.

S. ovàta Hill Skullcap

L. *ovatus*, egg-shaped, from *ovum*, egg—in allusion to the leaves.

Blue; June–July
Moist woods
Extreme east and Chautauqua County

Soft-hairy perennial producing slender rhizomes with occasional tuberous thickenings. Stems up to 7 dm tall, seldom branched below the inflorescence, covered with spreading glandular hairs. Leaves petioled, broadly lanceolate to cordate, 5–9 cm long and 2–6 cm wide, the apex acute, the base cordate, the margins crenate to coarsely serrate. Flowers blue, 15–25 mm long, borne in many-flowered terminal racemes with 1–3 branches. Calyx with spreading glandular hairs. Mericarps black or dark brown, about 1.5–2 mm long, covered with pointed tubercles.

S. incana Biehl., also known from Cherokee County, strongly resembles *S. ovata* but differs from it in having the calyx and numerous branches of the inflorescence densely white-hairy and at least the upper leaves with bases wedge-shaped to acute or rounded. It flowers from late July–August.

See plate *88*.

S. párvula Michx. — Small skullcap

L. *parvulus*, diminutive of *parvus*, small—applicable to the size of the plant and the length of the corolla.
Dark blue; late April–mid-June
Rocky prairie, pastures, and open woods
East two-fifths, primarily

Low, glandular or nonglandular perennial with 1 to several stems 1–3.5 dm tall, the rhizome alternately swollen and constricted, thus resembling a string of beads or sausages. Leaves more or less petioled below to sessile above, the blades cordate to broadly ovate or lanceolate, 8–16 mm long and 3–17 mm wide, the apex rounded or acute, the base cordate, truncate, or rounded, the margins entire or with a few teeth. Flowers borne singly in the leaf axils. Calyx about 3 mm long at anthesis. Corolla dark blue, 7–10 mm long, the lower lip with white markings. Mericarps dark brown or black, rather mushroom-shaped, 1–1.2 mm long, covered with peglike protuberances.

Three varieties of this skullcap are known for Kansas, including the plants recognized in earlier manuals as *S. australis*.

S. resinòsa Torr. — Resinous skullcap

L. *resinosus*, gummy, resinous, from *resina*, resin.
Blue; mid-May–early August
Dry, rocky, calcareous prairie hillsides and uplands
Blue Hills Upland of the north-central sixth, south to Hodgeman County, and west to Gove and Logan counties

Low, glandular, minutely hairy perennial with several to many stems 1–2 dm tall arising from a woody taproot and crown. Leaves short-petioled or nearly sessile, the blades of the main leaves mostly ovate to broadly elliptic, 8–18 mm long and 4–8 mm wide, the apex rounded, the base wedge-shaped or acuminate. Flowers borne singly in the leaf axils. Calyx 3.5–5 mm long at anthesis, densely covered with sessile glands and both glandular and nonglandular hairs. Corolla deep blue with white markings on the lower lip, 17–22 mm long. Mericarps black, very broadly obovoid to nearly globose, about 1.2–1.5 mm long, minutely tubercled.

STÁCHYS L. — Betony, hedge-nettle

Ancient Latin name of mouse-ear betony (*Stachys germanica*), from Gr. *stachys*, spike—the close succession of axillary flowers making the inflorescence as a whole spikelike. "Betony" is from L. *betonica*, the ancient name of the Linnaean *Betonica alopecuros*, the foxtail betony. Pliny says, "The Vettones, a people of Spain, were the original discoverers of the plant known as the 'Vettonica' in Gaul" (a spelling the Latins sometimes employed).

Annual, biennial, or perennial herbs. Leaves opposite, simple, sessile or petioled. Flowers in dense axillary clusters arranged in terminal, usually interrupted spikes. Calyx bell-shaped with 5 nearly equal, erect or spreading teeth. Corolla strongly 2-lipped, the upper lip erect, hoodlike, entire or notched, the lower lip spreading 3-lobed, the middle lobe largest and sometimes 2-lobed, the corolla tube elongate. Stamens 4, in 2 pairs, ascending under the upper corolla lip, the upper pair longest. Ovary deeply 4-lobed; style 2-lobed. Mericarps ovoid or oblong.

Two species in Kansas. *S. palustris* L., known from scattered sites in the northeast, strongly resembles the following species but has softly pubescent foliage and stems, the latter with hairs on the sides as well as on the angles.

459. *Scutellaria incana*

460. *Scutellaria ovata*

461. *Scutellaria parvula*

MINT FAMILY
326

S. tenuifòlia Willd.   Slenderleaf betony

NL., from L. *tenuis*, thin, + *folium*, leaf.
Purplish; July–August (–mid-October)
Moist woods, thickets, and roadsides
East third, primarily

Erect, glabrous or near-glabrous perennial. Stems up to 1 m tall, eventually branching above, entirely glabrous or with stiff, retrorse bristles on the angles only, not on the sides. Leaves petioled, the blades thin, mostly lanceolate to lance-oblong or elliptic, 3–10 cm long and 1.5–3.5 cm wide, the apex acute or acuminate, the base acute or rounded, the margins serrate. Calyx 6–9 mm long. Corolla purplish, 11–13 mm long. Mericarps brown, about 1.5–2 mm long, 3-angled in cross section, the outer face convex, very broadly elliptic, minutely wrinkled.

TEÙCRIUM L.   Germander

NL., from Gr. *teukrion*, ancient Greek name for a European germander. The name "germander" has an interesting history, beginning with the ancient Greek *chamaedrys*, name for a procumbent shrubby species of *Teucrium*, from Gr. *chamai*, low, on the ground, + *drys*, tree. In late Greek this becomes *chamandrya*; in Medieval Latin, *germandra*; and this in Old French, *germandree*, which by slurring the last syllable becomes, in English, "germander," a name now used by us for any species of *Teucrium*.

Annual or perennial herbs, often with rhizomes. Leaves opposite, simple, sessile or petioled. Flowers pinkish or purplish, in slender terminal spikes or in the axils of the upper leaves. Calyx 5-toothed or -lobed, irregular. Corolla irregular, the upper lip so deeply cleft that its lobes appear lateral, the lower lip much larger, with 2 small lateral lobes and a prominent middle lobe. Stamens 4, protruding from the cleft of the upper corolla lobe. Ovary shallowly 4-lobed; style terminal. Fruits roughened.

Two species in Kansas. *T. laciniatum* Torr., a low, tufted perennial with pinnately or bipinnately dissected leaves and white flowers usually streaked with purple, is known from Cheyenne, Hamilton, Stanton, and Morton counties only.

T. canadénse L.   American germander

L. *-ensis*, suffix added to nouns of place to make adjectives—"Canadian," in this instance.
Purplish to pink; mid-June–early September
Moist or wet soil
East two-thirds, primarily; scattered collections known from the west third

Slender, erect, soft-hairy perennial, up to 1 m tall, producing clones from horizontal rhizomes which give off clusters of fibrous roots at the nodes and terminate by growing upward to produce an upright shoot. Leaves petioled, the blades lanceolate to elliptic, 2–12 cm long and 2–6 cm wide, the apex acute, the base acute to rounded, the margins serrate, the lower surface more densely hairy and thus lighter green than the upper surface. Calyx 5–7 mm long. Corolla purplish to pink, 15–18 mm long, cleft to the calyx on the upper side, the 2 lobes of the upper lip erect and appearing to arise laterally, the lower lip with 2 small lateral lobes and a much larger, concave middle lobe. Stamens and style arching upward through the corolla cleft. Mericarps brown, about 1.8–2 mm long, wrinkled, especially toward the apex.

When the flower expands from the bud, the 2 lobes of the stigma are closed so that no pollen can be received. The 4 anthers are arched forward and downward, the introrse anthers opening on their lower sides, so that bees, perched on the lower lip and extracting nectar from inside the corolla tube at its base, will receive pollen on their backs. When the anthers have discharged their pollen, the stamens assume once more the erect position. The style arches so as to bring the stigma forward and downward, while the lobes of the stigma spread apart and become exposed to pollen brought by bees from freshly opened flowers.

In some states of higher rainfall than ours, the running rhizomes of this species can make it a troublesome weed. In all places of its occurrence, however, it is a valuable honey plant.

NOTE: Plants of this species are often confused with *Stachys palustris*.

SOLANÀCEAE   Nightshade Family

Herbaceous annuals or perennials, and, in warm or tropical climates, shrubs, woody vines, and trees. Leaves alternate, mostly simple, sessile or petiolate, estipulate.

462. *Scutellaria resinosa*

463. *Stachys tenuifolia*

464. *Teucrium canadense*

NIGHTSHADE FAMILY

327

465. *Chamaesaracha conioides*

Flowers bisexual, regular or slightly irregular, borne singly or in clusters. Calyx usually 5-lobed. Corolla of 5 petals fused part or all of their length. Stamens usually 5, inserted on the corolla tube, the anthers dehiscing via longitudinal slits or apical pores. Pistil 1; ovary superior, 2-celled (rarely 3- to 5-celled); style 1; stigma 2-lobed. Fruit a many-seeded berry, sometimes surrounded by the enlarged and persistent calyx, or a capsule.

A family of about 90 genera and more than 2,000 species, primarily of the American tropics. Economically important members include the tomato (*Lycopersicon esculentum* Mill. varieties), potato (*Solanum tuberosum* L.), eggplant (*Solanum melongena* L.), husk tomato (*Physalis* spp.), the red and green peppers (*Capsicum frutescens* L. varieties), tobacco (*Nicotiana tabacum* L.), drug plants such as belladonna and atropine (*Atropa belladonna* L.), and ornamentals such as *Petunia, Datura,* and *Salpiglossis.*

Kansas has 5 genera with species either native or widely naturalized, and a few others which are introduced but rarely reported.

Key to Genera:
1. Corolla funnelform, 4–10 cm long; fruit a spiny capsule .... *Datura*
1. Corolla funnelform to wheel-shaped, usually 1 cm long or less; fruit a berry (tightly enclosed by the spiny calyx in *Solanum rostratum* which has yellow flowers) 2
  2. Calyx usually not greatly enlarged in fruit, spiny if investing the fruit; anthers rather large and conspicuous, connivent, opening by terminal pores .... *Solanum*
  2. Calyx enlarging in fruit to invest the berry entirely or nearly so; anthers not connivent, opening by longitudinal slits 3
    3. Calyx closely investing the berry, not angled; corolla with hairy pads alternating with the filaments .... *Chamaesaracha*
    3. Calyx becoming much enlarged and inflated, permanently and entirely enclosing the berry; corolla (except in *P. lobata*) lacking hairy pads between the filaments ...... *Physalis*

CHAMAESARACHA Gray   Ground saracha, false nightshade

NL., from Gr. *chamai,* on the ground, + *Saracha,* a related tropical American genus named after Isador Saracha, a Spanish Benedictine monk and botanist.

Perennial herbs with rhizomes. Leaves simple, the margins entire or pinnatifid, the petioles winged. Flowers regular, axillary, borne singly or 1–5 together. Calyx bell-shaped, 5-lobed, thin, neither ribbed nor angled, persistent and enlarging somewhat in fruit, enclosing the berry snugly, but open at the top, not longer than the berry, and not inflated. Corolla wheel-shaped, 5-lobed, white or yellowish, often tinged with purple on the outside, the throat with white hairy appendages. Stamens 5, the anthers dehiscing by longitudinal slits. Fruit a many-seeded berry. Seeds reniform, more or less flattened, the surface rough-reticulate.

One species in Kansas.

C. conioides (Moric. *ex* Dunal.) Britt.   Ground saracha

NL., from Gr. *konia,* dust, + *-oides,* similar—in allusion to the fine pubescence covering the plant.
Whitish, sometimes purple-tinged; late April–early September
Dry, sandy soil of prairie hillsides, roadsides, and waste ground
West third, primarily; from Logan and Ellis counties south

Low plants with ascending or prostrate stems 8–20 cm long, the herbage densely covered with both glandular and spreading nonglandular hairs. Leaves oblanceolate and nearly entire below to elliptic or broadly lanceolate and toothed or pinnatifid above, 1–6 cm long and 5–20 mm wide. Calyx 4–5 mm long. Corolla whitish, sometimes purple-tinged, 7–9 mm long and 1–1.5 cm across. Fruit and enclosing calyx about 7 mm long, the pedicel much-elongated and somewhat sinuous. Seeds about 2 mm long, the surface so coarsely alveolate as to resemble a sponge.

DATURA L.   Datura

Hindu *dhatura,* name used by the Hindus for their *Datura metel.*

Poisonous annual or perennial herbs, or (in the tropics) trees or shrubs. Leaves large, simple, entire to toothed or lobed. Flowers large and showy, borne singly. Calyx cylindric or angled, generally abscissing horizontally above the base, leaving a flaring collar under the fruit. Stamens 5, included within the corolla tube. Ovary 2-celled or sometimes falsely 4-celled; stigma 2-lobed. Fruit a more or less prickly capsule, 4-celled below, 2-celled at the top, splitting irregularly or separating along 4 sutures. Seeds flat.

Two species in Kansas. *D. quercifolia* H. B. K. is known from Morton and Meade counties only. It is a native North American plant with leaves usually pinnately lobed, purplish flowers 4–7 cm long, and fruits with stout, flattened spines unequal in length, the longer spines more than 1 cm long.

D. stramònium L. — Jimsonweed

An old generic name, of unknown origin, used by the Venetians about the middle of the 16th century for *D. metel.* "Jimsonweed" is a corruption of "Jamestown-weed," the original common name.

White, sometimes tinged with purple; June–mid-October
Cultivated land, waste ground, hog-lots, barnyards
East three-fifths and the southwest corner; apparently intr. from Asia, now nearly cosmopolitan

466. *Datura stramonium*

Smooth, dark-green annual with erect, bifurcating stems 0.3–1 m tall. Leaves rank-smelling, the blades downward-inclined, ovate to lance-ovate, 6–25 cm long and 2.5–18 cm wide, the margins very coarsely and somewhat irregularly toothed; petioles 2.5–8 (–10) cm long. Calyx tubular, 3.5–5 cm long, the teeth lance-acuminate to triangular-acuminate. Corolla white, sometimes tinged with purple, 5-lobed, 7–10 cm long. Ovary ovoid, 2-celled at the apex and imperfectly 4-celled below; style slender, carrying the 2-lobed stigma to the height of the anthers. Capsule 3.5–5 cm long, erect, subtended by the collarlike base of the calyx, covered with many, relatively slender, round spines all about 1 cm long or less. Seeds dark brown, somewhat flattened, coarsely rugose, minutely pitted, about 3.5 mm long.

Individuals with purple stems and violet flowers are sometimes classed as *D. tatula* or *D. stramonium* var. *tatula* (Persian, from *tat*, to prick—in reference to the prickly fruit).

The flowers are adapted to cross-pollination by night-flying hawk moths especially, though hummingbirds and long-tongued bees may participate. The flowers open with a sweet narcotic odor in the evening, when nectar is held in the corolla tube to an appreciable distance up from its base. Standing in a patch of Jimsonweeds around 8 or 9 o'clock of an evening in June, one is greeted by the sight of hawk moths eagerly darting from flower to flower, never alighting but pausing in flight momentarily before each flower while sucking up through their long proboscises the nectar held in the corolla tube. As round and round they go, the moths are inevitably effecting cross-pollination.

In spite of the fact that the plants are unpleasant in odor and taste, poisoning of humans from ingesting the plant itself or decoctions made therefrom is not uncommon. Four to five grams of leaf or seed may be fatal to a child. Other cases of poisoning have resulted from overdoses of medication containing the plant or its alkaloids. Livestock apparently will not eat the plants unless forced to by lack of other forage. In some individuals, contact with flowers or foliage results in a dermatitis. For an extensive discussion of *Datura* poisoning in humans and animals see Kingsbury (21).

The same compounds which give the plant its toxic properties—the alkaloids hyocyamine, atropine, and hyoscine (scopalomine)—account for the use of this and other species of *Datura* in numerous medicinal and narcotic preparations. In Shakespeare's England the plants seem to have been used chiefly in the preparation of an unguent. About its "vertues" we find set down in Gerard's herbal (13) only this: "The juice of Thorn-apples boiled with hogs grease in the form of an unguent or salve, cures all inflammations whatsoever, all manner of burnings or scaldings, as well of fire, water, boiling lead, gunpowder, as that which comes by lightning, and that in very short time, as my selfe have found by my daily practice, to my great credit and profit. The first experience came from Colchester, where Mistresse Lobel a merchants wife there being most grievously burned by lightning and not finding ease or cure in any other thing, by this found helpe and was perfectly cured when all hope was past, by the report of Mr. William Ram publique Notarie of the said towne."

Among American Indians, several tribes used *Datura* as painkillers during surgery, while setting broken bones, or for wounds, ear infections, etc. The Zunis, Mohaves, Paiutes, and probably others as well, made use of the hallucinogenic properties of the plant, and its use in treating asthma was common to whites and some Indian groups. The leaves and flowering tops were listed officially in the *U.S. Pharmacopoeia* from 1820 to 1950 and in the *National Formulary* from 1950 to 1965. The seeds and roots were official in the *USP* from 1820 to 1905 and from 1842 to 1863, respectively. For an extensive survey of the literature of *stramonium*, as the drug is sometimes called, as well as numerous references to original early writings dealing with these plants and their uses, see Vogel (45) and Weiner (34). Because of the extreme potency of the *Datura* alkaloids, experimentation with these plants is definitely NOT RECOMMENDED.

See plate *89.*

PHÝSALIS L. — Groundcherry

NL., from Gr. *physallis*, for a plant with a bladdery husk.

Annual or perennial herbs with branched stems. Leaves alternate (sometimes a few internodes so short as to make some of the leaves appear to be opposite), entire or toothed, petioled. Flowers borne singly in the leaf axils, the peduncles usually slender and arched downward. Calyx bell-shaped, 5-lobed or -toothed. Corolla bell-shaped or funnelform (wheel-shaped in *C. lobata*), yellowish to blue or whitish, sometimes with a darker center; corolla tube more or less hairy, and sometimes glandular, at the base of the stamen filaments. Stamens separate, the anthers opening by longitudinal slits. Berry many-seeded, completely enclosed by the enlarged, persistent calyx which becomes bladderlike and 5-angled or prominently 10-ribbed. Seeds reniform, somewhat flattened, finely pitted.

Nine species (including *P. ixocarpa* Brot., a cultivar) reported for Kansas. This is a relatively difficult group to work with, partially because of the extreme variation in vegetative parts which occurs within many of the species.

Key to Species (67):

1. Corolla light purple, wheel-shaped; herbage with few to many inflated, crystal-like hairs sometimes giving the plant a scurfy appearance .... *P. lobata*
1. Corolla light yellow, with or without dark spots toward the throat, funnelform, the limb usually not reflexed even when the flower is fully opened; herbage glabrous or variously hairy, but lacking inflated hairs as described above 2
  2. Plants with jointed, 2- or 3-branched hairs .... *P. pumila*
  2. Plants nearly glabrous or variously hairy, but branched hairs (if present) very small, inconspicuous, and much less numerous than the short unbranched hairs mixed with them 3
    3. Annuals; anthers (0.5–) 1–2.3 (–2.8) mm long .... *P. angulata*
    3. Perennials; anthers 3–5 mm long (1.5–3 mm in our plants of *P. hederaefolia*) 4
      4. Flowering peduncles usually 3–8 mm long; corolla limb often reflexed when fully opened .... *P. hederaefolia*
      4. Flowering peduncles usually 10–15 mm long; corolla limb not usually reflexed when fully opened 5
        5. Hairs long, spreading, jointed .... *P. heterophylla*
        5. Hairs various, but not as above .... *P. virginiana*

P. angulàta L.

L. *angulatus*, with corners, from *angulus*, corner—in reference to the angled, bladderlike pod formed by the persistent, enlarged calyx.
Light yellow; mid-June–mid-October
Low, moist soil of woods, lake margins, river valleys, fields, and roadside ditches
Scattered in the east half and southwest fourth

Erect annual, glabrous or nearly so, 1.5–9 dm tall. Leaf blades lanceolate to elliptic, 2.5–12 cm long and 1–5 cm wide, the apex and base acute or acuminate, the margins usually coarsely toothed, less frequently sinuate or entire. Calyx and young herbage usually with a few minute, appressed nonglandular hairs. Flowering peduncles threadlike, 1.5–4 cm long. Corolla light yellow, not dark spotted, 4–10 mm long. Anthers bluish, 1–2.3 mm long. Fruiting calyx 2–3 cm long.

P. pubescens L., also an annual, is known from scattered sites in the east half. It has ovate or broadly elliptic leaves, and the herbage is covered with long, spreading hairs, some of which are glandular.

P. hederaefòlia Gray

NL., from the genus *Hedera*, English ivy, + L. *folium*, leaf—the leaf shape faintly resembling that of English ivy.
Light yellow to yellowish-green; late May–August
Prairie, roadsides, flood plains, usually in sandy soil
West two-thirds

Glandular-hairy perennial with erect to spreading stems 1–3 dm tall. Leaves petioled, the blades mostly cordate to deltoid or broadly ovate (occasionally nearly circular), 1.5–4 (–7) cm long and 1–3 (–5) cm wide, the apex acute or rounded, the base cordate to truncate or acute, the margins coarsely toothed to nearly entire. Herbage covered with short glandular hairs with relatively few longer, jointed nonglandular hairs intermixed. Flowering peduncles relatively stout (at least not threadlike), 3–8 mm long. Corolla light yellow to yellowish-green, usually darker toward the throat, funnelform with the limb usually reflexed when fully open, 10–17 mm long and about 15 mm across when fully open. Anthers yellow, 1.5–3 mm long. Fruiting calyx 2–3 cm long.

467. *Physalis angulata*

P. heterophýlla Nees.

Clammy groundcherry

NL., from Gr. *heteros*, different, + *phyllon*, leaf—in reference to the fact that the leaf margins are variously entire, sinuate, or sinuate-toothed.
Light yellow; mid-May–mid-September
Open or disturbed habitats such as roadsides, rocky prairie, overgrazed pastures, waste ground
East two-thirds

Erect, moderately to densely hairy perennial, 1.5–9 dm tall, arising from a deep-seated rhizome. Leaves distinctly petioled, the blades mostly cordate to ovate (the lowest few often circular), 2–10 cm long and (1–) 3–7.5 cm wide, the apex usually acute or acuminate, the base cordate to truncate or acute, often oblique, the margins coarsely toothed to merely sinuate. Herbage in ours covered with long, spreading, jointed glandular hairs, usually with shorter hairs intermixed. Flowering peduncles relatively stout (at least not threadlike), 8–20 mm long. Corolla light yellow with brownish or bluish spots toward the throat, 10–18 mm long. Anthers usually yellow, 3–4.5 mm long. Fruiting calyx 2.5–3 cm long, the ripe berry yellow.

468. *Physalis heterophylla*

P. lobàta Torr.

Plains chinese-lantern

NL., *lobatus*, provided with lobes, from L. *lobus*, Gr. *lobos*, a rounded projection—in reference to the leaf margins.
Light purple; late April–mid-October
Dry, sandy soil of roadsides, fields, prairie
West half, primarily; more common southward

Low, minutely scurfy perennial, increasing by deep-seated rhizomes which give off prostrate, much-branched, very leafy stems, forming a compact ground cover. Leaves elliptic to lance-elliptic, 2–8.5 cm long (including the winged petiole) and 0.5–3 cm wide, the apex acute to rounded, the base attenuate, the margins coarsely toothed to wavy or nearly entire. Herbage (especially the calyces and young leaves) with minute inflated hairs somewhat crystalline in appearance which flatten on drying. Flowers on threadlike pedicels 1–3 cm long. Calyx lobes lance-acuminate. Corolla light purple, wheel-shaped, 1.5–2.5 cm across. Stamens alternating with 5 hairy pads. Anthers yellow, 1.5–2 mm long. Fruiting calyx 5-angled, 1.5–2 cm long.

This plant has been listed as *Quincula lobata* (Torr.) Raf. in earlier manuals.

After seeing this brave little plant richly blooming in the dry, packed soil of roadsides and in other difficult situations of our western High Plains, and, after noting that it actually increased in frequency during the drought of the thirties, one must approve the statement in *Hortus* that it is suitable for dry, sunny places in the rock garden.

469. *Physalis lobata*

P. pùmila Nutt.

Prairie groundcherry

L. *pumilus*, dwarfish.
Yellow; late-May–September
Rocky soil of prairies, roadsides, and pastures
East two-fifths, primarily

Erect, more or less hairy perennial, 1.5–4.5 dm tall, with a horizontal rhizome. Leaves petioled, the blades mostly ovate to lanceolate or elliptic, 3–9 cm long and 1–4.5 cm wide, the apex usually acute, the base acuminate or attenuate, the margins entire or wavy. Pubescence of jointed, mostly 2- or 3-branched hairs, those on the stem spreading and 1–2 mm long. Flowering pedicels threadlike, 1.5–3 cm long. Corolla yellow, funnelform, 1–2 cm long and mostly 1.5–2 cm across. Anthers yellow, 2.5–3 mm long. Fruiting calyx 3–4 cm long and 1.5–2 cm across.

P. viscosa L. is known from a few scattered sites in southern counties. It has herbage covered with short stellate or branched hairs, and the stem does not appear bristly.

P. virginiàna Miller

Virginia groundcherry

"Virginian."
Light yellow; June–September
Roadsides, waste ground, river banks, pastures, prairie, less frequently in wooded areas
Throughout

Erect perennial, 3–6 dm tall, with a deep-seated rhizome. Leaf blades lanceolate to ovate or elliptic, mostly 4–9 cm long and 0.7–4.5 cm wide, the apex acute or slightly rounded, the base acute to acuminate, often oblique, the margins entire to sinuate or coarsely toothed. Herbage nearly glabrous to sparsely and minutely hairy. Flowers on threadlike pedicels 5–15 mm long. Corolla light yellow, dark spotted toward the throat,

470. *Physalis virginiana*

NIGHTSHADE FAMILY

331

funnelform, the limb not reflexed, 15–25 mm long. Anthers yellow, sometimes tinged with blue, about 3 mm long. Fruiting calyx 2.5–3.5 cm long, the ripe berry reddish.

SOLÁNUM L. Nightshade

Name used by Pliny for a member of the Nightshade Family, supposedly *Solanum nigrum*.

Annual or perennial herbs (ours), shrubs, or small trees, rarely vines, often prickly. Leaves petioled, entire to pinnatifid, alternate but sometimes accompanied by smaller lateral leaves (at points where lateral branches will develop) and therefore appearing to be opposite. Flowers in clusters, the peduncles arising in the internodes. Calyx bell-shaped to wheel-shaped, 5-lobed. Corolla wheel-shaped, more or less 5-lobed, the margins often reflexed. Stamen filaments short, the anthers rather large and conspicuous, connivent or united around the style, opening by means of apical pores (rarely by a longitudinal slit). Ovary 2-celled. Fruit a many-seeded berry subtended by and sometimes partially or completely enclosed by the persistent (but not bladdery) calyx.

Nine species reported for Kansas, several of which are introduced and probably have not persisted. The most commonly encountered species are treated here.

Many species of *Solanum* are known to be toxic, either when fresh or when dried. The poisonous principles are a variety of related glycoalkaloids, the concentration of which apparently varies considerably, depending on such factors as the part of the plant, the age of the plant, and the environment. Symptoms of *Solanum* poisoning include gastrointestinal irritation (minor to quite severe) and nervous effects such as stupefaction and loss of sensations in humans, and apathy, drowsiness, salivation, dyspnea, trembling, progressive weakness or paralysis, prostration and unconsciousness in animals (21). Poisoning may or may not be fatal.

Key to Species (68):

1. Flowers yellow; plants heavily armed with yellow prickles; fruits completely enclosed by the prickly, enlarged calyx .... *S. rostratum*
1. Flowers white to purple; plants moderately prickly to not at all, fruits not enclosed by the calyx 2
  2. Hairs of the stems and lower leaf surface simple or lacking; plants without spines or prickles 3
    3. Leaves entire to sinuate or coarsely toothed, 2–15 cm long; throughout Kansas .... *S. americanum*
    3. Leaves deeply pinnatifid, 1.5–3.5 cm long; west half primarily .... *S. triflorum*
  2. Hairs of the stems and lower leaf surface all or mostly stellate; plants usually with prickles 4
    4. Herbage silvery-canescent because of the dense scurflike pubescence, having a velvety appearance .... *S. elaeagnifolium*
    4. Stems and lower leaf surfaces not as above, the hairs typically yellowish and not so dense .... *S. carolinense*

S. americànum Mill.

"American."
White; mid-June–early October
Roadsides, pastures, river banks, prairie hillsides and ravines, and open woods
Throughout

Erect, branched annual, 1.7–6 dm tall. Herbage dark green, sparsely to moderately hairy with minute, curved, simple hairs. Leaf blades ovate to ovate-lanceolate, 2–15 cm long and 1–9 cm wide, the margins entire to sinuate or coarsely toothed. Calyx about 1.5 mm long, the lobes acute or rounded. Corolla white, 5–7 mm long and 8–10 (–15) mm across. Berries shiny black when ripe, 5–9 mm across, on drooping pedicels. Seeds yellowish, flattened, the body nearly round with a blunt beak at 1 end, about 1.5–1.8 mm long, the surface minutely reticulate (use high magnification).

This species was included in earlier Kansas works as *S. nigrum* L., which is actually a closely related European species introduced in eastern North America but not occurring in Kansas. A number of cases of livestock poisoning are reported for *S. nigrum* in the United States, but some of these perhaps are attributable to *S. americanum*. A cultivated plant known as wonderberry or sunberry has been considered a variety of *S. nigrum*, but recently authorities have given it the rank of a separate species, *S. intrusum* Soria, since it does not interbreed with wild plants of *S. nigrum*. The cultivar is apparently "dependably nontoxic" (21).

The fruit of horsenettle was official in the *National Formulary* from 1916 to 1936 as a sedative and antispasmodic. Krochmal and Krochmal (30) list numerous folk uses of various parts of the plant. The berries have been used to treat tetanus, to promote

471. *Solanum americanum*

472. *Solanum americanum* fruiting

urination, to relieve pain, and to ease nervous tension. The entire plant has been utilized in the preparation of a medication for epilepsy, asthma, bronchitis, and convulsions; and the root is believed to contain a sedative.

S. carolinénse L. — Carolina horsenettle

"Of Carolina."
Pale violet to white; June–early October
Open, disturbed habitats such as roadsides, pastures, flood plains, and waste ground
East half

Slender, branching, rather ragged perennial, 3–6 dm tall, with deep roots and slender rhizomes. Leaf blades ovate to elliptic, 3.5–17 cm long and 2–9 cm wide, usually with 1–3 pairs of coarse teeth or lobes, both surfaces covered with stellate hairs, those of the lower surface with 4–8 rays. Calyx bell-shaped, 5–7 mm long, the lobes lance-attenuate or -acuminate. Corolla pale violet to white, 1.5–2 cm long and 2–3 cm across. Berries yellowish, 1–2 cm across, on drooping pedicels. Seeds yellowish, lustrous, ovate to nearly circular, flattened, 1.8–2.8 mm long.

While the flowers and fruits are considered sightly by some people, the plant can be a troublesome weed in fields and flower beds, because the perennial underground parts persist in sending up new shoots after the tops have been cut away. Furthermore, fruits of this plant ingested by a six-year-old child in Pennsylvania were apparently responsible for his death in 1963, and the fruits have caused poisoning in cattle, sheep, and possibly deer (21). Since this species is so common and seems so frequently to attract the attention of small children, parents would be wise to learn to recognize it and to eradicate it from play areas and/or teach children to leave it alone, the former method probably being the more reliable.

473. *Solanum carolinense*

S. elaeagnifòlium Cav. — Silverleaf nightshade

NL., from the genus *Elaeagnus* + L. *folium*, leaf—because of the similarity between the silvery leaves of this *Solanum* and those of such kinds of *Elaeagnus* as the Russian olive and silverberry.
Purplish-blue; mid-May–early September
Roadsides, pastures, waste ground, rocky prairie
South two-thirds and Republic County

Erect, pale green perennial, 2–6 dm tall, with a long, stout taproot. Stems, lower leaf surface, and calyces covered with a silvery stellate scurf, frequently with some sharp prickles as well. Leaf blades lanceolate to oblong, 1–11 cm long and 0.5–3 cm wide, the apex and base acute or rounded, the margins entire to wavy or coarsely toothed. Calyx bell-shaped, 7–12 mm long, the lobes quite slender. Corolla purplish-blue, 1.3–2.2 cm long and 2–3 cm across. Berry yellowish or eventually black, up to about 13 mm across. Seeds brownish, lustrous, ovate to nearly circular, flattened, 2.5–3.5 mm long.

At the height of its bloom, the rare color combinations of this species of nightshade strike the eye: silvery foliage, purple corollas, golden anthers, and touches of green, purple, yellow, and black of the pendant fruits. It is not surprising, therefore, to find that *Hortus* includes this species among those that have been grown in the flower gardens of the United States and Canada.

The stellate nature of the hairs may be easily viewed with a 10X hand lens by scraping off some of them and placing them on a dark background.

Under certain conditions, as little as 0.1 percent of an animal's body weight of this species may be toxic (21).

474. *Solanum elaeagnifolium*

S. rostràtum Dunal. — Buffalo-bur nightshade

L. *rostratus*, having a beak—the lowest of the 5 anthers being elongated beyond the others, curved and beaklike.
Yellow; May–mid-October
Roadsides, waste ground, fields, overgrazed pastures
Throughout

Aggressive, drought-resistant annual, 1–5 dm tall, with erect or ascending branches, abundantly beset with formidable yellow prickles and minute stellate hairs. Leaf blades mostly ovate to broadly elliptic in general outline, 2–13 cm long and 1.5–8.5 cm wide, deeply pinnatifid to bipinnatifid, the lobes and sinuses rounded. Calyx 5–11 mm long, the lobes lance-linear to linear-attenuate, the tube prickly. Corolla yellow, 1.5–2.5 cm long and about 2.5 cm across. Berries completely enclosed by the enlarged calyx. Seeds brown, about 2–2.5 mm long, the surface wrinkled and minutely reticulate.

475. *Solanum rostratum*

NIGHTSHADE FAMILY

By November the plant has ripened and died, the coalesced calyx lobes have opened up, exposing the berry, now dried out to a thin baglike skin filled with tiny black seeds. The bag now is ruptured and the seeds are scattered. Sometimes the main stem breaks off near the ground, and the plant in tumbleweed fashion rolls before the wind, scattering the seeds as it goes.

The plant is to be admired for its ability to defend itself, even in the packed soil of cow-lots. However, it is a weed in the bad sense of that word and, as the original food plant of the Colorado potato beetle, it continues to foster that pest. Unfortunately, we find it in every county in the state.

Although the spininess of this plant would seem to make it unattractive to livestock, cases of poisoning in hogs are known from their having eaten either the above-ground parts or the roots.

S. triflòrum Nutt.

NL., from L. *tri-*, 3-, + *flos*, *floris*, flower, i.e., "3-flowered"—the blossoms often being borne in clusters of 3.
White; June–September
Open habitats such as roadsides and rocky prairie banks and breaks
West half, primarily

Erect to decumbent, branched annual, 1–4 dm tall. Stems and lower leaf surface with simple hairs. Leaf blades ovate to lanceolate in general outline, 1.5–3.5 cm long, deeply pinnatifid, the sinuses rounded but the lobes sharply pointed. Calyx 3–4 mm long, the lobes acute. Corolla white, 6–7 mm long and about 1 cm across. Berries green, about 1–1.5 cm across. Seeds yellowish, flattened, more or less ovate, 2–2.5 mm long, the surface minutely reticulate (use high magnification).

SCROPHULARIÀCEAE   Figwort Family

Mostly herbs, sometimes shrubs, trees, or woody vines. Leaves alternate or opposite (whorled in *Veronicastrum*), simple, sessile or petioled, estipulate. Flowers bisexual, irregular (sometimes only slightly so, as in *Verbascum*), frequently 2-lipped, borne singly in the leaf axils or in various types of clusters. Calyx 4- to 5-toothed or -lobed, sometimes the sepals nearly or quite separate. Corolla of 5 fused petals, although the number is sometimes obscured. Stamens 2 or 4, rarely 5, inserted on the corolla tube. Nectiferous disk often present at base of pistil. Pistil 1; ovary superior, 2-celled (rarely 1-celled); style 1; stigmas 2, or 1 with 2 lobes. Fruit usually a many-seeded capsule.

This large family is comprised of about 220 genera and 3,000 species, world-wide in distribution. Cultivated members include a number of ornamentals such as slipperwort (*Calceolaria* spp.), speedwell (*Veronica* spp.), beardtongue (*Penstemon* spp.), turtlehead (*Chelone* spp.), lousewort (*Pedicularis canadensis* L.), foxglove (*Digitalis* spp., especially *D. purpurea* L.), snapdragon (*Antirrhinum majus* L.), and toadflax or butter-and-eggs (*Linaria vulgaris* Mill.). Foxglove is also the source of the drug digitalis.

This family is represented in Kansas by 21 genera and 57 species.

Key to Genera (3):
1. Foliage leaves (not bracteal leaves subtending flowers) alternate 2
  2. Stamens 5 .... *Verbascum*
  2. Stamens 4 3
    3. Corolla with a distinct spur at base protruding between the 2 lower calyx lobes 4
      4. Flowers in terminal racemes .... *Linaria*
      4. Flowers solitary in the axils of the leaves .... *Chaenorrhinum*
    3. Corolla not spurred at base 5
      5. Corolla twice as long as the calyx or longer .... *Pedicularis*
      5. Corolla much less than twice as long as the calyx .... *Castilleja*
1. Foliage leaves opposite or whorled 6
  6. Corolla 5 mm long or less (these plants are also keyed under the next alternative; introduced here because the size of the flower makes observation difficult) 7
    7. Calyx lobes 4 .... *Veronica*
    7. Calyx lobes 5 8
      8. Leaves deeply pinnatifid .... *Leucospora*
      8. Leaves entire or merely toothed .... *Bacopa*
  6. Corolla 5 mm long or longer (including the small-flowered plants keyed out above) 9
    9. Stamens 2 10
      10. Calyx 4-lobed or sepals 4 11

11. Corolla lobes distinctly longer than the very short corolla tube .... *Veronica*
11. Corolla lobes much shorter than the corolla tube; leaves always whorled ....
.... *Veronicastrum*
10. Calyx 5-lobed, or sepals 5 12
    12. Fertile anthers with pollen sacs divergent; sterile filaments united with the corolla throat, forming 2 ridges terminated by projecting knobs; pedicels never bracted; ours with fruits narrowly ellipsoid .... *Lindernia*
    12. Fertile anthers with pollen sacs parallel; sterile stamens absent or minute, the corolla throat rounded or flattened but not ridged; pedicels with minute bractlets; ours with fruits globose .... *Gratiola*
9. Stamens 4 (rarely 2 or 3 in *Bacopa*) 13
13. Flowers greenish or brownish, numerous in a branched terminal cluster ....
.... *Scrophularia*
13. Not as above 14
    14. Flowers wholly or predominately yellow 15
        15. Corolla strongly 2-lipped, the lower lip arched upward or with longitudinal ridges which partly or wholly close the throat .... *Mimulus*
        15. Corolla weakly 2-lipped, the 5 lobes nearly alike .... *Seymeria*
    14. Flowers white to pink, purple, red, or blue 16
        16. Lower lip of the corolla with the median lobe folded downward between the lateral lobes, forming a pouch enclosing the stamens .... *Collinsia*
        16. Lower lip of the corolla flat or concave, or arched upward to form a palate 17
            17. Bracteal leaves greatly and abruptly reduced in comparison with the foliage leaves below them, producing a definite terminal inflorescence which may be a spike, a raceme, or a panicle 18
                18. Corolla tubular or bell-shaped, distinctly 2-lipped .... *Panstemon*
                18. Corolla salverform, nearly regular; inflorescence a spike ...... *Buchnera*
            17. Bracteal leaves gradually reduced, the flowers therefore appearing axillary or in a lateral raceme 19
                19. Sepals separate nearly or quite to the base; corolla 12 mm long or shorter 20
                    20. Corolla nearly regular, the lobes nearly equal, about as long as the tube .... *Bacopa*
                    20. Corolla distinctly irregular and 2-lipped, its lobes shorter than the tube .... *Leucospora*
                19. Sepals united into a distinct calyx tube 21
                    21. Corolla nearly or quite enclosed by the elevated palate of the lower lip .... *Mimulus*
                    21. Corolla open, the lower lip without a palate 22
                      22. Corolla yellow .... *Aureolaria*
                      22. Corolla pink or purplish 23
                        23. Leaf blades usually with a pair of sharp basal lobes or finely pinnatifid; calyx lobes longer than the tube; reticulations of outer seed coat raised .... *Tomanthera*
                        23. Leaf blades entire; calyx lobes usually shorter than the tube; reticulations of outer seed coat not raised .... *Agalinis*

AGALÌNIS Raf.   **Gerardia**

NL., from Gr. *aga-*, very much, very, + *linum*, flax—interpreted by Fernald (9) to mean "remarkable flax."

Annual or perennial herbs with slender stems. Leaves opposite (sometimes alternate on the branches), linear, sessile, entire. Flowers mostly pink to purplish, showy, borne singly or in few-flowered racemes in the upper leaf axils or appearing terminal. Calyx regular, bell-shaped or hemispheric, 5-toothed or -lobed. Corolla irregular, 2-lipped, the upper 2 lobes arched and spreading or somewhat recurved, the lower 3 spreading. Stamens 4, in 2 pairs, included in the corolla tube, the filaments more or less hairy. Fruit a loculicidal capsule.

Seven species in Kansas. This is a relatively difficult group to work with, since a number of the species are remarkably similar and are distinguished from one another on the basis of rather fine characters. Only our 3 most common species are included here.

All species of this genus appear to be facultative parasites on the roots of other plants.

Earlier manuals placed these plants, along with the species of *Aureolaria* and *Tomanthera*, in the genus *Gerardia*.

Key to Species (51):
1. Capsules oblong .... *A. aspera*
1. Capsules globose 2

2. Corolla with lobes all spreading, pubescent within at base of upper lobes, externally more or less pubescent .... *A. skinneriana*
2. Corolla with the upper lobes projecting over stamens and style, the lower spreading, glabrous or with slight pubescence within at the base of the upper lobes .... *A. tenuifolia*

A. áspera (Benth.) Britt.

L. *asper*, rough—descriptive of the leaves and stems.
Purple or purplish-pink; late August–early October
Prairie
East two-thirds

Erect annual, 2–7 dm tall, with many ascending branches. Leaves linear, 1.5–4.7 cm long and 0.5–2 mm wide, roughened with short, stiff hairs. Flowers borne singly in the upper leaf axils on relatively stout (at least not threadlike), strongly ascending pedicels 5–10 mm long. Calyx bell-shaped, 4–8 mm long, the lobes triangular-acuminate, shorter than the tube. Corolla purple or purplish-pink, 12–24 mm long. Capsules broadly ellipsoid, 6–10 mm long. Seeds black, irregular in shape, 1.2–1.5 mm long, the outer seed coat loose, transparent except for a black reticulate ornamentation.

Pollination is effected by long-tongued bees and by butterflies.

A. skinneriàna (Wood) Britt.

Named after Dr. A. G. Skinner, owner of the land on which Wood discovered the species in 1847.
Pink; September–mid-October
Open woods, usually in shallow, rocky soil over sandstone
East third, south of the Kansas River

Delicate, light green annual, 2–6 dm tall, roughened on the angles, usually with a few stiffly ascending or spreading branches. Leaves linear, 1–3.5 cm long and 0.5–1.2 mm wide, rough on the upper surface. Flowers several per branch, not appearing as if terminal. Calyx bell-shaped in flower, 4–5 mm long, the teeth triangular or triangular-acuminate, much shorter than the tube. Corolla pink, 1–1.5 cm long. Capsules globose, 4–5 mm long. Seeds brown, 0.8–1 mm long, reticulate.

A. tenuifòlia (Vahl.) Raf.

NL., from L. *tenuis*, thin + *folium*, leaf.
Pink; August–mid-October
Moist soil of low prairies, flood plains, roadside ditches, and open woods
East two-thirds, primarily; scattered records from the west third

Erect annual, 1–6 dm tall, the stems simple or with ascending branches. Stem smooth. Leaves linear, 0.8–4 cm long and 0.5–5 mm wide, roughened with short, stiff hairs. Flowers borne singly in the upper leaf axils on glabrous, widely spreading to upward-curving, threadlike pedicels 6–15 mm long. Calyx bell-shaped in flower, the tube 3–4 mm long, the teeth abruptly acuminate, 0.5–1 (–1.5) mm long. Corolla pink, 9–15 mm long. Fruit globose, 4–8 mm long. Seeds brown, irregular in shape, about 1–1.1 mm long, the outer seed coat loose, transparent except for the brown, reticulate ornamentation.

According to Pennell (51), this species is parasitic on a wide variety of hosts including *Anemone virginiana*, *Fragaria virginiana* (wild strawberry), *Solidago nemoralis*, *Achillea millifolium*, *Antennaria plantaginifolia* (pussytoes), and several grasses.

AUREOLÀRIA Raf.   **False foxglove**

NL., from L. *aureolus*, golden, glittering, splendid—in reference to the flowers.

Large, leafy-stemmed annual or perennial herbs. Leaves opposite below, becoming subopposite or alternate above, entire to twice-pinnatifid. Flowers yellow, large, showy, in terminal leafy-bracted racemes or panicles. Calyx bell-shaped, 5-lobed, the lobes entire or toothed. Corolla bell-shaped, somewhat irregular, with 5 broadly rounded, spreading lobes. Stamens 4, the lower pair longer; filaments and anther cells hairy, the latter awned at the base. Style slender; stigma 1, ovoid-capitate. Fruit an ovoid to ellipsoid, several-seeded loculicidal capsule beaked by the persistent style.

One species in Kansas.

Members of this genus are root-parasites of oaks (*Quercus* spp.).

A. grandiflòra (Benth.) Penn.

NL., from L. *grandis*, large, noble, magnificent, + *flos, floris*, flower—the yellow flowers being quite showy.
Yellow; August–mid-October

Moist, rocky woods
Cherokee County

Erect, soft-hairy, branched perennial, up to 1.2 m tall, resembling our more common *Seymeria macrophylla* but with showier flowers. Lower leaves ovate in general outline, more or less pinnatifid, the upper ones becoming gradually smaller and merely laciniate or serrate, the uppermost in the inflorescence sometimes entire. Entire plant usually turns black upon drying. Calyx 1–1.5 cm long, the lobes narrowly triangular to subulate. Corolla yellow, 3.5–4.5 cm long. Capsules 1.5–2 cm long.

Bumblebees are the most frequent insect visitors.

BACÒPA Aubl.   Water hyssop

According to Fernald (9), an aboriginal South American name.

476. *Aureolaria grandiflora*

Low, mostly succulent perennial herbs inhabiting wet places. Leaves opposite, sessile. Flowers borne singly or in pairs in the leaf axils. Calyx 5-lobed, somewhat irregular. Corolla deeply 5-lobed and only slightly irregular to definitely 2-lipped with the upper lip entire, notched, or 2-cleft and the lower lip 3-lobed. Stamens 4, included in the corolla tube. Style 2-lobed or merely enlarged at the apex; stigmas 2. Fruit a many-seeded septicidal capsule.

Two species in Kansas. *B. acuminata* (Walt.) Robins is known from one site only in southeastern Cherokee County.

B. rotundifòlia (Michx.) Wettst.

NL., from L. *rotundus*, circular, round, + *folium*, leaf.
White; late June–September
Wet soil
Scattered across the state

Plants creeping on mud or attached to the bottom and floating in shallow water. Stems hairy, 0.6–4 dm long. Main leaves broadly obovate to nearly round, palmately veined, 1.5–3.5 cm long and 1–3 cm wide, the bases clasping the stem. Calyx 3–4 mm long, the lobes ovate. Corolla white with a yellow throat, 5–7 mm long. Capsules globose or nearly so, nearly as long as the sepals. Seeds light tan, oblong to ellipsoid, about 0.5 mm long, the surface finely reticulate, the alveoli mostly square to rectangular (use high magnification and side illumination).

BUCHNÈRA L.   Bluehearts

Named after Dr. Johann Gottfried Buchner (1695–1749), German botanist.

477. *Buchnera americana*

Rough-hairy biennial or perennial herbs which blacken on drying. Leaves opposite (the uppermost sometimes alternate), sessile, usually toothed. Flowers white to blue or purple, borne in pairs in elongate terminal spikes. Calyx tubular, 5-lobed or -toothed, obscurely 5- to 10-ribbed. Corolla with a curved tube and 5 slightly unequal, spreading lobes. Stamens 4, in 2 pairs of unequal length, included in the corolla tube; anthers 1-celled. Style club-shaped. Fruit a short, 2-valved, many-seeded, loculicidal capsule.

One species in Kansas.

B. americàna L.   Bluehearts

"American."
Light purple; mid-June–October
Prairie
East fourth, mostly south of the Kansas River, and more common in the southeast

Erect, slender biennial or perennial herb, 3–8 dm tall. Main leaves inclined stiffly upward, lanceolate to elliptic or lance-linear, 3.5–10 cm long and 0.5–3 cm wide, the apex slightly rounded on the lower leaves to acute or attenuate above, the margins mostly serrate to dentate, the lower surface prominently veined. Herbage with stiff spreading or appressed hairs, those of the leaves often arising from siliceous disks. Spikes 4–15 cm long, elevated on a naked or small-bracted peduncle, each flower subtended by a small lanceolate to lance-linear bract. Calyx 5–9 mm long. Corolla light purple, the slender tube about 10 mm long, the throat hairy, the lobes 7–8 mm long. Capsules 6–10 mm long. Seeds black, about 0.6 mm long.

This species has an air of confidence and thrift, standing erect among the prairie grasses, yet occurring rarely and only singly, or too dispersed to form a society.

See plate *90*.

478. *Castilleja coccinea*

479. *Castilleja purpurea*

CASTILLÈJA L.f.   Indian paintbrush, painted-cup

Named (1781) in honor of the Spanish botanist Juan Castillejo of Cadiz.

Annual, biennial, or perennial herbs with roots partly parasitic on nearby plants. Leaves alternate, sessile, entire to toothed or pinnately lobed. Flowers yellow or reddish, in terminal spikes, each blossom subtended by a conspicuously colored, leafy bract which attributes most of the color to the inflorescence. Calyx tubular, deeply divided into 2 lateral halves, laterally flattened. Corolla 2-lipped, laterally flattened, with a slender tube; upper lip slender, keeled, arched, the lower lip much shorter than the upper, 3-lobed. Stamens 4, in 2 pairs of unequal length, ascending under the upper corolla lip; anther sacs unequal, the outer attached by the middle, the inner by its apex. Stigma 2-lobed or entire. Fruit an ovoid, 2-celled, many-seeded loculicidal capsule. Seeds with a loose, transparent outer coat with honeycombed ornamentation.

Three species in Kansas.

Key to Species:

1. Root biennial; calyx lobes expanding distally to a broadly rounded, sometimes notched apex; bracts and calyces normally red; extreme east-central and southeast Kansas .. *C. coccinea*
1. Root perennial; calyx lobes not expanding distally; bracts and calyces yellow or greenish 2
   2. Corolla 4–6 cm long, yellow, conspicuously protruding beyond the calyx; bracts and calyx green .... *C. sessiliflora*
   2. Corolla 1.8–4 cm long, greenish, about as long as the calyx; bracts and calyces yellow .... *C. purpurea* var. *citrina*

C. coccínea (L.) Spreng.   Indian paintbrush

L. *coccineus*, scarlet-colored.
Flowering spike bright red; mid-April–mid-June
Low prairies
Linn and Neosho counties, south to the state line

Erect, soft-hairy biennial with 1 to few stems 1.5–5.5 dm tall. Leaves of the basal rosette entire, oblong-elliptic to obovate, 1–3 cm long; stem leaves with the main portion of the blade linear or nearly so, parallel-veined, usually with 1–4 pairs of linear or thread-like lobes. Floral bracts 3- to 5-lobed, 1.5–3 cm long, the lobes more or less oblong, bright red, rarely yellow to cream-colored. Calyx 2–3 cm long, membranous, constricted above the maturing fruit, the lobes broader toward the apex and entire or shallowly notched. Corolla greenish-yellow, about as long as or slightly longer than the bracts. Stigma capsules 8–12 mm long. Seeds 1–1.5 mm long.

See plate *91*.

C. purpùrea (Nutt.) G. Don. var. *citrina* (Penn.) Shinners   Lemon painted-cup

L. *purpureus*, purple—descriptive of the inflorescence color of the specimen upon which the name is based. However, a great deal of variation in bract color occurs from plant to plant. In Kansas most of our plants have yellowish flowers and bracts and are placed in var. *citrina*, from L. *citrinus*, lemon-yellow, from L. *citrus*, citron tree (*Citrus medica* of Linnaeus), which bears fruit of the form and color of a lemon.
Flowering spike light yellow to greenish; late April–mid-June
Prairie, usually in sandy soil
In an area bounded by a line from Harper and Kingman counties to Rush, Scott, and Meade counties

Light-green, soft-hairy herbaceous perennial with few to several erect to ascending stems 1–3.5 dm tall. Leaf blades with the main portion oblong to narrowly elliptic or linear, 3–5 cm long and 4–8 mm wide, parallel-veined, with 1–2 pairs of linear lobes. Floral bracts 3-lobed (rarely 5-lobed), 2–3 cm long, the lateral lobes usually linear, the main lobe oblong, the tips yellow. Calyx 1.3–5 cm long, membranous, constricted above the maturing fruit, the lobes linear. Corolla light yellow or greenish, sometimes tinged with red, 3–5.5 cm long. Capsules 9–17 mm long. Seeds 1.2–2 mm long.

C. sessiliflòra Pursh   Downy painted-cup

NL., from L. *sessilis*, low, dwarf, + *flos*, *floris*, flower. In botanical terms, those flowers are sessile that are attached to the stem directly, without an intervening peduncle or pedicel.
Flowers yellow, bracts green; late April–mid-June
Rocky prairie hillsides
West two-thirds, primarily; more common northward; scattered records from the east third

Erect, soft-hairy perennial with 1 to few stems 0.9–3.5 dm tall. Leaf blades with the main portion linear or lance-linear, 2–8.5 cm long and 4–9 mm wide, parallel-veined, the lowermost entire, the uppermost with 1 pair of linear lobes. Floral bracts 3-lobed, mostly 2–4 cm long, green and leaflike. Calyx 2.5–3.5 cm long, greenish, the lobes linear-attenuate. Corolla yellow, 4–6 cm long, definitely longer than the calyx. Capsules 1–1.6 cm long. Seeds 1.7–2 mm long.

CHAENORRHÌNUM Reichenb.   Dwarf snapdragon

NL., from Gr. *chaino*, gape, + *rhis, rhinos*, nose, snout, beak—perhaps in reference to the gaping corolla throat and the spur.

Annual or perennial herbs. Leaves alternate or opposite, narrow, entire. Flowers small, borne singly in the leaf axils. Calyx somewhat irregular, of 5 separate or slightly united petals. Corolla strongly irregular and 2-lipped, spurred at the base, the lower lip with a well-developed palate, but the throat not closed. Stamens 4. Capsule globose or nearly so, oblique, each cell dehiscing through a large, irregular terminal pore.

One species in Kansas.

C. mìnus

L. *minus*, less, Gr. *minys*, little, small, short.
Pale lavender; late May–mid-October
Railroad ballast
East third, south of the Kansas River; intr. from Europe

Erect, glandular-hairy annual, 0.7–4 dm tall, eventually producing numerous ascending branches. Leaves opposite and narrowly elliptic below, alternate and linear above, 7–23 mm long and 1–3 mm wide. Sepals 3–6.5 mm long. Corolla pale lavender with a yellow throat, 4.7 mm long. Capsules 3–6 mm long. Seeds dark brown, ovoid to oblong-ovoid, 0.5–0.8 mm long, sharply longitudinally ridged.

This relatively recent introduction to the state is rapidly becoming more common along railroad embankments.

COLLÍNSIA Nutt.   Collinsia

Named after Zaccheus Collins (1764–1831), Philadelphia botanist and philanthropist, praised by Asa Gray for his accuracy in botanical work.

Slender, leafy-stemmed annual or biennial herbs. Leaves opposite or whorled, mostly sessile, narrow, entire or toothed. Flowers showy, parti-colored, white with violet or pink, solitary or clustered in the upper leaf axils. Calyx bell-shaped, deeply 5-lobed, the lower 2 lobes shorter and wider than the upper 3. Corolla 2-lipped, with a short, deflexed tube, the base conspicuously pouchlike on the upper side; the upper lip erect, 2-lobed, the lower lip deflexed, with the median lobe shorter than the lateral lobes, with a pouchlike fold enclosing the stamens. Fertile stamens 4, in 2 pairs of unequal length, declined, accompanied by a glandlike staminode near the base of the corolla. Fruit an ovoid or globose, few-seeded septicidal capsule separating into 2 2-toothed or -cleft segments.

Two species reported for Kansas. *C. verna* Nutt. is reported from only a few localities in extreme east and southeast Kansas. It inhabits moist woodlands and has the lower corolla lip bright blue, the corolla lobes notched 0.5–1 mm.

C. violàcea Nutt.   Violet collinsia

L. *violaceus*, violet-colored, from *viola*, violet.
Flower white, violet and purple; April–mid-May
Open woods, roadsides, occasionally in prairie
Southeast eighth, and Miami County

Erect, minutely hairy winter annuals, mostly 0.5–3 dm tall, with slender, simple or branched stems. Leaves opposite, entire or with a few shallow teeth, the lowermost much smaller than the others, petioled, with blades round to oblong or elliptic, main leaves mostly oblong to lance-oblong, 0.6–4 cm long and 1–13 mm wide. Calyx 4–8 mm long. Corolla 10–16 mm long, the upper lip white tipped with pale violet, the lower lip violet or purple, with stripes of white extending inward from the intervals between the lobes, the lobes notched 2–3 mm, the throat yellowish, sometimes with transverse orange lines. Capsules 4–7 mm long, partially enclosed by the persistent calyx. Seeds dark brown, broadly ellipsoid to oblong, 1.5–1.9 mm long, with a reticulate surface and a deep funicular scar.

See plate *92*.

GRATÌOLA L.   Hedge hyssop

NL., diminutive of L. *gratia*, grace, flavor, and, in the theological sense, grace of God to man—because a preparation from the Eurasian species, Gr. *officinalis*, was reputed to cure dropsy, scrofula, jaundice, and enlargement of the spleen, the active principles in the plant being 2 bitter glucosides—poisonous when taken in large doses, as is true of many other beneficial drugs such as morphine and aconite.

Small annual or perennial herbs of wet or damp places. Leaves opposite, sessile, entire or toothed, with palmate venation. Flowers small, yellow, white, or purplish, borne singly in the leaf axils. Sepals 5, united but slightly at the base, usually subtended by 2 small, sepal-like bracts. Corolla tubular or narrowly bell-shaped, somewhat irregular or 2-lipped, the upper lip shallowly notched, pubescent at the base within, the lobes of the lower lip rounded. Stamens 2, included within the corolla tube, a second pair of stamens sometimes represented by 2 scalelike or threadlike staminodes. Connective between the anther sacs in most species expanded and saucer-shaped, longer and wider than the transverse anther sacs. Fruit a globose or ovoid, many-seeded capsule which separates into 4 segments.

Two species in Kansas. *G. virginiana* L., a fleshy perennial with white flowers 8–12 mm long borne on stout pedicels 1–4 (–10) mm long and capsules 5–9 mm long, is known from scattered sites in the east half. It usually occurs in shallow water of ditches and pond margins.

G. neglécta Torr.

L. *neglectus*, to overlook, slight.
White and yellow; May–early August
Wet soil of low prairies, prairie ravines, and stream banks
East fourth, south of the Kansas River

Widely branched annual, 0.7–3.5 dm tall. Stems with minute glandular hairs. Leaves thin, mostly elliptic to narrowly oblanceolate or oblong-elliptic, 2–5 cm long and 3.5–11 mm wide, the apex acute or somewhat rounded, the margins entire to shallowly serrate. Flowers borne on threadlike pedicels 10–25 mm long (to 35 mm in fruit). Sepals at anthesis 2.5–4.5 mm long, lance-linear, subtended by a pair of inconspicuous bracts as long as to somewhat longer than the sepals. Corolla white with a yellow, pubescent throat, 7–9 mm long. Capsules broadly ovoid, 3–4 mm long, subtended by the persistent sepals and bracts. Seeds pale, about 0.5–0.6 mm long.

Compare with *Bacopa* and *Lindernia*.

LEUCÓSPORA Nutt.

NL., from Gr. *leukos*, white, + *spora*, seed—in reference to the pale seeds.

A genus of 1 species with characteristics as given below.

L. multífida (Michx.) Nutt.

L. *multifidus*, many-cleft—in reference to the leaves.
Lavender; late June–October
Moist soil of creek banks, lake and pond margins, ditches, and prairie ravines, less frequently in drier habitats
East three-fifths

Erect, much-branched, glandular-hairy annual herb, 1–2 dm tall. Leaves opposite, sessile, deeply pinnately lobed, 1–3 cm long and 10–15 mm wide, the lobes linear or oblong, usually with 1 or 2 teeth or lobes. Flowers borne singly or in pairs in the leaf axils, on threadlike pedicels. Sepals linear-attenuate, barely united at the base, about 3 mm long. Corolla pale lavender, tubular, about 4 mm long, 2-lipped, the lips short, the upper 2-lobed, the lower 3-lobed and shorter than the upper. Stamens 4, in 2 pairs, included within the corolla tube. Fruit an ovoid, septicidal, many-seeded capsule about 3 mm long. Seeds pale yellow, minute (about 0.3 mm long).

LINÀRIA Mill.   Toadflax

NL., from L. *linum*, flax, the plant resembling the flax plant in growth habit, stems, and leaves, but by no means in the shape and structure of the flowers.

Erect annual, biennial, or perennial herbs, usually glabrous. Leaves alternate or opposite, sessile, narrow. Flowers purple or yellow, showy, in terminal racemes. Calyx regular, deeply 5-parted or of 5 separate sepals. Corolla strongly 2-lipped, the tube spurred at the base, the upper lip erect, 2-lobed, the lower 3-lobed. Stamens 4, in 2 pairs of unequal length, included within the corolla tube. Stigma one. Fruit a short, 2-celled ovoid or globose capsule opening irregularly at the top.

One native species in Kansas. In addition, *L. vulgaris* Mill., butter-and-eggs, and *L. dalmatica* (L.) Mill., both yellow-flowered perennials, are found escaped from cultivation or possibly naturalized in some areas.

L. canadénsis (L.) Dum.   Oldfield toadflax

L. -ensis, suffix added to nouns of place to make adjectives—"Canadian," in this instance.
Pale blue or white; mid-April–June
Sandy soil of prairie and roadsides, occasionally in open woods
East three-fifths, primarily south of the Saline and Kansas rivers

Slender, erect annual or biennial, 2–5.5 dm tall, often with a basal rosette of short, vegetative branches. Leaves alternate, those of the flowering stems linear to narrowly elliptic, 1–3.7 cm long and 0.5–3 mm wide, those of the vegetative basal branches broadly elliptic to linear and usually much smaller. Flowers on short pedicels subtended by small bracts, in racemes 2–17 cm long. Sepals lanceolate to lance-oblong, slightly united at the base, 3–4 mm long, with membranous margins. Corolla pale blue or white, 8–16 mm long (excluding the spur). Capsules 3–4 mm long, subtended by the persistent sepals. Seeds black, less than 0.5 mm long, truncate-pyramidal, minutely tubercled.

See plate *93*.

480. *Linaria canadensis*

LINDÉRNIA All.   False pimpernel

Named for F. B. Lindern (1682–1755), botanist and physician in Strassburg, Germany.

Small, glabrous annual or biennial herbs. Leaves opposite, sessile in ours, entire or toothed, with 3–5 veins. Flowers small, white or purplish, borne singly in the leaf axils. Calyx regular, of 5 separate or slightly united sepals. Corolla 2-lipped, the upper lip erect, shallowly 2-lobed, the lower lip wider, somewhat deflexed, 3-lobed. Fertile stamens usually 2, included within the corolla tube, accompanied by 2 staminodes forming hairy ventral ridges on the throat of the corolla. Stigma 2-lobed. Fruit a many-seeded, globular to ovoid or ellipsoid septicidal capsule.

Two species in Kansas. While it is possible to find plants which key nicely to one species or the other, there are also numerous plants which appear to share characters of both species.

Key to Species:
1. Flower and fruit pedicels threadlike, most of them conspicuously surpassing the subtending leaves; seeds brownish yellow .... *L. anagallidea*
1. Flower pedicels threadlike or often somewhat thicker, most of them shorter than to slightly longer than the subtending leaves; seeds pale yellow .... *L. dubia*

481. *Lindernia anagallidea*

L. anagallídea (Michx.) Penn.

NL., from the genus *Anagallis* + Gr. *-eides*, like, i.e., "resembling *Anagallis*."
White to pale lavender; mid-July–September
Wet, sandy or muddy soil
East half

Slender, diffusely branched annuals, 0.5–4 dm tall. Leaves ovate to lanceolate or elliptic, 7–20 mm long and 2–9 mm wide, the apex acute or somewhat rounded, the margins entire or with a few shallow teeth. Flower and fruit pedicels threadlike, all much longer than the subtending leaves. Calyx 3.5–5 mm long, the lobes linear. Corolla white to pale lavender, 6–9 mm long. Capsules narrowly ellipsoid, 4–5 mm long, usually longer than the sepals, the slender pedicels often rigidly spreading or deflexed as the capsules mature. Seeds brownish-yellow, about 0.3 mm long.

L. dùbia (L.) Penn.   False pimpernel

L. dubius, uncertain, vacillating—in this species the flowering peduncles being sometimes shorter and sometimes longer than their subtending leaves.
Pale lavender; mid-June–mid-September
Moist soil or mud of pond margins, creek banks, and fields
East half

Plants 0.6–2.5 m tall, more robust than the previous species. Leaves mostly ovate, 1–3 cm long and 4–15 mm wide, the apex acute or rounded, the margins entire or with a few shallow serrations. Flower and fruit pedicels threadlike or often somewhat thicker,

482. *Lindernia dubia*

FIGWORT FAMILY

341

at least the lower ones shorter than the subtending leaves. Corolla lavender, 7–9 mm long. Capsules narrowly ellipsoid, 4–6 mm long. Seeds pale yellow, 0.3–0.4 mm long.

483. *Lindernia dubia*, close-up of flower

MÍMULUS L.   Monkeyflower

L. *mimulus*, diminutive of *mimus*, mimic or buffoon—from fancying that the face of the corolla is mimicking or grinning. The popular name may be similarly explained.

Ours perennial herbs with stolons or rhizomes. Leaves opposite, sessile or petioled. Flowers yellow, blue, red, or pink, mostly showy, solitary in the leaf axils. Calyx tubular, angled, 5-lobed or -toothed, regular or irregular. Corolla 2-lipped, the upper lip 2-lobed, erect or somewhat reflexed, the lower lip 3-lobed, spreading or deflexed. Stamens 4, in 2 pairs of unequal length, inserted near the middle of the corolla tube. Style elongate; stigmas 2, flattened, circular. Fruit a cylindric, many-seeded loculicidal capsule.

Three species in Kansas.

Key to Species:
1. Flowers yellow .... *M. glabratus*
1. Flowers blue or bluish-purple (rarely pink or white) 2
     2. Leaves petioled; flowering pedicels up to 1 cm long .... *M. alatus*
     2. Leaves sessile; flowering pedicels 1–3.5 cm long .... *M. ringens*

M. alàtus Ait.   Winged monkeyflower

L. *alatus*, winged, from *ala*, bird's wing—the stem having thin, longitudinal extensions of the angles.
Violet, with a yellow palate; late July–mid-October
Wet soil
East third

Glabrous, erect perennial herb, up to 8 dm tall, with narrowly winged stems and spreading or ascending branches. Leaf blades usually lanceolate, occasionally elliptic or nearly ovate, 3–12 cm long and 1–5.2 cm wide, the apex acuminate, the base acuminate or wedge-shaped, the margins serrate or rarely crenate; petioles 0.5–3 cm long. Flowering pedicels up to 1 cm long, ascending, much shorter than the subtending leaf. Calyx regular or nearly so, 10–15 mm long, the teeth abruptly constricted to bristlelike tips. Corolla violet with a yellow palate, 18–30 mm long. Capsules 8–12 mm long, completely enclosed by the persistent calyx. Seeds orangish, more or less oblong or ellipsoid, 0.3–0.4 mm long.

484. *Mimulus alatus*

M. glabràtus H. B. K.   Yellow monkeyflower

NL., *glabratus*, smooth, from L. *glaber*, destitute of hair.
Yellow; late April–mid-September
Shallow water of springs, marshes, creeks, and pond margins, usually in sandy soil
Scattered in the west four-fifths, primarily in the west half

Plants glabrous or nearly so, the stems erect to ascending or sometimes decumbent, 0.5–3.5 dm tall. Leaves petioled below, sessile above, the blades deltoid or nearly round below to broadly ovate, cordate, or reniform above, 1–2.6 cm long and 0.7–2.8 cm wide, the apex usually rounded, the base usually rounded or cordate, the margins shallowly toothed or nearly entire. Herbage often tinged with red. Pedicels slender, 2–4 mm long, the flowers nodding. Calyx irregular, 5–9 mm long (–12 mm in fruit). Corolla yellow, 8–18 mm long, the throat bearded. Capsules oblong, enclosed by the persistent calyx. Seeds brown, broadly elliptic to oblong, about 0.5 mm long.

485. *Mimulus glabratus*

FIGWORT FAMILY
342

M. ríngens L.   Monkeyflower

L. *ringens*, present participle of *ringi*, to open wide the mouth—in reference to the spreading apart of the 2 lips of the corolla.
Blue or bluish-purple; late June–September
Wet soil of stream banks, river banks, and low woods
Northeast fourth

Glabrous, erect perennial herbs with rhizomes. Stems eventually branched, up to about 1 m tall. Leaves sessile, elliptic to narrowly lanceolate or lance-oblong, 4–15 cm long and 1–3.5 cm wide, the apex acuminate to attenuate, the base rounded or auricled, clasping the stem, the margins serrate. Flowering pedicels ascending or widely spreading, 1–3.5 cm long (1–4.5 in fruit). Calyx regular or nearly so, 13–19 mm long, the teeth abruptly or gradually narrowed to bristlelike teeth. Corolla blue or bluish-purple, 2–3 cm long. Capsules 8–10 mm long, completely enclosed by the persistent calyx.

PEDICULÀRIS L.   Lousewort, wood betony

L. *pedicularis*, pertaining to lice, from L. *pediculus*, louse. There are various legends about the acquisition of this name: the plant was supposed to breed lice; it was formerly used to keep lice away; the seeds looked like lice; the puckered leaves looked as though they were full of lice; and, according to the famous herbal of John Gerard (13), "It filleth sheep and other cattle, that feed in meadows where this groweth, full of lice."

Annual, biennial, or perennial (ours) herbs. Leaves alternate or opposite, sessile or petioled, crenate or sharply toothed to once- or twice-pinnatifid. Flowers showy, yellow, red, pink, or purple, in leafy-bracted terminal spikes or racemes. Calyx bell-shaped to tubular, usually longer on the upper side, cleft on the lower side. Corolla strongly 2-lipped, the upper lip laterally compressed, arched, the lower lip with 3, mostly spreading lobes. Stamens 4, in 2 pairs of unequal length, ascending under the upper lip. Capsule ovate to oblong, beaked, much compressed, several- to many-seeded, loculicidal.

One species in Kansas.

486. *Mimulus glabratus,* front view of flower

P. canadénsis L.   Early pedicularis

L. *-ensis*, suffix added to nouns of place to make adjectives—"Canadian," in this instance.
Light yellow; mid-April–May
Prairie, less frequently in open woods
Primarily within the eastern 3 columns of counties

Erect, unbranched perennial with 1 or more stems 1–3 dm tall arising from a slender rhizome. Leaves alternate, petioled, the blades mostly oblong to elliptic or narrowly oblanceolate in general outline, pinnately lobed or twice-pinnatifid, 2–9 cm long and 0.5–2.5 cm wide, the margins of the lobes more or less crenate and often siliceous. Stems, petioles, lower leaf surface, and calyx more or less hairy. Flowers on short pedicels in crowded racemes 3–10 cm long. Calyx 7–11 mm long. Corolla light yellow, 18–24 mm long. Capsule oblong, oblique, 15–20 mm long. Seeds brown, more or less oblong, truncate at one end, 1.5–2 mm long, the surface minutely reticulate, the areoli in longitudinal rows.

This striking plant is among the earliest in bloom in the spring, its dense spikelike racemes shooting up ahead of the grasses, in full view of the eager bumblebees that flock to gather the nectar. When the plants are growing in colonies, as they frequently do, this festival is a sight worth seeing.

But more is going on than this superficial view reveals. The intention in the design of a flower is to get its eggs fertilized—preferably cross-fertilized—by the sperm contained in the pollen from another individual. The flower of *Pedicularis* is designed to promote this process. Before the flower opens, the stigma at the apex of the slender style and the 4 anthers on threadlike filaments are ensconced, facing downward, within the sharply down-curved apex of the upper lip. With the parting of the lips of the corolla, the style is extended so as to bring the stigma in front of the corolla entrance. At about the same time, the anthers split open in a manner to let the pollen fall upon the back of a bumblebee when, having alighted on the lower lip, it thrusts its head into the throat of the corolla and sucks nectar from the bottom of the tube. It would seem that the stigma of the same flower could pick off some of this pollen as the bee backs away, but the stigmas of other flowers would be cross-pollinated when the bee visits them.

487. *Mimulus ringens*

Sprengel in *Das entdeckte Geheimniss der Natur* gives his observations on the European lousewort, *P. sylvatica* (L. *silvaticus*, of or belonging to a forest or to trees, from *silva*, wood or forest), whose flower structure is fundamentally like that of our species. He notes that the pollen is protected from rain by the cowl-like form of the upper lip and that rain is kept from diluting the nectar by tufts of hairs on the filaments near the bottom of the corolla tube. He observed a well-defined white spot on the lower lip in front of the entrance to the tube and believed it to be a nectar guide. He was puzzled that the pollen is held so far above the lower lip that an insect standing there could not make contact with it. Remarking that it is of first importance to find out for what insect a flower is intended, he concluded with: "However, I have never yet found an insect on it." It was simply Sprengel's bad luck that bumblebees never happened around when he was watching these flowers. Other observers have been more fortunate.

See plate *94*.

488. *Pedicularis canadensis*

PÉNSTEMON Mitch.   Penstemon, beardtongue

NL., from Gr. *pente*, 5, + *stemon*, thread hanging from a distaff or thread of a spider web, or, botanically speaking, the stamen of a flower, often with a threadlike filament.

Erect perennial herbs with short rhizomes bearing numerous thickened fibrous

FIGWORT FAMILY

343

489. *Penstemon albidus*

490. *Penstemon buckleyi*

roots, or shrubs. Leaves opposite, often petioled below, sessile and clasping the stem above, mostly entire or toothed. Flowers showy, white to blue, purplish, pink or red, in terminal clusters. Calyx deeply 5-lobed or -toothed, usually but slightly irregular. Corolla tubular or trumpet-shaped, 2-lipped, the upper lip 2-lobed, usually erect, the lower lip 3-lobed, as long as or longer than the upper. Fertile stamens 4, nearly equal in length, about as long as the corolla tube, accompanied by a threadlike or spatulate, naked or bearded staminode. Stigma headlike. Fruit an ovoid to conical, many-seeded septicidal capsule. Seeds irregular in shape, quite angular, rough, often winged along the angles.

Eleven species in Kansas. Our 6 most common species are included in the following key.

Nectar is secreted in most species at the base of the upper 2 filaments. Long-tongued bees, mostly bumblebees, and some kinds of flies are the principal pollinators. The common name "beard tongue" refers to the bearded staminode of many of the species.

Flower color of most *Penstemons* changes upon drying, so it is well to note the color while the material is fresh.

Key to Species:
1. Herbage light green, glaucous, the leaves thick and somewhat fleshy; seeds 2 mm long or larger 2
  2. Corolla 40–50 mm long .... *P. grandiflora*
  2. Corolla 14–22 mm long .... *P. buckleyi*
1. Herbage and leaves not as above; seeds 1.5 mm long or smaller 3
  3. Herbage entirely glabrous or at most very sparsely and minutely hairy 4
    4. Corolla tube abruptly expanding about a third of the distance up from the base toward the throat, the lobes projecting forward; lowermost peduncles usually 3 cm or longer (occasionally as short as 1.5 cm) .... *P. digitalis*
    4. Corolla tube gradually flaring from the base to the throat, the lobes usually reflexed; lowermost peduncles usually about 1 cm long (seldom longer than 2 cm) .. *P. tubaeflorus*
  3. Herbage definitely hairy 5
    5. Corolla 35–60 mm long; east three-fifths .... *P. cobaea*
    5. Corolla 10–30 mm long; west half .... *P. albidus*

P. álbidus Nutt.

L. *albidus, albus,* white.
White to lavender; mid-May–mid-June
Dry, rocky or sandy prairie hillsides and uplands
West half

Stems 1 to few per plant, simple, 1–5.6 dm tall. Leaves 2–10 cm long and 0.3–2.7 cm wide, entire or shallowly serrate, the lower ones mostly spatulate to narrowly oblanceolate, the middle and upper ones gradually becoming narrowly lanceolate to nearly linear. Herbage covered with minute stiff hairs which give it the texture of fine sandpaper. Flowers borne on short pedicels in spikelike panicles 4–25 cm long. Calyx 5–10 mm long. Corolla white to lavender (drying to dull faded pink) with dark purple lines on the corolla throat, 10–30 mm long, with short glandular hairs inside and outside. Staminode sparsely hairy. Capsules ovoid-acuminate, 10–12 mm long. Seeds black, irregular in shape, narrowly winged along the angles, 2–3 mm long.

See plate *95*.

P. búckleyi Penn.

Named in honor of Samuel Botsford Buckley (1809–1884), naturalist and one-time state geologist of Texas who collected throughout much of the southern and southeastern United States and who first collected and described this species.
Pale lavender to pale pink; April–early June
Sandy prairie hillsides and banks
Primarily in the southwest fourth

Stems simple, 2.5–5.5 dm tall, arising from short rhizomes bearing numerous thickened fibrous roots. Herbage glabrous, light green, glaucous. Leaves thickened, becoming prominently veined with age, spatulate or oblanceolate below, gradually becoming oblong or elliptic near the middle of the stem and lanceolate to ovate or cordate above, 2–8 cm long and 0.7–2 cm wide, the apex acute or mucronate on the lower leaves, acuminate on the upper; the base of the lower leaves tapered gradually, that of the upper leaves rounded to cordate and clasping the stem; margins entire. Flowers on short pedicels in spikelike panicles (4–) 10–25 cm long. Calyx 4–7 mm long. Staminode linear, bearded with short yellow or purplish hairs. Capsules ovoid-acuminate, 11–15 mm long, beaked by the persistent style-base. Seeds brown, irregular in shape, winged along the angles, 2.5–3.5 mm long.

Individuals which appear somewhat intermediate between *P. buckleyi* and a more northern species, *P. angustifolia*, have been segregated by some authorities as a separate species, *P. caudatus*, or as a subspecies of *P. angustifolia*. In Kansas, however, the distribution of these intermediates appears to overlap completely with that of *P. buckleyi*, and they will key out with the latter species in this book. Additional field studies of this group are needed.

See plate *96*.

P. cobaèa Nutt.   Cobaea penstemon

Named after the genus *Cobaea* by the botanist Thomas Nuttall, who discovered this species in
    Arkansas more than a century ago and saw a resemblance between its flowers and those
    of the Mexican *Cobaea scandens*.
White to very pale purple, with dark purple or reddish stripes; mid-May–mid-June
Rocky prairie hillsides and banks
East three-fifths

Plants with 1 to few stems 1.7–7 dm tall. Leaves 4–14 cm long and 1.5–6.5 cm wide, at least those of the middle and upper stem serrate, the lower ones mostly spatulate to narrowly oblanceolate, the middle and upper ones becoming lance-oblong to lanceolate or ovate with rounded or cordate, somewhat clasping bases. Herbage glandular-pubescent, at least in the inflorescence, the leaves often glabrous or nearly so. Flowers borne on stout pedicels often longer than the subtending bracts, in spikelike panicles 1.5–3 dm long. Calyx 8–15 mm long. Corolla white to very pale purple, with darker purple lines (drying to dull dusty pink), 3.5–6 cm long, glandular-hairy except for the inner surface of the lower lip. Staminode bearded. Capsules 13–19 mm long. Seeds sharply angled but scarcely winged on the angles, 2.5–3 mm long.

This and *P. grandiflorus* are the most spectacular of our Penstemons.

See plate *97*.

P. digitàlis (Sweet) Nutt.

Named for the genus *Digitalis*, foxglove, because of the similarity in the flowers.
White; May–June
Prairie, roadsides
East fifth and Republic and Butler counties

Plants 3.5–8.5 dm tall, entirely glabrous or with the inflorescence glandular-hairy. Main stem leaves narrowly oblanceolate to lance-oblong or broadly lanceolate, 7–16 cm long and 1.5–4 cm wide, entire or shallowly serrate. Flowers borne in rather open panicles, the lowermost peduncles within the inflorescence seldom shorter than about 1.5 cm and usually 3 cm or longer. Calyx 5–8 mm long. Corolla white, 20–33 mm long, the tube dilating abruptly about a third of the way from the base to the throat, the lobes projecting forward.

This species tends to occur in weedier habitats than does the rather similar *P. tubaeflorus*.

P. grandiflòrus Nutt.   Large-flowered beardtongue

NL., from L. *grandis*, large, noble, magnificent, + *flos, floris*, flower—the flowers indeed being
    quite large and attractive.
Pale purple to pink; mid-May–June
Rocky prairie hillsides and uplands
Primarily in the north-central sixth, eastward to Wabaunsee and Shawnee counties, and also
    known from Cowley County

Handsome plants, 2.8–7 dm tall. Herbage glabrous, light green, glaucous. Basal leaves petioled, mostly spatulate to oblanceolate, 3–13 cm long and 1–3 cm wide; middle leaves elliptic to oblong or lanceolate, 5–8 cm long and 2–4 cm wide; uppermost leaves broadly ovate to nearly round or cordate, gradually reduced in size, the bases rounded or cordate and clasping the stem. Flowers borne on stout pedicels in spikelike panicles 1.4–4 dm long. Calyx 7–9 mm long. Corolla pale purple to pink (drying bluish), 40–50 mm long. Staminode linear, bearded with short yellow hairs. Capsules ovoid-acuminate, 1.8–2 cm long, beaked by the persistent style-base. Seeds brown, irregular in shape, winged on the angles, 3–3.5 mm long.

See plate *98*.

491. *Penstemon cobaea*

492. *Penstemon digitalis*

493. *Penstemon tubaeflorus*

494. *Penstemon tubaeflorus;* note variation in peduncle lengths between this specimen and that in the previous figure

495. *Scrophularia marilandica*

FIGWORT FAMILY

P. tubaeflòrus Nutt.

NL., from L. *tubae*, trumpet, + *flos, floris*, flower—in reference to the corolla shape.
White; late May–June
Prairie
East half, primarily; more abundant southeastward

Plants 3.5–11 dm tall, entirely glabrous or with the inflorescence glandular-hairy. Main stem leaves narrowly oblanceolate to lance-oblong, lance-linear, or occasionally lanceolate, 4.5–11 cm and 1–3 (–4) cm wide, entire or shallowly serrate. Flowers borne in spikelike panicles, the lowermost peduncles within the inflorescence seldom longer than 2 cm and usually about 1 cm long. Calyx 2–5 mm long. Corolla white, 18–30 mm long, flaring gradually from the base to the throat, at least the upper lobes somewhat reflexed, the lower reflexed or projecting forward. Staminode sparsely bearded. Capsules 5–8 mm long. Seeds about 1 mm long.

SCROPHULÀRIA L.   Figwort

NL., from L. *scrofulae*, scrofula, a swelling of the glands of the neck—a southern European species of this genus, *S. nodosa*, having been employed since the early Middle Ages as a remedy for this condition. "Figwort," the old English name for the European species, comes through OF. *figue*, from L. *ficus*, fig or pile, and AS. *wyrt*, herb or root, and is therefore equivalent to "pilewort."

Erect perennial herbs with 4-angled stems. Leaves opposite, usually petioled, toothed or incised. Flowers small, yellowish, greenish, or brownish, in large, openly branched terminal clusters. Calyx regular, 5-lobed or -toothed. Corolla 2-lipped, the upper lip directed forward, 2-lobed, the lower lip 3-lobed, with the middle lobe deflexed, the others directed forward. Stamens 4, short, in 2 pairs of unequal length, usually included within the corolla tube, accompanied by a spatulate, scalelike staminode which ascends under the upper lip; anther sacs transverse. Stigma entire or apically notched. Fruit an ovoid to globose, many-seeded septicidal capsule separating into 2 segments.

Two species in Kansas. For a brief description of *S. lanceolata*, see comments under the following species.

S. marilándica L.   Maryland figwort

"Of Maryland."
Corolla dull green outside, brownish-purple and glossy within; mid-July–early September
Woods and thickets
East two-fifths

Plants erect, up to 2.5 m tall, usually unbranched below the inflorescence. Leaves mostly ovate to lanceolate, 4.5–17 cm long and 1.5–8.5 cm wide, the apex acuminate, the base broadly acute to broadly rounded or truncate, often oblique, the margins coarsely serrate or double-serrate. Lower leaf surface and inflorescence branches minutely glandular-hairy. Calyx bowl-shaped, 3–4 mm long, the lobes ovate. Corolla 6–10 mm long, dull green outside, brownish-purple and glossy inside. Staminode dark purple or brown. Capsules 4–7 mm long, smooth. Seeds brown to gray, variable and angular in shape but more or less ovoid to ellipsoid, 0.8–1.2 mm long, longitudinally grooved, the surface also minutely reticulate.

Plants of this species yield enough nectar in some seasons and localities to add appreciably to the honey crop, and among apiarists the common name for the species is Simpson honey plant.

According to Krochmal and Krochmal (30), Maryland figwort has had a multitude of medicinal uses, including the treatment of nearly everything from cradle cap to tuberculosis and hemorrhoids.

S. lanceolata Pursh is reported from scattered sites in the east half. It strongly resembles *S. marilandica* and can be most easily distinguished from it on the basis of staminode color—greenish-yellow in *S. lanceolata*, dark purplish-brown in *S. marilandica*. In addition, the leaves of *S. lanceolata* tend to be deeply serrate or laciniate with narrowly winged petioles, the capsules tend to be more narrowly acuminate, and the plants bloom earlier.

SEYMÈRIA Pursh

In honor of English naturalist Henry Seymer (1745–1800).

Erect, branched annual or perennial herbs. Leaves opposite, once- or twice-pinnatifid or variously dissected. Flowers yellow, often marked with purple, borne singly in the upper leaf axils or in elongate, interrupted, leafy-bracted terminal racemes. Calyx regular or nearly so, 5-lobed or -toothed. Corolla somewhat 2-lipped, the tube short and broad, the upper lip with 2 rounded, spreading lobes, the lower with 3. Stamens 4, the filaments short and equal in length and hairy at the base; anthers linear-oblong, opening (in ours) by terminal clefts. Stigma minute. Capsule ovoid, many-seeded, loculicidal, beaked by the persistent style-base.

One species in Kansas.

S. macrophýlla Nutt.

NL., from Gr. *makros*, long, + *phyllon*, leaf, i.e., "long leafed" or "large leafed."
Yellow; June–August
Woods and thickets
East fourth

Robust, sparsely hairy annual, up to about 2 m tall. Lower leaves triangular in general outline, deeply pinnatifid, 14–30 cm long and 9–22 cm wide, the middle and upper leaves gradually becoming smaller, lance-oblong, with entire or crenate margins. Flowers on short pedicels in terminal, simple or branched, leafy-bracted racemes. Calyx bell-shaped, slightly irregular, 5–9 mm long, the 4 lower lobes ovate to oblong, the upper lobe shorter and narrower. Corolla yellow, about 2 cm long, the throat densely hairy. Capsules quite hard, 8–10 mm long. Seeds brown or black, variable in shape but generally truncate-pyramidal with a reticulate surface, 1.8–2.2 mm long.

According to Pennell (51), this species is parasitic on the roots of *Aesculus* (buckeye).

Pollination is by bumblebees which force their way down the throat, past the dense hairs, to the nectar at the base of the corolla.

TOMANTHÉRA Raf.

NL., from Gr. *tome*, a cut or section, + *anthera*, anther—in reference to the anthers which are deeply cleft between the pollen sacs.

Erect annual herbs, similar to *Agalinis*, with retrorsely barbed stems. Leaves opposite, entire or pinnatifid. Flowers large, pink to purple or white, borne in the axils of the upper leaves in ours. Calyx regular or nearly so, bell-shaped to hemispheric, 5-lobed. Corolla irregular, 5-lobed, the upper lip arched and spreading or slightly recurved, the lower lip spreading. Stamens 4, the lower pair the longer, the filaments and anthers hairy, the pollen sacs of each anther united less than half their length. Stigma linear. Fruit a loculicidal capsule. Seeds resembling those of *Agalinis* but with the reticulate ornamentation raised and honeycomblike.

Two species in Kansas.

T. auriculàta (Michx.) Raf.

L. *auriculatus*, having little ears or lobes, from *auricula*, lobe of the ear, little ear, dim. of *auris*, ear—in reference to the pair of spreading lobes at the base of each leaf.
Purple; August–September
Prairie, open upland woods
East fourth and Elk County, uncommon

Slender, rough-hairy plants, 2.5–8 dm tall, sparingly branched, if at all. Leaves sessile, narrowly lanceolate with a pair of sharply pointed basal lobes, 2–5 cm long and 1.5 cm wide. Flowers showy, borne in the axils of the upper leaves. Calyx 12–16 mm long, the lobes lanceolate, longer than the tube. Corolla purple, 2–2.5 cm long. Capsules broadly ovoid, 1–1.5 cm long. Seeds about 1.3 mm long.

T. densiflora (Benth.) Penn., which also blooms in August and September, is known primarily from rocky prairie hillsides in the Flint Hills. It has leaves deeply dissected into 3–7 linear, sharply pointed lobes 0.5–2 mm wide and purple flowers somewhat larger (2.3–3.5 cm long).

VERBÁSCUM L.
Mullein

Name used by Pliny for the great mullein.

Biennial herbs producing a basal rosette of leaves the first year and an erect flowering stem the second year. Leaves alternate, sessile, often decurrent, entire to crenate or rarely deeply toothed. Flowers yellow, white, or blue, each blossom lasting but a few hours, borne in elongate terminal spikes, racemes, or panicles. Calyx deeply 5-lobed. Corolla wheel- or saucer-shaped with a very short tube, slightly irregular, the lower 3 lobes larger than the upper 2. Stamens 5, unequal in size, the anther sacs of each fused into 1;

496. *Seymeria macrophylla*

497. *Seymeria macrophylla*, close-up of flower

498. *Seymeria macrophylla* leaves

FIGWORT FAMILY

347

499. *Verbascum blattaria*

500. *Verbascum blattaria* leaves

501. *Verbascum thapsus*

FIGWORT FAMILY
348

all the filaments hairy, or 3 hairy and 2 smooth, the anthers often of 2 kinds. Stigma flattened. Fruit an ovoid to globose, many-seeded, septicidal capsule.

The number of species is estimated to be 125–250, all natives of the Old World, at least 6 of them now naturalized in the United States, 2 in Kansas.

Key to Species:

1. Plants slender, glabrous except for the glandular hairs of the loose inflorescence; bases of the upper leaves clasping the stem; 5 filaments hairy .... *V. blatteria*
1. Plants stout, densely covered with branched nonglandular hairs; inflorescence crowded; bases of the upper leaves decurrent on the stem; 3 filaments hairy .... *V. thapsus*

V. blattària L. — Moth mullein

Name employed by Pliny for this species; from L. *blatta*, moth or cockroach—because, according to some, the flower resembles a moth with wings spread out, the hairy filaments looking like the feathered antennae of moths. There is also a fable that the flowers attract moths but repel cockroaches.
Yellow; mid-May–October
Disturbed areas such as roadsides, overgrazed pastures, yards, fields, and waste ground
East half; intr. from Europe or Eurasia

Slender, smooth, erect plants, 0.3–1.5 m tall. Lowermost leaves spatulate to narrowly oblanceolate or elliptic, with gradually tapered bases, 6–11 cm long and (1–) 2–4.5 cm wide, the margins both pinnately lobed and crenate or dentate, or merely crenate, or doubly or singly dentate; middle and upper leaves grading from oblong to elliptic or lanceolate above, 1.5–11 cm long and 1.5–3 cm wide, the bases truncate to cordate-clasping, the margins becoming more finely dentate on the upper leaves. Raceme 1–8 dm long, glandular-hairy, simple or with a few branches. Calyx 4–8 mm long. Corolla yellow (rarely white) with a purple base, often turning brown or pinkish with age or on drying, 2–3 cm across. All 5 filaments hairy. Capsules globose, glandular-hairy, 6–9 mm long. Seeds dark brown, about 0.8 mm long, somewhat irregular in shape but basically a truncated, 4-sided pyramid with corrugated sides.

Only 2 or 3 flowers in a receme bloom at a time. After a few hours, they cast off the corolla, carrying with it the 5 stamens inserted around its shallow tube, while the ovary is left behind, ensconced in the cup of the persisting calyx with the slender style protruding. As the flower opens in the early morning the following features are displayed: a violet border around the corolla tube; the stamens with retracted, violet-bearded filaments; the anthers fat with pollen; and the style, thrust forward beyond the anthers, inviting early cross-pollination by bees and beetles that have been gathering pollen from older flowers, very little if any nectar being provided. Some time thereafter the stamens extend forward, presenting pollen behind the stigma but without touching it. Thus cross-pollination is favored at first, yet self-pollination may follow by contact of anthers with the stigma as the corolla falls off. By one method or the other plenty of seeds are produced, so that seedling offspring fill the places of plants dying the year of their flowering.

See plate *99*.

V. thápsus L. — Flannel mullein, great mullein

NL., from L. *thapsia*, ancient name of a plant yielding a yellow dye. "Mullein" comes through OF. *moleine*, from L. *mollis*, soft—in reference to the dense coating of hairs.
Yellow; June–September
Rocky hills, roadside banks, fallow fields, pastures, waste ground
Throughout except for the southwest; intr. from Eurasia

Stout, pale green, densely woolly plant, 0.3–1 m tall, with a deep taproot and fibrous laterals. Leaves oblong-elliptic, 6–40 cm long and 2.5–12 cm wide, the bases decurrent along the stems, the margins crenate or entire. Herbage and outer surface of the corolla densely covered with branched hairs. Flowers crowded in wandlike spikes 0.5–5 dm long. Calyx 7–10 mm long, densely hairy but not glandular. Corolla yellow, 1.5–3 cm across. Three of the 5 filaments hairy. Capsules globose, 7–9 mm long. Seeds brown, about 0.5–0.9 mm long, short-cylindrical to short-pyramidal, with corrugated sides.

In its first year, the seedling produces a substantial crown, a deep taproot with a network of fibrous laterals, and a rosette of large, oblong-elliptic, densely hairy leaves which endure through the winter and provide food for storage in crown and roots. The second spring the crown gives rise to a stiffly upright, foliated stem terminating with the densely flowered spike.

Viewed through a lens, the long, close-set, branched hairs covering the herbage seem like a Lilliputian forest, breaking the force of the wind across the surface, and so making the plant drought resistant by preventing too great a loss of water through evaporation.

In the flowers, provision is made for both self- and cross-pollination. The anthers of the lower stamens are so close to the stigma as to effect self-pollination, resulting in self-fertilization and seed-production, according to the observations of Darwin. Honeybees, bumblebees, and a variety of smaller wild bees collect the vermilion-colored pollen, and in working over flower after flower they may be effecting cross-pollination also. No nectar glands are to be seen, and pollen seems to be the sole attraction to insects.

The leaves yield a greenish-yellow dye in wool mordanted either with potassium dichromate or with alum and cream of tartar. Concerning the employment of the flowers as a hair dye, really dating back to the 4th and 5th centuries B.C., Parkinson specifies in his *Theatrum Botanicum*, 1640, that "the golden flowers of Mullein boyled in lye dyeth the hairs of the head yellow and maketh them faire and smooth."

Numerous medicinal applications of the plant are recorded. In Europe, decoctions of the leaves and roots were employed for respiratory and alimentary ailments. In North America, Indians apparently adopted some of these practices, especially for treating colds, bronchitis, asthma, etc., from white settlers, though the exact method of use varied from group to group. The Forest Potawatomis, Mohegans, and Penobscots smoked the dried leaves. The Catawbas prepared a cough syrup from the boiled roots and also made a poultice of mashed leaves to alleviate pain and swelling associated with bruises, sprains, and wounds. The Choctaws also made a headache poultice from the leaves, while the Menominees smoked the pulverized dried root for pulmonary ailments. The leaves and flowers were listed officially in the *National Formulary* from 1916 to 1936, with the leaves classified as demulcent and emollient and the flowers as demulcent and pectoral.

See plate *100*.

VERÓNICA L.   Speedwell

After St. Veronica who, according to an early Christian legend, wiped the sweat and blood from the face of Jesus as he staggered under the weight of the cross. "Speedwell," in the sense of a parting salute, is the name given in England to the most common British Veronica, whose flower, on being picked or jarred, speedily drops its corolla.

Erect to spreading or creeping annual, biennial, or perennial herbs. Leaves opposite, at least below, sessile or petioled, mostly entire or toothed, less frequently deeply pinnatifid. Flowers small, white to blue or purple, borne singly in the leaf axils or in terminal or axillary racemes. Sepals 4, separate or slightly united at the base, more or less alike. Corolla wheel-shaped, with 4 lobes and a very short tube, very weakly 2-lipped. Stamens 2, protruding. Ovary short and flattened, the style protruding beyond the corolla tube. Fruit a flattened, few- to several-seeded, often heart-shaped capsule which is loculicidal or both loculicidal and septicidal, subtended by the persistent calyx.

Nine species in Kansas. Our 4 commonly encountered species are included in the following key.

Key to Species (3):
1. Main axis of the plant producing well-developed racemes from the axils of some opposite leaves .... *V. anagallis-aquatica*
1. Main axis of the plant (including its leafy branches if present) bearing flowers singly in the axils of the upper (alternate) leaves, the inflorescence therefore composed of axillary flowers or forming a terminal spike or raceme 2
  2. Pedicels longer than the sepals, 4 mm long or longer .... *V. agrestis*
  2. Pedicels shorter than the sepals, 0.5–2 mm long 3
    3. Flowers white; foliage leaves somewhat succulent, narrowly oblong to oblanceolate, neither palmately veined nor pinnatifid; style very short, the stigma nearly sessile in the apical notch of the capsule .... *V. peregrina*
    3. Flowers blue; foliage leaves not succulent, either palmately veined or pinnatifid; style elongate .... *V. arvensis*

V. agréstis L.

L. *agrestis*, of the land, rural, wild.
Blue, or white with blue veins; mid-April–early June, again in the fall
Lawns, fields, roadsides, waste ground
East six-sevenths, perhaps throughout; intr. from Eurasia

Small annual with prostrate or ascending stems 0.5–3 dm long, usually branched from the base. Leaves petioled, opposite below, becoming alternate above, the blades variable in shape, mostly broadly ovate to nearly round in outline, occasionally deltoid, 4–15 mm long and 3–11 mm wide, the margins deeply crenate. Flowers borne on threadlike pedicels in the axils of the upper alternate leaves, the pedicels longer than the subtending leaves. Calyx 2–4 mm long in flower, the lobes ovate. Corolla blue or white with blue lines, about 3–4 mm long. Capsules apically notched, 3–4 mm long and 4–5 mm wide. Seeds pale yellow, more or less ovate, strongly concavo-convex, transversely wrinkled on the back, about 1.2 mm long.

This species includes variants recognized by some authorities as separate species, *V. didyma* or *V. polita*.

502. *Veronica anagallis-aquatica*

V. anagállis-aquática L.   Water pimpernel, water speedwell, brooklime

NL., from the genus name *Anagallis*, pimpernel (in the Primrose Family), + L. *aquaticus*, living
    in the water.
Blue; May–early October
Shallow water or mud of stream and pond margins, ditches, and other wet places
Primarily in the southwest and south-central sixth

Glabrous or near-glabrous, somewhat succulent perennial with stems 2–10 dm tall, rooting at the lower nodes. Leaves opposite, mostly sessile (the lowermost or those produced late in the season sometimes short-petioled), quite variable in shape, oblanceolate to lanceolate, elliptic, or oblong, 2–9 cm long and 0.4–3 cm wide, the apex rounded or acute, the base tapered, rounded, or auricled, more or less clasping the stem, the margins entire or shallowly serrate. Flowers borne in terminal and axillary racemes 2–15 cm long, the pedicels threadlike, ascending in flower, gradually spreading widely or becoming deflexed in fruit. Calyx 2–3 mm long, the lobes lanceolate to ovate-acuminate. Corolla blue, 4.5–5.5 mm long. Style elongate. Capsule about 3 mm long and 3–4 mm wide, shallowly notched at the apex. Seeds round or broadly elliptic, about 0.5 mm long.

According to Fernald and Kinsey (52), this plant has been used as a potherb and in salads and was recommended by numerous European writers as a valuable source of Vitamin C, hence the common name "brooklime."

503. *Veronica peregrina*

V. arvénsis L.   Common speedwell

L. *arvensis*, pertaining to a cultivated field, from *arvum*, cultivated field, plowed land.
Blue; March–May (–September)
Lawns, fields, waste ground
East four-fifths; intr. from Europe

Small, hairy annual weed, 6–26 cm tall, the stems simple or several times branched from the base. Leaves opposite, the lowermost petioled, the uppermost sessile, the blades mostly ovate to deltoid, 4–14 mm long and 3–10 mm wide, the apex acute or rounded, the margins serrate or crenate. Flowers sessile or nearly so in the axils of reduced, leaflike bracts arranged alternately on the stem, mostly elliptic or oblong, and entire. Calyx 3–4 mm long. Corolla blue, 4–6 mm long. Capsule obcordate, 3–4 mm across and nearly as long. Seeds brownish-yellow, ovate, about 0.8–1.2 mm long.

V. peregrìna L.   Purslane speedwell

L. *peregrinus*, foreign, coming from foreign parts—Linnaeus having named the plant from specimens that came to him from America.
White; late April–May (–August)
Lawns, fields, waste ground
Probably throughout

Erect annual or winter annual, glabrous or nearly so, the stems simple or with a few branches, 0.5–4 dm tall. Leaves somewhat succulent, sessile, opposite below, alternate above, mostly oblanceolate to oblong or linear, the main leaves 1–4 cm long and 3–10 mm wide, the apex rounded, the margins entire or shallowly toothed. Flowers sessile or on short pedicels in the axils of the upper reduced leaves. Calyx 3–5 mm long, the lobes oblanceolate. Corolla white, slightly longer than the calyx. Capsules obcordate, 3–4 mm long and about 4 mm wide. Seeds yellowish, elliptic, plano-convex, 0.5–1 mm long, with a short keel or ridge along the back.

Glabrous individuals are recognized as var. *peregrina*, while those with short glandular hairs on the foliage and capsules are placed in var. *xalapensis* (H. B. K.) Penn.

VERONICÁSTRUM Fabr.   Culver's physic, Culver's root

NL., from the genus *Veronica* + L. suffix *-aster*, indicating partial resemblance or similarity, the
    similarity between this species and the Asiatic *Veronica spicata*, for instance, being quite
    evident.

Tall, slender perennial herbs. Leaves whorled, sessile, narrow, toothed. Flowers white or pink, in slender terminal racemes. Sepals 4 or 5, slightly united at the base, somewhat unequal in size. Corolla tubular, nearly regular, with 4 short lobes, somewhat 2-lipped. Stamens 2, the anthers protruding conspicuously on slender filaments. Fruit a narrowly ovoid or ellipsoid, many-seeded capsule, opening by 4 short terminal slits.

A genus composed of 2 species, 1 in North America and 1 in Siberia.

V. virgínicum L.   Culver's physic

"Of Virginia." An infusion of the rhizome containing an active crystalline glucoside was much
    used by Dr. Culver and other physicians of the 19th century as a mild cathartic.
White; mid-June–mid-August
Prairie, rarely in open woods
Northeast eighth and Cherokee County

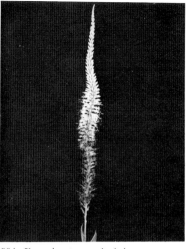

504. *Veronicastrum virginicum*

Glabrous to sparsely hairy, rigidly erect plants, 0.5–1.5 m tall, with a segmented rhizome bearing thickened fibrous roots. Leaves usually in whorls of 3–7, the uppermost sometimes in pairs, the blades usually glabrous, rarely with the lower surface minutely hairy, narrowly elliptic to lance-elliptic, 2.5–11 cm long and 0.5–3.5 cm wide, the apex long-acuminate to attenuate, the base wedge-shaped, the margins finely serrate. Racemes 0.7–3 dm long, simple or fascicled, terminating the main stem. Calyx 1.5–2.5 (–3.5) mm long, the lobes mostly lance-acuminate to lance-attenuate. Corolla white, 4–7 mm long. Anthers red. Capsules 4–5 mm long, capped by the persistent threadlike style. Seeds tan, elliptic, about 0.6 mm long.

Nectar occurring at the bottom of the corolla tube is sought after principally by bees, butterflies, and flies. The behavior of the flowers at first promotes cross-pollination but later self-pollination is possible, as follows: The flowers open progressively from the base towards the apex of the raceme, and as the corolla lobes spread apart, the anthers are thrust out beyond the corolla in advance of the style, presenting pollen to insect visitors. After a time, the style advances in position to receive pollen brought from flowers freshly opened. Since bees in going over a raceme are accustomed to begin at the base and work upward, they would as a matter of routine carry pollen from the anthers of young upper flowers to the advanced stigmas of lower older flowers of the raceme next visited.

Digging up a plant in midsummer, we discover that the underground system consists of a series of similar horizontal rhizome segments provided with stout fibrous roots, the youngest segment bearing at its front end the upright foliar and flowering shoot of the season and being connected at its rear end with the segment of last year that now shows the scar where the shoot it bore last year came away. This segment of last year is in turn connected at its rear end with the segment of year-before-last, which still shows the scar of its upright shoot. These segments are all alive, and since none older than 3 years is present we realize that each segment dies soon after the completion of its third year of life. While it is the terminal bud of a segment that produces the foliar-flowering shoot, it is a lateral bud that sometime in the fall gives rise to a new segment. The best time to transplant Culver's physic, therefore, is when the leaves begin to ripen in late summer, for then new segments that are to produce the foliar-flowering shoots of the following summer can grow forth and send out their roots without subsequent molestation. It can be seen that when a segment gives off but 1 lateral bud it simply reproduces itself without increasing the number of plants; however, it frequently happens that 2 or more lateral shoots arise from a segment, resulting at first in a clump of upright stems and eventually in the formation of a colony.

The underground parts contain leptandrin, a substance with powerful emetic and cathartic properties, and were used by the Senecas, Meskwakis, and Menominees for various medicinal purposes. The roots and rhizomes were listed in the *U.S. Pharmacopoeia* at one time but were eventually excluded as being possibly unsafe due to rather irregular and severe effects.

MARTYNIÀCEAE   Unicorn-plant Family

From the genus *Martynia*, which we do not have in our flora. Named in honor of John Martyn
    (1693–1768), professor of botany, Cambridge, England.

Coarse-stemmed, glandular-hairy annual or perennial herbs. Leaves mostly opposite, simple, with undulate or lobed margins, petiolate, estipulate. Flowers bisexual, irreg-

505. *Proboscidia louisianica*

ular, borne in terminal racemes. Sepals 5 and free from one another or fused, the calyx then 4- to 5-lobed, sometimes split along the under side. Petals 4–5, fused, the corolla irregular and somewhat 2-lipped, 4- to 5-lobed. Fertile stamens 2 or 4; when 2, then usually accompanied by 2 or 3 staminodes (sterile stamens); when 4, then 2 long and 2 short, and 1 staminode sometimes present. Pistil 1; ovary superior, 1-celled but with 2 broad, intruding parietal placentae bearing numerous or few ovules; style 1; stigma 2-lobed. Fruit a large woody capsule with a prominently protruding beak which separates into 2 curved claws when the capsule explosively opens longitudinally.

The family includes about 5 genera and 16 species, primarily of the American tropics and subtropics. A few species are grown as ornamentals.

Kansas has but a single genus and species.

PROBOSCÍDEA Schmid.   Devil's claw, unicorn plant

According to Fernald (9), the Greek name for the proboscislike beak of some fruit, from Gr. *proboskis*, snout, trunk of an elephant.

Annual or perennial herbs with thick, branching stems. Leaves opposite to alternate, broad, entire to palmately or pinnately lobed. Flowers large and showy, few to many in terminal racemes. Calyx 5-lobed, more or less irregular, split to the base along the lower side. Corolla bell-shaped to funnelform or tubular, the tube pouchlike on 1 side at the base, the limb oblique, 5-lobed. Fertile stamens 4, accompanied by 1 staminode, or 2 and then with 2 or 3 staminodes; filaments arched, the anthers united along their edges. Ovary ovate to lanceolate with a slender style and obovate to oblanceolate stigma lobes. Capsule with a ventral crest and sometimes also a dorsal one.

One species in Kansas.

P. louisiánica (Mill.) Thell.   Common devil's claw

"Of Louisiana."
Pink with yellow and red markings; late June–September
Fields, roadsides, pastures, waste ground, usually in sandy soil
West two-thirds, primarily; scattered in the east third

Erect, diffusely branched, thick-stemmed annuals, 1.5–10 dm tall, covered with ill-smelling, clammy glandular hairs. Leaves long-petioled, opposite below to more or less alternate above; blades cordate to reniform, 3.8–18 cm long and 4–21 cm wide, with palmate venation and somewhat uneven margins. Calyx rather membranous, oblique, 1–2 (–2.5) cm long on the upper (longer) side, subtended by a pair of bracts. Corolla funnelform, 3–4.5 cm long, usually pale pink, the throat and inner surface of the corolla tube marked with yellow lines and dark red or purplish spots. Style gradually thickened toward the stigma, protruding conspicuously beyond the calyx after the corolla has fallen. Capsule up to about 10 cm long and 3 cm thick, with a beak longer than the seed-bearing body. Seeds black, quite rough, narrowly ovate, somewhat flattened, 8–11 mm long.

The flowers resemble those of the Catalpa tree and, like those of Catalpa, are proterandrous. The 2 stigma lobes spread apart only after the pollen has been shed, and they close together again after having been pollinated by bees coming from younger flowers.

The fruits are adapted for seed dispersal by clasping the legs of deer, coyotes, rabbits, etc., by being caught in the hooves of cloven-footed animals, or by becoming hooked in the wool of sheep. The tender young fruits are sometimes pickled and eaten, while the dried mature fruits are often used to make a variety of ornaments.

OROBANCHÀCEAE   Broomrape Family

Herbaceous perennials with alternate scale-leaves but lacking foliage-leaves and chlorophyll, parasitic on the roots of other plants. Flowers bisexual, irregular and 2-lipped, borne singly or in few- to many-flowered terminal racemes or spikes. Calyx 2- to 5-lobed. Corolla of 5 petals fused together at least part of their length. Stamens 4, 2 long and 2 short, borne on the corolla tube. Pistil 1; ovary superior, 1-celled, usually with 2 placentae; style and stigma 1. Fruit a many-seeded capsule splitting into 2 segments.

A family of about 14 genera and 180 species, all of the Northern Hemisphere and primarily in the Old World.

Kansas has but a single genus.

OROBÁNCHE L. Broomrape

After L. *orobanche*, from Gr. *orobus*, vetch, + *anchein*, to choke—the name borrowed by Linnaeus for the type species, *O. major*, from Pliny, who seems to have used it not for a broomrape but for a European *Cuscuta*, since he says in his *Natural History*: "There is a certain plant, too, which kills the chick-pea and the vetch, by twining around them, the name of it is orobanche."

506. *Orobanche multiflora*

Small, inconspicuous, glandular-hairy plants, usually turning rusty-brown upon drying. Flowers purplish, reddish, or yellowish, in inflorescence types as for the family. Calyx (in ours) with 5 lanceolate to long-acuminate lobes. Corolla tubular, curved downward somewhat, more or less 2-lipped, the upper lip entire, notched, or 2-lobed, the lower lip 3-lobed. Ovary 1-celled with 2 or 4 parietal placentae; style elongate; stigma large, capitate or lobed.

Four species in Kansas. *O. multiflora* Nutt., reported only from a dry, rocky, exposed prairie hillside in Scott County State Park, is similar to *O. ludoviciana* but differs primarily in having rounded, rather than acute, corolla lobes and in being a lighter-colored yellowish plant. *O. fasciculata* Nutt., known from Cheyenne and Barber counties, resembles *O. uniflora* but has longer stems and shorter peduncles than that species, usually has more than 3 flowers, and has hairy scale-leaves.

For other examples of colorless, parasitic flowering plants without foliage, see *Corallorhiza* (Orchidaceae) and *Monotropa* (Ericaceae).

Key to Species (3):

1. Flowers solitary or few, the pedicels always much longer than the calyx; calyx not subtended by bractlets .... *O. uniflora*
1. Flowers several to many, sessile or mostly so in a dense spike; calyx commonly subtended by 1–3 bractlets inserted at its very base .... *O. ludoviciana*

507. *Orobanche multiflora* with underground parts exposed

O. ludoviciàna Nutt. Louisiana broomrape

NL., "pertaining to Louisiana," in this case referring to the Louisiana Territory.
Purplish; (June–) July–early September
Sandy soil of prairie and river flood plains
Scattered nearly throughout, relatively rare everywhere but more common in the west half

Plants parasitic on the roots of *Artemisia*, *Xanthium*, and other Compositae. Stems quite thickened, occasionally with 2 or 3 branches, the aboveground portion up to 2 dm tall, with few to several scale-leaves. Flowers several to many, borne in dense terminal spikelike clusters, the lowermost flowers sometimes on pedicels about 1 cm long. Calyx 1.3–1.7 cm long, the lobes long-attenuate. Corolla purplish, about 2 cm long, 5-lobed and definitely 2-lipped. Anthers glabrous or only slightly hairy. Capsules ellipsoid, 9–14 mm long. Seeds minute, about 0.3–0.5 mm long.

O. uniflòra L. One-flowered broomrape, cancer-root

NL., from L. *unus*, one, + *flos, floris*, flower—each plant having but a single flower.
Cream-colored, tinged with purple; late April–May (–June)
Rich, moist soil of oak-hickory woods
East fourth, primarily; rather rare

Plants parasitic (in our area) on the roots of trees. True stems 2–5.5 cm long, mostly belowground, sometimes with 2 or 3 short branches at ground level, bearing several glabrous scale-leaves at or near the surface. Flowers borne singly on 1–3 naked, more or less hairy peduncles (4–) 6–20 cm tall. Calyx 0.6–1 cm long, the lobes triangular-acuminate to attenuate. Corolla cream-colored, tinged with purple, 1.5–2 cm long, 5-lobed and slightly 2-lipped. Ovary ellipsoid or ovoid; stigma discoid, 2-lobed. Capsules 7–10 mm long. Seeds minute, about 0.2–0.4 mm long.

Although the anthers are situated close behind the stigma, and both anthers and stigma mature at the same time, self-pollination is prevented by the forward facing of the receptive surface of the stigma. Cross-pollination may occur when insects coming for nectar secreted by a gland at the base of the ovary get pollen on the upper surface of their proboscises, heads, and bodies and deliver it to the stigmas of other flowers. The anthers open by longitudinal slits, but it appears that jarring is required to loosen the pollen; this is provided for by sharp points projecting downward from the under side of the anthers and engaging the insects while entering and retiring.

LENTIBULARIÀCEAE   Bladderwort Family

Annual and perennial herbs, ours aquatic or strand plants, mostly carnivorous. Leaves alternate or in a basal rosette, simple, or dissected. Flowers bisexual, irregular, borne singly or 2–3 together at the tip of an erect scape. Calyx 2- to 5-lobed. Corolla of 5 fused petals, 2-lipped, the lower lip spurred. Stamens 2, inserted on the base of the corolla tube. Pistil 1; ovary superior, 1-celled; style 1 or absent; stigma 2-lobed. Fruit a many-seeded capsule.

A family of approximately 5 genera and 260 species which are widespread in distribution but of little economic importance.

Kansas has 2 species of *Utricularia*.

UTRICULÀRIA L.   Bladderwort

NL., from L. *utriculus*, little bladder, in reference to the small traps on the leaves.

Perennial herbs, entirely rootless, growing in water or wet soil. Leaves alternate, linear and entire or variously dissected (as in ours) and bearing tiny inflated bladders which actively trap minute aquatic insects, such as water fleas and mosquito larvae, as well as very small fish larvae and tadpoles. Flowers yellow or purple. Calyx 2-lipped, cleft to the base, the lobes entire. Corolla 2-lipped, the throat often closed by a conspicuous, rounded, often bearded projection (the palate) from the lower lip; upper lip entire or 2-lobed; lower lip entire or 3-lobed, spurred. Capsule ovoid to globose, opening irregularly or separating into 2 or 4 segments. Seeds minute.

The carnivorous traps found in this genus represent highly modified leaf divisions. The structural detail of the traps varies from species to species, but the general morphology and mode of operation is quite similar for all species. In size they range from about 0.2–3 (–5) mm. In *U. vulgaris* and *U. gibba*, to which the following somewhat simplified description pertains, the hollow traps are pear-shaped to obliquely ovate in side view, slightly flattened laterally, with the opening at the narrow end, toward the apex of the shoot. The entrance to each trap is closed by 2 flexible, roughly semicircular, inward-opening flaps—an outer one which is attached at the top of the opening and a thinner inner one which is attached along the bottom. The external surfaces of the trap and of the outer flap or door, and in fact of the entire plant, are studded with small mucilage-secreting cells. The surface cells of the door secrete sugar as well, which, along with the mucilage, is thought to serve as an attractant for prey. Above the entrance is a pair of branched, antennaelike projections. These, together with several elongate rigid bristles extending forward from along the sides and bottom of the entrance are thought to channel small swimming creatures toward the orifice.

By some means not well understood, the cell walls of the traps are capable of pumping water out of the closed trap and creating a negative pressure within, so that when a trap is mechanically triggered by a victim brushing against the 4 bristles at the base of the trap opening, water quite suddenly rushes into the trap, carrying the animal with it. This movement is usually too rapid to be seen with the unaided eye, but it has been recorded on movie film. After the trap has been sprung, the entrapped creature may be observed inside the vesicle, and the side walls will have changed from a concave to a flat or convex position. Once the trap is filled with water, the doors are forced tightly against the threshold again, and the evacuation process begins, thus resetting the trap. As long as the traps remain submersed and undamaged, they can be triggered and reset again and again. The minimum time for resetting required for *U. vulgaris* is 15–30 minutes, but some species take as long as 2 hours. A period of about 48 hours or less is required to digest the prey. After performing laboratory experiments with extracts from *Utricularia* plants and traps, some researchers have concluded that digestion is achieved by enzymes secreted by the plant, while others attribute it to bacterial action or to a combination of the two. Large victims, such as fish larvae and tadpoles, may not fit completely into the trap. In such a case, if the trap can close sufficiently around the animal to build up a vacuum again, the carcass is gradually digested as it is slowly sucked farther and farther into the trap. Persons wishing to read in greater detail about the structure and behavior of *Utricularia* traps are referred to 2 works by F. E. Lloyd (53, 54).

Key to Species:
1. Leafy stems floating beneath the surface of water; scapes quite stout, usually bearing more than 8 flowers; pedicels subtended by a membranous bract 3–6 mm long .... *U. vulgaris*
1. Leafy stems creeping at the bottom of shallow water; scapes threadlike, usually bearing only 1–3 flowers; pedicels subtended by a minute bract about 1 mm (–3 mm) long .... *U. gibba*

U. gíbba L.   **Bladderwort**

L. *gibbus*, humped, the reference here being unclear.
Yellow; late July–mid-October
Shallow water or mud along margins of marshes and small ponds
Southeast fourth, especially in the southeast fourth of that area

Delicate creeping plants with stems seldom over 1 m long. Leaves mostly 3–15 mm long, 2-parted at the base, each of these parts then with a few hairlike subdivisions, bearing relatively few bladders. Flowers usually fewer than 8, borne on slender, erect scapes mostly 0.2–1 dm tall, each pedicel subtended by a minute bract about 1 mm long. Calyx 1.5–3 mm long. Corolla yellow (sometimes drying pinkish), 5–8 mm long, the lower lip with a prominent palate. Capsules 2.5–3 mm long, subtended by the persistent calyx, the pedicels remaining erect or ascending as the fruits mature.

U. vulgàris L.   **Bladderwort**

L. *vulgaris*, common, ordinary.
Yellow; late June–August
Shallow water of oxbow lakes, small ponds, and wet ditches
Scattered in the east three-fourths

Rather coarse, submersed, free-floating, plumelike plants with stems up to 2 m long. Leaves 1–5 cm long, 2-parted at the base, each of these parts then finely divided, the ultimate subdivisions threadlike, usually bearing numerous tiny bladders. Flowers usually 8 or more, borne on stout, erect scapes mostly 0.5–3 dm tall, each pedicel subtended by a membranous bract 3–6 mm long. Calyx 3–5 mm long (or slightly longer in fruit). Corolla yellow, 10–15 mm long, the upper lip erect, the lower lip shallowly 3-lobed and with a prominent palate. Capsules 3–6 mm long, subtended by the persistent calyx, the peduncles elongating and arching downward as the fruits mature.

508. *Utricularia vulgaris*

ACANTHÀCEAE   **Acanthus Family**

Named after the genus *Acanthus*, which we do not have, from Gr. *akantha*, thorn—some species of the genus having leaves with spiny margins.

Annual or perennial herbs, shrubs, or (rarely) trees. Stems usually square in cross section and swollen at the nodes. Leaves opposite, simple, sessile or petiolate, estipulate. Flowers bisexual, irregular to nearly regular, borne singly in the leaf axils or in terminal or axillary clusters. Sepals fused to one another, sometimes only slightly at the base, the calyx 2- to 5-lobed. Petals 5, fused at least part of their length, the corolla usually 5-lobed and often 2-lipped. Stamens 4 (2 long and 2 short) or only 2, inserted on the corolla tube, often accompanied by 1–3 staminodes. Pistil 1; ovary superior, 2-celled, each cell 2- to many ovuled; style 1, threadlike; stigmas 2. Nectiferous disk present at the base of the ovary. Fruit sometimes a drupe, but usually a loculicidal capsule which dehisces elastically. Seeds borne on persistent, curved or hooked projections from the placentae.

A large family of about 250 genera and 2,500 species, widely distributed, especially in the tropics. Species of Bears-breech (*Acanthus*), clock-vine (*Thunbergia*), *Aphelandra*, *Ruellia*, *Justicia*, and a few other genera are grown as ornamentals.

Kansas has 3 genera.

509. *Utricularia vulgaris* leaves bearing insectivorous traps

Key to Genera:

1. Calyx 12–25 mm long; corolla almost equally 5-lobed; stamens 4; fruit club-shaped, 1–2 cm long, equally thickened on all sides, 6- to 8-seeded .... *Ruellia*
1. Calyx less than 10 mm long; corolla strongly 2-lipped; stamens 2; fruit 2- or 4-seeded 2
    2. Each flower subtended by 3 minute triangular bracts; lower lip deeply 3-lobed; fruit club-shaped, about 12 mm long, the walls equally thickened on all sides .... *Justicia*
    2. Each flower subtended by 2 or 4 conspicuous, obovate bracts; lower lip entire or merely 3-toothed; fruit ovoid, 5–6 mm long, the walls much thickened in the vicinity of the placentae .... *Dicliptera*

DICLÍPTERA Juss.

NL., from Gr. *diclis*, double-folding, + *pteron*, wing—in reference to the involucel.

Erect annual or perennial herbs. Leaves entire or rarely toothed, petioled. Flowers small, red to pink or purple, borne in few-flowered clusters subtended by small, leaflike bracts and arranged in interrupted spikes or panicles. Calyx membranous, with 5 trans-

510. *Dicliptera brachiata*

511. *Justicia americana*

lucent lobes. Corolla 2-lipped, the upper lip erect, entire or notched, the lower lip spreading, entire or 3-toothed. Stamens 2; 1 sac of each anther terminal, the other apparently lateral. Capsule ovoid to nearly globose, stipitate, 1- to 4-seeded, dehiscing elastically.

One species in Kansas.

D. brachiàta (Pursh) Spreng.

L. *brachiatus*, with arms or branches.
Purple or pink; July–mid-October
Moist, rich soil of wooded flood plains
East third, south of the Kansas River

Erect, minutely hairy, branched annuals, 3–7 dm tall. Leaf blades membranous, mostly broadly lanceolate to lanceolate, 2–13 cm long and 1–7 cm wide, the apex acuminate, the base abruptly acuminate, the petioles slender, 1–6 cm long. Flower clusters borne on the upper branches, the flowers subtended by oblong or obovate, leaflike bracts 5–9 mm long. Calyx membranous, about 3 mm long, the lobes linear-lanceolate. Corolla purple or pink, 11–17 mm long. Stamens protruding from the corolla tube. Capsules ovoid, 5–6 mm long. Seeds dark brown, nearly round in outline, flattened, 3–3.5 mm long, sparsely covered with rust-colored, papulose hairs (use high magnification).

JUSTÍCIA L.   Water willow

In honor of James Justice, 18th-century Scotch horticulturist and botanist.

Perennial herbs or shrubs of wet places. Leaves sessile or nearly so, entire. Flowers white to pink or purple, each subtended by 3 minute triangular bracts, borne singly or usually in stalked axillary clusters. Calyx regular, the 5 (rarely 4) sepals linear to lanceolate, united slightly at the base. Corolla 2-lipped, the upper lip 2-lobed, the lower lip 3-lobed, the tube usually narrow. Stamens 2, protruding beyond the corolla tube, the anther sacs of each stamen separate. Capsules club-shaped or cylindric, 4-seeded, contracted at the base into a short stalk.

One species in Kansas.

J. americàna (L.) Vahl.   American water willow

"American."
Light purple to white, marked with purple; June–August
Mud or shallow water
East two-fifths

Glabrous perennial herbs with branching rhizomes which produce numerous erect stems which rise to various heights according to the depth of the water, but mostly 5–10 dm tall. Leaves mostly narrowly elliptic to lance-linear, 4–14 cm long and (4–) 8–25 mm wide, the margins entire or shallowly indented. Flowers borne in crowded spikes 1.5–3 cm long on naked peduncles much longer than the subtending leaves. Calyx 5–6 mm long, herbaceous, the lobes linear-lanceolate or subulate. Corolla light purple to white, 10–12 mm long ,with a short tube and 4 prominent lobes, the lower marked with purple blotches. Anthers dark purple, the sac separated by a prominently forked connective. Capsules about 12 mm long, the sterile stipe about as long as the seed-bearing portion. Seeds brownish, nearly round in outline, flattened, covered with minute tubercles or wartlike bumps (use high magnification), about 3 mm long and 2 mm wide.

The slender style at first extends behind and out above the anthers but later arches forward to a position more in front of the mouth of the corolla tube. Nectar is secreted inside the base of the tube, and both long-tongued and short-tongued wild bees, in getting at the nectar in newly opened flowers, receive pollen on their backs, and, after going to older flowers, leave pollen adhering to stigmas that are in position to receive it.

It is interesting to see this plant's habit of growth and follow its method of vegetative reproduction. Invariably described as a perennial, it is in a sense biennial, since no part of a colony is more than 2 years old. Horizontal rhizomes, a few inches to a foot or more in length, grow forth from the under-mud bases of the foliage-bearing upright stems. The following spring the terminal buds of these rhizomes produce upright shoots with adventitious roots at their bases. These shoots begin blooming in June and new horizontal rhizomes arise at their bases. All older parts will have died by the end of their second year. Since the rhizomes radiate in all horizontal directions, the colony forms in the mud an interlaced network which protects against erosion.

RUÉLLIA L.   Ruellia

Named in honor of Jean de la Ruelle, French physician and botanist (1474–1537).

Erect perennial herbs or shrubs. Stems in ours square in cross section. Leaves petioled or nearly sessile, entire or rarely toothed. Flowers usually large and showy, blue to purple, solitary in the leaf axils or in axillary or terminal clusters. Small, reduced, cleistogamous flowers sometimes present also. Sepals 5, narrow, united at the base. Corolla funnelform, nearly regular, with a slender tube dilated toward the throat and 5 large, spreading, subequal lobes. Stamens 4, in 2 pairs of unequal length, usually accompanied by 1 staminode, the members of each pair weakly united at the base. Stigma lobes unequal, one much larger than the other and recurved. Capsule club-shaped to oblong or obovate, compressed, the lower portion sterile and contracted into a stalklike base, dehiscing elastically along 2 sutures.

Two species in Kansas.

Key to Species:

1. Plants pale green, conspicuously hairy, the leaves rather tough; calyx lobes linear, 0.5–1 mm wide .... *R. humilis*
1. Plants darker green, glabrous or nearly so, the leaves rather thin; calyx lobes linear-lanceolate, about (2–) 3.5 mm wide .... *R. strepens*

R. hùmilis Nutt. Ruellia

L. *humilis*, low, on the ground.
Lavender; late May–early September
Dry prairies, roadsides, pastures, occasionally in open woods
East half

512. *Ruellia humilis*

Pale green perennial herbs with 1 to few erect or decumbent, branched stems 1–5 dm tall arising from a rhizome with many coarse fibrous roots. Herbage sparsely covered with long, spreading white hairs. Leaves very short-petioled, strongly upwardly inclined, the blades of the main leaves ovate to narrowly lanceolate, 2–6 cm long and 1–4 cm wide, the apex acute or rounded, the base acute or rounded, the margins entire. Flowers mostly solitary in the leaf axils, sessile or nearly so. Calyx lobes linear-attenuate, ciliate, 12–25 mm long and 1 mm wide or less. Corolla lavender, 4–8 cm long, the lobes rounded, rather wrinkled in appearance. Capsules hard, club-shaped, few-seeded, 10–12 mm long, subtended by the persistent calyx. Seeds rusty brown, nearly round in outline, flattened, about 3.5 mm long, covered with inconspicuous hairs which swell when the seed is wetted.

The corolla is ephemeral, opening during the night and falling off without fading, carrying the stamens with it the following night, or sooner if the plant is shaken. However, if in the evening branches with buds ready to open are cut and kept in a vase of water indoors, the buds will open normally and the flowers will last longer than they do in the blustering out-of-doors, and younger buds will bloom in succession. If the flowers are timid, not so is the plant, for how valiantly it will grow and bloom and spread its seeds abroad under adverse conditions was demonstrated in the long drought period of the thirties when it could be seen advancing in rundown meadows along with that tough contender, *Aster ericoides*.

513. *Ruellia strepens*

R. strèpens L. Woodland ruellia

NL., loud, noisy, rustling, rattling, from L. *strepo*, *strepitus*, to make a great noise—in reference to the explosive capsules.
Pale blue-violet; mid-May–August (–October)
Moist, rich soil of woodlands
East two-fifths, more common eastward

Perennial herbs, glabrous or nearly so, with branched or unbranched stems 3.5–10.5 dm tall arising from a short, knobby rhizome with many thick fibrous roots. Leaf blades rather thin, mostly ovate to lanceolate, 5–16 cm long and 1.5–7 cm wide, the apex acute or acuminate (rounded on the small basal leaves), the base acute or gradually tapering to the petiole, the margins entire or with very shallow indentations. Flowers nearly sessile in leafy-bracted axillary clusters or terminal on short lateral branches, blooming 1 or 2 at a time and lasting but a day. Calyx lobes linear-lanceolate, ciliate and sometimes hairy on the back as well, 15–25 mm long and about 3.5 mm wide. Corolla pale blue-violet, 4–6.5 cm long, the lobes rounded, rather wrinkled in appearance. Capsules hard, club-shaped, few-seeded, 12–17 mm long, subtended by the persistent calyx. Seeds rusty-brown, nearly round in outline, flattened, about 3.5 mm long, covered with inconspicuous hairs which swell when the seed is wetted.

Recalling that *Ruellia humilis* thrives best in open, dry situations, we are interested to find *R. strepens* occurring in woods and thickets, whether in well-drained, rocky soil, or on flat land bordering streams, in both situations thriving through the worst of droughts and always fresh and upstanding when its companions of other species are drooping. Yet it seems to require partial shade.

ACANTHUS FAMILY
357

One feature of the blooming habits of this plant we should not overlook—the succession of cleistogamous flowers through the last half of summer, when one sees flower buds coming on and showing a whitish corolla that never opens, always falling off before it is half grown, the ovary remaining and becoming a capsule filled with seeds. Indeed, that there was such a thing as cleistogamy in flowers (Gr. *kleistos*, closed, hidden, + *gamos*, marriage, union for reproduction) was discovered by the German botanist Dillenius (1687–1747), when he was examining a plant named later by Linnaeus *R. clandestina* (*L. clandestinus*, hidden, secret). Since then many examples of cleistogamy have come to light in genera of other families, such as *Viola* and *Triodanis*. Pollination in the usual sense of the word, i.e., transference of pollen from the anther to the stigma, does not take place in cleistogamous flowers, but the pollen grains germinate while enclosed in the anthers, the pollen tubes enveloping and penetrating the style. By this method no pollen grains get lost, and in consonance with this fact few are produced.

514. *Phryma leptostachya*

PHRYMÀCEAE   Lopseed Family

Perennial herbs. Stems simple, rarely branching except in the inflorescence. Leaves opposite, simple, petiolate, estipulate. Flowers bisexual, irregular, bracted, borne in slender terminal spikes. Sepals 5, fused most of their length, the calyx 2-lipped with 2 shorter lobes or teeth below and 3 longer ones above. Petals 5, but the number somewhat obscured by their fusion into a 2-lipped corolla. Stamens 4, arising at 2 different heights on the corolla tube. Pistil 1; ovary superior, 1-celled; style 1, slender; stigma 2-lobed. Fruit a dry, indehiscent nutlet, enclosed by the persistent calyx and strongly reflexed at maturity.

The family contains but a single genus, *Phryma*, known only from northeastern Asia and eastern North America. Some authors divide the genus into 3 or 4 species, others treat the Asian and American plants as a single species.

PHRỲMA L.   Lopseed

Derivation of the name unknown.

Characters of the family.

P. leptostàchya L.

Early generic name, from Gr. *leptos*, fine, small, thin, delicate, + *stachys*, spike—in reference to the slender inflorescence.
Pale purple to white; late June–mid-August
Moist, rich soil of woodlands
East half, primarily, especially in the east third

Erect, glabrous or minutely hairy perennial, 3–8 (–10) dm tall, with a cluster of thickened roots. Leaves usually long-petioled below, becoming shorter-petioled above, the blades thin, mostly lanceolate to ovate or deltoid, 3.5–15 cm long and 2.2–12 cm long, pinnately veined, the apex acute or acuminate, the base varying from wedge-shaped to rounded to somewhat truncate, often oblique, the margins coarsely serrate or double-serrate. Spikes 2.5–15 cm long (excluding the peduncle), or longer in fruit. Calyx tubular, strongly ribbed, 4–6 mm long (9–10 mm long in fruit), the short lobes triangular-acuminate, the long lobes attenuate and recurved at the tip. Corolla pale purple to white, 5–9 mm long. Fruit tightly enclosed by the persistent calyx.

See plate *101*.

PLANTAGINÀCEAE   Plantago Family, Plantain Family

Mostly annual or perennial acaulescent herbs. Leaves mostly basal, simple, venation frequently appearing to be parallel, sessile or more or less narrowed to a petiole, estipulate. Flowers small and inconspicuous, bisexual, or unisexual and then monoecious or dioecious, more or less regular, borne in terminal spikes or rarely solitary. Sepals 4, fused much of their length or free nearly to the base, the margins often membranous. Petals 4, membranous, fused part of their length but free above. Stamens 4, rarely fewer, arising from the corolla and usually protruding at maturity. Pistil 1; ovary superior, usually 2-celled; style 1, threadlike, hairy, receptive to pollen most of its length. Fruit in ours a circumscissile pyxis, i.e., a capsule splitting apart transversely all around and casting off the top part.

There are 3 genera and about 270 species of world-wide distribution in this family. The group is of little economic importance.

Kansas has 1 genus.

PLANTÀGO L.   Plantain

L. *plantago*, the ancient name for the plant, from L. *planta*, sole of the foot, + *ago*, to put forth or extend—a name suggested by the shape and posture of the leaves of the type species, *P. major*.

Acaulescent annual or perennial herbs. Leaves all basal, often broad and prominently ribbed. Flowers bisexual in all our species except *P. virginiana*, in 1 to several bracted spikes or heads borne on naked peduncles. Corolla constricted at the throat, the lobes reflexed, spreading, or erect. Stamens included within the corolla tube or protruding, the filaments weak, the anthers soon falling away. Capsule 2-celled, 2- to several-seeded, more or less membranous, at least partly enclosed by the persistent calyx and corolla. Seeds usually with a gelatinous outer coat which swells when the seed is wetted.

Nine species in Kansas.

Plantago seeds are fed to cage birds, as well as being eaten by wild game birds, song birds, and rodents. The plants are eaten by rabbits, white-tailed deer, and small rodents.

Several medicinal uses for plantain are described by Krochmal and Krochmal (30). The crushed leaves have been applied to bruises and swellings, and a tea made from the roots and/or leaves has been used as a tonic or to treat dysentery, constipation, and blood disorders. The seeds, apparently because of the gelatinous outer coat described above, function as bulk laxatives and have been used in lotions and wave-setting solutions.

At least some species of *Plantago*, including *P. lanceolata*, *P. major*, and *P. regelii*, are known to cause hayfever.

515. *Plantago aristata*

Key to Species:

1. Bracts and sepals glabrous or inconspicuously ciliate 2
   2. Delicate plants up to about 1 dm tall; leaves linear, 0.4–1 mm wide .... *P. pusilla*
   2. Coarse plants up to 6 dm tall; leaves broadly ovate to lanceolate, narrowly elliptic, or linear, 0.8–15 cm wide 3
     3. Leaf blades broadly ovate to lanceolate, the apex acute or rounded; spikes at flowering time narrowly cylindrical, rather loosely flowered (at least toward the base), 8–40 cm long; capsules 4- to 9-seeded .... *P. rugelii*
     3. Leaf blades oblanceolate to narrowly elliptic or nearly linear, the apex acute or attenuate; spikes at flowering time broader at the base, densely flowered, 1–5 cm long; capsules 2-seeded .... *P. lanceolata*
1. Bracts and/or sepals distinctly hairy 4
   4. Leaves oblanceolate to obovate; corolla lobes erect after anthesis 5
     5. Anterior sepals rounded, the green midrib not extended beyond the scarious margin at the tip; seeds yellowish, narrowly ovate to oblong, 1.4–1.8 mm long .... *P. virginica*
     5. Anterior sepals acuminate, the green midrib extended beyond the scarious margin; seeds usually bright red or reddish-black, ovate to narrowly ovate, 2–3 mm long ...... *P. rhodosperma*
   4. Leaves linear; corolla lobes after anthesis widely spreading or reflexed 6
     6. Bracts, at least the lower ones, twice as long as the flowers or longer; eastern Kansas primarily .... *P. aristata*
     6. Bracts little if any exceeding the flowers and not at all conspicuous; throughout ...... *P. patagonica*

P. aristàta Michx.   Bracted plantain

NL., from L. *aristosus*, full of awns or bristles—descriptive of the spikes.
Greenish or whitish; June–July (–August or later)
Dry, rocky or sandy prairie hillsides and banks, roadsides, and waste ground
East half, primarily

Hairy, dark green annual, 1–4 dm tall. Leaves linear, 1–2.4 dm long and 2–8 mm wide, the apex attenuate, the margins entire. Spikes densely flowered, 1.5–15 cm long, the bracts linear, leaflike, rather rigid, up to 3 cm long, protruding conspicuously from the spike. Calyx about 3 mm long, the lobes oblong to narrowly obovate, with a rounded apex and membranous margins. Corolla about 4–5 mm long, the lobes spreading or reflexed, about 2 mm long. Capsules ellipsoid, 2-seeded, dehiscing slightly below the middle, always concealed by the persistent corolla. Seeds reddish-tan, boat-shaped 2–3 mm long, the surface minutely reticulate (use high magnification and side illumination).

When the seeds are wetted, fine gelatinous scalelike structures emerge from the areoli of the netlike surface. The function of these structures is open to speculation.

P. lanceolàta L.   Ribwort, ribgrass

L. *lanceolatus*, possessed of a little spear—in reference to the leaf shape. "Ribwort is the old English name, suggested by its ribbed leaves.
Flowers greenish; May–November
Disturbed areas such as yards, roadsides, waste ground
East half, primarily, especially in the east fourth; intr. from Europe

516. *Plantago lanceolata*

Coarse, glabrous to sparsely hairy biennial or perennial, 2–7 dm tall. Leaves 0.8–3 dm long (including the petiole) and 0.8–4.5 cm wide, oblanceolate to narrowly elliptic or nearly linear, strongly ribbed, the apex acute or attenuate, the base tapered very gradually to the petiole, the margins entire or shallowly and remotely toothed. Spikes densely flowered, at flowering broader at the base, 1–5 cm long, becoming more cylindric in fruit, the bracts ovate-acuminate, membranous with a narrow, dark, sharply keeled midvein, longer than the calyx. Calyx lobes about 4 mm long, much like the bracts, but the outer 2 united. Corolla about 3.5–5 mm long, the lobes spreading, about 2 mm long. Capsules ellipsoid, 2-seeded, 3–4 mm long, dehiscing near the base. Seeds golden brown to blackish, boat-shaped, 2–3 mm long, with a homogeneous gelatinous coat which swells up when the seed coat is wetted.

Looking at a spike before it begins to bloom, we see only the bracts that tightly cover the tiny flowers; later, beginning at the base of the spike and gradually proceeding toward the apex the single slender style of a flower is thrust out beyond the scale, ready to receive pollen borne on the wind.

When the style appears, the 4 stamens of the flower are tucked inside the tiny corolla; and when the style withers, the yellowish-white anthers are brought out beyond their covering bract on slender filaments and deliver their pollen to the wind. By this simple process, self-pollination is prevented and the impartial wind carrying the pollen would quite certainly drift over all the styles, leaving none unpollinated. Not being required to make a show before insect pollinators, the minute corolla with its short tube and 4-lobed limb remains in retirement. In fact, if we are not searching for floral details the stamens are all we are likely to notice.

Since ancient time, even to this day, the species has been used medicinally for a variety of ailments. Dioscorides and Pliny (1st century) reported its use by soothsayers and as a love charm; and a tea made of its leaves and seeds was reputed to be efficacious in cases of hemorrhoids, fever, cramps, stomach and liver troubles, headache, asthma, toothache, earache, whooping cough, diarrhea, wounds, and poisonous bites of serpents and insects.

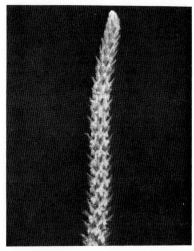

517. *Plantago patagonica*

P. patagònica Jacq.   Woolly Indianwheat

"Of Patagonia.
Whitish; mid-May–July
Prairie, usually in sandy soil
Throughout

Hairy, gray-green annual, 1–3.6 dm tall. Leaves linear, 5.5–17 cm long and 2–8 mm wide, the apex attenuate, the margins entire. Spikes 1–12 cm long, the bracts lanceolate to linear-attenuate, shorter than to definitely longer than the calyx, up to 9 mm long. Calyx about 1.5–2.5 mm long, the lobes more or less oblong. Corolla lobes spreading or reflexed, about 2 mm long. Capsules ellipsoid, 2-seeded, dehiscing slightly below the middle, always concealed by the persistent corolla. Seeds about 2–2.5 mm long, essentially like those of *P. aristata*.

P. pùsilla Nutt.

L. *pusillus*, very small.
Flowers greenish; mid-April–June
Sandy, often rocky soil of prairies or open woods
Southeast eighth, primarily; scattered collections reported as far north and west as Republic and Ellis counties

Delicate, glabrous or hairy annual, 4–11 cm tall. Leaves linear, usually entire, 2–6 cm long and 0.4–1 mm wide. Spikes rather loosely flowered, 1.5–6 cm long, the bracts triangular-ovate, about as long as the sepals, with membranous margins. Calyx 1–1.5 mm long, the lobes obovate, with an oblong green midregion and membranous margins. Corolla 1.5–2 mm long, the lobes about 0.5 mm long, usually erect and forming a closed

518. *Plantago pusilla*

PLANTAGO FAMILY

beak over the capsule. Capsules ovoid, 4-seeded, about 2 mm long, dehiscing slightly below the middle. Seeds dark brown, elliptic, minutely pitted, about 0.75–1.8 mm long, when wetted displaying gelatinous structures as described for *P. aristata*.

    P. elongata Pursh, a small annual closely resembling *P. pusilla*, is known from several counties, mostly in the central third of the state. It tends to be somewhat larger than the latter species, has corolla lobes which are spreading or reflexed in age, and has capsules with (3–) 4–5 (–6) seeds 1.75–2.5 mm long.

519. *Plantago rugellii*

P. rhodospérma Dcne.   Red-seeded plantain

NL., from Gr. *rhodon*, red, + *sperma*, seed.
Flowers greenish or whitish; May–June
Open sites in prairie, along roadsides, and in open woods, usually in rocky or sandy soil
East three-fourths

    Coarse, hairy, gray-green annual, 0.8–4.5 dm tall. Leaves oblanceolate to elliptic, 4.5–22 cm long and 1–4 cm wide, the apex usually acute, the margins entire or remotely shallowly toothed. Spikes 2–20 cm long, the bracts lance-linear, sharply keeled, about as long as the calyx. Calyx about 3 mm long, the lobes with a green keel which at the apex extends beyond the scarious margins. Corolla 5–8 mm long, the lobes erect after anthesis. Capsules ovoid, 2-seeded, 5–7 mm long, dehiscing slightly below the middle. Seeds usually reddish, ovate to narrowly ovate, concavo-convex, 2–3 mm long; when wetted producing gelatinous protuberances as described for *P. aristata*, although in this case the seed surface is not visibly reticulate.

P. rugèlii Dcne.   Blackseed plantain

Named for its discoverer, Ferdinand Rugel (1806–1879).
Flowers greenish; June–September
Disturbed areas such as roadsides, pastures, creek banks, yards, waste ground
East three-fifths, primarily

    Coarse, glabrous perennial, 1–6 dm tall. Leaf blades lanceolate to broadly ovate, 4–23 cm long and 2–15 cm wide, prominently ribbed from the base, the apex acute or rounded, the base gradually or rather abruptly tapering to the petiole, the margins often undulate, entire to coarsely toothed; petioles (1–) 3–13 cm long. Spikes 8–40 cm long, often loosely flowered, at least toward the base, the bracts resembling the calyx lobes but smaller. Calyx 3–3.5 mm long, the lobes lanceolate, with a prominent green keel and broad membranous margins. Corolla 6–7 mm long, the lobes about 1 mm long. Capsules narrowly ellipsoid, 4- to 9-seeded, 3–6 mm long, dehiscing far below the middle. Seeds black, elliptic but truncate at one end, concavo-convex, about 1.8–2.2 mm long, producing gelatinous protuberances when wetted.

    P. major L., an introduced European weed, is recorded from scattered localities, mostly in the eastern part of the state. It differs from *P. rugelii* in having the sepals and bracts ovate with rounded tips and capsules ovoid to stoutly elliptic or nearly globose, 6- to 16-seeded, and dehiscing near the middle (just below the sepal tips). The seeds are brown, smaller and more irregularly angled than those of *P. rugelii*, and have a minutely wrinkled or reticulate surface (use high magnification).

P. virgínica L.   Paleseed plantain

"Of Virginia."
Flowers greenish or whitish; late April–June
Exposed sites in prairie, open woods, along roadsides, usually in shallow or sandy soil
East two-thirds

    Coarse, hairy annual, 0.8–4 dm tall, much resembling *P. rhodosperma*. Leaves spatulate, 2–11 cm long and 8–32 mm wide. Spikes 2–15 cm long, the bracts oblong to lance-oblong, about 1.5 mm long. Calyx 2–2.5 mm long, the lobes ovate to broadly elliptic or oblong with membranous margins. Corolla 4–6 mm long, the lobes always erect. Capsules broadly ellipsoid, 2-seeded, dehiscing just below the middle. Seeds yellowish-brown, narrowly ovate to oblong, concavo-convex, 1.4–1.8 mm long, when wetted producing gelatinous protuberances as described for *P. aristata*, although in this case the seed surface is not visibly reticulate.

RUBIÁCEAE   **Madder Family**

Named after the Eurasian genus *Rubia*, from L. *ruber*, red—in reference to the red dyestuff madder, obtained from its roots.

520. *Diodia teres*

Annual or perennial herbs, or shrubs, trees, or woody vines. Leaves opposite or appearing whorled, simple, entire, stipulate, the stipules sometimes sheathing the stem, sometimes large and indistinguishable from the leaves. Flowers bisexual or unisexual, often of 2 or 3 kinds with regard to relative lengths of stamens and styles, regular, borne singly or in clusters. Sepals 4–5, fused to one another, sometimes much reduced or apparently absent as in the genus *Galium*. Petals usually 4–5, rarely 3, fused at least part of their length. Stamens of the same number as the corolla lobes, inserted on the corolla tube. Pistil 1; ovary inferior (rarely superior or half-inferior), usually 2-celled; style 1, often with 2 branches; stigmas 2 or 1 with 2 lobes. Fruit a capsule, a pair of small, indehiscent, 1-seeded burrs or nutlets, or a fleshy berry.

A large family with approximately 500 genera and 6,000 species, distributed around the world, primarily in tropical and subtropical zones. Economically important members of the family include coffee (*Coffea arabica* L.) and quinine (*Cinchona* spp.), plus a number of ornamentals such as *Gardenia*, woodruff (*Asperula* spp.), madder (*Rubia* spp.), partridgeberry (*Mitchella repens* L.), crosswort (*Crucianella stylosa* Trin.), and others. Sweet woodruff (*Asperula odorata* L.) is used as a flavoring in May wine.

Kansas has 5 genera. *Cephalanthus occidentalis* L. is a shrub with firm, shiny leaves and spherical heads of fragrant white tubular flowers and is found commonly along borders of streams, ponds, and other wet places in the eastern two-thirds of the state. *Spermacoce glabra* Michx., known from Montgomery, Labette, Cherokee, and Linn counties, is a glabrous perennial herb with elliptic to oblanceolate leaves and tiny white or purplish flowers crowded into dense, rounded axillary heads.

Key to Genera:
1. Principal leaves appearing whorled .... *Galium*
1. Principal leaves opposite 2
   2. Flowers sessile or nearly so .... *Diodia*
   2. Flowers pediceled .... *Hedyotis*

DIÓDIA L.   Buttonweed

NL., from Gr. *diodos*, thoroughfare—these plants often growing by the wayside.

Ascending to spreading or prostrate annual or perennial herbs. Leaves opposite, entire, narrow, mostly sessile, the stipules with long bristles, sheathing the stem. Flowers small, white or pale purple, bisexual, sessile, borne 1–3 in the leaf axils. Calyx teeth 2–5, often unequal. Corolla funnelform or salverform with 4 (rarely 3) short lobes. Stamens included within or protruding from the corolla tube. Ovary inferior, 2-celled; style threadlike; stigma 1 and 2-lobed or 2 and linear. Fruit a pair of small, 1-seeded, indehiscent nutlets.

One species in Kansas.

D. téres Walt

L. *teres*, cylindrical.
White to pink or pale purple; July–early October
Prairies, open woods, roadsides, usually in sandy soil
Primarily in the east half, south of the Kansas River

Rough annual with stems branched, erect to ascending or spreading, 1–5 dm tall. Leaves narrowly lanceolate to lance-linear, 1.5–4 cm long and 3–8 mm wide, the apex acute or attenuate, the base rounded or clasping the stem, the margins usually scabrous, revolute. Sepals ovate to lanceolate, 1.5–3 mm long. Corolla white to pink or pale purple, funnelform, 4–6 mm long. Nutlets 3–4 mm long (excluding the persistent sepals).

GÁLIUM L.   Bedstraw, cleavers

Though Gr. *galion* was the name of a plant, Linnaeus may have had in mind Gr. *gala*, milk— plants of this genus having been used formerly for curdling milk, as we learn from Gerard concerning the Eurasian G. *verum*: "The people in Cheshire, especially about Nantwich where the best cheese is made, do use it in their Rennet, esteeming greatly of that cheese above other made without it." Also, the fluffy straw of this species was used for filling mattresses—hence the vernacular name "bedstraw."

Slender, annual or perennial herbs with 4-angled stems, often with a red coloring matter in the roots. Leaves appearing whorled (actually opposite with large, leaflike stipules), entire, sessile or but short-petioled. Flowers small, mostly white (rarely red or brownish), mostly bisexual (dioecious in a few species), usually in clusters. Sepals apparently absent. Corolla wheel-shaped, 4- (rarely 3-) lobed. Stamens usually shorter than the corolla. Ovary inferior, 2-celled; styles 2. Fruit separating into 2, globose, 1-seeded, indehiscent, usually dry, often bristly segments (mericarps).

Seven species in Kansas.

Key to Species:
1. Fruit smooth 2
   2. Leaves abruptly short-acuminate or mucronate at the tip; inflorescences several- to many-flowered .... *G. concinnum*
   2. Leaves rounded to acute at the tip; never short-acuminate or mucronate; inflorescences 2- to 4-flowered .... *G. obtusum*
1. Fruit bristly or hairy 3
   3. Principal stem leaves in whorls of 6 or 8; stems reclining or prostrate, glabrous to more or less retrorsely roughened on the angles 4
      4. Annual; leaves linear to narrowly oblanceolate, mostly in whorls of 8; stems retrorsely roughened .... *G. aparine*
      4. Perennial; leaves oblanceolate to elliptic; mostly in whorls of 6; stems glabrous or retrorsely roughened .... *G. triflorum*
   3. Principal stem leaves in whorls of 4; stems erect or ascending, never retrorsely roughened on the angles 5
      5. Flowers all pediceled, terminating the branches of the inflorescence .... *G. pilosum*
      5. Flowers not as above 6
         6. Flowers sessile in the axils of foliage leaves; leaves 4–9 mm long .... *G. virgatum*
         6. Flowers, or some of them, sessile or nearly so along the side of the inflorescence; leaves 15–45 mm long .... *G. ciraezans*

521. *Galium aparine*

G. aparìne L.   Catchweed bedstraw

According to Fernald (9), an old generic name interpreted by botanists for centuries to mean "to catch, cling, or scratch."
White; May–June
Woods, thickets, prairie ravines, waste ground, usually in moist soil
Throughout except for the southwest tenth; intr. from Europe

Weak, sprawling or scrambling annual. Stems brittle, 1–15 dm long, retrorsely roughened. Leaves and stipules in whorls of 6 or 8, narrowly oblanceolate, mostly 1–5 cm long and 1–8 mm wide, the apex short-acuminate or mucronate, the base wedge-shaped. Flowers small and inconspicuous, in few-flowered axillary or terminal clusters. Corolla white, 1–2 mm long. Fruit 1–3 mm long, covered with hooked bristles.

This species was spoken of as having high medicinal value by both Dioscorides and Pliny in the 1st century, and the herbals of the Middle Ages and later times praise its curative powers. Thus Hieronymus Bock (1577) said of it that it was highly prized by the ancients for both external and internal use and that the expressed juice of the herbage drunk with wine kept the poison of snakes and scorpions from reaching the heart; a decoction in hot water was good for biliousness and when put into the ears relieved earache; the herbage pounded up in lard and spread over the neck dispelled goiter; and leaves applied to wounds stopped bleeding; it was good for scrofula and scurvy, and for St. Vitus's dance, epilepsy, hysteria, etc.

The herbalist Gerard tells us for the 16th century that "women do usually make pottage of Clevers [cleavers; an old popular name referring to the plant's cleaving to clothing, etc.] with a little mutton and otemeale, to cause lanknesse, and keepe them from fatnes." And Parkinson in the 17th century relates that "the herbe serveth well the Country people in stead of a strainer, to cleare their milke from strawes, haires, or any other thing that falleth into it."

522. *Galium circaezans*

According to Krochmal (30), the entire plant, harvested while in bloom, has been used to increase urine flow, stimulate the appetite, reduce fever, remedy vitamin C deficiency, and as a wash to reduce freckles.

The mature fruits, dried and roasted, provide a good coffee substitute, though collecting enough of them is a tedious task.

G. circaèzans Michx.

After the genus *Circaea*, enchanter's nightshade, in the Evening Primrose Family.
White; mid-May–June (–August)
Woodlands
East two-fifths

Erect to ascending perennial, 1.6–8 dm tall. Leaves and stipules in whorls of 4, ovate to lanceolate or broadly elliptic, usually 3-nerved, 1.5–4.5 cm long and 7–23 mm wide, the apex acute, the base acute or acuminate. Flowers small and inconspicuous, at least some of them sessile along the side of the branched, few-flowered axillary and terminal clusters. Corolla white, 1–2.5 mm long. Fruit 1–4 mm long, covered with hooked bristles.

G. concínnum T. & G.

L. *concinnus*, well arranged, skillfully joined, beautiful, striking.
White; June–July (–September)
Woods and thickets, usually in moist soil
East third, especially the northeast sixth

Delicate, glabrous, spreading or ascending perennial, 2–6.5 dm tall, with slender rhizomes. Leaves and stipules in whorls of 4 or 6, narrowly elliptic to linear, 0.7–2.5 cm long and 1.5–4 mm wide, the apex acute or rounded and minutely mucronate, the base wedge-shaped. Flowers small and inconspicuous, borne on threadlike pedicels and peduncles in several- to many-flowered, branched terminal and axillary clusters. Corolla white, 1–2 mm long. Fruits smooth, 2 mm long.

A tea made from the leaves has been used to treat dropsy and kidney disorders.

523. *Galium concinnum*

G. obtùsum Bigel.

L. *obtusus*, blunt, dull.
White; May–June
Moist soil of woods and prairie thickets
Scattered primarily in the east third

Delicate, erect perennial, 3–7 dm tall, closely resembling *G. concinnum*. Leaves and stipules in whorls of 4 or 6, narrowly elliptic to nearly linear, 0.5–2 cm long and 1–4 mm wide, the apex rounded or acute, but never mucronate. Flowers small and inconspicuous, in short, 2- to 4-flowered, mostly terminal clusters. Corolla white, 4-lobed, 1.5–2 mm long. Fruit about 3 mm long, smooth.

G. pilòsum Ait.   Hairy bedstraw

L. *pilosus*, hairy, from *pilus*, hair.
White to reddish-purple; mid-June–July
Open woods over rocky sandstone soil
Southeast eighth

Rough-hairy, erect to ascending perennial, 1–6.5 dm tall, with slender rhizomes. Leaves and stipules in whorls of 4, broadly elliptic, 1–2 cm long and 4–12 mm wide, the apex usually rounded and mucronate, the base acute or acuminate. Flowers small and inconspicuous, on short pedicels terminating the branches of the inflorescence. Corolla white to reddish-purple, 1–2 mm long. Fruit about 3 mm long, covered with hooked bristles.

The roots have been used as a source of red dye.

524. *Galium pilosum*

G. triflòrum Michx.

NL., from L. *tri-*, 3, + *flos*, *floris*, flower—the flowers (as in some of the other species as well) often being in 3's.
White; June–July
Woods and thickets
East third

Weak, prostrate or reclining perennial. Stems glabrous to retrorsely roughened, 2–10 dm long. Leaves and stipules in whorls of 6, oblanceolate to elliptic, 1–3.5 cm long and 6–12 mm wide, with a single prominent vein, the apex abruptly short-acuminate to merely mucronate, the base wedge-shaped or acuminate, the margins rough. Flowers small and inconspicuous, pediceled, in few- to several-flowered axillary and terminal clusters. Corolla white, 1–2 mm long. Fruit about 2 mm long, covered with hooked bristles.

This species, like sweet woodruff, may be used to flavor wine.

G. virgàtum Nutt.

L. *virgatus*, of twigs, twiglike from *virga*, twig, branch.
White; May–mid-June
Prairies and pastures, usually in shallow soil over limestone
Southeast corner, from Bourbon County to Cowley County

Slender, erect, glabrous to sparsely hairy annual. Stems simple, sparingly branched above and/or several times branched at the base, 0.8–3.3 dm tall. Leaves and stipules in whorls of 4, mostly elliptic, 4–9 mm long and 1–2 mm wide, the apex acute. Flowers sessile or nearly so, usually borne singly in the leaf axils. Corolla white, about 0.5 mm long. Fruit 2–2.5 mm long, covered with hooked bristles, reflexed at maturity.

HEDYOTIS

Bluets

NL., from Gr. *hedys*, sweet, pleasant, + *ous, otos*, ear—the reference unclear.

Prostrate to erect annual or perennial herbs, rarely low shrubs. Leaves opposite, sessile or petioled, often linear. Flowers white to blue or purple, small (in ours), bisexual, often of 2 or 3 kinds, terminal, borne singly or in clusters. Calyx 4-lobed, persistent. Corolla funnelform or salverform, 4-lobed, the throat often hairy. Ovary half-inferior, 2-celled; style 1; stigmas 2, narrow. Fruit a several-seeded, 2-celled capsule, more or less 2-lobed and loculicidal above.

Three species in Kansas. *H. longifolia* (Gaertn.) Hook. is known from extreme southeastern Cherokee County only and is not included here.

Key to Species:
1. Delicate annual, 2–11 cm tall; flowers blue or purple (rarely white), borne singly .... *H. crassifolia*
1. Bushy perennial, 1–5 dm tall; flowers white, in many-flowered clusters .... *H. nigricans*

525. *Hedyotis crassifolia*

H. crassifòlia Raf.

Tiny bluets

NL., from L. *crassus*, thick, + *folium*, leaf.
Blue or purplish; late March–early May
Prairies, open woods, pastures, occasionally in lawns
East two-fifths, primarily

Low, delicate annual, 2–11 cm tall, with a few spreading branches. Lower leaves mostly ovate to broadly spatulate, 4–10 mm long (including the petiole) and 2–7 mm wide; middle and upper leaves smaller, sessile, mostly elliptic. Herbage roughened with minute hairs. Flowers borne singly, 6–11 mm long. Sepals lanceolate to ovate-acuminate, about 2 mm long. Corolla salverform, blue or purplish (rarely white) with dark nectar guides radiating from the throat. Capsules thin-walled, about 2 mm long and 4 mm wide. Seeds brown, about 1 mm long, slightly flattened, nearly round to somewhat angular in outline, with a prominent pit on the funicular face, the rest of the surface with a netlike ornamentation (use high magnification).

This plant has been listed in earlier manuals as *Houstonia minima* and *Hedyotis minima*.

The stamens and style are included in the corolla tube, and while the flowers look alike, they are really of 2 kinds, some having styles extending above the anthers, others having the anthers standing above the styles, so that bees seeking nectar at the base of the corolla tube would get pollen on their proboscises from the long stamens at the right height to deposit it on the stigmas of other flowers with long styles, thus effecting cross-pollination. This would occur as well between flowers with short stamens and those with short styles.

526. *Hedyotis nigricans*

H. nìgricans (Lam.) Fosb.

Narrowleaf bluet

L. *nigricans*, blackish, dark.
White or pinkish; mid-June–July
Rocky prairie hillsides and banks and roadsides
Nearly throughout

Glabrous or nearly glabrous perennial with several to many stems 1–5 dm tall arising from a woody taproot and crown. Leaves linear to very narrowly elliptic, 1.5–4 cm long and 1–4 mm wide, the apex acute, the margins minutely roughened, often rolled under, fascicles of smaller leaves usually present in the axils. Flowers many, in terminal clusters, 5–9 mm long. Sepals lanceolate to lance-attenuate, about 1.5–2.5 mm long, usually keeled or at least with a prominent midrib. Corolla white or pinkish, funnelform, papillose. Anthers blue. Capsules relatively thick-walled, about 3–4 mm long and 2–2.5 mm wide.

This graceful, starry-flowered perennial shows hardiness and adaptability to a wide range of conditions by its occurrence in all parts of the state, which vary in amount of rainfall and character of soil. We find it, for instance, in the glacial soil of Doniphan and Brown counties, where the annual rainfall averages around 33 inches; in the residual soil from shale in Cherokee and Crawford counties, where the rainfall averages more than 42 inches; and in extreme western Hamilton County, with an average rainfall below 17 inches and a variety of soils—wind deposits; outwash, both heavy and light; shallow and poor marly soils; and dune sand.

Certainly this widespread, hardy species deserves a place in garden borders and rock gardens, yet it apparently is not cultivated anywhere.

527. *Triosteum perfoliatum*

CAPRIFOLIÀCEAE — Honeysuckle Family

After *caprifolium*, ML. name for woodbine (*Lonicera periclymenum*), "the honisuckle that groweth wilde in every hedge" (12). Caprifolium is NL., from L. *caper*, billy goat, or *capra*, nanny goat, + *folium*, leaf—the leaves on twining stems seeming to be capering in and about the trees and shrubs of hedgerows. Wittstein's *Etymologishbotanisches Handworterbuch* says "because the plants climb like a goat."

Shrubs, trees, woody vines, or perennial herbs. Leaves opposite, simple or pinnately compound, sessile or petiolate, stipules usually lacking (present in *Sambucus*, present as nectiferous glands in *Viburnum*). Flowers bisexual, regular or irregular, borne in axillary or terminal clusters of 2 to many. Sepals 5, usually fused at least part of their length. Petals 5, united, the corolla wheel-shaped, tubular, funnelform, bell-shaped, or urn-shaped, often 2-lipped, mostly 5-lobed. Stamens 5, inserted on the corolla tube. Pistil 1; ovary inferior, 1- to 5-celled; style 1 or absent; stigma with 2–5 lobes. Fruit a berry, drupe, or capsule.

The family includes about 12 genera and 450 species, primarily distributed in the Northern Hemisphere. Cultivated members of the family include the honeysuckles (*Lonicera* spp.), elder or elderberry (*Sambucus* spp.), *Virburnum*, coralberry (*Symphoricarpos orbiculatus* Moench.), beauty bush (*Kolkwitzia amabilis* Graebn.), bush honeysuckle (*Diervilla* spp.), and *Weigela*.

Kansas has 5 genera and 13 species, most of which are woody shrubs or vines, i.e., *Lonicera*, *Sambucus*, *Symphoricarpos*, and *Viburnum*. The only herbaceous genus represented in Kansas is *Triosteum*.

TRIÓSTEUM L. — Feverwort, horse gentian

NL., from Gr. *treis*, 3, + *osteon*, bone—in reference to the bony nutlets in each fruit.

Coarse, erect, unbranched perennial herbs. Leaves simple, sessile or sometimes perfoliate, entire. Flowers greenish-yellow to dull red, each subtended by 2 small bracts, borne singly or few together in the leaf axils. Calyx lobes linear to linear-lanceolate, as long (or nearly so) as the corolla. Corolla tubular, slightly irregular, pouchlike on 1 side at the base, curved outward, the 5 lobes erect. Stamen filaments quite short, the linear anthers included within the corolla tube. Ovary usually 3-(5-)celled, 1 ovule per cell; style slender; stigma capitate, included within or protruding from the corolla tube. Fruit a dry berry containing 3 hard ribbed seeds.

Three species in Kansas. *T. angustifolium* L. and *T. aurantiacum* Bickn. are quite rare in eastern Kansas and are not treated here.

T. perfoliàtum L. — Horse gentian

NL., from L. *per*, through, + *folium*, leaf—the united bases of some lower pairs of leaves appearing to be perforated by the stem.
Purplish-brown; May–early June
Woods and thickets
East third

Herbaceous perennial with a rugged, irregular rhizome producing 1 or more stout stems 0.4–1.3 m tall. Herbage more or less densely covered with soft nonglandular hairs intermixed with glandular hairs. Leaves mostly 8–22 cm long and 3.5–8 cm wide, the distal portion of each blade ovate or lanceolate with an acuminate apex, constricted below the middle, those of the lower leaves then widening toward the cordate-perfoliate bases, those of the upper leaves tapering gradually to more or less perfoliate bases. Flowers 1–4 per leaf axil. Calyx lobes linear, 10–18 mm long, uniformly covered on the back with glandular and nonglandular hairs. Corolla purplish-brown, 11–18 mm long, hairy on the outside. Fruit orange, 7–11 mm long, crowned by the persistent calyx. Seeds bony, longitudinally grooved, 7–8 mm long.

Several features in the design of the flower promote cross-pollination by bees: (1) Nectar from glands around the base of the style is protected from rain and from various small-fry insects useless in pollination by corolla lobes overlapping above the throat of the tube, and by blocking the way down the tube by means of fine hairs on the style and filaments; (2) correlation of the length of the corolla tube with the length of tongue of various kinds of bees; (3) prevention of self-pollination by early extension of the style and presentation of the stigma out in front of the corolla, while (4) the anthers remain behind and discharge their pollen inside the tube where the mouth parts of bees sucking nectar from the bottom of the tube would pick it up, and then transfer it to the stigma positioned before the entrance of another flower.

Various vernacular names have been given to this species: horse gentian, wild or wood ipecac, Tinker's weed, wild coffee, horse ginseng, white gentian, genson, and fever-wort, and, for a wonder, all of them can be rationalized: wild coffee, because the German settlers in Pennsylvania used the dried and roasted fruits as a substitute for coffee; fever-wort, because in the early practice of medicine in this country an infusion of the leaves was used to increase secretions and induce sweating in cases of intermittent fever; wild or wood ipecac, because a decoction of the plant was found to have the well-known emetic effect of ipecac; horse gentian, white gentian, and the corruptions of gentian—ginseng, genson—because *Triosteum* has sometimes been used in this country as a tonic, substituting for the officinal European yellow gentian; finally Tinker's weed, because a certain Dr. Tinker used the plant in his medical practice.

528. *Valerianella radiata*

VALERIANÀCEAE — Valerian Family

Annual, biennial, or perennial herbs. Leaves opposite or in basal rosettes, simple but sometimes deeply dissected or lobed, sessile or petiolate, estipulate. Flowers bisexual, or unisexual and dioecious, irregular to nearly regular, arranged in terminal clusters. Sepals minute or absent or developing late as the fruits mature and then present as filamentous appendages on the fruit. Petals 5, fused to one another part of their length. Stamens 1–4 (3 in ours), inserted on the corolla tube. Pistil 1; ovary inferior, 3-celled but 2 of the cells sterile and not well developed; style 1; stigma 1, simple or with 2–3 branches or lobes. Fruit dry and indehiscent.

The family comprises about 13 genera and 400 species, mostly of the Northern Hemisphere but also occurring in the South American Andes. Cultivated members of the family include ornamentals such as the common valerian or garden heliotrope (*Valeriana officinalis* L.), corn salad (*Valerianella olitoria* Poll.), African valerian (*Fedia cornucopiae* DC.), and the red valerian or Jupiter's beard (*Centranthus ruber* DC.).

Kansas has but a single genus and species.

VALERIANÉLLA Mill. — Corn salad

NL., from the genus *Valerian* + the L. diminutive suffix *-ella*.

Erect annuals or short-lived biennials with dichotomously branching stems. Leaves simple, the lowermost somewhat united around the stem, the uppermost sometimes with a few teeth near the base. Flowers small, white or bluish, regular, bracted, in dense terminal clusters. Calyx teeth 5 or not evident at all. Corolla funnelform to narrowly bell-shaped or tubular, pouchlike on 1 side at the base, with 5 more or less equal lobes. Stamens and style protruding from the corolla tube. Style 3-lobed.

One species in Kansas.

V. radiàta (L.) Dufr.

L. *radiatus*, having rays, from *radius*, ray—in reference to the radiating branches of the inflorescence.
White; late April–June
Low, moist areas in prairies or open woods, less frequently along roadsides or railroads
East fourth, south of the Kansas River; more common southward

Plants glabrous or sparsely hairy, 0.8–5 dm tall, somewhat malodorous when dried. Leaves spatulate below, spatulate to oblong or lanceolate above, 1–8 cm long and 0.4–2.2 cm wide, the apex rounded, the bases united around the stem, the margins minutely ciliate, entire or with a few teeth below the middle of the leaf. Flower clusters 8–12 mm across, subtended by lanceolate to lance-oblong or oblanceolate bracts. Corolla funnelform, white, 1.5–2 mm long. Fruit yellowish, ovoid, 1.5–2 mm long, glabrous or hairy, strongly indented between the 3 cells, only 1 of which contains a seed.

DIPSACÀCEAE — Teasel Family

Annual, biennial, or perennial herbs, rarely shrubs. Leaves mostly opposite,

529. *Dipsacus laciniatus*

simple, sessile or more or less narrowed to a petiole, estipulate. Flowers bisexual, or bisexual and unisexual flowers on the same plant, more or less irregular, borne in heads or dense spikes, each flower enveloped by a whorl of minute bracts, referred to as the involucel or epicalyx and sometimes also subtended by a longer, rigid bract of the receptacle. Sepals minute, sometimes similar to the pappus found in the Composite Family. Petals 4–5, fused to one another most of their length. Stamens usually 4, inserted on the corolla tube. Pistil 1; ovary inferior, 1-celled; style 1; stigma 1, simple or with 2 branches or lobes. Fruit an achene, often topped by the persistent calyx and enclosed by the involucel or epicalyx.

The family includes approximately 9 genera and 160 species, all of which are native to the Old World. Cultivated members of the family include such ornamentals as teasel (*Dipsacus* spp.), whorl-flower (*Morina longifolia* Wall.), *Scabiosa*, and a few others.

DÍPSACUS L. Teasel

Ancient Greek name for teasel, from *dipsa*, thirst, because the united, cuplike leaf bases characteristic of some species hold water.

Tall, coarse biennial or perennial herbs with prickly stems. Leaves opposite, entire to coarsely toothed or pinnatifid, the uppermost often united around the stem. Flowers small, white to blue or purplish, bisexual, long-tubular, slightly irregular, each subtended by a conspicuous, rigid, long-attenuate receptacular bract which persists long after the flowers and fruits have fallen, and also by the 4 minute united bracts of the epicalyx which completely envelops the inferior ovary, all this crowded into dense, many-flowered, ovoid or oblong-ellipsoid heads which are further subtended by an involucre of long, linear, rigid ascending bracts. Calyx cuplike, 4-angled or 4-lobed. Corolla long-tubular, unequally 4-lobed. Achenes tightly enclosed by the persistent epicalyx and crowned by the calyx.

Two species in Kansas, both introduced from Europe. *D. laciniatus* L., known from Crawford County only, differs from our more common species in having leaves with margins laciniate to variously once or twice pinnately lobed, the leaf bases united and forming a cup around the stem.

D. sylvéstris Huds. Wild teasel

L. *sylvestris*, of the woods, from *sylva*, *silva*, woods.
White to pale purple; July–early September
Roadside ditches, pastures, and waste ground, usually in moist soil
East fourth, plus Cheyenne and Rush counties; intr. from Europe

Stout, prickly-stemmed biennial, up to 2 m tall, with a taproot. Leaves prickly, at least along the midvein, the blades of lower leaves oblanceolate to elliptic, 1–4 dm long and 2–11 cm wide, with crenate margins, those of the upper leaves smaller, lanceolate to linear-lanceolate, with entire margins. Heads at flowering 3.5–8 cm long. Bracts of the epicalyx united about two-thirds of their length. Calyx 4-lobed, hairy, about 1 mm long. Corolla white to pale purple, minutely hairy, 7–10 mm long. Stigma 2-lobed. Achenes and epicalyx 4-angled, 3–3.5 mm long.

Compare with *Eryngium leavenworthii*.

The leaves and flower heads of *D. sylvestris* produce a yellow dye when mordanted with alum. The dried bristly heads of *D. fullonum* L., fuller's teasel, another European species, are still used in some parts of the world for raising the nap on woolen textiles, and the long-stalked heads of any of the teasels make attractive additions to dried flower arrangements.

See plate *102*.

CUCURBITÀCEAE Gourd Family

Herbaceous annual or perennial vines, usually climbing by tendrils, sometimes extending over the ground. Leaves alternate, simple or compound, mostly palmately lobed or divided, long-petioled, estipulate. Flowers unisexual, monoecious or dioecious, regular, solitary or in racemes. Perianth borne on the hypanthium, which, in female flowers, is united with the inferior ovary, sometimes extending beyond it. Calyx and corolla usually 4- or 5-lobed (6-lobed in *Echinocystis*). Stamens appearing to be 3 (actually 5 with 2 pairs united) or appearing to be but 1 (in *Cyclanthera*). Pistil 1; ovary inferior, with 1–3 cells; styles united into 1; stigmas thick and more or less united; a rudimentary pistil sometimes present in the male flowers. Fruit fleshy (a pepo) or a more or less dry berry.

This family of about 100 genera and 850 species includes our cultivated pumpkins and squashes (*Cucurbita* spp.), watermelon (*Citrullus vulgaris* Schrad.), cucumber and gherkin (*Cucumis* spp.), and muskmelon and cantaloupe (*Cucumis melo* L. varieties).

Five species in 5 different genera occur in the state. *Melothria pendula* L. is known from Cowley County only and is not included here.

Key to Genera:
1. Coarse, rough, prostrate, gray-green vine with thick stems lacking tendrils; corollas yellow, 6–12 cm long; fruit globose, smooth, hard, green striped with yellow or orange, many-seeded, 5–10 cm long .... *Cucurbita*
1. Vines with slender stems, climbing by means of tendrils; corollas much smaller and white to greenish or yellowish; fruit prickly, 1- to several-seeded 2
   2. Female flowers and fruits borne in dense headlike clusters, the fruits 1-seeded, indehiscent, the prickles minutely retrorsely barbed .... *Sicyos*
   2. Female flowers and fruits borne singly, the fruits with more than 1 seed, dehiscent, the prickles smooth 3
      3. Leaves palmately divided nearly to the midrib; male flowers with pollen sacs united and forming a single, horizontal ringlike sac around the anther column; ours with fruit opening irregularly and seeds 6.5–7.5 mm long .... *Cyclanthera*
      3. Leaves with 3–7 triangular or lanceolate lobes reaching about one-third to two-thirds of the way to the midrib; pollen sacs not as above; ours with fruit dehiscing through 2 pores at the apex and seeds 1.5–2 cm long .... *Echinocystis*

530. *Cucurbita foetidissima*

CUCÚRBITA L.   Gourd, squash, pumpkin

L. *cucurbita*, gourd.

Mostly trailing, annual or perennial herbs with forked tendrils. Leaves large, merely toothed to angled or palmately lobed. Flowers yellow, large and showy, male and female flowers on the same plant, borne singly in the leaf axils. Calyx lobes 5. Corolla bell-shaped or funnelform, the 5 lobes recurved at the tip. Ovary 1-celled with 3–5 placentae; stigmas 3–5, fleshy, each 2-lobed or -branched. Fruit hard, smooth, fleshy, indehiscent, many-seeded.

One species in Kansas.

C. foetidíssima H. B. K.   Buffalo gourd

L. *foetidus*, ill-smelling, + the suffix *-issimus*, most, i.e., most ill-smelling gourd—the name given to this plant in 1817 by Humboldt, Bonpland, and Kuntz after they had collected it in Mexico and noticed the fetid odor of the crushed leaves.
Yellow; June–mid-August
Rocky or sandy soil of disturbed prairie, roadsides, railroad embankments, and waste ground
Nearly throughout, but more common westward

Coarse, rough, prostrate, tendril-bearing perennial vine with stems 1.5–7.5 m long radiating from the crown of a strong, deep taproot which sometimes attains a diameter of 30 cm and a depth of 1.3–2 meters. Leaf blades triangular to cordate, 9–28 cm long and 5–20 cm wide, palmately veined, sometimes with shallow, angular lobes, the apex acute or acuminate, the base usually cordate to truncate, the margins minutely dentate. Calyx lobes subulate to linear-attenuate, 7–25 mm long. Corolla yellow, 6.5–12 cm long. Stamens of the male flowers with linear, coherent anthers and short free filaments surrounding a nectar gland. Fruit green with yellow or orange stripes, up to about 7 cm long, inedible, and so saponaceous that when crushed it can be used as a substitute for soap. Seeds white, smooth, 6–10 mm long, flattened, ovate, the narrow end acute.

On the dry plains, the plants are noteworthy for their constantly fresh appearance. Yet when a branch is cut off, its leaves immediately wilt, showing that the deep root system must be remarkably efficient in extracting water from an apparently dry soil.

Some tribes of prairie Indians attributed powers of magic to the plant. While they used the root medicinally, they took great care not to multilate it in digging it up, lest evil befall them.

See plate *103*.

CYCLANTHÉRA Schrad.   Cyclanthera

NL., from Gr. *kyklos*, ring or circle, + NL. *anthera*, anther—in reference to the unique feature in the flowers of this species.

Herbaceous, annual or perennial climbing vines with simple or 2- to 3-forked tendrils. Leaves deeply palmately lobed or compound. Flowers small, white to greenish or yellowish, male and female flowers borne in the same leaf axils, the male flowers in branched clusters, the female flowers single. Calyx lobes 5–6, subulate to threadlike, or absent. Corolla wheel-shaped, 5- to 6-lobed. Stamen filaments united into a central column, the anthers so coalesced as to form 1 ring-shaped pollen sac around the top of the column. Ovary 1- to 3-celled, each cell with a few ovules. Fruit prickly, several-seeded, elastically and irregularly dehiscent.

One species in Kansas.

531. *Cyclanthera dissecta*

C. dissécta (T. & G.) Arn. Cutleaf cyclanthera

L. *dissectus*, cut asunder—in reference to the deeply lobed leaves.
White or greenish; July–August
Rocky, brushy banks and thickets
Scattered primarily in the northwest fourth

Slender, glabrous annual vine with stems 1–3 m long climbing on other plants or sprawling where there is nothing to climb. Leaf blades mostly 3–9 cm long and 3–9 cm wide, divided nearly to the midrib into 3, 5, or 7 elliptic or lanceolate, serrate lobes, the apex of each lobe abruptly constricted to a threadlike tip. Flowers white or greenish, 3–5 mm across. Calyx united entirely with the corolla, scarcely discernible. Corolla broadly campanulate with triangular lobes, 2–3 mm long. Stigma 1, thick and fleshy. Fruit obliquely ovoid, 1.8–3 cm long, the prickles smooth. Seeds brown, ovate-acuminate, flattened, 6.5–7.5 mm long, usually with warty protuberances.

See plate *104.*

ECHINOCÝSTIS T. & G.

NL., from Gr. *echinos*, hedgehog, + *kystis*, bladder—because of the prickly, bladderlike fruit.

Herbaceous annual or perennial vines with simple or 2- or 3-forked tendrils. Leaves simple, palmately 5- to 7-lobed. Flowers small, white or greenish, monoecious, the female flowers solitary or in small clusters, the male flowers in conspicuous racemes or panicles. Calyx 5- or 6-lobed. Corolla wheel-shaped, 5- or 6-lobed. Male flowers with a bell-shaped hypanthium and 2–3 united stamens. Female flowers with an urn-shaped hypanthium and 2–3 more or less prominent staminodes. Ovary 2- or 3-parted or -lobed and 2- or 3-celled, each cell with 2 ovules; style 1; stigma broad, lobed. Fruit bladdery, covered with weak prickles, few-seeded, eventually becoming dry, spongy and fibrous within, dehiscing explosively either irregularly or through 2 pores at the apex.

One species in Kansas.

532. *Echynocistis lobata*

E. lobàta (Michx.) T. & G. Wild mock cucumber, wild balsam apple

NL., *lobatus*, having rounded projections or divisions, from Gr. *lobos*, rounded part or division of an organ—in reference to the leaves (which actually have sharply pointed lobes).
White or greenish; August–September
Moist soil of prairie ravines, thickets, creek banks, and pond margins, sometimes along roadsides or in waste places
Primarily north of a line drawn from Labette County to Butler and Sherman counties

Nearly glabrous, annual vine with angled stems, climbing by means of branched tendrils when occurring in thickets, or sprawling over the ground in waste places. Leaf blades with 3–7 triangular or lanceolate lobes, 4.5–10 cm long and 4.8–13 cm wide, the margins usually finely and remotely serrate. Male flowers in narrow axillary panicles 6–18 cm long; female flowers usually solitary and, when present, borne on a short peduncle at the same node as the male. Calyx barely discernible except for the free, threadlike tips. Corolla lobes 6, lanceolate- to linear-attenuate, spreading widely, 5–10 mm long. Fruit ovoid, 3–5 cm long, opening at the apex, the prickles smooth. Seeds dull brown, quite hard, elliptic, flattened, 1.5–2 cm long, usually roughened along the margins.

The vines are easily propagated by seeds, grow rapidly, bloom abundantly for a long period, and make a sightly coverage for trellises, arbors, low walls, and wire fences.

SÍCYOS L. Bur-cucumber

NL., from Gr. *sikyos*, cucumber.

Annual vines with coiling, forked tendrils. Leaves simple, 3- to 5-angled or lobed. Flowers small, white to yellowish or greenish, monoecious, male and female flowers usually from the same leaf axils, the female flowers in small, long-stalked headlike clusters, the male flowers in racemes or branched clusters. Calyx teeth 5, small. Corolla wheel-shaped or bell-shaped, 5-lobed. Male flowers with stamens united by filaments and anthers. Pistils of female flowers 1-celled, 1-ovuled; style slender; stigmas 3. Fruit ovoid, dry, indehiscent, completely filled by the 1 seed, and covered by minutely retrorsely barbed prickles which are readily detached.

One species in Kansas.

S. angulàtus L. Wall bur-cucumber

L. *angulatus*, angled—applicable to the stems, as well as to the form of the leaves.
Whitish; (late June–) August–September (–October)
Moist soil of wooded creek and river banks, occasionally in ditches or prairie ravines
East two-fifths, primarily

Annual vine with sticky stems climbing 4.5–7.5 m or trailing or clambering where there is nothing to climb. Leaf blades rough-hairy, 3–14 cm long and 4–16 cm wide, with 3 or 5 shallow, acute or acuminate lobes, the margins minutely serrate or dentate. Male flowers few and loose, 5–10 mm across, on a long peduncle; female flowers several, smaller than the male, in a headlike cluster on a short peduncle. Calyx barely discernible except for the free, threadlike tips. Corolla lobes triangular or lance-triangular. Fruit ovoid-acuminate, 1–1.5 cm long. Seeds dull brown, ovate to nearly round in outline, flattened, about 8–10 mm long, very minutely glandular-dotted but otherwise unornamented.

CAMPANULÀCEAE Bellflower Family

Annual, biennial, or perennial herbs with a milky sap. Leaves alternate, simple, sessile or petiolate, estipulate. Flowers bisexual, regular, borne singly or in 2's or 3's in the leaf axils or in various types of terminal or axillary clusters. Sepals or calyx lobes 3–5. Petals 5, fused to one another part of their length. Stamens as many as the corolla lobes, completely free or inserted on the base of the corolla tube. Pistil 1; ovary inferior, 3- to 5-celled; style 1, sometimes with 2–5 branches; stigma 1 and simple, or 2–5, or 1 with 2–5 lobes. A nectiferous disk is frequently present at the base of the style. Fruit a many-seeded capsule.

The family includes about 40 genera and 1,000 species, widely distributed but primarily in temperate climates. The group is closely allied to the family Lobeliaceae, and some authors do not separate the latter family from the Campanulaceae. Cultivated members of the family include the bluebells or bellflowers (some 120 species of *Campanula*), Venus's looking glass (*Triodanis* spp.), and balloon flower (*Platycodon grandiflorum* DC.).

Kansas has 2 genera.

Key to Genera:
1. Annuals or perennials; flowers all alike, borne in long terminal spikes or racemes .. *Campanula*
1. Low, slender, erect or reclining annuals; flowers sessile or nearly so, borne 1 to several together in the axils of the middle and upper leaves, the earliest and lowermost flowers cleistogamous, the later and upper flowers chasmogamous .... *Triodanis*

CAMPÁNULA L. Bellflower

NL., from L. *campana*, bell, + the diminutive suffix *-ula*—"little bell."

Mostly perennial herbs, less frequently annuals or biennials. Leaves usually more or less toothed. Flowers usually blue, showy, all alike, borne in racemes or panicles (rarely solitary). Sepals 5, narrow, often alternating with small, reflexed herbaceous appendages. Corolla usually bell-shaped, less frequently wheel- or bowl-shaped or tubular, 5-lobed. Stamens 5, inserted on the base of the corolla tube, free from one another, the filament bases membranous, dilated, anthers linear, coiling after dehiscence. Ovary 3- to 5-celled; style elongate; stigmas 3–5. Capsule short, usually strongly ribbed, dehiscing by as many pores as there are cells, the pores located either near the base or the apex.

Two species in Kansas. *C. rapunculoides* L., creeping bellflower, a perennial spreading from rhizomes, is a cultivar, native to Eurasia, which has become naturalized in eastern North America and is now reported from Scott County.

C. americàna L. American bellflower

"American."
Blue; late June–September
Moist soil of woods, thickets, shaded roadside banks, and creek banks
East third

533. *Campanula americana*

Erect, usually unbranched annual, 0.6–2 m tall. Leaves thin, ovate to lanceolate or elliptic, mostly 8–20 cm long (including the winged petiole) and 1.2–6 cm wide, the apex acuminate or attenuate, the margins serrate. Flowers borne in narrow terminal racemes 1–6 dm long. Sepals 10–17 mm long. Corolla blue, wheel- or bowl-shaped, 15–20 mm long, the lobes ovate-lanceolate. Capsules narrowly obconic, sometimes curved slightly toward the stem, 8–15 mm long, crowned by the persistent sepals, dehiscing through round pores near the top. Seeds reddish-brown, shiny, mostly short-oblong to broadly elliptic, somewhat flattened, about 0.9–1.3 mm long, with a narrow rib along the margin and a thickened collar around the funicular scar.

The stamens in this and most other species of *Campanula* are proterandrous—that is, they mature and shed their pollen long before the pistil of the same flower is capable of being pollinated, in this case even before the flower opens and before the style has elongated or the stigmatic lobes spread. The anthers open inwardly, depositing their pinkish pollen onto the upper and middle part of the style, which is covered with tiny bristles and papillae. The style elongates, carrying pollen upward, but for a time yet the stigma lobes remain close together so self-pollination is impossible. Bumblebees, honeybees, and a variety of wild bees, as well as some Diptera and butterflies, sucking nectar from between the broadened bases of the filaments, may incidentally transfer pollen from younger flowers to the stigmas of older flowers. Eventually the stigma lobes open and apparently become receptive to pollen from either the same or a different flower. The stigma lobes continue to grow, curving backward until ultimately they come into contact with any pollen remaining on the upper part of the style, thus bringing about self-pollination.

This species is seldom cultivated, but it would be an interesting feature in a partly shaded section of a wildflower garden, especially for those who like to watch the behavior of flowers at their different stages of blooming and that of insect pollinators.

See plate *105*.

TRIODÁNIS Raf.   Venus's looking-glass

NL., from Gr. *tri-*, 3, + *odous, odon, odontos*, tooth—in reference to the unequal teeth of the calyx.

Mostly low, erect or reclining annuals. Stems angled, minutely and retrorsely barbed, simple or with a few branches. Leaves (except sometimes the lowermost) sessile, often clasping the stem, the margins usually toothed and ciliate, the lower surfaces more or less hairy on the veins. Flowers mostly blue or purplish, sessile or nearly so, borne 1 to several in the middle and upper leaf axils. Sepals 5. Corolla wheel-shaped with a short funnelform tube and 5 elliptic lobes. Stamens 5, free from one another, the filaments short, dilated, and hairy near the base. Ovary usually 3-celled; stigmas 3. Capsules linear to ellipsoid, club-shaped, or obovoid, angled or round in cross section, many-seeded, usually with each cell dehiscing through a pore near the top or the middle.

Five species in Kansas. *T. lamprosperma* McVaugh is known from Montgomery and Cherokee counties only. It closely resembles *T. perfoliata* and *T. biflora* but differs from both in having seeds which are larger (0.8–1 mm long) and more flattened. The capsules dehisce through pores near the apex.

In addition to the normal flowers described above, all species of *Triodanis*, early in the season or during periods of climatic stress, produce small cleistogamous flowers which have reduced parts and which never open. In addition, vegetative parts of the plants vary tremendously in size, depending upon the vigor of the plants and the favorability of the habitat.

The common name comes from the genus *Specularia*, with which *Triodanis* in the past has been combined. The name of the type species of *Specularia*, *S. speculum-veneris*, which means "Venus's looking-glass," comes from the medieval name of that plant and, according to Prior's *Popular Names of British Plants* (23), refers to "the resemblance of its flowers set upon their cylindrical ovary to an ancient round mirror at the end of a straight handle."

Key to Species (55):

1. Leaves or bracts subtending the flowers lanceolate to linear; capsules of the cleistogamous flowers round in cross section, subulate, curved and with more or less spreading tips, 8–12 (–20) mm long, dehiscing through longitudinal apical fractures or a single apical pore .... *T. leptocarpa*
1. Leaves or bracts subtending the flowers ovate to cordate or reniform; capsules of the cleistogamous flowers round or flattened in cross section, oblong to ellipsoid or ovoid, straight and more or less appressed, usually less than 8 mm long, dehiscing through 2 or 3 apical or lateral pores 2

2. Openings of the capsules linear, located about midway down the capsule; seeds with longitudinal rows of minute rounded tubercles (use 20X magnification) .... *T. holzingeri*
2. Openings of the capsules broadly elliptic or rounded, located about midway down the capsule or near the apex; seeds not as above 3
  3. Pores at or very near the apex of the capsule; seeds smooth and glossy; leaves not prominently veined beneath .... *T. biflora*
  3. Pores well below the apex of the capsule, usually about midway between the base and the apex; seeds smooth but merely lustrous, or with minute sharp-pointed protuberances (use 20X magnification); leaves with midrib and 1 or 2 pairs of lateral veins prominent beneath .... *T. perfoliata*

T. biflòra (R. & P.) Greene

NL., from L. *bi-*, 2, + *flos, floris*, flower—referring to the 2 types of flowers.
Purple or blue; mid-May–June
Roadsides, waste ground, fields, prairie, open woods, usually in sandy or rocky soil
Southeast eighth and 4 central counties

Slender plants 1–6 dm tall. Leaves mostly ovate or elliptic, 7–20 mm long and 4–10 mm wide. Sepals lanceolate to lance-attenuate, 1.5–5 mm long. Corolla purple or blue (rarely to pinkish or white with purple lines), 5–10 mm long. Capsules mostly ellipsoid and 5–8.5 mm long (excluding the persistent sepals), dehiscing through broadly elliptic pores near the apex. Seeds brown, smooth, glossy, broadly elliptic, somewhat flattened, about 0.5–0.65 mm long.

534. *Triodanis biflora*

T. hòlzingeri McVaugh

Named after John M. Holzinger (1853–1929), who seems to have been the first botanist to suspect that these plants were unlike those of the typical *T. perfoliata*.
Purple; June
Sandy prairie, railroad right-of-ways
Primarily south and west of the Arkansas River and in sandy areas just north of the river; also in east-central Kansas, where it may have been introduced from farther west

Plants 2.5–6 dm tall. Leaves ovate to cordate or less frequently elliptic, 9–25 mm long and 4–15 mm wide. Sepals long-triangular to lance-attenuate, 3–5 mm long. Corolla purple or bluish-purple, 5–11 mm long. Capsules mostly ellipsoid and 6–12 mm long (excluding the persistent sepals), dehiscing through linear pores located about midway down the capsule. Seeds brown, lustrous, broadly elliptic, somewhat flattened, about 0.4–0.7 mm long, ornamented with minute tubercles in longitudinal rows (use 20X magnification and side illumination).

535. *Triodanis leptocarpa*

T. leptocárpa (Mutt.) Nieuw.

NL., from Gr. *leptos*, fine, thin, slender, + *karpos*, fruit—slender-fruited.
Purple; late May–June
Prairie, roadsides, pastures, open woods, frequently on rocky or sandy soil
East four-fifths and Kearny County

Slender plants 1.8–6 dm tall. Leaves elliptic or oblanceolate below to lance-oblong or -linear above, 1–3 cm long and 1.5–7 mm wide, the margins entire, ciliate. Sepals linear-attenuate, (2–) 4–9 mm long. Corolla purple, 7–9 mm long. Capsules linear to subulate, round or somewhat flattened in cross section, 8–25 mm long, dehiscing through slits alternating with the sepals, or less frequently through a single apical pore, those of the cleistogamous flowers often spirally twisted or appressed and smaller than those produced by the chasmogamous flowers. Seeds light brown, smooth, lustrous, broadly elliptic, about 0.7–1 mm long.

T. perfoliàta (L.) Nieuw.   Venus's looking-glass

NL., from L. *per*, through, + *folium*, leaf—the clasping leaf bases making it appear that the stems pass through the leaves.
Purple or blue; late May–June
Sandy or rocky soil, usually in disturbed habitats
East three-fourths, primarily; several records from the west fourth

Plants up to 1 m tall. Leaves ovate or the upper ones, especially, cordate with bases clasping the stem, 10–25 mm long and 10–27 mm wide. Sepals triangular-attenuate, 3–8 mm long. Corolla purple or blue, 5–12 mm long. Capsules mostly ellipsoid or narrowly top-shaped, 4–10 mm long (excluding the persistent sepals), largely concealed by the leaves, dehiscing through broadly elliptic or rounded pores about midway down the capsule. Seeds brown, lustrous, broadly elliptic, somewhat flattened, about 0.5 mm long,

536. *Triodanis perfoliata*

BELLFLOWER FAMILY

373

usually (in our area) ornamented with minute sharp-pointed projections (use 20X magnification and side illumination).

According to Trent (56), 47.5 percent of the seed produced by normal (i.e., chasmogamous) flowers are viable, while only 0.7 percent of the seeds produced by cleistogamous flowers will germinate. Ungerminated seeds are apparently normal in most respects but lack embryos.

See plate *106*.

LOBELIÀCEAE Lobelia Family

Mostly annual or perennial herbs with acrid, often poisonous, milky or colored juice; some shrubs or trees in the tropics. Leaves alternate, simple, sessile or petiolate, estipulate. Flowers bisexual, irregular, in ours borne in terminal spikes or racemes. Calyx lobes 5. Petals 5, fused most of their length, the corolla 2-lipped, often split along the upper side. Stamens 5, arising low on the corolla tube, the anthers and upper part of the filaments united to form a cylinder around the style. Pistil 1; ovary half-inferior, 2-celled; style 1; stigma 2-lobed. A ring of nectiferous tissue is present around the base of the style, more or less at the region of conjuncture between the corolla tube, stamen filaments, and the exposed portion of the ovary. Fruit a many-seeded capsule or a berry.

The family contains approximately 24 genera and 700 species, widespread but especially well represented in the tropics. The group is closely related to and often included in the Campanulaceae. Cultivated members of the family include the cardinal flower (*Lobelia cardinalis* L.) and other species of *Lobelia*, *Downingia*, and *Pratia*.

Kansas has 1 genus.

LOBÈLIA L. Lobelia

Named after Mathieu de Lobel (1538–1616), Flemish physician, herbalist, and writer on botanical subjects.

Annual, biennial, or perennial herbs. Leaves often lanceolate with toothed margins. Flowers red, blue, pale purple, or white in ours, usually showy, borne in bracted spikes, racemes, or panicles. Calyx cleft all the way to the ovary or with a short tube above the ovary. Corolla tube split along the upper side to or nearly to the base, the upper lip with 2 narrow, spreading or reflexed lobes, the lower lip 3-lobed, broader than the upper lip. Stamens protruding through the slit in the corolla tube, the anthers usually colored, the upper 2 with a tuft of white hairs at the tip. Ovary fused most of its length with the hypanthium. Fruit a 2-celled, many-seeded capsule topped by the persistent calyx, corolla, and stamens, opening at the top, the 2 halves recurving.

Four species in Kansas.

Our lobelias contain several closely related pyridine alkaloids, including lobeline, which render the plants or potions made therefrom toxic when ingested in large enough quantity (21). These same compounds have accounted for their various uses as medicinals, and most cases of human poisoning have resulted from overdoses of homemade remedies. Although livestock poisoning has been reported, this is not a common problem.

Key to Species:
1. Flowers bright blue or red (rarely white), mostly 2–4 cm long; corolla tube with a slit on each side near the base; plants of wet ground 2
   2. Flowers red .... *L. cardinalis*
   2. Flowers bright blue (rarely white) .... *L. siphilitica*
1. Flowers pale blue or pale purple, 7–18 mm long; corolla tube lacking a pair of slits near the base; plants of prairies or open woods 3
   3. Hypanthium obovoid to nearly globose, nearly as long as the corolla, becoming much inflated in fruit; plants of open woods .... *L. inflata*
   3. Hypanthium obconic, shorter than the corolla, not becoming inflated in fruit; plants usually of prairies .... *L. spicata*

L. cardinàlis L. Cardinal flower, scarlet lobelia

L. *cardinalis*, chief, principal, as a cardinal is ecclesiastically—the flower being so named because its color is nearly that of a cardinal's hat and cloak.
Red; August–mid-October
Wet ground
Nearly throughout

Erect perennial, glabrous or nearly so, with strong fibrous roots. Stems 2–11 dm tall, simple or with a few branches. Leaves thin, mostly lanceolate to elliptic or oblanceolate, 6–18 cm long and 1–3.5 cm wide, the margins irregularly serrate. Flowers red, 2.5–4 cm long, borne in racemes 1–4 dm long, the bracts subtending each flower more or less leaflike, lanceolate below to linear above. Calyx lobes linear attenuate, 7–17 mm long. Corolla with a slit on each side near the base. Anther tube bluish-gray. Capsule 6–9 mm long. Seeds orangish or tan, ovoid to ellipsoid, minutely tubercled, 0.6–0.8 mm long.

The bright scarlet flowers are perhaps unmatched in brilliance by any other of our wildflowers, though they are approached in this respect by the inflorescence bracts of Indian paintbrush, *Castilleja coccinea.*

While we are charmed by the color, we find the plan of the flower for securing cross-pollination no less fascinating. We notice that the corolla is tubular below and 2-lipped above, the upper lip 2-lobed and the lower lip 3-lobed, and that the tube has a slit along its upper side, running from between the 2 lobes of the upper lip to the base of the tube. The 5 stamens, in the bud and in the freshly opened flower, are united into a tube around the stiff style, the short anthers forming a cap over the globose stigma, discharging their pollen inwardly upon a ruff of papillae surrounding the base of the stigma. As the flower opens, the cylinder of stamens and the enclosed style extend beyond the corolla tube, approaching the upper lip but bending downward near the end, so that the tip stands before the entrance to the tube. Soon the style elongates, bringing the stigma and the pollen-covered ruff beyond the anthers. At this stage, the 2 lobes of the stigma have not separated and no pollen can get on the inner stigmatic surfaces; hummingbirds or hawk moths arriving at this time can receive pollen on their proboscises from the ruff but cannot leave any of it on the stigma. In flowers a little older, the stigmatic lobes roll back, and then visitors coming with pollen from younger flowers can pollinate them. Nectar is secreted around the base of the style inside the staminal tube and can be reached through gaps between the filaments near their bases. Our ruby-throated hummingbirds are the principal visitors, and the encounter between such beautiful birds and flowers is a sight to be remembered.

This species has long been in cultivation in American gardens, and it was used medicinally, along with *L. siphilitica* and *L. inflata,* by the Indians and the European colonists. Cherokee Indians used a decoction of the roots as a remedy for worms (45). The Meskwakis used roots of *L. cardinalis* and *L. siphilitica,* finely ground and mixed together, as a "love medicine." When secretly placed in the food of an arguing couple, the preparation was believed to "avert divorce and make the pair love each other again" (34). Eighteenth- and nineteenth-century physicians in Europe and the United States recognized this species as a cathartic and emetic, though not so strong as *L. inflata.*

See plate *107.*

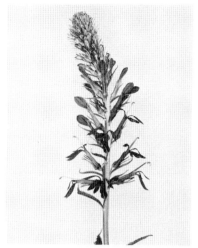

537. *Lobelia cardinalis*

L. inflàta L.   Indian tobacco

L. inflatus, puffed up, swollen—in reference to the enlarged hypanthium.
Light purple; mid-July–August (–October)
Moist, open woods
Scattered in the east fourth

538. *Lobelia inflata*

Erect, sparsely hairy annual with a cluster of weak fibrous roots. Stems up to 8 dm tall, simple or with a few branches. Leaves elliptic to lanceolate above or oblanceolate below, 3–9 cm long and 1–3.5 cm wide, the margins irregularly dentate or serrate. Flowers small and inconspicuous, light purple, 7–9 mm long, the racemes mostly 3–15 cm long, terminating the main stem and its branches, the bract subtending each flower more or less leaflike, lanceolate. Calyx lobes linear-attenuate, 2.5–5 mm long. Corolla tube split only along the upper side, not along the sides near the base. Hypanthium becoming much inflated as the fruit matures, 6–8 mm long, the walls thin, net-veined. Seeds orangish to tan, narrowly ellipsoid, minutely reticulate, about 0.6 mm long.

According to Kingsbury (21), at least 14 pyridine alkaloids, which have structures similar to that of nicotine, have been isolated from this plant. This species was reportedly used by some Indians as a purgative and to treat dysentery, but most documented instances of its use as a medicinal have involved whites. The dried leaves and tops were listed in the *U.S. Pharmacopoeia* from 1820 to 1936 and in the *National Formulary* from 1936 to 1960 as expectorant and emetic. As late as 1854, *Good's Family Flora* strongly recommended the free use of *L. inflata* for all nervous diseases, fits, convulsions, spasms, asthma, tetanus, St. Vitus's dance, hydrophobia, etc. According to Krochmal (30 and 57), the plant also has been used to treat whooping cough, bronchitis, laryngitis, pneumonia, and hysteria, and currently serves as a commercial source of lobeline, which is utilized in antismoking preparations. Over and above the therapeutic potency attributed to this plant, it was reputed among the Pawnees to exert a favorable influence when proffered as a love charm (28).

LOBELIA FAMILY
375

539. *Lobelia siphilitica*

540. *Lobelia spicata*

Overdoses of the plant or its extracts produce vomiting, sweating, pain, paralysis, depressed temperature, rapid but feeble pulse, collapse, coma, and death in humans (21).

L. siphilítica L.   Big blue lobelia

NL., *siphiliticus, syphiliticus*, pertaining to syphilis—this species reputedly having been used by the Indians in the treatment of syphilis.
Blue (rarely white); August–early October
Wet places
East three-fifths, primarily

Erect, sparsely hairy perennial with strong fibrous roots. Stems 2–9.5 dm tall, simple or with a few branches. Leaves thin, mostly oblanceolate to elliptic, 3–18 cm long and 1–6 cm wide, the margins finely serrate to crenate or nearly entire. Flowers bright blue (rarely white), 2–3 cm long, borne in racemes 1–5 dm long, the bracts subtending each flower more or less leaflike, mostly lanceolate to ovate. Calyx lobes usually lance-attenuate, 8–17 mm long, with basal auricles and ciliate, crisped or undulate margins. Corolla tube with a slit on each side near the base. Capsule 8–10 mm long. Seeds orangish or tan, narrowly ellipsoid, minutely mucronate, about 0.6–0.8 mm long.

The structure of the flower and the provision for cross-pollination are essentially as indicated for *L. cardinalis.* The relatively large size of the corolla tube would seem to favor bumblebees especially.

Benjamin Smith Barton observed that this plant had been used by the Iroquois, the Cherokees, and some other tribes to treat venereal disease. He recognized it as an effective diuretic and believed it useful in treating gonorrhea, but doubted its power to cure syphilis (45). Although some reports indicate that whites in North America obtained successful results by using a decoction of *L. siphilitica* to treat syphilis, European physicians apparently could not repeat these results. Millspaugh (43) suggests that this failure may have been due to the deterioration of the active principle in the dried herb or the fact that the Europeans were not using the lobelia in combination with mayapple roots (*Podophyllum peltatum*), wild cherry bark (*Prunus virginiana*), and powdered New Jersey tea bark (*Ceanothus americanus*), as the Indians did.

See plate *108.*

L. spicàta Lam.   Palespike lobelia

L. *spicatus*, pointed, furnished with spikes or ears, as of wheat—in reference to the slender inflorescence.
Light blue; mid-May–July
Prairies, occasionally in open woods
East fourth

Slender, more or less hairy, erect perennials with fibrous roots. Stems 1.5–8.5 dm tall, usually single and unbranched. Leaves mostly oblanceolate below to elliptic or narrowly lanceolate above, 2–8 cm long and 6–24 mm wide, the margins entire to shallowly serrate or crenate. Flowers light blue to nearly white, 7–12 mm long, the racemes 0.7–3 dm long, the bract subtending each flower linear-attenuate, quite distinct from the leaves. Calyx lobes linear-attenuate, 3.5–5 mm long, usually with basal auricles varying from longer to shorter than the obconic hypanthium. Corolla tube split only along the upper side, not along the sides near the base. Hypanthium not becoming inflated in fruit, the capsule 8–10 mm long. Seeds orangish-brown, ellipsoid, minutely tubercled, 0.6–0.8 mm long.

COMPÓSITAE   Composite Family

NL., from L. *compositus*, made up of parts, because a group of small flowers or florets is made to look like a single flower by being bound together by 1 or more series of bracts known as the involucre, as in the sunflower and dandelion.

Mostly herbs, with some shrubs, trees, and vines, sometimes with milky sap. Leaves usually alternate or opposite (rarely whorled), simple or pinnately or palmately lobed or compounded, sessile or petioled, estipulate. Calyx (in this family called the *pappus*) highly modified and present in the form of teeth, scales, bristles, or awns, or a ring of elevated tissue, or absent entirely. The pappus usually must be examined under magnification. Petals 5, fused into either a radially symmetrical, tubular corolla or a bilaterally symmetrical, ligulate (strap-shaped) corolla, the tubular form (4-)5-lobed, the ligulate form entire or toothed at the apex and short-tubular at the base. Stamens mostly 5, inserted on the corolla tube, the anthers nearly always united or at least coherent around (but free from) the style. Pistil 1; ovary inferior, 1-celled; style branches 2. Fruit an achene, often crowned by the persistent pappus. In some cases, the achene body is contracted at the apex into a beak, which may be either stout or quite slender and of various lengths, to which the pappus is attached. Mature achenes are frequently necessary for using the keys to genera and species.

The inflorescence type in this family is a compact head composed of few to many (rarely only 1 or 2) flowers closely embraced by 1 or more series of involucral bracts (sometimes called phyllaries). This head is often mistaken by the uninitiated for a single flower, but close examination of its structure quickly reveals its true nature. Within the head, the individual flowers are situated on a common receptacle which may be more or less flattened (as in *Taraxacum*), conical (as in *Echinacea*), or columnar (as in *Ratibida columnifera*). This arrangement often prompts beginning botanists to assume that the pistils are superior; however, the perianth of each individual flower is fused to and arises from the top of the ovary, the latter therefore being inferior. In some cases, the individual flowers are subtended by minute scales, bracts, or bristles which arise from the upper surface of the receptacle and which may or may not be deciduous. When such structures are absent, the receptacle is said to be naked; when such bracts, scales, etc., are present, the receptacle is said to be chaffy.

Though the head is the true inflorescence in this family, the term "inflorescence" is also used in a more general sense to refer to the arrangement of heads on a plant.

In this family, flowers with tubular corollas are called disk flowers, and when only such flowers are present, the head is said to be discoid, as in *Liatris*. In a discoid head, all the flowers are usually bisexual and fertile. Exceptions include *Gnaphalium*, which has the outermost flowers of the head female and the inner ones bisexual, and *Artemisia* and *Erechtites*, which also have the outer flowers female, but the inner flowers bisexual or sometimes neutral.

Flowers with ligulate corollas are called rays or ray flowers. When these flowers are present in the head, associated with disk flowers as in *Helianthus*, the heads are said to be radiate. In this type of head, usually the disk flowers are bisexual and the ray flowers, which are restricted to the periphery of the head, are either female and fertile, or neutral and infertile, perhaps with a rudimentary pistil present. Variations on this theme include *Grindelia* and *Xanthocephalum*, which may have some disk flowers neutral; *Erigeron*, which has some species with rayless female flowers situated between the typical rays and the disk flowers; and *Polymnia*, *Silphium*, and *Parthenium*, which have male or neutral disk flowers and female (or perfect) ray flowers.

When only ligulate flowers are present, as in *Taraxacum*, the heads are said to be ligulate. In this case, all the flowers are bisexual.

In 4 of our genera which have heads more or less of the discoid type, although few-flowered and much reduced in size, male and female flowers are found in separate heads. In *Antennaria*, in fact, there are separate male and female plants. In *Iva*, *Ambrosia*, and *Xanthium*, the male heads usually occur higher on the plant than the female heads, and their flowers may contain a rudimentary pistil with unopened style branches. In *Ambrosia*, the involucral bracts of the female flowers are fused around the ovary to form a nutlike involucre with a series of short, erect spines or tubercles at the apex. In *Xanthium*, the cocklebur, this modification is even more extreme, with the involucre of the female heads completely enclosing the 2 flowers and forming a conspicuous burr with rigid, hooked bristles.

This is the largest of all families of vascular plants and contains about 950 genera and 20,000 species. It is sometimes subdivided into smaller families, subfamilies, or tribes (a category of classification slightly lower in rank than subfamily).

For convenience in comparing similar plants, the genera are arranged in this book according to tribes. However, for ease in keying, an artificial key (one in which closely related genera do not necessarily "key out" adjacent to one another), rather than a natural key (which reveals affinities among genera), is used. The artificial key makes use of the most obvious characters possible, and therefore is usually easier to use, but these characters are not always ones which are useful in determining true relationships among the Composite genera. To give the reader some idea of the characters which have been used at times to separate one tribe from another, a brief synopsis is given at the beginning of each subsection.

Members of this family used as food by man include the sunflower (*Helianthus* spp.), Jerusalem artichoke (*Helianthus tuberosus* L.), globe artichoke (*Cynara scolymus* L.), lettuce (*Lactuca sativa* L. and varieties), endive (*Cichorium endivia* L.), chicory

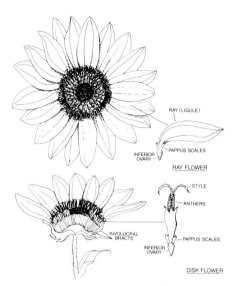

541. Flower structure in *Helianthus*

(*Cichorium intybus* L.), and salsify (*Tragopogon porrifolius* L.). The contact insecticide pyrethrum is derived from the African *Chrysanthemum coccineum* Willd. The safflower (*Carthamus tinctorius* L.) produces both a cooking oil and a red dye. A number of ornamentals also belong to this group: perennial asters (*Aster* spp.), garden or China asters (*Callistephus* spp.), *Chrysanthemum* spp., *Coreopsis* spp., *Cosmos* spp., *Dahlia* spp., sunflower (*Helianthus* spp.), marigolds (*Tagetes* spp.), and *Zinnia* spp., to name a few.

Kansas has 86 genera, making it our largest family also.

Key to Genera (69):
1. Heads with ray flowers only, the flowers all perfect, the rays 5-toothed; juice milky or colored .... Key A, p. 378
1. Heads with both ray and disk flowers, or with disk flowers only; juice watery, rarely milky 2
  2. Heads with both ray and disk flowers; rays restricted to the margin of the heads, never bisexual 3
    3. Rays yellow or orange, occasionally marked with purple or reddish-brown at the base 4
      4. Receptacle with small, straw-colored or membranous bracts subtending each individual flower; pappus chaffy, of awns, or none .... Key B, p. 378
      4. Receptacle naked (lacking bracts as described above); pappus various .. Key C, p. 379
    3. Rays some color other than yellow or orange .... Key D, p. 380
  2. Heads with disk flowers only (some plants with very small and inconspicuous rays are keyed here as well as with the radiate group) 5
    5. Receptacle with bristles or small straw-colored bracts subtending each individual flower; plants often spiny .... Key E, p. 380
    5. Receptacle naked (lacking bristles or bracts as described above) .... Key F, p. 381

Key A:
1. Pappus of simple, hairlike bristles only 2
  2. Achenes more or less strongly flattened (not very flattened in a few species of *Lactuca*) 3
    3. Achenes beaked or beakless, in either case somewhat enlarged at the summit where the pappus is attached; heads relatively few-flowered (about 5–56 flowers in our species) .... *Lactuca*, p. 384
    3. Achenes beakless, without any enlarged pappiferous disk at the summit; heads many-flowered (about 85–250 flowers in our species) .... *Sonchus*, p. 389
  2. Achenes round or triangular in cross section, only slightly flattened 4
    4. Achenes minutely spiny, or with some short processes near the summit of the body, tipped by a slender beak; flowers yellow .... *Taraxacum*, p. 390
    4. Achenes smooth or nearly so, not evidently spiny or rough 5
      5. Flowers pink or purple to sometimes white or creamy; heads several to many; cauline leaves present or reduced to mere scales; achenes beakless 6
        6. Flowers cream-colored in ours; cauline leaves well developed, mostly well over 1 cm wide .... *Prenanthes*, p. 388
        6. Flowers pale pink or purplish; cauline leaves less than 1 cm wide, often reduced to scales .... *Lygodesmia*, p. 386
      5. Flowers bright yellow to orange or orange-red 7
        7. Perennial from an elongate or very short rhizome, with numerous fibrous roots; taproot wanting; achenes truncate or sometimes narrowed upwards, beakless .... *Hieracium*, p. 382
        7. Annual, biennial, or perennial, from a taproot or several strong roots, without rhizomes; achenes beaked .... *Pyrrhopappus*, p. 388
1. Pappus of plumose bristles, or bristles and scales, or scales only, or none, the scales sometimes minute and inconspicuous in species of *Krigia*, or very slender and bristlelike in *Microseris* 8
  8. Pappus of plumose bristles, at least in part .... *Tragopogon*, p. 391
  8. Pappus of scales, or bristles and scales, or none, not plumose 9
    9. Flowers blue (white); pappus of minute narrow scales only .... *Cichorium*, p. 378
    9. Flowers yellow or orange; pappus various, but not as above 10
      10. Pappus none or vestigial; plant annual .... *Krigia*, p. 383
      10. Pappus well developed; annual or perennial plants 11
        11. Pappus of hairlike bristles intermixed with long, very slender, gradually attenuate scales or flattened hairs, involucre 17–25 mm high; perennial with vertical rhizomes .... *Microseris*, p. 387
        11. Pappus of 5 or more prominent or sometimes very inconspicuous scales and generally 5–40 longer hairlike bristles; involucre 3.5–14 mm; annual or perennial, tending to be fibrous-rooted (definitely so when perennial) .... *Krigia*, p. 383

Key B:
1. Disk flowers sterile, with undivided styles, their ovaries much smaller than those of the fertile ray flowers .... *Silphium*, p. 454

11. Pappus bristles, at least those of the disk flowers, falling very early, often as a unit; ours glabrous plants with stiff leathery leaves .... *Prionopsis*, p. 433
11. Pappus bristles more persistent, not falling as a unit; ours mostly hairy 12
    12. Leaf blades pinnatifid .... *Machaeranthera*, p. 433
    12. Leaf blades entire or merely toothed 13
        13. Pappus in 2 distinct series in the disk flowers, the outer series much shorter than the inner; ours perennials .... *Heterotheca*, p. 430
        13. Pappus of disk flowers not differentiated into 2 distinct series; ours annuals with rather soft glandular foliage .... *Croptilon*, p. 427

Key D:
1. Receptacle chaffy; pappus chaffy, or of awns or a crown, or none 2
  2. Leaves, or many of them, opposite 3
    3. Disk flowers infertile, with undivided style; pappus none .... *Melampodium*, p. 450
    3. Disk flowers fertile, with divided style; pappus present or lacking .... *Eclipta*, p. 444
  2. Leaves alternate 4
    4. Involucral bracts dry, membranous-margined 5
      5. Heads large, terminating the branches; rays elongate, 5–13 mm .... *Anthemis*, p. 407
      5. Heads small, the rays short and broad, 1–5 mm; inflorescence a more or less flattened cluster .... *Achillea*, p. 406
    4. Involucral bracts more or less herbaceous, not membranous-margined 6
      6. Leaves pinnatifid .... *Ratibida*, p. 451
      6. Leaves entire or merely toothed 7
        7. Heads small, the disk less than 1 cm wide in flower; receptacular bracts not at all spinescent .... *Verbesina*, p. 456
        7. Heads larger, the disk 1.5–3.5 cm wide; receptacular bracts spinescent, longer than the disk flowers .... *Echinacea*, p. 443
1. Receptacle naked (or bristly in *Gaillardia*); pappus various 8
  8. Pappus chaffy, or of awns, or none 9
    9. Receptacle bristly, the bristles about as long as the achenes .... *Gaillardia*, p. 415
    9. Receptacle not bristly 10
      10. Leaves pinnatifid in ours; pappus a short crown or none; heads solitary or few .... *Chrysanthemum*, p. 409
      10. Leaves entire; pappus not as above; heads several to many 11
        11. Leaves opposite; pappus of about 7–10 lance-attenuate scales with prominent midribs and membranous margins; achenes 4-angled .... *Palafoxia*, p. 419
        11. Leaves alternate; pappus of several minute teeth or scales and 2–4 longer scales or awns; achenes 2-(3-)winged .... *Boltonia*, p. 425
  8. Pappus of capillary bristles 12
    12. Rays very numerous, short, narrow, and inconspicuous; disk flowers about 20 or generally fewer; annual weeds blooming in late summer and early fall .. *Conyza*, p. 426
    12. Rays few to many, more or less well developed, or if small and numerous, then the heads with about 30 or more disk flowers 13
      13. Plants blooming in spring and early summer, occasionally continuing until fall; involucral bracts neither definitely leafy nor with a stiff papery base and herbaceous tip .... *Erigeron*, p. 427
      13. Plants blooming in late summer and fall; involucral bracts in most species either leafy or with a stiff papery herbaceous tip 14
        14. Leaves pinnatifid .... *Machaeranthera*, p. 433
        14. Leaves entire or merely toothed .... *Aster*, p. 420

Key E:
1. Pappus of hairlike bristles 2
  2. Pappus bristles plumose .... *Cirsium*, p. 394
  2. Pappus bristles merely minutely barbed, not plumose .... *Carduus*, p. 393
1. Pappus of scales, or awns, or none 3
  3. Pappus present 4
    4. Leaves alternate; involucral bracts distinctly hooked at the tip .... *Arctium*, p. 392
    4. Leaves, or most of them, opposite 5
      5. Principal (inner) involucral bracts partially united .... *Thelesperma*, p. 456
      5. Principal involucral bracts separate .... *Bidens*, p. 439
  3. Pappus absent 6
    6. Flowers either all bisexual and fertile with divided style, or the central ones thus and the marginal ones female or neutral .... *Eclipta*, p. 444
    6. Flowers of 2 kinds, some female and fertile, others bisexual and infertile with undivided styles 7
      7. Male (infertile bisexual) and female flowers in the same head; involucre not tuberculate or spiny .... *Iva*, p. 460
      7. Male and female flowers in separate heads, the male generally uppermost; involucre of the female heads nutlike or burlike 8
        8. Involucre of the female heads a bur with many hooked prickles .. *Xanthium*, p. 461
        8. Involucre of the female heads with one row of tubercles or spines .. *Ambrosia*, p. 458

Key F:

1. Pappus chaffy, or of awns, or none ... 2
  2. Leaves, at least the lower, opposite ... *Dyssodia*, p. 414
  2. Leaves alternate or all basal ... 3
    3. Pappus of 2 to several firm, long awns ... *Grindelia*, p. 429
    3. Pappus of a short cup or crown, or of short, rounded scales, or none ... 4
      4. Involucral bracts nearly equal, colored, petal-like at least distally ... *Hymenopappus*, p. 417
      4. Involucral bracts not at all petal-like, evidently imbricate ... 5
        5. Inflorescence spiciform, racemiform, or paniculiform ... *Artemisia*, p. 407
        5. Inflorescence corymbiform or the heads solitary ... *Chrysanthemum*, p. 409
1. Pappus of hairlike bristles ... 6
  6. Outer flowers, or all the flowers of some heads female ... 7
    7. Plants more or less white-woolly at least on the lower leaf surface ... 8
      8. Plants dioecious or nearly so ... *Antennaria*, p. 404
      8. Plants not dioecious; flowers all fertile, the outer numerous and female, the inner few and bisexual ... *Gnaphalium*, p. 405
    7. Plants not at all white-woolly, though sometimes otherwise pubescent ... 9
      9. Outer (female) flowers essentially lacking a pappus ... *Erigeron*, p. 427
      9. Flowers all with well-developed pappus ... 10
        10. Involucral bracts herbaceous-tipped ... *Aster*, p. 420
        10. Involucral bracts sometimes greenish, but scarcely herbaceous ... *Conyza*, p. 426
  6. Flowers all bisexual ... 11
    11. Flowers bright yellow or orange ... *Senecio*, p. 412
    11. Flowers white, pink, purple, or pale creamy ... 12
      12. Heads few-flowered, aggregated into secondary heads ... *Elephantopus*, p. 395
      12. Heads separate ... 13
        13. Pappus double, the inner of long bristles, the outer of very short bristles ... *Vernonia*, p. 396
        13. Pappus simple, of similar bristles ... 14
          14. Achenes 3- to 5-angled; leaves opposite or whorled in ours ... *Eupatorium*, p. 398
          14. Achenes 6- to 10-ribbed, not angled; leaves alternate (opposite in some species of *Brickellia*), entire to broadly lobed ... 15
            15. Involucral bracts in 1 or few series, all nearly equal in length ... *Cacalia*, p. 410
            15. Involucral bracts definitely graduated in length and in several series ... 16
              16. Inflorescence simple, spikelike or racemelike in ours, the heads sessile or pedunculate on the main axis of the plant; flowers rose-purple (rarely white) ... *Liatris*, p. 401
              16. Inflorescence more branching; flowers cream-colored in ours ... 17
                17. Pappus bristles nearly smooth, easily detached .. *Brickellia*, p. 397
                17. Pappus bristles plumose, persistent ... *Kuhnia*, p. 400

CICHÒRIEAE — Chicory Tribe

Sap of plants milky or colored. Flower heads all alike, with ray flowers only, the flowers usually all bisexual and fertile. Receptacle usually naked. Anthers with a thin, short appendage at the apex, sagittate at the base but not tailed. Style branches stigmatic the entire length of the inner faces. Pappus of bristles and/or slender hairlike scales.

CICHÒRIUM L. — Chicory

L. *cichorium*, Gr. *kichora*. According to Pliny, the ancient name was Egyptian in origin.

Branched biennial or perennial herbs with milky juice. Leaves alternate, entire or pinnatifid. Flowers blue (rarely white), all ligulate and bisexual, the heads several-flowered, sessile, 2–3 together in the leaf axils or at the tips of branches or solitary on thick, peduncle-like branches. Involucre cylindric, the bracts of 2 kinds, the inner longer, narrower, and thinner than the outer reflexed ones, and often with membranous margins. Receptacle convex, with a few chaffy scales. Achenes glabrous, longitudinally ribbed, approximately 5-angled in cross section, the outer ones somewhat flattened, with a pappus of minute scales.

A European genus with 1 species naturalized in the United States. Another species, *C. endivia* L., endive, is cultivated as a green vegetable.

C. íntybus L. — Common chicory

Intybus, intibus, intibum were the ancient Latin names used for this plant.

Blue, rarely white; late June–October
Roadsides, waste places, fields
East fourth, primarily, with a few scattered collections from farther west; intr. from Europe

Perennial herb with a thickened taproot. Stems 0.3–1 m tall, with rigid, ascending or widely spreading branches. Leaves of the basal rosette spatulate, 8–25 cm long, with incised, lobed, and toothed and often undulate margins, much like those of the dandelion; the leaves higher up oblanceolate to lanceolate, becoming progressively smaller and less deeply lobed, the margins coarsely toothed to entire and bractlike where they subtend flower heads. Flowers usually blue, the corollas mostly 10–18 mm long, sparsely hairy on the under side, with 5 apical teeth. Bracts of the involucre with conspicuous glandular hairs along the margins and/or on the back near the apex; basal portions of the bracts become hardened and persist after the achenes have fallen. The heads open in the early morning and close about noon on bright days. Achenes light brown mottled with dark brown, obconic with a truncated base, minutely roughened, about 2.5–3 mm long.

Chicory was anciently used as a potherb and salad in Egypt and in Europe. Pliny tells of its medicinal value also: "Applied by way of liniment it disperses abscesses. . . . It is also very beneficial to the liver, kidneys and stomach . . . taken in honied wine, it is a cure for the jaundice, if unattended by fever. . . . This plant, in consequence of its numerous salutary virtues, has been called by some persons 'chreston' [the useful] and 'pancration' [the all-powerful]." Herbals of the Middle Ages proclaim the virtues of chicory no less extravagantly, but modern materia medicas simply say that a decoction of the root provides a bitter tonic.

Parkinson (12) speaks only of the use of chicory as a potherb and salad, but in his century the roots began to be used as an adulterant and substitute for coffee, a practice that has grown to enormous proportions in Europe, and in some years millions of pounds of the root have been imported into the United States for this purpose. The roots are processed by slicing, kiln-drying, roasting, and grinding.

The dried cultivated root when harvested in the fall contains around 60 percent inulin, 7 percent fructose, 6 percent nitrogenous substance, 0.4 percent fat, and 0.15 percent bitter principle. In the process of roasting, a part of the inulin is converted into fructose and some of the fructose becomes caramel. Evidently the fructose, caramel, and bitter principle are essentials in the taste of the chicory brew.

542. *Cichorium intybus*

543. *Hieracium longipilum*

HIERÀCIUM L. Hawkweed

NL., from Gr. *hierakon*, for a kind of plant, from Gr. *hierax*, hawk. Pliny relates that, according to an ancient notion, hawks used a species of this genus (*H. pilosella*) to sharpen their eyesight.

Erect perennial herbs with milky juice and short or elongate rhizomes with fibrous roots. Leaves alternate, usually sessile and entire. Herbage usually with long, spreading tawny hairs, often glandular also, and usually with some stellate hairs. Flowers yellowish-white to yellow or red-orange, all ligulate and bisexual, the heads 12- to many-flowered, arranged in a branched cluster or in a narrow elongate cluster. Involucre cylindric or hemispheric, the bracts of 2 different kinds: 2 inner row of longer, green, hairy, linear-lanceolate or linear bracts and an outer set of shorter bracts. Receptacle slightly convex, naked. Achenes slender, cylindric or tapered but not beaked, 10-ribbed, round or 3-angled in cross section, with a pappus of many tawny-white or brownish bristles.

Two species in Kansas. *Hieracium gronovii* L. is occasionally found in the southeastern sixth, especially in Woodson and Cherokee counties, and in Franklin and Douglas counties. It is similar in aspect to the following species, but is more delicate and seldom taller than 1 m and has hairs about 5 mm long and heads with only 20–40 flowers.

H. longipìlum Torr. Long-bearded hawkweed

NL., from L. *longus*, long, + *pilus*, hair—the leaves and stems being covered with long, coarse hairs.
Yellow; July–early September
Prairie, less frequently in open oak woods
East two-fifths

A wandlike perennial, 0.6–2 m tall, with a short vertical rhizome and coarse fibrous roots. Stems unbranched except within the inflorescence. Leaves crowded toward the base of the stem, the lower ones oblanceolate, 9–22 cm long and 1.5–4 cm wide, becoming smaller and elliptic to linear and bractlike above. Leaves and lower parts of the stem with brownish, spreading hairs 1–2 cm long. Flowers yellow, the corollas about 7–8 mm long. Heads mostly 40- to 90-flowered; involucre 8–12 mm long. Involucral bracts and branches of the inflorescence sparsely to densely glandular-hairy. Achene body black, tapered slightly toward both ends, 3.5–4 mm long; pappus 5–7 mm long.

NOTE: Compare with *Prenanthes aspera*.

After the achenes have matured and sailed away, the stem breaks off near the ground. In the fall, the plant produces a rosette of overwintering leaves that remain green, photosynthesizing when the temperature permits, thereby sustaining the new stem which begins growth in the spring in advance of the grasses.

Ancient and medieval herbalists recommended the hawkweed for a variety of ailments. Dioscorides (1st century) affirmed that "the juice of it in wine helps digestion" and that "it is good against the bite of venomous serpents"; and he recommended the application of bruised leaves, with a little salt, to burns, erysipelas, and all kinds of eruptions.

544. *Hieracium longipilum* leaves

KRĪGIA Schreb.   Dwarf dandelion

In honor of David Krig or Krieg, a German physician who was one of the first to collect plants in Maryland.

Small annual or perennial herbs with milky juice and without aboveground stems. Leaves sessile, entire or pinnatifid, alternate to not quite opposite, or all basal. Flowers yellow or orange, all ligulate and bisexual, in 1 or a few, several- to many-flowered heads. Involucre bell-shaped, of 2 rows of similar bracts. Receptacle naked, initially flat or concave, becoming convex in fruit. Achenes top-shaped or cylindric, round or angled in cross section, the pappus absent, or of a few membranous scales, or with an outer row of scales and an inner row of rough bristles.

Four species in Kansas. Only the 2 most common are included in the following key. *K. biflora* (Walt.) Blake, known only from Cherokee County, is a caulescent, branched perennial, 2–8 dm tall, with heads close in size to those of *K. dandelion*. *K. occidentalis* Nutt., known from a few southeastern counties, is a caespitose, tufted annual 2.5–10 cm tall with spatulate or narrowly oblanceolate leaves with 1 to many small scapose heads.

Key to Species:

1. Caespitose perennials with solitary scapose heads; pappus of 20–40 bristles; involucre mostly (7–) 10–15 mm long .... *K. dandelion*
1. Branched annuals with several to many heads; pappus absent or a minute scaly crown; involucre mostly 3–5 (–7) mm long .... *K. oppositifolia*

K. dándelion (L.) Nutt.   Potato dandelion

NL., from its resemblance to the common dandelion, *Taraxacum officinalis*.
Yellow; late April–early June
Prairie meadows, roadsides, open woods, often on rocky soil
Primarily in the southeast, from Bourbon and Allen counties southward; scattered records west to Reno County

Smooth, perennial herbs, 1.5–3.5 dm tall, with slender rhizomes producing fibrous roots and small tubers up to 2 cm long, lacking aboveground stems except for the single, naked, much-elongated flowering peduncle bearing a single head of yellow flowers. Leaves all basal, spatulate to oblanceolate or linear, 4–14 cm long and 2.5–20 mm wide, the margins entire, with a few shallow teeth, or sharply pinnately lobed. Flowers yellow, the corollas about 1.5–2.3 cm long. Heads 15- to 20-flowered; involucre 10–15 mm long, the bracts linear-lanceolate, linear-oblong, or subulate, thin and with membranous margins, the apex rounded or acute. Achene body reddish-brown, cylindrical, 10- to 15-ribbed, about 2.5 mm long, with appressed hairs; pappus in 2 series, the inner of many rough bristles 5–8 mm long, the outer of 10 membranous scales about 0.6–1 mm long.

NOTE: Compare with *Pyrrhopappus grandiflorus*.

545. *Krigia dandelion*

K. oppositifòlia Raf.

NL., from L. *oppositus*, opposite, + *-folius*, -leaved.
Yellow; May–June
Prairie or open woods, usually on moist soil
East two-fifths and Barber County

Smooth, slender annuals or winter annuals with fibrous roots. Stems 1 to several, erect or ascending, eventually branched, 0.7–3.5 dm tall. Basal leaves spatulate to linear-oblanceolate, 1.5–10 cm long and 3–10 (–25) mm wide, the margins entire, with a few teeth, or sharply pinnately lobed; stem leaves spatulate to linear or narrowly lanceolate, 1.3–10 cm long and 2–6 mm wide, alternate below but appearing opposite at the upper

546. *Krigia oppositifolia*

CHICORY TRIBE
383

1 or 2 nodes. Flowers yellow, the corollas 2–4 mm long. Heads several to many, on wiry, simple or branched, terminal or axillary peduncles (0.7–) 1.5–7 (–10) cm long; involucre 3–5 mm long, the bracts ovate or lanceolate with acute or acuminate apexes, the bases becoming keeled along the midrib in fruit. Achene body reddish-brown, narrowly ob-ovoid, about 1.25–1.5 mm long, 10- to 15-ribbed, with minute transverse ridges (use high magnification); pappus absent or scarcely visible under high magnification as a low, scaly crown.

LACTÙCA L.   Lettuce

L. *lactuca*, the ancient Latin name for garden lettuce, from *lac*, milk—in reference to the milky sap of the plant.

Erect annual, biennial, or perennial herbs with milky juice. Stems usually un-branched except in the inflorescence. Leaves alternate, entire or pinnatifid, often with spiny-toothed margins. Flowers yellow, white, pink, purplish, or blue, all ligulate and bisexual, the heads 5- to many-flowered in branched, usually terminal clusters. Involucre cylindric or urn-shaped, the bracts green with a white margin, gradually becoming longer from the outside toward the inside and overlapping like shingles, strongly reflexed after the achenes are gone. Receptacle flat, naked. Achenes flattened, winged or laterally ribbed, with or without a beak, enlarged again at the point where the pappus is attached; pappus of many soft, white hairlike bristles which are persistent or which eventually fall away one at a time.

Six species in Kansas. NOTE: Mature achenes are usually necessary for positive identification. For *L. oblongifolius*, which has relatively large showy blue flowers and also fewer heads than our other species, mature achenes usually are not available on a plant with flowers at their prime.

Garden lettuce (*L. sativa* L., from Latin *sativus*, that which is sown or planted) belongs to this genus, and several wild species have edible parts. The seeds are eaten by some birds, including pheasants and goldfinches, and the plants are eaten by white-tailed prairie dogs, antelope, and deer, at least in some regions of the United States.

Since ancient times the milk, or latex, of wild lettuces has been used for medicinal purposes. The emperor Augustus, being convinced that he had been cured of a dangerous illness by the use of lettuce, erected an altar and a statue in its honor. Gerard's herbal tells us that "there be according to the opinion of the Antients, of Lettuce two sorts, the one wilde or of the field, the other tame or of the garden," and of the latter he says: "Lettuce cooleth the heat of the stomacke, called the heart-burning; and helpeth it when it is troubled with choler: it quencheth thirst, and causeth sleepe . . . being taken before meat it doth many times stir up appetite; and eaten after supper it keepeth away drunk-ennesse which cometh by the wine; and that is by reason that it staieth the vapours from rising up into the head."

The process of collecting the wild milk in quantity for the drug trade, as long employed in some sections of Europe and Britain, was to cut off the top of the plant a short distance below its apex, as the flowers began to appear, and scrape off the exuding milk into a china cup, repeating several times daily and from day to day the cutting back and scraping off, until the milk stopped flowing. The milk on drying becomes a hard, brown mass of narcotic odor and bitter taste called lactucarium. This was thought to have anodyne and sedative properties that at one time or another made it acceptable to all the pharmacopoeias.

Key to Species (3):

1. Achenes with only a median nerve on each face, occasionally with an additional pair of very obscure ones 2
   2. Heads relatively small, the involucre mostly (8–) 10–15 mm high; achene body mostly 5–6 mm, the beak 2 mm long, the pappus mostly 5–7 mm long .... *L. canadensis*
   2. Heads larger, the fruiting involucres mostly 15–22 mm high; achene body mostly (3.5–) 4–5 mm long, the beak 4 mm, the pappus mostly 7 (–12) mm long .... *L. ludoviciana*
1. Achenes evidently to very prominently several-nerved on each face 3
   3. Perennial from a taproot with a deep-seated crown (collected specimens are often broken off at point of attachment of this crown); heads relatively large and very showy, the involucre 14–18 mm high; flowers blue; leaves usually not sagittate .... *L. oblongifolia*
   3. Annual or biennial; heads relatively small, the fruiting involucre mostly 9–15 mm long (to 18 mm in *L. saligna*, which has distinctly sagittate leaves); flowers yellow, blue, or whitish 4
      4. Achenes with a short, stout beak less than half as long as the body, or beakless ......
       .... *L. floridana*
      4. Achenes with a filiform beak nearly as long to twice as long as the body 5

547. *Lactuca*

L. canadénsis L.   Canada wild lettuce

L. *-ensis*, suffix added to nouns of place to make adjectives—"Canadian," in this instance.
Yellow to pinkish-orange; July–September
Roadsides, fields, waste ground, woodlands
East half, primarily

548. *Lactuca ludoviciana*

Annual or biennial, up to 2.5 m tall, with coarse fibrous roots. Leaves quite variable, the lower ones mostly 1.5–3.5 dm long and 4–14 cm wide, oblanceolate in general outline, coarsely pinnatifid or sometimes merely toothed, especially on smaller (younger?) plants, the upper leaves gradually smaller toward the top of the plant, mostly oblanceolate to lanceolate, lance-oblong, or nearly linear and tapering to a long, thin point, more frequently tending to be entire with toothed margins but sometimes as deeply lobed as the lower leaves; bases narrow and somewhat petiolelike or, especially on upper leaves, sessile with sagittate lobes; margins and midvein of lower surface sometimes spiny. Herbage usually glabrous and more or less glaucous. Flowers yellow to pinkish-orange, the corollas 7–8.5 mm long. Heads many, mostly in a branched terminal cluster; involucre 8–12 mm long, the bracts grading from narrowly ovate or lanceolate on the outside to awl-shaped or linear in the innermost series. Achene body brown, often mottled with darker brown, elliptic or obovate, flattened, marginally winged, minutely and horizontally rugulose (use magnification), 3–4 mm long (excluding the threadlike 2 mm beak), with a single median rib along each face; pappus about 6 mm long.

Canada wild lettuce contains a bitter substance which taints the milk of cattle which graze on it, but the young leaves (before the bitter taste is strongly developed) may be used in salads or as a cooked vegetable.

The milky juice has been used as a sedative to treat coughs, and Menominee Indians used the juice of a freshly picked plant to cure poison ivy.

L. floridàna (L.) Gaertn.

"Of Florida."
Blue; mid-August–mid-September
Moist soil of creek banks and open woods, occasionally in ditches and prairie ravines
East two-fifths and a few counties farther west

Annual or biennial, up to 2 m tall, with coarse fibrous roots. Leaves mostly oblanceolate in general outline, sometimes broadly so, mostly 0.9–3 dm long and 3–18 cm wide, usually tapering to a petiolelike base lacking clasping basal auricles, and with 2–4 pinnatifid lobes and a prominent, triangular terminal lobe, the margins toothed, the upper leaves sometimes entire or merely toothed, sessile, somewhat smaller than the lower leaves. Flowers blue (rarely white), the corollas about 8–10 mm long. Heads many, mostly in a branched terminal cluster; involucre (8–) 10–12 mm long, the bracts grading from triangular or lanceolate on the outside to lance-linear or awl-shaped on the inside. Achene body brown mottled with darker brown, narrowly elliptic or fusiform, more or less flattened, slightly curved, lacking a prominent marginal wing, minutely rugulose (use magnification), 5–6 mm long, with 2–3 prominent ribs on each face, usually beakless in ours but sometimes narrowed to a thick, 1 mm beak; pappus about 5–7 mm long.

L. ludoviciàna (Nutt.) Ridd.

NL., from latinized form of Ludwig (Louis); "of Louisiana," in this case, probably referring to the Louisiana Territory.
Yellow (rarely blue); late July–mid-September
Roadsides, prairie, waste places
Reported from throughout the state, but infrequently encountered

Biennial or short-lived perennial, up to 1.5 m tall, quite similar to *L. canadensis*. Leaves variable, oblanceolate in outline below to lanceolate above, the lower ones 1.2–2.5 dm long and 4–7 cm wide, the upper ones gradually smaller, the margins merely toothed and prickly or coarsely pinnatifid, the bases lobed, even on the oblanceolate leaves; midvein on lower surface prickly. Flowers yellow (drying bluish), rarely blue when fresh, the corollas 11–14 mm long. Heads many, in a branched terminal cluster; involucre (15–) 17–22 mm long, the bracts grading from ovate or lanceolate on the outside to lance-attenuate or awl-shaped in the innermost series. Achene body brown mottled with darker brown, elliptic or oblanceolate, flattened, marginally winged, minutely horizontally rugulose (use magnification), (3.5–) 4–5 mm long (excluding the threadlike 4 mm beak), with a single median rib along each face; pappus mostly about 7 (–12) mm long.

L. oblongifòlia Nutt.

NL., from L. *oblongus*, oblong, + *folium*, leaf.
Blue or purplish; late June–August
Roadsides, creek banks, moist prairie ravines
North and west of a line drawn from Atchison County to Ford County

Perennials, up to 1 m tall, arising from taproots with deep-seated crowns. Leaves varying from oblanceolate to lanceolate, lance-oblong, narrowly elliptic, or linear-oblong, the lower ones mostly 8–20 cm long and 1–2.5 (–3) cm wide, the upper ones somewhat smaller; margins, at least on the upper leaves, quite entire, the lower leaves entire to more or less pinnatifid; bases of upper leaves sometimes slightly lobed and clasping the stem. Flowers blue or purplish, quite showy, the corollas 14–24 mm long. Heads few to many in a branched terminal cluster; involucre 14–18 mm long, the bracts grading from lanceolate on the outside to awl-shaped or lance-linear in the innermost series. Achene body gray, oblanceolate or narrowly elliptic, flattened or slightly 3-angled in cross section, very narrowly winged, about 3.5–4 mm long, tapering gradually into the stout, 2–2.5 mm beak, each face with 5–6 longitudinally ribs or veins; pappus about 8–9 mm long.

L. salígna L.   Willow-leaved lettuce

NL., from the genus *Salix*, willow—some of the leaves, i.e., the linear, entire ones, resembling those of willows.
Pale yellow; August–early October
Roadsides, weedy places
East fifth and a few central counties farther west; intr. from Europe

Smooth, slender annuals or biennials, up to 1 m tall, with fibrous roots. Leaves sessile, with conspicuous sagittate basal lobes and a prominent white midvein, never spiny, linear and entire or oblanceolate with narrow, widely spaced, few-toothed pinnatifid lobes, mostly (2.5–) 4–12 (–15) cm long and 2–9 mm wide (the pinnatifid leaves up to 25 mm wide), the upper leaves somewhat smaller than the lower ones. Flowers pale yellow (drying blue), the corollas about 8–12 mm long. Heads many, in a slender, usually branched terminal cluster; involucre 6–7 mm long in flower, up to 13 (–18) mm long in fruit, the bracts grading from scalelike or lanceolate on the outside to lance-attenuate or linear in the inner series. Achene body dark brown or grayish, oblanceolate or fusiform, somewhat flattened, lacking a marginal wing, 3–3.5 mm long (excluding the threadlike 5 mm beak), with 6–8 ribs on each face; pappus about 3–4 mm long.

L. serriòla L.   Prickly lettuce

NL., from L. *serra*, saw, + L. diminutive suffix *-olus*—the margins of the leaves and back of the midrib being minutely prickly.
Pale yellow; July–early October
Roadsides, disturbed prairie, alluvial flood plains, lake shores
Throughout; intr. from Europe

Biennial or winter annual, up to 1.5 m tall, with a large taproot. Leaves variable, oblanceolate to broadly elliptic, oblong, or lanceolate in outline, 0.5–3 dm long and 1–10 cm wide, gradually smaller upward, either entire or pinnatifid, the base with rounded or sagittate lobes, margins prickly toothed, midvein of lower surface and sometimes lower portions of the stem spiny. Flowers yellow (drying blue), the corollas about 9–10 mm long. Heads many, in a branched terminal cluster; involucre 6–14 mm long, the bracts grading from lanceolate or narrowly oblong on the outside to linear in the innermost series. Achene body brown, narrowly oblanceolate, flattened, minutely spiny toward the apex, 3–4 mm long (excluding the threadlike 4 mm beak), with 5–7 ribs on each face; pappus about 4–5 mm long.

The leaves of this species have a compass-plant tendency to align themselves in a vertical plane with the tips pointing north and south.

Garden lettuce is supposed to have originated from the wild *L. serriola*. It is more palatable than its ancestor because its milk contains a much smaller amount of the bitter principles to which the medicinal potency of the various wild species is attributed.

LYGODÉSMIA (Pursh) D. Don   Skeleton plant

NL., from Gr. *lygos*, willow twig, + *desma*, bundle—in reference to the much-branched, slender, twiggy, nearly leafless stems.

Annual or perennial herbs with much-branched, green ribbed stems and milky juice, the leaves mostly linear or subulate or reduced to small bracts. Flowers pink or purplish (rarely white), all ligulate and bisexual, in solitary, few-flowered heads at the ends of branches. Involucre cylindric, the bracts with membranous margins, 4–8 long bracts subtended by a few much-shorter ones. Receptacle slightly convex, naked. Achenes narrowly cylindrical or spindle-shaped, glabrous, with or without prominent ribs, with a pappus of hairlike bristles.

Two species in Kansas. *L. rostrata* Gray is known in Kansas from Hamilton, Finney, Morton, Stevens, Seward, Stafford, and Republic counties. It is an annual, more slender in aspect than the following species, and has its lower leaves opposite.

L. júncea (Pursh) Hooker   Rush skeleton plant

L. *junceus*, similar to *Juncus*, the rush—applicable to the stems and branches.
Pink; late May–August (–September)
Prairie hillsides and uplands, on loamy, usually sandy soil
West two-thirds and Doniphan County

Erect, gray-green perennial herb with a vertical rootlike rhizome that extends approximately 1.5–6 m down, giving off adventitious roots and shoots at various levels. Stems 2.5–5 dm tall, with numerous ascending branches. Leaves alternate, distantly spaced, the lower linear-lanceolate, the upper much reduced and awl-shaped or scalelike. Flowers pink to pale lavender, the corollas 14–20 mm long. Heads 5-flowered; involucre 12–16 mm long, the main bracts linear to linear-subulate, rather thin, and with membranous margins. Achenes prominently ribbed, the pappus off-white.

One frequently finds plants with globose, fruitlike bodies along the stem appearing to be normal features of the plant. These are galls caused by the ovipositing of a gall wasp, *Anistrophus pisum.*

A tea brewed from the steeped leaves and perhaps other parts of this plant was used by Pawnee, Ponca, Omaha, and Blackfoot Indians to increase the milk flow of nursing mothers, and the Cheyennes called it milk medicine. According to W. C. Stevens, some of the prairie Indians reportedly used an infusion of the plant as a wash for sore eyes and hardened exudations of the juice for chewing gum.

MICRÓSERIS D. Don.

NL., from L. *micro*, Gr. *mikro*, small, little, + *seris*, for a kind of endive—the cluster of basal leaves with wavy margins being reminiscent of the cluster of leaves produced by a variety of chicory (*Cichorium intybus*) which is sometimes called endive, as is *C. endivia*, the true endive.

Annual or perennial herbs with thickened, sometimes branched vertical rhizomes and milky juice. Leaves in ours all basal, the margins entire and usually undulate. Flowers yellow, all ligulate and bisexual. Heads many-flowered, scapose; involucre bell-shaped, the bracts either all of the same length or gradually longer toward the inside, or of 2 distinct lengths. Receptacle slightly convex, naked. Achenes cylindric, or fusiform, longitudinally ribbed, with a pappus of white, flattened, hairlike bristles, some of which have a broader membranous base (use magnification).

One species in Kansas.

M. cuspidàta (Pursh) Sch.-Bip.   Wavyleaf microseris

L. *cuspidatus*, having a sharp, rigid point—in reference to the leaves.
Yellow; late April–May (–June)
Prairie hillsides and uplands, usually on sandy or rocky soil
Scattered nearly throughout, but uncommon in the southwest fourth and not known from the
    extreme southeast counties

Acaulescent perennial. Leaves linear or lance-linear, grasslike, often folded along the midvein, 6–25 cm long and 0.7–2 cm wide, the margins usually undulate. Herbage, especially the leaf margins, with short, white kinky hairs. Flowers yellow (drying white with a purple stripe down the center of the ligule or ray), the corollas 2–3 cm long. Heads 1–4, borne singly on naked peduncles mostly 0.5–3 dm tall; involucres 2–2.5 (–3) cm long, the bracts lanceolate with an attenuate apex or awl-shaped, with broad membranous margins, the outer bracts somewhat shorter than the innermost and grading into the longer ones. Achene body straw-colored, slightly tapered toward each end, glabrous, about 6.5–9 mm long; pappus 10–15 mm long.

NOTE: Compare with *Pyrrhopappus grandiflorus* and *Krigia dandelion.*
See plate *109.*

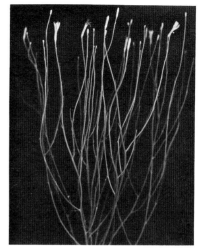

549. *Lygodesmia juncea*

550. *Lygodesmia juncea,* close-up of flower

551. *Lygodesmia juncea,* insect galls on stem

CHICORY TRIBE

387

552. *Microseris cuspidata*

553. *Prenanthes aspera*

PRENÁNTHES L.   Rattlesnake root

NL., from Gr. *prenes*, prone, or bent forward, drooping, + *anthes*, the suffix form of *anthos*, flower—the flower heads of the type species, *P. trifoliata*, and several other species with open panicles being bent forward. "Rattlesnake root" because the milky juice and an infusion of the leaves were formerly used in the South to treat snakebites.

Erect perennial herbs with thick, bitter taproots or corms and milky juice. Leaves alternate, at least some deeply lobed. Flowers purple to pink, white, or yellowish, all ligulate and bisexual, the heads 5- to 12-(35-)flowered, usually nodding or drooping (though not in ours) and in branched clusters. Involucre cylindrical, with 2 inner rows of long linear bracts subtended by a few much-shorter bracts, or sometimes the outer ones grading into the longer ones. Receptacle slightly convex, naked. Achenes narrowly cylindric, glabrous, ribbed, with a pappus of many deciduous, hairlike bristles.

One species in Kansas.

P. áspera Michx.   Rough rattlesnake root

L. *asper*, rough.
Cream-colored; mid-August–mid-October
Prairie, usually in rocky soil, less frequently in dry, rocky, open woods
East third, primarily

Wandlike, rough-hairy perennial with a short, thick rhizome and fibrous roots. Stems 0.6–1.3 m tall, unbranched except perhaps in the inflorescence. Lower leaves 4–10 cm long and 1–4 cm wide upright, more or less oblanceolate, tapering into a narrow, clasping, petiolelike base, coarsely toothed; upper leaves lanceolate to elliptic, gradually becoming shorter, definitely sessile, the margins often entire. Flowers cream-colored, fragrant, the corollas 11–18 mm long. Involucre 12–17 mm long, the bracts definitely of 2 different lengths. Heads 12- to 16-flowered, mostly ascending or spreading (rather than drooping, as is typical for the genus), in a long, strict, compound spike or pyramidal panicle. Body of the achenes reddish-brown, about 6–7 mm long, the pappus light tan, about 8–10 mm long.

The plants begin growth early in the spring, producing first of all a rosette of leaves, followed by others that are long and upright and at first able to compete for light with the grasses growing up around them but which die and fall away as the season progresses. When blooming in late summer and fall, the plants are tall enough to be easily distinguished from other vegetation by their flowers and habit of growth.

NOTE: Compare with *Hieracium longipilum*.

PYRRHOPÁPPUS DC.   False dandelion

NL., from Gr. *pyrrhos*, red or tawny, + *pappos*, old man, botanically used for the series of bristles, scales, teeth, etc. which represents the calyx in the Composite Family—the pappus in this genus being red or rust-colored.

Mostly annual or biennial herbs with taproots and milky juice. Leaves alternate on the stem or all basal, entire or pinnatifid. Flowers yellow, all ligulate and bisexual, with a 5-toothed corolla, the heads many-flowered in ours, solitary and only a few per plant. Involucre cylindric to somewhat conical or bell-shaped, the many bracts lanceolate to subulate, thin, with membranous margins, the inner ones longer with a glandular crestlike apex, the outer ones much shorter, subulate, and lacking an apical crest. Receptacle slightly convex, naked. Achenes fusiform, nearly round in cross section, with 5 longitudinal grooves and rounded ribs; pappus elevated on a long, slender beak, dirty white or light tan, of many hairlike bristles which are persistent or fall as a unit if the beak breaks, usually subtended by a ring of minute woolly white hairs.

Two species in Kansas.

Key to Species:
1. Annual or biennial with leafy aboveground stems and eventually several flower heads; plants mostly of the east third of Kansas .... *P. carolinianus*
1. Acaulescent perennial with a deep-seated tuber, leaves all basal, and 1 (–4) scapose flower heads; plants mostly of the central third of Kansas .... *P. grandiflorus*

P. caroliniànus (Walt.) DC.   False dandelion

"Of Carolina."
Yellow; late May–October
Prairies, roadsides, less frequently in open woods
East two-fifths plus Barber County, more common eastward

Erect, leafy-stemmed annual or biennial, up to 1.6 m tall (much shorter in mowed areas). Lower leaves oblanceolate to almost linear, (6–) 10–25 cm long and 1–9 cm wide, tapering to a petiolelike base, the margins entire, coarsely toothed, or pinnatifid; upper leaves gradually becoming sessile with hastate or clasping bases and usually smaller than the lower leaves. Flowers yellow, the corollas about 1.5–2 cm long. Heads 1 to several on long axillary or terminal peduncles; involucre (1–) 1.5–2.3 cm long, bell-shaped, the inner bracts linear and minutely hairy. Achene body brown, minutely roughened, about 6 mm long; pappus about 10 mm long (excluding the 10 mm beak).

P. grandiflòrus (Nutt.) Nutt.   Rough false dandelion

NL., from L. *grandis*, large, great, + *flos, floris*, flower.
Yellow; May–June
Prairies and roadsides
In a roughly triangular area from Seward County to Rooks, Republic, and Washington counties and southward to Chautauqua County

Acaulescent perennials arising from ovoid or ellipsoid tubers up to 3.5 cm long situated 2–18 cm below ground level. Leaves all basal, oblanceolate or narrowly so, mostly 2.5–17 (–22) cm long and 1.5–5 (–7.5) cm wide, the margins merely with a few widely spaced teeth to pinnatifid with toothed lobes. Herbage glabrous or sparsely hairy. Flowers yellow, the corollas about 2.5–3 cm long. Heads 1–4, each borne singly on a scape-like stalk 1–3.5 dm tall which is naked or has 1 or 2 small bracts (rarely with a second head arising in the axil of 1 of the bracts); involucre bell-shaped, 1.7–2.2 cm long, the inner bracts linear or awl-shaped, minutely hairy, some on each head with and some without an apical crest as in *P. carolinianus*. Achene body fusiform, straight or slightly curved, 4–5 mm long, initially light tan with irregular transverse ridges, eventually becoming dark brown and minutely scaly as the epidermis peels loose from the ridges; pappus 10–13 mm long (excluding the 5–7 mm beak).

NOTE: Compare with *Krigia dandelion* and *Microseris cuspidata*.

554. *Pyrrhopappus grandiflorus*

SÒNCHUS L.   Sowthistle

L. *sonchus*, from Gr. *sonchos*, for any sowthistle. "Sowthistle" is the old English name, from AS. *sugethistel*, the first component being from AS. *suga*, sow. Plants of this genus were reputed to be relished by pigs.

Erect, leafy-stemmed annual or perennial herbs with hollow stems and milky juice. Leaves alternate or all basal, entire to variously pinnatifid, the bases frequently auricled and clasping the stem, the margins usually spiny-toothed. Herbage mostly smooth and glaucous. Flowers yellow, all ligulate and bisexual, the rays or ligules linear and quite narrow, 5-toothed, the heads many-flowered, solitary or several to many in branched clusters terminating the main stem or lateral branches. Involucres mostly ovoid or bell-shaped, the bracts of 2 sizes, the longer inner ones lanceolate to subulate, green but with broad membranous margins, with or without 1–3 capitate, stalked glandular structures along the midrib, the short outer bracts of various lengths, overlapping like shingles. Receptacle flat or slightly convex, naked. Achenes elliptic to narrowly ovate or lanceolate, flattened 6- to 10-ribbed, marginally winged, glabrous, the outer row straw-colored, the others reddish-brown, the pappus of many very fine, white, persistent, hairlike bristles surrounding 6–10 flattened, early-deciduous bristles.

555. *Sonchus asper*

Two species in Kansas. *S. oleraceus* L., an introduced annual weed similar to the following species, is known from several counties, mostly in the east third of the state. It differs from *S. asper* primarily in having the basal lobes of the middle and upper leaves pointed rather than rounded, and in having achenes with firm walls distinctly transversely roughened.

S. àsper (L.) All.   Prickly sowthistle

L. *asper*, rough, prickly—in reference to the leaf margins.
Yellow; late May–July, again in late September–October
Fields, roadsides, waste places, usually in moist soil
East half, primarily; intr. from Europe

Annuals, up to 2 m tall, with only a few branches, if any. Lower leaves oblanceolate with an attenuate base, mostly 0.4–2 dm long and 1.5–8 cm wide, the margins finely toothed and prickly to coarsely toothed or deeply pinnatifid with toothed lobes; upper leaves gradually becoming smaller and sessile with rounded, clasping basal lobes, the margins usually only finely to coarsely toothed and prickly. Upper parts of the stem, at least within the inflorescence, with conspicuous glandular hairs. Flowers yellow (but often drying white with a blue tinge), the corollas about 17–21 mm long. Heads several to many; involucres bell-shaped, 11–15 mm long, the inner bracts subulate or lance-linear; bases, at least of the outer bracts, becoming whitish and hardened and persisting after the achenes have blown away. Achene body elliptic or lanceolate, about 2.5–2.8 mm long; pappus about 8 mm long.

TARAXACUM Zinn   Dandelion

NL. *taraxacum*, from *tarakhshaquq, -aqun*, of Persian origin.

Biennial or perennial herbs with long taproots and milky juice. Leaves sessile, all basal, entire or mostly variously pinnatifid. Flowers yellow, all ligulate and bisexual, the heads many-flowered, borne singly on naked, hollow peduncles arising from the rosette of leaves. Involucre bell-shaped, with an inner row of linear or subulate, erect herbaceous bracts which have glandular structures near the apex and 2 outer rows of shorter, reflexed ones. Receptacle slightly convex, naked. Achenes cylindric or broadly spindle-shaped, round or 4- to 5-angled in cross section, longitudinally grooved or ribbed, the upper half usually roughened, usually with a long, slender beak topped with a pappus of many long, white, hairlike bristles.

One species in Kansas. Occasional red-seeded plants have been recognized by some authorities as a distinct species, *T. erythrospermum* Andrz. (from Gr. *erythros*, red, + *spermum*, seed), but in our area these plants appear to be mere variants which occur mixed in populations of the common *T. officinalis*, and one never encounters a colony composed even predominantly of red-seeded individuals (R. L. McGregor, personal communication).

T. koksaghis Rodin., a native of central and western Asia and a close relative of our dandelion, was used by Russia and the United States during World War II as a commercial source of latex.

556. *Taraxacum officinale*

T. officinàle Weber   Dandelion

L. *officinalis*, pertaining to drugs and medicine, from *officina*, laboratory—the name indicating that the plant has been used as a medicinal. The dandelion is a native of Eurasia, but none of the Greek and Latin writers gives an account of any such plant. The first unmistakable identification of it is found in the writings of Arab physicians of the early Middle Ages, and thenceforward it appears with consistency in the herbals of southern and central Europe. "Dandelion" is the English version of Fr. *dent de lion*, lion's tooth, from L. *dens lionis*. The German version is *Löwenzahn*, with the same meaning. Indeed, the vernacular names in practically all European countries have this meaning. The commonly accepted (and possibly correct) explanation of this name is that it refers to the prominent dentation of the leaf margin. However, in *Hortus Sanitatus*, a renowned herbal first published in 1485 in Mainz, Germany, we read that a certain Master William prized this plant highly because of its therapeutic potency "and therefore likened it to a lion's tooth, called in Latin *dens leonis*."

Yellow; April–June, mid-August–October, sparingly through the winter
Lawns, roadsides, waste places
Throughout; intr. from Europe

Acaulescent perennial with a long, thickened, contractile taproot and a crown giving rise to a rosette of oblanceolate, mostly strongly toothed to deeply and coarsely pinnatifid, usually runcinate leaves 0.5–3 dm long and (1–) 2.5–6 cm wide. Flowers yellow, the corollas 15–22 mm long, the rays or ligules linear. Heads 1 to several, the flowering and fruiting stalks mostly 0.5–3 dm tall; involucre broadly cylindric or bell-shaped, 1.5–2.5 (–3) cm long, the inner bracts subulate, glaucous, with a purple glandular area at the apex. Achene body gray to light brown or rarely red, slightly flattened, 3–4.5 mm long, antrorsely barbed toward the slender, 6–8 mm beak; pappus 4–5 mm long.

Though this is one of our most common and troublesome weeds, it is also one of the most interesting plants in the state, for several reasons. Dandelions bloom only when the day's length is 12 hours or less. The flower buds develop near the ground, in the center of the basal rosette of leaves. This takes about 1 week, and near the end of this week, over a period of about 48 hours, each bud is elevated on a naked scape and the bracts of the involucre reflex, allowing the flowers to spread wide apart. At this time, one may observe various insects gathering pollen from the blossoms and might assume that cross-pollination is occurring. However, dandelions are of hybrid origin and have 3 sets of chromosomes, rather than the usual 2 sets necessary for normal meiosis to take place, and most of the pollen grains are infertile. Each head remains open but 1 day; then it closes and the scape and head flatten to the ground. The ovules and seeds develop apomictically —that is, without being fertilized. This is made possible by an error or abnormality in the process of meiosis, which normally produces an egg cell with half the number of chromosomes as in the parent plant (24, in this case). The first stages of meiosis proceed properly, but at some early point a cell and its nucleus fail to divide completely and a nucleus with 24 chromosomes results. From this nucleus, the embryo and seed develop. Maturation of the achenes takes several days, after which time the scape again becomes erect, the bracts recurve, and the receptacle becomes permanently convex, spreading the achenes apart and allowing them to be carried away by the wind.

Apparently equipped to overcome all adversity, dandelions have fleshy taproots, up to 3 dm long and 3 cm wide, which have several interesting capabilities: First, they are contractile and can pull the crown very close against the ground or even below the surface, making it more difficult for the plant to be cut off or dug up. Secondly, the tissues of this root, when wounded (as when one attempts to pull the plant out by grasping the leaves), produce undifferentiated callus tissue, similar to that in the center of the root, and from each wound usually 2–5 new shoots and eventually new plants develop!

We would be wronging this stubborn, tenacious invader, however, to call attention only to those characters that now are in disrepute. Its introduction into this country was probably intentional, and there are indications that seeds were brought on the May-flower. Its use for greens in this and other countries is well known, and according to the U.S. Department of Agriculture's publication, *The Composition of Foods*, the young leaves are a good source of iron, calcium, phosphorous, potassium, and vitamin A and have some vitamin C. Most authorities recommend blanching the leaves—that is, covering the plants with an opaque box, flower pot, etc., so that the leaves remain whitish or yellowish green—to reduce the bitterness that otherwise tends to develop. Large plants may be transplanted into bushel baskets or other sizable containers, covered with leaf litter, and kept in the basement for use as a fresh vegetable during winter months. The blossoms are used in making dandelion wine; and the roots, which contain a bitter principle, taraxin, may be dried, browned, ground, and used as a coffee substitute or adulterant. In spite of the bitterness, the roots were eaten by the Digger Indians of Colorado and the Apaches of Arizona. According to W. C. Stevens, when a cloud of grasshoppers descended on the island of Minorca and devoured most of its vegetation, the inhabitants were able to eke out an existence with dandelion roots.

All parts of the dandelion have been used as medicinals, and the dried roots were listed officially in the *U.S. Pharmacopoeia* from 1831 to 1926. Herbals of the Middle Ages and later times recommended infusions and extracts of its several parts for promoting the appetite and stimulating the organs of digestion and secretion, and it has been recognized as a benefactor by purifying the blood and dispelling the lethargy known as spring fever when used as greens and salads and in medicinal preparations. Various Indian groups of North America adopted its use soon after its introduction into this country. Kiowa women used a tea made from dandelion flowers and pennyroyal (*Hedeoma pulegioides*) leaves to relieve menstrual cramps, while a tea made from the roots was used by the Pillager Ojibwas to treat heartburn.

In addition, the foliage is eaten by small mammals and browsers and the seeds by birds and small rodents.

TRAGOPÒGON L.   Salsify, goat's beard, oyster plant

NL., from Gr. *tragopogon*, for a kind of plant, from Gr. *tragos*, goat, + *pogon*, beard, in reference to the pappus.

Glabrous biennial herbs with fleshy taproots and milky juice. Leaves alternate, grasslike, sessile and clasping the stem. Flowers yellow or purple, all ligulate and bisexual, the rays 5-toothed, the heads many-flowered and solitary at the ends of the branches. Involucre more or less obconical to broadly cylindric or bell-shaped, the bracts narrowly lanceolate to subulate, tapered to a fine point, in 2 rows, of about the same length. Receptacle flat, rough but not chaffy. Achenes linear or fusiform, round or angled in cross section, 5- to 10-ribbed, at least all but the outer ones with a slender beak, the pappus of long, silky, plumose bristles.

Two species in Kansas. *T. pratensis* L., an introduced European weed established in the eastern United States, has been reported for Kansas but is very rare if present at all.

Key to Species:
1. Flowers yellow .... *T. dubius*
1. Flowers purple .... *T. porrifolius*

T. dùbius Scop.   Meadow salsify, goat's beard

L. *dubius*, doubtful, uncertain—this species being so nearly like *T. pratensis* L., which was named first.

557. *Tragopogon dubius*

558. *Tragopogon dubius* fruiting

Yellow; mid-May–June
Roadsides, waste places, prairie
Throughout; intr. from Europe

Plants erect, 0.3–1 m tall, sometimes branched from the base. Leaves lance-linear and gradually tapering to a long, fine point, mostly 1–2.5 (–4) dm long and 3–14 mm wide. Flowers yellow, the corollas about 2–2.5 (–3) cm long, the ligules or rays linear. Heads 1 to several, the peduncles becoming broader toward the base of each head; involucres 3–4 cm in flower (to 7 cm in fruit), the 13 bracts appressed or ascending around the flowers and immature fruits, but strongly reflexed after the achenes have blown away. Achene body light brown or straw-colored, narrowly fusiform, straight or slightly curved, obscurely 5-ribbed and angled in cross section, antrorsely roughened, 13–17 mm long, tapering gradually into a slender beak 25–35 mm long; pappus bristles silvery or grayish, finely plumose, 30–35 mm long.

The large fruiting heads (fig. 558) are quite conspicuous and attractive in the morning sunlight.

The very young leaves and stems may be used in spring for salads or as a cooked green vegetable. The roots should be gathered in fall to earliest spring (before the flowering stems develop) for use as a vegetable which has, for centuries, enjoyed a reputation better than that of the cultivated *T. porrifolius.*

T. porrifòlius L. Vegetable-oyster salsify

NL., from L. *porrum,* leek or chives (*Allium porrum*), + *folium,* leaf—the leaves of the 2 species
being similar. The roots, properly cooked, are said to have the flavor of oysters.
Purple; mostly May–June
Fields, roadsides, waste places
Scattered localities, mostly in the east half; intr. from Europe

Plants erect, 0.5–1.3 m tall, similar in size and aspect to *T. dubius.* Flowers purple; involucres 2–4 cm long in flower, with 8–11 bracts. Achene body orange-brown or straw-colored, plumply fusiform, 5-angled, antrorsely roughened, about 13 mm long, rather abruptly narrowed to a slender beak 20–23 mm long; pappus bristles finely plumose, 20–25 mm long, eventually becoming brownish.

This is the only species of *Tragopogon* in cultivation because its roots are the largest—up to 3 dm in length and 5 cm across at the top—though for esculent quality the roots of the other species growing wild are superior. Parkinson (12) in his *Paradisus Terrestris* puts it this way: "If the rootes of any of these kindes being young, be boyled and dressed as a Parsnep, they make a pleasant dish of meate, farre passing the Parsnep in many mens judgements, and that with yellow flowers to be the best."

CYNÀREAE Thistle Tribe

Sap of plants not milky. Stems, leaves, and/or bracts of involucre usually spiny. Bracts of involucre usually prickly or finely fringed. Flower heads all alike, with disk flowers only, though in *Centaurea* the outer disk flowers are larger and irregularly cut with elongate corolla lobes resembling rays. Receptacle usually furnished with bristly chaff, sometimes naked. Anthers with tail-like projections at the base. Styles with a hairy or thickened ring below their branches. Pappus of bristles or slender, hairlike scales.

ÀRCTIUM L. Burdock

NL., from Gr. *arktos,* bear—the rough bracts of the involucre suggesting a bear's coarse fur.

Biennial herbs. Leaves alternate, simple, petioled, the blades mostly cordate, the margins entire to toothed or rarely laciniate. Heads small, many-flowered, discoid, borne singly or many together in branched terminal clusters. Involucre globose or nearly so, the many bracts in several series, graduated in length, leathery, narrow, the lower part appressed, the tip attenuate, spreading, and inwardly hooked. Receptacle flat, bristly. Pappus of many short, rough, separately deciduous bristles. Flowers all tubular, bisexual, fertile, the corolla purple (rarely white), with a slender tube. Anthers with tail-like bases. Style branches with a hairy ring below. Achenes oblong, somewhat compressed, 3-angled, many-nerved, the top truncate.

One species in Kansas.

A. minùs Schkuhr Common burdock

L. *minus,* less, lesser—in comparison with the great burdock, *A. lappa.*

Rose-purple; July–November
Partially shaded waste places
East two-fifths, primarily, with scattered records from the northwest fourth; intr. from Eurasia

Coarse, erect plants, up to 1.5 m tall. Leaf blades broadly to narrowly ovate, those of the main leaves mostly 1–5 dm long and 0.7–4 dm wide, the apex rounded and usually mucronate, the base abruptly acuminate or sagittate, with rounded basal lobes, the margins entire or coarsely toothed, frequently undulate. Heads several to many, 1–2 cm long and 1.2–2.3 cm across. Corollas rose-purple (rarely white), 7–9 mm long. Achene body 4.5–5.5 mm long, the pappus 1–1.5 mm long.

By the time the fruits have ripened, the slender, hooked bracts of the involucre become quite rigid. These get hooked to clothing or the hair of animals and hold so fast that the mature heads are pulled off and carried away, and the seedlike achenes eventually become widely scattered.

We know from the writings of Dioscorides and Galen (1st and 2nd century) that the root of the burdock was used medicinally in their time, and down through the centuries herbals have made extravagant claims for its curative value.

Matthiolus in the 16th century wrote that it was an excellent medicine for "hawking up phlegm and blood," and the green leaves in summer when spread over sprained members were beneficial "even though a bone was broken"; crushed leaves mixed with salt and applied to the wound were "very good for those who have been bitten by vipers, mad dogs and other poisonous animals," and "the root even drives away goiter when mixed with melted lard and applied thereto."

After the burdock got started in this country some tribes of our Plains Indians used its root as a remedy for pleurisy. The Japanese eat the root of the burdock as a vegetable, and its high content of inulin, as in the case of artichoke tubers, would certainly insure its nutritive value. Weiner (34) recommends peeling the roots and stems and cooking them with two changes of water before eating them.

See plate *110*.

559. *Arctium minus*

CÁRDUUS L.   **Thistle**

L. *carduus*, name used by Pliny, Virgil, and others for a common thistle.

Stout, spiny annual, biennial, or perennial herbs. Leaves alternate, simple, sessile, pinnately lobed or toothed with spiny margins, the leaf bases forming spiny wings down the stem. Heads many-flowered, discoid, borne singly or in clusters at the ends of the stems. Involucre hemispheric, the bracts many, in several series, graduated in length, mostly spine-tipped and eventually recurved. Receptacle flat or convex, covered with soft hairlike bristles. Pappus of many firm, rough hairlike bristles united in a ring and falling as a group. Flowers all tubular, bisexual, fertile, the corollas pink or purplish (rarely white or yellow), deeply and somewhat unevenly 5-lobed. Stamens with separate hairy filaments and sagittate anthers. Style branches threadlike, the pair subtended by a thickened, sometimes hairy ring of tissue. Achenes obovate, 4-angled or somewhat flattened, glabrous.

560. *Carduus nutans*

Two species in Kansas. *C. acanthoides* L. is known from Nemaha County only. It is a plant of bushier habit than the common *C. nutans* and has numerous small, erect heads, mostly 2–3 cm long in flower, with very narrow involucral bracts 0.5–1 (–2) mm wide.

C. nùtans L.   **Nodding thistle, musk thistle**

L. *nutans*, nodding—in reference to the pose eventually taken by the flower heads.
Rose-purple to white; late May–early August (–October)
Roadsides, overgrazed pastures, waste places, and other disturbed habitats.
East three-fifths, primarily; intr. from Europe

Coarse, spiny biennial or winter annual, 0.3–2 m tall, the stems simple, or branched above. Leaves glabrous, 5–22 cm long and 1.5–9 cm wide, mostly lanceolate to elliptic or oblong in general outline, coarsely serrate to pinnately lobed about halfway to the midrib, the main veins extended beyond the margins as rigid spines. Heads mostly nodding, the middle and outer involucral bracts (2–) 5–7 mm wide. Corolla rose-purple to white, 2–3.3 cm long. Achene body smooth, the pappus about 2 cm long.

In 1932, this plant was known only from Washington County, and by 1948 it was still limited to Washington, Shawnee, Riley, Greenwood, Atchison, Republic, and Saline counties. Today it is well established in more than half of the state and is officially considered a noxious weed. According to Steyermark (4), the dried flowers have been used as rennet to curdle milk, and the pith of the stem is edible when boiled.

561. *Carduus nutans*, close-up of head

THISTLE TRIBE
393

The nodding thistle is closely related to our species of *Cirsium*, but even in the vegetative stage *Carduus nutans* can be distinguished from our species of *Cirsium* on the basis of its entirely hairless leaves.

CÍRSIUM Mill.   Thistle

NL., from Gr. *kirsian,* ancient name of a kind of thistle, possibly from *Kirsus,* varicose veins—
    plants of the genus having been used for this and various other afflictions.

Spiny annual, biennial, or perennial herbs. Leaves alternate, simple, sessile, entire to toothed or lobed, the margins spiny. Heads many-flowered, discoid, borne singly on peduncielike branches, usually on the upper part of the plant. Involucre globose to ovoid, bell-shaped, or nearly cylindric, opening more broadly as the achenes mature, the many bracts in several series, graduated in length, sometimes with a thickened, resinous keel, some or all usually spiny-tipped. Receptacle flat, densely covered with soft hairs. Pappus of many plumose bristles united basally and falling away as a group. Flowers all tubular, bisexual, fertile, the corollas lavender to rose-purple or pink, rarely white or yellow, deeply and unequally 5-lobed. Stamens usually with separate filaments, the anthers with long or short, tail-like basal appendages. Style branches subtended by a thickened, hairy ring and an abrupt change in texture. Achenes obovate to oblong or somewhat elliptic, somewhat flattened or 4-angled, glabrous.

Five species in Kansas.

Key to Species:
1. Stem conspicuously winged by the decurrent leaf bases, the wings extending (1.5–) 2–5 cm below the point of departure of the leaf blade; upper leaf surface armed with yellowish prickles .... *C. vulgare*
1. Stem not winged at all or winged no more than 1–2 cm below the leaf bases; upper leaf surface lacking prickles 2
    2. Stems slightly hairy to glabrous; leaves entire to toothed or pinnatifid, the upper surface usually dark green and glabrous .... *C. altissimum*
    2. Stems densely covered with matted white hairs; leaves mostly pinnatifid, the upper surface more or less densely covered with matted hairs, the lower surface quite densely covered 3
       3. Leaf bases not decurrent at all or for less than 5 mm; upper and lower leaf surfaces usually equally hairy; flowering heads mostly 4–7 cm long .... *C. undulatum*
       3. Leaf bases decurrent for 1–2 cm; upper leaf surface frequently less hairy than the lower; flowering heads mostly 3.5–5 cm long .... *C. ochrocentrum*

C. altíssimum (L.) Spreng.   Tall thistle

L. *altissimus,* tallest, from *altus,* tall.
Purple to rosy-lavender; August–October
Roadsides, waste ground, pastures
East half, primarily; scattered records as far west as Sheridan County

Robust biennial with a thickened root and rosette of large leaves from the crown the first year, growing 1–2 m tall the second year. Stems sparsely hairy to nearly glabrous, not winged. Leaves mostly oblanceolate to narrowly elliptic, varying from entire to toothed or pinnatifid, the upper surface dark green and glabrous, the lower surface densely covered with matted white hairs. Heads mostly 3–4.5 cm long. Involucral bracts tipped with small, outward-pointing prickles. Corollas purple to rosy-lavender, 2.5–3.7 cm long. Achene body 5–6 mm long, the pappus 1.8–2.6 cm long.

This is our tallest species of *Cirsium* and the least formidable to handle. Though rank and weedy in habit, it is handsome in flower, as well as when the ripe, seedlike achenes are poised with outspread, plumose bristles for drifting away with the wind.

Individuals with deeply pinnatifid, spinier leaves have been segregated by some authorities as *C. discolor* (Muhl.) Spreng.

C. arvense L., Canada thistle, is known from scattered localities across the state. It tends to be nearly glabrous and has numerous small heads mostly 1.5–2 cm long. It is officially considered a noxious weed in the state of Kansas.

See plate *111*.

C. ochrocéntrum A. Gray   Yellow-spined thistle

NL., from Gr. *ochros,* light yellow, + *kentron,* prickle.
Purple to rosy-lavender; late June–August
Roadsides, pastures, dry prairie
West three-fifths

Erect perennial strongly resembling *C. undulatum* and apparently sometimes inter-

562. *Cirsium altissimum*

grading with that species. Stems densely covered with matted white hairs, winged by the decurrent leaf bases for about 1–2 cm below the point of departure of each leaf. Leaves much like those of *C. undulatum* but more heavily armed. Mature flowering heads mostly 3.5–5 cm long. Involucral bracts tipped with rigid prickles about 1 cm long. Corollas purple to rosy-lavender (rarely white), 3–3.7 cm long. Achene body about 5 mm long, the pappus 2.5–3.5 cm long.

This plant was an outstanding figure amidst the desolation of overgrazed pastures during the great drought of the thirties.

The flowers of this species, boiled in water, were utilized by the Kiowa Indians in a preparation to treat burns and skin sores, while the Zunis made a tea of the entire plant to treat syphilis (34).

C. undulàtum (Nutt.) Spreng. — Wavyleaf thistle

L. *undulatus*, wavy—in reference to the leaf margin.
Purple to rosy-lavender; June–September
Roadsides, overgrazed pastures, dry prairies
Throughout

Stout biennial, up to 1.2 m tall, more or less densely covered with matted white hairs, which are especially thick on the stems and under surface of the leaves. Stems not winged by decurrent leaf bases. Leaves mostly pinnatifid, the margins typically strongly undulate, the sinuses between the major lobes rounded or, less frequently, rectangular, mostly 0.5–1.5 cm wide at the base of the sinus, the upper surface densely covered. Mature flowering heads mostly 4–7 cm long. Involucral bracts tipped with prominent, outward-pointing prickles. Corollas purple to rosy-lavender (rarely white), 3.3–5.2 cm long. Achene body 5–6.5 mm long, the pappus 2.7–4 cm long.

This is our most widely distributed thistle. The plants tend to increase in overgrazed or drought-afflicted pastures, but they are so formidably armed with prickles that cattle are discouraged from eating them. In spite of the plant's unfriendly demeanor, the large purple heads are quite attractive against the silvery foliage, and handsome specimens occasionally are allowed to persist in yards as ornamentals. The cut flowers or, later on, the fruiting heads are sometimes used in fresh or dried floral arrangements.

Cross-pollination is effected in this and other species of *Cirsium* chiefly by bees and butterflies.

C. vulgàre (Savi) Ten. — Common thistle, bull thistle

L. *vulgaris*, common, ordinary.
Purple to rosy-lavender; June–October
Pastures, fields, roadsides, waste ground
East three-fifths; intr. from Eurasia

Coarse, dark green biennial, up to 1.5 m tall. Stem conspicuously winged by the decurrent leaf bases, usually also with kinky brownish hairs. Leaves deeply pinnatifid, the sinuses between the major lobes rectangular, mostly 1.5–3 cm wide, the upper surface of the blade covered with appressed yellowish prickles, the lower surface glabrous or sparsely covered with kinky white hairs. Heads 4–7 cm long and 4–5.5 cm across, the involucral bracts all spine-tipped. Corollas purple to rosy-lavender, 2.8–3.6 cm long. Achene body 3–4 mm long, the pappus 1.5–2 cm long.

Earlier manuals have listed this plant erroneously as *C. lanceolatum*.

VERNÒNIEAE — Ironweed Tribe

Sap of plants not milky. Flower heads all alike, with disk flowers only, the flowers all bisexual and fertile. Receptacle naked. Anthers sagittate at the base but not caudate (tailed). Style branches threadlike or hairlike, round in cross section, minutely hairy. Pappus of bristles or of slender, bristlelike scales.

ELEPHÀNTOPUS L. — Elephant's foot

NL., from Gr. *elephos, elephantos*, elephant, + *pous*, foot—from the shape of the stem leaves, or perhaps, according to Fernald (9), a translation of some aboriginal name.

Perennial herbs, usually coarsely hairy and branched above. Leaves alternate or mostly basal, simple with toothed margins, sessile, mostly ovate to obovate or elliptic. Heads small, with (1–) 3–4 (–5) flowers, these clustered in larger, headlike glomerules

563. *Cirsium ochrocentrum*

564. *Cirsium undulatum*

subtended by 1–3 leaflike, cordate or deltoid bracts. Involucral bracts of the true heads stramineous, in 4 pairs, the outer 2 pairs shorter than the inner 2. Receptacle flat. Flowers all bisexual and fertile. Pappus of 5 (–8 or 10) flattened, rigid bristles or awns which are dilated and hardened at the base. Corollas all regular and tubular, pale pink or purplish. Anthers sagittate at the base. Style branches long, slender, round in cross section, minutely hairy all over. Achenes columnar, 10-ribbed, truncate, antrorsely barbed, crowned by the persistent pappus.

One species in Kansas.

565. *Elephantopus carolinianus*

E. caroliniànus Willd.   Elephant's foot

"Of Carolina."
Lilac-purple; mid-August–October
Moist woodlands
Bourbon to Wilson counties and south, and Osage County

Perennial from a fibrous-rooted rhizome. Stem 2.5–7 dm tall, with a few widely ascending branches. Lower leaves with the main portion of the blade broadly elliptic to ovate or obovate and more or less abruptly constricted to a long, tapering base, 11–21 cm long and 5–9 cm wide, the apex acute or acuminate, the margins crenate; leaves within the inflorescence much smaller and mostly elliptic. True involucre 6–11 mm long, the bracts strawlike in color and texture toward the base, green and minutely resin-dotted toward the apex. Corolla lilac-purple, about 6 mm long. Achene body 4–4.5 mm long, the pappus about 5 mm long.

VERNÒNIA Schreb.   Ironweed

Named for William Vernon, early English botanist who made excursions into North America.

Coarse, erect perennial herbs. Stems usually unbranched below the inflorescence. Leaves alternate, simple, sessile or nearly so, mostly narrow, the margins entire or toothed. Heads 11- to many-flowered, in branched terminal clusters. Involucres hemispheric to oblong-cylindric or top-shaped, the many bracts herbaceous, appressed, imbricate (i.e., overlapping like shingles) in several series. Receptacle flat. Flowers all bisexual and fertile. Pappus purplish or somewhat rusty in color, of many coarse bristles in 2 series, the inner long and slender, the outer short and narrow, somewhat scalelike. Corollas all regular and tubular, reddish-purple (rarely white). Anthers narrowly sagittate at the base. Style branches long and slender, round in cross section, minutely hairy all over. Achenes (6-) 8- to 10-ribbed, usually hairy and with resin dots between the ribs, truncate at the apex.

566. *Elephantopus carolinianus*, close-up of head

Six species in Kansas. Naturally occurring hybrids are common in this genus, and one may even encounter colonies which exhibit the effects of genetic exchange among 3 or even 4 species. Our most common species are included in the following key.

Key to Species:
1. Main involucral bracts prolonged into a slender, threadlike tip, usually curled inward
.... *V. arkansana*
1. Main involucral bracts rounded or acuminate 2
 2. Lower surface of leaves glabrous .... *V. fasciculata*
 2. Lower surface of leaves hairy .... *V. baldwini*

V. arkansàna DC.

"Of Arkansas."
Reddish-purple; July–October
Low, moist soil
Southeast eighth

Stems 0.4–3 m tall, usually unbranched below the inflorescence, arising from rhizomes bearing thickened fibrous roots. Leaves minutely rough-hairy and gland-dotted, narrowly elliptic, to linear-elliptic, 5–15 cm long and 5–17 (–25) mm wide, the apex attenuate, the base usually attenuate or wedge-shaped, the margins shallowly glandular-toothed, often revolute. Flowers reddish-purple, the corollas 0.8–1.2 cm long. Heads several to many with 55–89 flowers, in branched terminal clusters; involucres hemispheric, 7–10 mm long, the bracts narrowly lanceolate to linear with a very long-attenuate tip which is usually curled inward, the inner bracts broader than the outer bracts, often purple-tinged. Achene body more or less oblong to oblong-elliptic, strongly ribbed, resin-dotted, 4–6 mm long, the pappus somewhat purplish, 6–7 mm long.

This species is known to hybridize in nature with *V. baldwini*, *V. fasciculata*, and *V. missurica*.

567. *Vernonia arkansana*

V. báldwini Torr.   Ironweed

Named after its discoverer, William Baldwin (1779–1819).
Reddish-purple; mid-July–October
Pastures, roadsides, prairie, flood plains
Throughout

Stems 0.4–1 m tall, arising from rhizomes with thickened fibrous roots. Leaves gland-dotted, soft-hairy when young, becoming rough with age, broadly to narrowly lanceolate, 5–18 cm long and 2–7.5 cm wide, the apex acuminate, the base slightly rounded to acute or acuminate, the margins glandular-toothed. Flowers reddish-purple, the corollas 0.8–1.1 cm long. Heads usually many, with 18–34 flowers, in branched terminal clusters; involucres bell-shaped, 4–9 mm long, the bracts strongly imbricated, resin-dotted, usually with a dark midvein, the outermost shortest, lanceolate, grading into the longer, narrowly oblong inner bracts, the apex usually acuminate but grading from acute and often mucronate to long-acuminate and recurved, the margins ciliate or arachnoid. Achene body 3–4 mm long, the pappus 5–7 mm.

This is one of our most robust and drought-resistant perennials, coming especially into prominence in dried-out, overgrazed pastures, where it is preserved by its extremely bitter taste from being browsed by cattle, though sheep devour it.

V. gigantea and *V. missurica*, each known from a few sites in extreme eastern Kansas, will key out here with *V. baldwini*. They resemble that species in most characters but have the involucral bracts rounded and seldom conspicuously resinous. *V. gigantea* has heads with 18–29 flowers, and lower leaf surface with short straight hairs, while *V. missurica* has heads with 34–55 flowers and lower leaf surface with long crooked hairs.

568. *Vernonia baldwini*

V. fasciculàta Michx.   Western ironweed

NL., from L. *fasciculus*, small bundle, bunch of flowers, or nosegay—in reference to the many crowded heads on ascending peduncles.
Reddish-purple; mid-July–September
Low, moist soil
East two-thirds, especially the east two-fifths

Stems 0.4–1.2 m tall, from elongate rhizomes bearing thickened fibrous roots. Leaves narrowly linear to lanceolate, 5.5–14 cm long and 3–22 (–45) mm wide, the apex and base attenuate or acuminate, the margins shallowly to coarsely glandular-toothed, often revolute, the upper surface minutely rough-hairy and gland-dotted, the lower surface glabrous and gland-dotted. Flowers reddish-purple, the corollas 11–13 mm long. Heads usually many, with 18–21 flowers, crowded in more or less flat-topped terminal clusters; involucres narrowly bell-shaped, 6–9 mm long, the bracts strongly imbricated (overlapping like shingles), the outermost shortest, ovate, grading into the longer, oblong inner bracts, all with the apex acute and sometimes mucronate as well, the margins usually ciliate. Achene body 3–3.5 mm long, glabrous or nearly so, the pappus 6–7 mm long.

V. marginata (Torr.) Raf., plains ironweed, is known from Morton, Seward, and Meade counties. It is very closely related to *V. fasciculata* and differs from it primarily in having the involucral bracts acuminate rather than acute.

569. *Vernonia fasciculata*

EUPATÒRIEAE   Thoroughwort Tribe

Sap of plants not milky. Heads all the same, with disk flowers only. Receptacle naked. All the flowers bisexual. Anthers rounded at base. Branches of style thickened upward, minutely and uniformly papillate, rounded at tip. Pappus of plumose or minutely barbed capillary bristles.

BRICKÈLLIA Ell.

In honor of Dr. John Brickell (1749–1809) of Savannah, Georgia, amateur botanist and helpful correspondent of Muhlenberg, Fraser, and other professional botanists.

Perennial herbs or shrubs. Leaves alternate or opposite. Heads 3- to many-flowered, usually in branched terminal clusters. Involucres bell-shaped or oblong, the many bracts graduated in length, rather dry, and longitudinally striated. Receptable flat or convex, naked. Flowers all bisexual and fertile. Pappus of many, usually minutely roughened bristles. Corollas all regular, tubular, with 5 small, erect teeth, slender, white to yellowish or pink. Anthers rounded at the base, united around the style. Achenes 10-

570. *Brickellia grandiflora*

ribbed (5-ribbed in a few species, not ours), cylindric, usually quite hairy, crowned by the more or less persistent pappus of 10–80 bristles.

One species in Kansas.

B. grandiflòra (Hook.) Nutt.

NL., from L. *grandis*, large, great, noble, magnificent, + *flos, floris*, flower.
Cream-colored; mid-July–October
In a variety of habitats ranging from moist, shaded streambanks to brushy prairie ravines or exposed prairie hillsides
Northwest fourth and Stanton County

Erect herbaceous perennial, 1.4–6.5 dm tall, with a fusiform taproot or sometimes a fascicle of fleshy storage roots. Stems and petioles minutely soft-hairy. Leaves opposite below, becoming alternate above, the blades deltoid, 2–9 cm long and 2–6 cm wide, the apex acuminate or short attenuate, the base more or less deeply sagittate, the margins serrate or crenate, the upper surface usually glabrous and darker green than the gland-dotted, minutely hairy lower surface. Heads nodding, in branched terminal clusters; involucre bell-shaped, 9–10 mm long. Corollas cream-colored (often drying rusty brown), 6–6.5 mm long, the teeth gland-dotted. Achene body retrorsely barbed, 3.5–5 mm long, the pappus also minutely barbed, 3.5–6 mm long, the bristles tending to fall away individually.

Plants of moister habitats tend to be more luxuriant in growth than plants growing on drier, more exposed sites.

EUPATÒRIUM L.   Eupatorium, thoroughwort

NL., from Gr. *eupatorion*, after Mithradates Eupator, king of Pontus (134 B.C.–63 B.C.), who, according to Pliny, was the first to use a plant of this genus for liver complaint.

Ours erect herbaceous perennials with rhizomes, the herbage often gland-dotted. Leaves usually opposite (sometimes whorled or alternate), entire to toothed or rarely dissected. Heads few- to many-flowered, in branched terminal clusters. Involucres cylindrical to bell-shaped or hemispheric, the bracts 4 to many in 2–6 series, usually herbaceous, appressed, nearly equal or strongly graduated in length. Receptacle flat, naked. Flowers all bisexual and fertile. Pappus of slender bristles, usually in 1 series. Corollas all regular and tubular, whitish to pale pink, purplish, or bluish. Anthers rounded at the base. Style branches thickened upward, rounded at the apex, very minutely and uniformly hairy or nearly glabrous, stigmatic at the base. Achenes 5-angled and -ribbed, cylindric or narrowed slightly toward the base, truncate at the apex, glabrous or slightly hairy or gland-dotted, crowned by the persistent pappus.

Seven species in Kansas. *E. coelestinum* L., mist-flower, occurs in moist alluvial woods in Crawford, Cherokee, Labette, Montgomery, and Chautauqua counties. It is our only eupatorium with blue flowers and resembles somewhat, especially in flower color, the cultivated species of the closely related genus *Ageratum* which are used as annual bedding plants.

Key to Species (3):
1. Leaves whorled (merely opposite in occasional depauperate plants), relatively broad, mostly 2–15 cm wide; flowers purplish, sometimes very pale, or rarely white ...... *E. purpureum*
1. Leaves opposite or alternate, except in occasional unusual individuals; flowers mostly white 2
  2. Flowers mostly 5 in each head, rarely 3–7 .... *E. altissimum*
  2. Flowers mostly 9–70 in each head 3
    3. Leaves sessile .... *E. perfoliatum*
    3. Leaves evidently petiolate 4
      4. Involucre bracts definitely graduated in length and imbricate (overlapping like shingles) .... *E. serotinum*
      4. Involucre bracts scarcely if at all imbricate, the outer mostly nearly or quite as long as the inner .... *E. rugosum*

E. altíssimum L.   Tall eupatorium

L. *altissimus*, tallest, from *altus*, tall.
Whitish; mid-August–mid-October
Prairies, rocky hills and banks, and thickets
East two-fifths, primarily

Erect herbaceous perennials with rhizomes bearing fibrous roots and several stems (0.5–) 1–2 m tall. Leaves sessile, opposite, mostly narrowly elliptic to lance-linear, 4–11 cm long and 0.5–2 cm wide, the apex acute or attenuate, the base wedge-shaped, the margins usually toothed toward the apex, the lower surface with 3 prominent veins. Herbage, including involucral bracts, minutely hairy and gland-dotted. Heads with 5 flowers; involucre cylindrical, 3–6 mm long, the bracts graduated in length, the outermost ovate or short-oblong, the innermost oblong to linear-oblong. Corollas whitish, gland-dotted, 4–5 mm long. Achene body dark brown or black, 2.5–3 mm long, the pappus minutely barbed, 5–6 mm long.

This is one of our most frequently encountered eupatoriums and is often seen beside farm fences and along railroad right-of-ways. It is often confused with *Kuhnia eupatorioides*.

E. perfoliàtum L.   **Boneset**

NL., from L. *per*, through, + *folium*, leaf—since the leaves, opposite in pairs, unite at their
    bases so that the stem seems to grow through them.
White; July–September
Moist or wet soil
East half, primarily; west to Kiowa and Meade counties in the south

Stems 0.5–1.5 m tall, densely covered with spreading kinky hairs. Leaves opposite, narrowly triangular to lanceolate, 5.5–16 cm long and 1.5–5 cm wide, the apex acuminate or attenuate, the bases perfoliate, the margins crenate or serrate, the upper surface sparsely rough-hairy, the lower surface gland-dotted and more or less densely covered with soft kinky hairs. Heads mostly with 2–23 flowers; involucre obconic, 3–6 mm long, the inner bracts attenuate, much longer than the outer ones. Corollas white, 2.6–4 mm long. Achene body 1.5–2 mm long, the pappus 2–3 mm long.

This plant, which ranges from Nova Scotia and Quebec south to Florida and west to Minnesota, Nebraska, Kansas, Oklahoma, Texas, and Louisiana, is one of the best known and most widely used of the medicinal plants native to North America. White settlers apparently learned of its use from various Indian groups who utilized it in the treatment of a number of ailments including colds, fevers, body pains, stomachache, and epilepsy. Settlers employed it as a tonic in dyspepsia and general debility and a remedy for influenza, muscular rheumatism, and fevers of whatever source. It was particularly preferred in the treatment of intermittent fevers (possibly malaria) called dengue or breakbone fever. It has also been used as a laxative and a vermifuge. The flowers and dried leaves were listed officially in the *U.S. Pharmacopoeia* from 1820 until 1916 and in the *National Formulary* from 1926 to 1950.

The chopped flower tops may also be used to prepare a dye bath which produces a yellow color in wool mordanted with alum and cream of tartar or with potassium dichromate.

E. purpùreum L.   **Joe-Pye Weed**

L. *purpureus*, purple—in reference to the flowers. As to Joe Pye, New England tradition has it
    that he was an Indian who used species of *Eupatorium* in the treatment of spotted fever.
Pink or purple, sometimes whitish; late June–August
Open woods
East fourth

Erect herbaceous perennials with rhizomes bearing fibrous roots. Stems 6–17 dm tall, glaucous, simple below the inflorescence, often purplish at the nodes. Leaves whorled, 3–4 per node, broadly to narrowly lanceolate, abruptly narrowed to a wedge-shaped base, 8.5–25 cm long and 3–10 cm wide, the apex usually acuminate, the margins serrate, the upper surface glabrous or nearly so, the lower surface minutely soft-hairy and gland-dotted. Heads mostly 5- to 7-flowered; involucres cylindrical to narrowly obconic, 5–9 mm long, the bracts rather dry and strawlike in texture and sheen, except for the green veins, the outermost ovate to short oblong, minutely hairy, grading into the longer, glabrous, linear-elliptic inner bracts. Corollas pink or purple, 4.5–7.5 mm long. Achene body 3–5 mm long, the pappus bristles minutely barbed, 5–7 mm long.

This species is listed in some earlier manuals as *E. falcatum*.

E. maculatum L., which also has whorled leaves, is known from Douglas and Atchison counties. It very closely resembles *E. purpureum* but has heads with 9–22 flowers.

In addition to its use in the treatment of typhus, *E. purpureum* was also employed as a diuretic. Weiner (34) reports that the Meskwaki Indians "considered it a love medicine to be nibbled when speaking to women when they are in the wooing mood."
    See plate *112*.

571. *Eupatorium altissimum*

572. *Eupatorium perfoliatum*

573. *Eupatorium purpureum*

Thoroughwort Tribe

574. *Eupatorium rugosum*

575. *Eupatorium serotinum*

E. rugòsum Houtt.

White snakeroot

L. *rugosus*, wrinkled, shriveled.
White; August–October
Open woods, frequently in moist, alluvial soil
East two-fifths and Ellsworth County

Plants 0.3–1 (–1.5) m tall, the stems openly branched above, minutely hairy. Leaves petioled, opposite, the blades thin, lanceolate to ovate, occasionally deltoid or cordate, 4–13 cm long and 1.5–9 cm long, the apex acuminate to attenuate, the base acute to broadly rounded, truncate, or somewhat cordate, the margins serrate, the lower surface minutely hairy on the veins. Heads with 12–24 flowers; involucre obconic, 3–4 mm long, the bracts sparsely hairy, scarcely graduated in length. Corolla white, 3–4 mm long. Achene body 2–2.5 mm long, the pappus 3–3.5 mm long.

The plants form clumps of upright stems by the production of adventitious buds on the stubby subterranean base of a parent stem. Each of these stems gives off a cluster of fibrous roots just below the soil surface. These upright stems eventually separate from the parent stock or may easily be pulled away without injury, each with its set of roots, and be transplanted without serious setback. These facts are worth knowing because the plant is a favorite for shady places of the home grounds, where it produces its cluster of small white flower heads from July into November, yielding through the season a great crop of wind-borne, seedlike achenes which soon start new plants in shady places round about. However, when growing in wooded or shaded areas of pastures the plants if browsed by livestock may cause the sickness known as trembles or milk sickness, owing to their containing a poisonous principle called tremetol. This is soluble in milk fat and may be transferred to human beings, causing nausea, vomiting, feeble pulse, slow respiration, weakness, and sometimes collapse.

E. seròtinum Michx.

L. *serotinus*, happening late—in reference to the blooming period.
White; late August–October
Moist soil, usually in woods and thickets, less frequently in pastures and along roadsides
East fourth, primarily south of the Kansas River

Plants 0.6–1.5 m tall, the stems openly branched above, densely covered with minute incurved hairs. Leaves petioled, opposite, the blades lanceolate, 4.5–11 cm long and 1–4 cm wide, the apex acute to attenuate, the base acute to rounded or truncate, often oblique, the margins serrate, the upper surface glabrous or nearly so, the lower surface minutely hairy and gland-dotted and with 3 prominent veins. Heads mostly with (9–) 13 (–15) flowers; involucre obconic, 3–4 mm long, the bracts graduated in length, at least some of them densely hairy. Corollas white, 2.5–3.5 mm long. Achene body about 1.5–2 mm long, the pappus 2–3 mm long.

KŪHNIA L.

Kuhnia, false boneset

After Dr. Adam Kuhn (1741–1817) of Pennsylvania who traveled in America, about the middle of the 18th century, and later took to Linnaeus a living plant of the type species, *K. eupatorioides*, transplanted from Pennsylvania.

Erect to decumbent perennial herbs with stout, woody taproots. Leaves unevenly alternate (less frequently opposite or nearly so), sessile. Heads 10- to 25-flowered, borne singly or few to many in branched terminal clusters. Involucres more or less cylindrical to top- or bell-shaped, the bracts rather dry, longitudinally striate-nerved, in 4–7 series, those of the outer several series different in shape and more markedly graduated in size than those of the inner series. Receptacle flat or somewhat convex, naked. Flowers all bisexual and fertile. Pappus of 10, 15, or 20 plumose bristles in 1 series, about the same length as the corolla. Corollas all regular and tubular, cream-colored to dull yellow or reddish. Anthers rounded at the base, tending to separate by anthesis. Style branches elongate, flattened, club-shaped, papillate, stigmatic only at the base. Achenes 10-ribbed (smaller intermediate ribs occasionally present also), cylindric or becoming wider toward the apex, topped by the persistent, plumose pappus.

One species in Kansas.

K. eupatorioìdes L.

False boneset

NL., from the genus *Eupatorium*, in the Composite Family, + the Gr. suffix *-oides*, like, resembling.
Rusty- to yellowish-white; mid-August–early October
Prairie, pastures, roadsides, rarely in dry, open woods
Throughout

A sturdy, erect, grayish-green perennial herb, 2–11 dm tall, blooming the first year from seed and forming a stubby rhizome that produces a cluster of stems and many adventitious, deep roots as well as a strong taproot that may reach a depth of 16 or more feet. Herbage rough-hairy. Main leaves mostly lanceolate to lance-linear, 3–6 cm long and 8–20 mm wide, usually serrate or at least with 1 or 2 pairs of teeth, the lower surface gland-dotted and prominently veined, the upper leaves smaller, lance-linear to linear, and often entire. Heads several to many, in clusters terminating the main stem and the ascending branches, involucre 7–13 mm long. Corollas rusty- to yellowish-white, 5–8 mm long, the teeth gland-dotted. Achene body 3–5 mm long, the pappus 4–8 mm long.

Our plants are recognized by authorities as belonging to *K. eupatorioides* var. *corymbulosa* T. & G. and have been listed in earlier manuals as *K. suaveolens*.

This species of *Kuhnia* might be confused with *Eupatorium altissimum* if one has not noticed that the former has prevailingly alternate leaves, while those of the latter are opposite. Closer examination shows further that the 2 far-protruding styles of *Kuhnia* flowers are enlarged and flattened at the end, while in *Eupatorium* they are more slender and the tips are merely rounded. There is also a distinction between the capillary pappus of the 2 species—that of *Kuhnia* being tawny and plumose while the *Eupatorium* pappus is merely roughened. Furthermore, the anthers of *Kuhnia* flowers are nearly free from one another, and those of *Eupatorium* preserve the family feature of laterally united anthers. Finally, as to height, *Kuhnia* seldom exceeds 2 feet and *Eupatorium* reaches 3–6 feet.

Though the flowers of *Kuhnia* are unpretentious, their structure is admirable and their neutral color affords a blending element in the design of bouquets. The dry achenes, bound in a cluster by the bracts of the involucre and crowned with their plumose pappus, are valuable in the make-up of durable winter bouquets.

576. *Kuhnia eupatorioides*

LIÁTRIS Schreb.   Gayfeather, blazing star, button snakeroot

Von Schreber, the German botanist who gave this euphonious name, apparently invented it.

Erect perennial herbs, mostly with woody, more or less globose corms, less frequently with a more elongate rootstock or rhizome, in some species almost a woody, branched taproot. Leaves alternate, simple, sessile, narrow, entire, linear to ovate-lanceolate. Herbage usually minutely glandular-punctate. Heads few- to many-flowered, usually in spikes or racemes. Involucres oblong to hemispheric, the many bracts graduated in length, often with membranous petaloid margins. Receptacle essentially flat, naked. Flowers all bisexual and fertile. Pappus of 1 or 2 series of plumose or minutely barbed bristles. Corollas all regular, tubular, rose-purple (rarely white), the 5 narrow lobes erect or spreading. Anthers oblong, about half as long as the filaments, style stiff, exserted late in anthesis, the style branches club-shaped, stigmatose only at the base. Achenes somewhat cylindrical but tapering basally, about 10-ribbed, more or less hairy (use magnification), crowned by the persistent pappus.

Seven species in Kansas. This is a relatively difficult group to work with. Several of the species are rather variable and include some individuals with diploid and some with tetraploid chromosome numbers. Plants intermediate in some characters between *L. mucronata* and *L. punctata* or between *L. pycnostachya* and *L. lancifolia* are not uncommon and may cause problems in identification.

Key to Species:
1. Pappus bristles minutely barbed, the length of the lateral hairs 3–6 times as great as the thickness of the central axis 2
   2. Heads about as wide as long, 15- to 40-flowered, the involucral bracts broad and bullate (appearing blistered), the tips rounded .... *L. aspera*
   2. Heads longer than wide, 3- to 20-flowered, the involucral bracts relatively narrow, not bullate, the tips acute and reflexed .... *L. pycnostachya*
1. Pappus bristles plumose, the length of the lateral hairs 15 times or more as great as the thickness of the central axis 3
   3. Heads 4- to 14-flowered, narrowly cylindrical or obconic, many, arranged in densely crowded spikes 4
      4. Plants with a long underground rootstock; leaf margins ciliate .... *L. punctata*
      4. Plants with a more or less globular corm; leaf margins not ciliate .... *L. mucronata*
   3. Heads 15- to 60-flowered, more broadly obconic or oblong, (1-) few to several (-many) in relatively open spikes or slender racemes 5
      5. Stems with widely spreading hairs; middle and lower involucral bracts of each head with strongly recurved tips .... *L. squarrosa* var. *hirsuta*
      5. Stems glabrous; middle and lower involucral bracts of each head with tips ascending or spreading (rarely recurved and then usually in old age) .... *L. glabrata*

L. áspera Michx.

L. *asper*, rough, harsh.
Rose-purple; August–mid-October
Rocky prairie, occasionally in dry, open woods
East two-fifths

577. *Liatris aspera*

578. *Liatris glabrata*

579. *Liatris hirsuta*

Plants 2.8–12 dm tall, with a rugged, sometimes branched corm which gives off one to few stems above and strong fibrous roots reaching far into the subsoil. Leaves narrowly oblanceolate to linear, 3.5–20 cm long and 3.5–22 mm wide, much reduced above, usually minutely rough-hairy, the apex acute or mucronate. Heads with 25–40 flowers, more or less distantly spaced in slender spikes or spikelike racemes; involucres broadly obconic to bowl-shaped, 10–18 mm long and 11–20 mm wide, the bracts broad, rounded, often purplish, appearing "blistered," the margins membranous and erose. Corollas rose-purple (rarely white), 1–1.5 cm long. Achene body 5–6 mm long, the pappus 6–8 mm long, barbed.

This plant has been listed in earlier manuals for this region as *L. scariosa*, but the latter species, although resembling *L. aspera*, is restricted in its distribution to the Appalachian Mountains.

This blazing star is as effective as the gayfeather in enlivening the prairie scene, and both are equally esteemed for winter bouquets. With its flower heads relatively large and distantly spaced, the involucral bracts so distinct, and the individual flowers displayed with especial clarity, this species is superior in its presentation of the essential floral character of the genus.

This species is good also for observing the characters of flowers of different ages. When we open a bud we find the anthers cohering and forming a tube enclosing a rodlike style. In a later stage we find the style elongating and with its papillose surface brushing out the pollen discharged during the bud stage, the style later spreading its 2 slender divisions apart and exposing the inner stigmatic surfaces, which as yet have not been pollinated.

The flowers offer to insect visitors pollen adhering to the outer surfaces of the style branches and nectar secreted around the style base and held in the lower part of the corolla tube. Bees, butterflies, and flies in getting what they want are certain to transfer pollen from the papillae of the style branches to the stigmatic surfaces, sometimes effecting self-pollination, and, going from plant to plant, causing cross-pollination also.

The Plains Indians relate that they boiled the leaves and corms together and gave the decoction to children with diarrhea. Some tribes claimed to increase the speed of their horses by chewing the corm and blowing it into the nostrils, and by mixing corms with the corn fed to the horses.

Also, we read in Barton's *Medical Botany* (Philadelphia, 1818) that "all the tuberous-rooted species of the genus *Liatris* are active plants and seem to be uniformly diuretic . . . and Schoepf describes the *L. scariosa* as an acid, subbitter plant, possessed of diuretic virtues . . . Pursh says the same plant and *L. squarrosa* are known among the inhabitants of Virginia, Kentucky, and Carolina by the name of 'Rattle Snake's Master'; and tells us, that when bitten by that animal, they bruise the bulbs of these plants, and apply them to the wounds, while at the same time they make a decoction of them in milk, which is taken inwardly in the same manner as *Prenanthes serpentaria*."

See plate *113*.

L. glabràta Rydb.

L. *glabratus*, from *glaber*, smooth—this species being hairless or nearly so.
Rose-purple; July–mid-September
Sandy prairie
Primarily in the southwest fourth, but north and east to Washington and Pottawatomie counties

Glabrous plants, 3–8 dm tall, with pale green foliage, simple stems, and more or less globose corms 1–4 cm across. Leaves linear, 5.5–18 cm long and 2–6 mm wide, the apex attenuate, the margins white and slightly thickened. Heads with 17–28 (usually about 22) flowers; involucre bell-shaped to obconic, 10–18 mm long, the bracts sometimes dark reddish-purple, with abruptly or gradually long-acuminate tips usually ascending, sometimes recurving slightly with age, the margins smooth. Corollas rose-purple, 8–12 mm long. Achene body 5–6 mm long, hairy, the pappus 6–8 mm, plumose.

Some manuals treat this species as a variety of *L. squarrosa*.

See plate *114*.

L. mucronàta DC.   Gayfeather

L. *mucronatus*, sharply pointed, from *mucro*, sharp point—descriptive of the leaves and involucral
bracts.
Rose-purple; mid-August–early October
Rocky prairie hillsides and banks
East half and Morton County

Glabrous (usually) plants, 3–7 dm tall, with 1 to several stems arising from a
globose corm. Leaves linear, 4–15 cm long and 1–4 mm wide, gradually smaller above, the
apex usually sharply pointed, the margins callous. Heads with 4–6 flowers, borne in
crowded spikes 6.5–37 cm long; involucre cylindrical or narrowly obconical, 0.7–1.7 cm
long, the outer bracts abruptly mucronate or acuminate, the middle and inner ones more
gradually so, the margins membranous, sometimes finely ciliate also. Corolla rose-purple
(rarely white), 1–2 cm long. Achene body 5–8 mm long, the pappus 6–9 mm long,
plumose.

Our plants have been listed as L. *angustifolia* in earlier manuals. Some authorities
recognize L. *angustifolia* as a species separate from L. *mucronata*, but most treat the 2 as
composing a single variable species, in which case the latter name is applied to the entire
group. This species is also closely related to L. *punctata* and plants (hybrids?) intermediate in characters between the 2 species are sometimes found.

580. *Liatris punctata*

L. punctàta Hook.   Blazing star

NL., from L. *punctum*, pricked in, sharp point, or small hole—the herbage of this species being
conspicuously gland-dotted.
Rose-purple; mid-August–early October
Dry prairie
Nearly throughout, but more common westward

Glabrous plants with pale green herbage and 1 to several stems 1.5–8 dm tall
arising from a thick, short rootstock subtended by a stout taproot that extends 1.3 m to
about 5 m, depending on the depth of available water. Leaves linear, rather rigidly
ascending or spreading, 3.5–14 cm long and 1.5–5 mm wide, the apex with a small, sharp
mucro, the margins callous and coarsely ciliate. Heads with 4–6 flowers, arranged in more
or less densely crowded spikes; involucre cylindric to narrowly obconic, 10–17 mm long,
the bracts appressed, the tips more or less abruptly acuminate or mucronate, the margins
ciliate. Corollas rose-purple, 1–1.5 cm long. Achene body 6–7.5 mm long, the pappus
6–10 mm long, plumose.

The taproot, with laterals reaching out for water at various levels, is not present in
our other species of *Liatris*, and we may look upon it as the principal factor enabling this
species to occur in all quarters of the state, even in all our westernmost counties of least
rainfall, least atmospheric humidity, and highest evaporation.

Note also comments under the closely related species L. *mucronata*.

581. *Liatris pycnostachya*

L. pycnostàchya Michx.   Kansas gayfeather

Gr. *pyknos*, dense, crowded, + *stachys*, ear or spike of grain—the flower heads being crowded
into an elongate spike or narrow raceme.
Rose-purple; July–mid-August
Prairies, frequently in low, moist prairie
East third, primarily

Plants 0.6–1.5 dm tall, with 1 to several simple stems arising from rather irregularly
shaped corms 2.5–8 cm across, usually crowned by the fibrous remnants of old leaves.
Leaves linear, 1.5–20 cm long and 1.5–14 mm wide, usually becoming much reduced
toward the inflorescence. Herbage usually glabrous except within the inflorescence. Heads
with 5–12 flowers, borne in densely crowded spikes 6–30 cm long and 2–3 cm wide,
involucre short-oblong to somewhat bell-shaped, 7–10 mm long, the bracts resin-dotted,
frequently purplish, with acute, usually recurved tips and membranous, usually ciliate
margins. Corollas rose-purple (rarely white), 7–9 mm long, glabrous or with a few hairs
within. Achene body 3.5–7 mm long, the pappus 6–7 mm long, barbed.

L. *lancifolia* (Greene) Kittell, a closely related species occurring in the western
plains from South Dakota to New Mexico, is recorded from several counties in the west
half of Kansas, primarily in the southwest fourth. Plants of this species are smaller than
the more eastern L. *pycnostachya*, usually being no more than 0.6 m tall, and have
involucral bracts with rounded, appressed tips.

L. *pycnostachya*, the stateliest of our species of *Liatris*, occurs singly or in groups,
rarely in extensive societies. Needing more moisture than the others, it is at its best in

low prairies or near the bases of slopes. It is one of the best native species for planting into the wildflower garden. Transplanting should be done early in the fall, and its seeds should be sown out of doors at the same period. The flower spikes make handsome winter bouquets. For this purpose cut the stems at the desired length when the spikes are well in bloom and hang them upside down to dry where dust cannot accumulate and cloud their brilliance. An occasional going over with the blower of the vacuum cleaner should act as a freshener.

See plate *115*.

L. squarròsa (L.) Michx. var. *hirsuta* (Rydb.) Gaiser   Blazing star, colic root

L. *squarrosus*, roughened by spreading scales—the recurved tips of the involucral bracts giving that effect; L. *hirsutus*, beset with stiff hairs—as is the herbage.
Rose-purple; mid-July–mid-September
Rocky, open, wooded hillsides, glades, and banks, less frequently in prairie
East fourth

Plants more or less densely rough-hairy, with few to many stems 3–7.5 dm tall arising from globose or somewhat irregular corms 2–5 cm across. Leaves linear, 5–17 cm long and 2.5–6 (–8) mm wide, the apex attenuate, the margins white and slightly thickened. Heads with (13–) 20–29 flowers; involucre oblong, 12–20 mm long, the bracts with tips usually abruptly short-acuminate and recurved, the margins ciliate, often membranous also. Corollas rose-purple, 9–14 mm long. Achene body 5–7.8 (averaging 6) mm long, the pappus 7.5–9 mm long, plumose.

While collecting transplants in Polk County, Missouri, in August 1966, I visited with local residents who volunteered information about their use of this plant to treat colic and gas on the stomach. They utilize the strongly aromatic corms, first peeling them, then chewing them.

INÙLEAE   Elecampane Tribe

After the genus *Inula*, the ancient Latin name for the Linnaean *Inula helenium*, the elecampane.

Sap of plants not milky. Leaves alternate. Flower heads in ours lacking ray flowers. Receptacle usually naked. Flowers either all bisexual and mostly fertile, or unisexual and dioecious (male and female flowers on separate plants), only the female flowers producing fruit, occasionally with a few bisexual flowers present also. Corollas of disk flowers thread-like or hairlike and truncate at summit. Anthers with tail-like appendages at base. Branches of style threadlike, receptive their entire length. Pappus usually of threadlike bristles.

ANTENNÀRIA Gaertn.   Pussytoes, ladies' tobacco

NL., from NL. *antenna*, feelers of an insect—the pappus bristles of the male flowers being enlarged at the ends, much like the antennae of butterflies, for instance.

Perennial, silky or woolly herbs of small stature, with slender, branching rhizomes or stolons. Leaves alternate, entire, sessile, mostly spatulate to narrowly obovate, the upper surface dull- or gray-green and more or less covered with arachnoid hairs or eventually glabrous, the lower surface gray-woolly, the basal leaves much larger than the upper leaves and often forming rosettes. Heads many-flowered, 1 to many, usually in crowded terminal clusters. Ray flowers absent. Involucres hemispheric to bell- or top-shaped, the bracts linear or oblong, dry and membranous (at least near the tip), silvery or colored, graduated in length. Receptacle flat or convex, naked, deeply pitted. Flowers unisexual, male and female flowers on separate plants, the male plants usually fewer in number and even unknown for some species. Male flowers with a scanty pappus of white bristles which are thickened, somewhat club-shaped, flattened, and minutely barbed toward the apex and somewhat united at the base; corolla tubular, regular, the teeth spreading; anthers with slender, tail-like basal appendages; a rudimentary pistil and undivided style usually present. Female flowers with a pappus of many hairlike, minutely barbed bristles united at the base; corolla tubular but much more slender than those of the male flowers, truncate at the apex; style 2-branched. Achenes quite small (less than 1 mm long), oblong, round in cross section or somewhat flattened.

Two species in Kansas. Many species of pussytoes are entirely or partially apomictic—that is, all or some of the female flowers produce fruit without being fertilized.

Key to Species (3):

1. Basal leaves and those at the ends of the stolons relatively small, less than 1.5 cm wide, 1-nerved or obscurely 3-nerved .... *A. neglecta*
1. Basal leaves and those at the ends of the stolons relatively large, prominently 3–5 nerved, the larger ones 1.5 cm large or more .... *A. plantaginifolia*

A. neglécta Greene

L. *neglectus*, to pay no attention, disregard, slight, overlook.
White; April–mid-May
Rocky or sandy prairie
East two-thirds, primarily; west to Sheridan County in the northwest

Herbaceous perennial, 0.4–3.5 dm tall, closely resembling *A. plantaginifolia* but smaller in most respects and with leaves 1- or obscurely 3-veined.

A. plantaginifòlia (L.) Richards — Plantainleaf pussytoes

NL., from the genus *Plantago*, in the Plantain Family, + *folium*, leaf—because the leaves of this species of *Antennaria* resemble the leaves of some *Plantagos*, such as those of *P. media*, a European species that Linnaeus probably had in mind when he named this *Antennaria*. Other common names used for this plant in various sections of the country include the following: cat's foot, dog's toes, spring everlasting, cudweed, Indian tobacco, ladies' tobacco, and poverty weed.
White; mid-April–mid-May
Rocky, open oak woods, less frequently in prairie
East third, primarily

582. *Antennaria plantaginifolia,* female flowers

Herbaceous perennial forming dense patches by means of underground rhizomes and short, decumbent aboveground stolons which strike root here and there. Leaves mostly basal or borne on the stolons, narrowed to petiolelike bases, the blades spatulate to ovate, elliptic, or obovate, 3- to 5-veined, 2–5 cm long and 0.4–5.5 cm wide, the apex broadly acute or rounded, mucronate. Heads in clusters 1.5–3 cm across, borne on erect stems 0.5–4 dm tall bearing narrow, reduced leaves or leaflike bracts. Involucre 3–7 mm long, the bracts green and hairy at the base, white and membranous toward the tip. Corollas whitish, about 4–6 mm long. Achene body about 1.5 mm long, the pappus about 5 mm long, deciduous.

The leaves remain green through the winter and are eaten to some extent by deer, rabbits, grouse, and bobwhite quail (4).

GNAPHÀLIUM L. — Cudweed, everlasting, pearly everlasting

NL., from Gr. *gnaphalion*, name used by Pliny for a woolly plant, of which he says "its white soft leaves are used as a flock [waste cotton or wool], and, indeed, there is no perceptible difference."

583. *Antennaria plantaginifolia,* male flowers

Mostly erect, woolly, sometimes glandular annual herbs, less frequently biennial or perennial, with taproots. Leaves alternate, simple, sessile, narrow, entire. Heads several- to many-flowered, in more or less densely crowded terminal clusters. Involucres ovoid to bell-shaped or hemispheric, the bracts dry and membranous with a green midrib, more or less graduated in length. Receptacle flat, naked, sometimes pitted. Flowers all fertile, mostly bisexual, but those toward the center of the head female. Pappus of a single series of minute hairlike bristles, sometimes thickened toward the tip, united at the base, deciduous. Corollas threadlike, with 5 minute teeth (use high magnification). Anthers with slender, tail-like basal appendages. Style branches without elongate appendages protruding beyond the stigmatic portion. Achenes small, round in cross section or somewhat flattened.

Two species in Kansas. *G. purpureum* L. is known from scattered sites in the east half, primarily in the southeast eighth. It is a more slender plant than the more common *G. obtusifolium*; has spatulate leaves, heads borne in narrow, spikelike clusters, involucral bracts light brown to pink or purplish; and blooms much earlier in the season (May–June). It lacks the fragrance of our sweet everlasting.

G. obtusifòlium L. — Fragrant cudweed, sweet everlasting

NL., from L. *obtusus*, blunt, + *folium*, leaf.
Whitish; mid-August–mid-October
Prairie, occasionally in open woods, frequently on sandy soil
East three-fifths

Fragrant, erect annual or winter annual, 2–9 dm tall. Stems simple below, branched above, covered with matted white hairs. Leaves linear-elliptic, 2–6 cm long and

584. *Gnaphalium obtusifolium*

1.5–5 mm wide, the apex attenuate, the upper surface glandular (at least on young leaves), the lower surface woolly. Heads borne in branched terminal clusters mostly (4–) 6–20 cm across. Involucre ovoid to bell-shaped, 3–6 mm long, the bracts whitish, strongly reflexed after the achenes have blown away. Corollas white, very slender and inconspicuous, about 4 mm long. Achene body smooth, oblong to ellipsoid, about 0.5 mm long, the pappus bristles 3–4 mm long, readily deciduous.

The dried flowering stalks have an aroma resembling that of brown sugar and make attractive additions to winter bouquets. The plants reportedly impart a yellow dye to wool mordanted with alum and cream of tartar or with potassium dichromate.

The Creek Indians prepared a tea or infusion of this plant together with *Monarda punctata* to treat delirium.

ANTHEMÍDEAE   Camomile Tribe

Sap of plants not milky. Heads all the same, usually with both ray flowers and disk flowers present or with disk flowers only. Leaves alternate. Stems, leaves, and/or bracts of involucre not spiny. Bracts of involucre with thin, dry or membranous margins. Receptacle chaffy or naked (rarely hairy). Marginal flowers of radiate heads usually female and fertile with definite rays, with narrow, threadlike rays, or (in *Artemisia*) with the ligules of the ray flowers obsolete. Disk flowers usually bisexual and fertile. Anthers not tailed (caudate) at the base. Style branches not prolonged beyond the pollen-receptive portion, the tips truncate and with a small tuft of hairs. Pappus a short crown or lacking altogether.

ACHILLÈA L.   Yarrow

NL., from the name of a plant said to have been named from Achilles, hero of Homer's *Iliad*, who supposedly used this plant to heal a soldier's wounds.

Erect herbaceous perennials. Leaves alternate, simple, sessile or petioled, toothed to pinnately dissected. Heads many-flowered, radiate (rarely discoid), in branched, mostly terminal clusters. Involucre bell-shaped to hemispheric, the bracts dry with membranous translucent margins and often a green midrib, graduated in length. Receptacle slightly convex to conic, chaffy throughout with flat membranous scales similar to the involucral bracts. Pappus absent. Ray flowers 5–12, female, fertile (rarely neutral and infertile), the corollas white to pink (rarely yellow). Disk flowers 10 to many, bisexual, fertile; corollas yellowish-white to straw-colored, 5-toothed. Anthers with ovate, scalelike appendages at the tips, entire at the base. Style branches flattened, the tips truncate and with a small tuft of hairs. Achenes oblong to obovate, compressed parallel to the involucral bracts, glabrous, with a thickened margin (use high magnification).

One species in Kansas.

A. millefòlium L.   Common yarrow, milfoil

Latin name used for this plant by Pliny and others, compounded from *mille*, 1,000, + *folium*, leaf—because of the many fine dissections of the leaf. "Yarrow" is an old English name and "milfoil" an old French name for this plant.
White; May–mid-July (–October)
Prairie, pastures, and roadsides
Nearly throughout, but more common eastward

Plants aromatic, mostly 2–10 dm tall, spreading by means of slender rhizomes. Leaves finely pinnately dissected, 3–10 cm long and 0.5–2 cm wide, the lowermost petioled, spatulate to narrowly elliptic in general outline, those on the stem sessile, elliptic to oblong or lance-oblong. Herbage more or less densely covered with long, often matted hairs. Heads many, mostly 4–5 mm across, in flat-topped or somewhat rounded terminal clusters. Involucre 4–5.5 mm long. Rays white (rarely pink), usually 5, 3–5 mm long, 3-lobed, relatively broad. Disk flowers whitish. Achene smooth, straw-colored, about 1.5 mm long.

Our plants are native to North America, not introduced as was once thought, and all belong to subspecies *lanulosa*.

Our *Achilleas* are hardy, persistent through the heat and aridity of summer, even through long drought periods, and may succeed as a ground cover where grasses fail, being enabled to do this by their habit of spreading by underground offshoots.

There are volatile, aromatic substances in the *Achillea*, very evident to the smell when the leaves are rubbed between thumb and fingers. These substances, together with bitter principles and tannins, must be thanked for the medicinal value of the plant observed from ancient times, but doubtless too much vaunted in the herbals of the Middle Ages, according to which ailments of the liver and kidneys, gout, influenza, and rheumatism were alike amenable to its curative power. Various Indian tribes of North America utilized preparations to treat nearly every ailment imaginable. During the Civil War, it was so widely used to treat wounds that it became known as soldier's woundwort ("wort" from AS. *wyrt*, plant). It was listed officially in the *U.S. Pharmacopoeia* from 1863 to 1882.

See plate *116*.

585. *Achillea millefolium*

ÁNTHEMIS L.   Camomile, Chamomile

L. and Gr. *anthemis*, name used by Theophrastus and Pliny for the Roman camomile, *A. nobilis*.

Branching annual or perennial herbs, usually strong-scented. Leaves alternate, simple, pinnately dissected. Heads many-flowered, radiate, borne singly on terminal peduncles. Involucre hemispheric to bell-shaped, the bracts many, small, usually in 2 series and graduated in length, dry, with more or less membranous margins. Receptacle merely convex to conic or hemispheric, usually chaffy, at least toward the apex. Pappus a minute crown or absent. Ray flowers 10–20, female or neutral, infertile, the corollas white or yellow. Disk flowers many, bisexual, fertile, the corollas yellow. Anther tips with ovate, scalelike appendages, the bases entire. Style branches flattened, the apexes truncate and with a small tuft of hairs. Achenes round in cross section or 4- or 5-angled, occasionally slightly compressed, glabrous, the top truncated.

One species in Kansas.

A. cótula (L.) DC.   Dog fennel, Mayweed camomile

From Gr. *kotyle*, for anything hollow, or a little cup—probably suggested by the fact that when the flower heads are approaching the stage of opening, the rays are erect and, together with the involucre, give the appearance of a cup. "Dog fennel" is meant to be uncomplimentary because, though the leafage might suggest the sweet-smelling fennel, *Foeniculum vulgare*, its odor is repellent.
Rays white, disk yellow; mid-May–early October
Pastures, roadsides, wasteground, and other disturbed habitats
East two-fifths, primarily; intr. from Europe

586. *Anthemis cotula*

Erect, odorous annuals, 1–6.5 dm tall. Leaves mostly obovate to ovate in general outline, 1.5–5 cm long, very finely 2 or 3 times pinnately dissected into segments about 0.5 mm wide. Heads several to many, 1.5–2.5 cm across. Involucre 3–4 mm long, the bracts with pointed or somewhat rounded tips. Chaff of the receptacle narrowly linear and translucent. Ray flowers usually neutral, the rays white, 6–9 mm long, the tip entire or very slightly 2 or 3 toothed. Achenes brown, about 1.5 mm long, with longitudinal rows of tubercles, lacking a pappus.

The seeds of this plant are eaten by birds.

Children who have played around in patches of dog fennel have complained of irritated skin and smarting lips and eyelids; but the odor, which Gerard in his herbal calls "a naughty smell," does not seem to bother them.

That pungent smell is due to an ethereal oil that, together with several alcohols, organic acids, and bitter principles, have made this and other species of its genus important as a medicine. So it has been in Europe for 2,000 years and more, and so it has been in this country, infusions of flowers, leaves, and stems being employed as a tonic after fevers, as an astringent and emetic; and the crushed fresh plant being applied externally as a blister and counterirritant. We do not find it today on the pharmacist's preferred list of medicinal plants. Though it has spiciness, it is out of favor in the herb gardens today; and even cattle find it unpalatable, though sheep will crop its flowering tops. If these statements seem uncomplimentary to the yarrow, we counterbalance them by mentioning 2 of its virtues: it is good to look at as it stands growing in meadows and pastures and, when put in a vase by itself or with other flowers, it holds its freshness and spiciness a long time.

The Roman camomile, a low, pleasantly scented perennial herb with creeping and rooting branches, is the source of camomile tea.

ARTEMÍSIA L.   Wormweed, sagebrush

L. *artemisia*, ancient name of the mudwort, explained by Pliny as being in honor of Artemisia, wife of Mausolus, king of Caria, a providence of Asia Minor, for whom after his death she built the renowned Mausoleum, about 350 B.C., one of the Seven Wonders of the World.

Annual, biennial, or perennial herbs or subshrubs, or shrubs, usually aromatic.

Leaves alternate, simple, sessile or petioled, entire to lobed or finely pinnately dissected. Heads quite small, mostly few-flowered, apparently discoid, usually many together in elongate, branched, sometimes leafy terminal clusters. Involucre at first urn-shaped, becoming bell-shaped or hemispheric in fruit, the bracts in 2–4 series, at least the innermost dry and membranous, often hairy, graduated in length. Receptacle convex or conic, naked or rarely hairy. Pappus absent. Ray flowers 1 to few, female, fertile or infertile, the corolla cylindric or nearly so and with a V-shaped notch down the inner side, lacking a ligule. Disk flowers 3–20, bisexual, fertile or infertile; corollas yellowish-white, minutely 5-toothed. Anthers rounded or somewhat cordate at the base, with lanceolate or subulate, scalelike appendages at the tips. Style branches 2, flattened, the branches recurved, the apexes truncate and with a minute tuft of hairs, or the style not cleft. Achenes minute (less than 1 mm long), more or less ellipsoid to obovoid or rarely cylindrical or angled, glabrous.

Nine species in Kansas. The 5 most commonly encountered species are treated here.

Sundry medicinal values were attributed to the European species, *Artemisia vulgaris*, by Hippocrates, Greek physician (c. 460–359 B.C., called Father of Medicine), as well as by Pliny, Dioscorides, and Galen (of the 1st and 2nd centuries). Even magical powers were claimed for it, a reputation that persisted through the Middle Ages into more recent times. Thus Leonhart Fuchs writes in his herbal (1543): "To him who has this herb with him no poisonous animal nor other harmful thing can bring ill or injury. If one who is traveling across country but has this herb with him it will prevent fatigue." Another early German herbalist, Hieronymous Bock, 1539, stresses the more credible medicinal value of the herb: "A decoction of it prepared with honey and sugar mitigates coughing, dispels goiter, pulverizes stone, clears out the lungs, kidneys and bladder." An old German name for the plant was *Sonnwendgürtel* ("solstice girdle"), from the fact that a girdle of it was worn by Celts, Germans, Balts, and Slavs for the benefit of its magic at the time of their solstice celebrations, just as nowadays some people hang a horseshoe over the door or carry a rabbit's foot for good luck. However, we know that not everyone in that far-distant time was ridden by credulity, for Hieronymus Bock sets it down for his day that "this monkey business and ritual does not in the least count in Paris, in France."

In the New World, as in the Old, the lives of the natives were intimately and vitally related to the plant population, and it need not surprise us that our Indians put the indigenous *Artemisias* to much the same medicinal uses as the early Europeans and Asiatics did theirs; but that our Indians should have, as they did, the same kind of superstitions about the *Artemisias* and use them in similar rites and ceremonies, with confidence in their magic powers, is amazing.

Key to Species:
1. Leaves linear and threadlike or deeply dissected into threadlike segments; disk flowers infertile; ray flowers fertile but achenes maturing late 2
   2. Plant subshrubby; leaves densely covered with long whitish or grayish hairs (at least when young, the upper surface often becoming glabrous with age) .... *A. filifolia*
   2. Plant mostly herbaceous; leaves glabrous .... *A. campestris*
1. Leaves linear-attenuate or with 2 or 3 (–5) linear-attenuate segments; flowers variously fertile 3
   3. Peripheral flowers of each head fertile, inner flowers infertile .... *A. dracunculus*
   3. Peripheral flowers of each head rarely fertile, inner flowers fertile 4
      4. Leaves 0.5–2 cm long, either entire and linear or pinnately dissected into a few linear lobes 1 mm wide or narrower .... *A. carruthii*
      4. Leaves 2–10 cm long, if dissected the lobes 2–12 mm wide .... *A. ludoviciana*

A. campéstris L.

L. *campestris*, of fields, from campus, field.
Flowers yellowish-white, inconspicuous; late July–October
Dry prairie
West three-fifths

Glabrous biennial or short-lived perennial with a stout taproot and numerous lateral roots. Stems 1 to few, stiff, 3.5–8 dm tall. Leaves 1.5–6 cm long, divided into linear segments 0.5–1 mm wide. Heads 1–2 mm long. Peripheral flowers of each head fertile, but the achenes quite late in maturing. Central flowers infertile, the ovary abortive. Achenes ribbed, about 0.8 mm long, lacking a pappus.

A. carrùthii Wood

In honor of its discoverer, James A. Carruth (1807–1896).

Flowers yellowish-white, inconspicuous; August–October
Dry prairie
West third, primarily; especially southward

Densely woolly perennial herb with several stiff stems 1–6 dm tall arising from a rhizome. Leaves mostly 1–2 cm long, linear or pinnately dissected into a few linear lobes 0.5–1 mm wide. Heads 2–3 mm long. Peripheral flowers of each head sterile (rarely fertile). Central flowers fertile.

A. dracúnculus L.

L. *dracunculus*, diminutive of *draco*, dragon—the reference unclear.
Flowers yellowish-white, inconspicuous; August–early October
Prairie
Scattered, mostly in the northern half

Perennial herb with stems 5–15 dm tall arising from a thick rhizome. Leaves 3–8 cm long and 1–6 mm wide, mostly linear-attenuate but occasionally 2- or 3-cleft. Herbage entirely glabrous or somewhat woolly, at least in the inflorescence. Heads about 2 mm long. Peripheral flowers of each head female and fertile. Inner flowers infertile.

The cooking herb tarragon is the leaves of this species.

587. *Artemisia carruthii*

A. filifòlia Torr.　　　　　　　　Sandhill sage

NL., from L. *filum*, filament, thread, + *folium*, leaf.
Flowers yellowish-white, inconspicuous; August–October
Prairie
West half, especially westward and southward

Subshrubby perennial with freely branching stems 3–10 dm tall. Leaves entire and linear or dissected into 3 linear lobes about 0.5 mm wide and 0.5 mm thick, the margins often rolled under, densely woolly when young, becoming less so with age. Heads about 2 mm long, woolly. Peripheral flowers of each head female, fertile, but the achenes maturing late. Inner flowers infertile, the styles not cleft. Mature achenes about 0.7–1 mm long.

A. ludoviciàna Nutt.　　　　Mugwort wormwood, white sagebrush

NL., "belonging to Louisiana"—not the present state of that name but that vast territory obtained
from France in 1803, extending from the Mississippi to the Rocky Mountains and from
the Gulf of Mexico to British America.
Flowers yellowish-white, inconspicuous; August–September
Weedy prairie, roadsides, and pastures
Nearly throughout, but uncommon in the southeast

588. *Artemisia filifolia*

Highly variable perennial herb with several to many rigid stems 3–10 dm tall arising from a short rhizome. Leaves 2–10 cm long, varying from narrowly elliptic and entire to deeply pinnately lobed, the blades or subdivisions 2–12 mm wide, both upper and lower surfaces densely woolly in var. *ludoviciana*, the upper surface glabrous or nearly so in var. *mexicana*. Heads 2–4 mm long, more or less woolly. Peripheral flowers usually sterile. Central flowers fertile.

This species is closely related to and perhaps even conspecific with the European *Artemisia vulgaris*.

The Kiowa Indians of the Plains chewed the leaves of var. *mexicana* to soothe sore throats (34).

CHRYSÁNTHEMUM L.　　　　　　　Chrysanthemum

L., from Gr. *chrysanthemon*, from Gr. *chrysos*, gold, + *anthemon*, flower—the type species, C.
coronarium (L. *coronarius*, suitable for garlands, from *corona*, garland, wreath, chaplet),
having lemon-yellow flowers.

Annual or perennial herbs. Leaves alternate, simple, sessile or petioled, entire to toothed or pinnatifid. Heads many-flowered, radiate (rarely discoid), borne singly or in branched clusters. Involucre broad and flat, the bracts in several series, graduated in length, dry and membranous throughout or just along the margins. Receptacle flat or convex, naked. Pappus a short crown or absent. Ray flowers many (rarely absent or deformed), female, fertile; corollas usually white (ours), less frequently yellow or pink, the ligules well developed. Disk flowers many, bisexual, fertile, the corollas yellow, tubular, (4-) 5-toothed. Anthers with scalelike appendages at the apex, the bases entire. Style branches of ray flowers oblong, protruding from the corolla tube, those of the disk flowers short, truncate, and with a small tuft of hairs at the apex, included within the

589. *Artemisia ludoviciana*

CAMOMILE TRIBE
409

590. *Artemisia ludoviciana;* note in this specimen and the previous one the variation which may exist among individuals

591. *Chrysanthemum leucanthemum*

corolla tube. Achenes cylindric, nearly round to somewhat angled in cross section, 5- to 10-ribbed or -angled, those of the ray flowers sometimes with 2 or 3 of the ribs somewhat winglike.

One species in Kansas.

C. leucánthemum L. Oxeye daisy

L., from Gr. *leukos,* white, + *anthemon,* flower—the ray flowers being white.
Rays white, disk yellow; mid-May–June (July)
Prairie meadows, pastures, and roadsides
East two-fifths, primarily; intr. from Eurasia

Glabrous, erect herbaceous perennial with 1 to several stems 2–7 dm tall arising from a short vertical rhizome which gives off horizontal branches, offsets, and roots. Leaves 2.5–7 cm long and 4–16 mm wide, the lowermost long-petioled with blades nearly round to obovate, oblanceolate or spatulate, the margins crenate and/or pinnately lobed, the middle and upper leaves sessile and somewhat clasping the stem, smaller, mostly with widely spaced teeth or pinnate lobes. Heads 2.5–4.5 cm across, borne singly at the ends of the stem or its few branches. Involucre 4–7 mm long, the margins of the bracts membranous toward the tip. Rays white, 11–22 mm long, the tips with 1 or 2 very shallow notches. Achenes dark brown with prominent, lighter ribs, 1–2 mm long, lacking a pappus.

The disk flowers provide both nectar and pollen, and the insect visitors comprise various kinds of bees, flies, butterflies, and beetles. The disk flowers bloom in succession from the circumference towards the center of the disk, displaying at once all stages of blossoming, from very young buds to fading flowers. With the aid of a lens, we can easily see how the opening of a bud begins with the upthrust of the style from within the tube of united anthers, the 2 branches of the style standing stiffly upright and so appressed together that pollen swept out by the tuft of hairs at their tips is kept from their inner stigmatic surfaces. At a later stage, the style branches spread apart, exposing the stigmatic surfaces to contact with insects that have come with pollen from younger flowers adhering to them, thus affording the chance of cross-pollination.

When meadows or pastures are run-down, because of drought or for other reasons, the oxeye daisy sometimes gets started and becomes widespread. Horses, sheep, and goats will browse the plants. Cattle usually are repelled by their pungent, bitter taste, though cows do sometimes nibble them by chance or from scarcity of other forage, giving the milk an unpleasant taste.

In the Eurasian homeland of the oxeye daisy, people find the plants of common occurrence, and in olden times interpreted their bitter taste and aggressive growth-habit as indicating medicinal potency. They made use of an infusion of the flowers, as well as a drink, prepared by boiling together leaves, stems, and flowers and sweetening with honey, to relieve chronic coughs and bronchial catarrh; and a decoction of the fresh herb in ale was made for curing jaundice.

We now esteem the oxeye daisy most for the charm of its flowers, and we welcome its abundance in the fields when we are wanting a multitude of flowers for Memorial Day.

See plate *117.*

SENECIÓNEAE Groundsel Tribe

Sap of plants usually not milky. Heads all alike, either radiate or discoid. Disk flowers bisexual and fertile or occasionally the marginal flowers female and with slightly irregular tubular corollas; ray flowers, if present, female and infertile. Receptacle naked. Anthers rounded or sagittate at the base, never caudate (tailed). Style branches usually truncate. Pappus of many very slender bristles.

CACÁLIA L. Indian plantain

L. *cacalia,* name used by Pliny for the coltsfoot (now assigned the name *Tussilago farfara* L.), a garden vegetable used for greens.

Tall, glabrous perennial herbs (ours), less frequently shrubs or rarely trees, sometimes with milky juice. Stems stout, unbranched below the inflorescence. Leaves alternate, simple, petioled at least below, entire or coarsely toothed. Heads 4- to 6-flowered, discoid, borne in large, rather flat-topped, branched terminal clusters. Involucre urn-shaped or cylindric, the bracts 5 in a single series, herbaceous or essentially so, sometimes with dry, membranous margins, sometimes subtended by a few smaller bractlets. Receptacle naked, flat with a minute conical projection in the center. Pappus of many white hairlike bristles arising from a flat disk. Flowers all tubular, bisexual, fertile; corollas white or yellowish, deeply 5-lobed. Anthers rounded at the base. Style branches flattened, the apexes truncate and with a minute tuft of hairs. Achenes cylindric, smooth, several-ribbed, the pappus eventually deciduous.

Two species in Kansas.

Key to Species:
1. Leaves entire or shallowly serrate, the main veins parallel, converging at the apex; stems
  markedly longitudinally angled and grooved; involucral bracts with a winged keel ....
  .... *C. tuberosa*
1. Leaves very coarsely toothed, incised or lobed, the main veins palmate, diverging toward
  the margins; stems smooth; involucral bracts lacking a winged keel .... *C. atriplicifolia*

C. atriplicifòlia L.   Pale Indian plantain

NL., from L. *atriplex*, ancient name for the garden vegetable orache (*Atriplex hortensis*), +
  folium, leaf—there being a similarity between the leaves of the 2 species.
Rusty-white; July–September
Open woods and thickets, less frequently along roadsides
East third

Perennial herbs 1–2 m tall, the stem round in cross section. Leaves petioled, the blades thin, green above and glaucous beneath, coarsely toothed, incised, or lobed, with veins diverging palmately toward the margins, the lower leaves reniform to broadly triangular, 10–15 cm long and 15–19 cm wide, those farther up the stem becoming shorter-petioled, smaller, and round or ovate in general outline, the uppermost usually oblanceolate to elliptic and often entire. Heads many, the involucres 8–10 mm long, the bracts oblong to lance-oblong, with acute apex and membranous margins. Corolla rusty-white (drying yellowish), 7–8.5 mm long. Achene body about 5 mm long, the pappus 5–7 mm long.

C. tuberòsa Nutt.   Tuberous Indian plantain

L. *tuberosus*, knobby, from *tuber*, knob or swelling—in reference to the thick, tuberous root.
Rusty-white; June–July (–August)
Prairies
East two-fifths

Perennial herb, 4.5–18 dm tall, the stem prominently longitudinally angled and grooved. Leaf blades thick and leathery, green on both sides, mostly ovate to elliptic, with parallel veins which converge toward the apex and margins entire or shallowly serrate, the lower leaves (5–) 8–20 cm long and (2.5–) 3–11 cm wide, the middle and upper leaves smaller and shorter-petioled. Heads many, the involucres 8–10 mm long, the bracts with a prominently winged keel. Corolla rusty-white (drying yellowish), 8–10 mm long. Achene body about 5 mm long, the pappus 7–8 mm long.

ERECHTÌTES Raf.   Fireweed

An ancient name used by Dioscorides for some species of groundsel (*Senecio*, in the Composite
  Family).

Erect annual (ours) or perennial herbs. Leaves alternate, simple, entire to toothed or pinnately dissected. Heads many-flowered, discoid, in branched terminal clusters. Involucre cylindric or urn-shaped at flowering time, becoming bell-shaped in fruit, the bracts about 13–14, linear, sharply pointed, in a single series, sometimes subtended by a few smaller bractlets. Receptacle flat, naked. Pappus of many soft, white hairlike bristles. Outer flowers of each head female, fertile, the corollas yellowish, tubular but very slender and with the 3 outer teeth much longer than the 2 inner ones. Inner flowers many, bisexual, usually fertile; corolla yellowish, tubular, the 5 teeth all alike. Anthers entire or somewhat sagittate at the base. Style branches elongate, flattened, the apexes truncate. Achenes cylindric or tapering slightly to the base, 5-angled and 10- o 20-ribbed.

One species in Kansas.

E. hieracifòlia (L.) Raf.   Pilewort

NL., from the genus *Hieracium*, hawkweed, in the Composite Family + L. *folium*, leaf—having
  leaves resembling those of *Hieracium*.
Yellowish-white; mid-August–mid-September
Moist woods and thickets
East fourth

592. *Cacalia tuberosa*

593. *Cacalia tuberosa*, close-up of heads

Glabrous to sparsely hairy, rank-smelling annuals, 0.3–2 m tall. Leaves thin, oblanceolate to elliptic or oblong in general outline, 6–20 cm long and 1.3–6 cm wide, shallowly toothed to coarsely toothed and/or pinnatifid, the lower leaves with wedge-shaped bases, the upper leaves with auricled bases. Heads several to many on long, slender peduncles in an openly branched terminal cluster; involucre 6–8 mm long, subtended by a few threadlike bracts. Corollas yellowish-white, 10–15 mm long. Achene body 2.5–3 mm long, the pappus 9–12 mm long, falling away as a unit.

In Asia, the young shoots and leaves of this plant are eaten either cooked or raw.

SENÈCIO L. — Groundsel, ragwort

L. name for a plant, groundsel, from L. *senesco*, to grow hoary—in allusion to the hoary pappus and tomentum of some species. "Groundsel" is from AS. *grundeswelge*, possibly "ground-swallower," in reference to the rapid growth of the European *Senecio vulgaris*; but an earlier form, *grundeswelge* or *grundeswylige*, variant of *grundaeswelgae*, is explained as "pus-absorber" (AS. *gund*, pus), a name which would allude to the application of the herb to all kinds of wounds, as recommended by Dioscorides as long ago as the 1st century.

Annual, biennial, or perennial herbs or subshrubs, in some parts of the world vines, shrubs, or even trees. Leaves alternate or all or mostly basal, simple, sessile or petioled, entire to toothed or pinnatifid. Heads many-flowered, radiate (rarely discoid), usually in branched terminal clusters. Involucre urn-shaped to more or less bell-shaped, the main bracts in 1 series, 12–25, linear, usually sharply pointed, equal in length, sometimes subtended by a few much-smaller bractlets. Receptacle flat or slightly convex, naked but often pitted. Pappus of many soft, white hairlike bristles. Ray flowers female, fertile; corollas yellow or orange, the ligules linear to somewhat elliptic, 3-toothed. Disk flowers bisexual, fertile; corollas yellow to orange or reddish, tubular, with 5 equal teeth. Anthers rounded at the base or somewhat sagittate. Style branches flattened or nearly round in cross section, the apexes truncate and with a tuft of minute hairs. Achenes cylindric, round in cross section or slightly flattened, 5- to 10-ribbed, the pappus deciduous.

Ten species reported for Kansas. The following are the most commonly encountered species.

Key to Species:
1. Subshrubby perennials with leafy, woody stems; leaves with linear lobes; west half .... *S. riddellii*
1. Annual or perennial herbs; leaves not as above 2
  2. Annuals with fibrous roots; leaves well distributed along the stem and only gradually reduced upward, very deeply pinnately lobed .... *S. glabellus*
  2. Perennials with rhizomes; leaves mostly crowded near the base, usually becoming much-reduced and remote toward the middle of the stem 3
    3. Plants with persistent, matted woolly hairs, at least in patches near the nodes; nearly throughout, usually in prairies .... *S. plattensis*
    3. Plants glabrous at maturity or with very small spots of wool near the base of each leaf; east third, usually in woods .... *S. obovatus*

S. glabéllus Poir. — Butterweed

L. *glabellus*, nearly smooth, diminutive of *glaber*, smooth.
Yellow; April–June
Low, wet ground, usually shaded
East fifth, primarily

Glabrous, erect annual with fibrous roots and a thick stem (1.5–) 3.5–7.5 dm tall, unbranched below the inflorescence. Leaves long-petioled below, becoming sessile above but not reduced much in size, the blades deeply pinnately lobed, 5.5–14 (–20) cm long and 2–7 cm wide, the terminal lobe more or less rounded, the others becoming more oblong and reduced toward the base of the blade, the margins coarsely toothed and often undulate. Heads many, mostly (1–) 1.5–2 cm across; involucre broadly bell-shaped, 5–7 mm long. Rays yellow, 5–9 mm long. Achene body minutely hairy or glabrous, about 1.5–1.8 mm long, the pappus 4–6 mm long.

This species is much less common than the other species included here.

S. obovàtus Muhl. — Golden groundsel

NL. *obovatus*, reverse egg-shaped, that is, egg-shaped, but with the broadest point toward the apex rather than the base, from L. *ovatus*, egg-shaped.
Yellow; April–May
Moist woods
East third

Erect, glabrous perennial, 1–6 dm tall, producing long, slender stolons of rhizomes, the stem unbranched below the inflorescence. Leaves mostly basal, the lowermost long-petioled with blades mostly round to broadly ovate or elliptic, 2.5–6 cm long and 2–5 cm wide, usually merely serrate but sometimes with 1 or 2 pairs of small, pinnate basal lobes and a large terminal lobe; middle and upper leaves becoming short-petioled or sessile, the blades oblanceolate to elliptic, lanceolate, or lance-linear, 3–8 cm long and 0.7–4 cm wide, usually pinnately lobed at least halfway to the midrib. Heads few to many, mostly 1.5–2 cm across. Involucre broadly bell-shaped, 4–8 mm long. Rays yellow, 5–10 mm long. Achene body glabrous, 2–2.5 mm long, retrorsely barbed on the ribs, the pappus 4–5 mm long, deciduous.

The rhizome is at first simple, but as time goes on, it produces short branches, each giving rise to an erect flowering stem. The achenes are scattered widely by the wind, but many settle down in the vicinity of the parent, so that at first a patch and then a sizable colony may appear whenever conditions are favorable. The wild sweet william phlox has the same habit of spreading, and when both are planted in a garden of wild-flowers they form a prosperous association, with overlapping blooming periods and a harmonious combination of contrasting colors.

See plate *118*.

594. *Senecio plattensis*

S. platténsis Nutt.   Prairie groundsel

NL. from Platte, the species having been just collected along the Platte River in Nebraska + the L. suffix -*ensis*, indicating the place where that plant occurs.
Yellow; mid-April–early June
Prairie
Nearly throughout, but infrequently collected in the southwest eighth

Erect perennial, 0.7–6 dm tall, with a short vertical rhizome, the stem unbranched below the inflorescence. Leaves mostly basal, the lowermost long-petioled with blades lanceolate to elliptic, oblong, or oblanceolate in general outline, merely serrate to deeply pinnatifid, 1.5–8 cm long and 0.7–3 (–3.5) cm wide; middle and upper leaves abruptly reduced in size and becoming short-petioled or sessile, pinnately lobed half or more of the distance to the midrib. Heads several to many, mostly 2–3 cm across, on long, slender peduncles in branched terminal clusters; involucre broadly bell-shaped, 5–7 mm long, the bracts mostly linear-attenuate. Rays yellow, 6–11 mm long. Achene body glabrous, 2–3 mm long, the ribs not barbed, the pappus 4–4.5 mm long, deciduous.

This is our most widely distributed groundsel.

S. riddéllii T. & G.

Named after John L. Riddell (1807–1865), professor of chemistry in the Medical College of Louisiana (1836–1865), author of *Catalogus florae ludovicianae* (1852), and inventor of the binocular microscope.
Yellow; September–October
Dry, sandy or rocky prairie, pastures, or roadsides
West half

595. *Senecio plattensis*, close-up of heads

Erect, glabrous subshrub, 3.5–8.5 (–10) m tall, the stems branched, woody near the base. Leaves alternate, 2.5–6.5 cm long, pinnately divided into linear lobes 1–1.5 (–2) mm wide. Heads many, 2–3 cm across, in branched terminal clusters; involucre obconic, 7–11 mm long and 6–10 mm wide (when pressed), the bracts usually about 12 (–14), subtended by a few small bractlets. Rays light yellow, 10–15 mm long. Achene body 3–5 mm long, hairy, the pappus 7–10 mm long.

S. douglasii, another subshrub resembling *S. riddellii*, is known from a few south-western counties. It differs from *S. riddellii* in having stems and young leaves with long, matted white hairs and in having the involucre broadly bell-shaped, 7–9 mm long and 12–15 mm across (when pressed), with about 20 bracts.

Both species are reputedly toxic to livestock.

HELÉNIEAE   Sneezewort Tribe

Sap of plants not milky. Heads all alike, usually radiate (occasionally obscurely so or entirely discoid). Ray flowers female and fertile, the rays broadly obtriangular and deeply 3-lobed. Disk flowers bisexual and fertile. Receptacle naked or sometimes with poorly developed scales (or bristles). Anthers rounded at the base to somewhat auricled or sagittate and with lanceolate or ovate scales at the tip. Style branches slender, receptive only near the base. Pappus of membranous scales.

596. *Senecio riddellii*

597. *Dyssodia papposa*

598. *Gaillardia fastigiata*

DYSSÒDIA Cav. Dogweed, fetid marigold

NL., from Gr. *dysodia*, ill smell—in allusion to the plant as a whole.

Strong-scented annual, biennial, or perennial herbs with rather large, conspicuous, translucent, orange or brown oil glands. Leaves mostly opposite, simple, entire or toothed to deeply pinnately dissected. Heads many-flowered, usually radiate (rarely discoid). Involucre hemispheric to narrowly top-shaped or cylindric, the bracts in 1 or 2 series, the innermost joined at the base. Receptacle usually flat, pitted, naked or with minute scales. Pappus of 10–20 scales, each usually divided at least to the middle into several awns or bristles. Ray flowers usually 8–13, yellow or orange, female, fertile. Disk flowers several to many, bisexual, fertile, the corollas dull yellow, tubular, with 5 triangular lobes. Anthers rounded at the base. Style branches elongate (more so in disk flowers than ray flowers), flattened. Achenes narrow, glabrous, and more or less striate to stout, obpyramidal, and silky-hairy, the pappus persistent.

Two species in Kansas. *Dyssodia aurea* (Gray) A. Nels., a short, much-branched annual with bright yellow flowers and conspicuous rays, is known from Hamilton County only. *Pectis angustifolia* Torr., another short (up to 2 dm) annual with bright yellow flowers and conspicuous rays, also belongs to the Sneezewort Tribe and is known from scattered sites in the western half of the state. Its glandular herbage, when crushed, smells of lemon oil.

D. pappòsa (Vent.) Hitchc. Dogweed

NL., *papposus*, crowned with a pappus—applicable to the achenes of both the disk flowers and ray flowers.
Dull yellow; (late May–) July–October
Usually in disturbed sites such as roadsides, pastures, farm lots, waste ground, etc.
Throughout

Erect, diffusely branched annual, 0.6–5.5 dm tall. Leaves gland-dotted, mostly 1.5–4.5 cm long, once or twice pinnately parted into slender, entire or toothed segments 0.5–2 mm wide. Heads many, radiate, but the ray flowers with very short ligules and not greatly distinguished from the disk flowers, the heads therefore appearing discoid. Involucre broadly ellipsoid, 5–10 mm long, the bracts with prominent, blisterlike oil glands, the inner row of bracts erect, elliptic to oblanceolate, prominently glandular, and thin-margined, the outer row shorter, narrower, and more leaflike in texture. Receptacle pitted and beset with minute chaffy scales (use high magnification). Corolla dull yellow. Pappus of slender scales apically divided into slender, minutely barbed bristles. Achene body black, hairy, 3–4 mm long, the pappus 3–4 mm long, persistent.

The strong-smelling volatile oil and resin secretions are irritating to the mucous membranes and may cause dermatitis in susceptible persons. Livestock will not browse the plants, leaving them to multiply and fruit abundantly in barnyards and cow lots.

Several medicinal uses of this plant are recorded in the literature. The Sioux Indians took snuff of the powdered leaves and tops to induce nosebleed for relief of headache. According to Krochmal and Krochmal (30), the fresh leaves, either chewed or made into a tea, have been used to treat stomachache and diarrhea; and an infusion of the achenes steeped in warm water has been given to colicky babies.

GAILLÁRDIA Foug. Gaillardia, Indian blanket, blanketflower

Named in honor of Gaillard de Marentonneau, French botanist of the 18th century.

Erect annual or perennial herbs, rarely woody at the base, with or without above-ground stems. Leaves alternate or all basal, sessile or petioled, entire to toothed or pinnatifid. Heads many-flowered, usually radiate but occasionally discoid, borne singly on short or elongate peduncles. Involucre saucer- or wheel-shaped, the bracts entirely herbaceous, sometimes rather stiff or hardened and discolored at the base, ovate to oblong or lanceolate, in 2–3 series, strongly reflexed in fruit. Receptacle convex to globose, pitted, chaffy with long bristles or shorter subulate scales. Pappus of 5–10 long, membranous, awn-tipped scales. Ray flowers usually neutral or female and infertile (occasionally absent), the ray itself yellow, or yellow and red or purplish-red, broad, wedge-shaped or fan-shaped, the apex deeply 3-toothed or -cleft. Disk flowers bisexual, fertile, the corollas yellow or purplish-red, the tips of the 5 teeth covered with moniliform hairs (from L. *monile*, necklace, because the multicellular hairs resemble strings of beads). Anther sacs auricled at the base. Style branches usually long and minutely hairy (sometimes short and glabrous), stigmatic only at the base. Achenes broadly obpyramidal, partly or entirely covered by long, stiff, ascending hairs, crowned by the persistent pappus.

Three species in Kansas.

Key to Species:
1. Heads discoid (rarely with inconspicuous rays); leaves all basal .... *G. suavis*
1. Heads radiate; leaves borne along a branched stem 2
   2. Receptacle with long, stiff bristles; rays usually red tipped with yellow .... *G. pulchella*
   2. Receptacle with small, lacerated scales; rays usually all yellow .... *G. fastigiata*

G. fastigiàta Greene Gaillardia

NL., from L. *fastigium*, gable, slope. In botany, the term "fastigiate" is applied to branches that
   are erect and near together, the juncture of their bases often forming a downward point.
Rays yellow, disk flowers purple; July–September
Dry, rocky or sandy prairie
South-central sixth, primarily

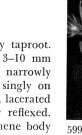

    Erect, branched herbaceous perennial, 2–6 dm tall, with a deep, woody taproot.
Herbage minutely gland-dotted and rough-hairy. Leaves 1.5–6 cm long and 3–10 mm
wide, the lowermost spatulate, entire or with a few teeth, the uppermost narrowly
oblanceolate to linear or linear-oblong, entire. Heads 3–5 cm across, borne singly on
axillary and terminal peduncles, the receptacle globose, covered with small, soft, lacerated
scales, the involucral bracts lanceolate to lance-oblong, not ciliate, eventually reflexed.
Rays yellow, 8–17 mm long; disk flowers dark purple, the teeth attenuate. Achene body
2–3 mm long, bearing a basal whorl of long, ascending hairs, the pappus scales 6–8 mm
long, lance-linear, the midrib extended into a long awn.

599. *Gaillardia pulchella*

G. pulchélla Foug. Annual gaillardia, Indian blanket

L. *pulchellus*, diminutive of *pulcher*, beautiful—in reference to the small flowers. "Indian
   blanket" is an appropriate common name suggested by a colony of these plants in bloom,
   often widespread over the plains.
Rays yellow apically, red to rose-purple basally; mid-May–September
Dry, rocky or sandy prairie and roadsides
West four-fifths and Wilson County; more common westward

    Erect annual with a taproot and branching stems 1–6 dm tall. Leaves 1.5–6.5 cm
long and 4–28 mm wide, the lowermost oblanceolate and merely serrate to pinnately
lobed, the middle and upper leaves mostly lanceolate to oblong, entire (rarely pinnately
lobed). Herbage minutely gland-dotted and rough-hairy. Heads 3–5 cm across, borne
singly on long peduncles, the receptacle globose, covered with long bristles, the involucral
bracts lance- to linear-attenuate, ciliate, strongly reflexed. Rays red to reddish-purple at
the base, yellow toward the tip (often fading to rose-purple and white when dried), 1–2
cm long; disk flowers dark purple, the teeth attenuate. Achene body 2–2.5 mm long,
bearing a basal tuft of long, ascending hairs, the pappus scales 4–7 mm long, the midrib
extended into a long awn.

    As the flowers wither, the receptacle becomes more rounded, presenting over its
surface a mass of achenes, each tipped by the long-attenuate, rigid, grayish awns of the
persistent pappus. The variety *picta* of this species (L. *pictus*, painted or embellished),
with larger flower heads of various colors, is the annual gaillardia most commonly
cultivated.

    See plate *119*.

G. suàvis (Gray & Engelm.) Britt. & Rusby

L. *sauvis*, sweet.
Rays absent or inconspicuous, disk flowers reddish-brown; May–mid-June
Dry, rocky or sandy prairie and roadsides
Primarily south of a line from Reno County to Meade and Sumner counties

    Erect herbaceous perennial. Leaves glabrous, all basal, spatulate to oblanceolate
and tapered to a petiolelike base, 5–12 cm long and 1–4 cm wide, the margins entire to
toothed or pinnately lobed. Heads 1–4 per plant, 2–2.5 cm across, each borne singly on
a naked peduncle 3–6 cm tall, the receptacle globose, the involucral bracts lance-linear to
linear-attenuate, eventually reflexed. Rays absent in ours; disk flowers reddish-brown, the
teeth triangular. Achene body 2–3 mm long, bearing a basal whorl of long, ascending
hairs, the pappus scales 7–9 mm long, narrowly lanceolate to elliptic, the midrib extended
into a long awn.

HELÉNIUM L. Sneezeweed

Gr. *Helenion*, name used by Hippocrates (died 4th century B.C.) and Theophrastus (died c.
   285 B.C.) for the plant now called *Inula helenium*. "Sneezeweed" is appropriate because
   the powdered disk flowers used as snuff cause violent and long-continued sneezing.

SNEEZEWORT TRIBE

600. *Helenium amarum*

601. *Helenium autumnale*

Erect, often branched annual, biennial, or perennial herbs, usually with taproots. Leaves alternate, simple, sessile or nearly so, entire to pinnately lobed, gland-dotted. Heads many-flowered, usually radiate, 1 to many, borne singly or in pairs. Involucre small, wheel-shaped, the bracts about 16, in 2 series, lanceolate to subulate or linear, herbaceous, gland-dotted, about equal in length or the inner ones shorter, spreading at first but eventually reflexed. Receptacle globose to ovoid or conic, pitted, naked or sometimes with a few bracts between the ray and disk flowers. Pappus of 5–10 transparent scales with or without awns, sometimes the scales entirely slender and awnlike or absent altogether. Ray flowers usually present, about 8, female, and fertile, in some species neutral and infertile; rays yellow (less frequently reddish-brown with a yellow tip), large and conspicuous, glandular-hairy, often reflexed, the apex 3-toothed or -lobed. Disk flowers many, bisexual, fertile; corollas mostly yellow, with 4–5 yellow or red-brown, triangular teeth. Style branches slender, each with a minute apical tuft of hairs. Achenes obpyramidal, 4- to 5-angled with as many intervening ribs, usually hairy, crowned by the persistent pappus.

Three species in Kansas. *H. flexuosum* Raf., a perennial resembling *H. autumnale* but with the disk flowers and sometimes also the rays reddish-brown or reddish-purple, is known from the extreme southeast only.

All of our species are poisonous to livestock, but are apparently grazed only when other forage is scarce and seem to be eaten more readily by sheep than by other stock. A toxic, alcohol-soluble glycoside named dugaldin has been isolated from *H. hoopesii* (not of our flora), and the same or similar compounds must be responsible for the toxicity of the other species. A bitter taste is imparted to milk and cream produced by animals grazing on sneezeweed. For a detailed description of the symptoms of sneezeweed poisoning, see Kingsbury (21).

The flowers and leaves of various species of *Helenium* have been used as a snuff to clear congested nasal passages, as fish poisons, and in the treatment of intestinal worms (30).

Key to Species:
1. Annual with linear leaves 1–2.5 mm wide; stems not winged .... *H. amarum*
1. Perennial with narrowly elliptic leaves 8–30 mm wide; stems winged .... *H. autumnale*

H. ámarum (Raf.) Rock   Bitter sneezeweed

L. *amarus*, bitter.
Yellow; July–October
Weedy, overgrazed pastures, fields, and waste places
East half, primarily the southeast eighth

Gland-dotted annual, 1–8 dm tall, the stems simple below and with many ascending branches above. Leaves narrowly linear, 2.8–6 cm long and 1–2.5 mm wide. Heads 1–2.5 cm across, borne singly on elongate peduncles, the involucral bracts strongly reflexed and obscured by the rays, the receptacle globose. Rays yellow, deeply 3-lobed, 5–13 mm long. Disk flowers yellow. Achene body brown, broadly pyramidal, long-hairy, 0.5–1 mm long, the pappus scales with slender arms.

This species is listed in earlier manuals for our area as *H. tenuifolium*.

Bitter sneezeweed comes and goes. It may be present in a locality one year and gone the next, depending on whether conditions are favorable for seed germination. It becomes abundant primarily in pastures where too many cattle are turned in too early for the grass to get a good start.

H. autumnàle L.   Sneezeweed

L. *autumnalis*, belonging to autumn—in reference to the late blooming period.
Yellow; August–October
Moist soil
East fourth, primarily

Erect, much-branched herbaceous perennial, up to 1 m tall, with a short rhizome. Stems winged by the decurrent leaf bases. Leaves 3–10 cm long and 0.8–3 cm wide, mostly narrowly elliptic to somewhat oblanceolate, the apex acuminate, the margins shallowly and remotely serrate. Herbage gland-dotted and minutely hairy. Heads 1.5–3 cm across, borne singly or in pairs on long axillary peduncles, the involucral bracts strongly reflexed and obscured by the rays, the receptable globose. Rays yellow, deeply 3-lobed, 8–16 mm long. Disk flowers yellow. Achene body reddish-brown, very narrowly pyramidal, hairy (at least on the angles), about 1.5 mm long, the pappus scales long-acuminate and sometimes lacerated, about 0.5 mm long.

HYMENOPÁPPUS L'Her

NL., from Gr. *hymen*, membrane, + *pappos*, old man—in reference to the hoary bristles or pappus that usually represents the calyx in this family, and which in this genus is present in the form of membranous scales.

Erect, more or less woolly herbs, either biennials with obconic taproots or perennials with woody, branched roots and several crowns. Stems longitudinally grooved, more or less densely covered with matted white hairs. Leaves alternate, simple, sessile or petioled, entire, lobed, or finely dissected, becoming smaller upward, minutely gland-dotted. Heads many-flowered, usually discoid (always, in our species), several to many in branched terminal clusters, rarely borne singly. Involucre hemispheric to bell- or top-shaped, the bracts 6–12 in 2–3 series, appressed, rather broad and rounded, at least the tips or margins somewhat petaloid and membranous. Receptacle convex to nearly flat, naked. Pappus of 10–20 broadly rounded or truncate, membranous scales. Ray flowers absent in our species. Disk flowers all bisexual and fertile; corolla yellowish or white, regular, tubular, with a flaring throat and 5 reflexed lobes. Anther tips with ovate, scale-like appendages, bases entire or nearly so. Style branches flattened, with short, conic, minutely hairy appendages at the tips. Achenes obpyramidal, 4- to 5-angled, each face with no rib or 1–3 ribs, the pappus persistent.

602. *Hymenopappus scabiosaeus*

Four species in Kansas. *H. filifolius* Hook., our only perennial species, is reported from scattered sites in the west third. It has roots bearing several to many crowns or stems, relatively few stem leaves (0–12) and heads 1–20 (rarely to –50) per stem, and yellow corollas with the throat 2–8 times as long as the lobes. All of our other species are biennials and usually have but a single stem and unbranched taproot, though they may occasionally produce, through injury to the root crown, several stems.

Key to Species (58):
1. Corolla tube 2–3 mm long; basal rosette leaves simple to bipinnate with broad ultimate segments 5–20 mm wide; flowers white or purplish, never yellow, corolla throat funnelform (rarely campanulate) .... *H. scabiosaeus*
1. Corolla tube 1.5–2 (–2.2) mm long; basal rosette leaves bipinnately dissected, with mostly linear ultimate segments, 0.5–6 mm wide; flowers yellow or white; corolla throat campanulate (very rarely funnelform) 2
  2. Flowers white; ultimate leaf segments narrowly linear, 0.5–1.5 mm wide .... *H. tenuifolius*
  2. Flowers yellow; ultimate leaf segments short, narrow to broad, 1–6 mm wide .... *H. flavescens*

H. flavéscens Gray

L. *flavescens*, turning yellow, present participle of *flavescre*, to turn yellow, inceptive of *flavus*, to be yellow, from *flavus*, yellow, golden yellow.
Yellow; mid-May–late June
Dry prairie
Southwest twelfth

Plants biennial, 3–6 (–9) dm tall. Leaf blades 5–10 cm long, bipinnately dissected into linear segments 1–5 mm wide. Heads many, discoid, 30- to 70-flowered. Involucre 4–7 mm long, the bracts yellowish, more or less hairy, scarcely or not at all glandular. Disk corollas bright yellow, 34 mm long (measured with the lobes reflexed), gland-dotted, the tube gradually expanded into a funnelform throat about as long as the lobes. Achene body black, hairy on the angles, glandular, 3–3.5 mm long, the pappus about 0.5–1 mm long.

H. scabiosaèus L'Her var. corymbosus (T. & G.) B. L. Turner   Old plainsman

NL., from the genus *Scabiosa*, in the Teasel Family, + the suffix -*us*, having the nature of, i.e., "resembling *Scabiosa*"; *corymbosus* from Gr. *korymbos*, flower cluster—the flower heads in this species being borne in the plan of a "corymb"—the name that in botanical parlance is applied to a flower cluster with a flat or somewhat rounded top and in which the oldest flowers are at the periphery and the younger open progressively toward the center. In "old plainsman" we have a common name rich in fancy and atmosphere, but notice that the scientific name has the virtue of precise description.
Whitish to cream-colored; May–early July
Prairie, frequently on rocky hillsides and banks
East two-thirds

Plants biennial, (1.2–) 2–10 dm tall. Leaves broadly lanceolate in general outline, 4–15 cm long, finely pinnately or bipinnately divided into segments 0.5–5 mm wide, the basal leaves long-petioled and with relatively broad, rounded subdivisions, the middle and upper leaves gradually becoming shorter-petioled, smaller, and more finely dissected. Heads many, discoid, 20- to 60-flowered. Involucre 5–8 mm long, the bracts whitish, not glandular at all or with only a few glands near the base. Disk corollas white to cream-colored, 2.5–3 mm long (measured with the lobes reflexed), gland-dotted, fragrant, gradually expanded into a funnelform throat about as long as the reflexed lobes. Achene body black, hairy, 3–4 mm long, the pappus about 0.5 mm long.

A hardy, drought-resistant species, often occurring with species of *Gaillardia*, *Monarda*, *Opuntia*, and *Yucca*.

603. *Hymenopappus scabiosaeus*

H. tenuifòlius Pursh   Old plainsman

NL., from L. *tenuis*, thin, + *folium*, leaf.
White; June–August
Dry prairie hillsides and uplands, frequently in rocky, calcareous soil
West two-thirds, primarily

Plants biennial, mostly 2–7 dm tall. Leaf blades 2–13 cm long, finely pinnately dissected into linear segments 0.5–1 mm wide, even on the basal leaves. Heads many, discoid, 25- to 50-flowered. Involucre 4–8 mm long, the bracts yellowish, definitely glandular, usually with matted white hairs, at least near the base. Disk corollas white, 2.5–3 mm long (measured with the lobes reflexed), gland-dotted, the throat abruptly expanded and bell-shaped or somewhat urn-shaped, definitely longer than the reflexed lobes. Achene body black, hairy on the angles, 3.5–4.5 mm long, the pappus about 1–1.5 mm long.

HYMENÓXYS Cass.   Bitterweed

NL., from Gr. *hymen*, membrane, + *oxys*, sharp—in reference to the pappus scales.

Erect, more or less long-hairy or silky, minutely gland-dotted annual or perennial herbs with taproots. Leaves all basal or some alternate on the stem above, simple, usually entire, less frequently toothed or pinnately lobed or cleft. Heads many-flowered, usually radiate (rarely discoid), borne singly on naked scapes or few together in a branched terminal cluster. Involucre hemispheric to bell-shaped or cylindric, the outer bracts rigid and either united below and quite thickened at the base or neither united nor strongly thickened below, the inner bracts either leathery or more or less herbaceous or membranous, all more or less equal in length. Receptacle hemispheric or conical, naked but pitted. Pappus of 5–8 thin translucent scales. Ray flowers (when present) 10–30, female, fertile; the corolla yellow (often becoming whitish with age or upon drying), the ray oblong, broad, with 4 veins and 3 apical teeth or lobes, strongly reflexed. Disk flowers many, bisexual, fertile; corolla yellow, with 5 short, hairy teeth; anthers with membranous, triangular or ovate, scalelike appendages at the apex and sagittate bases. Style branches flattened, the apex somewhat enlarged, truncate, with a minute tuft of hairs. Achenes obpyramidal or top-shaped, 5-angled, densely hairy, crowned by the persistent pappus.

604. *Hymenoxys acaulis*

Four species in Kansas. *H. linearifolia* Hook. is known from a few counties only, and it is not included here.

Key to Species:
1. Annuals with well-developed, branched, leafy aboveground stems bearing leaves their entire length .... *H. odorata*
1. Perennials lacking well-developed aboveground stems, the leaves all borne at or near the base 2
   2. Leaves densely covered with silky hairs, our populations mostly with discoid (rarely radiate) heads .... *H. acaulis*
   2. Leaves glabrous or sparsely hairy but not silky, heads always radiate .... *H. scaposus*

H. acaùlis (Pursh) Parker

NL., from Gr. *akaulos*, without a stem—the plant having no evident aboveground stem.
Yellow; mid-May–August
Rocky prairie hillsides, usually in calcareous soil
Scattered in the west half

Perennial herb with branched, woody caudexes. Leaves all basal, spatulate, entire, densely covered with silky hairs, 1.5–8 cm long and 2.5–8 mm wide. Heads usually discoid, rarely radiate, borne singly on naked peduncles 8–30 cm tall. Involucre 5–8 mm long and 12–15 mm across, the bracts silky, rather thin. Corollas yellow, the rays (when present) 5–8 mm long.

H. odoràta DC.   Bitterweed

L. *odoratus*, having an odor, from *odor, odoris*, smell, scent, + the suffix -*atus*, indicating possession.

Yellow; May–early July
Overgrazed prairie pastures, roadsides
Southwest eighth, primarily

Erect, bushy-branched annual 1.5–2.5 dm tall bearing numerous finely dissected leaves 1.5–4 cm long along the stems and branches. Heads radiate, borne on terminal or axillary peduncles 2–5 cm long. Involucre 4–5 mm long and 4–9 mm across, the bracts united at the base and becoming quite hard. Corollas yellow, the rays 6–8 mm long. Achene body 2–2.5 mm long, the pappus 1–1.5 mm long.

These plants are poisonous to sheep. The toxic agent is unknown but remains active in dried plants (21).

H. scapòsa (DC.) Parker

L. *scaposus*, having a stem, from L. *scapus*, Gr. *skapos*, stem, staff—each head being borne above
    the leaves on a leafless stalk or scape.
Yellow; April–July (–October)
Dry prairie hillsides and uplands, prairie pastures and roadsides
West half

605. *Hymenoxys scaposa*

Perennial herb, usually with branched, woody caudexes. Leaves all borne at or near the base, linear to linear-oblanceolate, glabrous or sparsely hairy (never silky), 1.5–7.5 cm long and 0.5–3 mm wide. Heads radiate, borne singly on naked peduncles 4–27 cm tall. Involucre 5–10 mm long and 11–18 mm wide, the bracts silky, rather thin. Corollas yellow, the rays with 4 veins becoming dark and prominent as the flowers dry, hence an older name for the genus, *Tetraneuris*, from Gr. *tetra-*, 4, + *neuron*, nerve. Achene body 2.5–3 mm long, the pappus also 2.5–3 mm long.

This plant is listed in earlier manuals for Kansas as *Tetraneuris fastigiata*.

PALAFÓXIA Lag.

Named for a Spanish general, José de Palafox y Melzi (1780–1847).

Erect, glandular annuals with taproots. Leaves opposite, simple, petioled, entire. Heads few-flowered, radiate (ours) or discoid, in branched terminal clusters. Involucre hemispheric or bell-shaped to top-shaped or cylindric, the bracts 8–10 in (1?–) 2–3 series, appressed, narrow, mostly herbaceous but the tip sometimes reddish. Receptacle flat, naked. Pappus in ours present, of (6–) 7–10 (–12) lance-attenuate scales with prominent midribs and membranous margins. Ray flowers (when present) female, fertile; corollas pink or purplish, broad, 3-toothed or -lobed. Disk flowers bisexual, fertile; corollas pink or purplish, tubular, with 5 usually linear teeth or lobes. Anthers entire or notched at the base. Style branches threadlike, slightly flattened, minutely hairy throughout. Achenes 4-angled, hairy.

One species in Kansas.

P. sphacelàta (Nutt. ex Torr.) Cory

NL., from Gr. *sphakelos*, gangrene, mortification. The application to this species is uncertain.
Pink or purplish; June–early October
Loose, sandy soil of dunes, prairie, pasture, river banks, and dry stream beds
West fourth, primarily the southwest eighth of the state

Plants 1.5–5.5 dm tall, the stems eventually much branched, covered with appressed hairs intermingled with spreading hairs tipped with dark glands. Leaves narrowly lanceolate to nearly linear, 3–6.5 cm long and 0.3–2 cm wide, the apex acute to attenuate, both surfaces covered with stiff appressed hairs. Involucres narrowly obconic, 9–13 mm long, the bracts equal in length, in 2 series, at least the outer ones glandular-hairy. Rays pink or purplish, deeply 3-lobed, 8–11 mm long. Disk corollas pink or purplish, about 9–10 mm long. Achene body black, quite slender and tapered to the base, 6–8 mm long, the pappus 6–9 mm long, persistent.

This plant is listed in earlier manuals for our area as *Othake sphacelata*.

PSILÓSTROPHE DC.   Paperflower

NL., from Gr. *psilos*, bare, stripped of, + *trophe*, nourishment—in allusion to the slender achenes.

Erect, several-stemmed biennial or perennial herbs with taproots (ours) or low shrubs. Leaves alternate, simple, petioled below to sessile above, entire or lobed. Heads several-flowered, radiate, crowded in branched terminal clusters. Involucre bell-shaped or cylindric, the bracts mostly in 2 series, the outer ones lanceolate to linear-oblong, hairy, the inner ones (when present) narrow, dry, and membranous. Receptacle more or less

606. *Psilostrophe villosa*

flat, naked. Pappus of 4–6 membranous scales. Ray flowers 3–7, female, fertile; corolla yellow, the ligule broad, 3-lobed, eventually becoming papery, persistent on the achenes. Disk flowers 5–12, bisexual, fertile; corolla yellow, glandular, with 5 teeth or lobes. Anther bases sagittate. Achenes small, linear, longitudinally ribbed, round in cross section or somewhat angled, glabrous or long-hairy.

One species in Kansas.

P. villòsa Rydb.   Paperflower

L. *villosus*, hairy, shaggy.
Yellow; mid-May–mid-September
Sandy or rocky prairie
Mostly in southern counties from Harper County west, scattered records north to Gove County

Woolly, gray-green plants, 1.5–5.5 dm tall, with branches slender, stiff, and ascending or widely spreading. Leaves 1.5–8 cm long and 0.4–2 cm wide, the lowermost often 3- or 5-lobed, the others entire and spatulate to elliptic. Heads many. Involucre 5–8 mm long, the bracts lance-linear to linear, densely woolly. Rays yellow (white in age), conspicuous, 3–5 mm long. Disk corollas yellow, 4–5 mm long, constricted near the base, the teeth glandular-hairy. Achene body glabrous, about 2.5 mm long, the pappus scales about 3 mm long, linear-attenuate, persistent.

607. *Psilostrophe villosa,* close-up of heads

ASTÉREAE   Aster Tribe

Sap of plants not milky. Heads all the same, either with only disk flowers present or with both disk and ray flowers present, some or all of them often yellow. Leaves alternate. Stems, leaves, and/or bracts of involucre usually not spiny; bracts of involucre usually not prickly or finely fringed; receptacle naked; anthers rounded at base; style lacking a hairy or thickened ring below its branches; branches of the style of the bisexual flowers flat and smooth within, but with triangular or lanceolate terminal appendages more or less hairy on the outside. Pappus of scales, awns, or bristles, or absent.

ÁSTER L.   Aster

Gr. and L. *aster*, star—because of the radiating ray flowers.

Annual or perennial herbs. Leaves alternate, simple, entire or toothed, often petioled below and sessile above or all sessile. Heads many-flowered, radiate, the heads usually several to many in branched terminal clusters. Involucral bracts in 2 or more series, about equal or graduated in length, herbaceous or more or less papery. Receptacle flat or slightly convex, naked. Pappus of many hairlike bristles, usually in a single series. Ray flowers several to many, female, fertile, the corollas white to pale blue, pink, or purple. Disk flowers many, bisexual, fertile; corolla tubular below, funnelform above, 5-lobed, yellowish or sometimes tinged with pink, blue, or purple; anthers rounded at the base; style branches flattened, the tips acute or acuminate, hairy. Achenes more or less flattened, with several ribs crowned by the persistent pappus.

About 25 species in Kansas. Sixteen of our most common or most conspicuous species are included here.

Key to Species (4):
1. Plants annual, inhabiting low, moist, and frequently saline soils .... *A. subulatus*
1. Plants perennial, habitats various 2
    2. Most of the stem-leaves, but at least the basal and lower leaves, on noticeable petioles and also cordate or somewhat so at the base 3
        3. Leaves nearly or quite entire or with only shallow inconspicuous teeth; both surfaces of leaves rough-hairy; involucral bracts with diamond-shaped green tips .... *A. azureus*
        3. At least the lower leaves toothed; upper surface of leaves more or less rough-hairy, the lower surface soft-hairy; involucral bracts with elongated green tips .. *A. drummondii*
    2. Basal leaves not both petioled and cordate at the base, the other leaves sessile 4
        4. Both sides of leaves covered with silvery or gray silky hairs .... *A. sericeus*
        4. Leaves glabrous (without hairs) or more or less hairy, but not silvery-silky 5
            5. Stem glabrous or nearly so, or sometimes the upper portion more or less hairy in lines 6
                6. Base of the stem-leaves rounded or heart-shaped, strongly clasping or partly encircling the stem 7
                    7. Stem and leaves often or mostly glaucous (with a white waxy "bloom" which can be rubbed off), sometimes green or greenish; bracts of involucre acute .... *A. laevis*

A. azùreus Lindl.   **Aster**

L. *azureus*, bright blue.
Rays blue, disk yellow; mid-September–early November

Open woods, roadside banks, and prairie
East fourth

Erect herbaceous perennial with a short rhizome. Stem up to 1 m tall, simple below, with several ascending or spreading branches above. Leaves rather thick and firm, rough-hairy, the margins entire or sometimes inconspicuously serrate, the lowermost long-petioled, mostly with lanceolate blades 5–15 cm long and 1–4.5 cm wide, the upper leaves gradually becoming elliptic or linear-elliptic and tapered to a sessile base. Heads 1.5–2 cm across, the inflorescence branches with many tiny, subulate bracts. Involucre 6–10 mm long, the bracts with diamond-shaped green tips. Rays blue (rarely white or pink), 7–13 mm long. Achene body pale, 3- to 5-nerved, 2–2.5 mm long, the pappus 5–6 mm long.

A. drummóndii Lindl.   Drummond aster

Named after James Drummond, English botanist who, in the second quarter of the 19th century, explored the Hudson highlands, Texas, and Louisiana.
Rays blue to purplish, disk yellow; August–October
Woods and thickets
East two-fifths

Herbaceous perennial with one to several stems 3–13 dm tall arising from a fibrous-rooted rhizome. Leaf blades 4–14 cm long and 2–6.5 cm wide, the margins serrate (sometimes entire on the uppermost), the upper surface more or less roughened with short, stiff hairs, the lower surface with soft or sometimes rough hairs, the lowermost blades cordate and on slender petioles, those higher up with shorter winged petioles, varying to lanceolate and sessile above. Heads 10–15 mm across. Involucre 3–5 mm long, the bracts with elongate green tips. Rays blue, 6–9 mm long. Achene body straw-colored, glabrous to sparsely hairy, about 2 mm long, the pappus about 5–6 mm long.

A. ericòides L.   Heath aster

NL., from L. *erice*, a name which was used by Pliny for the heath plant, and from which *Erica*, the name of the heath genus, is derived, + -*oides*, resembling—in reference to the spreading habit of growth, the many slender flowering branchlets with small to almost scalelike leaves, often racemose along one side of the stem like a heath raceme.
Rays white, disk yellow; mid-September–October
Prairie, pastures, roadsides
Throughout

Herbaceous perennial with well-developed creeping rhizomes. Stems slender, often clustered, ascending to nearly prostrate, 2–10 dm tall, covered with appressed hairs, at least above. Lower and middle stem leaves usually gone by flowering time, the upper ones linear or subulate, more or less rough-hairy, 5–20 (–30) mm long and 1–5 mm wide, the apex bristle-tipped, the bases straight or slightly rounded, never clasping. Heads 8–17 mm across. Involucre 3–5 mm long, the bracts linear-acute to linear-acuminate, with an elliptic or oblanceolate green area at the tip, bristly-hairy along the margins but never glandular. Rays white (rarely pink), 5–8 mm long. Achene body silky, about 1–1.5 mm long, the pappus 3–5 mm long.

This is the most common, hardiest, and smallest-flowered of our asters. In addition to the fibrous roots along the rhizome, additional longer roots go down 1–3 m, depending on the character and water content of the soil. Far from being suppressed by the drought of the thirties, the heath aster became more widespread. Once established, it is difficult to eradicate because of the prolific branching of its persistent horizontal rhizomes.

A. féndleri A. Gray

Named in honor of Augustus Fendler (1813–1883), a German immigrant who collected plants in New Mexico, Kansas, and Nebraska for Asa Gray.
Rays blue or purplish, disk yellow; late June–mid-October
Dry, rocky prairie banks and hillsides
West half, primarily

Perennial from a gnarled, woody caudex. Stems several, erect to ascending or decumbent, hairy (at least in the upper parts), 0.7–3.5 dm tall. Leaves firm, linear, 6–55 mm long and 6–11 mm wide, the bases straight, the margins conspicuously ciliate. Heads 1.5–3 cm across. Involucre 5–8 mm long, the bracts rather thick, glandular-hairy, the outer ones largely herbaceous, often recurved. Rays blue or purplish, 8–13 mm long. Achene body 2–3 mm long, the pappus off-white, 3–4.5 mm long.

See plate *120.*

608. *Aster drummondii*

609. *Aster ericoides*

610. *Aster ericoides;* note in this specimen and the previous one the amount of variation which may exist among individuals

ASTER TRIBE

A. nòvae-ángliae L.   New England aster

NL., from *nova*, new, + *Anglia*, England.
Rays purple to rose-purple; September–October
Low, moist ground
Scattered in the east fourth, mostly in the extreme northeast

Stout, erect, coarsely hairy perennial, 5–12 dm tall, with a stout rhizome. Lower leaves usually gone by flowering time, those of the middle and upper portions of the stem numerous and crowded, lanceolate to lance-oblong or lance-linear, mostly 4–9 cm long and 7–17 mm wide, the bases auricled and clasping the stem, the upper surface usually darker than the lower. Heads quite showy, about 2.5–3.5 cm across, the individual peduncles usually with 1–3 small bracts. Involucre 7–10 mm long, the bracts mostly linear-attenuate, at least the outermost glandular-hairy and herbaceous for most of their length. Rays purple to rose-purple, 1–2 cm long. Achene body densely silky, about 2 mm long, the pappus off-white or purplish, about 5 mm long.

This, our handsomest aster, we now find most frequently in the grass strips beside roads, especially where the group is low or where seepage occurs from ponds and drainage ditches. Although it grows wild here, it is also grown in nurseries to supply a demand of landscape designers and gardeners in this country and abroad.

611. *Aster fendleri*

A. laèvis L.

L. *laevis*, smooth—descriptive of the stems and leaf surfaces.
Rays blue to purple, disk yellow or reddish; September–mid-October
Rocky banks and hillsides
East third, more common northward

Erect perennial, glabrous or nearly so, with stems 3–15 dm tall arising from a rhizome. Leaves rather thick and firm, smooth except along the margins, glaucous, sessile (those of the new vegetative rosettes tapered to the winged, petiolelike bases), the blades variable in shape but generally narrowly elliptic to lance-linear, less frequently lanceolate or oblanceolate, 3.5–16 cm long and 0.5–3.5 cm wide, the bases strongly clasping the stem above, the margins entire or very shallowly serrate. Heads 2–2.5 cm across. Involucre 5–10 mm long, the bracts linear-acute, with diamond-shaped green areas at the tip. Rays blue to purple, 8–15 mm long. Disk flowers yellow or reddish. Achene body glabrous, 2–2.5 mm long, the pappus 4–5 mm long, off-white or sometimes reddish.

A. oblongifòlius Nutt.   Aromatic aster

NL., from L. *oblongus*, rather long, + *folium*, leaf.
Bluish-purple; September–October
Prairie, prairie pastures, and roadsides
East two-thirds and a few counties in the north half of the west third

612. *Aster paludosus* ssp. *hemisphericus*

Erect, slender-branched perennial, glandular-hairy and aromatic above, 1–8 dm tall, with long rhizomes. Lower and middle leaves usually gone by flowering time, the upper ones oblong to linear-elliptic, 7–48 mm long and 1.5–8 mm wide, the bases rounded or sometimes slightly clasping the stem. Heads 2–3 cm across. Involucre 5–6 mm long, the bracts with green, glandular herbaceous tips, spreading or recurved. Rays bluish-purple (rarely rose), 9–14 mm long. Achene body more or less hairy, about 2 mm long, the pappus off-white or purplish, about 5 mm long.

A. paludòsus Ait.

L. *paludosus*, boggy, marshy—in reference to the low, moist habitat in which these plants are
     sometimes found.
Rays blue, disk yellow; mid-August–mid-October
Usually in low, moist prairie
Crawford, Cherokee, and Labette counties

Herbaceous perennial with a scarcely branched stem 2–8 dm tall arising from a thickened creeping rhizome. Leaves relatively thick and firm, rough, usually strongly ascending, linear to linear-elliptic, 5–11 cm long and 3–10 mm wide, the apex acute to attenuate, the margins entire. Heads 3–5 cm across, borne singly in the upper leaf axils. Involucre 9–12 (–15) mm long, the bracts relatively broad, straw-colored, and rather hard below the middle, definitely herbaceous toward the tip, ascending or sometimes recurved. Rays blue, 16–20 mm long. Achene body glabrous or minutely hairy, 3–3.5 mm long, the pappus brownish-white, 7–9 mm long.

Our plants are recognized as subspecies *hemisphaericus* (Alex.) Cronq. Some authorities place them in a separate genus as *Heleastrum hemisphaericum* (Alex.) Shinners.

A. párviceps (Burgess) Mack. & Bush.

NL., from L. *parvus*, little, + the suffix *-ceps*, -headed.
Rays white; September–October
Open woodlands, prairie, pastures, and roadsides
Southeast eighth and Douglas County

Slender, erect perennial, 2–12 dm tall, with a short caudex and fibrous roots. Stems hairy in lines. Lower leaves mostly gone by flowering time, the middle and upper leaves linear-elliptic to linear or awl-like, 1–6 cm long and 1–3.5 mm long, usually glabrous except for the ciliate margins. Heads 8–10 mm across. Involucre 3–5 mm long, the bracts linear-attenuate, the margins of the tips often inrolled. Rays white, 3–6 (–7) mm long. Achene body minutely hairy, the pappus off-white, about 3 mm long.

A. pàtens Ait.

L. *patens*, open, exposed.
Rays blue; mid-August–October
Rocky, open woods over sandstone, less frequently in prairie
Southeast eighth

Slender, erect perennial, 2.5–10 dm tall, with a short, erect rhizome or rootstock and sometimes slender, creeping horizontal rhizomes also. Stem covered with appressed hairs. Middle and upper leaves lanceolate to lance-oblong, 1.5–5.5 cm long and 1–2 cm wide, the bases auricled and clasping the stem. Heads 2–3 cm across. The peduncles with several to many small bracts. Involucre 6–12 mm long, the bracts grading from linear-acute to linear-acuminate, hairy but not glandular. Rays blue (rarely white or pink), 9–15 mm long. Achene body 2–3 mm long, covered with short, silky hairs, the pappus off-white or purplish, 4–5 mm long.

A. pilòsus Willd.

L. *pilosus*, hairy, from L. *pilus*, Gr. *pilos*, hair.
Rays white, disk yellow; September–October
Rocky prairie and roadsides, occasionally in open woods
East fourth, primarily

Erect herbaceous perennial, 2–15 dm tall. Stems rough-hairy. Leaves lance-linear to linear-elliptic, 2–10 cm long and 1.5–20 mm wide, those of the stem below the inflorescence often dried or gone by flowering time. Heads 12–20 cm across, the inflorescence branches with many tiny, subulate bracts. Involucre 4–6 mm long, the bracts linear-acuminate, green at the tip, at least the outermost with minute spinelike tips. Rays white (rarely pink or purple), 6–11 mm long. Achene body straw-colored, sparsely and minutely hairy, 1–1.5 mm long, the pappus white, 3–3.5 mm long.

A. praeáltus Poir.

L. *praealtus*, very tall.
Rays bluish-purple, disk yellow; mid-September–October
Moist soil of creek banks, roadside ditches
East fourth, primarily

Erect herbaceous perennial, often forming dense colonies by means of its creeping rhizomes. Stems 3–13 dm tall, simple below but eventually becoming much branched above, usually with lines of minute hairs. Leaves rather thick and firm but not glaucous, smooth except along the margins, lance-linear to narrowly elliptic, mostly 2–11 cm long and 2–25 mm wide, the veins of the lower surface conspicuous, demarking areas about as long as wide (use magnification), the margins entire. Heads 1.5–2 cm across, the inflorescence leafy-bracted. Involucre 4–7 mm long, the bracts narrowly linear with an elongate green area near the acute or acuminate tips. Rays bluish-purple (rarely white), 5–10 mm long. Achene body glabrous or sparsely hairy, about 1.5 mm long, the pappus 1.5–2 mm long, white, longer than the disk corollas.

See plate *121*.

A. seríceus Vent.

L. *sericeus*, silken, in reference to the silky hairs on the leaves and bracts.
Rays reddish-purple to blue, disk yellow; late August–October
Prairie hillsides and uplands
East two-fifths, but absent from the southeast and extreme northeast and east-central counties

Erect herbaceous perennial with wiry stems 1–7 dm tall arising from a woody caudex. Leaves of the main stem usually gone by flowering time, those of the inflorescence

613. *Aster patens*

614. *Aster pilosus*

615. *Aster praealtus*

branches lance-oblong to elliptic, 1–3 cm long and 2–7 mm wide, densely covered with silky silvery hairs. Heads 1.5–3 cm across. Involucre 6–11 mm long, the bracts relatively broad, silky, rather hard, and straw-colored below the middle, definitely herbaceous toward the tip, appressed or sometimes recurved. Rays reddish-purple to blue, 10–15 mm long. Achene body glabrous, 5-veined on each face, about 3 mm long, the pappus bristles brownish-white, about 6 mm long.

A. símplex Willd.

L. *simplex*, onefold, unmixed, single, simple.
Rays white, disk yellow or reddish; September–October
Moist soil of wooded creek banks and roadside ditches
East two-thirds, primarily; scattered in the west third

Erect perennial with stems 0.4–1.5 m tall, forming colonies by means of its long creeping rhizomes. Stem glabrous below, with lines of minute hairs above (use magnification). Leaves lanceolate to elliptic or nearly linear, 2–16 cm long and 3–21 mm wide, the margins entire or shallowly serrate, sometimes ciliate, the upper leaves definitely sessile, the lowermost tapered to a winged, petiolelike base. Heads 10–20 mm across. Involucre 4–7 mm long, the bracts narrow, sharply pointed, with an elongate green area near the tip. Rays white (occasionally purplish or blue), 9–12 mm long. Disk corollas yellow or occasionally reddish or purplish, 3–4 mm long, the lobes about 1–1.5 mm long. Achene body hairy, the pappus 3–6 mm long.

Plants listed in earlier Kansas manuals as *A. paniculatus* belong to this species.

A. subulàtus Michx.

L. *subulatus*, awl-shaped, pointed.
Rays pale blue or purplish; late July–October
Low, wet or moist soil of river banks, marshes, ditches, and swales
South-central sixth, primarily; but scattered east, north, and west to Lyon, Republic, and Finney
    counties, respectively

Erect, glabrous annual, 2.5–10 dm tall, with a short taproot. Leaves linear-elliptic to linear, 2–13 cm long and 2–12 mm wide. Heads 12–18 mm across. Involucre 5–6 mm long, the bracts linear-attenuate, with a green midrib and tip, the margins of the tips mostly inrolled. Rays pale blue or purplish, 4–8 mm long. Achene body sparsely hairy, 2–3 mm long, the pappus 3–4 mm long.

BOLTÒNIA L'Her.   Boltonia

Named after James Bolton, 18th-century English botanist.

Erect, short-lived perennials with taproots and slender rhizomes. Leaves alternate, simple, entire, mostly sessile. Heads small, many-flowered, radiate, several to many, borne singly at the ends of the inflorescence branches. Involucre hemispheric, the bracts in about 3 series, about equal or graduated in length, with dark midribs and membranous margins. Receptacle hemispheric or conic, naked. Pappus of several minute teeth or scales and 2–4 longer scales or awns. Ray flowers many, female, fertile, the corollas pale purple or white (sometimes drying pink). Disk flowers many, bisexual, fertile; corollas funnelform, shallowly 5-toothed, yellow; anthers rounded at the base; style branches flattened, the tips short-lanceolate, hairy. Achenes more or less obovate, strongly flattened, 2-winged (those of the ray flowers 3-winged).

One species in Kansas.

B. asteroìdes (L.) L'Her.   White boltonia

NL., from the genus *Aster*, in the Composite Family + the Gr. suffix *-oides*, similar to.
Rays white to pinkish, disk flowers yellow; August–October
Moist or wet soil of marshes, ditches, stream banks, and low meadows
East third, primarily; west to Barton and Stafford counties in the central part of the state

Glabrous plants, 3–15 dm tall, with longitudinally ridged stems much branched above. Leaves mostly very narrowly elliptic to linear, 4.5–12 cm long and 4–13 mm wide, the apex usually acute or attenuate, the margins minutely roughened, the lower leaves often gone by flowering time. Heads many, mostly 1.5–3 cm across when fully expanded, the involucres 3–6 mm long, the bracts narrow and acute or gradually acuminate in var. *recognita* or broadly spatulate with an abruptly short-acuminate tip in var. *latisquama* (NL., from L. *latus*, broad, + *squama*, scale). Rays white to pale purple or pink, 10–17 mm long. Disk corollas yellow. Achene body 2–2.5 mm long, the pappus of 2 well-developed awns and a few much smaller bristles, 1–1.5 mm long.

616. *Aster sericeus*

617. *Aster simplex*

618. *Aster subulatus*

ASTER TRIBE

425

619. *Conyza canadensis*

The chief difference between the *Asters* and the *Boltinias* is in the pappus, that of the former consisting of many long, hairlike bristles.

CONỲZA L.   Horseweed

L. *conyza*, Gr. *konyza*, ancient name for the fleabane.

Coarse annual herbs. Leaves alternate, simple, entire or toothed in our species. Heads small, many-flowered, functionally radiate (though the ray flowers may sometimes lack ligules and differ from the disk flowers only in being female rather than bisexual), several to many, often borne in terminal clusters. Involucre cylindric to top-shaped, the bracts in 2–3 series, narrow, about equal in length, thin and herbaceous, ours with membranous margins. Receptacle flat or nearly so, naked. Pappus of fragile, hairlike bristles in ours. Ray flowers many, in 1 row in ours, female, fertile, the corolla cream-colored, with a threadlike tube and a minute ligule. Disk flowers several, bisexual, fertile; corolla cream-colored, tubular, 5-toothed; style branches with short appendages. Achenes flattened and 2-ribbed, or the outer ones 3-angled and 3-ribbed.

Two species in Kansas.

Key to Species:

1. Plants 0.4–2.5 m tall, with a stout stem unbranched below the well-defined, many-headed terminal inflorescence .... *C. canadensis*
1. Plants 0.5–3 dm tall, with a slender, much-branched stem, the heads mostly borne singly in the leaf axils towards the ends of the branches .... *C. ramosissima*

C. canadénsis (L.) Cronquist   Horseweed fleabane

L. *-ensis*, suffix added to nouns of place to make adjectives—"Canadian," in this instance.
White; late July–September
Fields, roadsides, overgrazed pastures, and waste places
Throughout

Erect, coarsely pubescent plants with an erect stem 0.4–2.5 m tall, unbranched below but with many slender, ascending branches in the terminal inflorescence. Leaves very narrowly elliptic to linear-attenuate, more or less densely crowded on the stem, 3–10 cm long and 2–10 mm wide, entire or with a few teeth. Heads many, in branched terminal clusters. Involucre 2–3 mm long, the bracts linear-attenuate. Flowers white, the rays about 0.5 mm long, very inconspicuous. Achene body straw-colored, 1–1.5 mm long, the pappus of fragile, hairlike bristles about 2 mm long.

The herbage and inflorescence of this plant have a strong, pungent smell caused by a volatile terpene oil which is secreted by countless dotlike glands. The oil may afflict particularly sensitive persons with irritation, or even inflammation, of the skin and of the mucous membranes of the mouth, nose, and eyes. The plant also contains a bitter principle and tannic and gallic acids, giving to infusions and decoctions of the roots and herbage astringent and styptic properties which the Indians made use of. Later the white settlers used the plant for the treatment of internal hemorrhage, diarrhea, and dysentery.

This species is listed in earlier manuals for this area as *Erigeron canadensis*. Under that name, the leaves and flowers were listed officially in the *U.S. Pharmacopoeia* from 1820 to 1882. Erigeron oil was official from 1863 to 1916.

C. ramosíssima Cronq.   Low horseweed

L. *ramosissimus*, with the most branches, from L. *ramosus*, full of branches, + the intensive suffix *-issimus*.
Pale purple; June–mid-October
Fields, roadsides, overgrazed pastures, and waste places
East two-thirds, primarily; several records in the south half of the west third

Low, much-branched, weedy annual, 5–30 cm tall. Leaves mostly linear, 8–35 mm long and 0.5–2 mm wide. Herbage sparsely covered with long white hairs. Heads borne on threadlike axillary peduncles near the ends of the branches. Involucre 3–4 mm long, the bracts linear-attenuate. Rays pale purple, scarcely longer than the pappus. Achene body minutely hairy, 1–1.5 mm long, the pappus 2–2.5 mm long.

This plant is listed in earlier manuals for our area as *Erigeron divaricatus*.

In sandstone areas of the Chautauqua Hills and in Cherokee County, one may encounter occasionally *Chaetopappa asteroides* (Nutt.) D.C., a small taprooted annual resembling low horseweed. Our *Chaetopappa* differs from low horseweed in having conspicuous white rays and yellow disk flowers, oblong involucral bracts with ciliate tips, and a pappus with an outer whorl of white scales and an inner whorl of stiff bristles.

CROPTILON Raf.   Scratch-daisy

NL., possibly from Gr. *chroa*, skin color, or *chroo*, to stain or color, + *ptilon*, down, in reference to the brownish pappus bristles.

Small annuals, often hairy or glandular-hairy. Leaves alternate, simple, oblanceolate, sessile. Heads several- to many-flowered, radiate, borne singly or in branched clusters. Involucre narrowly top-shaped, the bracts in several series, graduated in length, linear, each with a herbaceous median strip, dry, membranous margins, and an acute or acuminate tip. Receptacle flat or slightly convex, naked, very rough. Pappus of 1 series of off-white or reddish-brown bristles. Ray flowers about 5–30, female, fertile, the corollas yellow. Disk flowers 10–40, bisexual, fertile; corolla tubular, 5-toothed, yellow; anther sacs rounded basally; style branches with elongate, linear-acute appendages. Achenes varying from linear to narrowly top-shaped, sometimes somewhat angled, crowned by the persistent pappus bristles.

One species in Kansas.

620. *Croptilon divaricatum*

C. divaricàtum (Nutt.) Raf.

L. *divaricatus*, spread apart, probably in reference to the open branching habit of the plant.
Yellow; July–October
Sandy soil of prairie hillsides, dunes, flood plains, and roadsides
Central third, primarily, especially southward

Glandular-hairy, taprooted annuals, 1.5–8 dm tall. Leaves narrowly oblanceolate or elliptic below, becoming elliptic or narrowly oblong to lance-linear above, 3–9.5 cm long and 0.5–1.7 cm wide, the margins coarsely and distantly serrate, at least toward the tip, the lower surface whitish and more or less glandular-hairy, the upper surface usually glabrous and smooth. Heads 12–25 mm across, few to many in an open, widely branched terminal inflorescence. Involucre 5–10 mm long. Rays yellow, 9–15 mm long. Achene body 2–3 mm long, covered with silky hairs, the pappus light orangish-brown, 3–4 mm long.

ERÍGERON L.   Fleabane

L. and Gr. name for the groundsel, *Senecio vulgaris*, in the composite family, and applied to the fleabanes by Linnaeus; from Gr. *eri*, early, + *geron*, old man—the hoary pappus being conspicuous soon after the flowers fade. "Fleabane" is an old English name applied to *E. acris* because of its supposed power to destroy or banish fleas.

Annual, biennial, or perennial herbs. Leaves alternate, simple, entire to toothed or lobed, sessile or narrowed to petiolelike bases. Heads many-flowered, radiate, borne singly or in terminal clusters. Involucre hemispheric, the bracts in 2–3 series, about equal in length, mostly very narrowly oblanceolate to linear, the tips acute to acuminate or attenuate, with a narrow or wide membranous margin. Receptacle flat or convex, naked. Pappus of unequal, hairlike bristles in 2 series, or of short scales and hairlike bristles, or short scales only, or absent. Ray flowers many, in about 2 series, female, fertile, the ligules linear, white to pink or pale blue. Disk flowers many, bisexual, fertile; corolla short-tubular, 5-lobed, yellow; anthers rounded at the base; style branches flattened, the extensions beyond the pollen-receptive portion short, lanceolate, and acute to broadly triangular and rounded, sometimes absent. Achenes flattened, 2-ribbed (or the outermost 3-ribbed), crowned by the persistent pappus.

Eight species in Kansas. Our 5 most common species are included in the following key. *E. pumilus* Nutt., our only long-lived perennial with a branched, definitely woody caudex, is known from scattered sites in the northwest fourth.

Key to Species:
1. Pappus of ray and disk flowers unlike, that of the disk flowers composed of bristles and
   short outer setae, that of ray flowers lacking the bristles 2
   2. Foliage ample; plant mostly 6–15 dm tall; main leaves mostly (6–) 8–45 mm wide .. *E. annuus*
   2. Foliage sparse; plant mostly 3–7 dm tall; main leaves mostly 2–15 mm wide .... *E. strigosus*
1. Pappus of the ray flowers and disk flowers alike, of bristles, sometimes also with an outer
   whorl of bristles or scales 3
3. Pappus double .... *E. modestus*
3. Pappus simple 4
      4. Moderately to diffusely branched annual with linear or linear-oblong leaves; hairs of
         the stem characteristically short and incurved .... *E. bellidiastrum*
      4. Simple or moderately branched biennials or short-lived perennials (very rarely
         annual) with relatively broad leaves; hairs of the stem not characteristically short
         and incurved .... *E. philadelphicus*

621. *Erigeron annuus*

622. *Erigeron philadelphicus*

E. ánnuus (L.) Pers.   Annual fleabane

L. *annuus*, of a year's duration, from *annus*, year—indicating the supposed length of life of the plant.
Rays white, disk yellow; mid-May–October
Disturbed habitats such as roadsides, fields, pastures, and flood plains, also invading prairies and woodlands
East two-fifths

Erect biennial, the first year producing a rosette of long-petioled, elliptic to ovate, coarsely serrate leaves; the second year, an upright stem 1.5–12 dm tall and branched above. Leaves 3–13 cm long and 0.5–4.5 cm wide, the lower ones long-petioled, with blades mostly ovate to elliptic or oblanceolate, coarsely serrate, at least toward the tip, the upper ones gradually becoming sessile and smaller, narrowly oblanceolate to lance-linear, entire or serrate. Herbage sparsely to moderately hairy. Heads several to many, mostly 1.5–2 cm across. Involucres 3–5 mm long. Rays white (often turning blue on drying), 5–8 mm long. Achene body hairy, 2-ribbed, about 1 mm long, the pappus 1–2 mm long, those of the disk flowers with an outer series of very slender, very short scales and an inner series of fragile bristles, those of the ray flowers with scales only.

E. bellidiástrum Nutt.

NL., from L. *bellidis*, daisy, + *-astrum*, diminutive suffix with derogatory implications, wild, resembling but inferior to a cultivated daisy.
Rays white or pink, disk yellow; mid-May–early October
Sandy soil of prairies, roadsides, pastures, and flood plains
Southwest fourth

Erect, moderately- to much-branched annual, 0.5–5 dm tall. Leaves 1–4.5 cm long and 1–10 mm wide, oblanceolate to spatulate or linear, entire. Heads several to many, mostly 1.5–2 cm across when fully expanded. Involucre 2–5 mm long, the bracts hairy. Herbage with incurved hairs. Rays white or pink, 6–9 mm long. Achene body short-hairy, about 1.5 mm long, the pappus with a single series of fragile bristles about 2 mm long.

E. modéstus Gray

L. *modestus*, moderate, gentle, unassuming, modest.
Rays white or pink, disk yellow; May–June (–July)
Dry, rocky or sandy prairies and pastures
West two-fifths, especially southward

Erect, hairy annual, biennial, or short-lived perennial with a small, very short, branched caudex and several stems 1–5 dm tall. Leaves mostly 1–5 cm long and 1–5 mm wide, the basal leaves long-petioled, oblanceolate to spatulate, the middle and upper leaves smaller, mostly linear or nearly so and 1–2 (–3) mm wide. Heads 1 to several, 1.5–2 (–2.5) cm across. Involucre 4–5 mm long, the bracts hairy. Rays white or pink, 6–9 mm long and less than 1 mm wide. Achene body sparsely hairy, about 1 mm long, the pappus about 2 mm long, with an outer series of short bristles or narrow scales and an inner series of fragile bristles.

E. tenuis T. & G., known from a few counties in the extreme southeast, will key out here. It differs from *E. modestus* in having broader leaves (those of the stem 3–6 mm wide) and shorter, broader rays.

E. philadélphicus L.   Philadelphia fleabane

"Of Philadelphia."
Rays white to pink or pale purple, disk yellow; mid-April–early June, sometimes again in late summer or fall
Fields, roadsides, waste ground, and other disturbed habitats
East third, primarily

Erect biennial or short-lived perennial with 1 to few stems 1.5–7.5 dm tall, branched above, lacking slender rhizomes. Leaves 2.5–15.5 cm long and 0.8–4 cm wide, the lower ones more or less long-petioled, the blades oblanceolate to spatulate, mostly coarsely serrate or crenate, the middle and upper leaves sessile with clasping bases, oblanceolate to oblong or lance-oblong, the margins coarsely serrate or entire. Heads several to many, mostly 1.5–2.5 cm across. Involucres 3–6 mm long, the bracts linear, with a ciliate midrib. Herbage more or less hairy. Rays white to pink or pale purple, 8–10 mm long and 0.5 mm wide or less. Achene body sparsely hairy, about 1 mm long, 2-ribbed, the pappus of a single series of fragile bristles 1–3 mm long.

E. pulchellus Michx., known only from Cherokee County, will key out here with

E. philadelphicus. The former tends to have larger heads (mostly 2–3 cm across) with rays 1 mm wide or more, has most of the leaves in a basal rosette, and spreads by means of slender rhizomes.

Philadelphia fleabane was listed officially in the *U.S. Pharmacopoeia* from 1831 to 1882 as a stimulant and diuretic.

E. strigòsus Muhl.   Daisy fleabane

L. *strigosus*, thin, lean, meager—in reference to the skimpy leaves.
Rays white, disk yellow; May–September
Upland prairie and weedy pastures, frequently in rocky soil
East three-fourths, primarily; scattered records in the west fourth

Slender, erect annual, 2–10 dm tall, strongly resembling *E. annuus* but typically with narrower, and often fewer, leaves. Leaves 3–12 cm long and 2–15 mm wide, the lower ones long-petioled, with blades narrowly oblanceolate to nearly linear, the margins entire or with a few teeth, the upper ones linear or narrowly elliptic, gradually becoming sessile and smaller, the margins usually entire. Flowers and achenes as for *E. annuus*.

This plant has been listed in earlier Kansas manuals as *E. ramosus*.

GRINDÈLIA Willd.   Gumweed

623. *Erigeron strigosus*

Named after David Hieronymus Grindel (1776–1836), professor of chemistry and pharmacy in Dorpat, Estonia, and writer on pharmacal and botanical subjects.

Annual, biennial, or perennial herbs with taproots, sometimes woody at the base. Leaves alternate, simple, often toothed and gland-dotted, sessile. Heads many-flowered, radiate (rarely discoid), borne at the ends of the branches. Involucre hemispheric, the many bracts in 4–8 series, graduated or nearly equal in length, linear, ascending, spreading, or strongly reflexed. Receptacle flat or convex, naked but rather rough. Pappus of 2–10 deciduous awns or scales. Ray flowers mostly 15–45 (rarely absent), female, fertile, the ligules yellow. Disk flowers many, bisexual, the innermost and sometimes the outermost infertile; corolla yellow, funnelform, 5-lobed; anther sac rounded at the base; style branches flattened, the tips lance-linear or sometimes very short, hairy. Achenes oblong to obovoid, slightly compressed, smooth.

Two species in Kansas.

NOTE: See also *Prionopsis.*

Key to Species:
1. Involucral bracts not noticeably graduated in length, the tips not recurved; leaf margins with bristle-tipped teeth; plants of southeast Kansas .... *G. lanceolata*
1. Involucral bracts noticeably graduated in length, the tips recurved; teeth of the leaf margins not bristle-tipped; plants primarily of the west half and northeast fourth ....
.... *G. squarrosa*

G. lanceolàta Nutt.   Narrowleaf gumweed

L. *lanceolatus*, armed with a little lance, from *lanceola*, little lance—in reference to the leaf shape.
Yellow; August–mid-October
Dry soil of prairies, pastures, and roadsides
Southeast eighth

Glabrous, erect biennial or short-lived perennial, 2.5–11 dm tall, with a deep root, the main stems with long, ascending branches. Leaves narrowly oblanceolate below to narrowly elliptic, oblong, linear or lance-linear above, 3–8.5 cm long and 3–16 mm wide, the margins serrate, the teeth minutely bristle-tipped. Heads 3–4 cm across, the involucre 9–16 mm long, the bracts gland-dotted, rather loose, not conspicuously graduated in length, mostly with the tips erect or spreading somewhat, the outermost occasionally recurved. Rays yellow, 10–15 mm long. Disk flowers yellow. Achene body 4–6 mm long, the pappus of 2 slender, stiff, entire awns 4–7 mm long which soon drop off.

G. squarròsa (Pursh) Dunal   Curlycup gumweed

L. *squarrosus*, scabby, scaly, or roughened by the spreading tips of scales—in reference to the reflexed bracts of the involucre.
Yellow; late July–early October
Dry soil of prairies, pastures, and roadsides
Primarily in the west half and northeast fourth, scattered records from the southeast fourth

Glabrous biennial or short-lived perennial with a short vertical rhizome which gives off smooth, erect stems 1.8–8 dm tall, and a vertical, branching root which descends

624. *Grindelia squarrosa*

625. *Heterotheca latifolia*

to about 2 m. Leaves conspicuously gland-dotted, spatulate or oblanceolate below to mostly oblong or lance-oblong above, 1.5–6.5 cm long and 6–17 mm wide, the apex rounded or acute, the bases of the upper leaves cordate and clasping the stem, the margins serrate or crenate, the teeth gland-tipped. Heads 2.5–4 cm across, the involucre 6–10 mm long, the bracts resinous, conspicuously graduated in length, the tips strongly recurved. Rays absent in var. *nuda*, present, yellow, and 9–14 mm long in var. *squarrosa*. Disk flowers yellow. Achene body 2–3 mm long, ribbed, the pappus of 2–8 slender, entire or very minutely toothed awns 3–5 mm long which soon fall away.

The flower heads are very sticky because of resinous excretions from the several rows of glandular-dotted involucral bracts with strongly reflexed or hooked tips.

The species is quite drought-resistant, owing to its deep root and resinous secretions. Because of its unpalatability to cattle and sheep it often takes possession of run-down pastures during protracted periods of drought. Then it is that in western Kansas we may see it in societies miles wide over the plains—an impressive sight.

The unpleasant taste of plants is usually due to their content of tannins, volatile oils, resins, bitter alkaloids and glucosides, some of which are valuable to us as stimulants, sedatives, astringents, purgatives, emetics, diuretics, antiseptics, disinfectants, etc. The *Grindelias*, having secretions of tannin, volatile oil, a bitter saponin, and 3 kinds of resin, were used by our native Indians for asthma and bronchitis and for colic in children. The Pawnees boiled the flowering tops and leaves and used the decoction for bathing saddle sores and other rawness of the skin. Our own medical practice, following the lead of the Indians, has employed fluid extracts of the leaves and buds for bronchial spasm, whooping cough, and asthma, and the powdered drug for making asthma cigarettes.

It is worthy of note that the bees, both wild and domesticated, have their own use for the *Grindelia*, storing the comb with its nectar and pollen, untroubled by the fact that we find honey from it too strong in taste and too prone to granulate. Then, in addition to the values for man and bees related above, our *G. squarrosa* is desirable in the flower garden for the show of golden flowers it provides over a long period, without requiring solicitous attention, even when the soil is poor and dry.

Also of interest is the fact that the plants are facultative selenium absorbers—that is, they absorb selenium if it is present but do not absolutely require it.

HETEROTHÈCA Cass.   Heterotheca

NL., from Gr. *heteros*, di–erent, + *theke*, box or case—the achenes of the ray flowers being thickened while those of the disk flowers are flattened.

Hairy, aromatic annual or perennial herbs. Leaves alternate, simple, entire or toothed, usually petioled below and sessile or clasping above. Heads many-flowered, radiate (rarely discoid), borne in branched terminal clusters. Involucre broadly bell-shaped to narrowly top-shaped, the bracts in 3–9 series, graduated in length, the inner ones with dry, membranous margins. Receptacle flat or slightly convex, naked. Pappus in ours species with an inner series of many stiff, barbed, hairlike bristles and an outer series of few to many short bristles or scales, absent in the ray flowers of *H. latifolia*. Ray flowers 10–35, female, usually fertile, the corollas yellow. Disk flowers many, bisexual, fertile; corolla tubular, 5-lobed, yellow; anther sacs rounded basally; style branches flattened, the appendages hairy. Achenes glabrous or variously hairy, sometimes glandular.

Five species in Kansas. Two of our most common species are included here. Our other species strongly resemble *H. stenophylla* but are not treated here for lack of sufficient material to examine.

Key to Species:
1. Aromatic annual; middle and upper leaves 8–48 mm wide, clasping the stem; ray achenes glabrous and lacking a pappus .... *H. latifolia*
1. Tufted perennial with several to many stems; middle and upper leaves linear-elliptic to narrowly spatulate, 1–7 mm wide; all achenes silky .... *H. stenophylla*

H. latifòlia Buckl.

NL., from L. *latus*, broad, + *folium*, leaf, i.e., "wide-leaved."
Yellow; mid-July–mid-October
Prairies, pastures, and roadsides, usually in sandy or rocky soil
West two-thirds, primarily

Erect, glandular, aromatic annual or biennial. Stems simple below, branched above, 0.2–1 m tall, more or less densely covered with long, spreading nonglandular hairs intermixed with shorter glandular hairs. Leaves 1.5–9 cm long and 0.8–4.8 cm wide, the lowermost petioled or narrowed to a winged, petiolelike base, often gone by flowering time, the middle and upper leaves becoming definitely sessile and clasping, oblanceolate to lance-oblong, elliptic or ovate, the apex acute or mucronate, the margins entire or with a few teeth toward the apex, both surfaces conspicuously glandular. Heads several to many, 1.5–2.5 cm across, mostly in a more or less flat-topped terminal cluster. Involucre 5–9 mm long, the bracts linear-acute or linear-acuminate with a green midrib. Rays yellow, 8–13 mm long, lacking a pappus. Disk flowers yellow. Achene body 2–3 mm long, those of the ray flowers glabrous and lacking a pappus, those of the disk flowers covered with silky appressed hairs and with a pappus 4–5 mm long.

This species is listed in some earlier manuals for Kansas as *Heterotheca subaxillaris*.

H. stenophýlla (Gray) Shinners

NL., from Gr. *stenos*, narrow, + *phyllon*, leaf, i.e., "narrow-leaved."
Yellow; June–August
Dry prairies and plains
Primarily in the southwest and south-central counties

Herbaceous, glandular perennial with several to many slender, brittle, erect or ascending stems 2–4 dm tall arising from a thick woody taproot and crown. Lower leaves petioled, oblanceolate or spatulate, usually gone by flowering time, the upper leaves linear-elliptic or narrowly spatulate, 1–4 cm long and 1.7 mm wide, the apex acute, the margins entire, both surfaces hairy and glandular. Heads several to many, usually 1 to few together at or near the ends of the stems. Involucres 5–7 (–10) mm long, the bracts lance-linear. Rays yellow, 7–15 mm long. Disk flowers yellow. Achenes all alike, the body densely appressed-hairy, the pappus 5–7 mm long.

626. *Heterotheca stenophylla*

LEÙCELENE Greene

NL., from L. *leo*, lion, + Gr. *kelaines*, black, dark, murky.

Low, suffrutescent perennial herbs with 10 to many stems arising from a horizontal, rhizomelike root. Leaves alternate, simple, linear to spatulate, sessile, rather thick, ascending or appressed, those of the upper half much reduced in size. Heads small, few to many, radiate, borne singly at the ends of branches. Involucre hemispheric or top-shaped, the bracts in a few series, graduated in length, each with a green median strip and narrow white margins. Receptacle naked. Pappus of 20–30 hairlike bristles. Ray flowers many, female, fertile, the corollas white. Disk flowers many, bisexual, fertile, corollas tubular or somewhat funnelform, 5-toothed, yellow; anther sacs rounded basally; style branches with rounded appendages. Achenes long and slender, crowned by the persistent pappus.

The genus contains but a single species and in the past has been merged in the genus *Aster* by some authorities. It differs from most members of that genus in having (1) the creeping underground root and partly underground stems, rather than rhizomes or a woody caudex, and (2) involucral bracts without herbaceous tips, and also in blooming in both the spring and late summer or fall, rather than just late in the season.

627. *Leucelene ericoides*

L. ericòides (Torr.) Greene

NL., from L. *erice*, a name which was used by Pliny for the heath plant and from which *Erica*, the name of the heath genus, is derived, + Gr. suffix *-oides*, resembling—in respect to the spreading habit of growth, the many slender flowering branchlets with small or almost scalelike leaves, often racemose along 1 side of the stem (secund) like a heath raceme.
Rays white, disk yellow; late April–June, again in late summer or fall
Prairie
West half, primarily; east in the north to Clay County

Plants much branched from the base, 4.5–18 cm tall, the herbage more or less densely covered with long, appressed hairs. Leaves 3–13 mm long and 0.5–1 mm wide, rather thick and triangular in cross section, the margins with coarse, siliceous-based cilia. Heads 12–18 mm across, the involucre 5–7 mm long. Rays white, coiling under and often turning purplish or brown with age or on drying, 3–6 (–8) mm long. Disk yellow. Achene body hairy, the pappus 5–6 mm long, minutely barbed.

In sandy soils of the southeastern counties, one may occasionally find *Chaetopappa asteroides* (Nutt.) DC., a small-statured, taprooted annual superficially resembling *Aster ericoides*. Plants of the former have a basal rosette of broadly spatulate leaves and a few branches bearing widely spaced, more narrowly spatulate leaves. The pappus consists of 5 short membranous scales alternating with 5 much longer bristles.

MACHAERANTHÈRA Nees.

NL., from Gr. *machaira*, saber, + the NL. suffix *-anthera*, indicating anthers of a particular type—the anthers in plants of this genus being tipped with a sharp point.

Annual or perennial herbs. Leaves alternate, simple, entire to toothed or once or twice pinnately lobed, sessile. Heads many-flowered, radiate, borne singly or in terminal clusters. Involucre usually hemispheric, the bracts many, graduated in length, strawlike in color and texture, with a green midrib or green, elliptic or diamond-shaped area near the tip, erect, spreading, or recurved. Receptacle flat or slightly convex, naked, rough. Pappus of many slender, off-white, somewhat unequal bristles, sometimes absent in the ray flowers. Ray flowers many, female, fertile, the corollas usually yellow, white, or blue. Disk flowers many, bisexual, fertile; corollas funnelform, 5-toothed, yellow; anther sacs rounded at the base; achenes narrowly top-shaped to club-shaped or broadly linear, usually crowned by the persistent pappus of slender, subequal bristles.

Three species in Kansas as treated here. *M. annua* (DC.) Shinners is known from scattered sites, mostly in the southwest. It is a yellow-flowered annual, 2–9 dm tall, with remotely toothed, elliptic to oblanceolate leaves 1.5–7 cm long and 3–18 mm wide and involucral bracts with herbaceous, recurved tips.

628. *Machaeranthera pinnatifida*

Key to Species:
1. Leaves pinnately dissected; rays blue .... *M. tanacetifolia*
1. Leaves merely toothed; rays yellow .... *M. pinnatifida*

M. pinnatifida (Hook.) M. C. Johnst.   Cutleaf ironplant

NL., *spinulosus*, full of little thorns, from L. *spinula*, little thorn—in reference to the sharp tips of the divisions of the pinnatifid leaves.
Yellow; late May–October
Dry soil of prairies, pastures, and roadsides
West two-thirds

Glabrous perennial, 1–6 dm tall, with few to many erect or ascending stems arising from a woody crown and taproot frequently reaching depths of 1 m or more. Leaves crowded on the stem, pinnatifid, 1–4.5 cm long and 5–10 mm wide, the lobes bristle-tipped. Heads several to many, mostly 1.5–3 cm across (with the rays fully expanded). Involucre 4–7 mm long, the bracts linear, green near the appressed tips. Rays many, yellow, 7–13 mm long. Disk flowers yellow. Achene body minutely hairy, 2–2.5 mm long, the pappus 4–5 mm long, minutely barbed.

This plant has been listed in earlier manuals for our area as *Sideranthus spinulosus* and *Haplopappus spinulosus*.

629. *Machaeranthera tanacetifolia*

M. tanacetifòlia (H. B. K.) Nees.   Tansy aster

NL., from *Tanacetum*, name of the tansy genus, + *folium*, leaf—because the leaves resemble those of the tansy. The common name is suitable, since the flower heads are so much like those of *Aster* that this plant was first classified as *Aster tanacetifolius*.
Rays purplish-blue, disk yellow; late May–mid-October
Sandy soil of prairies, pastures, roadsides, and flood plains
West third, primarily

Erect, glandular-hairy annual, 1–4.5 dm tall, sustained by a taproot with many fine, widespreading laterals. Branches ascending, widely spreading, or sometimes reclining. Leaves close together, mostly oblanceolate to spatulate in general outline, 2–4.5 (–6) cm long and 5–20 mm wide, 2 or 3 times pinnately divided and incised, the segments 1–4 mm wide. Heads several to many, 3–4.5 cm across. Involucre 8–13 mm long, the bracts linear-attenuate, with herbaceous green tips, the outer ones recurved. Rays many, purplish-blue, 12–17 mm long. Disk flowers yellow. Achene body finely hairy, 2–3 mm long, the pappus 4–7 mm long, very minutely barbed.

This species occurs in most of the counties contained within its range, even to our western border, but it is of infrequent occurrence. During the years of the great drought of the thirties it went into retirement; then, when the drought was broken, the tansy aster promptly reappeared. Bailey's *Hortus* does not list it as ever having been planted in flower gardens, though it is altogether admirable in its wild estate.

M. annua (Rydb.) Shinners, a tall, glandular-hairy, yellow-flowered annual is known from scattered sites in the south and far west. It is 1–4 (–6) dm tall, branched above but rarely from the base, and has leaves mostly oblanceolate to oblong, 2–5 cm long and 3–14 mm wide, with 7–10 prominent teeth.

PRIONÓPSIS Nutt.

NL., from Gr. *prion*, saw, + *opsis*, appearance—in reference to the toothed leaf margin.

A small genus consisting of but a single species and having the characteristics of that species. It is closely related to *Grindelia*, and some authorities suggest combining the 2 genera.

630. *Prionopsis ciliata*

P. ciliàta (Nutt.) Nutt.   Ciliate goldenweed

NL., from L. *cilium*, eyelid, but the plural form *cilia* in our language means "eyelashes" or "a border or tuft of hairs." In *Prionopsis* the attenuate teeth of the leaf margin give the ciliate effect.
Yellow; late July–mid-October
Pastures, fields, roadsides, waste ground, and disturbed prairie
South two-thirds, primarily; seldom collected in the extreme east

Stout, glabrous, drought-resistant annual or biennial with stiff, upright stems 0.1–1 m tall, eventually with ascending branches above. Leaves stiff, 1.5–6.5 cm long and 0.4–3.2 cm wide, the lowermost spatulate or oblanceolate with tapered bases, the middle and upper leaves grading to oblong, lanceolate, or ovate, with clasping bases, all with bristle- or spiny-toothed margins. Heads mostly 3–4 cm across, borne singly or a few together at the ends of the branches. Involucre hemispheric, 10–22 mm long, the bracts loosely arranged in 3 or 4 series, graduated in length, lance-attenuate to linear-attenuate, the tip green and spreading or somewhat reflexed. Receptacle more or less flat, rough. Ray flowers many, yellow, female, fertile, the rays 1–2 cm long. Disk flowers many, yellow, bisexual, fertile. Achene body glabrous, somewhat flattened, oblong to ellipsoid, 3–5 mm long, the pappus of 1–3 series of slender, unequal, off-white bristles 6–7.5 mm long, often falling as a unit.

SOLIDÀGO L.   Goldenrod

NL., from L. *solidare*, to make whole or sound. Because of its reputed medicinal value, the herbals of the Middle Ages and even later recommended an extract of *Solidago* in wine for diseases of the kidney and bladder and for treatment of wounds.

Perennial herbs with short, erect rhizomes or longer, horizontal rhizomes which often produce slender, stoloniferous branches. Leaves alternate, simple, sessile or petioled, often toothed. Heads small, few- to many-flowered, radiate (rarely discoid), usually borne in many-headed racemes or panicles. Involucre narrowly bell-shaped to nearly cylindric, the bracts in several series, graduated in length, usually rather strawlike in color and texture and with a green midrib. Receptacle flat or slightly convex, naked, pitted. Pappus of many equal or unequal, usually hairlike, usually white bristles. Ray flowers 2–20, female, fertile, the corollas yellow. Disk flowers relatively few, bisexual, fertile; corollas tubular, 5-lobed, yellow; anther sacs rounded at the base; style branches flattened, the appendages lanceolate, hairy. Achenes many-ribbed, angled or nearly round in cross section.

Twelve species in Kansas, some with several varieties. *S. flexicaulis* L. and *S. radula* Nutt. are seldom encountered and are not included in the following key.

Various species of goldenrod have been utilized as medicinals by Indians and whites. The roots have been used in poultices on boils and in teas to treat burns and scalds. Flowers and/or leaves were used in teas for fevers and chest pains. Whites have utilized goldenrods as carminatives, antispasmodics, and astringents (45).

Key to Species (70):
1. Plants much branched from about halfway up the stem; leaves all linear-attenuate, 2–8 mm wide; heads borne in more or less flat-topped terminal clusters; involucral bracts striate-nerved .... *S. graminifolia*
1. Plants usually branched only in the inflorescence; leaves various, but not linear-attenuate, mostly wider than 10 mm; inflorescences various, but usually not flat-topped 2
    2. Inflorescence flat-topped or somewhat rounded, crowded, the heads terminal or nearly so on the branches; bracts striate-nerved .... *S. rigida*
    2. Inflorescence pyramid-shaped or widest near the middle or more or less cylindric, the heads often borne unilaterally along the branches; bracts not striate-nerved 3
        3. Inflorescence either of reduced axillary clusters scattered along the sides of the stem or a terminal, simple or branched cluster (narrow or cylindrical) with heads spirally arranged; flower clusters not in one-sided, racemelike branches, neither nodding at the tip nor with recurved branches 4
            4. Flower clusters arranged along the stem and arising from leaf axils .... *S. petiolaris*
            4. Flower clusters forming a narrow or cylindrical inflorescence (short or elongated) 5

631. *Solidago canadensis*

632. *Solidago gigantea*

5. Basal and lower cauline leaves the largest, usually persistent; cauline leaves progressively reduced upwards, often of different shape from the lowest ones 6
   6. Essentially glabrous or slightly rough (the inflorescence coarsely minutely hairy in *S. speciosa*) 7
      7. Leaves 3-veined; achenes hairy .... *S. missouriensis*
      7. Leaves pinnately veined; achenes glabrous .... *S. speciosa*
   6. Plants moderately to densely hairy throughout 8
      8. Leaf veins pinnate .... *S. petiolaris*
      8. Leaves mostly 3-veined .... *S. mollis*
5. Basal and lower cauline leaves smaller than the upper ones, soon falling away 9
   9. Plant essentially glabrous, 1-veined or pinnately veined .... *S. speciosa*
   9. Plant more or less densely hairy 10
      10. Leaves 1-veined or pinnately veined .... *S. petiolaris*
      10. Leaves 3-veined (sometimes obscurely so) .... *S. mollis*
3. Inflorescence terminal, broadest at the base or middle with at least the lower branches more or less strongly recurved and the heads borne unilaterally 11
   11. Stems glabrous or essentially so below the inflorescence 12
      12. Branches of the inflorescence glabrous or nearly so .... *S. missouriensis*
      12. Branches of the inflorescence more or less hairy 13
         13. Leaves 1-veined or pinnately veined or only faintly 3-veined .... *S. ulmifolia*
         13. Leaves distinctly 3-veined .... *S. gigantea*
   11. Stem more or less hairy below the inflorescence 14
      14. Leaves 1-veined, or not obviously 3-veined 15
         15. Leaves narrowly obovate, spatulate, or oblanceolate to linear; lower leaves usually more than 4 times as long as wide; basal leaves petioled; the upper usually with small axillary branches .... *S. memoralis*
         15. Leaves elliptic or oval to narrowly obovate, usually less than 4 times as long as wide; basal leaves sessile or nearly so; upper leaves without small axillary branches .... *S. mollis*
      14. Leaves obviously 3-veined 16
         16. Ray flowers of same number as to as many as 4 more than the disk flowers; leaves up to 5 times as long as wide, narrowly elliptic to obovate or oblanceolate (never narrowly lanceolate), blunt to merely acute; middle ones entire or with a few shallow teeth .... *S. mollis*
         16. Ray flowers usually twice or more as many as the disk flowers; leaves more than 5 times as long as wide and usually acuminate (if less than 5 and acute, then middle leaves obviously serrate) .... *S. canadensis*

S. *canadénsis* L.

"Canadian."
Yellow; September–October
Prairie
East three-fourths, primarily; scattered records in the west fourth

Stems 0.3–1.6 m tall, simple below the inflorescence, more or less short-hairy or rough-hairy. Middle and upper leaves very narrowly oblanceolate to lance-linear or linear-elliptic, almost always 3-veined, 4.5–14 cm long and 0.5–2.2 cm wide, the apex acute or wedge-shaped, the margins shallowly serrate (often entire within the inflorescence), the upper surface rough; the lower surface soft-hairy; lowermost leaves often gone by flowering time. Heads borne unilaterally on the more or less recurved branches of the panicle. Involucre 2–4 mm long, the bracts linear-oblong to linear, the tips rounded or acute (the innermost sometimes almost attenuate). Flowers yellow, the rays (7–) 10–17 per head, 1–2 mm long. Achene body short-hairy, 1–1.5 mm long, the pappus 2–3.5 mm long.

Several shades of dyes can be produced from goldenrod flowers and/or plants. The flowering heads alone give a yellowish-tan color to wool mordanted with alum; gold, to that mordanted with chrome. A dye bath prepared from the flowers, stalks, and leaves combined yields a light yellow to wool mordanted with alum and cream of tartar; yellow-orange, to that mordanted with chrome.

S. *gigantèa* Ait.   Giant goldenrod

NL., from L. *gigas, gigantis*, giant.
Yellow; (July-) August–October
Creekbanks, roadsides, prairie, usually in moist soil
Throughout

Stems both simple and glabrous below the inflorescence, 0.5–2 m tall. Leaves rather thin, smooth, narrowly lanceolate to narrowly elliptic, often (but not always) with 3 prominent veins, 5–14 cm long and 1–3.5 cm wide, the apex acuminate to attenuate, the base acute to narrowly wedge-shaped, the margins serrate (sometimes entire on the uppermost leaves) and rough, the lowermost leaves usually gone by flowering time. Heads borne unilaterally along the usually recurved branches of the panicle. Involucre 3–5 mm long, the bracts lance-linear to linear-elliptic, the tips acute to short-attenuate. Flowers yellow, the rays 10–17 per head, 4–6 mm long. Achene body short-hairy, 1–2 mm long, the pappus 2–3 mm long.

This species is listed in earlier manuals for this area as *S. serotina*.

The rubber content of the leaves is high, though some goldenrods have more, and the plant might sometime have commercial value as a source of rubber.

S. graminifòlia (L.) Salisb.

NL., from L. *gramen, graminis*, grass, + *folium*, leaf—in reference to the narrow leaves.
Yellow; (late July–) September–October
Dry, rocky prairies, pastures, and roadsides
East four-fifths, primarily; scattered records farther west

633. *Solidago missouriensis*

Stems branched above the middle, 0.4–1.5 m tall. Leaves sessile, glabrous, minutely gland-dotted, linear-attenuate, 2.5–15 cm long and 2–8 mm wide, those on the main stem usually gone by flowering time. Heads borne terminally or nearly so on the branches of flat-topped or slightly rounded terminal clusters. Involucre obconic, 4–6.5 mm long, the bracts oblong to linear-oblong, not striate, resinous near the tip. Flowers yellow, the rays 8–17 per head, 3.5–5.5 mm long. Achene body hairy, 1 mm long, the pappus 3–4 mm long.

NOTE: Compare with *Gutierrezia dracunculoides*.

Ours plants were listed in earlier manuals for Kansas as *Euthamia gymnospermoides* and *E. media*.

S. missouriénsis Nutt.   Common goldenrod

"Of Missouri."
Yellow; mid-July–October
Prairie and grassy roadsides
Throughout

634. *Solidago missouriensis*; note in this specimen and the previous one the amount of variation which may occur among individuals

Stems 1 to few, 1.5–10 dm tall, with roots which sometimes reach a depth of 2 m. Leaves glabrous, mostly linear-oblanceolate or narrowly oblanceolate below to narrowly elliptic or nearly linear above, 2.5–12.5 cm long and 4–15 mm wide, often (but not always) with 3 prominent parallel veins, the apex acute, the base long-tapered, the margins rough, those of the lower leaves serrate near the tip. Heads borne in branched terminal clusters with the branches more or less recurved and the heads more or less unilateral along them. Involucre 2–6 mm long, the bracts linear-oblong, the tips acute or rounded. Flowers yellow, the rays 7–13 per head, 4–5 mm long. Achene body glabrous or sparsely hairy, 1–1.5 (–2.2) mm long, the pappus 2.5–3 mm long.

This plant is listed in some earlier Kansas manuals as *S. glaberrima*.

This species is one of the commonest and earliest to bloom of our goldenrods. It distinguished itself for hardiness during the long drought period of the thirties when it colonized bare areas where grasses and other native plants had died out.

Since the heads are the objective of the flies, bees, butterflies, and beetles in quest of pollen and nectar, we pay attention to some features that are clearly to be seen with a hand lens. We find sterile female ray flowers and fertile bisexual disk flowers snugly bound together by the closely applied bracts of the cylindric involucre. The tubular disk flowers are at first tightly closed by the folding over of the lobes of the corolla. Within the corollas, the linear anthers are united into a tube around the 2-branched style, and the flowers are made to open up when the style elongates and carries the pollen out of the tube on the tips and sides of the closely applied style branches. Though the anthers are united, the filaments are free from one another, and through the gaps the proboscises of bees, butterflies, and flies are thrust to suck up nectar collected inside the corolla tube. While they are thus engaged, pollen adheres to them, some of it being later transferred to the stigmatic surfaces of the protruding, recurved style branches of older flowers. Certain species of little wild bees are especially active in this business, but the even more active honeybees add a significant amount of goldenrod honey to their stores, and bumblebees are frequent and eager visitors.

S. móllis Bartl.

L. *mollis*, soft.
Yellow; August–early October
Dry prairie uplands and hillsides, pastures, and roadsides, usually on sandy or rocky soil
West two-thirds, especially in the northwest fourth; not recorded from the southwest corner of
    the state

635. *Solidago rigida*

Stems 1 to several, 1.3–8 dm tall, usually unbranched below the inflorescence. Leaves mostly oblanceolate to narrowly oblanceolate or elliptic, rough-hairy, 2–8 cm long and 0.5–3.5 cm wide, 3-veined, the apex more or less broadly acute, often cuspidate, the margins entire or serrate toward the apex. Heads borne along the branches of terminal clusters. Involucre 4–5 mm long, the bracts linear-elliptic, the tips acute. Flowers yellow, the rays 6–10 per head, 4–7 mm long. Achene body hairy, 0.6–1.6 mm long, the pappus 3–4 mm long.

S. nemoràlis Ait.

L. *nemoralis*, woody, sylvan, of the woods, from L. *nemus*, Gr. *nemos*, forest or wood with pasture
        for cattle, grove, glade.
Yellow; (late July–) September–October
Dry, rocky and usually sandy soil of open woods, roadside banks, pastures, and prairie
East half, primarily

Stems 1 to several, simple below the inflorescence, 0.3–1 m tall. Leaves rough-hairy, mostly narrowly oblanceolate to nearly linear, 2.5–10 cm long and 0.6–2.5 cm wide, the apex acute, the lower leaves persistent, with bases long-tapered and margins shallowly serrate, the upper leaves less tapered at the base, smaller, and entire. Herbage minutely hairy throughout. Heads borne in branched terminal clusters, usually with the tip of the cluster nodding and/or the branches elongate and recurved, the heads borne along the upper sides of the branches. Involucre 3.5–5 mm long, the bracts linear-oblong, the tips usually rounded. Flowers yellow, the rays 5–9 per head, 3–5 mm long. Achene body hairy, the pappus 2–3 mm long.

The Houma Indians utilized a tea made from the roots to cure yellow jaundice.

S. rígida L. Stiff goldenrod

L. *rigidus*, stiff, firm—in reference to the leaves especially, but the stem and inflorescence have
        this characteristic as well.
Yellow; late August–October
Rocky, open ground of roadsides, pastures, waste ground, and prairie
East two-thirds, primarily; scattered records in the west third

Stems simple below the inflorescence, often clustered, up to 1.6 m tall. Leaves long-petioled below, sessile or nearly so above, the blades firm, rough-hairy, mostly ovate to lanceolate or elliptic, 2.5–16 cm long and 1–6 cm wide, the apex acute or rounded, the bases of upper leaves rounded, those of the lower leaves gradually tapered to the petiole, the margins usually shallowly crenate or serrate, sometimes entire. Heads borne terminally or nearly so on the branches of a flat-topped or slightly rounded terminal cluster. Involucre bell-shaped, 5–8 mm long, the bracts mostly oblong, more or less striate, not resinous, the tips rounded. Flowers yellow, the rays 7-14 per head, 5–8 mm long, the disk flowers 19–31 per head. Achene body usually glabrous, 2–2.5 mm long, the pappus persistent, 4–5 mm long.

In the fall, short branches arise from the rhizome, each with a cluster of ascending, long-petioled basal leaves, sometimes as much as a foot in length, which persist through the winter. Enclosed by the bases of these leaves is a growing point which produces an upright shoot with its panicle of flowers the following season. It should be realized that the over-wintering basal leaf clusters of such plants as this are ready to photosynthesize at any time when the temperature and light are favorable, thus, by contributing food, helping to stimulate an early start in growth when spring arrives. In this way the goldenrods prevent the grasses from overshadowing and starving them out. Of paramount importance also for this goldenrod are its abundant fibrous roots going down 2 meters or so to maintain a sufficient water supply among strongly competitive grasses.

See plate *122*.

S. petiolàris Ait.

NL., from L. *petiolus*, little foot, stalk, stem, + the suffix *-arum, -aris*, indicating possession.
Yellow; late August–October
Dry, rocky soil of roadside banks, prairie, and open woods
East two-thirds, primarily; especially in the north half of the central third

Stems 0.3–1.5 m tall, usually unbranched below the inflorescence. Leaves oblanceolate to narrowly elliptic, rarely linear-elliptic, mostly 3–11 cm long and 1–3 cm wide, usually rough-hairy (rarely glabrous in our area), the apex acute, the margins entire or serrate. Heads borne in usually narrow, elongate, branched clusters, often conspicuously leafy-bracted. Involucre 4–7 mm long, the bracts linear, often glandular, the tips acute or attenuate. Flowers yellow, the rays 5–11 per head, 5–8 mm long. Achene body glabrous, or nearly so, 2–3 mm long.

S. speciòsa Nutt.

L. *speciosus*, beautiful, splendid, showy.
Yellow; late August–October
Prairie, pastures, and open woods, usually in rocky and/or sandy soil
East two-fifths

Stems simple below the inflorescence, 0.4–1.5 m tall. Leaves smooth except for the rough margins, mostly narrowly oblanceolate to narrowly elliptic, 3.5–10 cm long and 0.5–3.5 cm wide, the apex acute, the base wedge-shaped, the lower leaves often gone by flowering time. Heads borne along the branches of a cylindric or somewhat pyramidal terminal cluster. Involucre 3–6 mm long, the bracts oblong, the tips rounded. Flowers yellow, the rays 4–11 per head, 4–6 mm long. Achene body glabrous, 2–2.5 mm long, the pappus 3–4 mm long.

The flowers of this species make an important contribution to the honey crop.

S. ulmifòlia Muhl.   Elmleaf goldenrod

636. *Solidago speciosa*

NL., from *Ulmus*, the elm genus, + L. *folium*, leaf.
Yellow; July–October
Rocky, wooded hillsides and banks
East fourth, primarily; more common toward the southeast

Stems 0.3–1 dm tall, simple below the inflorescence, arising from a rhizome producing slender, stoloniferous branches and fibrous roots. Middle leaves mostly broadly lanceolate to more narrowly lanceolate or elliptic or occasionally oblanceolate, mostly 3.5–14 cm long and 1–4.5 cm wide, the apex acuminate or attenuate, the base acute to acuminate or wedge-shaped, the margins serrate, the upper surface glabrous or rough-hairy, the lower surface rough-hairy; basal leaves smaller than upper leaves, usually gone by flowering time. Heads borne unilaterally along the slender, often recurved branches of the panicle. Involucre 3–5 mm long, the bracts lance-linear to linear-oblong or linear-elliptic, the tips nearly rounded to acute or attenuate. Flowers yellow, the rays 3–5 per head, about 1.5 mm long. Achene body short-hairy, about 1–1.5 mm long, the pappus about 2 mm long, deciduous.

This goldenrod would be good for planting in partly shaded places of the garden along with white boneset, scarlet lobelia, big blue lobelia, and the tall bellflower.

XANTHOCÈPHALUM Willd.   Broomweed, snakeweed

637. *Solidago ulmifolia*

NL., from Gr. *xanthos*, yellow, + *kephale*, head—in reference to the heads of yellow flowers.

Annual or perennial herbs or low shrubs, with taproots. Leaves alternate, simple, entire, usually resinous. Heads small, radiate, borne singly or in clusters. Involucre shaped variously, the bracts in about 3 series, graduated in length, closely appressed, usually firm, thick, straw-colored, often resinous. Receptacle naked, often deeply pitted. Ray flowers few, female, fertile, with the pappus a low, rough crown, or of small scales, or absent. Disk flowers few, bisexual, fertile or infertile, the corolla tubular, 5-lobed, yellow, the pappus a minute crown of scales, or of a few linear, basally-united scales, the anther sacs rounded basally, the style branches various. Achenes obpyramidal, usually densely hairy, crowned by the persistent pappus (when present).

Two species in Kansas. This is a complex genus which is sometimes divided into several genera, including *Gutierrezia*, *Amphiachrys*, *Xanthocephalum*, and others, but here it is considered in the broad sense.

Key to Species:

1. Annual; stem simple below, much branched above .... *X. dracunculoides*
1. Perennial; stems much branched from the base .... *X. sarothrae*

X. dracunculoìdes (DC.) Shinners   Annual broomweed

NL., from L. *dracunculus*, little serpent, + the Gr. suffix *-oides*, resembling, in the form of. In this instance, the application is not evident.
Yellow; August–mid-October
Pastures, roadsides, disturbed prairie, and other disturbed sites
East two-thirds, primarily, but west to Haskell and Morton counties in the south

Glabrous, erect annual, 1.8–7.5 dm tall, with stems simple below and with many slender, ascending branches above suggesting a whisk broom. Leaves linear to very narrowly elliptic, 1.8–7 cm long and 0.5–5 mm wide, quite resinous. Involucre more or less top-shaped, 3–4 mm long, the bracts firm and straw-colored, with herbaceous green tips. Ray flowers fertile, the ligule yellow, 3–4 mm long, the pappus consisting of a minute fringed crown. Disk flowers yellow, infertile, the pappus of a slender membranous cylinder bearing 5 linear projections about as long as the corolla. Achenes about 1.5 mm long.

This plant tends to increase in abundance during periods of drought and under severe grazing pressures. It has a definite odor and is distasteful to cattle, but they will eat it if other forage is not available and then suffer gastro-intestinal upsets. There is some toxic substance, in the herbage or the pollen or both, which causes an itchy dermatitis and symptoms resembling pinkeye in humans and livestock.

638. *Xanthocephalum sarothrae*

X. saròthrae (Pursh) Shinners   Perennial broomweed, broom snakeweed

NL., from Gr. *sarothron*, broom—suggested by the form and texture of the dried plant.
Yellow; mid-August–mid-October
Dry prairie hillsides and pastures
West half

Perennial with a strong taproot. Stems several to many, woody at the base, finely branched above, 1–4 dm tall. Leaves linear, 1–5 cm long and cylindric, 2.5–3 mm long, the bracts straw-colored below, green at the tip. Ray flowers fertile, yellow, the ligules inconspicuous, about 2 mm long, the pappus of fimbriate membranous scales. Disk flowers yellow, fertile. Achene body about 1.5 mm long.

Perennial broomweed is a secondary or facultative selenium absorber and is toxic to sheep, cattle, goats, pigs, chickens, and rabbits. Most losses result from abortion in pregnant animals or other reproductive irregularities, though death may result from ingesting 10–20 percent of the animal's body weight over a period of 3 or 4 days to 2 weeks. The plants are apparently most toxic when the young leaves are forming and may be more toxic on sandy soils than on harder soils (21).

Several medicinal uses for this plant are recorded in the literature. Navajos chewed a piece of the plant and placed the wad on bee and wasp stings and ant bites. The women of the tribe drank a tea prepared from the entire plant to promote expulsion of the placenta after childbirth (34). The Hopi Indians and also whites used the same brew to treat stomachache and other gastric upsets.

See plate *123*.

HELIÁNTHEAE   Sunflower Tribe

Sap of plants not milky. Flower heads all alike, radiate (rarely discoid). Ray flowers usually female and infertile; disk flowers usually bisexual and fertile. Receptacle chaffy. Bracts of the involucre foliaceous. Anthers rounded or sagittate at the base, never caudate (tailed). Style branches of the disk flowers usually with hairy, subulate, lanceolate, or conic appendages. Pappus of scales, awns, or a toothed crown, never of hairlike bristles.

BERLANDIÈRA DC.

In honor of Jean Louis Berlandier (1805–1851), a Swiss botanist who collected in Texas and Mexico from 1827 to 1830.

Hairy, herbaceous perennials, sometimes woody toward the base, with fleshy taproots. Leaves alternate, simple, sessile or petioled, toothed or pinnatifid. Heads many-flowered, radiate, borne singly or in branched terminal clusters. Involucre hemispheric or bowl-shaped, the bracts in 3 series, the outer ones green and leaflike in texture, the inner ones obovate, longer, thinner, chaffy toward the base. Receptacle flat or nearly so, chaffy. Pappus absent, or a minute crown of 2 or 3 small teeth. Ray flowers usually 5–12, in 1 series, female, fertile; corolla yellow or orange-yellow, the ligule with several green or reddish veins on the lower side, notched at the apex. Disk flowers many, bisexual, infertile; corollas red to maroon, gradually dilated upward and prominently veined; anthers entire or minutely toothed at the base and with ovate or lanceolate apical appendages; style undivided. Achenes obovate, strongly flattened parallel to the involucral bracts, neither winged nor notched at the apex, usually hairy on the inner face, usually more or less united to the subtending involucral bract and adjacent 2 or 3 flowers and their subtending receptacular bracts and all falling together as a unit.

Two species in Kansas. *B. lyrata* Benth., known from Morton County only, has most of its leaves pinnately lobed, some of them forming a persistent basal rosette, and has red veins on the under surface of the rays.

B. texàna DC.

"Of Texas."
Rays deep to orange-yellow, disk dark red; June–mid-September
Sandy prairie or sand dunes
South part of the central third, and Stevens and Meade counties

Plants up to about 1 m tall, woody near the base. Stems moderately to densely hairy, eventually branching above. Leaves usually crowded on the stem and stiffly ascending, petioled below to sessile above, the blades narrowly triangular to ovate, 3.5–9 cm long and 2.5–5 cm wide, the apex acute, the base truncate to cordate, the margins crenate, both surfaces hairy and gland-dotted, but the lower one especially so. Heads 1–20, 2.5–3.5 cm across. Involucre 8–15 mm long, the outer bracts more or less oblong to oblanceolate, acute, 1–6 mm wide, the inner ones obovate to broadly rhombic, 7–13 mm wide, the tips broadly acute to rounded. Rays deep- to yellow-orange with green veins beneath, usually 7–8, 8–13 mm long. Disk flowers dark red or maroon, strongly veined, densely resinous, the teeth hairy. Achenes obovate, densely hairy, strongly attached to the subtending bract and more loosely to the flanking disk flowers, 4.5–6 mm long (excluding the bract), the pappus scarcely evident.

639. *Berlandiera texana*

BÌDENS L.   Beggarticks, bur marigold

NL., from L. *bi-*, *bis*, 2, twice, + *dens*, tooth—the achenes of some species having a pappus of only 2 awns.

Mostly annuals, less frequently perennial herbs. Leaves opposite (sometimes alternate above), simple, entire to toothed, incised, or 1 or more times ternately or pinnately dissected. Heads many-flowered, radiate or discoid, terminal or axillary, borne singly or in clusters of 2–3. Involucre bell-shaped or nearly hemispheric, the bracts in 2 series, the outer ones herbaceous and leaflike, the inner ones usually membranous with yellow or translucent margins. Receptacle flat or slightly convex, chaffy, the receptacular bracts persistent, resembling the inner involucral bracts but narrower. Pappus usually of 1–8 barbed awns, less frequently of 2 small scales or teeth or absent. Ray flowers, when present, 3–8, usually yellow (rarely white or pink), usually neutral, infertile. Disk flowers usually many, bisexual, fertile, the corollas usually yellow, tubular, 4- or 5-toothed. Anthers rounded or minutely tailed at the base. Style branches flattened, the appendages short, hairy, and acute to acuminate. Achenes flattened parallel to the involucral bracts, or 3- to 4-angled, rarely nearly round in cross section, crowned by the persistent pappus, falling with the receptacular bracts.

Seven species in Kansas.

The seeds of various species are eaten to a limited extent by birds. The plants also yield yellow or orange dyes.

Key to Species:
1. Leaves entire (or at most with 2 or 4 divergent basal lobes in *B. connata*) 2
  2. Bases of most leaf pairs united around the stem; heads with conspicuous rays ...... *B. cernua*
  2. Leaf pairs not united at all or the petioles but slightly united; heads apparently discoid, the rays quite small and inconspicuous or absent altogether 3
    3. Leaves narrowed to a narrowly winged petiole 0.5–4 cm long; disk corollas orange-yellow, 5-lobed; achenes 4-angled and 4- to 6-awned .... *B. connata*
    3. Leaves sessile; disk corollas pale yellow, 4-lobed; achenes flattened, mostly 3-awned .... *B. comosa*
1. Leaves one or more times pinnately divided or dissected 4
  4. Heads with conspicuous rays; outer involucral bracts lance-linear to linear-attenuate, very coarsely ciliate, the margins quite wrinkled or "crisped"; achene body flattened, with an interrupted, membranous marginal wing .... *B. polylepis*
  4. Heads apparently discoid, the rays inconspicuous or absent altogether; involucral bracts not as above; achene body flattened or not, but if flattened then lacking a marginal wing 5
    5. Leaves fernlike, 2- to 3-times pinnately divided; inner involucral bracts mostly linear or subulate, 0.5–1.5 mm wide; achene body narrowly linear, longitudinally ribbed, 12–19 mm long, the awns retrorsely barbed .... *B. pinnata*
    5. Leaves 3- to 5-cleft or -foliolate; inner involucral bracts mostly lanceolate to lance-oblong or oblong, (1.5–) 2 mm wide or wider; achene body wedge-shaped, 6–10 (–12) mm long 6
      6. Outer involucral bracts (5–) 8 (–10); achenes black, slightly hairy .... *B. frondosa*

640. *Bidens bipinnata*

641. *Bidens cernua*

6. Outer involucral bracts (10–) 13 (–16); achenes greenish or brown, smooth or
tubercled .... *B. vulgata*

B. bipinnàta L.

Spanish needles

NL., from L. *bi-*, *bis*, 2, twice, + *pinnatus*, feathered, winged—the leaves being pinnately com-
pound, with the leaflets also deeply divided.
Rays yellow, inconspicuous, disk orange-yellow; late July–October
Yards, pastures, creekbanks, waste ground, and other disturbed sites, frequently in the shade
East three-fourths, primarily; more common eastward

Erect annual, 1–8 dm tall, glabrous or minutely hairy, the stems 4-angled, eventually branching. Leaves opposite, petioled, the blades mostly cordate in general outline
but 2- to 3-times pinnately divided, 3.5–15 cm long and 1.6–11 cm wide, the ultimate
subdivisions mostly triangular to ovate or lanceolate and also pinnately toothed or lobed.
Heads few to many, terminal or axillary, borne singly or in clusters of 2–3, the disk 4–6
mm across at anthesis. Outer involucral bracts mostly subulate to linear, shorter than to
about as long as or slightly longer than the inner; inner involucral bracts mostly linear
or subulate, sometimes elliptic, 6–10 mm long. Rays yellow, 4 or fewer, very short and
inconspicuous, or absent entirely. Disk corollas orange-yellow, 3–4.5 mm long. Achene
body narrowly linear, longitudinally ribbed, minutely roughened, 12–19 mm long, the
awns 4, prominently retrorsely barbed, 2–4 mm long, clinging tenaciously to clothing.

Though plants of this species are classed as weeds, they are the source of delicious
honey, and their fernlike foliage can be used effectively in floral arrangements. In west
Africa, the foliage is cooked and eaten as a green vegetable (4).

B. cérnua L.

Stick-tight

L. *cernuus*, drooping, nodding, facing earthward—in reference to the flowers.
Yellow; August–mid-October
Wet soil of ditches, lake margins, marshes, and other wet places
Scattered across the state

Erect or decumbent, branched annual, 1.5–40 dm tall, glabrous or nearly so, often
rooting at the lower nodes. Leaves sessile, narrowly oblanceolate to elliptic or lance-
linear, 3.5–23 cm long and 0.6–4.7 cm wide, the apex usually long-acuminate to attenuate,
the bases usually connate-perfoliate around the stem, the margins serrate. Heads radiate,
several to many, (2–) 2.5–5 (–6) cm across, borne singly or in clusters of 2–3 at the ends
of the stems and branches. Outer involucral bracts lance-attenuate to linear-elliptic or
linear-oblong, spreading or somewhat reflexed, at least some of them longer than the rays;
inner bracts lanceolate to lance-elliptic, 5–12 mm long. Rays yellow, 8–17 mm long.
Achene body narrowly cuneiform, 6–7 mm long, the angles tuberculate and retrorsely
barbed, the pappus awns usually 4, 3–4 mm long, retrorsely barbed.

B. comòsa (Gray) Wieg.

Beggarticks

L. *comosus*, hairy, since these plants are glabrous, one wonders whether Gray might have had
cymosus, "full of shoots or branches" in mind when he named this species.
Yellow; late July–mid-October
Lake margins, riverbanks, creekbanks, sloughs, and other wet places

Glabrous, branched annual herb, 1.6–10 dm or more tall, the stems erect or occasionally decumbent. Leaves opposite (occasionally subopposite), sessile, narrowly oblanceolate to narrowly elliptic, narrowly lanceolate, on the uppermost sometimes nearly
linear, 4–15 cm long and 0.8–4 cm wide, the apex usually acute to attenuate, the base
wedge-shaped, not connate-perfoliate, the margins serrate. Heads discoid, few to many,
usually borne singly, terminal or axillary, the disk 12–20 mm across. Outer involucral
bracts narrowly lanceolate to linear-elliptic or linear-oblong, longer than the disk, the
inner bracts narrowly lanceolate to narrowly elliptic, 7–10 mm long. Rays absent. Disk
corollas pale yellow, 3–4 mm long, 4-lobed. Achene body flattened, 6–7 mm long, the
awns usually 3–4 mm long, the margins and awns retrorsely barbed.

B. connàta Muhl.

Beggarticks, stick-tight

L. *connatus*, born together, joined—apparently in reference to the petioles which are but slightly
united around the stem.
Rays yellow, inconspicuous, the disk orange-yellow; mid-August–mid-October
Riverbanks, sand bars, lake margins, and other wet places
Northwest fourth, primarily

Glabrous, erect, widely branched annual, 2–15 dm tall. Leaves opposite, petioled, the blades lanceolate to elliptic or the uppermost lance-linear, 3–14 cm long and 0.5–5.5 cm wide, the apex acuminate, the base acute or acuminate, the margins coarsely serrate or dentate, the petioles narrowly winged, 0.5–4 cm long. Heads few to many, terminal or axillary, borne singly or in clusters of 2–3, the disk 7–15 mm across. Outer involucral bracts narrowly elliptic to narrowly oblong or spatulate, longer than the disk, often leaf-like and quite conspicuous, the inner bracts lanceolate to elliptic, 4.5–8 mm long. Rays yellow, quite small and inconspicuous, or absent altogether. Disk corollas orange-yellow, 5-lobed, 2.5–4 mm long. Achene body slender, mostly 4-angled, (3.5–) 6–9 mm long, the awns 4–6, (1.5–) 3–5 mm long, retrorsely barbed.

B. frondòsa L. — Devil's beggarticks

NL., full of leaves, from L. *frons, frondis*, leafy branch, bough, foliage, leaf, + the suffix *-osus*, in-
dicating abundance—in reference to the foliaceous bracts of the involucre.
Orange-yellow; (mid-August–) September–mid-October
Seeps, creekbanks, lake margins, ditches, and other wet places
Scattered nearly throughout, but less common westward

Erect, branched annual herb, 2–12 dm tall, glabrous or sparsely hairy. Leaves opposite, the slender petioles 0.5–6 cm long, the blades thin, mostly triangular in general outline, 3–11 cm long and (1–) 4–16 cm wide, so deeply cleft as to appear 3- (occasionally 5-) foliolate, the segments or leaflets lanceolate to elliptic with an acuminate apex and serrate margins, the terminal segment usually larger than the lateral 2. Heads discoid, several to many, terminal or axillary, borne singly or in clusters of 2–3, the disk 7–15 mm across. Outer involucral bracts (5–) 8 (–10), linear to linear-oblong or spatulate, longer than the disk, the inner bracts lanceolate to lance-oblong or oblong, 5–9 mm long. Rays minute or absent. Disk flowers orange-yellow, 2.5–4 mm long, 5-lobed. Achene body flat, wedge-shaped, 6–10 mm long, definitely appressed-hairy to nearly glabrous, antrorsely barbed along the angles, the 2 awns about 4 mm long, retrorsely barbed.

Plants of this species have been known to accumulate toxic concentrations of nitrates (21), and handling the plants causes local irritation of the skin in some individuals (4).

B. polylèpis Blake — Bur marigold

NL., from Gr. *polys*, many, + *lepis*, scale—perhaps in reference to the indented membranous margins of the achenes.
Yellow; August–October
Creekbanks, pond margins, ditches, and other wet places
East two-fifths, especially the east third

Glabrous, erect, branched annual, 2–15 dm tall. Leaves opposite, pinnately divided into 3–7 serrate, lance-linear or linear-elliptic segments 1.7–10 cm long and 3–22 mm wide. Heads radiate, several to many, quite showy, 3–7 cm across. Outer involucral bracts lance-linear to linear-attenuate, conspicuously and coarsely ciliate and hairy on the back, strongly wrinkled when dried, about as long as or slightly longer than the disk; inner bracts lanceolate to lance-oblong, often minutely hairy, 5–10 mm long. Achene body flattened, 5–7 mm long, with an interrupted wing, usually ciliate also, awns present or absent, long or short, antrorsely or retrorsely barbed.

This species has been listed in earlier manuals as *B. involucrata*.

We find this bur marigold most abundantly in the relatively moist soil along the drainage ditches of the highways and country roads in the eastern part of the state where it blooms abundantly over a long period in summer and fall, joining with various species of sunflower, *Silphium*, blazing star, and gayfeather in decorating the borders of our fields and grasslands.

B. vulgàta Greene — Beggarticks, stick-tight

L. *vulgatus*, to make common, spread.
Yellow; late August–September
Low, wet places
Primarily in the north two-thirds

Annual, 2.5–9 (–15) dm tall, strongly resembling *B. frondosa* but tending to be stouter, nearly glabrous to densely soft-hairy. Leaves opposite, petioled, the leaflets 3–5. Outer involucral bracts (10–) 13 (–16), ciliate, mostly linear to linear-elliptic or more or less broadly spatulate. Rays absent. Disk flowers yellow, 5-lobed, 2–3 mm long. Achene body flattened, smooth or tubercled, 6–10 (–12) mm long, antrorsely ciliate along the margins, the awns 2, 3–4 mm long, retrorsely barbed.

642. *Bidens polylepis*

COREÓPSIS L.   Coreopsis

NL., from Gr. *koris*, bug, + *opsis*, appearance—the achenes of some species somewhat resembling ticks.

Erect annual or perennial herbs. Leaves opposite or rarely alternate, simple and entire to toothed or dissected, or compound, sessile or petioled. Heads many-flowered, radiate, borne singly on long terminal or axillary peduncles. Involucre bell-shaped, the bracts in 2 series, the inner ones broad, orange or brown, dry and membranous or at least with membranous margins, the outer ones narrow, leaflike in color and texture. Receptacle flat or slightly convex, chaffy, the receptacular bracts flat, thin, narrow, and deciduous. Pappus usually of 2 persistent, barbless awns, sometimes a minute crown or absent. Ray flowers usually 8, female or neuter, usually infertile; ligules yellow, sometimes with a reddish-brown spot at the base, the apex 3-toothed. Disk flowers many, bisexual, fertile; corollas yellowish, regular, 5-toothed; anthers entire or sagittate at the base; style branches flattened, with short or elongate, hairy appendages. Achenes more or less broadly elliptic or oblong, flattened parallel to the involucral bracts, usually winged, the apex 2-toothed, or -awned or without such projections.

Six species reported for Kansas. Our 3 most common species are included here.

Key to Species:

1. Plants annual; rays yellow with a red-brown spot at the base; disk flowers red-brown; achenes wingless .... *C. tinctoria*
1. Plants perennial; rays entirely yellow; disk flowers yellow; achenes with at least a narrow wing 2
   2. Leaves lax, the subdivisions mostly 0.5–4 mm wide; outer involucral bracts lance-acuminate to lance-linear; achene body broadly winged, about 2.5 mm long .... *C. grandiflora*
   2. Leaves rigidly ascending, the subdivisions mostly 2.5–6 mm wide; outer involucral bracts mostly linear-oblong; achene body narrowly winged, 5–6.5 mm long .... *C. palmata*

C. grandiflòra Hogg.   Bigflower coreopsis

NL., from L. *grandis*, large, great, magnificent, + *flos, floris*, flower—this being an outstanding species in the size and abundance of its flowers.
Yellow; mid-May–early July
Prairie, usually in rocky or sandy soil, occasionally in open oak-bluestem woodlands
Primarily east and south of a line from Brown to Franklin and Sumner counties

Glabrous perennial, the stems simple or with a few branches, 2–8 dm tall, arising from a short rhizome. Leaves opposite, the blades 2.5–9 cm long, mostly with 3 or 5 lax, linear or linear-oblanceolate segments mostly 0.5–4 (–7) mm wide. Heads (2–) 3–5 cm across, the outer (lower) involucral bracts lance-acuminate to lance-linear, 6–10 mm long. Rays yellow, 11–30 mm long. Disk flowers yellow. Achene body black, winged, about 2.5 mm long, the pappus of 2 small scales.

This native has long been a favorite in American and foreign gardens.

See plate *124*.

C. palmàta Nutt.   Finger coreopsis

L. *palmatus*, shaped like a hand with spread fingers—in reference to the leaves.
Yellow; June–mid-July
Prairie, roadsides, and railroad right-of-ways
East fourth

Glabrous perennial, the stems simple or with a few branches, 4.5–9 dm tall, arising from a slender, branching rhizome with fibrous roots. Leaves opposite, sessile, rigidly ascending, 4–9 cm long, divided into 3 linear or linear-elliptic segments 2.5–6 mm wide, the segments sometimes with 1 or 2 lateral teeth or lobes. Heads 3–5 cm across, the outer (lower) involucral bracts mostly linear-oblong, 6–9 (–11) mm long. Disk flowers yellow. Achenes black, very narrowly winged, 5–6.5 mm long.

This species seems to be better adapted for association with the prairie grasses than are the others we have of its genus, owing, it would seem, to a combination of enabling characters, such as its perennial duration; strict, upright habit of stems and leaves; and early, rapid growth.

See plate *125*.

C. tinctòria Nutt.   Plains coreopsis

L. *tinctorius*, pertaining to dying, from *tingere*, to dye, *tinctor*, dyer—this plant being a source of yellow and red-brown dyes.

643. *Coreopsis grandiflora*

644. *Coreopsis palmata*

Rays yellow with reddish-brown bases; mid-June–mid-October
Moist soil
East two-thirds, primarily; scattered records in the west third, especially southward

Glabrous annual, up to about 1 m tall, with few to many branches. Leaves opposite, petioled, the blades 2.5–10 cm long, once or twice pinnately dissected into linear or linear-elliptic segments 1.5–3.5 mm wide, the terminal segment usually longer than the lateral ones. Heads 1.5–2.5 (–3) cm across, the outer (lower) involucral bracts ovate to lanceolate or subulate, 1–2 mm long. Rays yellow with a reddish-brown spot at the base, 8–15 mm long. Disk flowers reddish-brown. Achenes black, wingless, 1–4 mm long, lacking a pappus.

This species is often grown in gardens, where it puts out a grand show of flowers over a long period. Even without aid of cultivation, flourishing colonies of it are sometimes seen strung along country roadsides over long distances.

Various species of *Coreopsis* have been used to produce brown, yellow, orange, red, or purplish-red dyes, with *C. tinctoria* perhaps being the species used most frequently. Either the flowers alone or the entire plant may be chopped and soaked overnight to produce the dye bath. Dyers offer a variety of suggestions as to which mordants yield which colors, and allowing the dye bath to age 2 or 3 days may produce deeper colors. Adjusting the amount of mordant used also increases the color variations possible. This may well be one of our most versatile native dye plants.

645. *Coreopsis tinctoria*

ECHINĂCEA Moench.   Echinacea, purple coneflower

NL., from Gr. *echinos*, hedgehog—suggested by the spiny chaff among the disk flowers.

Erect perennial herbs. Leaves alternate, simple, entire or coarsely toothed, sessile above, petioled below. Heads many-flowered, radiate, terminal, borne singly. Involucre conical or hemispherical, the bracts in 3–4 series, herbaceous, more or less lanceolate, graduated in length. Receptacle conic, chaffy, the bracts consisting of conspicuous, persistent, blunt or sharp, conduplicate spines longer than the disk flowers. Pappus represented by a short, smooth or toothed crown. Ray flowers usually purplish-pink, infertile. Disk flowers many, bisexual, fertile, the corollas purplish or yellowish, tubular but somewhat enlarged and bulblike at the base, 5-toothed. Anthers sagittate at the base and with an ovate apical appendage. Style branches acuminate, hairy. Achenes 4-angled, glabrous.

Four species in Kansas. *E. purpurea* (L.) Moench, is known from one site in Cherokee County only and is not included here.

Key to Species:
1. Rays dark reddish-purple (rarely pink or white); stems with stiff, appressed hairs; plants mostly in a narrow band of counties from Wabaunsee, Shawnee, and Douglas counties south .... *E. atrorubens*
1. Rays pale pink to purplish-pink (rarely white); stems usually with coarse, spreading hairs, at least above; distribution not as above 2
    2. Rays purplish-pink, 18–35 mm long, spreading or strongly reflexed; plants of the west four-fifths .... *E. angustifolia*
    2. Rays pale pink to purplish-pink, 20–70 mm long, eventually drooping; plants mostly of the east fourth .... *E. pallida*

646. *Echinacea angustifolia*

E. angustifòlia DC.   Black Sampson echinacea

NL., from L. *angustus*, narrow, + *folium*, leaf, i.e., "narrow-leaved."
Ray flowers purplish-pink, disk purplish-brown; June–early July
Dry, upland prairies
West four-fifths

Stems stout, 1 to few, with coarse, spreading hairs (at least above), 2–7 dm tall, arising from a short vertical rhizome and a strong taproot extending 1.5–2 m into the soil. Leaf blades mostly spatulate to narrowly elliptic with 3 prominent veins, 5–19 cm long and 2–4 cm wide, the apex acute, the base wedge-shaped, both surfaces with rough, appressed hairs. Head terminal, 3–7 cm across. Rays purplish-pink (rarely white), 18–35 mm long, spreading or strongly reflexed. Achenes 4–5 mm long.

This most widespread of our species of *Echinacea* has had numerous medicinal uses. Among various tribes of Plains Indians it was one of the most frequently used plants. The underground parts were used in one form or another to treat snake and insect bites, toothache, swollen glands (as in mumps), sore throat, hydrophobia, fits, and stomach cramps (45).

By 1852, it was in use by whites as an aromatic and carminative. In 1885, Dr. H. C. F. Meyer of Pawnee City, Nebraska, introduced an extract of the rootstock as

647. *Echinacea atrorubens*

648. *Echinacea pallida*

649. *Eclipta alba*

"Meyer's Blood Purifier" which he claimed was an effective cure for tumors, syphilis, fever sores, ulcers, carbuncles, piles, gangrene, fever, typhus, bee stings, snakebite, and hydrophobia (59). In 1887, Lloyd Brothers Pharmaceutical Company in Cincinnati, Ohio, began marketing an extract of the plant which was much used for several decades. In recent years, interest in *Echinacea*, this time including *E. pallida*, has revived, and the dried roots have been purchased by pharmaceutical firms for as much as $1.25 per pound.

E. atrorùbens Nutt.

NL., from L. *ater, atra*, black, $+$ *rubens*, red.
Rays dark reddish-purple; June
Prairies, usually in rocky, calcareous soil
Primarily in the east fourth, south of the Kansas River and excluding the east column of counties

Stems 1 to few, stout, 4–10 dm tall, with stiff, appressed hairs (at least above), arising from a short vertical rhizome and a stout taproot. Leaf blades mostly lance-linear to linear-elliptic (occasionally broadly elliptic) with 3 or 5 prominent veins, glabrous except for the roughly ciliate margins, 9–19 cm long and 0.5–5 cm wide, the apex acute to attenuate, the base wedge-shaped. Heads terminal, 3–7 cm across. Rays dark reddish-purple (rarely pink or white), 17–30 mm long, spreading or strongly recurved. Achenes 4–5 mm long.

E. pállida (Nutt.) Nutt.   Pale echinacea

L. *pallidus*, pale—in reference to the ray flowers.
Rays pale pink to purplish-pink; late May–mid-July
Prairie
East fourth, west to Cowley and Butler counties in the south

Stems 1 to few, stout, 2–10 dm tall, more or less densely covered with coarse, spreading hairs, arising from a short vertical rhizome with a stout taproot. Leaf blades mostly lance-linear to linear-elliptic, usually with 3 prominent veins, 6–20 cm long and 0.8–3 cm wide, the apex acute, the base wedge-shaped, both surfaces with coarse, spreading hairs. Heads terminal, 6–12 cm across. Rays pale pink to purplish-pink (rarely white), 20–70 mm long, drooping. Achenes mostly 4–5 mm long.

ECLÍPTA L.

NL., from Gr. *ekleipsis*, lacking, obscure—in reference to the inconspicuous ray flowers.

Ours slender, branching annuals with taproots, sometimes sprawling and rooting at the nodes. Leaves opposite, simple, narrow, sessile or petioled. Heads many-flowered, radiate, borne singly. Involucre broadly bell-shaped, the bracts 10–12, in 2 series, the outer ones oblong to somewhat obovate, blunt, the inner ones shorter and narrower. Receptacle flat or slightly convex, the receptacular bracts linear and bristlelike. Pappus absent or consisting merely of minute points at the angles of the achenes. Ray flowers many, female, fertile, the corollas minute, white. Disk flowers many, bisexual, fertile, the corollas minute, white, tubular, usually 4-toothed.

Style branches flattened, the appendages short, rounded, hairy. Achenes 3- to 4-sided, or those of the disk flowers laterally flattened, roughened on the sides, hairy at the apex.

One species in Kansas.

E. álba (L.) Hassk.   Yerba de tajo

L. *albus*, white—indicating the flower color. Spanish *yerba*, herbe, $+$ *tajo*, ditch—suggested by the frequent occurrence of the plants along ditches and in wet places.
White; mid-July–early October
Wet places
East two-thirds, primarily; a few records from the west third

Rather nondescript plants, 0.1–1 m tall. Leaves short-petioled or sessile, the blades thin, lance-elliptic to linear-oblong, 1–10 cm long and 0.4–3 cm wide, the apex rounded on some of the lower or earlier leaves, but mostly acute to attenuate, the base variously tapered or, especially on the upper leaves, often with rounded auricles more or less clasping the stem, the margins entire or remotely serrate. Leaves and stems with stiff, appressed hairs. Heads about 5–10 mm across, on terminal and axillary peduncles. Involucre 5–8 mm long, the outer bracts ovate- to lance-acuminate. Rays white, about 2 mm long, quite inconspicuous. Achenes 2–2.5 mm long.

ENGELMÁNNIA T. & G.   Engelmannia

In honor of Dr. George Engelmann (1809–1884) of St. Louis, botanical explorer and writer.

The genus consists of but a single species, with characters as described below.

E. pinnatífida T. & G.   Engelmannia

NL., from L. *pinnatus*, feathered, from *pinna*, feather, + *fidus*, split, from *findere*, to split—in reference to the leaves.
Yellow; mid-May–mid-September
Dry prairie hillsides, and uplands and roadsides
West half and Sumner County

650. *Engelmannia pinnatifida*

Erect, coarsely hairy perennial with 1 to few stems 1.5–7.5 dm tall arising from a branched, woody caudex and a strong, deep taproot. Leaves alternate, simple, petioled below, sessile above, the blades deeply pinnatifid, varying in general outline from oblanceolate or ovate below to lanceolate or triangular above, 3–16 cm long and 1.5–7.5 cm wide, the lobes mostly oblong to linear, sometimes with a few teeth. Heads many-flowered, radiate, showy, 2.5–4 cm across, in branched terminal clusters. Involucre hemispheric, 7–12 mm long, the bracts in 2 or 3 series, the outermost narrow and green, the inner ones each dilated, whitish, and rather hard toward the base. Receptacle flat, chaffy. Pappus an irregular crown of minutely hairy, eventually deciduous scales or short awns. Ray flowers yellow, 8–10, female, fertile, the ligule 1–2 cm long, the style branches slender and elongated. Disk flowers many, bisexual, infertile, the corolla yellow, gradually dilated upward, prominently veined, 5-toothed; anthers with 2 basal teeth, style undivided. Achenes obovate, flattened, minutely hairy, each face with 1 prominent rib or keel.

HELIÁNTHUS L.   Sunflower

NL., from Gr. *helios*, sun, + *anthos*, flower.

Coarse, erect annual or perennial herbs. Leaves opposite or alternate, simple, at least the lower ones petioled. Heads large, many-flowered, radiate, borne singly on naked terminal or axillary peduncles or in few- to several-flowered clusters. Involucre saucer-shaped to hemispheric, the bracts many, in 2–4 series, usually green and somewhat leaf-like, graduated in length or all nearly equal in length. Receptacle flat or convex, chaffy, the receptacular bracts folded around the disk flowers. Pappus of 2 elongated, scalelike awns, sometimes accompanied by smaller scales, all deciduous. Ray flowers several to many, usually female, infertile, the corollas yellow, 3-toothed. Disk flowers many, bisexual, fertile, the corollas tubular, usually with a constriction just above the ovary, yellow below with 5 yellow, brown, red, or purplish teeth; anthers entire or with 2 minute basal lobes; style branches flattened, the appendages short or elongated. Achenes thick, slightly compressed at right angles to the involucral bracts, usually with 2 obvious angles and 2 obscure angles.

Eleven species in Kansas. *H. ciliatus* DC. and *H. strumosus* L. are infrequently encountered and are not included here.

We hardly need to be reminded that the sunflower became the state flower of Kansas by legislative enactment in 1903. When we read the statute conferring this distinction on the sunflower we make the happy discovery that, though cast in legalistic form, it has the eloquence of a panegyric:

"Whereas, Kansas has a native wild flower common throughout her borders, hardy and conspicuous, of definite, unvarying and striking shape, easily sketched, moulded, and carved, having armorial capacities, ideally adapted for artistic reproduction, with its strong, distinct disk and its golden circle of clear glowing rays—a flower that a child can draw on a slate, a woman can work in silk, or a man can carve on stone or fashion in clay; and

"Whereas, This flower has to all Kansans a historic symbolism which speaks of frontier days, winding trails, pathless prairies, and is full of the life and glory of the past, the pride of the present, and richly emblematic of the majesty of a golden future, and is a flower which has given Kansas the world-wide name, 'the sunflower state'; therefore, *Be it enacted by the Legislature of the State of Kansas:*

"That the helianthus or wild native sunflower is hereby made, designated and declared to be the state flower and floral emblem of the state of Kansas."

Although the legislature did not specify which of our species should be the state flower, *H. annuus* is the one often thought of in this respect.

Numerous records in the literature indicate that several species of sunflowers, though we cannot always be certain which ones, have been utilized by various North American Indian tribes as a food, as a medicine, or in ceremonial applications.

Do sunflowers "follow the sun," as the saying goes? The answer is that both the flower heads and the leaves have a tendency to keep faced towards the sun as long as they are capable of growth—the side of a peduncle, petiole, and leaf blade that gets the most light not growing as fast as the opposite darker side, with the result that a bending towards the light takes place. Such growth movements are common among plants.

Key to Species (3):

1. Annuals; receptacle flat or nearly so; disk nearly always red or purple 2
  2. Central receptacular bracts inconspicuously short-hairy, not at all bearded; involucral bracts chiefly ovate or ovate-oblong and abruptly contracted above the middle, usually ciliate and with some longer hairs on the back; lower leaves cordate in well-developed plants .... *H. annuus*
  2. Central receptacular bracts conspicuously white-bearded at the tip; involucral bracts lanceolate, tapering, not long-hairy, scarcely ciliate, leaves rarely cordate .... *H. petiolaris*
1. Perennials, perhaps occasionally biennial; receptacle generally convex or low conic; disk various 3
  3. Bracts of the involucre obviously graduated in length and overlapping like shingles, broad, firm, appressed, rounded to sharply acute; disk nearly always red or purple .... *H. rigidus*
  3. Bracts of the involucre not as above; disk yellow or purplish-brown 4
    4. Leaves sessile or short-petioled (petioles rarely over 15 mm long, scarcely winged), abruptly contracted or more commonly broadly rounded, truncate, or even subcordate at the base, firm, nearly entire, or shallowly toothed, the teeth seldom more than 1 mm deep 5
      5. Leaves sessile, subcordate, densely hairy with short, curved, spreading hairs on both surfaces, stem densely hairy; rays about 15–30 mm .... *H. mollis*
      5. Leaves evidently short petiolate (except sometimes in *H. divaricatus*), scabrous on the upper surface, more or less hairy on the lower, ordinarily not at all cordate; stem glabrous or hairy; rays about 5–15 .... *H. hirsutus*
    4. Leaves otherwise, either more gradually narrowed to the base or with longer or definitely winged petioles, and often with coarser teeth 6
      6. Leaves broadly lanceolate or broader, commonly at least one-third as wide as long, or over 4 cm wide, or both .... *H. tuberosus*
      6. Leaves lanceolate or narrower, at least 3 times as long as wide (generally much more) and seldom over 4 cm wide 7
        7. Leaves very narrow, not more than one-tenth as wide as long, rarely over 1 cm wide, or the lowermost occasionally proportionately and absolutely broader, disk red or purple .... *H. salicifolius*
        7. Leaves nearly always at least 1 cm wide and generally less than 10 times as long as wide, disk yellow 8
          8. Stem more or less hairy below the inflorescence, rarely nearly glabrous, leaves evidently rough-hairy on the upper surface .... *H. maximiliani*
          8. Stem glabrous below the inflorescence; leaves scarcely or not at all rough-hairy on the upper surface, the short, stiff hairs, if present, bent over and appressed .... *H. grosseserratus*

H. ánnuus L.   **Common sunflower**

L. *annuus*, annual, from *annus*, year.
Rays yellow, disk reddish- or purplish-brown; July–September
Fields, roadsides, and waste places
Throughout

A coarse, rough-hairy annual with a taproot, growing to various heights up to 5 m, giving off many spreading lateral branches when not too crowded. Leaves alternate, long-petioled, the blades cordate or deltoid below and becoming ovate or lanceolate above, 4–40 cm long and 1.5–35 cm wide, the apex acute or acuminate, the base cordate to truncate or acute upwards, the margins serrate. Heads 1 to many, mostly 6–12 cm across; involucre 1.5–3 cm long, the outer bracts mostly obovate to ovate or ovate-lanceolate, abruptly attenuate, and 5–10 mm wide, more or less hairy, the margins conspicuously ciliate. Rays yellow, 2–4.5 cm long. Disk corollas with teeth reddish or purplish. Pappus scales usually 2, readily deciduous. Bracts of the receptacle deeply 3-cleft, none of them conspicuously white-hairy at the tip. Achenes 3–5 mm long, variously colored, glabrous or nearly so.

The sunflower was held in high esteem by the Aztecs, as is shown by the fact that the Spanish conquerors found in their temples images of it wrought in pure gold; and apparently the North American Indians were cultivating the sunflower 1,000 or more years ago, for urns have been found in the habitations of the Bluff Dwellers of Arkansas and Missouri containing sunflower seeds identical with those of the monocephalic sunflowers sometimes grown in our own gardens and more or less extensively as a field crop in Russia, Japan, Manchuria, India, Egypt, Italy, Spain, France, and Germany.

The early French missionaries in North America reported that the Indians were fond of preparations comprising sunflower-seed meal, as well as the oil obtained by boiling the meal in water and skimming off the oil as it rose to the surface.

The Huron Indians used the oil for hair oil as well as for food. For a lengthy discussion of the common sunflower among North American Indians see Heiser (60, 61).

About 2 centuries later when the explorers of the Lewis and Clark Expedition were along the Missouri River in western Montana, the following entry was made in their journal, July 17, 1805:

"Along the bottoms, which have a covering of high grass, we observe the sunflower blooming in great abundance. The Indians of the Missouri, more especially those who do not cultivate maize, make great use of the seed of this plant for bread, or in thickening their soup. They first parch and then pound it between two stones, until it is reduced to a fine meal. Sometimes they add a portion of water, and drink it thus diluted; at other times they add a sufficient proportion of marrow-grease to reduce it to the consistency of common dough, and eat it in this manner. This last composition we preferred to all the rest, and thought it at that time a very palatable dish."

Nearly all parts of the plant have some measure of use. The seeds contain about 20 percent of palatable oil and 16 percent of protein. The expressed oil is good for table use, cooking, soap-making, and burning; and after the oil has been expressed the residue or "oil cake" makes a good food for cattle and poultry. The young flower heads while still in bud are boiled and eaten like French artichokes. The plants make good fodder, and sunflower silage has about 90 percent of the food value of corn silage. By retting the stalks, a floss is obtained which the Chinese make into cloth in combination with silk. The pith is lighter than cork and has been used for making life preservers.

See plate *126*.

651. *Helianthus grosseserratus*

H. grosseserràtus Martens   Sawtooth sunflower

NL., from *grossus*, coarse, + *serratus*, sawlike, from *serra*, saw—in reference to the leaf margins.
Rays yellow, disk yellow; (July–) mid-August–mid-October
Roadsides, fence rows, pastures, occasionally in open woods
East two-fifths

Perennial 2–3 (–5) m tall, with stems glabrous and glaucous below, often reddish, usually rough in the inflorescence. Leaves mostly alternate, although the lower ones may be opposite, the blades lanceolate to lance-linear or linear-elliptic, 6–23 (–32) cm long and 1–6 (–9) cm wide, the apex attenuate or less frequently acute, the base abruptly or gradually narrowed to the winged petiole, the margins serrate, the lower surface with soft, appressed hairs, the upper surface with some hairs and usually a bit rougher. Heads several to many, 4–8 cm across, mostly on elongate peduncles in branched clusters. Involucre 8–25 mm long, the bracts lance-attenuate, ciliate at least below the middle, otherwise glabrous or but sparsely appressed-hairy. Rays yellow, 1.5–4.5 cm long. Disk flowers yellow. Receptacular bracts mostly entire, with a ciliate, triangular or acuminate tip, the outermost ones sometimes with 2 small lateral teeth. Achenes glabrous, 3–4 mm long.

652. *Helianthus hirsutus*

H. hirsùtus Raf.

L. *hirsutus*, rough, hairy, shaggy—in reference to the rough leaves.
Rays and disk flowers yellow; late July–early October
Open woods and thickets, occasionally in prairie or along roadsides
East third and Republic County

Perennial, the stem 0.4–2 m tall, simple below the inflorescence and sparsely covered with coarse, spreading, siliceous hairs, arising from a rhizome bearing numerous slender, stoloniferous branches. Leaves opposite, short-petioled, the blades mostly lance-attenuate to long-triangular, usually with 3 prominent palmate veins, the base more or less abruptly narrowed to the winged petiole, the margins serrate or nearly entire, both surfaces rough-hairy. Heads 1 to few, mostly 4–8 cm across. Involucre 1–2 cm long, the outer bracts lance-linear to linear-attenuate, the margins ciliate, the backs more or less rough-hairy. Rays yellow, (1–) 2–4 (–5) cm long. Disk flowers yellow. Bracts of the receptacle 3-cleft and the middle tooth longest, acute or short-acuminate, more or less densely hairy. Achenes about 4.5–5 mm long.

See plate *127*.

H. maximiliàni Schrad.   Maximilian sunflower

Named after Maximilian Alexander Philip, prince of Neuwied, who undertook scientific explorations in South America (1813–1817) and in North America (1832–1834).

653. *Helianthus maximiliani*

654. *Helianthus mollis*

655. *Helianthus petiolaris*

Rays yellow, disk yellow; late July–mid-October
Prairie, roadsides, pastures, waste ground
Nearly throughout, but less common in the west fourth

Perennial with 1 or more stems up to 3 m tall arising from a stout rhizome. Stems moderately to densely covered with appressed hairs, at least above, glabrous or nearly so below, often reddish. Leaves opposite below, becoming alternate above, sessile or with a short, winged petiole, the blades lance-linear to linear-elliptic, 7–22 (–30) cm long and 1–2.5 (–5.5) cm wide, often light green and folded along the midrib, the apex attenuate, the margins entire or remotely and shallowly serrate, both surfaces with stiff, appressed hairs. Heads several to many, 4–8 cm across, usually on short peduncles in a racemelike inflorescence. Involucre 1.5–2.5 cm long, the outer bracts lance-attenuate, moderately to densely white-hairy. Rays yellow, 1.7–3.7 cm long. Disk flowers yellow. Receptacular bracts mostly entire, acuminate, hairy. Achenes 3–4 mm long.

H. *móllis* Lam.   Ashy sunflower

L. *mollis*, soft—referring to the dense coating of fine hairs on the under side of the leaf and the bracts and stems of the inflorescence.
Rays and disk flowers yellow; late July–mid-October
Prairie, pastures, roadsides, fields, and waste ground
Primarily in the east fourth, south of the Kansas River; scattered records west to Cloud and Kingman counties

Perennial with 1 or more stout stems more or less densely covered with long, spreading hairs, 0.5–1 m tall, simple below the inflorescence, arising from a thick, branched, creeping rhizome and fibrous roots. Leaves grayish-green, opposite, sessile, cordate to lanceolate, usually crowded on the stem, 5–11 cm long and 2.5–7.5 cm wide, becoming gradually smaller upward, the apex acute or acuminate, the base rounded or cordate, the margins shallowly serrate, both surfaces but especially the lower one with appressed white hairs. Heads 1 to several, 4–7 cm across, sessile or peduncled in a terminal cluster. Involucre 1–2 cm long, the outer bracts lance-acuminate to lance-linear, 2–4 mm wide, densely covered with long white hairs. Rays yellow, 1.5–3 cm long. Disk flowers yellow. Bracts of the receptacle entire, the tips acuminate and finely hairy. Achenes about 4 mm long.

The beauty of this species in its wild state has caused its introduction into the flower garden, but it is reported that cultivation induces a rankness of growth that detracts from its charm.

H. *petiolàris* Nutt.   Prairie sunflower

NL. *petiolaris*, petioled, from L. *petiolus*, little foot, stem of a fruit—in botany, coming to mean "stem of a leaf."
Rays yellow, disk purplish-brown; late May–September
Sandy soil of roadsides, fields, pastures, waste ground, and prairies
West three-fifths, primarily; scattered in the east two-fifths

Annual, much resembling *H. annuus* but tending to be shorter and smaller in all respects. Leaves alternate, petioled, the blades mostly lanceolate to lance-triangular, sometimes ovate or nearly cordate, 3–10 cm long and 0.8–7 cm wide, the apex variously sharp-pointed or somewhat rounded, the base acute to truncate or nearly cordate, the margins entire or shallowly serrate. Herbage more or less densely covered with stiff, appressed hairs, the under surface of the leaves usually more so than the upper. Heads 1 to several, 5–9 cm across; involucre 1–2 cm long, the outer bracts mostly 3–6 mm wide, lance-acuminate to lance- or linear-attenuate, covered with appressed hairs but not conspicuously ciliate. Rays yellow, 1.8–4.5 cm long. Disk corollas with teeth reddish-purple, the pappus of 2 thin, scalelike awns, falling easily. Bracts of the receptacle 3-cleft, those near the center of the head with conspicuous white hairs at the tip (easily seen without magnification). Achenes 3.5–4.5 mm long, covered with silky, appressed hairs.

H. *rígidus* (Cass.) Desf.   Stiff sunflower

L. *rigidus*, stiff, in reference to the leaves.
Rays yellow, disk reddish-purple or yellow; mid-May–September
Prairie, pastures, and roadsides
East half

Perennial with stems 0.5–2 m tall, simple or with a few ascending branches above, with a stout rhizome bearing fibrous roots and slender, stoloniferous rhizomes. Leaves opposite, the lowermost with winged petioles, the uppermost sessile, the blades rather thick, light green or gray-green, broadly lanceolate to narrowly elliptic or the uppermost nearly linear, 6–23 cm long and 1–7 cm wide, the apex acute or short-attenuate, the base more or less gradually tapered, the margins entire or serrate, both surfaces quite rough. Heads 1 to several on long terminal peduncles, 5–8 cm across. Involucre 1–1.5 cm long, the bracts definitely graduated in length, the outermost ovate to lanceolate with an acute (or occasionally rounded) tip and finely ciliate margins, but usually glabrous on the back. Rays yellow, 1.3–3.5 cm long. Disk corollas with teeth reddish-purple or occasionally entirely yellow. Bracts of the receptacle entire, with an acute or rounded tip, grading into the involucral bracts. Achenes 5–6 mm long, glabrous or sparsely hairy.

See plate *128.*

656. *Helianthus rigidus*

H. salicifòlius A. Dietr.   Willowleaf sunflower

NL., from L. *salicinus*, willowlike, from L. *salix*, willow, + *folium*, leaf.
Rays yellow, disk purplish-brown; (mid-July–) September–October
Rocky hillsides and banks
East third, primarily

Glabrous perennial, 1.5–3 m tall, with a large elongate rhizome and thickened roots. Leaves alternate, sessile, usually crowded on the stem, linear to narrowly linear-elliptic, 7–17 cm long and 1–6 mm wide, the apex attenuate, the margins entire or occasionally shallowly and remotely toothed, the lowermost leaves usually falling early in the season. Heads several to many, 3–6 cm across, in branched terminal clusters. Involucre 0.8–1.5 cm long, the outer bracts usually lance- or linear-attenuate, glabrous but with ciliate margins. Rays yellow, 1–2.5 cm long. Disk flowers purplish-brown. Receptacular bracts entire, with an acute apex, or shallowly 3-toothed. Achenes glabrous, 4.5–6 mm long.

657. *Helianthus salicifolius*

H. tuberòsus L.   Jerusalem-artichoke sunflower

L. *tuberosus*, full of swellings, from *tuber*, swelling, knob, or excrescence—here in reference to the tuberous swellings of the rhizomes. "Jerusalem" in this instance is merely a corruption of Italian *girasole*, turn-sun, in reference to the turning of the young flower heads toward the sun. "Artichoke" is a corruption of the Italian *articiocco*, the full Italian name for this species being *girasole articiocco*.
Rays yellow, disk yellow; mid-August–early October
Roadsides, pastures, prairie, waste ground, creekbanks, usually in low, moist ground

Perennial with 1 or more stems 1–2.8 m tall, rough or sometimes nearly glabrous, arising from a thickened rhizome which typically produces slender rhizomes with terminal tubers. Leaves opposite below, sometimes becoming alternate above, the blades mostly lanceolate to somewhat triangular, 8–21 (–30) cm long and 2.5–11 (–20) cm wide, the apex acuminate to attenuate, the base abruptly or gradually contracted to the winged petiole, the margins serrate or nearly entire, the upper surface roughened with short, stiff, siliceous hairs, the lower surface with softer, finer hairs. Heads several to many, 5–8 cm across, usually in a branched terminal cluster. Involucre 0.8–1.6 cm long, the outer bracts lanceolate to lance-linear, the margins ciliate, the backs more or less hairy. Rays yellow, 1.5–4 cm long. Disk flowers yellow. Receptacular bracts 3-toothed, the middle tooth triangular, the largest, and white-hairy. Achenes glabrous, 5–7 mm long.

The Indians supplemented their corn, beans, and squashes with Jerusalem artichoke tubers raw or cooked, harvesting the tubers from wild plants or plants cultivated outside their natural range.

The Jerusalem artichoke was introduced into England in 1617 and became popular for table use in some quarters, even being declared by the herbalist Parkinson "a dainty for a queen." The English used it simply as a vegetable, or made a stew of it with butter, wine, and spices, and even compounded it with marrow, dates, ginger, raisins, and sack for making pies. As time went on, however, the novelty of it wore off and it lost caste as a dainty and became instead a highly rated feed for cattle and hogs. In our own country at the present time the artichoke is grown mostly in the South and the Pacific Northwest for feeding livestock; but since it is relatively free from insects and disease, it succeeds, even without cultivation, in any warm, well-drained soil, and can be harvested over the long period from fall till spring. At one time at least, it was considered as a potential source of alcohol and synthetic rubber.

658. *Helianthus tuberosus*

HELIÓPSIS Pers.   Heliopsis, oxeye

NL., from Gr. *helios*, sun, + *opsis*, appearance—in reference to the radiate composite heads.

Erect perennial herbs. Leaves opposite, simple, petioled, toothed. Heads many-

659. *Heliopsis helianthoides*

660. *Melampodium leucanthum*

flowered, radiate, showy, borne singly or in few-flowered terminal clusters. Involucre bell-shaped or hemispheric, the bracts in 2 or 3 series, ovate to ovate-lanceolate, nearly equal in length, herbaceous or somewhat leathery. Receptacle convex or conic, often hollow, chaffy, the receptacular bracts linear-acute, about as long as the disk flowers and folded around them, persistent. Pappus absent, or merely a 1- to 4-toothed, ringlike border. Ray flowers 10 or more, female, fertile; corollas yellow, the ligules oblong, apically notched, more or less persistent on the achenes. Disk flowers many, bisexual, fertile; corolla yellowish, brownish, or purple, cylindric; anthers entire or nearly so at the base; style branched, flattened, with short, hairy appendages. Achenes smooth, thick, truncate, those of the disk flowers 4-angled, those of the ray flowers 3-angled.

One species in Kansas.

H. helianthoìdes (L.) Sweet · Sunflower heliopsis

NL., from the sunflower genus, *Helianthus*, + the Gr. suffix *-oides*, like, resembling—because of a general resemblance to some sunflowers with yellow disk flowers, such as *Helianthus tuberosus*.
Yellow; late July–October
Roadside ditches, low pastures, creekbanks, and thickets
East third, primarily

Perennial by its herbaceous stem which each spring produces erect, more or less rough-hairy stems 0.4–2 m tall. Leaf blades ovate or triangular to lanceolate, usually with 3 prominent veins arising from the base, 4–14 cm long and 1.5–8 cm wide, the apex acute, the margins coarsely serrate, both surfaces covered with rough, siliceous-based hairs or sometimes merely slightly rough. Heads mostly 3–7 cm across on stout, naked peduncles. Involucre 7–20 mm long, the bracts oblong with a rounded apex to broadly lanceolate with an acute apex, rather thick, rough-hairy. Rays dark, rich yellow, 1–3 cm long. Achenes dark brown, 4–5 mm long, the pappus scarcely evident. Disk flowers yellow.

Plants with very rough herbage have been recognized in the past as *H. scabra* and those with smooth herbage as *H. helianthoides*, but most manuals now include them in a single species. Both types occur throughout the range of the species in Kansas.

MELAMPÒDIUM L.

NL., from Gr. *melas*, black, + Gr. *podion*, footlike part. This name was drawn by Linnaeus from a list of ancient plant names and apparently applied to this genus without particular reason. The name was anciently used for *Helleborus niger* which has a short black rootstock.

Annual or perennial herbs. Leaves opposite, simple, sessile, entire to toothed or pinnately lobed. Heads many-flowered, radiate, many, but borne singly at the ends of stems and branches. Involucre bowl-shaped, the bracts with an outer series of 4–5 partly united, lance-ovate bracts and an inner series of concave bracts, each subtending a ray flower and falling away with the achene. Receptacle convex or conic. Pappus absent. Ray flowers female, fertile, the corollas yellow or white, the ligules 2- or 3-toothed, the style branches flattened along the inner face, stigmatic along the margins all the way to the tip. Disk flowers few to many, bisexual, infertile, each nearly enclosed by a membranous, deciduous bract with a conspicuous midrib and a toothed or laciniate tip; corollas yellow, more or less funnelform, 5-lobed; anthers entire at the base; ovaries linear, nonfunctional, the style undivided. Achenes of the ray flowers short and thick, broadened upward, not flattened but somewhat incurved.

One species in Kansas.

M. leucánthum T. & G. · Plains blackfoot

NL., from Gr. *leukos*, white, + *anthos*, flower.
Rays white, disk yellow; late May–early October
Dry prairie hillsides and uplands
Southwest fourth

A tufted perennial with many slender stems 0.8–3 dm tall arising from a woody crown above a strong taproot. Leaves narrowly oblanceolate to narrowly elliptic or linear-elliptic, 2–4 cm long and 1.5–9 mm wide, the apex acute or rounded, the base wedge-shaped, the margins entire or with a few shallow teeth. Herbage densely covered with stiff, appressed hairs and resin dots. Heads 12–25 mm across. Outer involucral bracts 4–7 mm long, broadly rounded at the tip. Rays white, broad, 7–11 mm long. Achenes completely enclosed, except at the very apex, by the inner series of involucral bracts, each bract rather thick and hardened and strangely sculptured, the lower portion covered with sharp protuberances, the apex usually extended into a hood- or cowl-like structure, the whole together about 3.5–5 mm long.

This is a handsome plant as we see its spreading branches covered with flowers. It is strongly resistant to drought on the plains of southwest Kansas; and when we dig one up, shake off its soil, and put it in a bowl of water, it keeps indefinitely. Sometimes this species occurs in groups intermingled with *Zinnia grandiflora*, blooming together in beautiful association over considerable areas.

See plate *129*.

RATÍBIDA Raf.   Prairie coneflower

Meaning of the name uncertain.

Erect biennial or perennial herbs. Stems longitudinally ribbed. Leaves alternate, simple but often deeply divided into pinnate or twice-pinnate segments, rough-hairy. Heads showy, many-flowered, radiate, borne singly on naked peduncles. Involucre short, obconic, the bracts few, small, in 2 series, more or less herbaceous, linear to linear-lanceolate. Receptacle columnar, chaffy, the receptacular bracts folded around the achenes and falling with them. Pappus absent or a minute crown with 2 awnlike teeth. Ray flowers few, neutral, infertile, the ligule yellow, red-brown, or yellow with a red-brown basal spot, strongly reflexed or drooping. Disk flowers many, bisexual, fertile; corollas short-cylindric, greenish to purplish-red-brown, 5-toothed; anther bases sagittate; style branches slightly flattened, with ovate to subulate, hairy appendages. Achenes flattened at right angles to the involucral bracts, 4-angled.

Three species in Kansas.

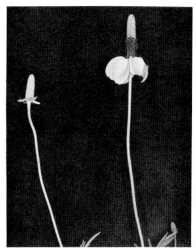

661. *Ratibida columnifera*

Key to Species:

1. Disk oblong-cylindric, 1–4 cm long; smallest subdivisions of the leaves linear-elliptic or oblong, 2–8 mm wide, occasionally with 1 tooth but never serrate; achenes with a pappus of 1 or 2 short, sharp awns; throughout the state but more common westward .... *R. columnifera*
1. Disk globose to ovoid or broadly ellipsoid, mostly 1–2 cm long; achenes and subdivisions of the leaves not as above 2
   2. Rays yellow, 1.5–6.5 cm long; achenes lacking a pappus entirely; east fourth of the state .... *R. pinnata*
   2. Rays yellow or entirely or partly purplish-brown, 3–10 mm long (occasionally absent); achenes with a crown of thickened teeth and a deeply cleft marginal wing; primarily in the southwest fourth .... *R. tagetes*

R. columnífera (Nutt.) Wooton & Standl.   Columnar prairie coneflower

NL., from L. *columna*, column, + *ferre*, to bear—indicating the columnar disk, a most distinctive feature of the head.
Rays yellow or basally or entirely purplish-brown, disk eventually purplish-brown; June–early September
Prairie and roadsides
Throughout; more common westward

Herbaceous perennial, 2–11 dm tall, usually with few to several strongly ascending branches. Leaf blades 2.5–11 cm long, deeply pinnately divided quite or nearly to the midrib into entire (occasionally with 1 tooth), linear-elliptic or oblong, sharply pointed segments 2–8 mm wide. Herbage gland-dotted and covered with stiff, appressed hairs. Flowering peduncles 6–30 cm long. Rays yellow in var. *columnifera*, purplish-brown or yellow with a purplish-brown base in var. *pulcherrima* (L. *pulcherrimus*, prettiest), mostly 1–3 cm long. Disk flowers greenish or purplish-brown, the disk itself becoming strongly columnar, 1–4 cm long and 7–10 mm wide. Tips of the style branches blunt. Bracts of the receptacle hairy at the tip. Achenes gray, strongly flattened, glabrous or with a fringe of hairs along one edge, about 2–2.5 mm long, the pappus of 1 or 2 prominent spinelike awns.

See plate *130*.

R. pinnàta (Vent.) Barnh.   Pinnate-leaf prairie coneflower

L. *pinnatus*, feathered, from *pinna*, feather—the leaves being divided transversely in the manner of a feather.
Rays yellow, disk eventually purplish-brown; late June–early August (sometimes again in the fall)
Prairie, roadsides, thickets, creekbanks, and woodland openings
East fourth

Herbaceous perennial, the stems 0.4–1.6 m tall, simple or with a few ascending branches, arising from a compact rhizome with many stout, deep fibrous roots. Leaf

662. *Ratibida pinnata*

blades 3.5–13 cm long, deeply pinnately divided nearly or quite to the midrib into usually 3 or 5 entire or serrate segments 0.5–3 cm wide, these often deeply 2- or 3-lobed, the ultimate segments linear-elliptic to lanceolate, the veins of the lower surface quite prominent. Herbage covered with stiff, mostly appressed hairs. Flowering peduncles 6–20 cm long. Rays yellow, 1.5–6.5 cm long, drooping. Disk flowers greenish or purplish-brown, the tips of the style branches subulate, the disk itself globose to broadly ellipsoid, 1–2 cm long and 12–15 mm wide. Bracts of the receptacle densely hairy at the tip. Achenes gray, glabrous, 2–3 mm long, lacking a pappus entirely.

R. tagètes (James) Barnh.   Short-rayed coneflower

Named after the Mexican genus *Tagetes*, which includes the horticultural French marigolds, some
    species of which superficially resemble this species of *Ratibida*.
Rays yellow and/or purplish-brown, disk eventually purplish-brown; mid-May–early August
Dry, rocky or sandy prairie or pastures
Southwest fourth, primarily; but north to Cheyenne and Sheridan counties and east to Saline
    County

Herbaceous perennial with 1 or more much-branched stems 1.5–5 dm tall arising from a short crown and strong, deep taproot. Leaf blades 2–7.5 cm long, the lowermost narrowly oblanceolate with 2 linear lateral lobes reaching about halfway to the midrib, the upper leaves more deeply divided and with 2–4 lateral lobes 1–2 mm wide. Flowering peduncles 0.5–4 cm long. Rays yellow, purplish-brown, or yellow with a purplish-brown base, 3–10 mm long, or sometimes absent altogether. Disk flowers greenish- or purplish-brown, the disk itself globose to ovoid or broadly ellipsoid, 9–13 mm long and 6–11 mm wide. Tips of the style branches blunt. Achenes gray, 2.5–3 mm long, with a crown of thick, rounded teeth and with a deeply cleft, fringed wing along 1 margin.

RUDBÉCKIA L.   Coneflower

Named in honor of Olaus Rudbeck (1630–1702) and his son, teachers of anatomy and botany at
    the University of Upsala.

Erect annual, biennial, or perennial herbs. Leaves alternate (sometimes opposite at the base), simple, entire to pinnatifid. Heads showy, many-flowered, radiate, borne singly at the end of the stem or branches. Involucre hemispheric, the bracts in 2–3 series, graduated in length or of about the same length, green and more or less herbaceous, spreading or reflexed. Receptacle conical or columnar, chaffy, the receptacular bracts narrow, membranous, acute. Pappus absent or a crown of 2–4 minute teeth. Ray flowers several to many, neutral, infertile, the corollas yellow or orange-yellow, sometimes with a brown basal spot. Disk flowers many, bisexual, fertile; corolla tubular, usually narrowed toward the base, 5-toothed, brown; anthers more or less truncate at the base; style branches flattened, with hairy, blunt or subulate tips. Achenes 4-angled in cross section, with a flat apex, smooth.

Six species in Kansas. *R. grandiflora* (D. Don.) Gmel. is known only from Cherokee, Crawford, and Labette counties where it occurs in prairie meadows. It usually has a single unbranched stem bearing a single large terminal head with drooping yellow rays and several long-petioled leaves, the latter being elliptic to lanceolate with 5 prominent parallel veins, and serrate margins, and rough-hairy on both surfaces.

Various species of *Rudbeckia* have been used to produce a yellow or green dye, usually with alum and cream of tartar as a mordant.

Key to Species:
1. Glabrous annual; leaf bases cordate and clasping the stem; style branches subulate;
    achenes round in cross section .... *R. amplexicaulis*
1. Plants perennial, usually hairy; leaf bases not clasping the stem; style branches blunt
    except in *R. hirta*; achenes square in cross section 2
    2. Leaves coarsely pubescent, rather thick, oblanceolate below to elliptic, lanceolate or
        lance-linear above, the margins entire or remotely toothed, never lobed; style branches
        attenuate; achenes lacking a pappus .... *R. hirta*
    2. Leaves glabrous or variously hairy, at least some of them deeply 3- or 5-lobed (some-
        times all of them entire in *R. triloba*); style branches blunt; pappus a low, toothed
        membranous crown 3
        3. Disk flowers yellow; receptacular bracts truncate, densely hairy; plants glabrous or
            nearly so .... *R. laciniata*
        3. Disk flowers purplish-brown; receptacular bracts not as above; plants sparsely to
            densely hairy 4
            4. Leaves thin, nearly glabrous, entire or the middle ones 3-lobed; receptacular
                bracts abruptly attenuate, glabrous .... *R. triloba*

4. Leaves rough on the upper surface, densely hairy on the under surface, some or all of them always 3- or 5-lobed; receptacular bracts broadly acute to somewhat rounded, densely hairy .... *R. subtomentosa*

R. amplexicaùlis Vahl.   Coneflower

NL., from L. *amplexus*, past participle of *amplecti*, to wind around, + *caulis*, stem—in reference to the leaf bases, which are cordate and clasp the stem.
Rays yellow or orange, sometimes with a purplish-brown base, disk purplish-brown; late May–July (rarely again in the fall)
Prairies, pastures, ditches, usually in moist soil
Primarily south of a line through Sumner, Greenwood, Woodson, and Crawford counties

Glabrous, glaucous annual, 3–7 dm tall, the stems eventually branched. Leaves 3–13 cm long and 1.5–4.5 cm wide, spatulate or oblanceolate below, becoming lanceolate or cordate above, the apex usually acute or acuminate, the base cordate and clasping the stem, the margins entire or with a few teeth. Heads 1 to many, 2–4.5 cm across, the disk becoming columnar. Involucre 6–11 mm long, the bracts lanceolate to lance-linear, glabrous. Rays 6–10, yellow or orange, sometimes purplish-brown near the base, 7–25 mm long. Receptacular bracts abruptly mucronate or short-acuminate, mostly glabrous except for the finely ciliate margins. Disk flowers purplish-brown, the style branches attenuate. Achenes about 2 mm long, round in cross section, lacking a pappus.

This species is sometimes placed in a separate genus, *Dracopsis*, by itself.

663. *Rudbeckia amplexicaulis*

R. hírta L.   Black-eyed Susan

L. *hirtus*, shaggy, rough, covered with short, stiff hairs.
Rays orange-yellow, disk purplish-brown; June–August (–September)
Prairie, pastures, brushy hillsides
East two-fifths, primarily; scattered records farther west

Erect, coarsely pubescent biennial with 1 to few stems 1–10 dm tall, usually with a few ascending branches. Leaves 1.5–11 cm long, the lower ones mostly oblanceolate, often tapered to a petiolelike base, the middle and upper leaves becoming elliptic to lanceolate or lance-linear and gradually reduced in size upwards, the apex usually acute, the margins entire or remotely serrate. Heads few to many, 2.5–5 (–8) cm across, the disk conical. Involucre 1–2.2 cm long, the bracts linear-oblong or subulate, reflexed in flower, conspicuously long-hairy on both surfaces. Rays 8–21, orange-yellow, 1–3 cm long. Disk flowers purplish-brown, the style branches elongate and sharply pointed. Bracts of the receptacle hairy near the tip and stiffly ciliate (use magnification). Achenes gray or brown, about 1.5 mm long and 0.3 mm across, 4-angled in cross section, lacking a pappus.

This is our commonest *Rudbeckia*. It is not imposing in size like the cutleaf coneflower (*R. laciniata*) nor so self-assertive as the sweet coneflower (*R. subtomentosa*), elbowing tall wayside weeds. However, as we find it in the prairie meadows, blooming as dispersed individual stems or in clumps, before being hidden by grasses, it seems one of the richest bits of the prairie mosaic.

664. *Rudbeckia hirta*

Bees gather both nectar and pollen from the disk flowers.

The Forest Potawatomis treated colds with a tea prepared from the roots of the black-eyed Susan.

See plate *131*.

R. laciniàta L.   Cutleaf coneflower

NL., from L. *lacinia*, fringe—in reference to the divided condition of the leaves.
Rays yellow, disk yellow; August–October
Moist soil of creekbanks, thickets, and open woods
East third, primarily

Herbaceous perennial, glabrous or nearly so, with branched stems up to 1.5 m tall. Main leaves broadly lanceolate to very broadly ovate in general outline, deeply 3- or 5-lobed, 7–25 cm long and 2.5–20 cm wide, the margins serrate and sometimes laciniate as well; upper leaves much smaller, lanceolate, the margins entire or merely serrate. Heads several to many, mostly 3–8 cm across, the disk hemispheric. Involucre 7–14 mm long, the bracts elliptic to oblong or linear with acute tips, reflexed, glabrous or sparsely hairy on the underneath (abaxial) side. Rays 6–13, yellow, drooping, 1–5 cm long. Receptacular bracts truncate, densely glandular-hairy (use magnification). Disk flowers yellow, the style branch tips ending bluntly with a globose tuft of hairs. Achenes about 4–5 mm long, square in cross section, the pappus a very low, membranous, toothed crown.

Cutleaf coneflower plants, steeped as a tea, have been used to increase urine flow and to treat kidney problems and Bright's disease. A tea made from this plant and the

665. *Rudbeckia laciniata*

666. *Rudbeckia subtomentosa*

667. *Rudbeckia triloba*

stem of stork's bill (*Erodium cicutarium*) has been utilized in the Southwest to treat gonorrhea or to induce menstruation (30).

R. subtomentòsa Pursh Sweet coneflower

L. *sub*, slightly, + *tomentum*, stuffing of wool, hair, or feathers—the stems and under surfaces of the leaves made ashen by a coating of downy hairs.
Rays yellow, disk purplish-brown; (mid-July–) August–September
Low, moist soil of thickets, brushy pastures, creekbanks, prairie ravines, and ditches
East column of counties, west in the north to Jackson and Osage counties

A vigorous, pubescent perennial with several stiff, erect stems 1–2 m tall arising from a thick, irregular rhizome which provides a circle of strong fibrous roots. Leaves similar to those of *R. laciniata*, but densely soft-hairy on the under surface and rough on the upper surface. Heads several to many, 4–8 cm across, the disk hemispheric to broadly ovoid. Involucre 7–17 mm long, the bracts lance-linear to linear-attenuate, densely covered with white hairs on the lower (abaxial) surface. Rays 12–21, yellow, 2–4 cm long. Receptacular bracts broadly acute to somewhat rounded, densely hairy (use magnification). Disk flowers purplish-brown, the tips of the style branches truncate and with a minute tuft of hairs. Achenes square in cross section, the pappus a minute toothed crown.

The attitude of the whole plant is one of self-confidence and vigor; yet it is not constituted to endure in dry situations, and we find it in drainage areas where moisture is available near the surface. As long as its fibrous root system can find water, it can rival the sunflower in show of flowers. As in all our Rudbeckias, a large variety of insects find in the disk flowers, according to their needs, nectar and/or pollen. The recipients of this bounty include the honeybee and many kinds of wild bees and other hymenoptera, butterflies, flies, bugs, and beetles.

R. tríloba L. Brown-eyed Susan

NL., from Gr. *tri-*, 3, + *lobos*, rounded projection or lobe—in reference to the 3-parted condition of some of the lower leaves.
Rays yellow or orange, sometimes with a dark-purplish base, the disk purplish-brown; mid-July–October
Moist soil, in sun or partial shade
East fifth, primarily

A much-branched, sparsely to moderately hairy biennial, up to 12 (–15) dm tall. Leaves thin, 2.5–11 cm long, the lower ones ovate to obovate and narrowed to petiolelike bases, the middle and upper ones elliptic or lanceolate becoming smaller and sessile above, the apexes mostly acuminate, the margins entire or with a few teeth, sometimes some of the middle leaves deeply 3-lobed. Heads several to many, 1.5–5 cm across, the disk hemispheric or ovoid, but not becoming columnar. Involucre 7–21 mm long, the bracts linear, reflexed in flower, glabrous above, sparsely hairy below. Rays 6–12, yellow or orange, or with a dark purplish spot at the base, 1–2.5 cm long. Receptacular bracts abruptly attenuate, glabrous, and bearing a pair of elongate, dark purple resin canals. Disk flowers purplish-brown, the style branches short and blunt. Achenes 2–3 mm long, 4-angled in cross section, the pappus a minute crown.

SÍLPHIUM L. Rosinweed

L. *silphium*, from Gr. *silphion*, for a member of the Carrot Family, used medicinally by the Greeks. In 1753 Linnaeus arbitrarily transferred the name to its present use.

Tall, coarse perennial herbs with rhizomes, often with resinous sap. Leaves alternate or opposite; simple; sessile, perfoliate, or petioled; entire to toothed or pinnatifid; usually rough. Heads many-flowered, radiate, rather large and showy, in branched terminal clusters. Involucre bowl-shaped or hemispheric, the bracts in 1 to several species, the outermost broad, firm and leaflike, the innermost becoming smaller and thinner and subtending the ray flowers. Receptacle flat, chaffy. Pappus absent, or consisting of 2 awns arising from the wings of the achenes. Ray flowers in 2 or 3 series instead of the more typical single series, female, fertile; corollas yellow (white in 1 species, not ours). Disk flowers bisexual, infertile; corollas yellow, cylindrical, 5-toothed; anthers entire or minutely tailed or toothed at the base; style undivided. Achenes broad, markedly flattened parallel to the involucral bracts, winged, with or without 2 apical awns.

Four species in Kansas. *S. integrifolium* Michx. is known from a few counties in the eastern fifth. It strongly resembles *S. speciosum* but differs in having rough-hairy stems and all the leaves entire and rough on both surfaces.

Key to Species:
1. Leaf bases united around the stem; stem definitely square in cross section .... *S. perfoliatum*
1. Leaf bases not united around the stem; stem round or but slightly 4-angled in cross section .. 1
   2. Leaves alternate, deeply pinnately lobed .... *S. laciniatum*
   2. Leaves opposite, entire or merely toothed .... *S. speciosum*

S. *laciniàtum* L.   Compass plant, rosinweed

NL., *laciniatus*, divided into small strips, from L. *lacinia*, small strip or part, lappet, fringe—in
   reference to the slender divisions of the leaf.
Yellow; late June–mid-August
Prairies
East half, primarily; a few records as far west as Ellis County

668. *Silphium laciniatum*

Herbaceous perennial, 1–3 m tall, with a massive rootstock and a taproot penetrating the soil as deep as 3 m. Stems round in cross section, sparsely covered with coarse,
spreading hairs. Leaves alternate, long-petioled below, sessile above, the blades 7–40 cm
long and 3.5–32 cm wide, progressively smaller upward, rough on both surfaces though
often more so on the lower surface, in general outline varying from ovate to lanceolate,
elliptic, oblanceolate or obovate, deeply divided into linear or subulate lobes 3–25 mm
wide, these sometimes also with a few lateral teeth or lobes. Heads 5–9 cm across. Involucre 1.5–3.5 cm long, the outer bracts ovate-acuminate to lance-acuminate, rough-hairy
and with conspicuously ciliate margins, the tips spreading or recurved. Rays yellow,
mostly 1.5–3.5 cm long.

This stately perennial is one of the most conspicuous features of the prairie flora.
Its presence in a given locality is an indication of a deep and moist subsoil. Although the
large flower heads resemble sunflowers, there are radical differences. For example, the
disk flowers of the *Silphiums* produce no seeds, this function belonging solely to the ray
flowers. In the sunflowers the situation is reversed, with only the disk flowers maturing
achenes.

Compass plants tend to align their rigid leaves vertically in a north-south plane,
hence the origin of that common name. "Rosinweed" comes from the fact that, when
wounded, the plants exude a balsamitic resin often used by children as a kind of
chewing gum.

See plate *132*.

669. *Silphium perfoliatum*

S. *perfoliàtum* L.   Cup rosinweed

NL., from *per*, through, + *folium*, leaf—the stem going through the fused bases of the paired
   leaves.
Yellow; mid-July–August (–September)
Low, moist, shaded places
East fifth, primarily; west to Washington County in the north, and also reported from Sedgwick
   County

Robust herbaceous perennial, 1–2.5 m tall, with several to many square, smooth
stems arising from the thickened rootstock, branching above and bearing numerous
flower heads. Leaves opposite, lanceolate, 12–33 cm long and 4–20 cm wide, the apex
usually acute or acuminate, the bases of the upper pairs of leaves or the broad petioles
of the lower leaves united around the stem, the margins coarsely serrate to nearly entire,
both surfaces rough. Heads 5–7 cm across. Involucre 1–2 cm long, the bracts glabrous
but with rough margins, the outermost broadly ovate to broadly lanceolate. Rays yellow,
15–35 mm long. Achene broadly obovate, about 10–12 mm long, each falling as a unit
with its subtending bract, lacking awns.

S. *speciòsum* Nutt.

L. *speciosus*, beautiful, splendid, showy.
Yellow; July–August (–September)
Weedy prairie, roadsides, creekbanks, frequently in moist, rocky soil
East two-thirds

Plants 0.5–1.2 m tall, the stems glabrous. Leaves opposite, sessile, the blades
lanceolate, 3.5–20 cm long, the apex acute or long-acuminate, the base rounded, the
margins coarsely serrate on the lower leaves, becoming entire above, the upper surface
rough, the lower one smooth. Heads 5–10 cm across. Involucre 1.5–3 cm long, the outer
bracts lance-acuminate to ovate-acuminate, smooth except for the minutely roughened
margins, the tips spreading or somewhat recurved. Rays yellow, 1.5–4 cm long. Achenes
about 10–15 mm long, apically notched.

670. *Silphium speciosum*

671. *Thelesperma filifolium*

672. *Thelesperma megapotamicum*

THELESPÉRMA Less.   Thelesperma

NL., from Gr. *thele*, nipple, $+$ *sperma*, seed—because of the conical projections or papillae on the achenes of some species.

Erect annual, biennial, or perennial herbs, sometimes somewhat woody at the base. Leaves opposite, simple, entire to finely pinnately or ternately dissected. Heads many-flowered, radiate (discoid in *T. megapotamicum*), borne singly on naked peduncles. Involucre bell-shaped or hemispheric, the bracts in 2 series, the inner ones broad, appressed, united below, with scarious margins, the outer ones shorter, narrow, more or less herbaceous, often spreading. Receptacle flat, chaffy, the receptacular bracts broad, membranous, white, 2-nerved. Pappus absent or of 2 stout, barbed awns. Ray flowers usually present, several, neutral, infertile; ligules yellow or yellow with a reddish-brown or purplish area at the base (rarely entirely reddish-brown or purplish). Disk flowers many, bisexual, fertile; corollas yellow or reddish-brown, funnelform, deeply toothed in ours; anthers rounded at the base; style branches flattened, the tips acute or subulate, hairy. Achenes more or less flattened parallel to the involucral bracts, linear to linear oblong, straight to slightly curved or hoodlike.

Two species in Kansas.

Key to Species:
1. Rays present; lower involucral bracts linear-attenuate .... *T. filifolium*
1. Rays absent; lower involucral bracts ovate to oblong with rounded tips ...... *T. megapotamicum*

T. filifòlium (Hook.) Gray   Fineleaf thelesperma

NL., from L. *filum*, thread, $+$ *folium*, leaf—in reference to the narrow pinnate segments of the leaves.
Rays yellow; mid-May–August (–mid-October)
Prairie hillsides, pastures, and roadsides, in sandy or rocky soil
Southwest corner, from Finney and Hamilton counties south

Slender, glabrous, more or less bushy-branched biennial, 1.5–5 (–10) dm tall, with a simple or branched taproot. Leaves 3.5–10 cm long, mostly pinnately or ternately divided into linear segments 0.5–2 mm wide, the uppermost especially occasionally simple. Heads radiate, 1.5–4 cm across, on peduncles 4–15 cm long, drooping in bud but becoming erect as the buds open. Outer (lower) involucral bracts linear-attenuate. Rays 8, yellow, 3-lobed, 7–16 mm long. Achenes dark gray, 4–6 mm long (including the awns), with a row of rough protuberances along the outer (abaxial) side, or the entire outer face rarely warty, the inner (adaxial) face smooth.

Our plants have been listed in some earlier manuals for this area as *T. trifidum*.

T. megapotàmicum (Spreng.) Kuntze   Rayless thelesperma

Apparently from Gr. *Mesopotamia*, the ancient country between the Tigris and Euphrates rivers, from Gr. *megas*, large, $+$ *potamikus*, pertaining to streams, although the reason for the application here is not clear.
Dull orangish-yellow; late May–August (–mid-October)
Dry prairie hillsides and uplands, pastures, and roadsides, in rocky, gravelly, frequently calcareous soil
West two-thirds, primarily; scattered records from the east third

Slender, glabrous, herbaceous perennial with 1 to few stems, 2.5–8 dm tall, branched above, arising from a short, woody crown and a slender taproot reaching down about a meter, giving off numerous laterals which sometimes go as deep as the principal root. Leaves 2.5–10 cm long, the linear segments 0.5–3 mm wide. Heads discoid, 1–2 cm across, on peduncles 8–23 cm long. Outer (lower) involucral bracts ovate or oblong, the tips rounded. Rays absent. Disk flowers dull orangish-yellow. Achenes reddish-brown or straw-colored, 6–9 mm long (including the awns), the entire outer (abaxial) face quite warty, the inner (adaxial) face usually also somewhat bumpy.

Our plants have been listed in some earlier manuals for this area as *T. gracile*.

The general aspect of the plant indicates adaptation to a dry soil and climate, and it is in just such circumstances that we meet with it over the western part of the state.

VERBESÌNA L.   Crownbeard

According to Fernald (9), the name was "metamorphosed" from the genus name *Verbena*.

Erect annual or perennial herbs. Leaves opposite or alternate, simple, petioled or at least narrowed to a petiolelike base. Heads showy, many-flowered, usually radiate, borne singly or in open terminal clusters. Involucre hemispheric to bell-shaped, the bracts in 2 to several series, usually about equal in length, the outer ones more or less herbaceous, the inner ones usually papery. Receptacle low and convex to high and conical or nearly globose, chaffy, the receptacular bracts folded at one edge of the achene. Pappus absent or of 2 deciduous or persistent awns. Ray flowers few to several (rarely absent), neutral and infertile or female and fertile, the corollas usually yellow (rarely white). Disk flowers many, bisexual, fertile; corollas tubular, 5-toothed, usually yellow; anthers rounded or minutely tailed at the base; style branches flattened, the appendages acute or attenuate, papillate or hairy. Achenes flattened at right angles to the involucral bracts, smooth or hairy, usually winged.

Four species in Kansas. *V. encelioides* (Cav.) Benth. & Hook., cowpen daisy, is known only from scattered sites across the state and is not included in the following key. It is a yellow-flowered annual with coarsely toothed, more or less triangular leaves which are white-hairy on the lower surface, and is usually found in disturbed, sandy soil.

Key to Species:

1. Flowers yellow .... *V. alternifolia*
1. Flowers white .... *V. virginica*

V. alternifòlia (L.) Britt.   Wingstem

NL., from L. *alternus*, alternate, + *folium*, leaf.
Yellow; mid-August to mid-October
Low, moist woods, frequently along stream banks
East third, primarily; a few records farther west

Rough-hairy perennial, 0.5–2.5 m tall, with narrowly winged stems. Leaves alternate, the blades narrowly lanceolate to elliptic or narrowly oblanceolate, 5.5–23 cm long and 1.5–7.5 cm wide, the apex acuminate or attenuate, the base narrowly wedge-shaped to acuminate, the margins serrate, both surfaces rough-hairy. Heads radiate, 3–6 cm across, in terminal clusters. Involucre 5–11 mm long, the bracts few, herbaceous, mostly lance-linear, glabrous or nearly so, soon reflexed. Ray flowers yellow, neutral and infertile, the rays 1.5–3 cm long. Disk globose or nearly so. Achenes smooth, broadly winged, 5–7 mm long (including the 2 stiff pappus awns).

This plant is listed in earlier manuals for Kansas as *Actinomeris alternifolia*.

V. helianthoides Michx., also a yellow-flowered perennial, is known from the southeast corner of the state. It differs from our more common wingstem in having a more developed involucre, the bracts overlapping and never reflexed, and leaves with the lower surface covered with soft, appressed hairs.

V. virgínica L.   Tickweed, frostweed

"Virginian."
White; late July–October
Open, wooded hillsides, creekbanks, pastures, and roadsides
Southeast eighth, primarily

Hairy perennial, 0.5–1.8 dm tall, with strongly winged stems. Leaves alternate, the blades more or less broadly lanceolate to narrowly elliptic, 7–23 cm long and 2.5–8 cm wide, the apex acuminate or attenuate, the base more or less abruptly narrowed to a petiolelike base, the margins coarsely serrate or merely wavy, the upper surface rough, the lower surface soft-hairy. Heads radiate, 10–16 mm across, in terminal clusters. Involucre bell-shaped, 3–8 mm long, the bracts lance-oblong to oblong or oblanceolate, hairy, the innermost membranous. Rays white, 4–8 mm long. Disk white, flat. Achenes sparsely hairy, broadly winged, 7–9 mm long (including the 2 stiff pappus awns).

ZÍNNIA L.   Zinnia

Named in honor of Johann Gottfried Zinn, professor of medicine in Göttingen and botanical writer of the mid-18th century.

Annuals or perennial herbs with woody taproots. Leaves opposite, simple, sessile, usually narrow (linear) and entire. Heads several- to many-flowered, radiate, many, but borne singly at the tips of the branches. Involucre bell-shaped to nearly cylindric, the bracts in 3–5 series, graduated in length, oblong to obovate, appressed, the outer ones herbaceous. Receptacle convex or conic, chaffy, the dry, membranous scales clasping the achenes. Pappus absent or of a few teeth or short awns. Ray flowers 3–8, female, fertile; corollas yellow (ours) or white, the ligules entire or 3-toothed, persistent on the achenes, the style branches slender and glabrous. Disk flowers several to many, bisexual, fertile; corollas white or yellowish, 5-toothed with 1 tooth longer than the others; anther bases

673. *Verbesina alternifolia*

674. *Verbesina virginica*

rounded; style branches minutely hairy. Achenes of the disk flowers flattened, those of the ray flowers 3-angled.

One species in Kansas.

Z. grandiflòra Nutt.

Rocky mountain zinnia

NL., from L. *grandis*, large, grand, magnificent, + *flos, floris*, flower.
Yellow; late May–mid-October
Dry prairie hillsides, uplands, and roadside banks
South half of the southwest fourth

Low, tufted perennial, woody toward the base, with a short, branched caudex giving off many slender, crowded branches 4–18 cm tall and a stout, deep taproot with laterals at different levels. Leaves lance-linear or narrowly elliptic, 8–42 mm long and 3–9 mm wide, resin-dotted, usually with 2 veins paralleling the midrib, the apex acute, the base more or less rounded, at least the margins with stiff, appressed hairs. Heads 10–25 cm across, the involucre 7–8 mm long, the bracts with a green crescent near the broadly rounded, membranous tip. Rays yellow, broadly ovate to nearly round, 9–15 mm long, the tip entire or notched. Achenes strongly ribbed, 4.5–6.5 mm long, the pappus of 2 short, ciliate awns.

How different this fine-textured perennial native species is from the annual *Zinnia elegans*, of Mexican origin, whose various sizes and colors of single- and double-flowered heads are commonly seen in our gardens; yet both in the distant past have had a common ancestry.

675. *Zinnia grandiflora*

AMBRÒSIEAE

Ragweed Tribe

Sap of plants not milky. Heads small, inconspicuous, apparently discoid. Flowers unisexual, male and female flowers in the same heads or on different heads on the same plant, the pappus and corolla rudimentary or lacking altogether. Female flowers, when in separate heads, usually enclosed in a burlike or nutlike structure formed by fusion of the involucral bracts.

AMBRÒSIA L.

Ragweed

Greek and Latin name for several kinds of plants, as well as for "the food of the gods," reputed to confer immortality on those who partook of it, from Gr. *ambrotos*, immortal.

Coarse, erect, aromatic annual or perennial herbs, with glandular hairs. Leaves alternate or opposite, simple, usually petioled, entire to toothed or pinnately or palmately lobed. Heads small, unisexual. Male heads 5- to 20-flowered, nodding, in narrow terminal clusters, the involucres saucer- or top-shaped, with 5–12 partly united bracts, the receptacle flat, chaffy, the flowers lacking a pappus, the corolla membranous, bell-shaped, usually 5-lobed, the stamens 5, the anthers more or less separated at anthesis, a rudimentary pistil present. Female heads 1- to few-flowered, sessile, inconspicuous axillary clusters of 1 to few further down on the same plant, the bracts of the involucre fused to form a hard, top-shaped or nearly globose, indehiscent, nutlike or burlike structure, the tips of the bracts represented by a series of tubercles or spines, the flower(s) lacking pappus, corolla, and stamens, and (except for the style) completely enclosed by the involucre.

Seven species in Kansas. Our 5 most common species are included in the following key.

Key to Species:
1. Fruiting involucres burlike, the spines scattered in several series over the body of the involucre .... *A. grayi*
1. Fruiting involucres nutlike, the tubercles or stout spines in 1–2 series near the apex, or absent 2
   2. Male heads sessile in a spikelike arrangement, the involucres very oblique and drawn out on one side into a lance-acuminate lobe .... *A. bidentata*
   2. Male heads on short stalks in racemelike arrangements, the involucre only slightly oblique and lacking a prominent lobe 3
      3. Perennial; main leaves of the stems sessile or nearly so .... *A. psilostachya*
      3. Annual; main leaves of the stem definitely petioled 4
         4. Leaves palmately 3- to 5-lobed (occasionally unlobed); male involucres with dark veins .... *A. trifida*
         4. Leaves pinnately lobed or dissected; male involucres lacking dark veins .... *A. artemisiifolia*

A. artemisiifòlia L. Pale ragweed

NL., from the genus *Artemisia* + L. *folium*, leaf—because of the similarity of the leaves of this
  species to those of *Artemisia*.
Flowers greenish, inconspicuous; August–October
Pastures, roadsides, creekbanks, waste ground, and other disturbed sites
East two-thirds and the north part of the west third

Erect, eventually much-branched annual, up to 1.3 m tall. Stems covered with
coarse, spreading hairs, at least in the upper parts of the plant. Leaves alternate, 2–12 cm
long and 1–8.5 cm wide, deeply pinnately or bipinnately cleft into oblong, elliptic, or
linear segments which are entire or have a few teeth. Male heads stalked, the involucres
shallowly toothed, lacking prominent veins. Fruiting involucre nutlike, minutely hairy,
obovoid, nearly round in cross section, 2.5–4.5 mm long, with a prominent conical apical
spine and usually a single row of 2–8 spines.

This species is listed in some earlier Kansas manuals as *A. elatior.*

Pale ragweed is an important hayfever plant and, when ingested, may cause
nausea in cattle (4). The seeds are a valuable source of winter food for birds (22).

A. bidentàta Southern ragweed

676. *Ambrosia bidentata*

NL., from L. *bi-*, combining form of *bis*, 2, + *dentatus*, toothed, from *dens*, tooth—in reference
  to the leaves.
Flowers greenish, inconspicuous; mid-August–mid-October
Pastures, roadsides, waste ground, and other disturbed sites
Southeast eighth

Rough-hairy annual, the stem 2.5–8 (–10) dm tall with several ascending branches.
Leaves alternate, the lower ones usually gone by flowering time, the upper ones sessile,
lance-linear with 2 spreading basal teeth or lobes, 1–6 cm long and 0.3–1 cm wide, the
apex acute to attenuate, the base rounded. Male heads sessile, oblique, drawn out on
one side into a lance-acuminate lobe. Fruiting involucre nutlike, about 6–8 mm long,
sparsely hairy (use magnification), more or less obovate, 4-angled in cross section, the
apex with a single central beak or spine, and 4 stout spines in a single series about 1 mm
long, 1 at each angle.

The pollen may cause hay fever. The seeds are eaten by wild turkeys (4).

A. gràyi (A. Nels.) Shinners

Named after Asa Gray (1810–1888), professor of botany at Harvard University, considered by
  many to be the Father of American Plant Taxonomy.
Flowers greenish, inconspicuous; mid-August–early October
Moist places of fields, prairie and roadsides
West half, primarily

677. *Ambrosia psilostachya*

Erect herbaceous perennial, 0.3–1 m tall, forming colonies by means of adventitious shoots arising from the roots. Leaves alternate, 3–11 cm long and 2–5.5 (–8) cm
wide, irregularly pinnately lobed, the lobes spreading widely, the lobes mostly narrowly
lanceolate to elliptic or oblanceolate with serrate margins, the lower surface densely
covered with silvery hairs, the upper surface silvery to nearly glabrous. Involucres at
flowering time 2–3 mm long. Fruiting heads up to 7 mm long and 4 mm wide, with 2
apical beaks, the other spines up to 13 in number, scattered, the tips slender, acuminate,
hooked at the tip.

A. acanthicarpa Hook. and *A. confertifolia* DC., both known primarily from the
southwest corner of the state, will key out here with *A. grayi* since they also have burlike
fruiting involucres. They differ from *A. grayi* in having fruiting involucres with 1 apical
beak and more finely divided leaves. *A. acanthicarpa* is an annual; *A. confertifolia*, a
perennial.

A. psilostáchya Western ragweed

NL., from Gr. *psilos*, bare, smooth, + *stachys*, ear of grain, spike.
Flowers greenish, inconspicuous; mid-July–October
Pastures, roadsides, waste ground, and disturbed places in prairie
Throughout

Erect herbaceous perennial, 3.5–8 dm tall, usually with numerous ascending or
spreading branches above, forming extensive colonies by means of horizontal, runnerlike
roots. Leaves opposite, the lower ones usually gone at flowering time, the upper ones
more or less lanceolate in general outline and sessile or nearly so, 2.5–9 cm long and
1–5 cm wide, more or less deeply pinnately lobed, the segments of the upper leaves acute

678. *Ambrosia trifida*

(those of the lowermost leaves, when present, more rounded), the margins usually with a few teeth also, both surfaces resin-dotted and covered with appressed hairs. Male heads stalked, the involucres nutlike, obovoid, about 2.5 mm long, minutely hairy, with 4–6 short, blunt or sharp, apical tubercles.

Several medicinal uses of western ragweed among American Indian groups are recorded in the literature (62). Pueblo Indian women at Acoma, New Mexico, drank a tea made from the plant in cases of difficult labor at childbirth, while the Kiowas of Oklahoma prepared a concoction to be rubbed on sores, either on humans or on horses. The Cheyennes of the central plains drank a ragweed infusion to treat bloody stools, colds, and bowel cramps. The Luiseno Indians of southern California used some species of *Artemisia*, probably this one, as an emetic.

A. trífida   Giant ragweed

L. *trifidus*, 3-cleft, from *tri-*, 3, + *fidus*, cleft—in reference to the 3-lobed leaves.
Flowers greenish, inconspicuous; late July–mid-October
Fields, roadsides, waste places, and other disturbed sites
Throughout

Tall, coarse, rough annual herb, up to 5 m tall. Leaves opposite, definitely petioled, the blades mostly deeply palmately 3- to 5-lobed, occasionally some of them entire, 5–22 cm long and 1.5–22 cm wide, the lobes more or less lance-acuminate, the margins serrate. Male heads stalked, the involucres with 3 (–4) dark veins. Fruiting involucre nutlike, glabrous, obovoid, angled, 5–8 mm long, with a prominent conical apical beak, and 4–8 shorter apical tubercles or spines in 1 or 2 rows, usually with purplish streaks or blotches.

This is one of the tallest and coarsest of all our weeds, one of the worst in our grain fields, and one of the most unsightly in waste ground. The pollen, widely distributed by the wind, becomes public enemy number one for people subject to hay fever. In its favor, it is recorded that Indians used fibers from the stems for making thread and cords, and some Indian tribes applied the bruised leaves to the abdomen for the relief of nausea. The Meskwaki Indians of Wisconsin reportedly chewed the root of giant ragweed to drive away fear at night (63). The leaves are eaten by white-tailed deer.

ÍVA L.   Marsh elder

NL., from Fr. *ive*, akin to OF. *if*, yew. Linnaeus is said to have given the name *iva* to a species of the Mint Family, *Ajuga iva*, and to this genus of the Composite Family because of a similarity of smell.

Annual or perennial herbs. Leaves opposite below, becoming alternate above, simple, entire to toothed, lobed, or dissected. Heads several-flowered, appearing discoid, although the marginal flowers are functionally different from the central ones, the heads arranged in spikes, racemes, or branched clusters. Involucre top-shaped or hemispheric, the 3–9 bracts in 1–3 series, free or somewhat united, sometimes graduated in length. Receptacle flat, chaffy. Pappus absent. Marginal flowers 1–8, female, usually more or less hidden by the larger central flowers, the corolla tubular and truncate, or sometimes merely rudimentary, the style branches linear. Disk flowers 3–20, male, the corolla funnelform, 5-lobed, the stamens united by their filaments, the anthers rounded at the base, only slightly united, the style undivided. Achenes obovate, flattened parallel to the involucral bracts.

Four species in Kansas. *I. angustifolia* Nutt. and *I. axillaris* Pursh are each known from only a few counties and are not included here.

Key to Species:
1. Inflorescence spikelike, the heads sessile in the axils of reduced, leaflike bracts; involucral bracts 3–5, merely subtending the achenes .... *I. annua*
1. Inflorescence branched, the individual heads not subtended by leaves or bracts, though some may be scattered throughout the inflorescence; involucral bracts 10, the inner 5 membranous and actually embracing the achenes .... *I. xanthifolia*

I. ánnua L.   Sumpweed

L. *annuus*, annual, from *annus*, year—describing the lifetime of this plant. "Sump" is akin to G. *sumpf*, "swamp" or "marsh."
Flowers greenish; August–early October
Moist soil
East three-fourths, primarily

Coarse, pale or gray green, erect annual, 2–12 dm tall, the stems simple or with several pairs of ascending branches. Leaves petioled, the blades mostly broadly ovate to lanceolate, occasionally elliptic or nearly deltoid with 3 prominent veins, 2–12 cm long and 0.8–7 cm wide, the apex acute or acuminate, the base acute to acuminate (rarely truncate), the margins serrate, both surfaces with stiff, appressed hairs, the petioles ciliate. Herbage resin-dotted. Heads about 3 mm across, each subtended by a conspicuous ciliate, leaflike bract and arranged in slender terminal spikes 2–10 (–15) cm long. Involucral bracts 3–5, obovate to obtriangular, 3–4 mm long, the margins and back of each bract with hollow, siliceous hairs. Individual flowers greenish and inconspicuous. Achenes black, 4–5 mm long.

This species is listed in earlier manuals for Kansas as *I. ciliata*.

I. xanthifòlia Nutt.

NL., from the genus *Xanthium* (cockleburr), in the Composite Family, + *folium*, leaf—in reference to the similarity of the leaves of these two plants.
Flowers greenish, inconspicuous; August–September
Fields, roadsides, waste ground, and other disturbed habitats
West half, primarily; but scattered records from the east half also

Erect, soft-hairy annual, 0.4–2 m tall, with a taproot. Stems simple, or branched above, glabrous below, becoming pubescent and glandular in the inflorescence. Leaves long-petioled below, becoming shorter-petioled above, the blades of the main leaves mostly cordate to deltoid or rhombic, the uppermost becoming lanceolate, or lance-linear, palmately veined, the apex acuminate, the base acute to more or less abruptly acuminate, the margins coarsely dentate or incised (rarely 3- to 5-lobed), the upper surface minutely hairy, the lower surface conspicuously white-hairy. Herbage resin-dotted. Heads many in branched terminal and axillary clusters, about 3–5 mm across, not subtended by a conspicuous leaflike bract, although some of the inflorescence branches are subtended by reduced leaves or linear bracts. Involucral bracts 10, the outer 5 herbaceous and obovate with an acuminate apex, lacking siliceous hairs or cilia, the innermost membranous with an erose, more or less truncate apex, embracing the achenes. Individual flowers greenish and inconspicuous. Achenes black, about 3 mm long.

According to Steyermark (4), this plant is a serious cause of hayfever, and the pollen and leaves can cause a dermatitis. Cows grazing on the plant produce bitter milk.

XÀNTHIUM L.   Cocklebur

NL., from Gr. *xanthion*, name used by Dioscorides for a plant from which an infusion was made for dying the hair yellow, from *xanthos*, yellow.

Coarse, weedy annuals with taproots. Leaves alternate, simple, petioled, toothed or lobed. Heads small, unisexual. Male heads many-flowered, borne in terminal spikes or racemes; involucre cup-shaped, the bracts herbaceous, free from one another, in 1–3 series; receptacle cylindric, chaffy; pappus absent; corolla tubular, 5-lobed; stamens united by their filaments, the anthers separate, rounded or acute at the base; pistil rudimentary. Female heads 2-flowered, borne in the leaf axils farther down on the same plant; involucre ovoid or ellipsoid, burlike, indehiscent, the bracts united and hardened, the tips of the bracts represented by spines; flowers nearly completely enclosed by the 2-chambered involucre, lacking pappus and corolla, with 2 stigmas. Achenes 2, retained within the spiny involucre.

Two species in Kansas. *X. spinosum* L., an introduced species, is known from scattered sites across the state. It differs from our more common cocklebur in having lanceolate leaves which taper to the base and which are provided with a conspicuous yellow 3-armed spine in the leaf axil.

X. strumàrium L.

NL., from L. *struma*, a scrofulous tumor, the plant having been used at one time for treating such tumors.
Flowers greenish, inconspicuous; June–September
Fields, roadsides, waste ground, and other disturbed sites
Throughout

Stems simple or branched, 1.5–20 dm tall. Leaves long-petioled, the blades mostly cordate to deltoid or broadly ovate, palmately veined, 3.5–16 cm long and 2–15 cm wide, the apex rounded or broadly acute, the base cordate or reniform, the margins more or less coarsely serrate and sometimes 3-lobed as well. Herbage covered with resinous dots and stiff, appressed, siliceous hairs, those on the stem and many of those on the leaves with bulbous bases (use high magnification). Male heads globose, 5–7 mm across. Female heads enlarging to form burs 2–3.5 cm long covered with hooked spines.

679. *Iva annua*

680. *Xanthium strumarium*

RAGWEED TRIBE
461

This species has been listed in earlier manuals for Kansas under several names including *X. commune, X. chinense, X. globosum, X. italicum, X. pennsylvanicum,* and *X. speciosum.*

The 2 stigmatic branches extend beyond the apex of the involucre through a temporary opening and for a brief period of time are exposed to wind-borne pollen. Thereafter the ovary in each cavity becomes an achene containing an embryo with elongate cotyledons which hold reserve food and a poisonous glucoside, hydroquinone, that bodes no good for any animal eating the achenes. The hard, spinose involucral bur deters most creatures from eating the achenes; but when germination takes place, the poisonous cotyledons come up above the ground and are sometimes eaten with the rest of the plant by livestock, with disastrous results. According to Kingsbury (21), toxicity decreases rapidly as the first true leaves develop, but when the seedlings are very young about 1.5 percent of the animal's body weight is enough to be lethal.

Preparations of the leaves have been used to treat tuberculosis of the neck glands, shingles (herpes), and skin and bladder infections, and also to stop the bleeding of skin cuts and abrasions (30).

Several dye colors have been reported for this plant. The entire plant is chopped and utilized to prepare the dye bath. Wool mordanted with alum and cream of tartar takes a lemon yellow dye. Adding one-fourth cup of alum to a 3-gallon dye bath gives unmordanted wool a purple color, while reducing the alum to 1 teaspoon produces a light tan color (57).

Appendixes

APPENDIX 1: NOTES ON ACCENTS, PRONUNCIATION, AND ABBREVIATIONS

To assist our readers in the pronunciation of scientific names, this book indicates syllables of principal accent by means of acute accent marks (´) if they are to receive the short English sound or a grave (`) accent mark if they are to be given the long sound.

It will be noted that the scientific names of the families have the ending -aceae. This is spoken as 3 syllables, a-ce-ae; the beginning a has the principal accent of the name and the long sound as in "ate"; the e of the second syllable has the long sound, but clipped as in the initial e of "event"; and the ae of the last syllable is given the long sound of e as in "eve."

When an initial o is followed by e, as in Oenothera, the o is silent and the e is given the long sound.

With such modifications as the above practices impose, the scientific names are usually spoken in this country as if they were in English.

Abbreviations used in the name derivations, mainly standard ones such as are given in dictionaries, are listed below for the reader's convenience:

Ar.	Arabic	L.	Latin
AS.	Anglo-Saxon or Old English	LL.	Late Latin
		ME.	Middle English
c.	circa (L., "about")	MF.	Middle French
E.	English	ML.	Medieval Latin
F.	French	NL.	New Latin
G.	German	OF.	Old French
Gr.	Greek	Sp.	Spanish

APPENDIX 2: ABBREVIATIONS OF AUTHORS' NAMES

Abbe	Ernst Cleveland Abbe (1905–), American botanist and phytogeographer
Adans.	Michel Adanson (1727–1806), French botanist and explorer; author of some 1,600 generic names
Aellen	Paul Aellen (1896–1973), professor in Basel, Switzerland
Agardh	Carl Adolph Agardh (1785–1859), Swedish algologist, professor at Lund from 1807–1835, later became Bishop of Karlstad
Ait.	William Aiton (1731–1793), English botanist, Royal Gardener at Kew
Alex.	Edward Johnston Alexander (1901–), New York Botanical Garden; authority on botany of the southeastern United States
All.	Carlo Allioni (1725?–1804), Italian physician and botanist, professor of botany at Turin
Ames	Oakes Ames (1874–1950), director of the Botanical Museum, Harvard University; eminent authority on orchids and economic botany
Anderson	Edgar Anderson (1897–1969), affiliated with the Missouri Botanical

Garden and George Washington University in St. Louis; renowned as a teacher and experimental taxonomist

Andrz. — Antoni Lukianowicz Andrzejowski (1784–1868), professor of botany at Vilna, Lithuania

A. Nels. — Aven Nelson (1859–1952), professor of botany and later president of the University of Wyoming, revisor of Coulter's *New Manual of Botany of the Central Rocky Mountains*

Ard. — Pietro Ardvino (1728–1805), Italian botanist and agriculturist at Padua

Arn. — George Arnold Walker Arnott (1799–1868), professor of botany at Glasgow and Edinburgh

A. S. Hitchc. — Albert Spear Hitchcock (1865–1935), botanist, Kansas State Agricultural College at Manhattan and later of the U.S. National Herbarium at the Smithsonian Institution; leading authority on grasses

Aubl. — Jean Baptiste Christophe Fusee Aublet (1720–1778), French botanical collector

Ave-Lall. — Julius Edward Leopold Ave-Lallemant

Bailey — Liberty Hyde Bailey (1858–1954), eminent botanist and author of numerous horticultural works including the *Standard Cyclopedia of Horticulture* and *Manual of Cultivated Plants*

Baill. — Henri Ernest Baillon (1827–1895), Paris botanist, physician, and author

Baldw. — William Baldwin (1779–1819), Pennsylvania physician and botanist who collected and studied plants of Georgia, Delaware, and South America

Barnh. — John Hendley Barnhart (1871–1949), American bibliographer

Bart. — Benjamin Smith Barton (1766–1815), professor of materia medica and natural history at the University of Pennsylvania; in 1808 engaged Thomas Nuttall to collect plants in western America

Bartl. — Friedrich Gottlibe Bartling (1798–1875), professor of botany at Göttingen, Germany

Beck — Lewis Caleb Beck (1798–1853), American botanist

Benth. — George Bentham (1800–1884), long-time president of the Linnaean Society; outstanding English taxonomist and author of numerous major botanical works

Bernh. — Johann Jacob Bernhardi (1774–1850), professor of botany at Erfurt, Germany

Berth. — Sabin Berthelot (1794–1880), French consul on Tenerife Island and co-author with Philip B. Webb of a natural history of the Canary Islands

Bess. — Wilibald Swibert Joseph Gottlieb von Besser (1784–1842), Austrian botanist

Biehl. — Johann Friedrich Theodor Biehler, German botanist

Blake — Sidney Fay Blake (1892–1959), American botanist and bibliographer

B. L. Turner — Billie Lee Turner (1925–   ), professor of botany, University of Texas, Austin

Blume — Carl Ludwig von Blume (1796–1862), Dutch botanist

Boenn. — Clemens Maria Friedrich von Boenninghausen

Bogin — Clifford Bogin (1920–   ), New York Botanical Garden and Woodmere Academy, Long Island, New York

Bor. — Alexandre Boreau (1803–1875), French botanist

Borbas — Vincze von Borbas (1844–1905), Hungarian botanist

Britt. — Nathaniel Lord Britton (1859–1934), American botanist, author, and one time director of the New York Botanical Garden (1896–1930)

Brongn. — Adolphe Theodore Brongniart (1801–1876), noted French paleobotanist and systematist

Brot. — Felix di Avelar Brotero (1744–1828), Portuguese professor of botany

B. S. P.	N. L. Britton, Emerson Ellick Sterns (1846–1926), Justus Ferdinand Poggenburg (1840–1893), American botanists
Buckl.	Samuel Botsford Buckley (1809–1884), naturalist and State Geologist of Texas
Bullock	Arthur A. Bullock (1906–), English botanist at Kew
Burgess	Edward S. Burgess (1855–1928), Hunter College, New York
Burm. f.	Nicolaas Laurens Burman, the son (1734?–1793), Dutch physician and botanist
Bush	Benjamin Franklin Bush (1858–1937), postmaster at Independence, Missouri, and amateur botanist
Butt.	Frederick King Butters (1878–1945), professor of botany at the University of Minnesota
Cass.	Alexandre-Henri Gabriel Comte de Cassini (1781–1832), French botanist
Cav.	Antonio Jose Cavanilles (1745–1804), Spanish botanist, professor of botany and director of the botanic gardens at Madrid
Cerv.	Vicente de Cervantes (1759?–1829), professor of botany and director of the botanic garden in Mexico City
Cham.	Ludolf Adalbert von Chamisso (1781–1838), also listed variously as Adelbert Ludwig von Chamisso or Louis Charles Adelaide Chamisso de Boncourt
Chât.	Jean Jacques Châtelain, published work in 1760 in Switzerland
Ch. des Moulins	Charles Robert Alexandre Des Moulins (1798–1875)
Chois.	Jacques Dennis Choisy (1799–1859), Swiss philosopher, clergyman, and botanist
C. H. Thompson	Charles Henry Thompson (1870–1931), botanist at the Missouri Botanical Garden, St. Louis
Clarke	Charles Baron Clarke (1832–1906), superintendent of the Royal Botanic Gardens at Calcutta
Clayt.	John Clayton (1685–1773), physician and plant collector in Virginia
Clem. & Clem.	Frederic Edwards Clements (1874–1945), American plant ecologist and climatologist; Edith Gertrude (Schwartz) Clements (1877–?), ecologist
C. L. Hitchcock	Charles Leo Hitchcock (1902–), professor of botany, University of Washington, Seattle, and senior author of *Vascular Plants of the Pacific Northwest* (1955–1973)
Conrad	Soloman White Conrad (1779–1831), American botanist
Const.	Lincoln Constance (1909–), professor of botany at the University of California at Berkeley
Cory	Victor Louis Cory (1880–1964), Texas botanist
Coult.	John Merle Coulter (1851–1928), eminent American botanist and author of manuals of Texas and the Rocky Mountains
Cov.	Frederick Vernon Coville (1867–1937), curator of U.S. National Herbarium (1893–1937)
Crantz	Heinrich Johann Nepomuk von Crantz (1722–1799), professor of medicine at Vienna and author of a flora of Austria
Croiz.	Leon Camille Marius Croizat (1894–?)
Cronquist	Arthur John Cronquist (1919–), curator, New York Botanical Garden, prominent American plant systematist and author of numerous botanical articles and manuals
Cyrillo	Domenico Cyrillo (1739–1799), Italian professor of botany
DC.	Augustin Pyramus de Candolle (1778–1841), Swiss botanist
Dcne.	Joseph Decaisne (1807–1882), Belgian botanist and director of the Jardin des Plantes in Paris
D. Don	David Don (1799–1841), British botanist and brother of George Don
Desf.	René Louiche Desfontaines (1750–1833), French botanist
Desr.	Louis Augusto Joseph Desrousseaux (1753–1838), contributor to *Lamarck's Encyclopedia*

Desv.	Nicaise Auguste Desvaux (1784–1856), French botanist
Dougl.	David Douglas (1798–1834), Scottish plant explorer who collected in the Pacific Northwest for the Royal Horticultural Society
Druce	George Claridge Druce (1850–1932), British naturalist, curator, and author
Drude	Carl Georg Oscar Drude (1852–1933), Germany
Duchesne	Antoine Nicolas Duchesne (1747–1827), French author of a work on useful plants
Dufr.	Pierre Dufresne (1786–1836), Geneva, Switzerland
Dum.	Count Barthélemy Charles Joseph Dumortier (1797–1878), Belgian botanist
Dumont	George Louise Marie Dumont de Courset (1746–1824), French agronomist and horticultural writer
Dunal	Michel Felix Dunal (1789–1856), French botanist and professor of botany
Durand	Elias Magloire Durand (1794–1873), Philadelphia pharmacist
Eat.	Amos Eaton (1776–1842), American botanist and author
Ell.	Stephen Elliott (1771–1830), American botanist
Engelm.	George Engelmann (1809–1884), German-born physician and botanist in St. Louis, Missouri
Fabricius	Philip Conrad Fabricius (1714–1774), German physician, botanist, and professor at the University of Helmstedt
Farwell	Oliver Atkins Farwell (1867–1944), consulting botanist for Parke, Davis & Co., Detroit, Michigan
Fassett	Norman Carter Fassett (1900–1954), professor of botany at the University of Wisconsin at Madison
F. B. Jones	Frederick Butler Jones (1909–), botanist with the Welder Wildlife Foundation, Sinton, Texas
Fedde	Friedrich Karl Georg Fedde (1873–1942), professor and editor in Berlin, Germany
Fern.	Merrit Lyndon Fernald (1873–1950), eminent plant systematist, one-time director of Gray Herbarium at Harvard (1937–1947), and author of numerous botanical works, including the 8th edition of *Gray's Manual of Botany*
Fisch.	Friedrich Ernst Ludwig von Fischer (1782–1854), director of the botanic garden at St. Petersburg from 1823–1850
Forsk.	Petrus Forsskål (also Pehr Forskal) (1732–1763), Danish student of Linnaeus who traveled to Arabia and wrote a flora of Egypt and Arabia
Fosb.	Francis Raymond Fosberg (1908–), U.S. Geological Survey, Washington; student of the Californian, South American, and Polynesian floras
Foug.	Auguste Denis Fougeroux de Bondaroy (1732–1789), of France
Franch.	Adrien R. Franchet (1834–1900), French botanist
Frém.	John Charles Frémont (1813–1890), U.S. Army officer and explorer
Gaertn.	Joseph Gaertner (1732–1791), German physician and botanist at Stuttgart
Gaiser	Lulu Odel Gaiser (1896–1965), Canadian botanist
Gaud.	Jean Francois Aimee (Gottlieb) Philippe Gaudin (1766–1833), Swiss clergyman and astrologist
G. Don.	George Don (1798–1856), British plant collector for the Royal Horticultural Society
G. F. W. Mey.	Georg Friedrich Wilhelm Meyer (1782–1856), German botanist and professor in Göttingen
Gilg	Ernst Friedrich Gilg (1867–1933), German botanist at the Botanical Museum, Berlin
Gilib.	Jean Emmanuel Gilibert (1741–1814), French botanist

Gill.	John Gillies (1747–1836), Scotch physician who collected in Chile
Gleason	Henry Allan Gleason (1882–), American botanist and plant geographer, head curator of the New York Botanical Garden and author of the *New Britton and Brown Illustrated Flora of the Northeastern United States and Adjacent Canada*
Gmel.	Johann Friedrich Gmelin (1748–1804), professor in Göttingen, Germany
G. N. Jones	George Neville Jones (1904–), professor of botany at the University of Illinois at Urbana and author of several floras
Godr.	Dominique Alexandre Godron (1807–1880), French botanist
G. Ownbey	Gerald Bruce Ownbey (1916–), brother of Marion Ownbey and professor of botany at the University of Minnesota
Graebn.	Karl Otto Robert Peter Paul Graebner (1871–1933)
Gray	Asa Gray (1810–1888), distinguished American botanist, professor of botany at Harvard University, author of numerous botanical works including the *Manual of Botany: A Handbook of the Flowering Plants and Ferns of the Central and Northeastern United States and Adjacent Canada*
Greene	Edward Lee Greene (1842–1915), American botanist and editor of several botanical journals
Grev.	Robert K. Greville (1794–1866), professor in Edinburgh, Scotland
Griseb.	August Heinrich Rudolf Grisebach (1814–1879), German botanist at Göttingen
H. & A.	W. J. Hooker (see) and G. A. Arnott; authors of *The Botany of Captain Beechey's Voyage* (1830–1841)
Hallier	Hans Gottfried Hallier (1868–1932), Dutch botanist
Hara	Hiroshi Hara (1911–), Japanese taxonomist, cytologist, and synantherologist
Hassk.	Justus Carl Hasskarl (1811–1894), German botanist
Haw.	Adrian Hardy Haworth (1767–1833), English gardener and entomologist
H. B. K.	Baron F. W. H. A. von Humboldt; A. J. A. Bonpland, the two forming the most famous scientific expedition to tropical America; and C. S. Kunth, who wrote the text of their descriptive work *Nova genera et species Plantarum* (1815–1825)
Heister	Lorenz Heister (1683–1758), professor in Helmstedt, Germany
Heller	Amos Arthur Heller (1867–1944), Pennsylvania botanist, collector of western American plants
Hill	John Hill (1716–1775), London apothecary and naturalist and author of herbals and nature books
Hochr.	Bénédict Pierre Georges Hochreutiner (1873–1959), director of the Conservatoire de Botanique, Geneva
Hoffm.	Georg Franz Hoffman (1760–1826), professor of botany at Göttingen and later in Moscow
Hogg	Thomas Hogg (1777–1855), of England
Hook.	Sir William Jackson Hooker (1785–1865), director of the Royal Botanic Gardens at Kew from 1841–1865, author and/or editor of many noted botanical works
Horkel	Johann Horkel (1769–1846), German botanist
Horn. or Hornem.	Jens Wilken Horneman (1770–1841), professor of botany in Copenhagen, Denmark
HouH.	Maarten Houhuyn (1720–1794), Dutch physician and naturalist
Huds.	William Hudson (1730–1793), London apothecary, botanist, and author
Jacq.	Nikolaus Joseph Baron von Jacquin (1727–1817), noted Austrian botanist and director of the botanic garden at Vienna
James	Edwin James (1797–1861), surgeon-naturalist who traveled with Major Stephen H. Long's expedition to the Rocky Mountains in 1819–1820

Jeps.	Willis Linn Jepson (1867–1946), author of several major manuals of the California flora and founder of the California Botanical Society
J. G. Smith	Jaud Gage Smith (1866–1925), American agrostologist
Johnst.	Ivan Murray Johnston (1898–1960), professor of botany at Harvard University, authority on world floras, and plant explorer and author
Juss.	Antoine Laurent de Jussieu, nephew of Bernard (1748–1836), professor in the Jardin des Plantes, Paris, France
Karst.	Gustav Karl Wilhelm Hermann Karsten (1817–1908), German botanist
Ker.	John Bellenden Ker or John Ker Bellenden or (before 1804) John Gawler (1764–1842), British botanist
Kittell	Sister Teresita Kittell (1892–), Catholic University of America, Washington, later of Holy Family College, Manitowoc, Wisconsin
Kl.	Johann Friedrich Klotzsch (1805–1860), curator of the Herbarium in Berlin, Germany
Knerr	Ellsworth Brownell Knerr (1861–?)
Koch	Wilhelm Daniel Joseph Koch (1771–1849), German botanist
Koehne	Bernard Adalbert Emil Koehne (1848–1918), professor in Berlin, Germany
Kunth	Carl Sigismund Kunth (1788–1850), German botanist
Kuntze	Otto Kuntze (1843–1907), German botanist and author
L.	Carl Linnaeus (1707–1778), Swedish botanist, a prodigious author, and founder of the sexual system of classification and author of *Species Plantarum* (1753) upon which modern botanical nomenclature is based
Lag.	Mariano Lagasca y Segura (1776–1839), professor and director of the botanic garden in Madrid, Spain
Lag. & Rodr.	M. Lagasca y Segura and José Démetrio Rodriguez (1780–1846)
Lam.	Jean Baptiste Antoine Pierre Monnet de Lamarck (1744–1829), famous French botanist and zoologist who propounded a theory of evolution by the transmission of acquired characters
Ledeb.	Carl Friedrich von Ledebour (1785–1851), German-born botanist, professor in Dorpat, Estonia, and author of 2 floras
Lehm.	Johann Georg Christian Lehmann (1792–1860), director of the botanic garden in Hamburg, Germany
Lem.	Charles Antoine Lemaire (1801–1871), French botanist, author, and professor in Ghent, Belgium
Less.	Christian Friedrich Lessing (1809–1862), German physician
Lévl.	Auguste Abel Hector Léveillé (1863–1918), of France
L. f.	Carl von Linne, the son (1741–1783), successor to his father in the professorship of botany in Upsala, Sweden
L'Her.	Charles-Louis l'Héritier de Brutelle (1746–1800), French magistrate and botanist
Lindl.	John Lindley (1799–1865), professor of botany in London, eminent orchidologist and textbook writer
Link	Johann Heinrich Friedrich Link (1767–1851), professor of natural science and director of the botanic garden in Berlin, Germany
Lodd.	Conrad Loddiges (1738–1826)
Loud.	John Claudius Loudon (1783–1843), English horticulturalist and prolific author and editor of garden books
Lous.	João de Loureiro (1710–1791), Portuguese missionary and naturalist
Mack.	Kenneth Kent Mackenzie (1877–1934), New York City corporation attorney and noted caricologist
MacM.	Conway MacMillan (1867–1929), State Botanist of Minnesota
Magnus	Paul Wilhelm Magnus (1844–1941), German taxonomist
Mart.	Karl Friedrich Philipp von Martius (1794–1868), German biologist, author, and traveler
Martin	Robert Franklin Martin (1910–), Division of Plant Exploration

	and Introduction, U.S. Department of Agriculture, Beltsville, Maryland
Math.	Mildred Esther Mathias (1906–), professor of botany at the University of California, Los Angeles
Maxim.	Carl Johann (Karl Ivanovich Maksimovich) Maximowicz (1827–1891), Russian botanist, director of the botanic garden at St. Petersburg
McVaugh	Rogers McVaugh (1909–), professor of botany at the University of Michigan, Ann Arbor
Medic.	Friedrich Casimir Medicus (1734–1814), Dutch gardener
Meissn.	Carl Friedrich Meissner (1800–1874), Swiss botanist in Basel, Switzerland
Merr.	Elmer Drew Merrill (1876–1956), once director of the New York Botanical Garden, administrator of botanical collections at Harvard University, and author of numerous botanical works
Mert.	Franz Carl Mertens (1764–1831), professor in Bremen, Germany
Michx.	André Michaux (1746–1802), French botanist and explorer of North America
Mill.	Phillip Miller (1691–1771), British gardener and author of *The Gardeners Dictionary* (1731), which went through 8 editions
Moench	Conrad Moench (1744–1805), German botanist and professor in Marburg, Germany
Moq.	Christian Horace Bénédict Alfred Moquin-Tandon (1804–1863), French botanist
Moric. *ex*. Dunal.	Moise Etienne Moricand (1779–1854), Swiss botanist in Geneva, and Michel Felix Dunal (1789–1856), French botanist and professor of botany in Montpellier
Morong	Rev. Thomas Morong (1827–1894), Massachusetts minister and amateur botanist
Muell. Arg. (or Muell.)	Jean (Argoviensis, i.e., of Aargau) Mueller (1828–1896), Swiss botanist
Muhl.	Gotthilf Heinrich Ernst Muehlenberg (1753–1815), German-educated Lutheran minister and pioneer botanist of Pennsylvania
Murr.	Johann Andreas Murray (1740–1791), Swedish professor of medicine and botany at Göttingen, Germany
Nees	Christian Gottfried Daniel Nees von Esenbeck (1776–1858), German botanist, author, and professor in Breslau
Nieuw.	Julius Aloysius Arthur Nieuwland (1878–1936), professor of botany and of organic chemistry at Notre Dame University, Indiana
Nort.	John Bittling Smith Norton (1872–1966), plant pathologist at the University of Maryland, College Park
Nutt.	Thomas Nuttall (1786–1859), English-American naturalist who collected in western America and later curator of the botanic garden at Harvard College, Cambridge, Massachusetts
Oeder	Georg Christian von Oeder (1728–1791), Danish botanist and professor in Copenhagen, Denmark
O. Ktze.	Carl Ernst Otto Kuntze (1843–1907), German traveler and botanist
Opiz	Philipp Maximilian Opiz (1787–1858), Bohemian botanist
Ort.	Casimiro Gomez Ortega (1740–1818), Spanish botanist and director of the botanical garden in Madrid, Spain
Pall.	Peter Simon Pallas (1741–1811), German botanist and student of the Russian and Siberian floras
Parker	Kittie Lucille (Fenley) Parker (1910–), George Washington University and later at the U.S. National Herbarium
Payson	Edwin Blake Payson (1893–1927), professor of botany at the University of Wyoming
P. Br.	Patrick Browne (1720–1790), Irish physician and naturalist who explored in Jamaica

Penn.	Francis Whittier Pennell (1886–1952), curator of botany, Academy of Natural Sciences of Philadelphia
Pers.	Christiann Hendrick Persoon (1761–1836), botanist, mycologist, and author, of South African birth, who lived in Paris from 1802 onward
Phil.	Rudolf Amandus Philippi (1808–1904), Chilean botanist, zoologist, and paleontologist
Piper	Charles Vancouver Piper (1867–1926), professor of botany and zoology at the Washington Agricultural College at Pullman, later agrostologist with the U.S. Department of Agriculture
Planch.	Jules Emile Planchon (1823–1888), French botanist
Poir.	Jean Louis Marie Poiret (1755–1834), French botanist and traveler in North Africa
Poll.	Johann Adam Pollich
Porter	Thomas Conrad Porter (1822–1901), professor of botany at Lafayette College, Pennsylvania
Prince	William Robert Prince (1795–1869), American nurseryman
Pursh	Fredrick Traugott Pursh (1774–1820), German author and botanical traveler in North America
Raf.	Constantine Samuel Rafinesque (1783–1840), eccentric pioneer naturalist and author, born in Turkey, later lived in Sicily and then Kentucky
Raven	Peter Hamilton Raven (1936–), Stanford University, Stanford, California, later of the Missouri Botanical Garden, St. Louis
R. Br.	Robert Browne (1773–1858), British botanist, librarian, and first Keeper of Botany at the British Museum
Regel	Eduard August von Regel (1815–1892), director of the botanic garden at St. Petersburg
Reichenb.	Heinrich Gottlieb Ludwig Reichenbach (1793–1879), German naturalist, author, and professor in Dresden
Rich.	Louis Claude Marie Richard (1754–1821), French botanist who collected plants in South America and the West Indies
Richards.	Sir John Richardson (1787–1865), Scotch botanist and zoologist who accompanied Sir John Franklin's expedition to arctic America
Ridd.	John Leonard Riddell (1807–1865)
Robins.	Benjamin Lincoln Robinson (1864–1935), curator of Gray Herbarium at Harvard University from 1892 to 1935
Rock	Howard Francis Leonard Rock (1925–1964), professor of biology at Vanderbilt University, Nashville, Tennessee
Rose	Joseph Nelson Rose (1862–1928), botanist with the U.S. Department of Agriculture and later with the U.S. National Herbarium
Rostk.	Friedrich Wilhelm Gottlieb Rostkovius (1770–1848), Polish physician and botanist
Roth	Albrecht Wilhelm Roth (1757–1834), German physician and botanist
Rottb.	Christen Friss Rottboell (1727–1797), professor of botany and director of the botanical garden at Copenhagen, Denmark
R. & P.	Hipolito Ruis Lobez (1754–1815), and José Pavon (1754–1844), Spanish explorers and botanists, co-authors of a flora of Peru and Chile
R. & S.	Johann Jakob Roemer and Joseph August Schultes
Rusby	Henry Hurd Rusby (1855–1940), one-time dean of the New York College of Pharmacy and an active collector in South America
Russell	Norman Hudson Russell (1921–), professor of botany at Central State College, Edmond, Oklahoma, and later at Buena Vista College, Storm Lake, Iowa
Rydb.	Per Axel Rydberg (1860–1931), Swedish-born American botanist, curator of the New York Botanical Garden, and author of numerous botanical works
Salisb.	Richard Anthony Salisbury (1761–1829), British botanist

Sav.	Paul Amedée Ludovic Savatier (1830–1891), French marine medical officer and botanist
Savi	Gaetano Savi (1769–1844), Italian botanist
Sch.-Bip.	Carl Heinrich Schultz (1805–1867), Latinized name *Bipontius*, German botanist
Scheele	Georg Heinrich Adolf Scheele (1808–1864), German botanist who described plants collected in Texas by Ferdinand Lindheimer and Ferdinand Roemer
Schkuhr	Christian Schkuhr (1741–1811), German botanist
Schlecht.	Diederich Franz Leonhard von Schlechtendal (1794–1866), German botanist and botanical editor
Schleiden	Matthias Jakob Schleiden (1804–1881), German author of botanical manuals
Schmid.	Casimir Christoph Schmidel (1718–1792), German physician and botanist at Erlangen
Schmidt	Franz Wilibald Schmidt (1764–1796)
Schott	Heinrich Wilhelm Schott (1794–1865), Austrian botanist and director of the Schönbrunn Botanical Gardens in Vienna
Schrad.	Heinrich Adolph Schrader (1767–1836), German botanist and professor in Göttingen
Schreb.	Johann Christian Daniel von Schreber (1739–1810), German botanist and professor in Erlangen
Schult.	Joseph August Schultes (1773–1831), Austrian botanist and professor
Schulz	Otto Eugen Schulz (1874–1936), German botanist
Schwein.	Lewis David von Schweinitz (1780–1834), German-born Pennsylvania clergyman and noted amateur mycologist
Scop.	Johann Anton (Giovanni Antonio) Scopoli (1723–1788), Austrian botanist, physician, and professor of natural history
Sheld.	Edmund Perry Sheldon (1869–?), American forester and author
Shinners	Lloyd Herbert Shinners (1918–1971), noted Canadian-born botanist, professor of botany at Southern Methodist University, and author of numerous botanical works
Short	Charles Wilkins Short (1794–1863), medical botanist and professor in Louisville, Kentucky
Sibth.	John Sibthorp (1758–1796), professor of botany at Oxford, England
Sieb.	Philipp Franz von Siebold (1796–1866), German botanist
Sims	John Sims (1749–1831), English botanist and editor
Small	John Kunkel Small (1869–1938), American botanist, one-time head curator of the New York Botanical Garden, and author of a flora of the Southeastern United States
Smith	Sir James Edward Smith (1759–1828), British botanist
Smyth	Bernard Bryan Smyth (1843–1913), Kansas Academy of Science, Topeka, Kansas; author of numerous early records of Kansas plants
Spach	Edouard Spach (1801–1879), French botanist
Spreng.	Curt Polykarp Joachim Sprengel (1766–1833), German botanist, author, and physician at Halle University
Standl.	Paul Carpenter Standley (1884–1963), prominent American botanist, curator, and author of numerous botanical works
Steud.	Ernst Gottlieb Steudel (1783–1856), German physician, botanical bibliographer, and authority on grasses
St.-Hil.	Auguste Francois César Prouvencal de Saint Hilaire (1779–1853), French botanist, explorer, and author who collected extensively in Brazil and Paraguay
Stuckey	Ronald Lewis Stuckey (1938–), Ohio State University, Columbus
S. V. Fraser	Samuel Victorian Fraser (1890–), student of the flora of Cloud County, Kansas
Sw.	Olof Peter Swartz (1760–1818), Swedish botanist, student of Linnaeus, professor in Stockholm, and author of a flora of the West Indies

Sweet	Robert Sweet (1783–1835), English horticulturist and ornithologist
Swezey	Goodwin Deloss Swezey (1851–1934), professor of astronomy, Nebraska
Ten.	Michele Tenore (1780–1861), Italian botanist
T. F. Forst.	Thomas Furly Forster
T. & G.	John Torrey and Asa Gray (see)
Thell.	Albert Thellung (1881–1928), keeper in the Botanical Institute at the University of Zurich
Thuill.	Jean Louis Thuiller (1757–1822), French gardener, professor, and author of a flora of Paris
Thunb.	Carl Peter (Pehr) Thunberg (1743–1828), student of Linnaeus, later professor of botany at Uppsala, Sweden, and author of botanical works based on his travels to Japan, Ceylon, and South Africa
Torr.	John Torrey (1796–1873), noted American physician and professor of botany and chemistry, who described numerous plants from western America
Trin.	Carl Bernhard von Trinuis (1778–1844), Russian physician, poet, and authority on grasses
Urban	Ignatz Urban (1848–1931), Botanical Museum, Berlin, Germany, and authority on the flora of tropical America
Vahl	Martin Hendrickson Vahl (1749–1804), student of Linnaeus, later professor of botany in Copenhagen, Denmark
Vail	Anna Murray Vail (1863–1955), librarian at the New York Botanical Garden
Vaniot	Eugene Vaniot
Vent.	Etienne Pierre Ventenat (1757–1808), horticulturist and professor of botany in Paris, France
V. Grant	Verne Edwin Grant (1917–), cytogeneticist, formerly at Rancho Santa Ana Botanical Garden, Anaheim, California, currently at Boyce Thompson Southwestern Arboretum, Arizona
Wall.	Nathanael (formerly Nathan Wolff) Wallich (1786–1854), superintendent of the Royal Botanic Garden, Calcutta, India, from 1815 to 1841
Wallr.	Carl Friedrich Wilhelm Wallroth (1792–1857), Danish physician and botanist
Walt.	Thomas Walter (1740–1789), British-American botanist
Wats.	Sereno Watson (1826–1892), botanist at Gray Herbarium, Harvard University, from 1888 to 1892, and authority on western American plants
Webb	Philip Barker Webb (1793–1854), British botanical explorer and author
Weber	Frederick Albert Constant Weber (1830–1903), French botanist and member of a French expedition to Mexico in 1865–1866
Wedd.	Hugh Algernon Weddell (1819–1877), British botanist and traveler who collected in South America
Wettst.	Richard Ritter von Westersheim Wettstein (1863–1931), director of the botanic garden at Vienna
Wendl.	Hermann Wendland (1823–1903), director of the Royal Garden at Herrenhausen, Hannover, Germany
Wheeler	Louis Cutter Wheeler (1910–), professor of botany, University of Southern California at Los Angeles
Wherry	Edgar Theodore Wherry (1885–), professor of botany, University of Pennsylvania, and author of numerous botanical works
Willd.	Carl Ludwig Willdenow (1765–1812), German botanist, author, and director of the Berlin Botanical Garden from 1801 to 1812
Wolf	Carl Brandt Wolf (1905–), botanist at Rancho Santa Ana Botanic Garden in Anaheim, California, from 1930 to 1945

Wood	Alphonso Wood (1810–1881), author of first American book to employ dichotomous keys (*Class Book of Botany*, 1845)
Woods.	Robert Everard Woodson (1904–1963), Missouri Botanical Garden, St. Louis
Woot.	Elmer Otis Wooton (1865–1945), professor of biology at New Mexico State College and author of botanical works about the New Mexico flora, later affiliated with the U.S. Department of Agriculture in Washington, D.C.
Zinn	Johann Gottfried Zinn (1727–1759), professor of medicine at Göttingen, Germany
Zucc.	Joseph Gerhard Zuccarini (1797–1848), German botanist and professor in Munich

APPENDIX 3: GLOSSARY

abaxial: the surface of the leaf (or other part) away from the stem, the lower or ventral surface.

abortive: failing to develop completely.

acaulescent: lacking a well-developed, aboveground stem, the leaves usually being borne at the base of a long or short flowering peduncle, as in the dandelion.

accessory fruit: a fruit composed of the ripened ovary and some other part such as the receptacle or hypanthium.

accrescent: persisting and enlarging with age.

achene: a dry, thin-walled, indehiscent, 1-seeded fruit.

acicular: needlelike.

acrid: harsh, sharp, or bitter in taste.

actinomorphic: radially symmetrical, *regular* (see).

acuminate: gradually narrowing to a point, the sides usually somewhat concave; see also *attenuate*.

acute: ending in a point, the margins straight or somewhat convex.

adaxial: the surface of the leaf (or other part) toward the stem; the upper or dorsal surface.

adnate: fused to an unlike part, as stamen filaments adnate to petals.

adventitious: arising from a position not considered typical in a very general sense, as adventitious roots arising along the sides of a stem.

adventive: a plant becoming naturalized outside of its native place.

aggregate fruit: a fruit composed of several combined ripened ovaries of 1 flower.

alate: winged.

alternate: in reference to leaf arrangement, having but one leaf per node.

alveolate: having a surface ornamented with alveoli, pitted like a honeycomb, often descriptive of seeds.

alveolus (pl. alveoli): a deep angular cavity.

anastomosing: joining together, fusing, usually used in reference to leaf venation.

androecium: a collective term referring to all of the stamens of a flower as a group (from Gr. *andros*, male, + *oikos*, house).

androgynophore: in the Passifloraceae, a compound stalk formed from the fusion of the gynophore, a stalk which elevates the pistil above the receptacle, and the stamen filament bases (from Gr. *andros*, male, + *gyne*, female, + the suffix, *-phor*, to bear).

androphore: a compound stalk formed by the fusion of the stamen bases.

angiosperm: a plant which bears its reproductive structures in flowers and its seeds in a true fruit, i.e., in a ripened ovary.

annual: a plant which completes its entire life cycle from germination to flowering and fruiting and then dies within a single growing season. A winter annual usually germinates in the fall, overwinters in a vegetative state, then produces flowers and seeds the next spring or summer before it dies.

annulus: a ring-shaped structure.

anodyne: a medicine that relieves pain.

anther: the pollen-bearing portion of a stamen.

anthesis: the point in time at which the pistil and/or stamens of a flower are mature and capable of participating in pollination and fertilization.

anthocarp: a structure formed by the union of the fruit proper with the perianth or receptacle.

antrorsely: directed upward or toward the apex.

apetalous: lacking petals.

apex: the top or distal end of an organ.

apical: at the apex.

apiculate: abruptly constricted at the apex into a tiny sharp point.

apomictic: maturing seed without sexual union having taken place.

appressed: lying flat or nearly so against the surface.

arachnoid: resembling spider web or cobweb (from Gr. *arachne*, spider, + the suffix *-oid*, like, resembling).

aril: a fleshy outgrowth from the stalk of the ovule, which finally more or less completely envelops the seed.

aristate: tipped with a rather long, stiff bristle or awn, much more extreme than mucronate.

aristulate: diminutive of "aristate."

ascending: directed obliquely forward or upward with respect to the plant part (stem, leaf, etc.) to which another plant part (leaf, hair, etc.) is attached.

asexual: sexless; produced by vegetative rather than sexual means.

asymmetrical: not symmetrical.

astringent: causing tissue to draw together.

attenuate: gradually narrowing to a long, slender point, the sides usually concave, more extreme than acuminate.

auricle: a basal lobe usually of a leaf or petal blade and usually rounded, as an earlobe.

auriculate, auricled: having 1 or more auricles.

awl: a pointed tool for making holes in leather.

awn: a long, slender, tapered bristle.

axil: the angle between the upper surface of a leaf or petiole and the stem.

axillary: arising from a leaf axil.

axis: that portion of the stem within the inflorescence on which the flowers or fruits are borne, usually in a radial arrangement.

barbed: having short, rigid, reflexed points along the margin or side.

basal: located near the base of an organ or plant.

bast: fibrous plant materials, especially that derived from the inner bark.

beak: a protracted point, usually on a fruit or on the keel of a leguminous flower.

bearded: bearing a tuft or patch of hairs.

berry: a fruit with a completely pulpy ovary wall.

biennial: a plant which lives for 2 growing seasons, usually germinating the first year, then flowering, fruiting, and dying the second year.

bifid: 2-cleft.

bifurcate: 2-forked (from L. *bi-*, 2, + *furca*, fork).

bilabiate: 2-lipped.

bipinnate: pinnately divided and then with those subdivisions again pinnately divided.

bisexual: having male and female parts.

biternate: twice ternate, i.e., divided into 3 principal segments, each of which is divided into 3 segments.

blade: the expanded portion of a leaf or petal.

bract: a structure, frequently small and leaflike or herbaceous, which subtends some other plant part such as a flower or inflorescence.

bristle: a stiff hair.

bud: a small, unexpanded branch system, containing embryonic leaves and branches and/or flowers, usually enveloped by protective scales (bud scales).

bulb: a perennating device which is essentially an underground bud consisting of a short segment of stem tissue, to which are attached a number of thick modified "leaves" (bulb scales) specialized for food storage, and from which new aboveground stems arise each growing season. Compare with corm, rhizome, and tuber. The term is sometimes used rather loosely; true bulbs are found only among monocotyledonous plants.

bur: a prickly or spiny fruit.

calcareous: containing or derived from some form of calcium carbonate, as limestone or soils developed from limestone parent materials.

calyx: a collective term referring to all the sepals of a flower as a group, whether or not they are united with one another.

campanulate: bell-shaped.

canaliculate: longitudinally grooved.

canescent: with a coating of fine, gray, soft hairs.

capillary: hairlike in form.

capitate: head-shaped; arranged in very dense, headlike clusters.

capitellate: diminutive of "capitate."

capsule: a dry, usually dehiscent fruit which develops from a 2- to many-celled pistil.

carminative: a drug or medication used to expel gas from the stomach or bowel.

caruncle: an outgrowth near the hilum of some seeds, as in the Spurge Family.

cathartic: a purgative, a substance intended to evacuate the bowels.

caudex: stem; often restricted to short, woody stems, near ground level, of herbaceous perennials, from which each new season's growth develops.

caulescent: having a definite aboveground stem.

caulicle: the rudimentary stem of the plant embryo, usually called the radicle (little root) because of an early misconception.

cauline: borne along the stem, rather than basally.

cespitose, caespitose: growing in dense clumps.

chaffy: said of a common receptacle (in some members of the Composite Family) which bears dry scales or bracts intermingled with or subtending the individual flowers of the head.

channeled: deeply grooved longitudinally.

chartaceous: papery in texture.

chasmogamous: having pollination occurring while the flower is open (from Gr. *chasmos,* open, + *gamy,* marriage).

chlorophyll: the green coloring matter in chloroplasts.

ciliate: having marginal hairs.

cilium (pl. cilia): hairs borne along the margin of a leaf or other part.

cinereous: the color of ashes.

circinate: coiled downward, with the apex of the stem in the center of the coil.

circumboreal: occurring around the globe at higher latitudes.

circumscissile: dehiscing around the middle so that the upper half comes off like a lid.

clammy-pubescent: covered with glandular hairs which cause the plant to feel somewhat moist and sticky to the touch.

clasping: partly surrounding or embracing another organ at the base.

clavate: club-shaped.

clawed: rather abruptly constricted toward the base into a petiolelike stalk; applied to some petals and sepals.

cleft: cut approximately halfway to the midrib.

cleistogamous flowers: flowers which never open and are self-pollinating and -fertilizing.

coalescence: the union of parts of the same kind.

coherent: having similar parts united, as the petals of a sympetalous corolla, as in honeysuckles and bellflowers.

columella: a small column.

column: a columnar or tubular structure formed by the fusion of several plant parts (in orchids, for example, formed from stamens, style, and stigmas).

coma: a tuft of silky hairs found on seeds in the Milkweed Family and Dogbane Family.

compound leaves: divided into separate leaflets which are attached directly to the petiole or to a rachis (see fig. 6).

concavo-convex: concave on one side and convex on the other.

conduplicate: folded together along the long axis.

confluent: joined together.

conic: cone-shaped, with the broad end at the base.

connate: fused to a like part, as petals fused to petals, or stamen filaments to stamen filaments.

connective: the tissue uniting the 2 pollen chambers of an anther.

connivent: coming together but not actually fused to one another.

contiguous: adjacent and touching.

convex: having a surface or edge which curves outward.

convolute: rolled up longitudinally.

coralloid: resembling coral.

cordate: heart-shaped, broadly ovate with rounded basal lobes.

coriaceous: leathery in texture.

corm: a short, stout, erect, perennial underground stem in which food is stored; a corm lacks the thickened storage leaves found in a bulb.

cormose: having corms.

corolla: collective term referring to all of the petals of a flower as a group, whether or not they are united with one another.

corona: an appendage or whorl of parts that is situated between the corolla and stamens, or on the corolla (as in daffodils or jonquils) or that is the outgrowth of the staminal parts (as in the milkweeds).

corpusculum: the clip joining together the 2 pollen masses of the pollinium in the Milkweed Family.

cortex: the bark of a stem; more technically, that portion of the stem between the epidermis and the vascular tissue.

corymb: a convex or flat-topped flower cluster of the raceme type with pedicels arising at different levels on the peduncle, the outer flowers blooming first.

corymbose: borne in corymbs, corymblike.

cotyledon: the first leaf or one of a pair of first leaves of a plant embryo.

creeping (stems): growing flat on or beneath the surface of the ground and rooting.

crenate: having shallow, rounded teeth.

crenulate: diminutive of "crenate."

crested: having an elevated, more or less irregular or toothed ridge.

crisped: extremely undulate.

cross-pollination: transfer of pollen from one flower to the stigma of another flower, usually on a different plant.

crown: the juncture of stem and root in a seed plant; the system of branches and foliage forming the top of a tree; a ring of scales or teeth; see also *corona.*

cruciform: arranged in the shape of a cross or X, as the petals of plants in the Mustard Family.

cultivars: cultivated varieties.

cuneate: wedge-shaped, triangular with the narrow end toward the point of attachment.

cuspidate: having the apex abruptly, sharply, concavely constricted into an elongated, sharp-pointed tip.

cyanogenetic: yielding hydrocyanic acid (HCN), a toxic substance, when hydrolyzed.

cyathium (pl. cyathia): in *Euphorbia,* the cuplike involucre containing the male and female flowers.

cyme: a convex or flat-topped flower cluster, with the central flower the first to open.

cymose: resembling a cyme, bearing cymes.

cystolith: a mineral concretion, usually of calcium carbonate.

dauciform (roots): thickened at the crown and tapering downward, like a carrot.

deciduous: falling away.

decompound: compounded more than once, as in a bipinnately compound leaf.

decumbent: lying prostrate on the ground, with the tip growing upward.

decurrent: extending down the stem below the point of departure.

decussate: with alternate pairs of leaves arranged at right angles to one another on the stem.

deflexed: bent or turned downward at a sharp angle; reflexed.

dehisce: to open, thereby dispersing seeds or pollen.

deltoid: triangular, shaped like the Greek letter delta (Δ).

dentate: with pointed, coarse teeth spreading at right angles to the margin (compare with serrate).

denticulate: diminutive of "dentate."

depressed: flattened on top.

di-: a prefix denoting "2" or "twice."

diadelphous: having the stamens united by their filaments into 2 groups.

diaphoretic: inducing perspiration.

dichotomous: repeatedly forking in pairs.

dicotyledonous: having 2 cotyledons.

didymous: twinlike.

didynamous: having 4 stamens with 2 longer than the others.

diffuse: loosely spreading.

digitate: with leaflets (usually 5) arising from one point, as in Virginia creeper.

dilated: expanded, widened.

dimorphic: having 2 forms.

dioecious: having male flowers on one plant and female flowers on a separate plant (from Gr. *di-,* 2, + *oikos,* house).

discoid: applied to flower heads of Compositae having only tubular flowers; flattened and circular like a disk.

disk: any flattened, circular structure, but frequently used botanically to refer to a nectary of that shape located beneath the ovary; in the Composite Family, that portion in the center of a radiate head occupied by flowers with tubular corollas.

dissected: divided into numerous segments.

distal: away from the main axis (of the plant).

diuretic: a drug or other agent used to increase the production of urine.

divaricate: widely divergent.

divergent: angled in different directions from a common point, spreading.

divided: parted to the base or to the midrib.

dorsal: along the outer or abaxial surface of an organ.

dorsal keel: a sharp, longitudinal, outward-projecting angle along the midvein of a petal or sepal.

drupe: a fleshy, indehiscent fruit with a single seed enclosed within the hard, inner wall (endocarp) of the fruit; this hard endocarp is sometimes called a stone or pit, as in a cherry, peach, plum, or olive, and is often mistaken for the seed itself.

echinate: prickly.

eglandular: lacking glands.

ellipsoid: a 3-dimensional ellipse, round in cross section, widest at the middle, and narrowed to rounded ends.

elliptic: a 2-dimensional figure widest at the middle and narrowed to rounded ends.

emarginate: having a shallow notch at the tip.

embryo: the rudimentary plant in a seed.

emergent: elevated above the water.

emetic: a medicinal substance used to induce vomiting.

endocarp: the inner layer of the ovary wall in the fruit stage.

endosperm: the nutritive tissue formed in the embryo sac, and surrounding the embryo or absorbed by the embryo in the development of the seed.

ensiform: sword-shaped.

entire: said of leaves which are not lobed or cleft in any manner, or of margins which are not toothed in any manner.

ephemeral: lasting for a day or less.

epicarp: the outer layer of the ovary wall in the fruit stage.

epidermis: the outer cell layer(s) of plant parts.

epigynous: arising from the top of the ovary (from Gr. *epi,* upon, + *gyne,* female), said of stamens, petals, and sepals of a flower which has an inferior ovary.

epiphyte: a plant growing upon another plant without taking nourishment from it.

equitant: applied to conduplicate leaves which bestride and overlap another on the opposite side, as in the Iris Family.

erect: standing upright.

erose: irregularly ragged.

esculent: edible.

estipulate: lacking stipules.

estuarian: inhabiting estuaries.

evanescent: vanishing early.

excrescence: a small irregular outgrowth.

excurrent: with a tip projecting beyond the main part of the organ.

exfoliate: to peel off in thin sheets or flakes.

exocarp: the same as epicarp.

expectorant: a substance which promotes the secretion of fluid from the respiratory tract.

exsert: protrude, project.

exserted: prolonged beyond surrounding organs.

extrorse (anthers): facing and dehiscing outward.

facultative: capable of an adaptive response to some environmental variable. Example: A facultative selenium absorber can take selenium when the latter is present in the soil, but does not absolutely require it. Compare with *obligate.*

falcate: sickle-shaped.

farinose: covered with a mealy powder.

fascicle: a bundle or dense cluster.

fastigiate: said of stems or branches that are nearly erect and close together, as in Lombardy poplar.

fertile: producing good seed.

fertilization: fusion of male and female gametes within an ovule; in plants pollination must occur before fertilization can take place, but fertilization may not always follow each act of pollination.

filament: the stalk of a stamen which bears the anther.

filiform: threadlike.

fimbriate: fringed.

flabellate: fanlike.

flaccid: limp.

fleshy: juicy, as the fruit of an apple or plum, as contrasted to a dry fruit such as a capsule or nut.

flexuous: bending back and forth.

floccose: covered with flocks or tufts of soft woolly hairs that rub off easily.

floret: one of the small flowers of the head of a composite; any small flower.

foliaceous: leaflike.

foliose: leafy.

follicle: a dry, dehiscent fruit, produced from a simple ovary, which opens along one suture.

fruit: the seed-bearing organ of flowering (angiosperm) plants, produced from a ripened ovary and sometimes other parts of the flower as well.

fugacious: lasting but a short time.

funiculus: the tiny stalk by which an ovule or seed is attached to the ovary wall; a funicular scar, the hilum, usually remains on the seed after it is detached from the funiculus.

funnelform: shaped like a funnel.

fuscous: dark brownish-gray.

fusiform: spindle-shaped; a 3-dimensional figure which is round in cross section, widest at the middle, and narrowed to a sharp point at each end.

galea: any helmet-shaped part of a calyx or corolla; the upper lip of a 2-lipped corolla.

gall: an abnormal growth of vegetative tissue caused by damage related to insect egg-laying, or by invasion of the tissue by fungi, bacteria, or other injurious agents.

gamopetalous: with united petals; same as sympetalous.

gamosepalous: having the sepals united; synsepalous.

genus (pl. genera): a group of closely related species.

gibbous: swollen on 1 side, usually at the base, as in a snapdragon corolla.

glabrous: smooth, without hairs.

gland: a cell or tissue which secretes special substances such as nectar, volatile oils, resin, odorous substances, or enzymes.

gland-dotted: with spherical, sessile glands.

glandular: having glands.

glandular-hairy: having gland-tipped hairs.

glaucous: covered with a whitish waxy substance which rubs off.

globose, globular: spherical.

glochid: small barbed hairs or bristles.

glochidiate: having glochids.

glomerule: a small, dense, headlike cluster of minute flowers.

granuliferous: bearing or covered with small granules.

gynandrous: having the gynostegium and androecium more or less united, as in orchids.

gynodioecious: dioecious, but having some bisexual flowers on plants which otherwise have mostly pistillate flowers.

gynoecium: a collective term referring to the pistil or all the pistils of a flower as a group.

gynophore: a stalk which elevates the pistil above the receptacle (from Gr. *gynos*, female, + the suffix *-phor*, to bear).

gynostegium: in the Milkweed Family, a complex, compound structure formed by modification and union of the stamens and pistils.

habit: the characteristic form of a plant, including the posture and texture of its parts.

habitat: the kind of place where a plant naturally grows, such as prairie, woods, desert, marsh, etc.

halophytic: growing in salty or alkaline soils.

hastate: more or less arrowhead-shaped, but with narrow, pointed basal lobes projecting perpendicularly to the midvein or petiole.

haustoria: the rootlike absorbing organs of parasitic plants, by which means these plants attach themselves to their host.

head: a dense cluster of sessile or nearly sessile flowers, as in sunflower, button bush, horsemint, etc.

hermaphrodite: having both stamens and pistils; bisexual, perfect (said of flowers).

hemiparasite: a parasitic plant possessed of chlorophyll and able to make part of its food, as the mistletoe.

hepatogenic: produced by or originating in the liver.

herb: a plant which dies down, at least to the ground, at the end of each growing season.

herbaceous: not woody; see also *herb*

herbage: all the green parts of a plant.

heteromorphic: dissimilar in shape, structure, or size.

hilum: on a seed, the scar where the funiculus was attached.

hirsute: covered with rough, coarse or shaggy hairs.

hirtellous: covered with soft or minute hairs.

hispid: with bristly stiff hairs.

hoary: grayish-white owing to a dense, fine pubescence.

hyaline: transparent or translucent.

hybrid: as used here, a plant resulting from a cross between two species.

hydrophyte: a plant which grows in water or in saturated soil.

hygroscopic: capable of expanding or contracting on presence or absence, respectively, of water or water vapor.

Hymenoptera: the order of insects which includes bees, wasps, ants, ichneuman flies, saw-flies, and their relatives.

hypanthium: a compound structure formed by the fusion of the basal parts of the calyx, corolla, and androecium. It may be free from or united with the ovary.

hypogynous: arising at the base of the ovary (from Gr. *hypo,* under, + *gyne,* female), said of stamens, petals, and sepals of a flower which has a superior ovary.

imbricate: overlapping, as the shingles of a roof.

immersed: growing wholly under water.

imperfect (flowers): lacking either stamens or pistils.

incised: cut sharply and irregularly.

included: not at all protruded from the surrounding envelope, usually applied to stamens and styles.

incomplete (flowers): with one or more sets of organs (calyx, corolla, stamens, pistil) lacking.

indehiscent: not opening to release seeds (or pollen, in the case of some anthers).

indeterminate: said of inflorescences which continue to elongate and to produce new flower buds apically through the blooming season.

inferior: said of an ovary which is imbedded in tissue derived from the receptacle and from bases of the sepals, petals, and stems and which has the other floral parts departing above it.

infertile: not producing seed.

inflex: to flex inward.

inflorescence: in general, a cluster of flowers or their resultant fruits; more specifically, the term refers to the arrangement of flowers on the plant and their order of blooming within a cluster.

inserted: attached to or growing out of.

internode: that portion of the stem between any two nodes, the latter being a region of the stem where one or more leaves are inserted.

intersepalar: arising between the sepals.

introrse: facing and dehiscing inward, applied to anthers.

involucel: a secondary involucre.

involucrate: with or resembling an involucre.

involucre: one or more whorls of small leaves or bracts standing close underneath a flower or flower cluster as in the Umbelliferae or Compositae.

irregular: bilaterally symmetrical; having some parts different in size and/or shape from other parts of the same kinds as in a 2-lipped or papilionaceous corolla.

keel: the protruding structure formed by the lower 2 petals of a papilionaceous flower; a sharp ridge along the midvein of a sepal or other part.

labellum: in the Orchid Family, the prominent lower petal of the corolla, often different from the other petals in form and color (from L. *labellum,* lip).

labiate: lipped; a member of the Mint Family.

lacerate: irregularly cleft, as though torn.

laciniate: cut into narrow lobes.

lamina: the blade of a leaf or petal.

laminate: in layers.

lanate: woolly.

lance-acuminate: lanceolate in general shape, with the apex acuminate.

lance-attenuate: lanceolate in general shape but with the apex attenuate, i.e., very gradually drawn out into a long, slender tip.

lanceolate: lance-shaped, longer than wide, widest below the middle, usually with slightly convex sides (margins).

lanose: densely woolly.

lateral: belonging to or borne on the side.

latex: the milky juice of some plants.

lax: loose, open, the opposite of congested.

leaflets: the subdivisions of compound leaves.

legume: a member of the Legume Family, a dry, dehiscent fruit developed from a simple pistil and usually breaking open down 2 sides.

lenticular: lens-shaped, i.e., flattened, convex on both sides.

ligulate: provided with or resembling a ligule.

ligule: a strap-shaped organ or part; as used here, a strap-shaped corolla, as in the ray flowers of composites.

limb: the expanded part of a sympetalous corolla.

linear: long and narrow, with the sides parallel or nearly so.

linear-lanceolate: intermediate in shape between linear and lanceolate.

lip: either the lower or upper principal division of a bilabiate corolla or calyx; the labellum of orchids.

lobe: a more or less rounded subdivision of a leaf, petal, or other plant part.

loculicidal capsule: a capsule the outer walls of which open longitudinally midway between the dividing partitions of the fruit.

loment: a jointed legume, constricted and breaking apart between the seeds.

lorate: strap-shaped, usually also flexuous or limp and with the apex rounded.

lunate: crescent shaped, like the crescent moon.

margin: edge.

membranous: thin and somewhat transparent or translucent.

menstrum: a solvent.

mericarp: a dry, indehiscent, 1-seeded section of a schizocarp, the latter a fruit type found in such families as the Umbelliferae and Malvaceae.

midrib, midvein: the main vein of a leaf or leaflike part.

molluscicide: a pesticide which poisons mollusks such as snails and slugs.

monadelphous: having the stamens united into a single unit, usually by fusion of the filaments.

moniliform: resembling a string of beads.

monocotyledonous (embryo): having only one cotyledon.

monoecious: having male and female parts in separate flowers but with both male and female flowers on the same plant (from Gr. *mono*, one, + *oikos*, house).

monogeneric: containing but a single genus.

mucilaginous: slimy.

mucro: a short, sharp point or spine at the apex of a leaf or leaflike part.

mucronate: having a mucro.

multiple fruit: a fruit composed of the closely associated ripened ovaries of several flowers.

napiform: turnip-shaped.

naturalized: not indigenous but introduced and become as if native.

nectar: a sweet liquid produced by the nectaries of plants, consisting of 25 percent to 75 percent sugar, with varying proportions of glucose, fructose, and saccharose, and serving as the chief source of honey for insect visitors.

nectary: a gland or tissue which secretes nectar, located usually at the base of petals, sometimes in special receptacles, as in larkspurs, violets, and milkweeds.

nectiferous: producing nectar.

nerve: vein.

neutral: without stamens or pistils.

nodes: points along the stem at which leaves are borne.

nutlet: a small, dry, nutlike seed or fruit, often applied to the subdivisions of the fruit in the Labiatae or Verbenaceae.

ob-: inversely; a prefix which indicates 180° inversion; example: *oblanceolate* = lanceolate in shape but with the narrowed end toward the base and the broadest part toward the distal end.

obcordate: heart-shaped but with the broad part toward the apex.

oblanceolate: see *ob-*.

obligate: dependent for survival upon some particular environmental variable. Example: An obligate selenium absorber requires the presence of selenium in the soil in order to live.

oblique: with unequal sides, asymmetrical.

obovate: egg-shaped, but with the broad end toward the apex.

obtuse: rounded.

ocrea: in the Polygonaceae, a more or less membranous sheath around the stem at the nodes, formed by the stipules.

operculum: a lid or covering flap.

opposite: said of leaves arranged with two at each node.

orbicular: round in outline.

orifice: opening, hole.

ovary: that portion of the pistil (the female part of the flower) in which the ovules are produced.

ovate: egg-shaped in outline, with the broadest part toward the base.

overwintering: living through the winter.

ovoid: 3-dimensionally egg-shaped.

ovule: the structure, within the ovary, which, after fertilization, matures into a seed.

palmately compound: having the leaflets all attached to the distal end of the petiole rather than along the sides of an extension of the petiole.

panduriform: fiddle-shaped, with the broad end toward the apex.

panicle: an openly branched flower cluster, the lowest flowers of which open first, allowing the axis of the cluster to elongate during the flowering period.

pantropical: occurring in tropical regions around the world.

papilionaceous: having a standard, wings, and keel, as in the corolla of a pea or bean flower (from L. *papilio, butterfly*).

papilla (pl. papillae): a minute pimplelike protuberance or projection, especially an epidermal cell with a minute conical projection.

papillate, papillose: bearing minute pimplelike projections.

pappus: in the Composite Family, the modified calyx, usually consisting of a series of bristles, scales, or teeth.

parasitic: growing on and taking nourishment from another plant.

parietal: having the ovules borne on the inner surface of the peripheral ovary wall.

parted: cleft nearly but not quite to the base or midrib.

pectinate: with narrow, closely set segments; comblike.

pedate: palmately divided or parted.

pedicel: the stalk attaching an individual flower to the stem or inflorescence axis or branch.

peduncle: the stalk of a flower cluster. NOTE: In some cases, a peduncle may support but a single flower (see Violaceae). This indicates evolutionary reduction or loss of flowers from an inflorescence type initially composed of more than 1 flower.

peltate: having the stalk attached to the under surface some distance from the margin.

pendant: hanging downward, drooping, pendulous.

pepo: a melonlike fruit or gourd, produced from an inferior ovary, with a hard rind, 1 cell, and many seeds.

perennial: living for more than 3 years, and usually not producing flowers until the 2nd or 3rd year.

perfect (flowers): having both male and female parts.

perfoliate: sessile and having the base united around the stem in such a manner that the latter appears to pass through the leaf (or other leaflike part).

perianth: a collective term referring to all the sepals and petals as a group.

pericarp: the outer wall of a fruit.

perigynous: borne around the ovary (from Gr. *peri*, around, + *gyne*, female), said of stamens, petals, and sepals which arise from the rim of a hypanthium in a flower with a superior ovary.

persistent: not falling away.

petal: one member of the corolla, typically (but not always) white or colored.

petaliferous: bearing petals.

petaloid: petal-like in color and texture.

petiolate: having a petiole.

petiole: the stalk of a leaf.

petiolule: the stalk of a leaflet.

phyllodia: a leaflike petiole with no expanded blade, as in some species of *Potamogeton* or *Sagittaria.*

pilose: covered with soft, slender hairs.

pinnate: with leaflets arranged along the sides of an extension of the petiole, either with a terminal unpaired leaflet (odd-pinnate) or with all leaflets paired (even-pinnate).

pinnatifid: pinnately cleft about halfway to the midrib.

pistil: the female part of the flower, typically consisting of an ovule-bearing part (ovary), style(s), and stigma(s).

pistillate: having a pistil but lacking functional stamens, i.e., functionally female.

placenta (pl. placentae): location or region where ovules are attached within the ovary.

placentation: arrangement of ovules within the ovary.

plano-convex: flat on one side and convex on the other.

plicate: in folds like a fan.

plumose: beset with hairs along each side, like the plumes or beard of a feather.

pod: a dry, dehiscent fruit.

pollen: the dustlike male spores in an anther.

pollinium (pl. pollinia): the mass of coherent pollen grains characteristic of the orchid and milkweed families.

poly-: a combining form meaning "many."

polygamodioecious: having bisexual flowers and unisexual flowers on different individuals.

polygamomonoecious: having both bisexual and unisexual flowers, but the flowers mostly unisexual and with male and female flowers borne on the same plant.

polygamous: having unisexual and bisexual flowers on the same plant.

pome: a fleshy fruit of the apple type.

prickle: small spinelike structure produced from the epidermis or bark of a plant; see also *spine.*

prismatic: shaped like a prism.

proboscis: in some insects, such as butterflies, an elongate, flexible feeding organ.

procumbent: lying on the ground but not rooting at the nodes.

prostrate: lying flat on the ground.

proterandrous: having the anthers maturing before the stigma(s) of the same flower.

proterogynous: having the stigmas maturing before the anthers in the same flower.

proximal: near the main axis of the plant.

pseudo-: a combining form indicating a deceptive resemblance, "falsely."

puberulent: minutely and softly hairy.

pubescence: the hairs on the epidermis of a plant.

pubescent: hairy.

punctate: dotted, either with minute pits or round, sessile glands.

pungent: hard and sharp pointed; prickly-pointed.

purgative: a medicine or other agent which causes evacuation of the bowels.

pustule: a minute blister or blisterlike projection.

pyxis: a pod opening round horizontally by a lid.

quarter-spheroidal: shaped like 1 part of a sphere which has been divided longitudinally into 4 identical parts.

raceme: an elongate, unbranched flower cluster with stalked flowers, the lowermost of which open first.

racemose: bearing racemes.

rachis: the axis of a raceme, spike, or compound leaf.

radiate: bearing ray flowers around the margin; spreading from a common center, as some stigmas do.

radical: belonging to the root or apparently coming from the root.

radicle: the stem part of the embryo, the lower end of which forms the root.

ray: in the Composite Family, the flat strap-shaped corolla of the marginal flowers of the head in such plants as the daisy, sunflower, or aster.

receptacle: the apex of the floral axis which bears the sepals, petals, stamens, and pistil.

reclined: turned or curved downward.

recurved: curved outwards or backward.

reduced: as used here, smaller in size than normal or than is typical or average or elsewhere on the same plant.

reflexed: abruptly recurved or bent outwards or backwards.

regular: radially symmetrical; having all the similar parts alike in size and shape (i.e., all the sepals alike, all the petals alike, etc.) and arranged around the pistil(s) like the spokes of a wheel.

remote: separate spatially.

reniform: kidney- or bean-shaped.

repand: wavy-margined.

resinous: producing any of numerous clear or translucent, yellow or brown, viscous substances such as resin or amber.

reticulate: in the form of a network.

retrorsely: bent or pointed downward, toward the base.

retuse: blunt and somewhat notched at the apex.

revolute: rolled under along the margin.

rhizome: a usually horizontal, underground, perennial stem. Short, erect rhizomes are frequently referred to as "rootstocks."

rhombate, rhombic: diamond-shaped; shaped like an equilateral parallelogram.

rhomboid: a 3-dimensional figure which is diamond-shaped in general outline.

ribbed: having prominent or protruding ribs or veins.

ringent: gaping open.

riparian: occurring along a riverbank.

rootstock: see *rhizome.*

rosette: a basal, usually rather crowded, whorl of leaves.

rostellum: beak.

rostrate: having a beak.

rotate: wheel-shaped; flat and circular in outline.

rotund: rounded in outline.

rudimentary: imperfectly developed.

rugose: wrinkled.

rugulose: minutely wrinkled.

runcinate: coarsely toothed, with the teeth pointing backwards toward the base.

runner: a slender prostrate stem, rooting at the nodes, as in the strawberry.

sac: a closed membrane, or a purse-shaped cavity.

saccate: like a sac or pouch.

sagittate: triangular, with the basal lobes pointing downward or toward the petiole or stem; arrowhead-shaped.

saline: containing common salt (sodium chloride) or other mineral salts, such as magnesium chloride, with the chemical characteristics thereof.

salverform (corolla): with a border spreading at right angles to a slender tube, as of Phlox.

saponaceous: slippery, soaplike.

saprophyte: a plant which takes its nourishment from dead organic matter.

scabrous: very rough to the touch.

scale: any thin membranous bract or appendage.

scandent: climbing.

scape: a leafless peduncle arising from the ground or a basal whorl of leaves and bearing 1 or more flowers.

scapose: bearing a flower or flowers on a scape; resembling a scape.

scarious: thin, dry, membranous, and somewhat translucent.

schizocarp: a dry fruit which separates into indehiscent, 1-seeded segments called mericarps.

scorpioid: curved at the end, like a scorpion's tail.

scurfy: covered with minute scales.

secund: having flowers or leaves all turning to one side.

seed: a ripened ovule, containing embryonic plant tissues and usually stored food materials.

self-pollination: the transfer to pollen from an anther to the stigma of a flower on the same plant, usually to the stigma of the same flower from which the pollen came (see cross-pollination).

semi-: a combining form meaning "half."

sepal: one member of the outermost series of flower parts, frequently green and herbaceous, but sometimes petal-like in color and texture.

septate: divided by partitions.

septicidal capsule: a capsule the seeds of which are released through openings in the radial walls of the cells, i.e., between the cells, rather than through the outer (peripheral) walls; compare with loculicidal.

septum (pl. septa): dividing walls between cells of an ovary.

sericeous: silky.

serrate: with sharp teeth pointing toward the apex.

sessile: attached directly to the stem without an intermediate flower or leafstalk.

setose: bristly.

sheath: a tubular plant part, usually surrounding the stem or an inflorescence.

shoot: a stem with its leaves newly sprung from a bud.

siliceous: containing or consisting of silica.

silicle: in the Mustard Family, a short pod; compare with *silique.*

silique: in the Mustard Family, a long, slender pod; compare with *silicle.*

simple: not branched; not compounded.

sinuate: as applied to leaf margins, strongly wavy, in the same plant as the blade, i.e., wavy in and out rather than up and down.

sinus: the space between 2 lobes.

spadix: in the Araceae, a thick, fleshy column on which the flowers are borne.

spathe: a leaflike or corollalike bract which, at least initially, surrounds a flower cluster.

spatulate: spoon-shaped.

species: in general, a group of like individuals, such as the bird's-foot violets, the black locusts, the red clovers.

spicate: spikelike.

spike: an elongate, unbranched inflorescence with sessile flowers which open from the bottom of the inflorescence toward the top.

spindle-shaped: tapering at each end; *fusiform* (see).

spine: a strong, sharp-pointed body mostly arising from the wood of a stem and derived evolutionarily from modified leaf tissue; compare with *prickle* and *thorn.*

spinose: bearing many spines.

spreading: positioned or growing outward at a wide angle, less erect than ascending but more erect than horizontal.

spur: any slender, spurlike projection of a flower as that of larkspurs, columbines, etc. Usually these are nectiferous.

squarrose: spreading sharply away from the axis (said of bracts of involucres, thickly set leaves, scales).

stamen: a male part of a flower, usually consisting of a slender stalk (the filament) and a pollen-containing portion (the anther).

staminate: provided with stamens but without functional pistils.

staminode, staminodium: an abortive, sterile stamen.

standard: the upper petal in a papilionaceous corolla in the Legume Family.

stellate: starlike; with branches radiating like those of a conventional star.

sterile (stamens, pistils): nonfunctional; incapable of contributing to reproduction.

stigma: the part of the pistil which is receptive to pollen.

stipe: in general, a small stalk; referring here to the stalk which in some plants elevates the pistil above the receptacle.

stipel: a stipulelike structure associated with a leaflet.

stipellate: having stipels.

stipitate: provided with a stipe.

stipular: like, pertaining to, or provided with stipules.

stipulate: furnished with stipules.

stipule: appendages, usually more or less leaflike, occurring in pairs, one on either side of the petiole base, in some plants.

stolon: a specialized aboveground stem which creeps and roots at the nodes or bends to the ground and then roots and produces a new plant at the tip.

stramineous: strawlike in texture and color.

striate: marked with slender longitudinal grooves or lines.

strict: narrow; with close, upright branches.

strigose: furnished with appressed, rigid or scalelike bristles.

style: that part of the pistil which bears the stigma.

sub-: a prefix meaning "nearly but not perfectly so," as in "subacute" or "suborbicular."

subtend: to be situated closely beneath a bud, a branch, a peduncle, a pedicel, etc.

subulate: awl-shaped; sharp-pointed from a broadened base.

succulent: juicy or pulpy.

sucker: a shoot arising from an underground part.

suffrutescent: slightly shrubby or woody at the base only.

sulcate: deeply grooved longitudinally.

super-, supra-: prefixes indicating "above," "many," "many times."

superior: said of an ovary or pistil which has the other parts of the flower, i.e., the sepals, petals, and stamens, arising at its base and free from it.

suture: a line of fusion (as between the subdivisions of a multicellular or compound ovary) or a line of dehiscence in a pod.

symmetrical (flowers): with sepals, petals, stamens, of the same number; (leaves) with the 2 longitudinal halves alike in size and form.

sympetalous: having the petals united with one another; gamopetalous.

syngenesious: with stamens united by their anthers.

synsepalous: having the sepals united with one another; gamosepalous.

tailed: constricted to a slender, tail-like appendage.

taproot: a stout vertical root continuing the main axis of the plant downward.

tendril: a thread-shaped structure used by some plants for climbing.

tepal: a term applied to sepals and petals which are similar in color and form and therefore not readily distinguishable from one another.

terete: circular in cross section; cylindrical.

terminal: borne at or belonging to the extremity or summit.

ternate: in 3's, as a leaf consisting of 3 leaflets.

tetra-: a prefix signifying "4."

tetradynamous: used of a flower with 6 stamens, 2 of them shorter than the other 4.

thallus: a flat leaflike body.

thorn: a strong, sharp-pointed body derived evolutionarily from modified stem tissue; compare with *prickle* and *spine.*

throat: the orifice at or a little below the juncture of the limb (the expanded portion) with the tube of a sympetalous corolla, or the equivalent region of a synsepalous calyx.

thyrsus: a compact pyramidal panicle.

tillers: lateral branches formed near the surface of the soil and rooting at the base.

tomentose: clothed with matted woolly hairs (tomentum).

torose, torulose: said of a cylindrical body swollen at intervals.

torus: the receptacle of a flower.

transverse: perpendicular to the longitudinal axis of a structure.

tri-: a prefix denoting "3" or "thrice."

tridentate: 3-toothed.

trifoliate: having 3 leaves.

trifoliolate: having 3 leaflets.

triternate: thrice ternate; divided into 3 principal segments which are also divided into 3 segments, and each of these into 3 segments.

truncate: as if cut off abruptly at the end.

tube: the portion of a gamopetalous corolla or gamosepalous calyx in which the petals or sepals are united.

tuber: a thickened portion of an underground stem, provided with buds or "eyes" at regions corresponding to stem nodes, as of the Irish potato or the Jerusalem artichoke.

tubercles: a small tuber or swelling; a nodule caused by nitrogen-fixing bacteria, especially common on the roots of leguminous plants.

tuberculate: with rounded projections.

tubular: shaped like a slender, more or less elongated, hollow cylinder.

tuft: cluster.

tumid: swollen.

tunic: bulb coat.

tunicate: having coats or tunics, as the bulbs of some onions.

turbinate: top-shaped.

turgid: rigid or swollen with water, i.e., not wilted or flaccid.

twice-pinnate: see bipinnate.

ultimate segments: the smallest subdivisions of a dissected leaf or leaflet.

umbel: a simple inflorescence with pedicels radiating from a common point, as in the Carrot Family.

umbellate: in umbels.

uncinate: hooked at the tip or in the form of a hook.

undulate: as applied to leaf margins, strongly wavy in a plane perpendicular to that of the leaf blade, i.e., wavy up and down rather than in and out.

uni-: prefix denoting "1."

unicellular: 1-celled.

unifoliate: having only one leaf.

unifoliolate: applied to a theoretically compound leaf, with all but 1 leaflet suppressed.

unilateral(ly): borne along 1 side of a stem or inflorescence axis.

unilocular: 1-celled.

unisexual: having stamens only or pistils only.

urceolate: urn-shaped.

utricle: a small, thin-walled, bladdery, 1-seeded fruit.

valvate, valvular: opening by valves; meeting by margins, not overlapping.

valve: one of the segments into which a longitudinally dehiscent fruit splits.

vein: 1 of the branches of the vascular system of plants.

velutinous: velvety to touch.

venation: the arrangement of the veins of a leaf.

ventral: on the inner or adaxial surface of an organ.

verrucose: warty; beset with little projections resembling warts.

versatile (anthers): attached at or near their middle to the filament.

verticil: a circle of similar parts; a whorl.

verticillate: arranged in a circle around the stem, like the spokes of a wheel.

vestigial: rudimentary.

villous: shaggy with long soft hairs.

viscid, viscous: glutinous, sticky.

warty: with wartlike bumps, *verrucose* (see).

whorl: an arrangement of parts in a circle around an axis, like the spokes of a wheel, the petals of a flower, etc.

wing: any membranous or foliaceous extension of a stem angle, seed margin, etc.

woolly: covered with long, entangled soft hairs.

xeromorphic: of a form and structure to endure dryness.

xerophyte: a plant adapted to a dry habitat.

zygomorphic: capable of division into symmetrical halves only by one plane passing longitudinally through the axis; said, for example, of papilionaceous and bilabiate corollas; *irregular* (see), bilaterally symmetrical.

Literature Cited

1. Schoewe, W. H. 1949. The geography of Kansas. *Trans. Kansas Acad. Sci.* 52: 261–331.
2. Flora, S. D. 1948. Climate of Kansas. *Report of the Kansas State Board of Agriculture* 58: 1–320.
3. Gleason, H. A. 1952. *The New Britton and Brown Illustrated Flora of the Northeastern United States and Adjacent Canada.* The New York Botanical Garden. Vols. I, II, and III. 1808 pp.
4. Steyermark, J. A. 1963. *Flora of Missouri.* Iowa State University Press. 1725 pp.
5. Anderson, Edgar, and Woodson, R. L. 1935. The species of *Tradescantia* indigenous to the United States. *Contr. Arnold Arboretum* 9: 1–132.
6. Robertson, K. R. 1966. The genus *Erythronium* (Liliaceae) in Kansas. *Ann. Missouri Bot. Gard.* 53: 197–204.
7. Riley, C. V. 1892. The *Yucca* moth and *Yucca* pollination. *Third Ann. Report Missouri Bot. Gard.:* 99–166 and plates.
8. Johnston, A. 1962. *Chenopodium album* as a food plant in Blackfoot Indian prehistory. *Ecology* 43: 129–130.
9. Fernald, M. L. 1950. *Gray's Manual of Botany* (8th ed.). American Book Company. 1632 pp.
10. Gibbons, Euell. 1962. *Stalking the Wild Asparagus.* D. McKay Company. 303 pp.
11. Grieve, Maude, and Leyel, C. F. 1971 (facsimile reprint of 1931 edition). *A Modern Herbal—The Medicinal, Culinary, Cosmetic and Economic Properties, Cultivation and Folklore of Herbs, Grasses, Fungi, Shrubs and Trees with All Their Modern Scientific Uses.* Dover Publ. Co. 2 vols. 888 pp.
12. Parkinson, John. 1629. *Paradisi in Sole, Paridisus Terrestris.* H. Lowness and R. Young, London.
13. Gerard, John. 1597. *The Herball or Generall History of Plantes.* Iohn Norton, London.
14. Robinson, Sara. 1856. *Kansas, Its Interior and Exterior Life.* Crosby, Crosby, Nichols and Company, Boston.
15. Claus, E. P. 1961. *Pharmacognosy.* Lea and Febiger. 565 pp.
16. Ownbey, G. B. 1958. Monograph of the genus *Argemone* for North America and the West Indies. *Mem. Torr. Bot. Club* 21: 1–159.
17. Waterfall, U. T. 1966. *Keys to the Flora of Oklahoma.* Publ. by author. 243 pp.
18. Stuckey, R. L. 1972. Taxonomy and distribution of the genus *Rorippa* (Cruciferae) in North America. *Sida* 4: 286–292.
19. Jonsell, Bengt. 1968. Studies in the northwest European species of *Rorippa*. *S. Str. Symb. Bot. Upsal.* 19: 1–221.
20. Hitchcock, C. L.; Cronquist, A.; Ownbey, M.; and Thompson, J. W. 1964. *Vascular Plants of the Pacific Northwest, Part 2: Salicaceae to Saxifragaceae.* University of Washington Press. 597 pp.
21. Kingsbury, J. M. 1964. *Poisonous Plants of the United States and Canada.* Prentice-Hall, Inc. 626 pp.
22. Martin, A. C.; Zim, H. S.; and Nelson, A. L. 1951. *American Wildlife and Plants—A Guide to Wildlife Food Habits.* Dover Publications, Inc. 500 pp.
23. Prior, R. 1870. *Popular Names of British Plants.* Williams and Norgate, London.
24. Isely, Duane. 1962. Leguminosae of the north-central states IV. Psoraleae. *Iowa State Journ. Sci.* 37: 103–162.
25. Tatschl, A. K. 1970. A Taxonomic and Life History Study of *Psoralea tenuiflora* and

Psoralea floribunda in Kansas. Unpubl. Ph.D. dissertation, University of Kansas, Lawrence.

26. Isely, Duane. 1950. The Leguminosae of the north-central United States: Loteae and Trifolieae. *Iowa State Journ. Sci.* 25: 439–482.

27. Küchler, A. W. 1974. A new vegetation map of Kansas. *Ecology* 55: 586–604, incl. 1:800,000 scale folded map.

28. Stevens, W. C. 1948. *Kansas Wild Flowers.* University of Kansas Press, Lawrence. 461 pp.

29. Gates, F. C. 1941. *Weeds in Kansas.* Kansas State Board of Agriculture. 360 pp.

30. Krochmal, Arnold, and Krochmal, Connie. 1973. *A Guide to the Medicinal Plants of the United States.* Quadrangle/New York Times Book Co. 259 pp.

31. Carroll, F. B. 1919. The development of the chasmogamous and the cleistogamous flowers of *Impatiens fulva. Bot. Contr. Univ. Pennsylvania* 4: 144–183.

32. Winters, H. F. 1970. Our hardy *Hibiscus* species as ornamentals. *Ec. Bot.* 24: 155–164.

33. Bates, D. M. 1967. A reconsideration of *Sidopsis* Rydberg and notes on *Malvastrum* A. Gray (Malvaceae). *Rhodora* 69: 9–28.

34. Weiner, M. A. 1972. *Earth Medicine—Earth Foods: Plant Remedies, Drugs and Natural Foods of the North American Indians.* Collier Books. 214 pp.

35. Sprengel, C. K. 1793. *Das Entdeckte Geheimnis der Natur im Bau und der Befruchtung der Bluten.* Fr. Vieweg, Berlin.

36. Bigelow, Jacob. 1822. *A Treatise on the Materia Medica.* Charles Ewer, Boston.

37. Russell, N. H. 1965. Violets (*Viola*) of central and eastern United States: an introductory survey. *Sida* 2: 1–113.

38. Hoffman, W. J. 1891. The Midé wiwin or "Grand Medicine Society" of the Ojibwa. *7th Ann. Report Bur. American Ethnol.* 1885:86: 149–300.

39. Ferrari, J. B. 1633. *De Florum Cultura.* Stephanus Paulino, Rome. 522 pp.

40. Ayensu, E. S. 1973. The religious, "stupendous" passionflowers. *Smithsonian Magazine* 3: 56–58, 60–61.

41. Stephens, H. A. 1962. The cacti of Kansas. *Kansas School Naturalist* 8: 1–15.

42. Hardin, J. W., and Arena, J. M. 1974. *Human Poisoning from Native and Cultivated Plants.* Duke University Press. 194 pp.

43. Millspaugh, C. F. 1887. *American Medicinal Plants, an Illustrated and Descriptive Guide to the American Plants Used as Homeopathic Remedies,* 2 vols. Brericke and Tafel, New York.

44. Woodson, R. E. 1954. The North American species of *Asclepias. Ann. Missouri Bot. Gard.* 41: 1–211.

45. Vogel, V. J. 1970. *American Indian Medicine.* Univ. Oklahoma Press. 578 pp.

46. Yunker, T. G. 1965. *Cuscuta. North American Flora,* Series II, Part 4: 1–51.

47. Krochmal, Arnold, and Krochmal, Connie. 1974. *The Complete Illustrated Book of Dyes from Natural Sources.* Doubleday and Company, Inc. 272 pp.

48. Gilmore, M. R. 1919. Uses of plants by Indians of the Missouri River region. *33rd Ann. Report Bur. American Ethnol.* 1911–12: 43–154.

49. Bogn, W. 1927. *Die Heilwerthe Heimischer Pflanzen* ("The Medicinal Value of Native Plants"). Leipzig.

50. Griffith, R. E. 1847. *Medical Botany, or Descriptions of the More Important Plants Used in Medicine, with Their History.* Lea and Blanchard.

51. Pennell, F. W. 1929. *Agalinus* and allies in North America. II. *Proc. Acad. Nat. Sci. Philadelphia* 81: 111–249.

52. Fernald, M. L., and Kinsey, A. L. 1943. *Edible Wild Plants of Eastern North America.* Harper and Row. 451 pp.

53. Lloyd, F. E. 1935. *Utricularia. Biol. Rev.* 10: 72–110.

54. Lloyd, F. E. 1942. *The Carnivorous Plants.* Chronica Botanica Co. 352 pp.

55. McVaugh, Rogers. 1945. The genus *Triodanis* Raf. and its relationships to *Specularia* and *Campanula. Wrightia* 1: 13–52.

56. Trent, J. A. 1942. Studies pertaining to the life history of *Specularia perfoliata* (L.) A. D. C., with special reference to cleistogamy. *Trans. Kansas Acad. Sci.* 45: 152–164.

57. Krochmal, A.; Wilken, L.; and Chien, M. 1970. Lobeline Content of *Lobelia inflata:* Structural, Environmental and Developmental Effects. *U.S.D.A. Forest Service Research Paper* NE Series (A13.78:NE178). 13 pp.

58. Turner, B. L. 1956. A cytotaxonomic study of the genus *Hymenopappus* (Compositae). *Rhodora* 58: 163–186, 208–242, 251–269, 295–308.

59. McGregor, R. L. 1968. The taxonomy of the genus *Echinacea* (Compositae). *Univ. Kansas Sci. Bull.* 48: 113–142.

60. Heiser, C. B., Jr. 1976. *The Sunflower*. Univ. Oklahoma Press, Norman. 198 pp.
61. Heiser, C. B., Jr. 1951. The sunflower among the North American Indians. *Proc. American Phil. Soc.* 95: 432–447.
62. Payne, W. E., and Jones, V. H. 1962. The taxonomic status and archeological significance of a giant ragweed from prehistoric bluff shelters in the Ozark plateau region. *Papers Michigan Acad. Sci., Arts and Letters* 47: 147–163.
63. Smith, H. H. 1928. Ethnobotany of the Meskwaki Indians. *Publ. Mus. City of Milwaukee, Bull.* 4: 175–326.
64. Reed, C. F.; Correll, D. S.; and Johnston, M. C. 1970. *Manual of the Vascular Plants of Texas*. Texas Research Foundation. pp. 578–580.
65. Rogers, C. M. 1968. Yellow flowered species of *Linum* in Central America and western North America. *Brittonia* 20: 107–135.
66. Shinners, L. H. 1963. *Gilia* and *Ipomopsis* (Polemoniaceae) in Texas. *Sida* 1: 171–179.
67. Waterfall, U. T. 1958. A taxonomic study of the genus *Physalis* in North America north of Mexico. *Rhodora* 60: 107–114, 128–143.
68. Stebbins, G. L., Jr., and Paddock, E. L. 1949. The *Solanum nigrum* complex in Pacific North America. *Madroño* 10: 70–81.
69. Gleason, H. A., and Cronquist, A. 1963. *Manual of the Vascular Plants of Northeastern United States and Adjacent Canada*. D. Van Nostrand Company, Inc. 810 pp.
70. Croat, T. B. 1962. The genus *Solidago* of the northcentral Great Plains. Unpubl. Ph.D. dissertation. University of Kansas, Lawrence.
71. Merriam, D. F. 1963. The Geologic History of Kansas. *State Geol. Surv. Kansas Bull.* 162. 317 pp.
72. Correll, D. S. 1950. *Native Orchids of North America North of Mexico*. Chronica Botanica. 399 pp.
73. Darwin, Charles. 1877. *The Various Contrivances by Which Orchids Are Fertilized*. Appleton and Company, New York. 300 pp.
74. Woodson, R. E. 1964. The geography of flower color in butterfly weed. *Evolution* 18: 143–163.

Index

Items that are italicized in the index are those found in italics in the text. Common names consisting of two or more words are listed in order by the first letter of the first word of the name. Page numbers in bold type indicate the location of an illustration. Dashes are used to indicate the repetition of the species name (or genus, under some entries such as Dye plants).